Deflections

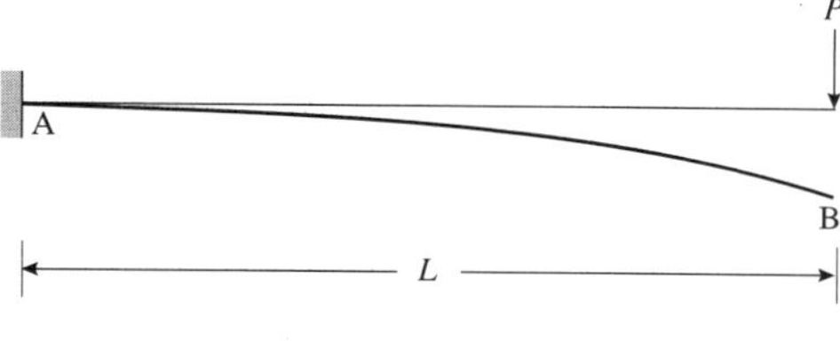

$$\theta_B = \frac{PL^2}{2EI} \;(\curvearrowright)$$

$$\Delta_B = \frac{PL^3}{3EI}(\downarrow)$$

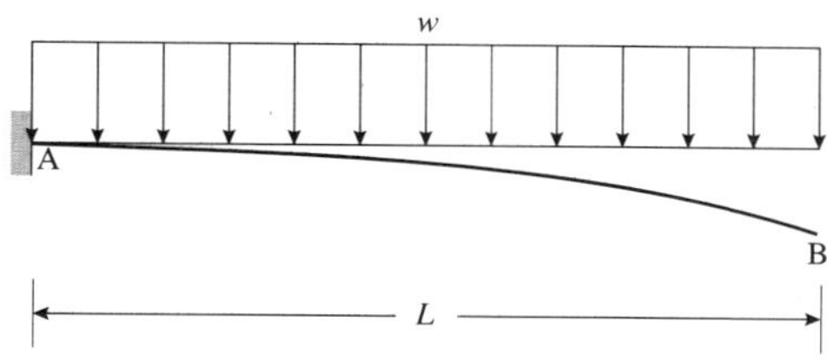

$$\theta_B = \frac{wL^2}{6EI} \;(\curvearrowright)$$

$$\Delta_B = \frac{wL^4}{8EI}(\downarrow)$$

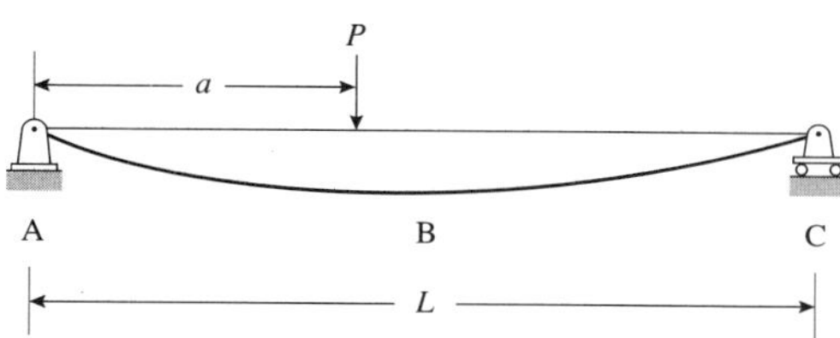

$$\theta_A = \frac{Pa(L-a)(2L-a)}{6EI} \;(\curvearrowright)$$

$$\theta_B = \frac{Pa(L-a)(2L+a)}{6EI} \;(\curvearrowleft)$$

$$v(x) = \frac{Pa(L-a)x}{6LEI}\left(-a^2 + 2aL - x^2\right)(\downarrow)$$

When $a = L/2$

$$\theta_{\max} = \frac{PL^2}{16EI} \text{ at A and C}$$

$$\Delta_{\max} = \frac{PL^3}{48EI}(\downarrow)$$

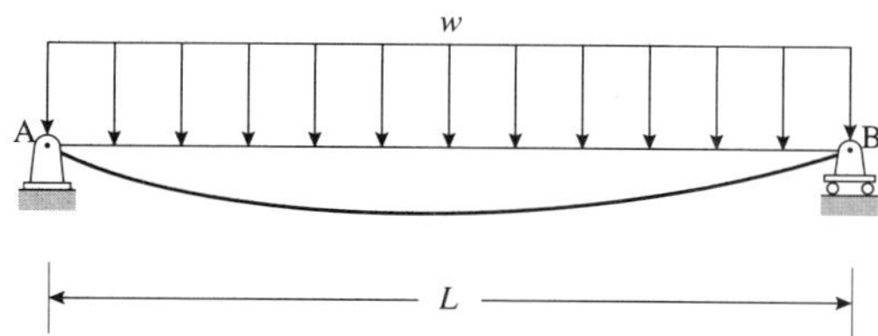

$$\theta_{\max} = \frac{wL^3}{24EI} \text{ at A and B}$$

$$\Delta_{\max} = \frac{5wL^4}{384EI}(\downarrow)$$

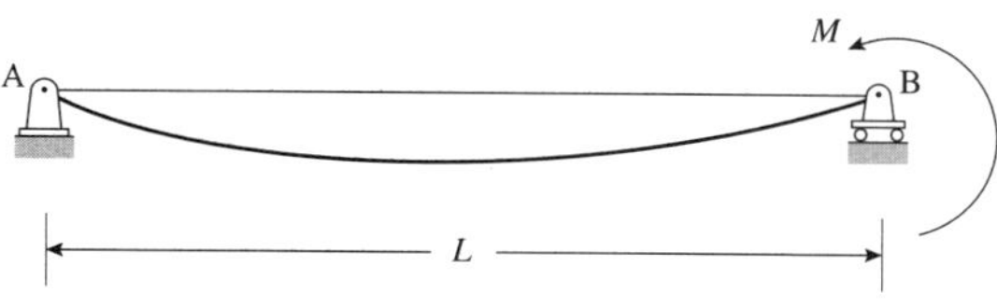

$$v(x) = \frac{Mx}{6EIL}\left(x^2 - 3Lx + 2L^2\right)(\downarrow)$$

$$\theta_A = \frac{ML}{6EI} \;(\curvearrowright)$$

$$\theta_B = \frac{ML}{3EI} \;(\curvearrowleft)$$

Introduction to Structural Analysis & Design

S. D. Rajan
Department of Civil and Environmental Engineering
Arizona State University

John Wiley & Sons, Inc.
New York / Chichester / Weinheim / Brisbane / Singapore / Toronto

ACQUISITIONS EDITOR	Wayne Anderson
MARKETING MANAGER	Katherine Hepburn
PRODUCTION SERVICES MANAGER	Jeanine Furino
PRODUCTION EDITOR	Sandra Russell
COVER DESIGNER	Madelyn Lesure
COVER PHOTO	Tim Trumble, ASU Sr. Photographer
PRODUCTION MANAGEMENT SERVICES	Argosy

This book was typeset in 10/12 Times by Argosy, and printed and bound by R. R. Donnelley & Sons (Willard). The cover was printed by Phoenix Color Corp.

The paper in this book was manufactured by a mill whose forest management programs include sustained yield harvesting of it timberlands. Sustained yield harvesting principles ensure that the number of trees cut each year does not exceed the amount of new growth.

This book is printed on acid-free paper. 

Library of Congress Cataloging in Publication Data:

Rajan, Subramaniam D.
Introduction to structural analysis & design / S.D. Rajan
p. cm.
ISBN 0-471-31997-X
1. Structural analysis (Engineering) 2. Structural design. I. Title: Introduction to structural analysis and design. II. Title.
TA645.R325 2000
624.1'71—dc21 00-021963

Printed in the United States of America

10 9 8 7 6 5 4 3 2 1

To my parents,
Subramaniam and Lakshmi,
who have always been my inspiration and guide.

Preface

Profound changes are taking place in every facet of society, including universities and the workplace. In a globally competitive world, it is necessary to develop skills that provide the edge to be a successful engineer. Today, such skills inevitably include not only a proper understanding of engineering principles but especially the *intelligent* use of computer-based tools. Such skills can be honed over time and with careful experience. On the other hand, curriculum changes in universities around the United States have made it necessary to reduce the number of credit hours in a typical civil engineering program. Instead of students taking several structures courses as a requirement of the program, courses such as concrete design or steel design are becoming elective. However, ABET (Accreditation Board for Engineering and Technology) requires students to be exposed to an adequate number of courses dealing with design principles. It is this background that provides the motivation for this book.

Why Structural Analysis and Design?

One of the most fascinating attractions of being a structural engineer is that the profession allows one to derive the satisfaction of designing structural systems on which society is so reliant—common dwellings such as homes and apartments, shopping plazas, bridges, retaining walls, commercial buildings, parking structures, hospitals, semiconductor fabs, aircraft hangars, sports stadiums, etc. To design correctly, one must understand structural analysis. While there are a multitude of structural systems and behaviors, the ones that are used most often are those that can be adequately modeled as trusses, beams, and frames exhibiting linear behavior. This book looks at these structural elements and systems in great detail. Key to understanding these structural elements is understanding and appreciating when they *cannot* be used as modeling elements. Engineering education is predominantly process-oriented, while engineering practice is predominantly system-oriented.[1] I hope that this text will bridge some of that gap.

The process of design inevitably involves three steps (among many others). First, the engineer must take the requirements stated by the client, go through a decision-making process, and finally arrive at a mathematical model of the structural system. This mathematical model can then be analyzed by one or more structural analysis techniques. Second, in evaluating the performance of the proposed structural system, the engineer looks for indications that the system is (i) safe, (ii) cost-effective, and (iii) acceptable to the client's and societal needs. Central to this evaluation is the required understanding of such issues as material behavior, theories of failure, and code requirements. Lastly, the process of redesign is almost always necessary. This step is perhaps one of the least discussed in any text.

Special Features

- Emphasis on mathematical modeling of structural systems for structural analysis.
- Early use of commercial-grade structural analysis software.
- Case studies to illustrate modeling techniques.

[1] W. F. Stoecker, Design of Thermal Systems, McGraw-Hill, 1989.

- Study of material behavior, structural performance considerations, and appropriate theories of failure.
- Emphasis on a limited number of powerful classical techniques.
- Easy transition from classical to numerical (direct stiffness/finite element) techniques.
- Formal design problem formulation to solve one-parameter and multiparameter design problems.
- Use of commercial-grade software to illustrate design techniques.
- Use of commercial-grade software to illustrate structural design techniques.
- Case studies involving simple but powerful design scenarios.
- Introduction to design of steel and concrete structures.
- Discussion on limitations of the techniques and principles discussed in the text—where and how to go forward to more powerful analysis and design techniques.

Motivation for the Changes

Compared to a conventional book on structural analysis, several topics have either been dropped or de-emphasized in this text. Some of these are compound and complex trusses, approximate analysis techniques, several less useful techniques for computing deflections, analysis of indeterminate structures using three-moment equations, moment distribution, and flexibility method. On the other hand, several new topics have been added: computer modeling of structural elements and systems, a second look at deformable solids and commonly encountered civil engineering materials, the theorem of minimum potential energy for deriving element equations for truss and frame elements, the mathematical programming approach to formulating and solving single- and multiparameter design problems, and computer-based structural design involving problems with non-unique (or multiple) solutions.

It should first be noted that the text is intended for a first course in structural analysis and design. The topics that are excluded deal with specialized structural systems (compound and complex trusses), are much less useful (approximate analysis techniques, three-moment equations, moment distribution) considering the fact that more powerful methods are available, are redundant (several less useful techniques for computing deflections), do not lead gracefully into future topics, and are less useful from a practical viewpoint (flexibility method).

The design-related material in this text is not intended to replace a course on steel, concrete, masonry and wood design, nor to replace a course on engineering design based on mathematical programming techniques. The motivation is to show and emphasize the intrinsic relationship between structural analysis and structural design. Hence the mathematical programming topics are handled in a condensed fashion. The structural design issues must be dealt with in so as to be comprehensible to students taking this course. Hence, in Section 3.7, the emphasis is twofold. First, design problem formulation is discussed. In a typical mechanics of materials course, design problems are one-parameter design problems (unique solutions). The task at hand here is to formulate the design problem more formally, identifying the major objective(s), the design parameters, and the requirements to be satisfied. A formal knowledge of optimization or mathematical programming is not needed. Second, simple solution techniques are used to solve the problems, such as graphical, intuitive, trial-and-error, etc. Once the reader is more adept and mature, Chapter 8 tackles the formal approach to solving the mathematical programming problem. Even in this chapter, the approach is to teach the student the dual task of formulating the design problem and finding a solution in an expeditious manner. Some design problems and ideas can be illustrated using unconstrained minimization techniques. This is the motivation for Section 8.2. Most structural design problems are constrained nonlinear programming (NLP) problems. In the author's opinion, it would be counterproductive to introduce mathematical programming background. Design problems that the students will comprehend and appreciate can be solved in a reasonable amount of time (1–5 minutes, if not seconds) on most personal computer systems. From a user perspective, genetic algorithms (GA) have the least amount of information needed and restrictions imposed in order to solve structural design problems. Traditional NLP solution techniques require far more information and are also far more restrictive. GA concepts (advantages, disadvantages, limitations, strengths) are introduced in Section 8.4. A point-and-click software system can be used to illustrate the GA solution methodology. Student programming, though desirable, is not required.

Structural modeling—taking structural systems and constructing the proper mathematical model for analysis and design—is a nice way of bridging the gap between analysis and design, education and practice. The modeling ideas can be illustrated with simple but powerful examples (e.g., design of industrial frames, roof trusses, simple bridges, retaining walls, etc.). Chapters 3, 7, and 8 deal with these ideas. They also point to future topics that interested students may explore—structural systems (space frames, plates and shells, cables and arches), modeling of material behavior (nonlinear, plastic), modeling of loads (earthquake, dynamic, thermal), and code-based design concepts (factored loads, probabilistic approach). From a structural mechanics (or solid mechanics) viewpoint, the energy-based approach can be more readily extended from truss and frame behavior than can the direct stiffness/flexibility approach.

Who Should Use This Book?

While one of the intents in structuring the contents of the book is to make the book self-sufficient, the reader is expected to have taken the following courses: Introduction to Engineering Design, Engineering Calculus, Differential Equation, Statics, Deformable Solids, and Introductory Linear Algebra and Numerical Analysis. These prerequisites will give the reader the necessary background to learn structural analysis techniques and design principles.

It is also anticipated that parts of the text will be used by practicing engineers, by those interested in studying for the PE exam and even in a second course on structural analysis.

	Typical 1-semester (4 semester-hours) curriculum	Typical 1-semester (3 semester-hours) curriculum	2-semester (6 semester-hours) curriculum[a]
Chapter	**Number of weeks**	**Number of weeks**	**Number of weeks**
1–3, 7	4	5	5 (1st semester)
4, 5	4	6 (select coverage)	8 (1st semester)
6, 7	4 (select coverage)	3 (select coverage)	6 (2nd semester)
8	2	2	2 (2nd semester)
9	2 (select coverage)	—	4 (2nd semester)

[a] Requires minimal supplemental material.

The appropriate location of Chapter 7, which deals with structural analysis computer programs, is left to the instructor. Chapter 7 is written almost as a standalone chapter, and the instructor of the course should provide guidance (and warnings against misuse). The author strongly believes that computer-based tools should be used early and carefully. As stated in Chapter 7, "Computers are wonderful tools to solve problems. They are a tool much like log tables, slide rules, and basic calculators that engineers have used in the past. But they are much more versatile and powerful. It is imperative that we use this powerful tool in an intelligent fashion. We must recognize its strengths and weaknesses (yes, weaknesses!). When used with caution, computer programs can provide an engineer with powerful tools to carry out mundane calculations very quickly and accurately, to understand structural behavior through visual examinations, to investigate several modeling alternatives, to improve and design better structural components and systems, etc."

This sentiment is echoed in different ways by those who have reviewed the text. Professor Spacone (University of Colorado) says, ". . . Students can, from the very beginning, check their results. In the book introduction, the author could suggest that instructors cover this chapter right after the Introduction chapter." Professor Bhatti (University of Iowa) comments, ". . . I really like the idea of introducing computer-based analysis early on in the course. This opens the possibility of assigning challenging group design projects relatively early in the semester." Professor Nowak (Gonzaga University) says, ". . . I am very familiar with the software. It has been used in several of my classes over the last seven years. I feel that the use of a structural software package would be imperative if this text were used."

Ancillaries

Two Windows 95/98/NT© based software systems will be used as a part of the instruction (and text). The first program is GS-USA© (Graphics-based System for Understanding Structural Analysis). This program, which has been used in a few universities over the last six to seven years, can be used for the analysis of planar truss and frames subjected to mechanical and thermal loads. In addition, the program has automated design capabilities and is based on genetic algorithms (GA) as the optimization engine. The second program is UNDO© (UNderstanding Design Optimization), which can be used for finding the optimal solution to unconstrained and constrained optimization problems without writing a single line of code (or program) (to be used in Chapter 8). The program also has a matrix toolbox that can be used for finite-element-based truss and frame analysis (Chapter 6). There are other equally useful programs on the CD-ROM.

A final thought on software. This book is not built around the capabilities of the aforementioned software. If you prefer another program with similar or even better capabilities, feel free to use it. The theories and problems discussed in this text lend themselves to any competent program.

Web Site

Additional resources, corrections, updates, etc. will be posted on the supporting Web site at `www.wiley.com/college/rajan`

Feedback

The process of continuous improvement should include feedback—from students, readers, and instructors. The author can be contacted through e-mail at

`s.rajan@asu.edu`

Your corrections, suggestions, and comments are welcome. The hope is to incorporate these ideas in future editions of the text.

Acknowledgments

Several individuals have shaped my thinking and career—Professor Swapan Majumdar at the Indian Institute of Technology, who introduced me to structural analysis; Professors Jasbir Arora, Ed Haug, and Asghar Bhatti at University of Iowa, who introduced me to design optimization; my peers, Professor Ashok Belegundu, Penn State University, and Professor

Nguyen T. Duc, Old Dominion University, who have provided me with motivation and strength to keep going; my colleagues at ASU, Professors Bob Hinks, Barzin Mobasher, Bill Bickford, Bob Rankin, and Hal Nelson, who have been a source of inspiration. A special thanks is owed to the graduate students who have tirelessly helped me debug and refine the computer programs—S-Y. Chen, Joanne Situ, Calvin Young, So Hui Yu, Gouri Sridevi, and Sambit Ghosh. Thanks to Dhaval Shah, Sandeep Mane and Kevin McCaughlin who meticulously read through the text and solved the problems. The arduous task of proofreading the text was accomplished by my son, Varun. I could certainly use his patience and attention to details. And finally, thanks to the undergraduate students who semester after semester have provided the major motivation for writing this text and whose feedback has made this text and computer programs better educational tools.

Several reviewers of the prospectus as well as the manuscript made invaluable suggestions— Professor Paul Nowak, Gonzaga University; Professor Enrico Spacone, University of Colorado; Professor Richard Gutkowski, Colorado State University; Professor M. A. Bhatti, University of Iowa; and Professor Graham Archer, Purdue University. Their insightful comments have shaped the treatment of the topics covered in this text.

This book would not have been possible without the assistance and guidance of the people at Wiley, especially my editor, Wayne Anderson. They have been most helpful and accommodating, as have the people at Argosy.

Finally, this book is a fruit of the countless discussions with my 'managers'—Vanitha, Varun, and Rohit—on the facts of life. They have made this endeavor possible and painless.

Subramaniam ("Subby") Rajan
Tempe, Arizona
July 2000

A Few Words About the Accompanying CD-ROM

The accompanying CD-ROM has several programs that the reader is urged to install and use early on. The following are the contents of the CD.

Software

GS-USA© Frame program: This program carries out the analysis and optimal design of planar truss and frame systems.

UNDO© program: This program carries out generic design optimization. It also has a matrix toolbox.

SlideTray© program: This program has several utilities such as a scientific calculator, an X–Y grapher, matrix toolbox, calculus toolbox, units converter, calendar, etc. It also has the AISC tables for some of the popular AISC cross-sections.

UCSD© program: This is a program for the design of concrete and steel components.

Adobe Acrobat© Reader: This program is needed to view all the program manuals, tutorials, and the extra problems and solutions. These are contained in files with the file extension *pdf*.

Reference Material

Problems and solutions: There are two directories. One (Extra Problems and Solutions) contains additional problems and solutions. The other (Examples) contains most of the program data files for the examples in the text.

Printing the manual: The install program creates two subdirectories under a program directory—*manual* and *samples*. For example, the program documentation for the GS-USA program is contained in the file *winframe.pdf* in the *manual* directory. Program data files for all the problems in the GS-USA tutorial can be found in the *samples* directory.

Contents

Chapter 1

Introduction

Modern society is increasingly reliant on lifelines such as electric transmission towers. Engineers design, construct and operate such systems.

"Engineers are problem solvers." Anon

"Few things are harder to put up with than a good example." Mark Twain

"I not only use all the brains that I have, but all that I can borrow." Woodrow Wilson

This book is about structural analysis and design. *Webster's Dictionary* defines *structure* as "1: the action of building: CONSTRUCTION . . . 4b: organization of parts as dominated by the general character of the whole . . . 5: the aggregate of elements of an entity in their relationships to each other." With this definition it is probably more appropriate to state that this book is about the analysis and design of structural components and systems as commonly used in the civil (and possibly, mechanical and aerospace) engineering industry.

Humans, for centuries, have designed and built homes, aqueducts, dams, bridges, city roads, freeways, office buildings, storage systems (e.g., water tanks), industrial buildings, towers (e.g., for communications, power transmission), sports facilities, domed stadiums, museums, amphitheaters, entertainment complexes, airports, runways, aircraft hangars, harbors, docks, cranes, launch pads, power generation plants, water and sewage treatment plants, refineries, and so on. An examination of texts dealing with statics and with mechanics of materials will also show that the principles of structural analysis and design can also be applied to the analysis and design of a very wide variety of systems—automobile frames, airframes, helicopter rotors, leads in a computer chip assembly, artificial limbs, robots, rotating machinery, exercise equipment, toys, chemical pipelines, manufacturing equipment, elevators, packaging that holds a computer monitor, and spring assembly in a computer keyboard, to name just a few!

The design of these systems is usually motivated by the need for such systems. The designer or the design team must go through a formal process that starts with constructing the specifications for the system. There are several intermediate steps—preliminary design(s), cost analysis, design selection, design refinements, detailed design, etc.—before the blueprints are made for the final design. One of the major pillars of structural design is structural analysis.

The overall objectives of this text are quite modest: (1) to demonstrate and practice structural analysis of determinate and indeterminate structures using classical and matrix methods; (2) to demonstrate and practice simple structural design concepts using structural analysis as a tool; (3) to learn and use general-purpose computer programs for structural analysis and design. Along the way we will take short detours and side trips to illustrate concepts and point out topics for future reference and study.

1.1 STRUCTURAL ENGINEERING

The history of structural engineering is replete with pioneers and innovators who with their contributions have changed the course of history and civilization. While very little is documented with respect to the design and construction methods of ancient civilizations, they were successful in building structural systems that have survived over the ages. For example, the Egyptians built pyramids, obelisks, and temples around 3000–2000 B.C., possibly using simple devices and tools such as wedges, mallets, chisels, straight levers, rollers, and the inclined plane. How exactly the structures were made remains a mystery today.

Archeological sites at Mohenjodaro and Harappa in India's Indus valley contain evidence of a sophisticated system of storing and distributing water, and collecting and disposing of waste water. The system, dating back to 2500 B.C., was built primarily of clay, bitumen, and bricks. The Greeks and the Romans enhanced the art of building. Stone and masonry were the two popular building materials used to construct aqueducts, trusses, domes, arches, vaults, etc. Greek architecture took on a refined form. Subtle use of proportions to design the taper of a column or the camber of steps at the Parthenon added artistic value to these structures. The Roman engineers developed efficient ways for masonry construction using locally available material to invent "cement"—lime mortar and volcanic ash. What knowledge they obtained from the Etruscans and the Greeks, they used to build grandiose structures—the vaults and domes of the Pantheon, the buildings at the Basilica of Constantine, arched aqueducts, bridges, roads, and harbors.

The period from 1800–1900 saw tremendous developments in the art of structural engineering. The early developments in the area of cement and concrete took place in Europe. In 1801, F. Ciognet published a paper on the principles of construction using concrete. In 1824, Joseph Aspdin in Leeds, England, invented the modern portland cement. J. L. Lambot in 1850 is credited with building the first cement boat for display at the 1855 World's Fair in Paris. However, people recognized that concrete was weak in tension. Surprisingly, a French gardener, Joseph Monier, patented a system of reinforcing concrete tree planters with steel wire. His work led to the widespread development and use of reinforced concrete, both in Europe and in the United States. To further improve the usage of concrete, the concept of prestressed concrete was successfully introduced and used by several individuals—W. H. Hewett in the early 1920s developed the principles of circular prestressing used in tanks and pipes; Eugene Freyssinet proposed methods to overcome prestress losses through the use of high-strength and high-ductility steels and, in 1940, introduced a widely followed system that bears his name.

The use of metal as a structural material began with the building of cast iron bridges in England in the late 1700s. The disadvantages of cast iron and its replacement, wrought iron, were overcome with better manufacturing processes. The development of the Bessemer process in the 1850s made it possible to produce building-grade iron and steel products.

Such developments led to revolutionary uses of building materials. John Roebling (1806–1869) was a pioneer in the design and construction of suspension bridges. As a designer of the Brooklyn Bridge, he was able to predict the effects of gravity loads, traffic loads, and wind loads on the design of bridges. He foresaw the effects of design and construction of bridges on mass transit, society, and art!

While manufacturing and construction methods improved, so did the analysis techniques. B. P. E. Clapeyron (1799–1864), a Frenchman, invented the three-moment theorem, a method for solving continuous beams. This method found widespread usage in the analysis of bridge structures. J. C. Maxwell (1831–1879) developed the method of consistent displacements that could be used to solve indeterminate trusses, beams, and frames. In Italy, Alberto Castigliano (1847–1884) developed a method (now known as Castigliano's theorem) for computing deflections in 1873, as did Charles Greene (1842–1903) in the United States (the moment-area theorem). In 1915, G. Maney (1888–1947) introduced the slope-deflection method in which the primary unknowns were displacements. The potential of the method was not realized since the methodology was more suited to a computer implementation. In fact, the slope-deflection method laid the foundation for the matrix-based methods and modern numerical methods. About a decade later, in 1924, the moment distribution method was invented by Hardy Cross (1885–1959). With this method it was possible to solve larger problems without having to deal with all the problem unknowns simultaneously. In other words, hand calculations could be used to solve larger problems in an iterative manner.

With the advent of modern computers in the 1950s, it was possible not only to analyze more complex structural systems accurately but also to analyze and design routine structures more efficiently. The finite element method is a good example of an analysis methodology that lends itself naturally to an implementation as a computer software system. The major formative ideas appeared in the 1940s. Courant's 1943 paper, "Variational Methods for the Solution of Problems of Equilibrium and Vibration," is a classic. However, the explicit use of computers in solving problems using the finite element method occurred in the 1950s through the work of Langefors and Argyris. The name "finite elements" first appeared in a paper by Clough in 1960. In a similar manner, the structural design optimization area has evolved into several numerical techniques that can be effectively implemented to solve design problems. The use of calculus-based ideas for minimization can be traced to Cauchy in 1847, Courant's paper on penalty functions in 1943, Dantzig's work on simplex method and linear programming in 1951, and Karush, Kuhn, and Tucker's theorem on solving constrained problems in 1951. Numerical techniques to solve small and large problems were pioneered by Rosen in 1960 (gradient projection method), Zoutendijk in 1960 (method of feasible directions), Abadie, Carpentier, and Hensgen in 1966 (reduced gradient method), Karmakar in 1984 (interior point methods), and several others. Other nongradient search techniques were developed by pioneers such as Rosenbrock in 1960 (method of orthogonal directions), Metropolis in 1953 (simulated annealing), and Holland in 1975 (genetic algorithm). The use of optimization techniques in structural design was pioneered by Schmit in 1960. Since then structural optimization techniques have made inroads into several commercially used computer programs and have been used to design civil, mechanical, aerospace, automotive, and several other structural systems.

Modern tools and construction methods made it possible to design and build systems that existed only in someone's imagination. Fazlur Khan (1929–1982) designed the first "tubular cantilever" buildings, the Sears Tower and John Hancock Center. Tubular systems opened up new economic possibilities in tall building design, and the idea is routinely used today to design tall buildings all over the world.

The tremendous growth took place also because of simultaneous developments in areas such as materials and construction practices. Developments in steel, concrete, polymers, aluminum, cement, timber, and composites make it possible to use these materials in designing a variety of structural systems—roads, bridges, dams, buildings, aqueducts, storage tanks, pipes, stadiums, and the like.

1.2 TYPES OF STRUCTURAL SYSTEMS

There is no unique way to classify structural elements and systems. If the geometry of the structural system or element is used as the basis for classification, structural elements are either discrete or continuous. Discrete elements are those that can be geometrically approximated as one-dimensional entities. Examples of discrete elements include trusses, frames, cables, and arches. Continuum structures are those in which the material distribution can be geometrically expressed in two-dimensional or three-dimensional space. Examples include plates and shells.

Truss

The simplest structural member is a short, prismatic, slender, and straight member. A *truss* is formed by connecting these members to one another with pin connections (Fig. 1.2.1). In two dimensions, the resulting structure usually consists of a pattern of triangles. When the structure is loaded, the members are subjected to only axial forces. The members are either in axial tension or compression. There are no shear forces or bending moments in a truss member. Trusses are introduced in Chapter 2 and discussed throughout the text.

Frame

Framed structures are formed by members connected to one another with rigid or semi-rigid connections. The horizontal members are usually referred to as *beams* (some of the other names used for beams are girder, joist, purlin, lintel), while the usually vertical members are *columns*. *Frames* can be very efficient in resisting gravity and lateral loads. Under the action of these loads, the beams are subjected primarily to bending moments, whereas the columns carry axial forces. As with truss members, the length of individual beams and columns is limited. The members are slender and prismatic. However, the members can be straight or curved (Fig. 1.2.2). Beams and frames are introduced in Chapter 2 and discussed throughout the text.

Cables

A *cable,* like a beam, is a slender member. However, the cross-sectional dimensions are very small compared to the length. As a consequence, the primary internal force is axial tension. The deflected shape of the cable is a function of the applied load. When the loading is discrete, the deformed shape is a straight line. When the loading is distributed, the deformed shape is a curve. For example, under self-weight, the curve is a catenary curve (Fig. 1.2.3).

Arches

An *arch* is a curved member. The cross-section is slender and prismatic, and the element spans two points. Modern buildings commonly have a rigid arch element in which the primary internal force is compression. Arches can be of different types depending on the manner in which they are supported—pin-pin, fixed-fixed, three-hinged, etc. Figure 1.2.4 shows a famous arch structure.

Fig. 1.2.1
Examples of truss structures (a) Truss bridge across Ohio River, St. Marys, WV (courtesy American Bridge Div. of U.S. Steel). (b) Space frame scaffold structure (courtesy The Statue of Liberty and Ellis Island Foundation, Inc.). (c) Example of open-web steel joist trusses (courtesy Bethlehem Steel Corporation). (d) Electrical transmission tower under erection—both tower and crane boom are truss-type structures (courtesy Bethlehem Steel Corporation).

Plates

The preceding structural elements were line members. On the other hand, a *plate* is a flat structural member (Fig. 1.2.5). The thickness of the plate is usually small compared to the lateral dimensions. The loading is usually normal to the flat surface. Under the action of this loading, the internal action is primarily bending. Under certain conditions, the plate can also be subjected to in-plane loading.

Shells

Similar to a plate, a *shell* is a structural member whose thickness is small compared to the lateral dimensions. However, the surface is not flat. Shells can be singly curved structures, such

Fig. 1.2.2 Examples of frame structures (a) Slant-leg frame bridge, Charlottesville, VA (courtesy Bethlehem Steel Corporation). (b) Continuous plate-girder bridge over Quinnipiac River, New Haven, CT (courtesy Steinman, Boynton, Gronquist, and Birdsall, Consulting Engineers). (c) Typical rigid frame construction (courtesy Lincoln Electric Company, Cleveland, OH). (d) Moment-resistant frame structure, Dresser Tower, Houston, TX (courtesy Bethlehem Steel Corporation).

Fig. 1.2.3 A suspended cable.

as cylindrical shells, or doubly curved, such as spherical shells or hyperbolic paraboloid surfaces (Fig. 1.2.6).

Those structural systems or components that do not fit the previous descriptions are usually treated as three-dimensional continuum structures. A dam and a thick bridge deck are examples of structural systems that are modeled using three-dimensional structural elements whose properties do not fit into any of the element categories discussed earlier.

Common Structural Systems

While the structural elements discussed earlier can be put together to form a system, often different structural elements are combined to create other structural systems. We discuss some commonly encountered systems below.

Fig. 1.2.4
The Jefferson National Gateway Arch, St. Louis, MO (courtesy Pittsburgh-Des Moines Steel Company).

Fig. 1.2.5
A floor-slab system (courtesy Portland Cement Association).

Fig. 1.2.6
St. Louis Priory Chapel; system of parabolic shells (courtesy Portland Cement Association).

Trussed Frame. Figure 1.2.7 shows an industrial building. The roof system is a truss while the structure that transmits the load to the ground is a frame. The truss assembly makes it possible to span long distances without intermediate supports.

Fig. 1.2.7
Industrial building built as a trussed frame.

Braced Systems. As we see in Chapter 3, structural loads are either gravity-induced or lateral. Examples of the latter include wind and earthquake loads. Figure 1.2.8 shows two similar systems. The unbraced frame on the left consists of girder and vertical columns. When the frame is subjected to lateral loads, large moments are induced in the columns. The frame on the right has cross-bracing members. These members provide lateral stability and reduce the overall deflection of the frame. By reducing the relative displacement between the bottom and top of the column, the column moments are reduced.

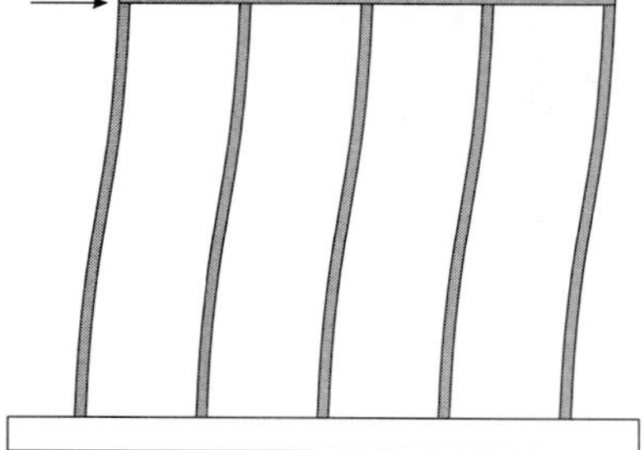

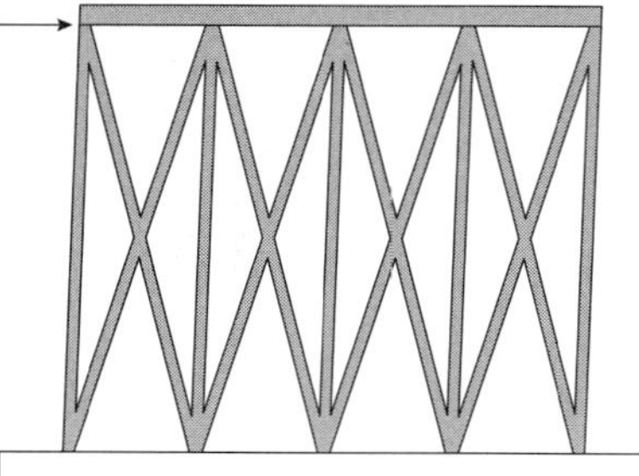

Fig. 1.2.8
Unbraced and braced frames.

Another approach to handling lateral loads is to use shear walls or diaphragms, as shown in Fig. 1.2.9. For small heights, a masonry wall can be used. For taller frames, typically, reinforced concrete is used. The basic action is similar to using cross-bracing members—to reduce the overall deflection by increasing the lateral stiffness and stability. The columns are subjected to primarily axial force and a much reduced bending moment.

Cable and Suspension Systems. For very long spans as required in some bridges, suspension systems can be used. Figure 1.2.10 shows a cable-stayed bridge. Multiple cables connect the tower to the main girder. Since the cables take the majority of the loading, the weight of the girder members can be effectively reduced.

A cable suspension bridge is shown in Fig. 1.2.11. The main cables stretch from one anchor to the other through the two towers. The main cables are connected to the stiffening truss and the bridge deck via multiple suspender cables. The structure is inherently flexible and some form of stiffening is required, for example, a stiffening truss or a diagonal cable system.

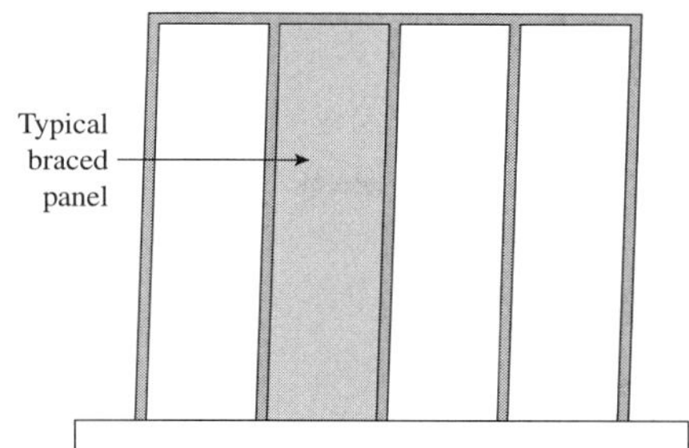

Fig. 1.2.9
Diaphragm-shear wall system.

Fig. 1.2.10
Sunshine Skyway Bridge, Tampa Bay, FL (courtesy Figg Engineering Group).

Fig. 1.2.11
Golden Gate suspension bridge, San Francisco, CA (courtesy Bethlehem Steel Corporation).

Slabs and Grids. Concrete slabs are very commonly found as floor slabs, roof slabs, or slabs on grade. In a one-way slab as shown on the left in Fig. 1.2.12, the basic load-transfer mechanism is along one direction. The beams supporting the one-way slabs also constitute a one-way load-transfer mechanism.

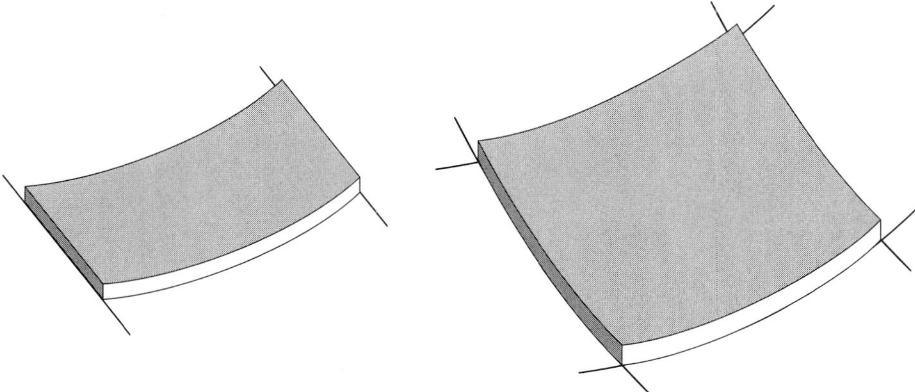

Fig. 1.2.12
One- and two-way slabs.

With a slab whose shape is more square than rectangular, the load-transfer mechanism is more complex. An effective design practice is to design the slab for flexure in the two orthogonal directions—part of the load is assumed to be carried in one direction and the rest in the other direction. Floor slabs are sometimes assumed to act as wide flat beams and are constructed in various shapes—T–beams, joist systems, waffle pan, precast planks, etc. It is also common to find wood flooring systems (Fig. 1.2.13) and steel flooring systems. The latter are sometimes constructed as beams with steel subfloors or concrete slabs (Fig. 1.2.13), or bar joists with steel subfloors.

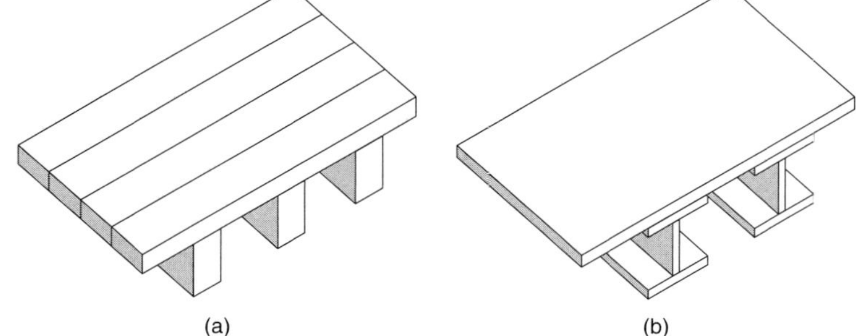

Fig. 1.2.13
(a) Wood flooring system. (b) Steel beam with concrete flooring.

Roof Structures. Conventional roofs consist of a slab and beam assembly—slabs spanning on secondary beams that then rest on main beams supported on columns. Hence the columns finally transmit the loads to the ground. As the column spacing increases, the sizes of all the components also increase, leading to an uneconomical and unaesthetic structure. An effective alternative is to use domed structures (Fig. 1.2.14). The curved surfaces carry the load primarily in direct compression or tension with very little bending or shear. The remarkable feature of such a structure is that even with very small shell thickness, a domed structure can be subjected to large loads over large column-free areas. The deflections are relatively small and the structure is quite economical.

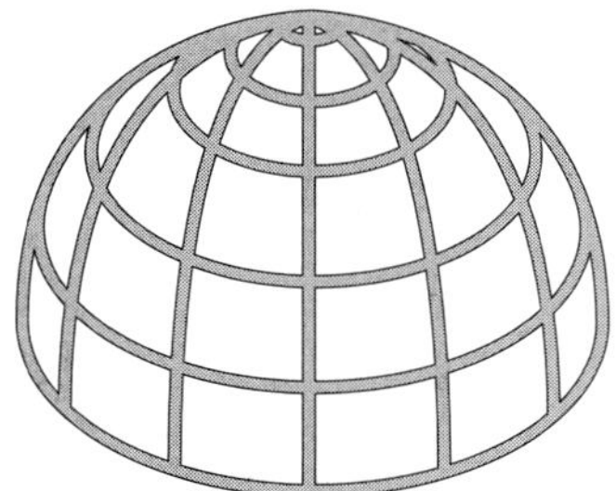

Fig. 1.2.14
Ribbed dome roof.

Not all domed structures are made of reinforced concrete. It is also possible to use fabric or cable nets, or fabric-inflated systems to construct roofs that span large unsupported areas.

The focus of this text is primarily on truss and frame structures, for at least two reasons. First, a wide variety of structural systems can be modeled (or, adequately approximated) as truss and frame structures. Second, among all the different systems, these are the simplest to visualize, approximate, and study in detail. The first course on structural analysis is unlikely to be the last course on structural analysis for those aspiring to be structural engineers. Advanced analysis courses such as finite element analysis cover the analysis of other types of structural systems and other types of structural analyses such as modal analysis, structural dynamics, nonlinear, etc. We will, however, cover some of the design issues in Chapters 3, 8, and 9.

1.3 STRUCTURAL ANALYSIS

Structural analysis is a systematic study of the relationship involving the structural material, members, the manner in which the structure is constructed and supported, the loads acting on the structure, and the resulting deflections and forces. The majority of this text deals with structural analysis.

Classical Techniques. Early on the evolutionary time line, sophisticated methods were invented to analyze a variety of structural systems. Based on the principles of engineering mechanics and suitable approximations, the classical methods have served the needs of the structural community for a very long time. However, with the advent of high-speed computing and user-friendly software, classical techniques are used primarily for quick but approximate analysis. The detailed analysis is done using structural analysis software. Classical techniques are invaluable in understanding the basic principles that govern structural analysis and form the basis of the advanced numerical techniques. We look at the classical structural analysis techniques in Chapters 2, 4, and 5.

Numerical Techniques. The easy availability, power, and cost-effectiveness of personal computers have revolutionized the workplace. They have given a new life to techniques that were considered unsuitable for hand-calculations and have spurred the development of even more powerful analysis techniques. Essential to understanding and harnessing the power of these numerical techniques is a proper knowledge of the fundamentals of numerical techniques and structural mechanics. The interaction between the classical techniques and numerical techniques cannot be overemphasized. In this text, a balance is struck between these two approaches. The classical techniques are introduced in the form of the analysis of determinate structures (Chapters 2 and 4)—static equilibrium, free-body diagrams, and computation of deflections are the three major ideas discussed in these chapters. In Chapter 7, we look at interacting with a computer program for structural analysis. A clear understanding of the ideas in Chapter 2 will facilitate this interaction. The classical techniques discussed in Chapter 5 are designed to solve statically and kinematically indeterminate structural systems. They build on the earlier ideas associated with equilibrium by tying compatibility to equilibrium. These ideas lay a firm foundation for the development and treatment of matrix-based numerical techniques in Chapter 6.

1.4 STRUCTURAL DESIGN

What is structural design? A conceptual view of the design process is shown in Fig. 1.4.1. A need (for a building, a bridge, etc.) initiates the design process. From that point onwards, the process is iterative, going through several stages—preliminary, intermediate, and final—where the various design options are discussed, refined, trimmed, and refined once again. The selection of certain design parameters establishes the governing design code. For example, if the decision is to use hot-rolled steel as the primary structural material for an office building, the governing design codes are likely to be the Uniform Building Code (UBC), which provides load-related data, and the American Institute of Steel Construction (AISC) code, which provides specifications for the limits on the structural response. The designer, at any stage, starts with his or her guess for what the design should be. For example, the design engineer may select W14 × 68 (an AISC wide-flange section about which we will learn more later) for a column. With values provided for all the structural parameters, an appropriate structural analysis is carried out. The structural response that the analysis provides, in the form of forces, displacements, reactions, etc., are used to check the validity of the design. If the design does not meet the specifications (e.g., if the displacements are too large), then the designer must refine the design with new values for some or all design parameters. On the other hand, if the design is too conservative, the designer has the option of finding a lower-cost design, possibly by reducing the size and dimensions of some of the design parameters.

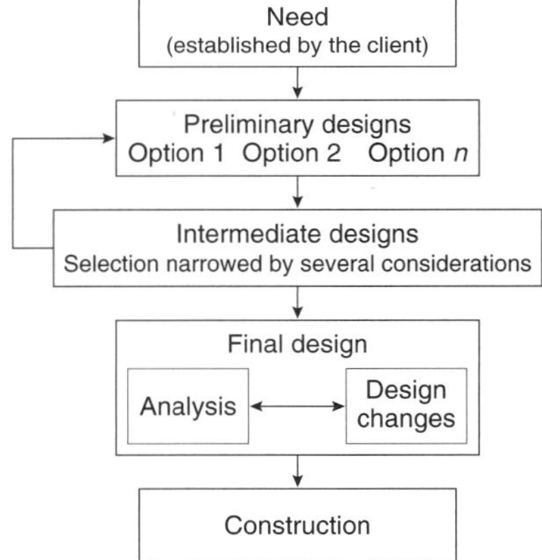

Fig. 1.4.1 Conceptual view of the design process.

In a typical design scenario, the structural engineer and his firm interact with the client and the regulating agency (e.g., the city building department). They may also interact with (a) an architect (or a transportation engineer or planner) who provides the structural form to meet the client's needs, (b) electrical engineers who may assist with meeting the electrical needs, (c) mechanical engineers designing the heating and cooling systems, (d) geotechnical engineers who will provide the soil properties and may design the foundation system, (e) hydraulic engineers concerned with the site's drainage issues, (f) surveyors who have the site topographic data, (g) CAD (computer-aided drafting) personnel who translate the design data into design details and "blueprints," and (h) the construction firm who will finally translate the structural drawings and details into the physical form. Smaller structural design consulting firms may deal with one or more firms over the course of a design project. Larger firms dealing with the "design and build" exercise may be self-sufficient by employing engineers and other personnel with diverse backgrounds to meet the different needs discussed above.

There are some key ingredients that the structural design engineer must be conversant with and account for in the various design stages. These include an intimate knowledge of the (a) material behavior and properties, (b) types of available cross-sections, (c) applicable design codes in computing the loads and checking the structural response and performance, and (d) process of designing or redesigning to meet the major design objectives.

Structural Material

In Chapter 3 we review the concepts associated with stress, strain, failure theories, and computation of stress distribution in truss, beam, and frame members. These concepts are linked to the most commonly used structural materials—concrete, steel, timber, and masonry.

Structural Sections

The truss and frame members are available in a variety of cross-sectional shapes. Different shapes provide different options to the designer to design the system effectively. Concrete cross-sections are typically rectangular, circular or T-sections (Fig. 1.4.2).

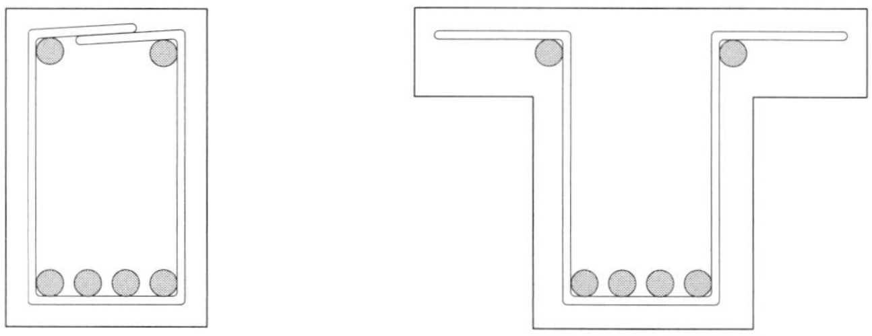

Fig. 1.4.2 Typical concrete cross-sections.

Steel and some other metallic cross-sections are available in a variety of shapes and dimensions. Some of these cross-sections are shown in Fig. 1.4.3.

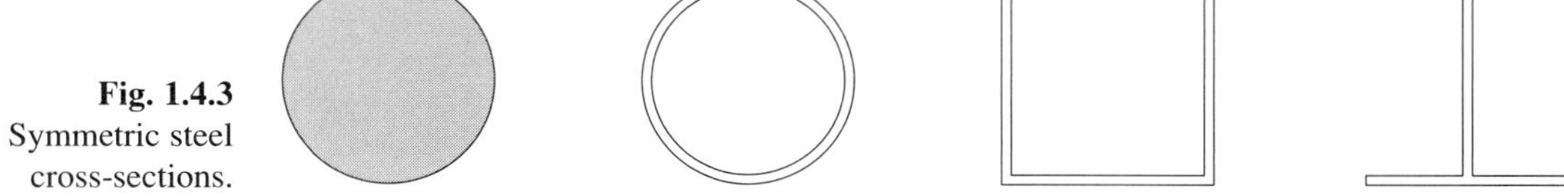

Fig. 1.4.3 Symmetric steel cross-sections.

Timber cross-sections are almost always rectangular. The computation of the cross-sectional properties is discussed in Chapter 3.

DESIGN CODES

Design codes provide such information as design loads, allowable stresses, material properties, cross-sectional dimensions and properties. Additional information provided in these documents includes design procedures and equations, sample calculations, discussions of the philosophy behind some of the procedures and equations, and tips for proper use of the procedures. The codes ensure that public health and safety are protected through the practice of safe engineering procedures. The provisions of the code are usually minimum acceptable standards. It is up to the design engineer to ensure that the specifications used for the final design are at least as good as if not better than the suggested specifications.

In Chapter 3 we discuss some of the design codes in the computation of loads. In Chapter 9 we discuss the design codes as applicable to the design of concrete and steel structural components.

OPTIMAL DESIGN

As discussed above, the design process is iterative. In a typical design, the designer starts with an initial guess. In other words, the designer assumes values for all the design parameters defining the structure. Examples include the initial shape, geometry and layout of the structure, the cross-sectional dimensions of the members, the manner in which the structure is supported, and the like. The results of the structural analysis make it possible to check whether the structure meets the design code specifications. One could ask a few pertinent questions. Given the design specifications and the structural needs, is the design unique? What should the designer do if the design does not meet the specifications? What should the designer do to reduce the overall cost of a component or the structural system? Can the design exercise be carried out in a systematic manner? An emerging methodology in answering these questions is called *design optimization*. The underlying driving force is to express the design problem in a mathematical form that can then be solved systematically to find the optimal solution.

An overview of problem formulation and optimization techniques in presented in Chapters 3 and 8, with several examples of structural optimization in Chapter 8.

1.5 UNITS AND VALUES

Problems in this text are solved using two different units—the SI and the U.S. customary (English) units. The International System of Units (SI) is now used throughout the world. Mass is expressed in kilograms (kg), length in meters (m), and time in seconds (s). Using these base units, one can express derived units. For example, force is a derived unit written as Newtons (N) and

$$1\ \text{N} = (1\ \text{kg})(1\ \text{m/s}^2) = 1\ \text{kg-m/s}^2$$

Sometimes it is more convenient to express the values in SI as multiples or fractions of 1000 or 10^3. The table below shows some of the most commonly used units.

Prefix	Notation	Factor	Prefix	Notation	Factor
nano-	n	10^{-9}	kilo-	k	10^3
micro-	μ	10^{-6}	mega-	M	10^6
milli-	m	10^{-3}	giga-	G	10^9

For example, it is more common to express modulus of elasticity in MPa than Pa.

The U.S. customary units are used primarily in the United States. Unlike the SI units, force is expressed in pounds (lb) or kips (k), length in feet (ft), and time in seconds (s). Using these base units, one can express derived units. For example, mass, a derived unit, is written as slug and

$$1\ \text{lb} = (1\ \text{slug})(1\ \text{ft/s}^2) \Rightarrow 1\ \text{slug} = 1\ \text{lb-s}^2/\text{ft}$$

An alternate form that is sometimes used is to express the mass in pound mass (lbm) and force as pound force (lbf). The former is defined as the mass having a weight of one pound at sea level.

$$1 \text{ lb} = (1 \text{ lbm})(32 \text{ ft/s}^2) \Rightarrow 1 \text{ lbm} = \frac{1}{32}\frac{\text{lb - s}^2}{\text{ft}} = \frac{1}{32} \text{ slug}$$

The table below shows some of the most commonly used units.

Quantity	U.S. Customary Units	SI Units
mass	slug	kg (kilogram)
length	ft (feet)	m (meters)
temperature	F (Fahrenheit)	C (Celsius)
stress	psi (pounds per sq. in)	Pa (pascal)
acceleration	ft/s^2	m/s^2
force	lb (pound)	N (Newton)
moment	lb-ft	N-m
displacement	ft	m
rotation or slope	rad	rad
time	s (second)	s (second)

Throughout the text, problem parameters are mostly expressed as numerical values. The number of significant digits represents the confidence with which values are presented. For example, the numbers –25.13, 0.02513, 0.001340, 4.150×10^3 all have four significant digits. As a matter of habit, we carry out the computations with the accuracy afforded by scientific calculators or computer programs. However, the final results are expressed with the appropriate number of significant digits.

1.6 TIPS AND AIDS

Finally, a few words about getting the most from this book and from a course on structural analysis and design. The presentation style used in this text has two emphases. First, the relevant theory and assumptions are presented that describe the method, technique, or theorem. Second, the theory is followed by several examples that illustrate different traits of the problem-solving abilities of the method, technique, or theorem. Every example has been chosen with care. You are encouraged to look at each and every example in detail. Ideas are explored and comments made so that we can appreciate the different nuances of the theory and its applications. Finally, every section is followed by several problems that you are encouraged to solve. Most have answers that can be used to check your solution. If you familiarize yourself with the computer programs accompanying this text, you have answers (but not necessarily the procedure) to almost all the problems.

I have accumulated over the years, some perspective on effective teaching and learning.

- Come prepared to class. Spend about 15–30 minutes reading the material for that day's lecture. Read it more than once if necessary, even if you do not understand most of the material. Familiarity with the "language of the lecture"—terminology, figures, equations, etc.—is a major advantage that you will carry with you to the lecture.
- At the end of the day (not the end of the week), carefully review the lecture material—derivations, solved examples, etc. Look at this task as changing the oil in an automobile—either you do it regularly or you take the car to a mechanic for a major repair job later.

Close the lecture notes and resolve the solved examples from start to finish. Go back to the lecture notes if you are unsuccessful. Repeat the exercise until you successfully solve the problem. Too often we are tempted to say "I understand the material but. . . ." There is simply no substitute for solving a problem from *start to finish.*

- Practice, practice, and practice. You will have a better understanding of the material by solving problems.
- Don't be afraid of asking questions. A correctly posed question can be a huge life- and time-saver. Your ability to ask the correctly posed question is directly proportional to the amount of "homework" you do.
- Use every conceivable resource—the instructor's knowledge and office hours; the TA's enthusiasm, knowledge, and office hours; the library; material available on the Internet; etc. This is perhaps the only time in your life when you will have the time, energy, motivation, atmosphere, and resources available simultaneously to meet your needs.
- Study in small groups, associating with other students who are as motivated, diligent, and capable, if not more so, than you are. Otherwise this is likely to be a waste of time. Put this study session into your weekly schedule.
- A student should be ethical before he or she becomes a practicing engineer. When it is not clear as to what is ethical, consult someone knowledgeable.
- Structural analysis (like statics, strength of materials, and so on) is one-half knowledge and one-half discipline. The former deals with principles and theories that have evolved over at least the last millennium. The good news is that there aren't too many of them. There are two types of students—sensor and intuitor.[1] Both have the ability to master the knowledge. The difficulty that most of us face is with the latter. Discipline is the trait that makes a successful engineer. Understand the problem, be systematic, and when required, don't ignore the details. The devil is in the details.

It is hoped that the material in this text encourages and helps you in being a better structural engineer and an even better student.

EXERCISES

Appetizers

1.1. Convert the following quantities.

(a) 1 lb to 1 N
(b) 1 m^3 to ft^3
(c) 1 kg/m^3 to $slug/in^3$
(d) 1 N/m to lb/in
(e) 1 m/m-°C to in/in-°F
(f) 1 N-m to lb-in
(g) 1 MPa to psi

Main Course

1.2. Estimate the following quantities.

(a) Weights of (i) 4" × 8" × 2" brick, (ii) 12" × 12" × 2" concrete block, (iii) a cubic foot of snow, (iv) a gallon of water.

(b) Floor heights in a typical office building.

(c) Width of a typical lane on a highway.

[1] Sensors learn through information that they capture through their senses; intuitors learn through reading, reasoning, and interpretation.

1.3. (a) A length is measured and reported as 1.21 m. What is the maximum percentage uncertainty in the measurement?

(b) A measure of a quantity is known to two significant digits. What is the maximum percentage uncertainty in the measurement?

Structural Concepts

1.4. Make a table showing quantities commonly encountered by a structural analyst or designer: forces, moments, cross-sectional dimensions, member lengths, mass density, modulus of elasticity, displacements, rotations, strain, and stress. For each of these quantities, write the precision that is adequate to express these quantities. For example, temperatures can be expressed to the nearest degree. Do your answers depend on the choice of the units?

1.5. In this chapter structures were classified based primarily on their geometry. There are other possible classifications. What are they?

Chapter 2

Determinate Structural Systems

Structural engineers model, analyze and design systems such as a concrete bridge over a desert wash

"Mechanics is the paradise of mathematical science because here we come to the fruits of mathematics." Leonardo da Vinci

"I must confess that my imagination . . . refuses to see any sort of submarine doing anything but suffocating its crew and floundering at sea." H. G. Wells

In earlier courses in statics and deformable solids (or strength of materials), most of the structural systems were statically determinate. The analyses of these structures were carried out using structural equilibrium concepts. In this chapter, the concepts associated with structural modeling, equilibrium, stability and instability, and static determinacy and indeterminacy are explored and reviewed. We will also look at one of the most important concepts in structural engineering—free-body diagrams (FBDs). These diagrams are used to analyze pin-connected structures, trusses, and frames.

This second look or review is necessary to lay the foundation for the subsequent chapters, especially with regard to structural analysis and analysis for design. Statically determinate structural systems constitute the majority of the systems that are built today. They are particularly useful for low-rise and small-span structures. That these structures are popular is partly due to their ease of analysis and design, and partly because of some of the advantages that determinate structures enjoy over indeterminate systems. The differences between the two systems are examined further in Chapter 5, where we analyze statically indeterminate systems.

OBJECTIVES

- To study and understand the concepts associated with structural connections, members, and supports.
- To understand how to draw FBDs.
- To recognize determinate, indeterminate, and unstable structural systems.
- To use the static equilibrium concepts with FBDs of determinate structures and solve for support reactions, pin forces, and member internal forces.
- To study and analyze determinate truss systems.
- To study and analyze determinate beams and frames. Use shear force and bending moment diagrams to track variations of these internal forces in beams and frames.

ASSUMPTIONS

- While deformations are not computed or used in this chapter, it is assumed that these values are small. Deformations are small (a) if their magnitudes are small in comparison to the dimensions of the structure, and (b) if the relationships between the applied loads and the reactions (supports and internal member forces) are not affected by the deformations of the structure.
- While strains and stresses are not computed or used in this chapter, it is assumed that the material behavior remains linear and elastic while the loads act on the structure.

At the end of this chapter, try to answer why it is essential that we state (and monitor) these assumptions.

MATHEMATICAL BACKGROUND

The reader is urged to review the mathematical background necessary for this chapter, presented in Appendix E.

2.1 COMPONENTS OF STRUCTURAL SYSTEMS

The structural systems discussed in this text are made up of at least three distinct components—members, connections, and supports. The nature or characteristics of the members, connections, and supports dictate how effectively the system will support the applied loads.

2.1.1 Structural Members

The primary load-bearing components of a structural system are the members. Examples include truss or frame members. The members are usually prismatic and made of a single material (see Fig. 2.1.1.1).

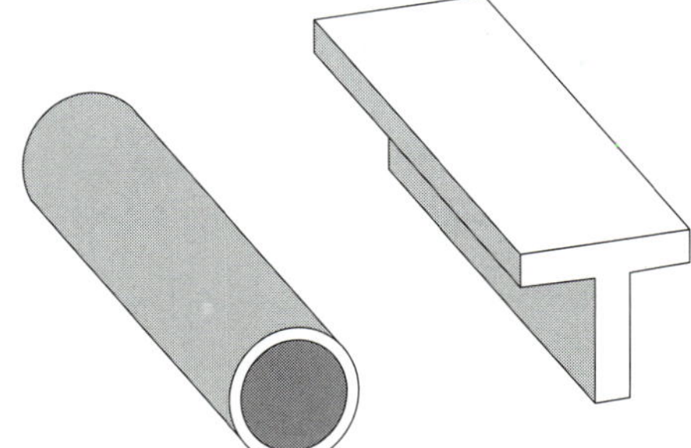

Fig. 2.1.1.1 Typical structural members—hollow circular and T sections.

2.1.2 Connections

Connections or joints are used to tie different structural members together. There are several types of connections, such as the more common ones following.

Rigid Connection. A rigid connection is one in which two or more members are connected to each other such that the angle between the members before and after deformation is the same. These connections are also known as moment-resisting connections. Axial force, shear force, and bending moments are transmitted across the connection.

Graphical notation: ┌

Figure 2.1.2.1 shows an end-plate moment connection. The joint is made by shop-welding a plate to the end of a beam and field-bolting to a column or another beam.

Fig. 2.1.2.1 Rigid connection (courtesy AISC).

Internal Hinge. The moment at an internal hinge is zero. The connection transmits shear and axial forces but not the moment. Hence, it is also known as a moment-release connection (or a frictionless pin). Fig. 2.1.2.2 shows an internal hinge in an elevated highway structure.

Fig. 2.1.2.2 Internal hinge (or moment-release connection) (courtesy W.G. Godden Structural Engineering Slide library).

Graphical notation: 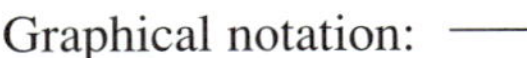

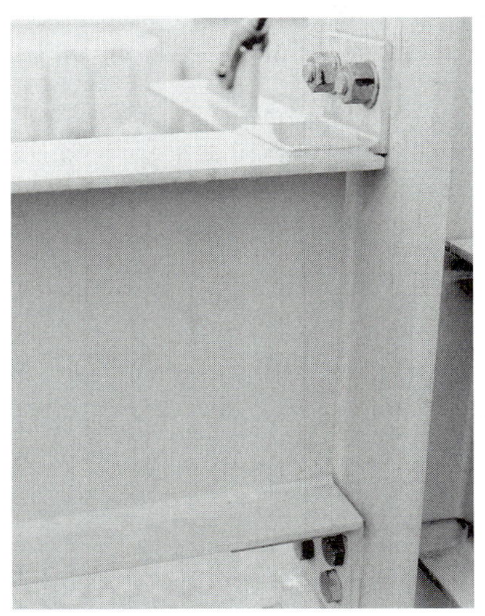

Fig. 2.1.2.3 More examples of structural connections (courtesy AISC).

Typical. In reality, connections are neither completely rigid nor internal hinges. The rigidity is a function of several parameters such as the stiffness and the geometry of the individual members, the bolt or screw layout, and the like. Estimating the rigidity of typical connections is covered in design texts.

Graphical notation: 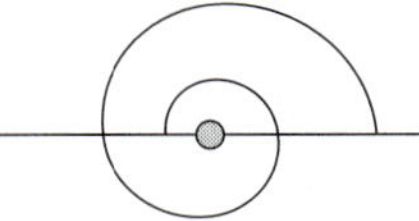

A torsional spring is used to depict the rigidity of a typical connection. If the torsional spring constant is infinity, then the connection behaves as a rigid connection. On the other hand, if the spring constant is zero, then the connection behaves as an internal hinge.

2.1.3 Supports

The applied loads on a structure are finally transmitted to the supports through the members and connections, as summarized in Table 2.1.3.1; see Fig. 2.1.3.1.

Table 2.1.3.1 Common Structural Supports

Support	Graphical Notation	Free-Body Diagram	Remarks
Rocker			There is a single reactive force at a roller or rocker support. The reaction acts normal to the support surface.
Roller		R_y	

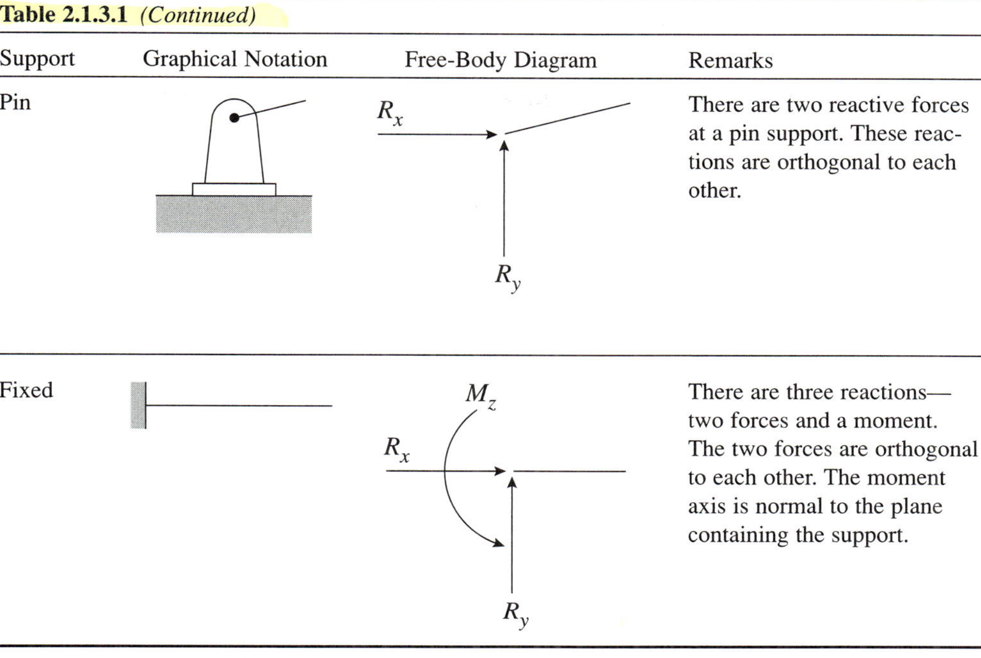

Table 2.1.3.1 *(Continued)*

Support	Graphical Notation	Free-Body Diagram	Remarks
Pin		R_x, R_y	There are two reactive forces at a pin support. These reactions are orthogonal to each other.
Fixed		M_z, R_x, R_y	There are three reactions—two forces and a moment. The two forces are orthogonal to each other. The moment axis is normal to the plane containing the support.

Fig. 2.1.3.1
Examples of structural supports (courtesy AISC).

The remarks concerning supports apply only to idealized conditions. Supports are affected by a variety of factors. For example, frictionless pins do not exist, or there may be friction affecting the ability of the roller or rocker supports to move or slide. Fixed support conditions assume that the medium into which the member is supported or embedded is infinitely stiff. Consider a column that is supported on the ground through an appropriate foundation. While we may assume that the support is a fixed support, soil properties and conditions (among other factors) determine how this support will behave when the column is loaded.

Observation: It is interesting to note that the member displacements and rotation at the supports along the direction of reactions are zero. For example, at a pin support there are two reactions. The displacements along those reactive forces are zero. This observation will be used when dealing with computer-based structural analysis in Chapters 6 and 7, where displacements rather than reactive forces need to be specified.

2.2 FREE-BODY DIAGRAMS

One of the most important concepts in structural analysis is the free-body diagram (FBD). An FBD is a figure of a structural component or system showing all the forces (external, internal, reactive) acting on it. We use FBDs throughout the text to compute structural response quantities using the concept of static equilibrium.

Consider the beam shown in Fig. 2.2.1. It is supported at A by a pin support and at B by a roller support. There are two concentrated forces acting on the beam.

Fig. 2.2.1
Simply supported beam.

In order to study the relationship between the applied loads and the support reactions, we must construct the FBD for the beam. First we remove the pin support at A and, in its place, introduce the two orthogonal reactions labeled A_x and A_y. Next we remove the roller support at B and replace it with the single reactive force B_y. The labeled and complete FBD is shown in Fig. 2.2.2.

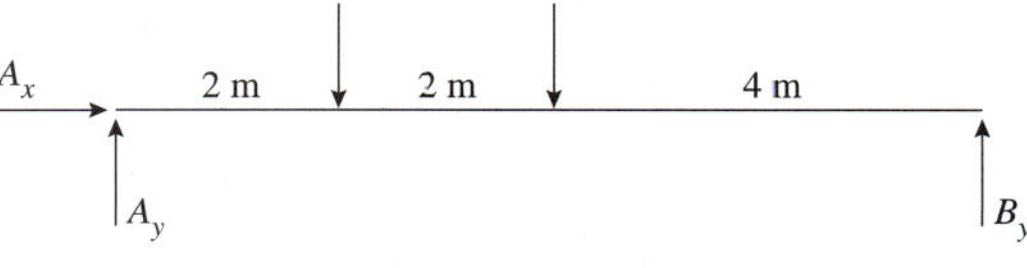

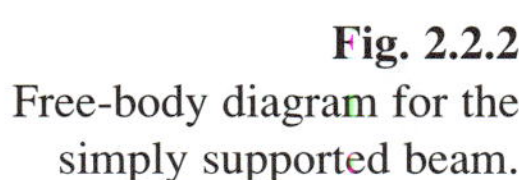

Fig. 2.2.2
Free-body diagram for the simply supported beam.

As is evident with this example, to draw the FBD of a structure, the supports are removed and replaced with the reactive forces that they provide. When drawing the reactive forces, their directions are assumed. Later, the structural analysis will show whether these assumptions are correct. When it is necessary to study the effects of the external loads and supports on the stress and strain distribution inside the structural members, the internal forces must be determined. We cut the member by passing an imaginary section through the member cross-section. In Fig. 2.2.3, a cut *a-a* is made through the beam to the right of the 5 kN force. This cuts the beam into two halves, and the resulting left and right FBDs are shown in Fig. 2.2.4. The internal forces at the cut are shown with their directions assumed.

However, note that since these forces are internal (not external), they are shown equal and opposite in the two FBDs. In other words, if the two FBDs are glued back together, the original FBD (Fig. 2.2.2) should result. This is possible only if the internal forces are equal and opposite in the two FBDs.

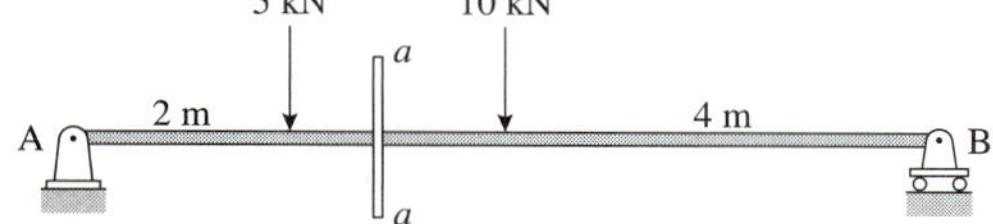

Fig. 2.2.3
A cut through the beam cross-section.

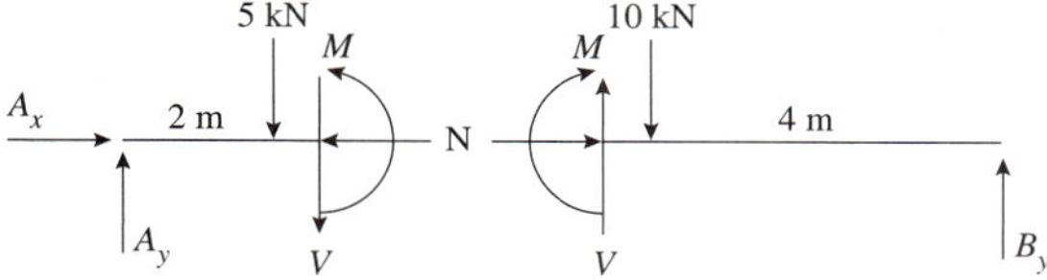

Fig. 2.2.4
Left and right FBDs.

The internal forces acting through the centroid of the cross-section can be computed using the left or the right FBD.

Structural analysis of systems acted on by distributed loads can be carried out after replacing the distributed loads with their equivalent force system. Once the equivalent force system is obtained, the FBD can be drawn. The next section lays the groundwork for computing the equivalent force system.

2.2.1 Resultant of Distributed Loading

Consider a distributed loading $w(x)$ acting on a span L, as in Fig. 2.2.1.1.

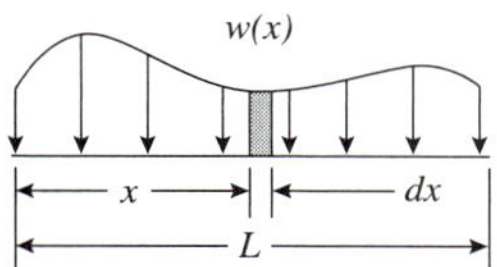

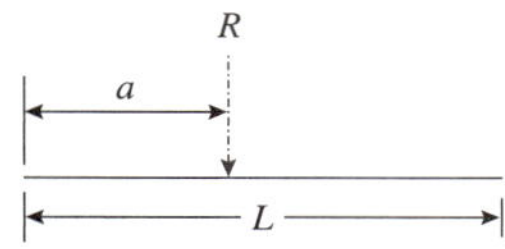

Fig. 2.2.1.1
Distributed loading and its equivalent force system.

An equivalent loading system involving a single force R acting at a distance a from the left end can be constructed as follows. Throughout the text we use a dashed line when replacing a distributed loading with its resultant. The resultant R represents the area under the loading $w(x)$. Hence

$$R = \int_0^L w(x)\,dx \tag{2.2.1.1}$$

Taking the moment of both the force systems about the left end, we have

$$(R)(a) = \int_0^L xw(x)\,dx \tag{2.2.1.2}$$

from which we find

$$a = \frac{\int_0^L xw(x)\,dx}{R} \tag{2.2.1.3}$$

Table 2.2.1.1 shows the resultants of distributed loading systems.

Table 2.2.1.1 Resultants of Distributed Loading Systems

Loading Type	$w(x)$	R	a
Uniform	q	$\int_0^L q\,dx = q[x]_0^L = qL$	$\dfrac{\int_0^L xq\,dx}{R} = \dfrac{q\left[\frac{x^2}{2}\right]_0^L}{qL} = \dfrac{L}{2}$
Triangular	$\dfrac{q_R}{L}x$	$\int_0^L \dfrac{q_R}{L}x\,dx = \dfrac{q_R}{L}\left[\dfrac{x^2}{2}\right]_0^L = \dfrac{q_R L}{2}$	$\dfrac{\int_0^L \frac{q_R}{L}x^2\,dx}{R} = \dfrac{\frac{q_R}{L}\left[\frac{x^3}{3}\right]_0^L}{\frac{q_R L}{2}} = \dfrac{2L}{3}$
Trapezoidal	$q_L + \dfrac{q_R - q_L}{L}x$	$\int_0^L \left[q_L + \dfrac{q_R - q_L}{L}x\right]dx = L\left[\dfrac{q_L + q_R}{2}\right]$	$\dfrac{\int_0^L \left[q_L + \frac{q_R - q_L}{L}x\right]x\,dx}{R} = \dfrac{L}{3}\left[\dfrac{q_L + 2q_R}{q_L + q_R}\right]$

The units for q, q_L, q_R are force per unit length (e.g., lb/in, N/m). It is common to encounter such loading in structural engineering. Examples of uniformly distributed loads include dead load and wind load; triangular loading includes hydrostatic load; and trapezoidal loading includes soil pressure and snow load.

Observation: In each of the above cases, the resultant is located within the span of the loading, a fact that is used in Section 2.8.2. The integration approach is extremely powerful. In this text, we emphasize the use of techniques that eliminate the need to memorize formulas.

EXAMPLE 2.2.1 ***Free-Body Diagrams***

Here are a few examples of complete and correct FBDs.

(a) A beam with an oblique force acting at its end is shown in Fig. E2.2.1(a). It is supported by a pin support at A and a roller support at B.

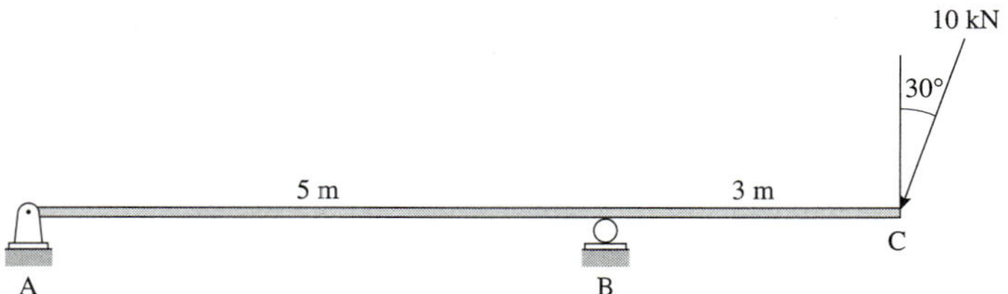

Fig. E2.2.1(a)

To draw the structural FBD in Fig. E2.2.1(b), we replace the pin support with its two orthogonal reactive forces. Similarly, we replace the roller support at B with a single reaction normal to the support surface. Note also that we have replaced the oblique force with its horizontal and vertical components. This is necessary so that the equilibrium equations can be used easily. In this text, we will also show force components with dashed lines.

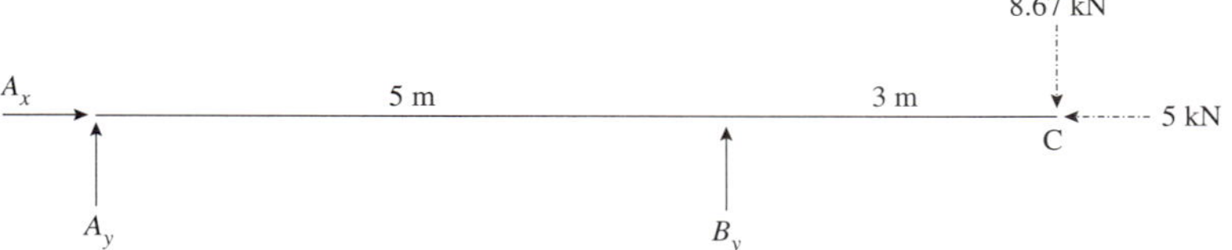

Fig. E2.2.1(b)

(b) Shown in Fig. E2.2.1(c) is a planar frame supporting a uniformly distributed load. Supports A and C are pin supports, B is a rigid connection, and D is a free end.

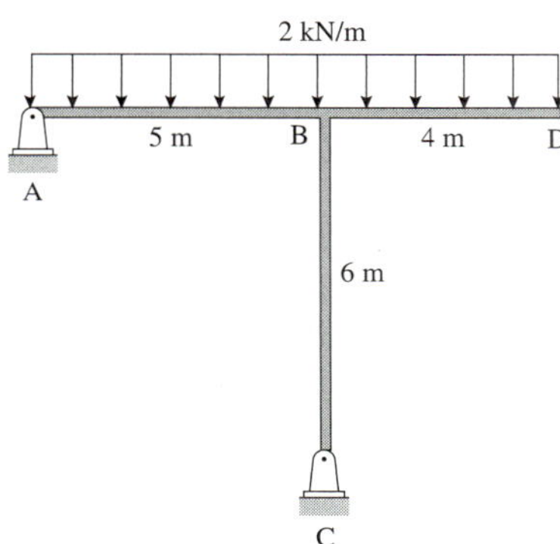

Fig. E2.2.1(c)

The structural FBD is shown in Fig. E2.2.1(d). Two support reactions each replace the pin supports at A and C. A single resultant R replaces the uniformly distributed load.

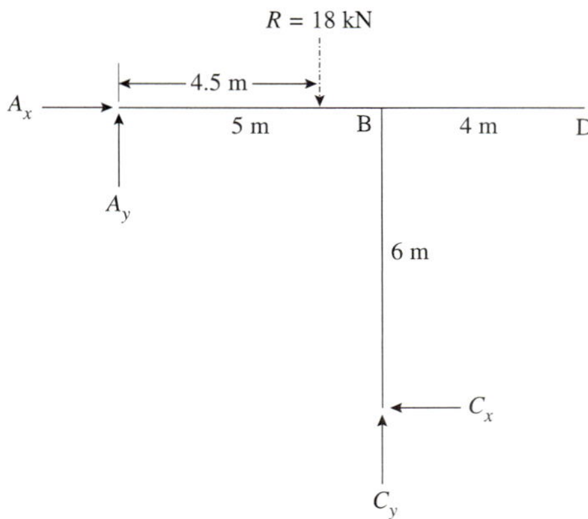

Fig. E2.2.1(d)

(c) A pin-connected frame is shown in Fig. E2.2.1(e). A is a fixed support, B and C are internal hinges, and D is a pin support. This frame can be broken into its three natural components—members AB, BC, and CD.

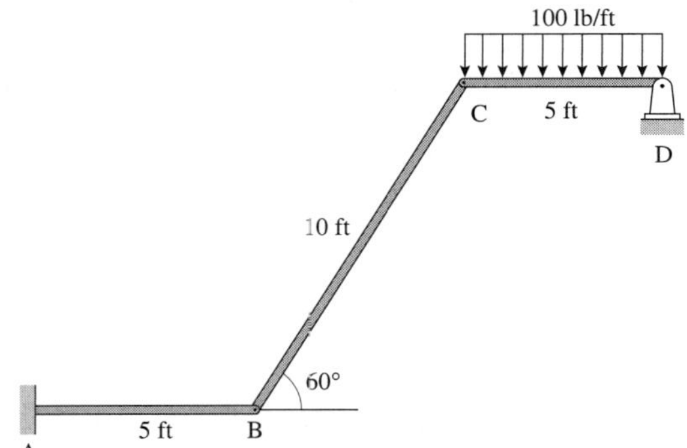

Fig. E2.2.1(e)

The resulting FBDs are shown in Fig. E2.2.1(f). The fixed support at A yields three reactions A_x, A_y, and M_A. The pin support at D yields D_x and D_y. As usual, we assume their directions when drawing the FBDs. An internal hinge when broken apart shows two pin forces that are orthogonal to each other. When showing these two pin forces, we must be careful to draw these forces consistently. First, we draw FBD 1. In this FBD, we assume the directions for B_x and B_y. Next, we draw FBD 2. However, the directions for B_x and B_y must be opposite to the directions shown in FBD 1. This way the pin forces in the two FBDs are equal and opposite. In a similar fashion, we assume the directions of C_x and C_y in FBD 2. Finally, when drawing FBD 3, the directions for C_x and C_y must be opposite to the directions shown in FBD 2.

Here are a few examples of incomplete and/or incorrect FBDs.

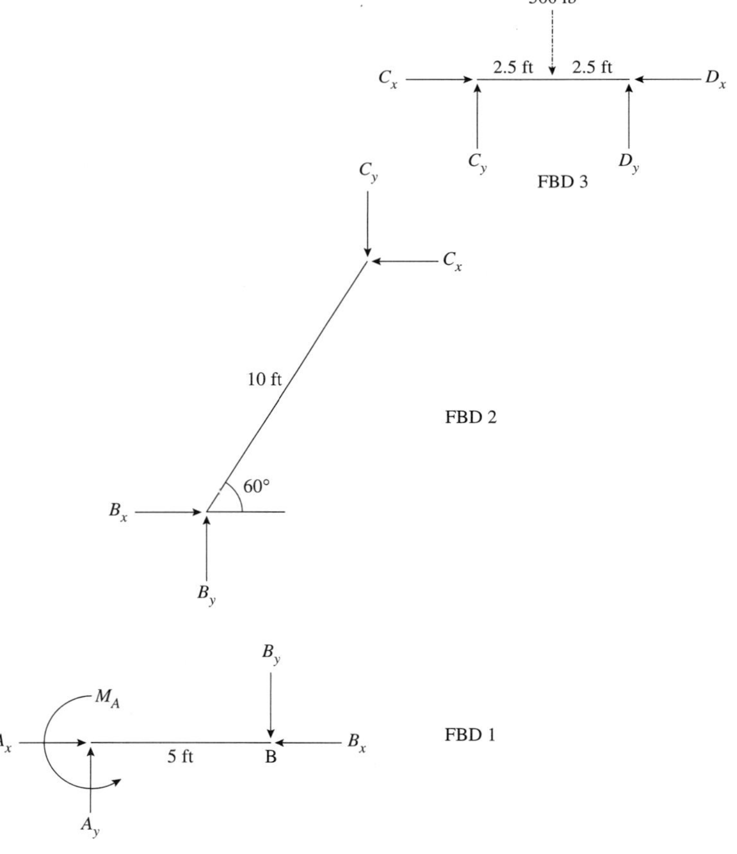

Fig. E2.2.1(f)

(d) A triangular loading acts on a propped cantilever beam (Fig. E2.2.1(g)).

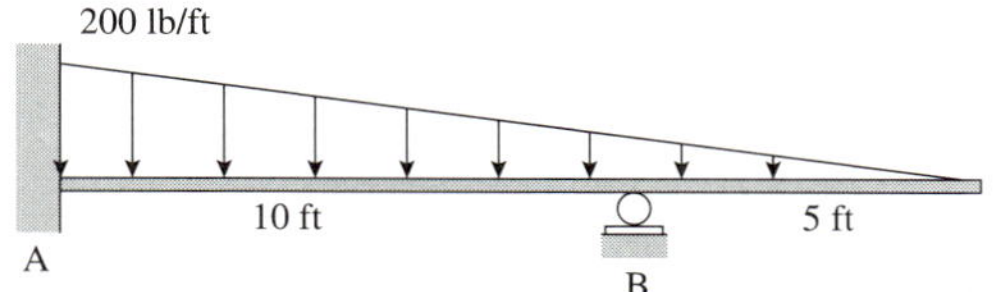

Fig. E2.2.1(g)

A suggested FBD for the beam is shown in Fig. E2.2.1(h).

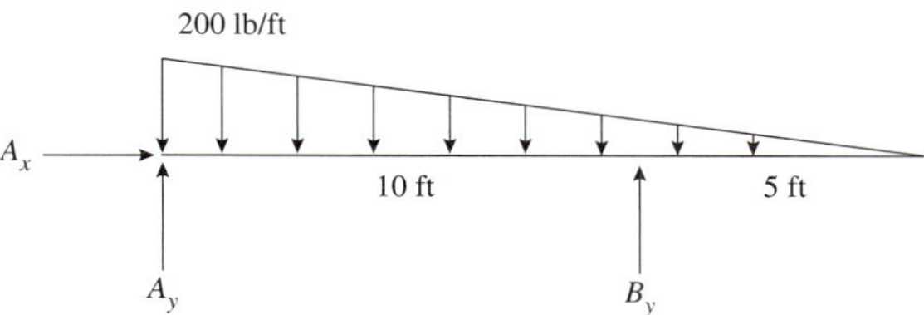

Fig. E2.2.1(h)
Incorrect FBD.

There are two problems with the FBD. First, the fixed support at A has three reactions, but the moment reaction is missing from the FBD. Second, one should show the resultant R of the distributed load. R will be needed to compute the values of the support reactions.

(e) We revisit Example (c) in a slightly different manner. The distributed load on CD is replaced with a vertical concentrated force that appears to act on joint C in Fig. E2.2.1(i).

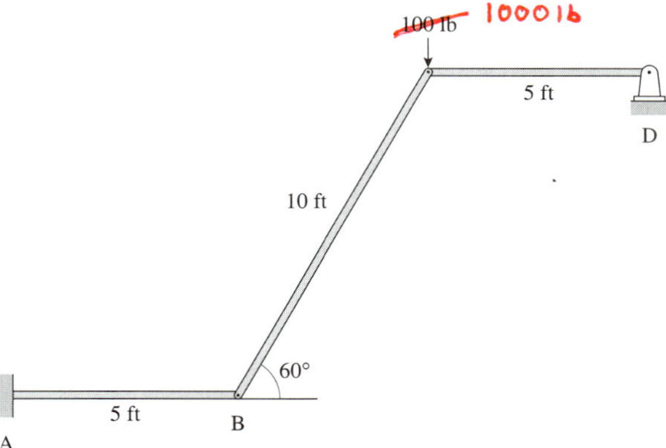

Fig. E2.2.1(i)

The three suggested FBDs are shown in Fig. E2.2.1(j).

There are several problems with the FBDs. First, pin force B_y in FBD 1 and FBD 2 must be equal and opposite. Assuming that we draw FBD 1 first, B_y must act upwards in FBD 2. Second, when the three FBDs are put back together, the net vertical force at C is 2000 lb. However, there is a hidden danger in the analysis if we do not know how and where the 1000 lb force acts. If the force represents a force that is transmitted from another member resting at the (C) end of member DC, then the force should be shown only in FBD 3. On the other hand, if the force acts directly on the internal hinge assembly, the details of the assembly must be known. Otherwise it is not possible to compute the fractional amount that should be shown in FBD 2 and FBD 3. Consider the following scenario: if the force is shown entirely in FBD 2, then from FBD 3, C_y and D_y are zero! The point of this example is that we should understand the details of the physical problem before constructing an idealized structural model.

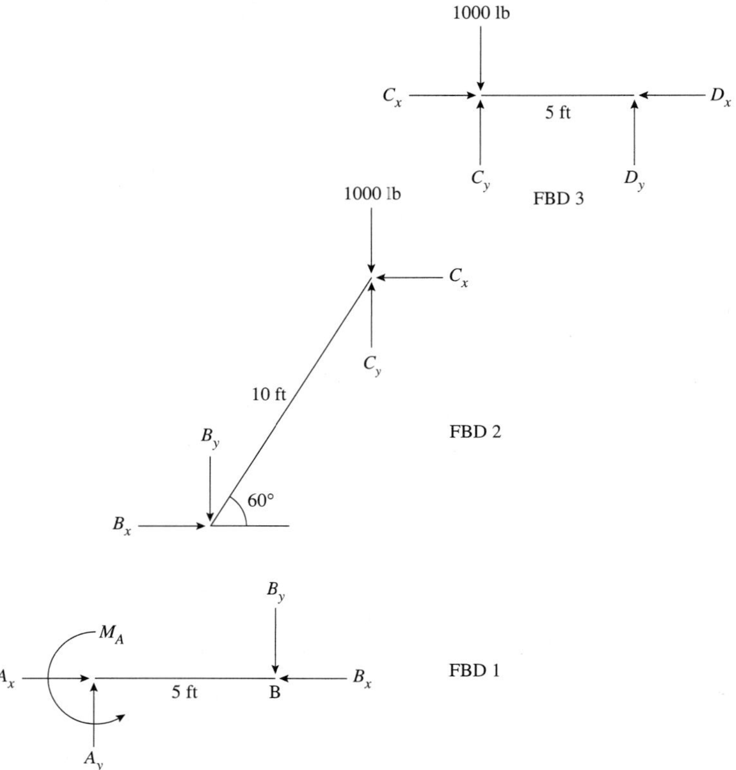

Fig. E2.2.1(j) Incorrect FBDs.

EXERCISES

Appetizers

Consider the beams and frames shown in Figs. P2.2.1–P2.2.3. For each structure draw the complete FBD. When forces are distributed or inclined, show only their resultant or components.

2.2.1

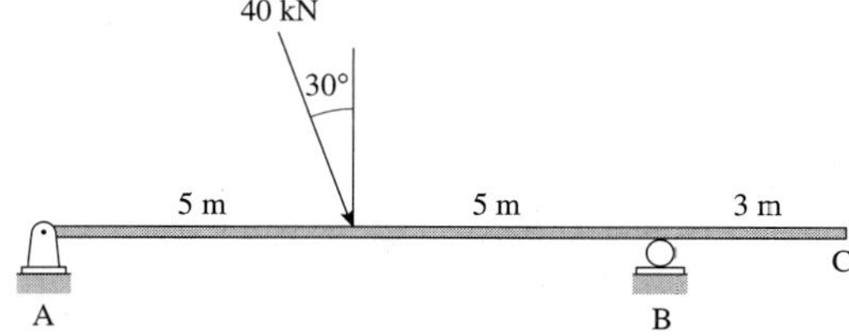

Fig. P2.2.1

2.2.2

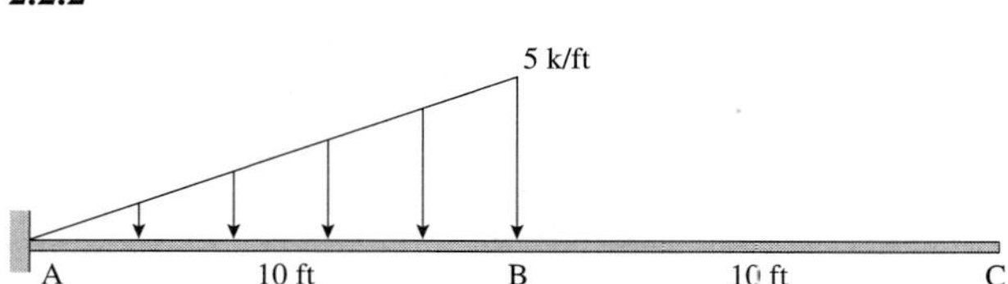

Fig. P2.2.2

2.2.3 A and D are pin supports, and B and C are rigid connections.

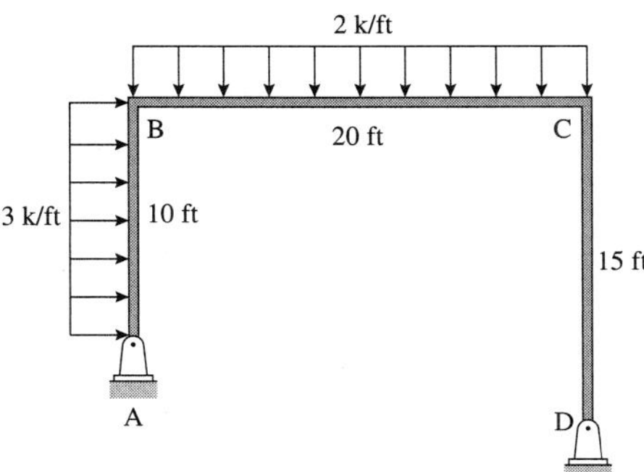

Fig. P2.2.3

Main Course

Consider the beams and frames shown in Figs. P2.2.4–P2.2.6. For each structure draw the complete FBD.

2.2.4 A is a fixed support, B is an internal hinge, and C is a roller support.

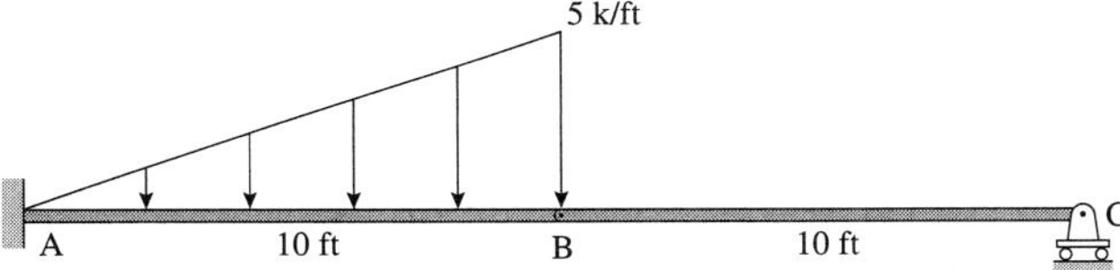

Fig. P2.2.4

2.2.5 A and D are fixed supports and B and C are internal hinges.

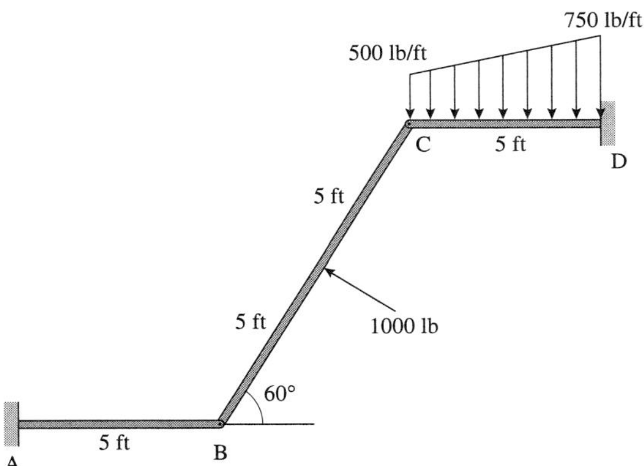

Fig. P2.2.5

2.2.6 A is a pin support, B is a rigid connection, and C is a roller support.

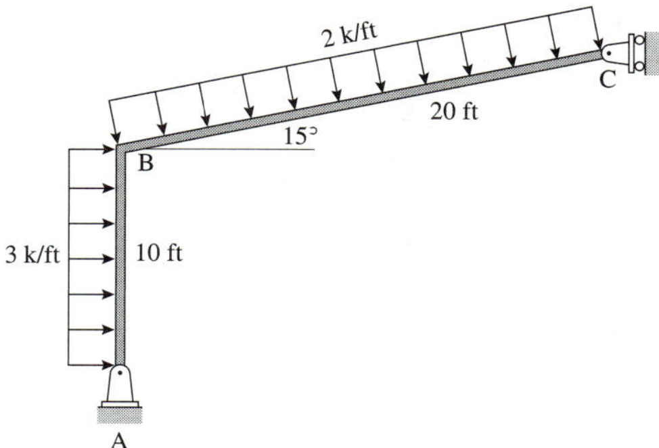

Fig. P2.2.6

Structural Concepts

2.2.7 Figure shows a simply supported beam that is loaded by a stack of bricks one layer deep. The profile of the stack can be approximated as a half sine curve. Given that the dimensions of a typical brick are 2 in by 8 in and that it weighs 2 lb, draw the FBD of the beam.

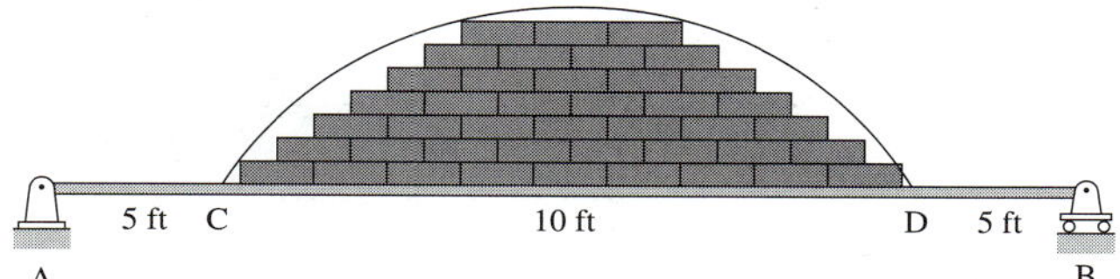

Fig. P2.2.7

2.2.8 Set up a few simple tests to verify the expressions for the resultant force and location of the trapezoidal loading shown in Table 2.2.1.1.

2.3 EQUILIBRIUM

The concept of static equilibrium is central in the analysis of structural systems subjected to static loads. Free-body diagrams are used with the concept of equilibrium to compute the support reactions. A system is in static equilibrium if it is initially at rest and continues to be at rest as it is acted on by external loads. This observation is according to Newton's Law, which states that for every action there is an equal and opposite reaction. Consider the FBD shown in Fig. 2.3.1 of a body that lies in the x-y plane.

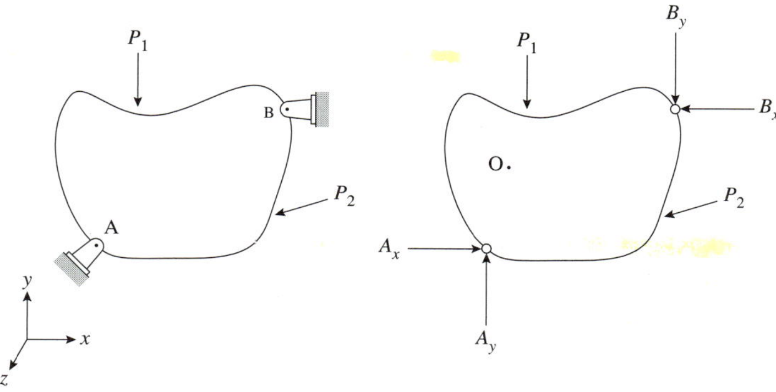

Fig. 2.3.1
Planar body and its FBD.

The structure is in equilibrium if it satisfies the following three equations:

$$\textit{Set 1:} \quad \overset{\rightarrow +}{\sum} F_x = 0 \quad \overset{\uparrow +}{\sum} F_y = 0 \quad \overset{\curvearrowleft +}{\sum} M_{zO} = 0 \qquad (2.3.1)$$

The moment equation is about the z axis and point O is an arbitrary point that need not be a point in the structure. As long as the three equations are linearly independent, they can be used to solve for up to three unknowns. Equations (2.3.1) are not the only equations available; we can use alternate sets of equations, whenever applicable:

$$\textit{Set 2:} \quad \overset{\rightarrow +}{\sum} F_x = 0 \quad \overset{\curvearrowleft +}{\sum} M_{zA} = 0 \quad \overset{\curvearrowleft +}{\sum} M_{zO} = 0 \qquad (2.3.2)$$

In set 2, A and B are distinct points and cannot lie on a line that is perpendicular to the x axis.

$$\textit{Set 3:} \quad \overset{\uparrow +}{\sum} F_y = 0 \quad \overset{\curvearrowleft +}{\sum} M_{zA} = 0 \quad \overset{\curvearrowleft +}{\sum} M_{zB} = 0 \qquad (2.3.3)$$

In set 3, A and B are distinct points and cannot lie on a line that is perpendicular to the y axis.

$$\textit{Set 4:} \quad \overset{\curvearrowleft +}{\sum} M_{zA} = 0 \quad \overset{\curvearrowleft +}{\sum} M_{zB} = 0 \quad \overset{\curvearrowleft +}{\sum} M_{zC} = 0 \qquad (2.3.4)$$

In set 4, A, B, and C are distinct points and cannot lie on a straight line.

Observation: If a structural system is in equilibrium, then any component of the system is also in equilibrium. These include individual joints, individual members, or certain parts (made up of several joints and members) of the structure.

2.4 DETERMINACY

For simple structures, if the number of unknowns (support reactions, pin forces, internal forces) is equal to the number of available equilibrium equations, then the structure is said to be statically determinate. Conversely, if the number of equations is less than the number of unknowns, the structure is statically indeterminate.

The reactions and internal forces in statically determinate structures (that are stable) can be found by using the equations of equilibrium alone. To ascertain whether a structure is statically determinate, identify and count the number of support reactions. For a planar structure, if the number of reactions is equal to three, the structure is (externally) determinate. When a structure has internal hinges, then it is known that the moment is zero at the hinge. Breaking the structure at the hinge yields an additional FBD with three equations at the cost of two additional unknowns (the two pin forces). Hence it is possible for a structure to be externally indeterminate but internally determinate, and hence a determinate structure (see the third problem in Example 2.4.1). The known condition at a hinge is recognized generally as an *equation of condition.*

To summarize, let m be the total number of unknowns and n the number of independent (planar) FBDs. Then,

if $m < 3n$, the structure is unstable,

if $m = 3n$, the structure is determinate if stable, and

if $m > 3n$, the structure is indeterminate if stable.

We address determinacy and stability issues with truss and frame structures later in this chapter.

EXAMPLE 2.4.1 ***[using Examples 2.2.1(a)–(c)]***

Example 2.2.1(a): The structural FBD shows three unknown forces: A_x, A_y, and B_y. Noting that each planar FBD yields three equations, we have as many equations as unknowns. Hence the beam is statically determinate.

Example 2.2.1(b): The structural FBD of the frame shows four unknown forces: A_x, A_y and C_x, C_y. With a single FBD, we have three equations. Hence the frame is statically indeterminate to the first degree (four unknowns—three equations).

Example 2.2.1(c): The pin-connected frame has five reactions that would show up in the structural FBD. We could call the frame *externally* indeterminate. However, we can generate three independent FBDs by breaking the frame into its three members. The three FBDs have a total of nine unknown forces (five reactions and four internal pin forces). The FBDs can generate nine equilibrium equations. Since the number of unknowns is equal to the number of equations, the structure is *internally* determinate and hence statically determinate.

Note that if we break the frame in Example 2.2.1(b) into its three components (by making cuts to the left, right, and below joint B), we will generate a total of four FBDs (FBDs of members AB, BD, and BC, and one of joint B) involving 13 unknowns (four support reactions and nine internal forces). The frame is (still) statically indeterminate to the first degree (13 unknowns—12 equations). Breaking a planar structure into two parts by making a cut through a member yields an additional FBD (hence three equations) but is offset by the introduction of three additional unknowns (the three internal forces at the cut).

2.5 SIMPLE DETERMINATE STRUCTURES

In this section, we use the concepts discussed in the earlier sections to compute support reactions and pin forces in simple structures.

General Procedure

Step 1: Identify the unknowns. Determine the number of independent FBDs. Are there enough FBDs to solve for all the unknowns? Note that each planar FBD yields three equations.

Step 2: For each FBD, remove the supports and replace them with their support reactions. Note that we guess the direction for each support reaction. Similarly, replace the internal hinges (or pin connections) with the pin forces.

Step 3: Replace distributed loads with their resultant. If necessary, resolve inclined loads and supports into their components.

Step 4: If there are multiple FBDs, identify the one with at most three unknowns. Can you identify which equilibrium equation will yield an equation involving one unknown (in most problems, a moment equilibrium equation can be used)? If so, use the equation to solve for the unknown. Use this idea to solve for the other unknowns. If it is not possible to obtain an equilibrium equation with a single unknown, you will have to solve the equilibrium equations simultaneously.

Step 5: Repeat step 4 for all the FBDs. If the structural FBD has not been used in the earlier steps, it can be used to check the results obtained. Note that if an unknown force appears in two FBDs, it must be shown equal and opposite in those FBDs.

Step 6: Summarize the results. If the values of the reactions and pin forces are positive, the assumed directions are correct; otherwise the directions are opposite to those assumed. If necessary, draw the corrected FBD.

EXAMPLE 2.5.1

Simply Supported Beam

Figure E2.5.1(a) shows a simply supported beam. Determine the support reactions.

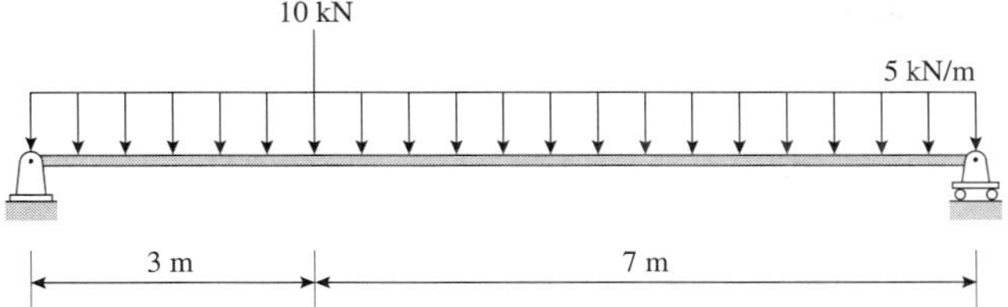

Fig. E2.5.1(a) Simply supported beam.

SOLUTION

Step 1: There are two reactions at the pin-support at A and one reaction at the roller-support at B (for a total of three reactions). The beam is statically determinate and the structural FBD can be used to compute these support reactions, A_x, A_y, and B_y.

Step 2: The resultant of the distributed load R is given as $R = 5$ kN/m × 10 m = 50 kN and is located at the center of the distributed loading (5 m from either A or B).

Step 3: The structural FBD is shown in Fig. E2.5.1(b).

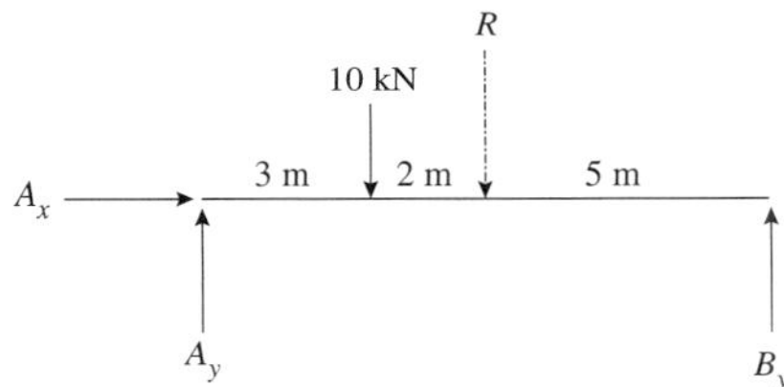

Fig. E2.5.1(b)

Step 4: Use the equations of equilibrium to determine the three reactions.

$$\overset{\curvearrowleft +}{\sum} M_A = 0 = B_y(10) - (10)(3) - R(5) \Rightarrow B_y = \frac{30 + 250}{10} = 28 \text{ kN}$$

$$\overset{\uparrow +}{\sum} F_y = 0 = A_y - 10 - R + B_y \Rightarrow A_y = 10 + 50 - 28 = 32 \text{ kN}$$

$$\overset{\rightarrow +}{\sum} F_x = 0 = A_x \Rightarrow A_x = 0 \text{ kN}$$

Note that we could have solved the problem in a variety of different ways: $\overset{\curvearrowleft +}{\sum} M_B = 0$, $\overset{\uparrow +}{\sum} F_y = 0$ followed by $\overset{\rightarrow +}{\sum} F_x = 0$, or $\overset{\curvearrowleft +}{\sum} M_B = 0$, $\overset{\curvearrowleft +}{\sum} M_A = 0$ followed by $\overset{\rightarrow +}{\sum} F_x = 0$.

Step 5: Check. Total load on beam = R + 10 = 60 kN
Total vertical reactions = $A_y + B_y$ = 28 + 32 = 60 kN

Answers: $A_x = 0$, $A_y = 32$ kN(↑), $B_y = 28$ kN(↑).

EXAMPLE 2.5.2

Simply Supported Beam with Inclined Roller Support

A simply supported beam with an inclined roller support at one end is shown in Fig. E2.5.2(a). Compute all the support reactions.

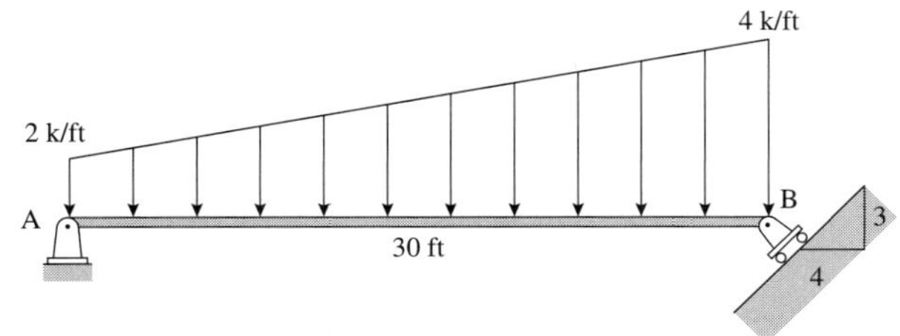

Fig. E2.5.2(a)

SOLUTION

Step 1: There are two reactions at the pin-support at A and one reaction at the inclined roller support at B (for a total of three reactions). The beam is statically determinate and the structural FBD can be used to compute these support reactions, A_x, A_y, and R_B.

Step 2: The resultant of the trapezoidal distributed load can be obtained by first breaking the load into a uniformly distributed load with its resultant R_1 and a triangular loading with its resultant R_2.

R_1: R_1 is given as $R_1 = 2$ k/ft × 30 ft = 60 k and is located 15 ft from either A or B.

R_2: R_2 is given as $R_2 = 1/2 \times 2$ k/ft × 30 ft = 30 k and is located 20 ft from A or 10 ft from B.

(Alternately, we could also have used the equations from Table 2.2.2.1 for the trapezoidal loading.) The geometry data involving the skew support includes $\theta = \tan^{-1}(3/4) = 36.87°$.

Step 3: The structural FBD is shown in Fig. E2.5.2(b).

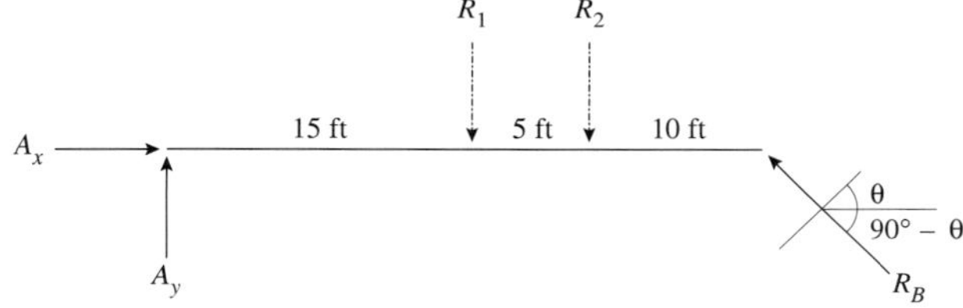

Fig. E2.5.2(b)

Step 4: Use the equations of equilibrium to determine the three reactions.

$$\overset{\curvearrowleft +}{\sum} M_A = 0 = (R_B \sin(90-\theta))(30) - R_1(15) - R_2(20) \Rightarrow R_B = \frac{900+600}{24} = 62.5 \text{ k}$$

$$\overset{\uparrow +}{\sum} F_y = 0 = A_y - R_1 - R_2 + R_B \sin(90-\theta) \Rightarrow A_y = 60 + 30 - 50 = 40 \text{ k}$$

$$\overset{\rightarrow +}{\sum} F_x = 0 = A_x - R_B \cos(90-\theta) \Rightarrow A_x = 37.5 \text{ k}$$

Answers: $A_x = 37.5$ k(→), $A_y = 40$ k(↑), $R_B = 62.5$ k.

EXAMPLE 2.5.3

Frame with Moment-Release Hinge

A planar frame is shown in Fig. E2.5.3(a) with a moment-release hinge at B. Compute all the support reactions and pin forces.

SOLUTION

Step 1: There are two reactions each at the pin-supports A and D, for a total of four reactions. The structural FBD cannot be used to compute these support reactions, A_x, A_y and D_x, D_y (four unknowns vs. three equations of equilibrium). We must break the structure at the internal hinge at B since it is known that the moment at the hinge is zero.

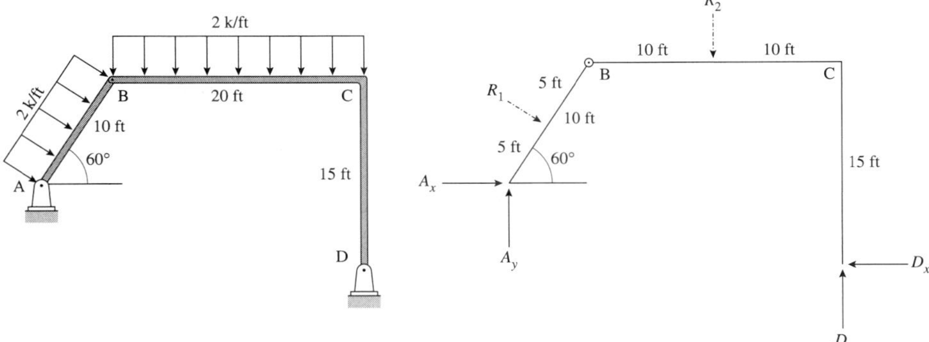

Fig. E2.5.3(a)
Frame with moment-release hinge and its structural FBD.

Step 2: The resultants of the distributed loads are as follows:

R_1: $R_1 = 2$ k/ft × 10 ft = 20 k and is located 5 ft from either A or B.

R_2: $R_2 = 2$ k/ft × 20 ft = 40 k and is located 10 ft from either B or C.

The rest of the geometry data are $d_1 = 10\cos(60^\circ) = 5$ ft and $d_2 = 10\sin(60^\circ) = 8.66$ ft.

Step 3: The relevant FBDs are shown in Fig. E2.5.3(b).

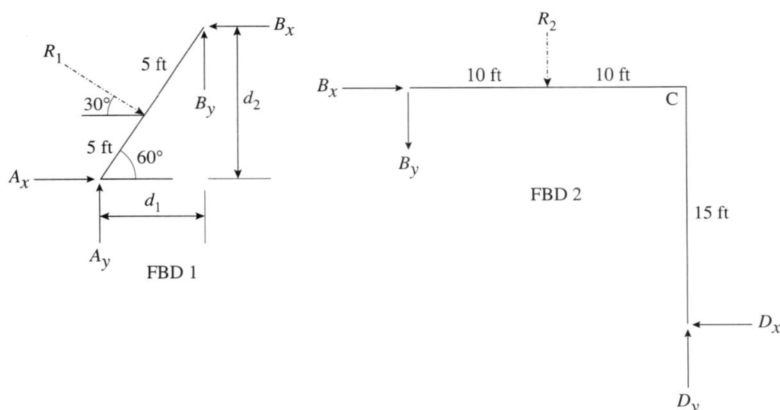

Fig. E2.5.3(b)

Note that the three FBDs are not independent of each other since the structural FBD can be constructed using (the superposition of) FBD 1 and FBD 2.

Step 4: With these FBDs, there are a total of six unknowns. FBD 1 and FBD 2 yield a total of six equilibrium equations. We cannot, however, use either of the FBDs to generate an equation involving a single unknown. One way of solving the problem is to generate an equation involving B_x, B_y from FBD 1 and another equation from FBD 2. We can then solve these equations simultaneously to find B_x, B_y.

$$\text{FBD 1:} \quad \overset{\curvearrowleft+}{\sum} M_A = 0 = -R_1(5) + B_y(d_1) + B_x(d_2) \Rightarrow 8.66B_x + 5B_y = 100 \tag{1}$$

$$\text{FBD 2:} \quad \overset{\curvearrowleft+}{\sum} M_D = 0 = R_2(10) + B_y(20) - B_x(15) \Rightarrow 15B_x - 20B_y = 400 \tag{2}$$

Multiplying (1) by 4 and adding it to (2) yields $49.64B_x = 800 \Rightarrow B_x = 16.1$k. Substituting the result in (1) or (2) yields $B_y = -7.9$k.

Now we can use FBD 1 followed by FBD 2 to find the four support reactions:

FBD 1: $\overset{\uparrow +}{\sum} F_y = 0 = A_y - R_1 \sin 30^\circ + B_y \Rightarrow A_y = 20(0.5) + 7.9 = 17.9 \text{ k}$

FBD 1: $\overset{\rightarrow +}{\sum} F_x = 0 = -A_x + R_1 \cos 30^\circ - B_x \Rightarrow A_x = 20(0.866) - 16.1 = 1.2 \text{ k}$

FBD 2: $\overset{\uparrow +}{\sum} F_y = 0 = -B_y - R_2 + D_y \Rightarrow D_y = -7.9 + 40 = 32.1 \text{ k}$

FBD 2: $\overset{\rightarrow +}{\sum} F_x = 0 = B_x - D_x \Rightarrow D_x = 16.1 \text{ k}$

Note that even though we assumed the direction for B_y incorrectly, we do not correct the original FBDs midway through the solution. We use the FBDs to generate the equilibrium equations and substitute –7.9 for B_y in all the equations. If B_y was a support reaction, we will note the corrected direction when the results are summarized.

Step 5: Resultant pin force at B: $R_B = \sqrt{B_x^2 + B_y^2} = \sqrt{(16.1)^2 + (-7.9)^2} = 17.9 \text{ k}$.

Check: Using the structural FBD, we find

$$\overset{\rightarrow +}{\sum} F_x = -A_x + R_1 \cos 30^\circ - D_x = -1.2 + 20(0.866) - 16.1 \approx 0 \qquad \text{OK.}$$

$$\overset{\uparrow +}{\sum} F_y = A_y - R_1 \sin 30^\circ - R_2 + D_y = 17.9 - 20(0.5) - 40 + 32.1 \approx 0 \qquad \text{OK.}$$

$$\overset{\curvearrowleft +}{\sum} M_A = -R_1(5) - R_2(d_1 + 10) + D_y(d_1 + 20)$$

$$- D_x(15 - d_2) = -100 - 40(15) + 32.1(25) - 16.1(6.34) \approx 0 \qquad \text{OK.}$$

Answers: $A_x = 1.2 \text{ k}(\leftarrow)$, $A_y = 17.9 \text{ k}(\uparrow)$, $D_x = 16.1 \text{ k}(\leftarrow)$, $D_y = 32.1 \text{ k}(\uparrow)$, $R_B = 17.9 \text{ k}$.

EXAMPLE 2.5.4

Beam with Internal Hinge and Internal Roller Support

Figure E2.5.4(a) shows a beam. Assume that the support at A is fixed, B is an internal hinge, and C is a roller support. Compute all the support reactions.

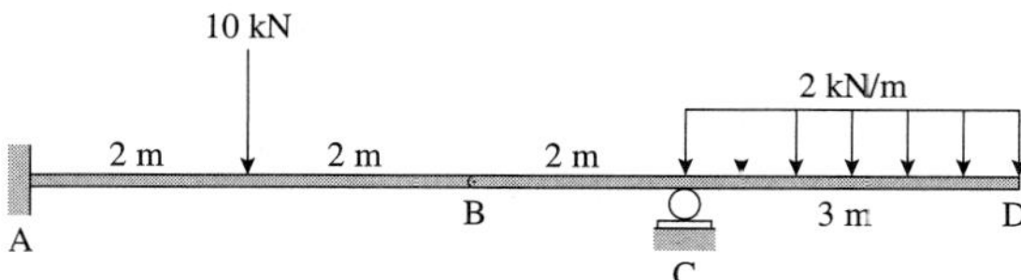

Fig. E2.5.4(a)

SOLUTION

Step 1: There are three reactions at the fixed support at A and one reaction at the roller support at C (for a total of four reactions). The structural FBD cannot be used to compute these support reactions, A_x, A_y, M_A, andC_y (four unknowns vs. three equations of equilibrium). As we showed in the previous example, we must break the structure at the internal hinge at B.

Step 2: The resultant of the distributed load R = 2 kN/m × 3m = 6 kN and is located 1.5 m from either C or D.

Step 3: The relevant FBDs are shown in Fig. E2.5.4(b).

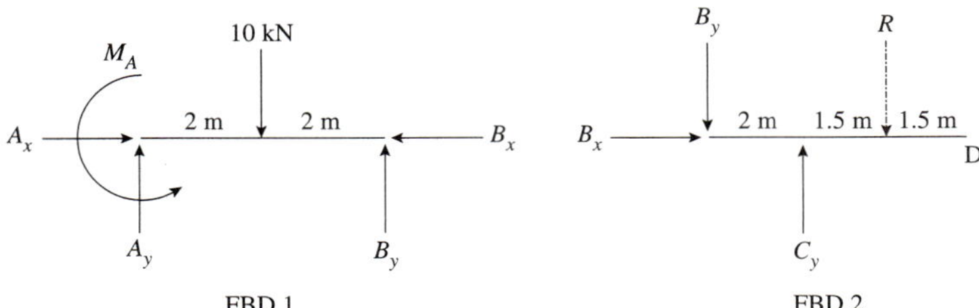

Fig. E2.5.4(b)

Step 4: With these two FBDs, there are a total of six unknowns. FBD 1 and FBD 2 yield a total of six equilibrium equations. An examination of FBD 1 shows too many unknowns for the use of any of the equilibrium equations. However, FBD 2 has only three unknowns. Using the moment equation about B yields an equation with one unknown, C_y. The other two unknowns can be found using the other equilibrium equations.

$$\text{FBD 2:}\quad \overset{\curvearrowleft +}{\sum} M_B = 0 = -R(3.5) + C_y(2) \Rightarrow C_y = 10.5\ \text{kN}$$

$$\text{FBD 2:}\quad \overset{\uparrow +}{\sum} F_y = 0 = -B_y + C_y - R \Rightarrow B_y = 4.5\ \text{kN}$$

$$\text{FBD 2:}\quad \overset{\rightarrow +}{\sum} F_x = 0 = B_x \Rightarrow B_x = 0$$

At this stage, FBD 1 has three unknowns, and solving for these we have

$$\text{FBD 1:}\quad \overset{\uparrow +}{\sum} F_y = 0 = A_y - 10 + B_y \Rightarrow A_y = 5.5\ \text{kN}$$

$$\text{FBD 1:}\quad \overset{\rightarrow +}{\sum} F_x = 0 = A_x - B_x \Rightarrow A_x = 0$$

$$\text{FBD 1:}\quad \overset{\curvearrowleft +}{\sum} M_A = 0 = M_A - (10)(2) + B_y(4) \Rightarrow M_A = 2\ \text{kN-m}$$

Check: Using the structural FBD, we find

$$\overset{\uparrow +}{\sum} F_y = A_y - 10 - R + C_y = 5.5 - 10 - 6 + 10.5 = 0 \qquad \text{OK.}$$

$$\curvearrowleft \overset{+}{\sum} M_A = M_A - (10)(2) + C_y(6) - R(7.5) = 2 - 20 + 63 - 45 = 0 \qquad \text{OK.}$$

Answers: $A_x = 0,\ A_y = 5.5\ \text{kN}(\uparrow),\ M_A = 2\ \text{kN} - \text{m}(\curvearrowright),\ C_y = 10.5\ \text{kN}(\uparrow).$

EXERCISES

Appetizers

Consider the beams and frames shown in Figs. P2.5.1–P2.5.4. Show whether the structure is determinate or indeterminate. If determinate, compute the pin forces, if any, and the support reactions.

2.5.1.

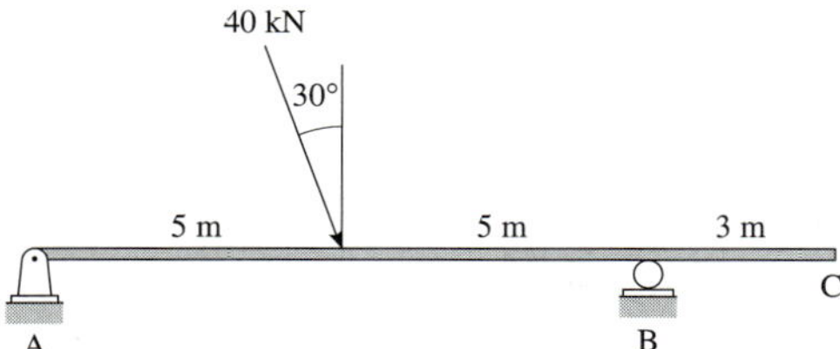

Fig. P2.5.1

2.5.2.

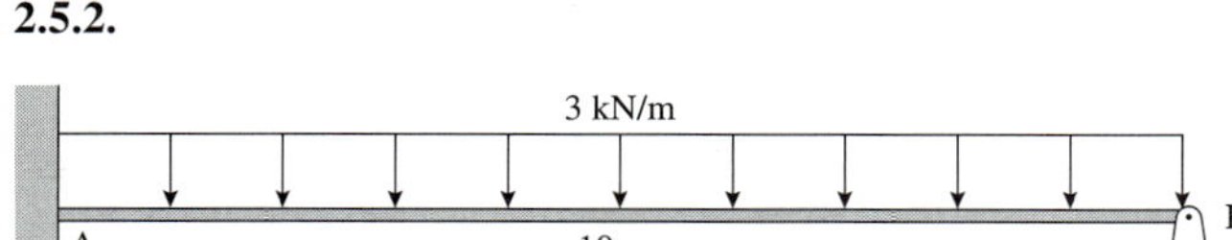

Fig. P2.5.2

2.5.3. A is a fixed support and the concentrated moment is applied at C.

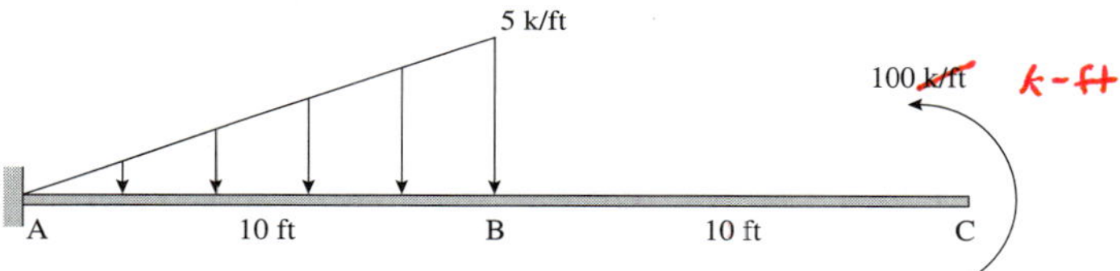

Fig. P2.5.3

2.5.4. The continuous piece ABC is supported by a pin support at B and a roller support at C.

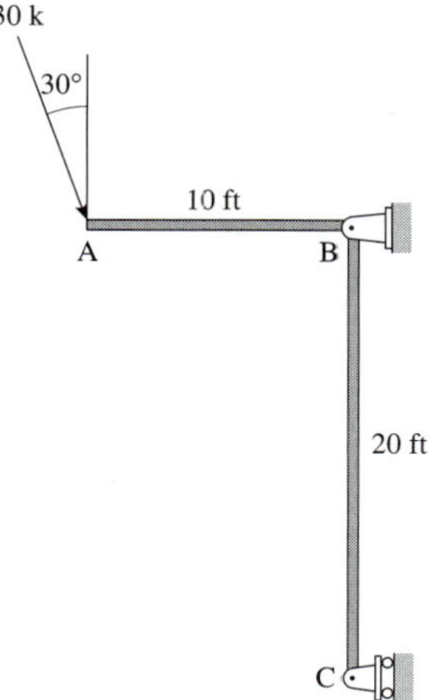

Fig. P2.5.4

Main Course

Consider the beams and frames shown in Figs. P2.5.5–P2.5.9. Show whether the structure is determinate or indeterminate. If determinate, compute the pin forces, if any, and the support reactions.

2.5.5. A is a fixed support, B is an internal hinge, and C is a roller support.

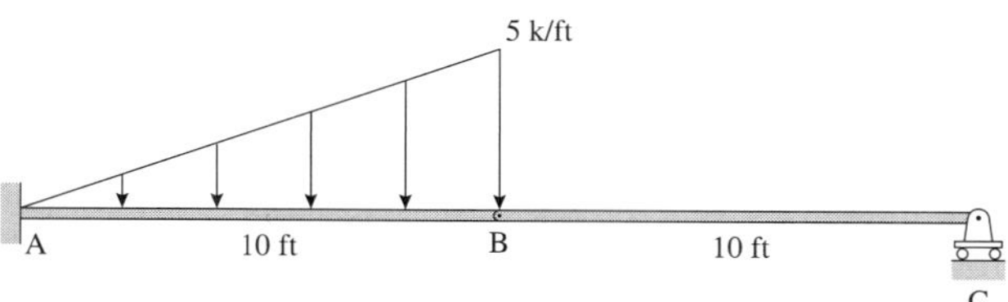

Fig. P2.5.5

2.5.6.

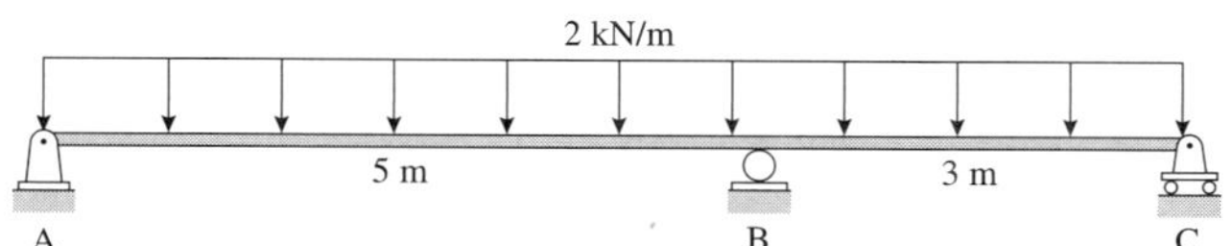

Fig. P2.5.6

2.5.7.

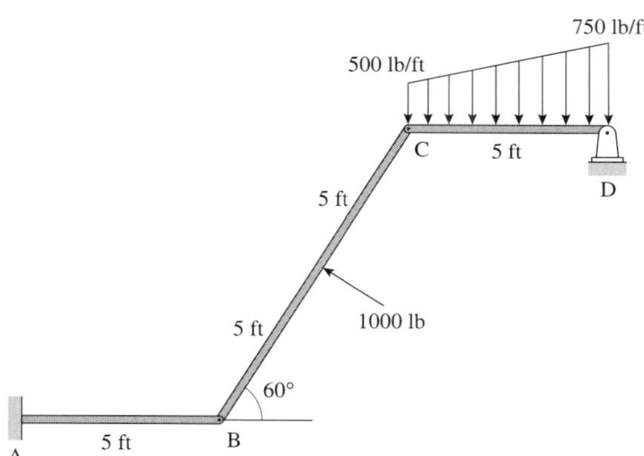

Fig. P2.5.7

2.5.8.

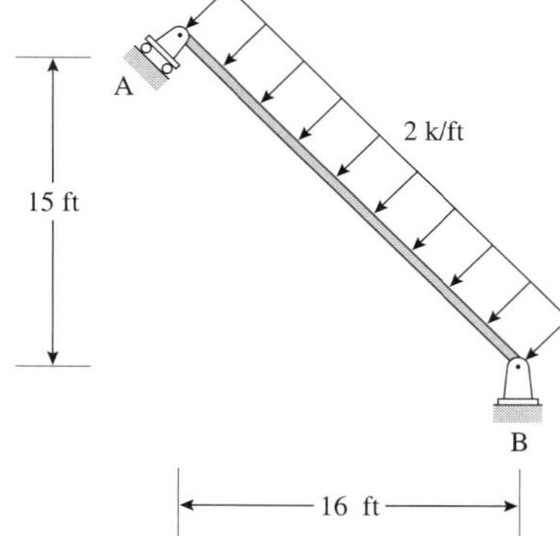

Fig. P2.5.8

2.5.9.

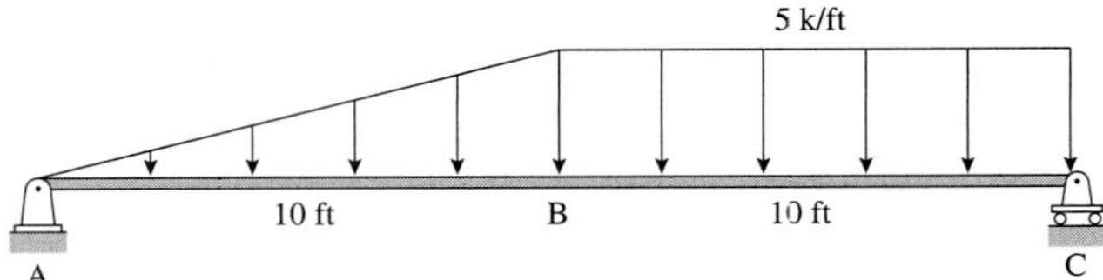

Fig. P2.5.9

Structural Concepts

2.5.10. Figure P2.5.10 shows a planar frame. D is a rigid connection and B is a pin connection (internal hinge). Is the problem data complete in order to calculate the pin forces at B and the support reactions? How would you classify the behavior of member BC?

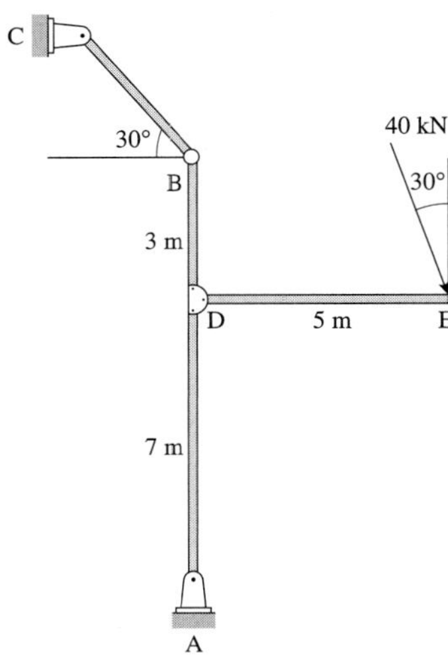

Fig. P2.5.10

2.5.11. Arches may be constructed with no hinges or with two or more hinges. Figure P2.5.11 below shows a three-hinged arch. Compute the pin forces and support reactions.

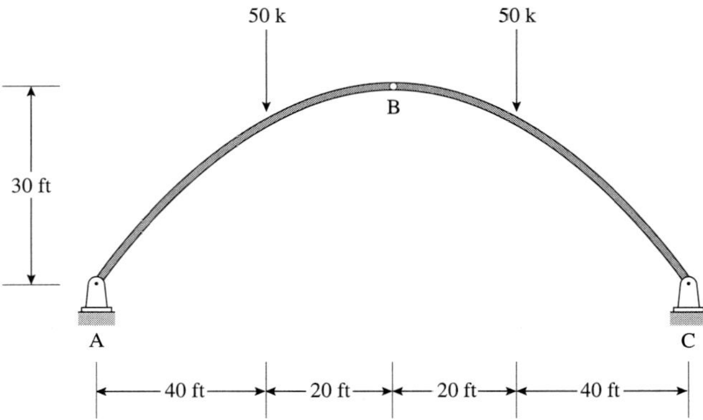

Fig. P2.5.11

2.5.12. A fluid at rest exerts hydrostatic pressure that is equal in all directions at a point (Pascal's Law). Consider the structure in Fig. P2.5.12, which is used to hold back water (density 62.5 lb/ft^3) as well as provide an access road on the top. The structure is modeled as a frame. Compute the support reactions.

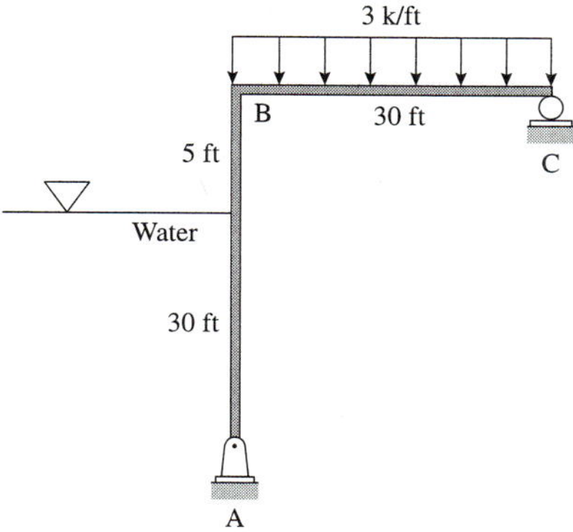

Fig. P2.5.12

2.6 STABILITY

When we talk about a structure being determinate or indeterminate, we assume that the structural system is stable. In the context of linear, small displacement analysis, the (geometric) stability of a system is an inherent property of the structure and not of the loads acting on the structure. In this section, a few examples of unstable structural systems are shown and discussed. In each case, we will see that the system fails to satisfy one or more of the equations of equilibrium. These structures will exhibit rigid body motions when loads are applied to them.

Consider the beam shown in Fig. 2.6.1.

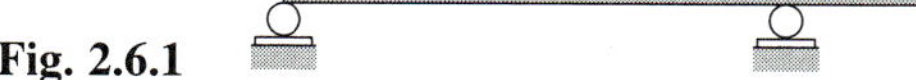

Fig. 2.6.1

The beam has three support reactions. However, there is no reaction (or restraint) in the horizontal direction and $\sum F_x = 0$ is not satisfied. Consider the two-member planar frame shown in Fig. 2.6.2. While the frame has three reactions, it does not satisfy the moment equilibrium equation, $\sum M_A = 0$. This is because all the three support reactions pass through a single point, A. Point A is known as the *center of instantaneous motion.*

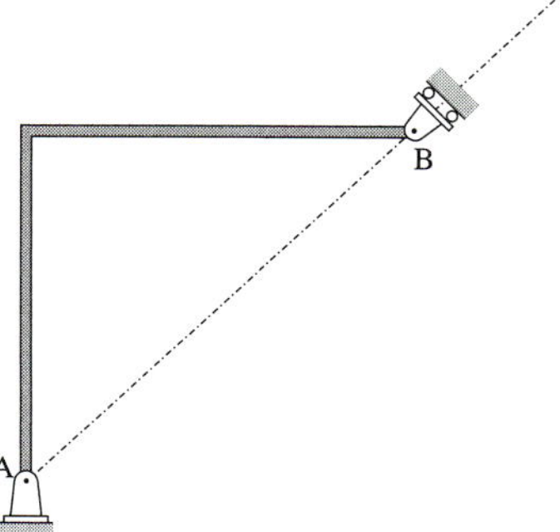

Fig. 2.6.2

Note that a planar structural system has three rigid-body modes—two orthogonal displacements and rotation about an axis normal to the plane of the system. In a similar manner, we can conclude that a space system has six rigid-body modes—three orthogonal displacements and rotations about the three orthogonal axes. Unless these rigid-body modes are adequately suppressed, the structure is unstable. A more detailed examination of these ideas takes place in the next few sections.

2.7 PLANAR TRUSS ANALYSIS

Truss structures abound all around us. Common examples include roof and bridge trusses. It is interesting to note that the first use of trusses by Romans involved wooden bridges and roofs. Activities in different European countries such as Italy (by the architect Palladio), Switzerland (by the carpenter Grubenmann), and others popularized their usage in the 1700s. In the United States, the advent of railroads spurred a systematic analysis and design of bridge trusses. The challenge was to span long distances. As the need for efficiency became more important, different truss configurations were invented and patented. Some of these systems are shown in Figs. 2.7.1 (a)–(c).

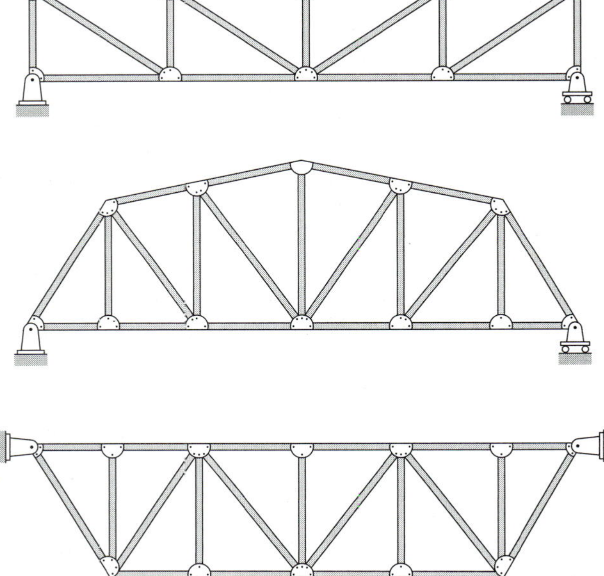

Fig. 2.7.1 (a) Flat Pratt. (b) Camelback. (c) Inverted Warren. *(Continued)*

Bridges are constructed today with a variety of materials and in several different forms. Construction of wooden bridges stopped about the end of nineteenth century. They were replaced by steel bridges, which can economically span tens to hundreds of feet. More and more bridges designed and constructed today are made of concrete, or prestressed concrete, or are cable-stayed bridges.

The ability to span long distances makes it possible to use trusses also as roof systems for residential homes, industrial buildings, sports arenas, etc. Some of the common forms are shown in Figs. 2.7.1 (d)–(f).

Only the very simplest of trusses exist as outlined in the figures. They interact with a variety of structural components. In a roof truss, the roof loads are transferred to the truss through purlins that span multiple trusses, or by a combination of rafters that are oriented parallel to trusses, and purlins. Some of these issues are dealt with in later chapters.

We now lay the groundwork for the rest of the chapter. A truss is a structural system that satisfies the following requirements.

a. The members are straight, slender, and prismatic. The cross-sectional dimensions are small in comparison to the member lengths. The weights of the members are small com-

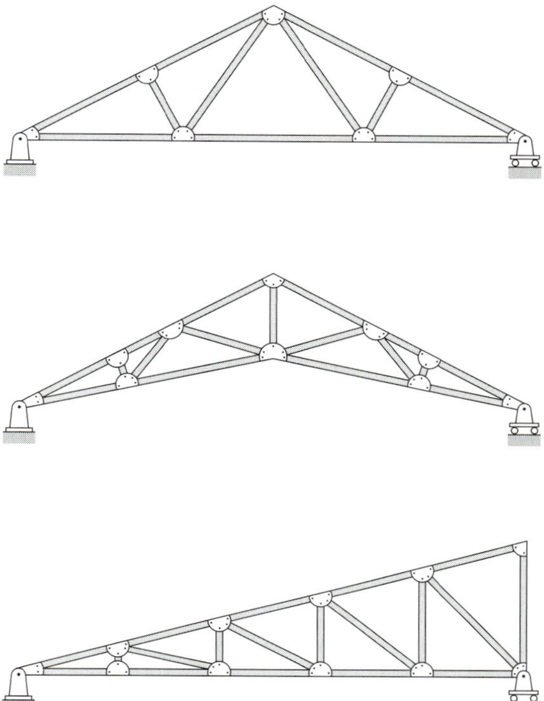

Fig. 2.7.1
(Continued) (d) Fink. (e) Scissors. (f) Sawtooth.

pared to the applied loads and can be neglected. Also, when constructing the truss model for analysis, we treat the members as a one-dimensional entity (having length and negligible cross-sectional dimensions).

b. The joints are assumed to be frictionless pins (or internal hinges).
c. The loads are applied only at the joints in the form of concentrated forces.

As a consequence of these assumptions, the members are two-force members, meaning that they carry only axial forces. (Shear forces, bending moments, and torsional moments do not exist in truss members.)

> **Observation:** In reality, connections or joints are not frictionless pins. Note also that certain types of loads, such as snow or rain loads, are truly distributed loads. The approximation of certain structural systems or components as trusses must be done with care.

The purpose of analyzing a truss is to determine the magnitude and nature of the axial force in the truss members. Members that are in axial compression behave differently from those in axial tension (see Fig. 2.7.2). Compressive members are prone to buckle (e.g., Euler buckling) and must be designed to resist buckling.

Trusses are determinate if the number of unknowns (member forces and support reactions) is equal to the number of equilibrium equations. Let m be the number of members, r the number of support reactions, and j the number of joints (including supports). The total number of unknowns is equal to $m + r$. The number of available equations is equal to $2j$ since every joint provides two equilibrium equations (equilibrium in the x and y directions).

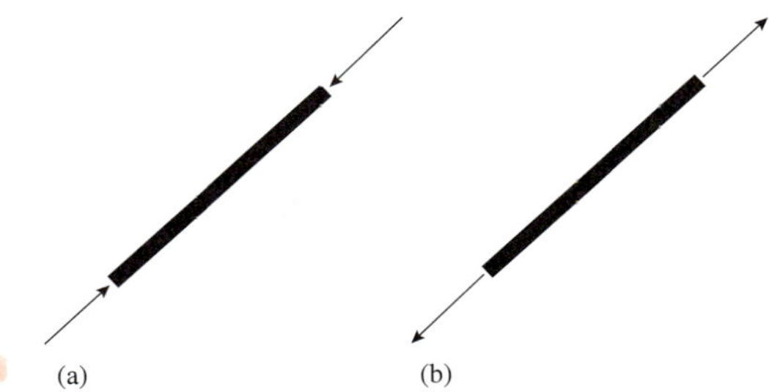

Fig. 2.7.2 Truss member in (a) compression, (b) tension.

If $(m + r) < 2j$, the truss is unstable. We examine this issue later.

If $(m + r) = 2j$, the truss is determinate, since the number of equations is equal to the number of unknowns.

If $(m + r) > 2j$, the truss is indeterminate. The degree of indeterminacy is equal to $(m + r - 2j)$.

2.7.1 Method of Joints

The method of joints is a useful technique if the force in every truss member is to be computed. The basic idea is to use the FBD of a truss joint and use the two equations of force equilibrium to solve for the unknowns. Consider a typical joint in a planar truss, as in Fig. 2.7.1.1.

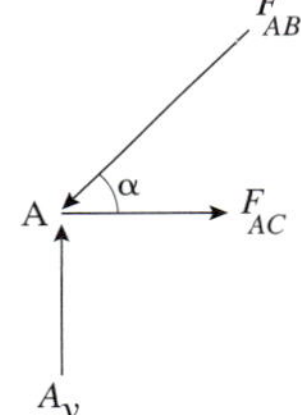

Fig. 2.7.1.1 FBD of a typical joint A.

Let A_y represent the support reaction at A and F_{AB} and F_{AC} the forces in members AB and AC, respectively. This joint is in static equilibrium if

$$\overset{\rightarrow +}{\sum} F_x = 0 = -F_{AB}\cos\alpha + F_{AC} \qquad \overset{\uparrow +}{\sum} F_y = 0 = A_y - F_{AB}\sin\alpha$$

The (third) moment equation does not exist since the FBD consists of forces that all pass through the same point, the joint. If A_y is known, F_{AB} and F_{AC} can be computed by using these two equations. Note that in Fig. 2.7.1.1, F_{AB} represents a member in compression and F_{AC} a member in tension.

2.7.2 General Procedure

Step 1: Use the structural FBD to solve for as many support reactions as possible. While this step is not always necessary, it does provide for checks at the end of the problem.

Step 2: Now start at a support/joint, where there are at most two unknowns. Draw the FBD of the joint. The FBD can be used with either $\overset{\rightarrow +}{\sum} F_x = 0$ or $\overset{\uparrow +}{\sum} F_y = 0$ so as to generate an equation with a single unknown. This is almost always possible if x and y are simply any convenient orthogonal directions. Now use the other equation to solve for the next unknown.

Tip: There are two approaches here in assuming the direction of the internal member force. In the first approach we reason out in sufficient detail so that we know the correct direction. This approach requires that we monitor the values of the member forces as we move to the next FBD containing that member. In the second approach, we assume that all the members are in tension (member force acts away from the joint). If the answer is positive, the member is in tension. Otherwise, a negative value indicates that the member is in compression (in the subsequent FBDs involving the member, use the negative values when writing the equilibrium equation; do not correct the FBDs).

Step 3: Repeat Step 2 for all the joints in the truss until all member forces are found. If the structural FBD is used to compute the support reactions (in step 1), use the extra equations as checks.

Step 4: Summarize the results showing the magnitude and nature (tension or compression) of the member force.

EXAMPLE 2.7.1

Cantilevered Truss

Figure E2.7.1(a) shows a planar truss. Compute the axial force in all the members.

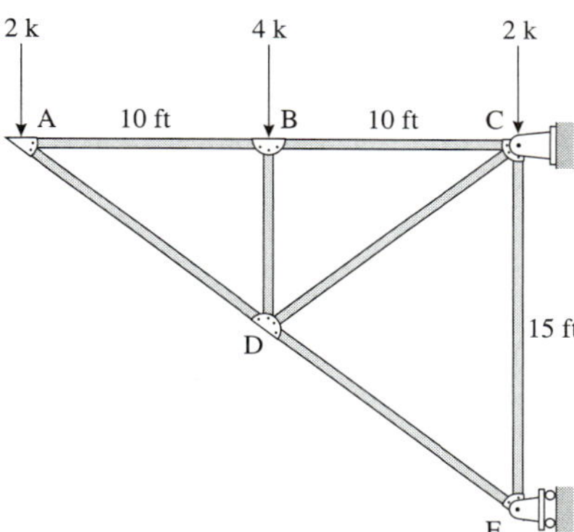

Fig. E2.7.1(a)

SOLUTION

Step 1: The truss is statically determinate ($j = 5$, $m = 7$, $r = 3$) since $2j = m + r$. Using the structural FBD (Fig. E2.7.1(b); $\alpha = \tan^{-1}(15/20) = 36.87^\circ$; $\beta = 90^\circ - \alpha$), we find

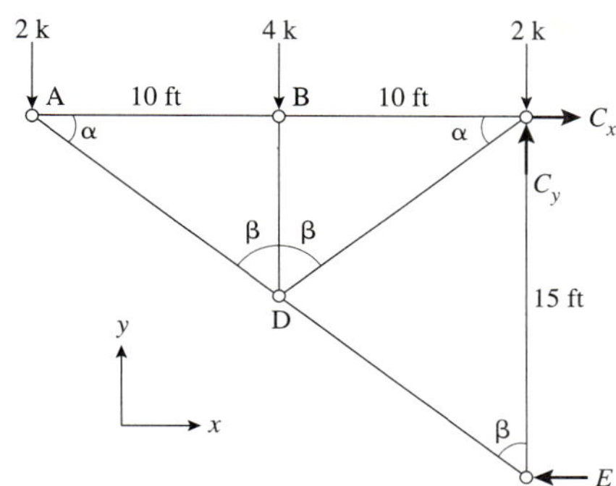

Fig. E2.7.1(b)

$$\overset{\curvearrowleft +}{\sum} M_C = 0 = 2(20) + 4(10) - E_x(15) \Rightarrow E_x = 5.33 \text{ k}$$

$$\overset{\uparrow +}{\sum} F_y = 0 = -2 - 4 - 2 + C_y \Rightarrow C_y = 8 \text{ k}$$

$$\overset{\rightarrow +}{\sum} F_x = 0 = C_x - E_x \Rightarrow C_x = 5.33 \text{ k}$$

Step 2: We now establish the sequence of the joints to process in order to compute the member forces.

Substep	Joint	Equation to Use	Solve Equation to Obtain	Equation to Use	Solve Equation to Obtain
1	A	$\overset{\uparrow +}{\sum} F_y = 0$	F_{AD}	$\overset{\rightarrow +}{\sum} F_x = 0$	F_{AB}
2	B	$\overset{\uparrow +}{\sum} F_y = 0$	F_{BD}	$\overset{\rightarrow +}{\sum} F_x = 0$	F_{BC}
3	C	$\overset{\rightarrow +}{\sum} F_x = 0$	F_{CD}	$\overset{\uparrow +}{\sum} F_y = 0$	F_{CE}
4	E	$\overset{\rightarrow +}{\sum} F_x = 0$	F_{ED}		

We explain how the joint sequence was set up while pointing out that the joint sequencing is *not* unique. As stated in the previous section, we should start at a support/joint where there are at most two unknowns. Then the key step is to use either $\overset{\rightarrow +}{\sum} F_x = 0$ or $\overset{\uparrow +}{\sum} F_y = 0$ so as to generate an equation with a single unknown. For this problem, having computed the reactions, we could start either at A or at E. We have chosen to start at A. Using $\overset{\rightarrow +}{\sum} F_x = 0$ first yields an equation involving both F_{AB} and F_{AD} as unknowns. However, $\overset{\uparrow +}{\sum} F_y = 0$ yields an equation with only F_{AD} as the unknown, since F_{AB} does not have a component in the y direction. Having computed F_{AD}, we can use $\overset{\rightarrow +}{\sum} F_x = 0$ to compute F_{AB}. The same thought process applies to other joints. Substep 3 (or Substep 4) could have involved joint D. However, with the regular x-y definition, $\overset{\rightarrow +}{\sum} F_x = 0$ and $\overset{\uparrow +}{\sum} F_y = 0$ will involve both F_{CD} and F_{ED} as unknowns, requiring us to solve simultaneous equations. We tackle this issue at the end of this problem.

Step 3: Use the joint FBDs to solve for the member forces (see Fig. E2.7.1(c)). When drawing the FBDs, if a member is assumed and drawn as if in compression, then in all subsequent FBDs its must be (consistently) shown in compression.

Substep 1: To satisfy $\overset{\uparrow +}{\sum} F_y = 0$, the vertical component of F_{AD} must act upwards to counteract the 2 k load. Similarly, from $\overset{\rightarrow +}{\sum} F_x = 0$, F_{AB} must act to the right.

$$\overset{\uparrow +}{\sum} F_y = 0 = -2 + F_{AD} \sin\alpha \Rightarrow F_{AD} = \frac{2}{\sin\alpha} = 3.33 \text{ k}$$

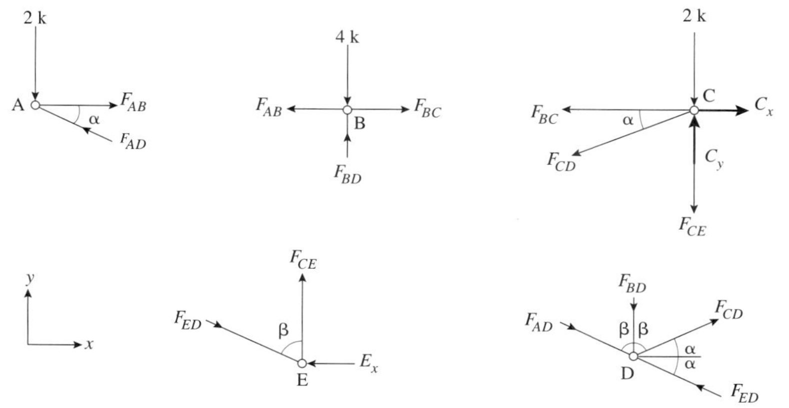

Fig. E2.7.1(c)

$$\overset{\rightarrow+}{\sum} F_x = 0 = F_{AB} - F_{AD}\cos\alpha \Rightarrow F_{AB} = F_{AD}\cos\alpha = 2.67 \text{ k}$$

Substep 2: To satisfy $\overset{\uparrow+}{\sum} F_y = 0$, F_{BD} must act upwards to counteract the 4 k load. Similarly, from $\overset{\rightarrow+}{\sum} F_x = 0$, F_{BC} must act to the right.

$$\overset{\uparrow+}{\sum} F_y = 0 = -4 + F_{BD} \Rightarrow F_{BD} = 4 \text{ k}$$

$$\overset{\rightarrow+}{\sum} F_x = 0 = -F_{AB} + F_{BC} \Rightarrow F_{BC} = 2.67 \text{ k}$$

Substep 3: To satisfy $\overset{\rightarrow+}{\sum} F_x = 0$, the horizontal component of F_{CD} must act to the left, since the resultant of C_x and F_{BC} acts to the right. Similarly, to satisfy $\overset{\uparrow+}{\sum} F_y = 0$, F_{CE} must act downwards since the resultant of C_y, the 2 k force, and the vertical component of F_{CD} acts upwards.

$$\overset{\rightarrow+}{\sum} F_x = 0 = -F_{BC} - F_{CD}\cos\alpha + C_x \Rightarrow F_{CD} = \frac{C_x - F_{BC}}{\cos\alpha} = 3.33 \text{ k}$$

$$\overset{\uparrow+}{\sum} F_y = 0 = -2 - F_{CD}\sin\alpha - F_{CE} + 8 \Rightarrow F_{CE} = 8 - 2 - F_{CD}\sin\alpha = 4\text{k}$$

Substep 4: To satisfy $\overset{\rightarrow+}{\sum} F_x = 0$, the horizontal component of F_{ED} must act to the right since the support reaction E_x acts to the left.

$$\overset{\rightarrow+}{\sum} F_x = 0 = -E_x + F_{ED}\sin\beta \Rightarrow F_{ED} = \frac{E_x}{\sin\beta} = 6.67 \text{ k}$$

Step 4: Checks:

Support E: $\overset{\uparrow+}{\sum} F_y = F_{CE} - F_{ED}\cos\beta = 4 - 6.67\cos\beta \approx 0$ OK.

Joint D: $\overset{\uparrow+}{\sum} F_y = -F_{AD}\sin\alpha - F_{BD} + F_{CE}\sin\alpha + F_{ED}\sin\alpha \approx 0$ OK.

$$\overset{\rightarrow +}{\sum} F_x = F_{AD}\cos\alpha + F_{CE}\cos\alpha - F_{ED}\cos\alpha \approx 0 \quad \text{OK.}$$

Answers: $F_{AB} = 2.67\text{ k }(T)$, $F_{AD} = 3.33\text{ k }(C)$, $F_{BC} = 2.67\text{ k }(T)$, $F_{BD} = 4\text{ k }(C)$, $F_{CE} = 4\text{ k }(T)$, $F_{CD} = 3.33\text{ k }(T)$, $F_{ED} = 6.67\text{ k }(C)$.

Defining a New Coordinate System. We did not solve joint D in substep 3 because the traditional x-y coordinate system would yield a set of simultaneous equations involving F_{CD} and F_{ED}. Now let us define a new coordinate system x'-y', as in Fig. E2.7.1(d). The advantage of this system is that we can generate an equation with just one unknown.

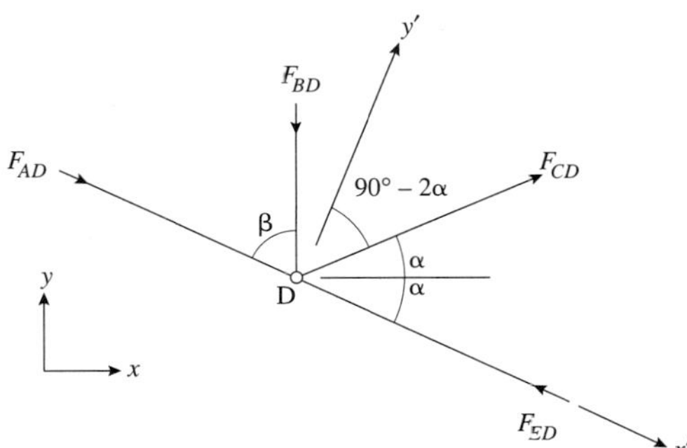

Fig. E2.7.1(d)

$$\overset{\uparrow +}{\sum} F_{y'} = 0 = -F_{BD}\sin\beta + F_{CD}\cos(90^\circ - 2\alpha) \Rightarrow F_{CD} = \frac{F_{BD}\sin\beta}{\cos(90^\circ - 2\alpha)} = 3.33\text{ k}$$

$$\overset{\rightarrow +}{\sum} F_{x'} = 0 = F_{AD} + F_{BD}\cos\beta + F_{CD}\cos(2\alpha) - F_{ED} \Rightarrow F_{ED} = 6.67\text{ k}$$

EXAMPLE 2.7.2

Simply Supported Truss

Figure E2.7.2(a) shows a planar truss. Compute the axial force in all the members.

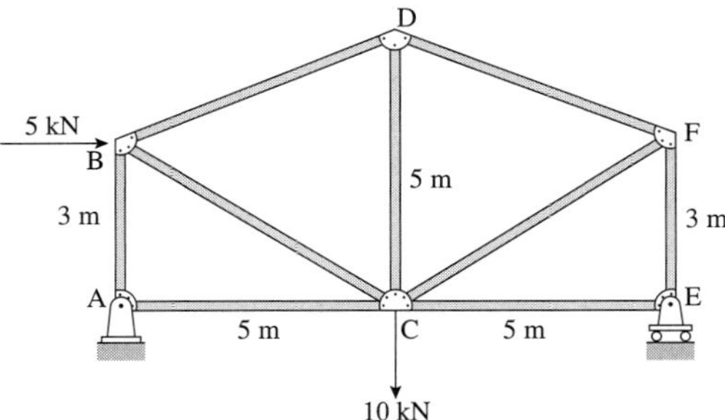

Fig. E2.7.2(a)

SOLUTION

Step 1: The truss is statically determinate ($j = 6$, $m = 9$, $r = 3$) since $2j = m + r$. Using the structural FBD (Fig. E2.7.2(b)), we find

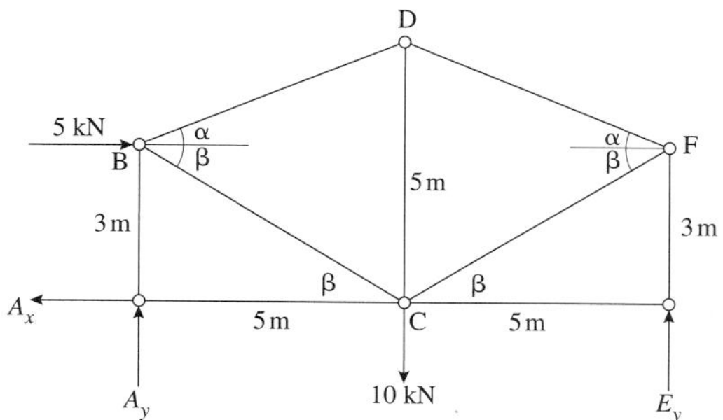

Fig. E2.7.2(b)

$$\overset{\curvearrowleft +}{\sum} M_A = 0 = -5(3) - 10(5) + E_y(10) \Rightarrow E_y = 6.5 \text{ kN}$$

$$\overset{\uparrow +}{\sum} F_y = 0 = A_y - 10 + E_y \Rightarrow A_y = 3.5 \text{ kN}$$

$$\overset{\rightarrow +}{\sum} F_x = 0 = 5 - A_x \Rightarrow A_x = 5 \text{ kN}$$

The rest of the geometry-related data are: $\beta = \tan^{-1}(3/5) = 30.96^\circ$, $\alpha = \tan^{-1}(2/5) = 21.8^\circ$, $\gamma = 90^\circ - (\alpha + \beta) = 37.24^\circ$.

Step 2: In this example (see Fig. E2.7.2(c)), we initially assume that all members are in tension. The following joint sequence yields the solution:

(a) A: $\overset{\uparrow +}{\sum} F_y = 0 \Rightarrow F_{AB}$, (b) A: $\overset{\rightarrow +}{\sum} F_x = 0 \Rightarrow F_{AC}$, (c) B: $\overset{\uparrow +}{\sum} F_{y'} = 0 \Rightarrow F_{BD}$,

(d) B: $\overset{\rightarrow +}{\sum} F_x = 0 \Rightarrow F_{BC}$, (e) D: $\overset{\rightarrow +}{\sum} F_x = 0 \Rightarrow F_{DF}$, (f) D: $\overset{\uparrow +}{\sum} F_y = 0 \Rightarrow F_{DC}$,

(g) F: $\overset{\rightarrow +}{\sum} F_x = 0 \Rightarrow F_{FC}$, (h) F: $\overset{\uparrow +}{\sum} F_y = 0 \Rightarrow F_{FE}$, (i) C: $\overset{\rightarrow +}{\sum} F_x = 0 \Rightarrow F_{CE}$.

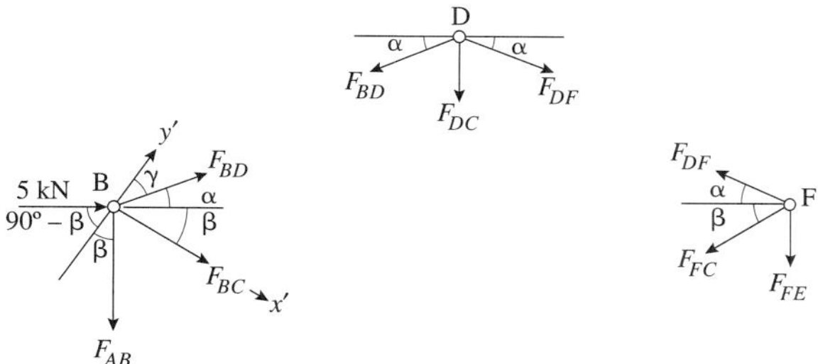

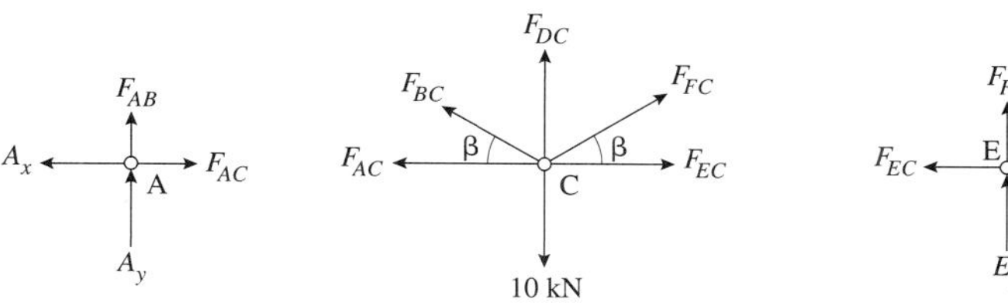

Fig. E2.7.2(c)

We find

$$\text{Joint A:}\quad \overset{\uparrow+}{\sum} F_y = 0 = A_y + F_{AB} \Rightarrow F_{AB} = -A_y = -3.5 \text{ kN}$$

The negative value indicates that our assumption for F_{AB} (being in tension) is incorrect. In the subsequent steps, we assume that F_{AB} is in tension but we substitute the negative value (–3.5) for F_{AB}. Do not attempt to rectify the FBDs; we state the correct sense of the member forces at the end of the problem.

$$\text{Joint A:}\quad \overset{\rightarrow+}{\sum} F_x = 0 = -A_x + F_{AC} \Rightarrow F_{AC} = A_x = 5 \text{ kN}$$

$$\text{Joint B:}\quad \overset{\uparrow+}{\sum} F_y = 0 = 5\cos(90^\circ - \beta) - F_{AB}\cos\beta + F_{BD}\cos\gamma$$

$$\text{or}\quad F_{BD} = \frac{F_{AB}\cos\beta - 5\sin\beta}{\cos\gamma} = \frac{-3.5(0.8575) - 5(0.5144)}{0.7961} = -7 \text{ kN}$$

$$\text{Joint B:}\quad \overset{\rightarrow+}{\sum} F_x = 0 = 5 + F_{BD}\cos\alpha + F_{BC}\cos\beta \Rightarrow F_{BC} = \frac{-5 - F_{BD}\cos\alpha}{\cos\beta} = 1.75 \text{ kN}$$

$$\text{Joint D:}\quad \overset{\rightarrow+}{\sum} F_x = 0 = -F_{BD}\cos\alpha + F_{DF}\cos\alpha \Rightarrow F_{DF} = -7 \text{ kN}$$

$$\text{Joint D:}\quad \overset{\uparrow+}{\sum} F_y = 0 = -F_{BD}\sin\alpha - F_{DF}\sin\alpha - F_{DC} \Rightarrow F_{DC} = 5.2 \text{ kN}$$

$$\text{Joint F:}\quad \overset{\rightarrow+}{\sum} F_x = 0 = -F_{DF}\cos\alpha - F_{FC}\cos\beta \Rightarrow F_{FC} = 7.58 \text{ kN}$$

$$\text{Joint F:}\quad \overset{\uparrow+}{\sum} F_y = 0 = F_{DF}\sin\alpha - F_{FC}\sin\beta - F_{FE} \Rightarrow F_{FE} = -6.5 \text{ kN}$$

$$\text{Joint C:}\quad \overset{\rightarrow+}{\sum} F_x = 0 = -F_{BC}\cos\beta - F_{AC} + F_{FC}\cos\beta + F_{EC} \Rightarrow F_{EC} = 0$$

Check: We now use the equations at joint C and support E to check our results.

$$\text{Joint C:}\quad \overset{\uparrow+}{\sum} F_y = F_{BC}\sin\beta + F_{DC} + F_{FC}\sin\beta - 10 = 0 \qquad \text{OK.}$$

$$\text{Joint E:}\quad \overset{\rightarrow+}{\sum} F_x = -F_{EC} = 0 \qquad \text{OK.}$$

$$\text{Joint E:}\quad \overset{\uparrow+}{\sum} F_y = E_y - F_{FE} = 0 \qquad \text{OK.}$$

Answers: The member forces are summarized below. Those that were computed to be negative are in compression.

$F_{AB} = 3.5$ kN (C), $F_{AC} = 5$ kN (T), $F_{BC} = 1.75$ kN (T), $F_{BD} = 7$ kN (C), $F_{DF} = 7$ kN (C), $F_{DC} = 5.2$ kN (T), $F_{FC} = 7.58$ kN (T), $F_{FE} = 6.5$ kN (C), $F_{EC} = 0$.

EXERCISES

Appetizers

For each of the trusses in Figs. P2.7.1–2.7.4 below, identify whether it is determinate or indeterminate. For determinate trusses, solve for the member forces using the method of joints. State whether the members are in tension or compression.

2.7.1.

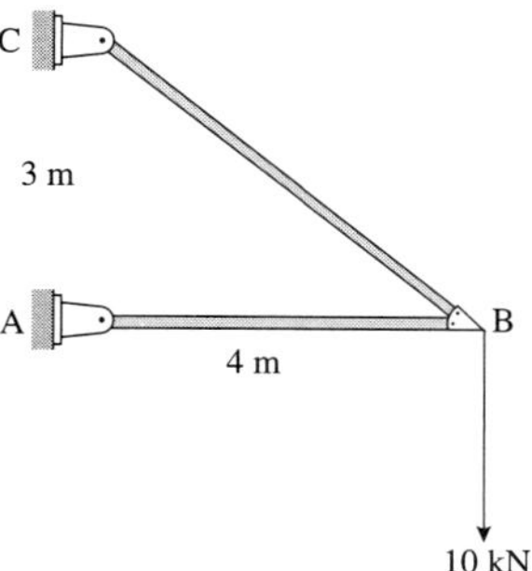

Fig. P2.7.1

2.7.2.

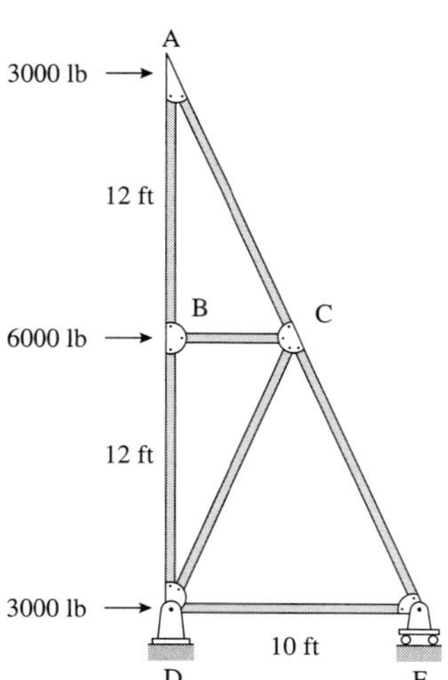

Fig. P2.7.2

2.7.3.

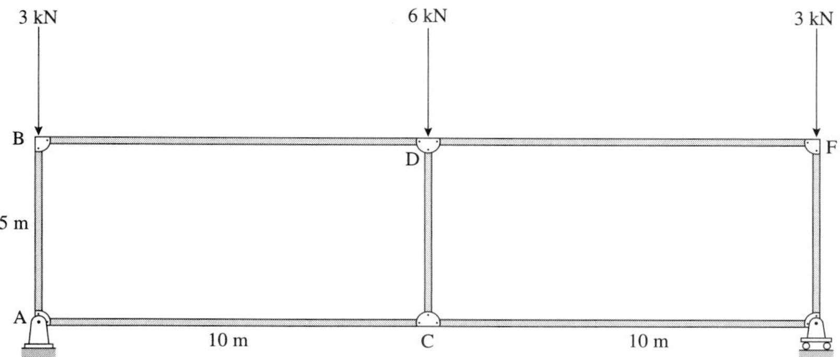

Fig. P2.7.3

2.7.4.

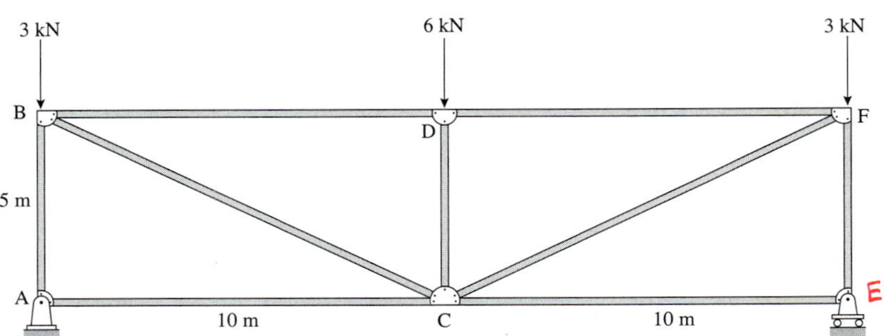

Fig. P2.7.4

Main Course

For each of the trusses in Figs. P2.7.5–P2.7.10, identify whether it is determinate or indeterminate. For determinate trusses, solve for the member forces using the method of joints.

2.7.5. Figure P2.7.5 shows a Warren truss. All triangles are equilateral triangles.

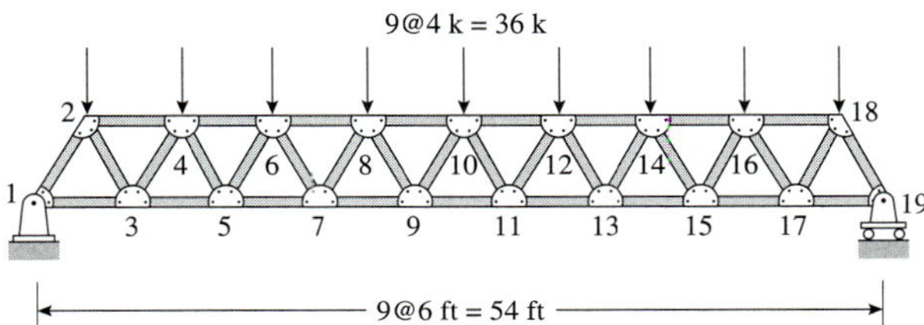

Fig. P2.7.5

2.7.6.

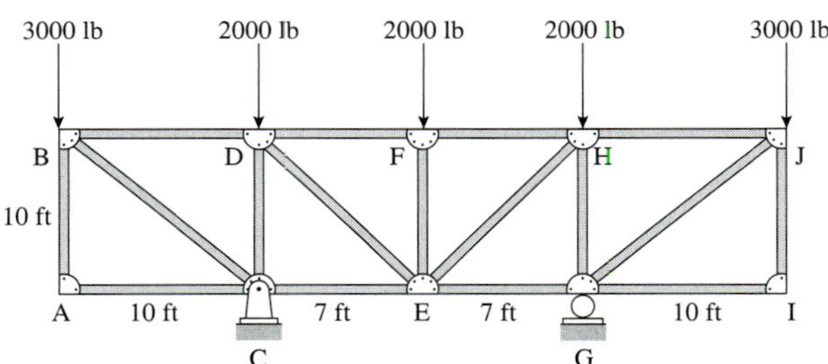

Fig. P2.7.6

2.7.7. One of the supporting frames in a billboard is modeled as a truss, as shown in Fig. P2.7.7. The loads approximate the effect of wind acting on the billboard.

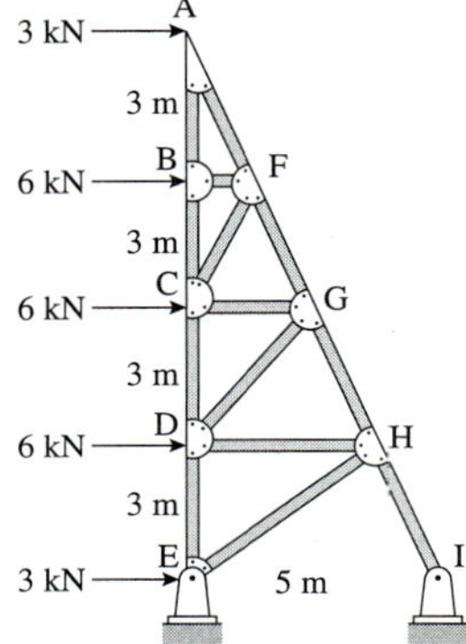

Fig. P2.7.7

2.7.8. Figure P2.7.8 shows a flat-bottom gable truss. Gable trusses are commonly used as roof trusses.

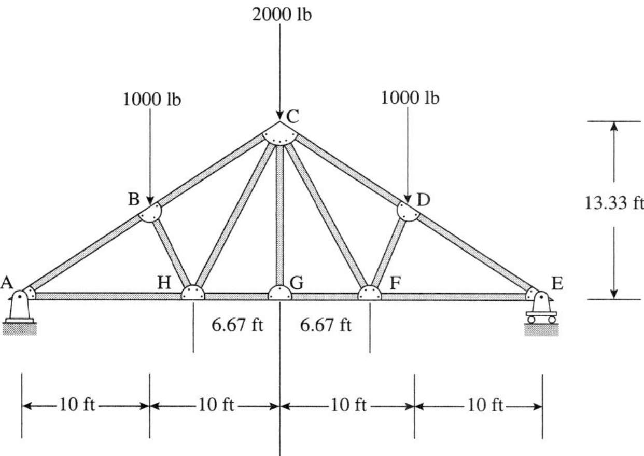

Fig. P2.7.8

2.7.9. Figure P2.7.9 shows a scissors gable truss. The orientation of the bottom members gives rise to a vaulted ceiling. Compare the performance of the two gable trusses.

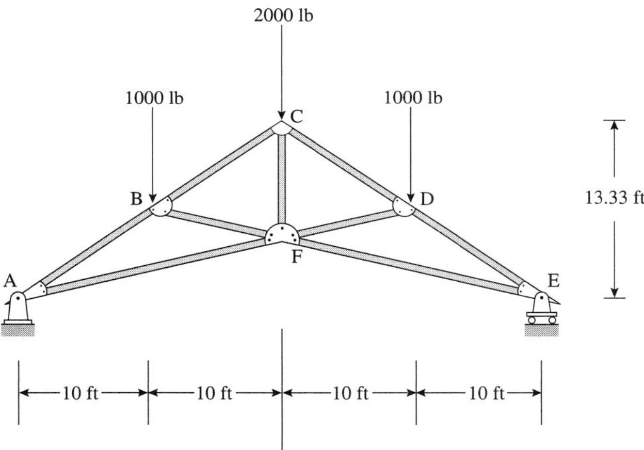

Fig. P2.7.9

2.7.10. Figure P2.7.10 shows a floor deck resting on floor beams, which in turn are supported by a truss. The loading on the floor is 2000 lb/ft.

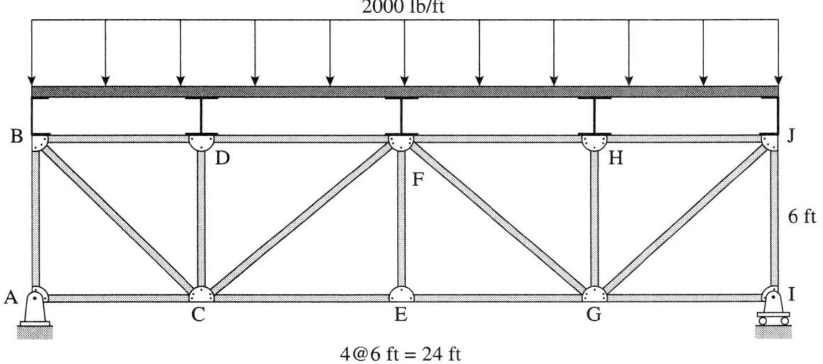

Fig. P2.7.10

Structural Concepts

2.7.11. Figure P2.7.11 shows the FBD of joint A in a planar truss. The nature of the member forces is not indicated in the FBD. However, it is known that members AB and AC are in tension and the forces in them are 1000 lb and 1732 lb, respectively. Given this information, can you compute the forces in the other two members?

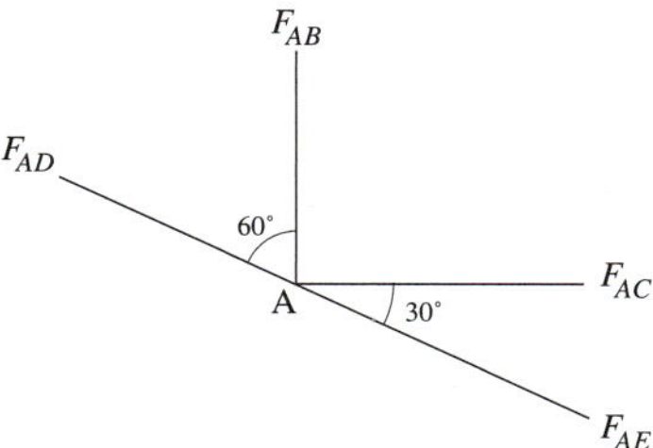

Fig. P2.7.11

2.7.12. How could you have reduced the computations in Problem 2.7.5?

2.7.3 Method of Sections

When the forces in a few truss members are needed, the method of sections is an easy way to compute them. Consider the (determinate planar) flat Howe truss shown in Fig. 2.7.3.1. Let us assume that we need to compute the forces in member BH and GH.

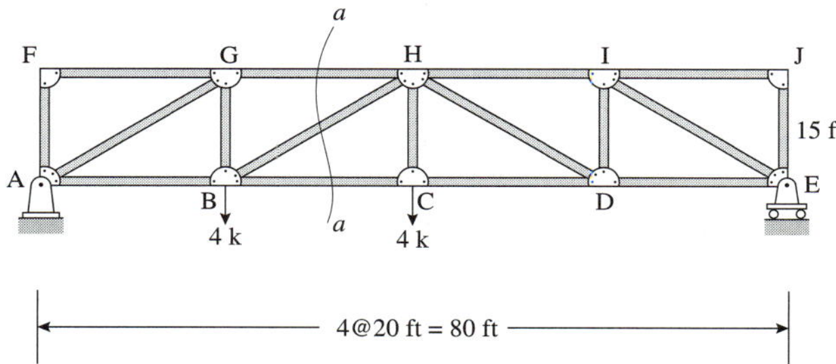

Fig. 2.7.3.1
A planar truss.

In the method of sections, a section is passed through the truss, breaking it into (typically) two parts or halves. If the section passes through a member, the member force must be included in the resulting FBD. Let section *a-a* be used so that it passes through GH, BH, and BC. The FBDs of the left and right half are shown in Fig. 2.7.3.2.

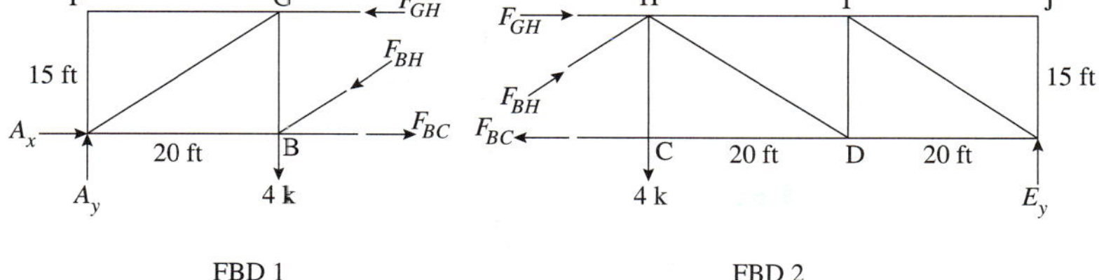

Fig. 2.7.3.2
FBDs of the left and right sides of section *a-a*.

If the truss is in equilibrium, then parts of the truss must also be in equilibrium. In other words, the equations of equilibrium must be satisfied by the forces acting on the two FBDs.

Note that in Fig. 2.7.3.2, members GH and BH are in compression and member BC is in tension.

2.7.4 General Procedure

Step 1: Use the structural FBD to solve for as many support reactions as possible. While this step is not always necessary, it does provide for checks at the end of the problem.

Step 2: Make a cut that passes through not more than three members[1] including the member(s) whose internal forces are to be computed. Draw the FBD of either the left or the right half of the truss.

Step 3: Now use one of the equations of the equilibrium to generate an equation in a single unknown. Typically this is the moment equation if you can find a point through which all but one of the unknown forces pass through. Solve for the unknown member force. Now use the other equations to solve for the other unknowns.

> **Tip:** When using the moment equation, remember to slide the force vector to the nearest joint and then take its x-y components. It is easier to compute the lever arm this way. For example, in Fig. 2.7.3.2, in order to compute the moment due to F_{BH} (in the right FBD) about point E, one would slide F_{BH} to joint H. Then the lever arm of the horizontal component of F_{BH} is 15 ft and for the vertical component, the lever arm is 40 ft.

Step 4: Repeat steps 2 and 3 if necessary.

EXAMPLE 2.7.3

Warren Truss

Figure E2.7.3(a) shows a Warren truss made up of equilateral triangles. Compute the axial force in members EF, BF, and BC.

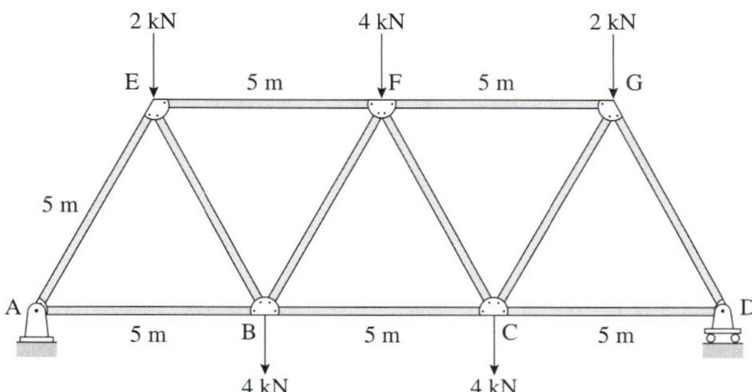

Fig. E2.7.3(a)

SOLUTION

Step 1: The structural FBD is shown in Fig. E2.7.3(b). All interior angles are 60°.

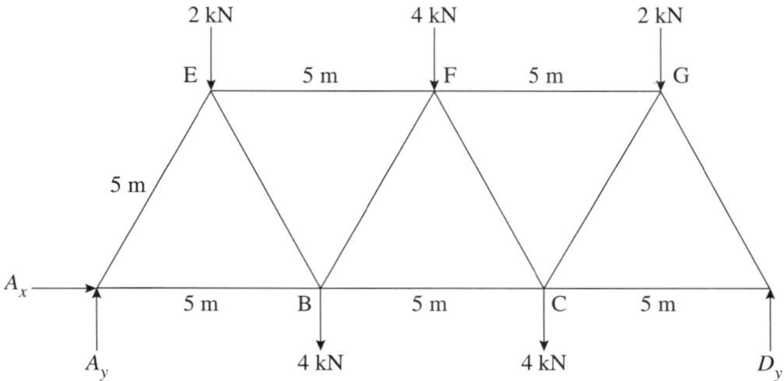

Fig. E2.7.3(b)

[1] See exception in Example 2.7.4.

Using the structural FBD, we find

$$\overset{\curvearrowleft +}{\sum} M_A = 0 = -2(2.5) - 4(5) - 4(7.5) - 4(10) - 2(12.5) + D_y(15) \Rightarrow D_y = 8 \text{ kN}$$

$$\overset{\uparrow +}{\sum} F_y = 0 = A_y - (2 + 4 + 4 + 4 + 2) + D_y \Rightarrow A_y = 8 \text{ kN}$$

$$\overset{\rightarrow +}{\sum} F_x = 0 = A_x \Rightarrow A_x = 0$$

Step 2: To compute the member forces in EF, BF, and BC, a cut through these three members is made. The resulting (left) FBD is shown in Fig. 2.7.3(c). There are three unknowns and three available equations of equilibrium.

The $\overset{\rightarrow +}{\sum} F_x = 0$ cannot be used, since it would result in an equation with three unknowns. We can use $\overset{\uparrow +}{\sum} F_y = 0$ to obtain F_{BF}. We could also use the moment equation $\overset{\curvearrowleft +}{\sum} M = 0$. If the moments are taken about point F, the unknown in the resulting equation is F_{BC}. Similarly, a moment equation about B yields F_{EF}. The moment equation usually provides a starting point for computing the unknowns.

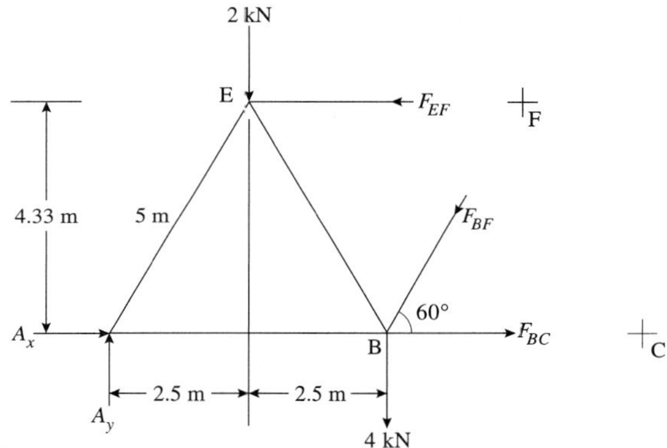

Fig. E2.7.3(c)

$$\overset{\curvearrowleft +}{\sum} M_B = 0 = -A_y(5) + 2(2.5) + F_{EF}(4.33) \Rightarrow F_{EF} = \frac{40 - 5}{4.33} = 8.08 \text{ kN}$$

$$\overset{\uparrow +}{\sum} F_y = 0 = A_y - (2 + 4) - F_{BF} \sin 60^\circ \Rightarrow F_{BF} = 2.31 \text{ kN}$$

$$\overset{\rightarrow +}{\sum} F_x = 0 = A_x - F_{EF} - F_{BF} \cos 60^\circ + F_{BC} \Rightarrow F_{BC} = 9.24 \text{ kN}$$

Answers: $F_{EF} = 8.08$ kN (C), $F_{BF} = 2.31$ kN (C), $F_{BC} = 9.24$ kN (T).

EXAMPLE 2.7.4

K-Truss

Figure E2.7.4(a) shows a K-truss. Compute the axial force in members TU, TN, FN, and FG.

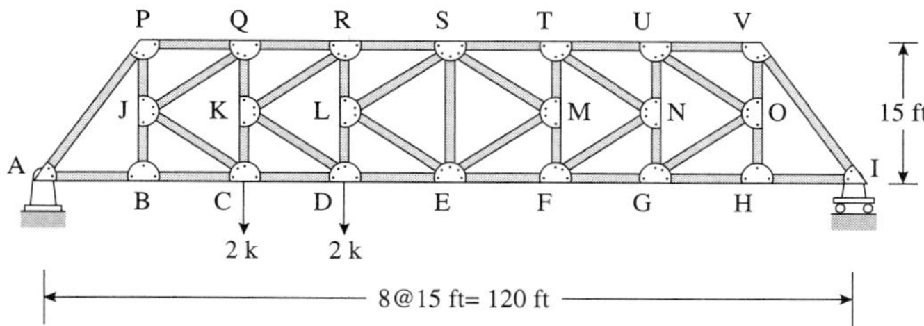

Fig. E2.7.4(a)

SOLUTION

Step 1: The structural FBD is shown in Fig. E2.7.4(b).

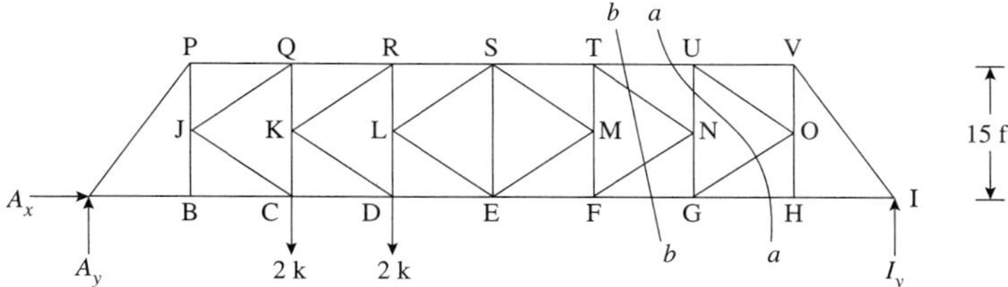

Fig. E2.7.4(b)

Using the structural FBD, we obtain

$$\overset{\curvearrowleft +}{\sum} M_A = 0 = -2(30) - 2(45) + I_y(120) \Rightarrow I_y = 1.25 \text{ k}$$

$$\overset{\uparrow +}{\sum} F_y = 0 = A_y - 2 - 2 + I_y \Rightarrow A_y = 2.75 \text{ k}$$

$$\overset{\rightarrow +}{\sum} F_x = 0 = A_x \Rightarrow A_x = 0$$

Step 2: A single meaningful cut will not yield the values of the four members. Hence we solve the problem in two stages. First we make the cut *a-a* through members TU and three additional members. The resulting FBD is shown in Fig. E2.7.4(c). While the FBD has more unknowns than available equations, we can use the moment equilibrium equation judiciously to compute one of the member forces, F_{TU}.

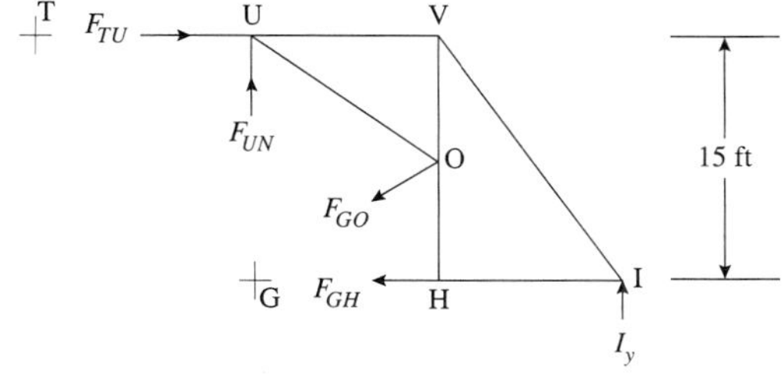

Fig. E2.7.4(c) FBD associated with cut *a-a*.

$$\overset{\curvearrowleft +}{\sum} M_G = 0 = -F_{TU}(15) + I_y(30) \Rightarrow F_{TU} = 2.5 \text{ k}$$

Step 3: Now to compute the other three unknowns, we make a second cut *b-b*. The resulting FBD is shown in Fig. 2.7.4(d) ($\alpha = \tan^{-1}(7.5/15) = 26.56^\circ$).

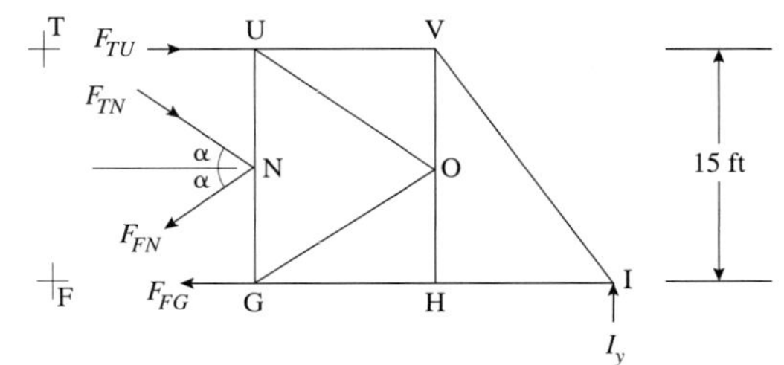

Fig. E2.7.4(d) FBD associated with cut *b-b*.

Again, we use a combination of moment and force equations to compute the three member forces, as follows.

$$\overset{\curvearrowleft +}{\sum} M_F = 0 = -F_{TU}(15) - F_{TN}\cos\alpha(15) + I_y(45) \Rightarrow F_{TN} = 1.4 \text{ k}$$

$$\overset{\uparrow +}{\sum} F_y = 0 = -F_{TN}\sin\alpha - F_{FN}\sin\alpha + I_y \Rightarrow F_{FN} = 1.4 \text{ k}$$

$$\overset{\curvearrowleft +}{\sum} M_N = 0 = -F_{TU}(7.5) - F_{FG}(7.5) + I_y(30) \Rightarrow F_{FG} = 2.5 \text{ k}$$

Answers: $F_{TU} = 2.5$ k (C), $F_{TN} = 1.4$ k (C), $F_{FN} = 1.4$ k (T), $F_{FG} = 2.5$ k (T).

EXERCISES

Appetizers

2.7.13. Compute the forces in members CD, DG, and GH in Fig. P2.7.13.

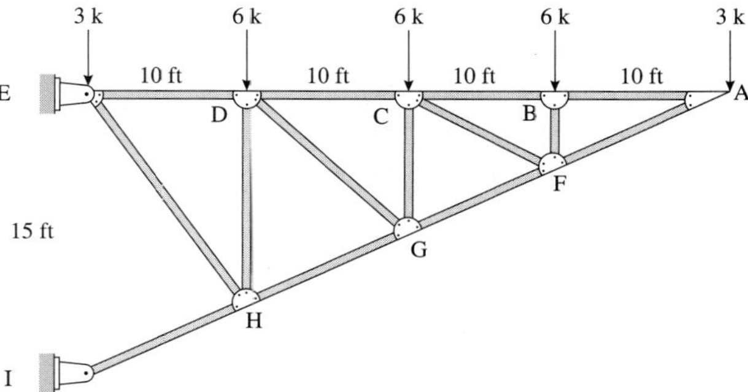

Fig. P2.7.13

2.7.14. Figure P2.7.14 shows a Fink truss. Compute the forces in the members that meet at joint C.

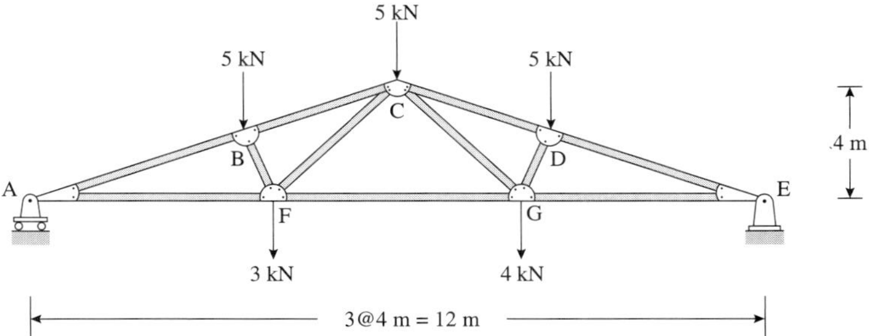

Fig. P2.7.14

2.7.15. Check your answers for members DF and DE in Problem 2.7.6 using the method of sections.

2.7.16. A Northlight truss as shown in Fig. P2.7.16 is loaded at joints D and E. Compute the forces in all the members that meet at joint B.

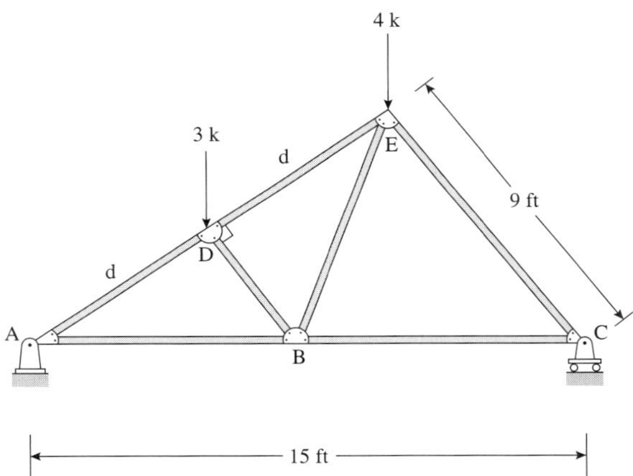

Fig. P2.7.16

Main Course

2.7.17. The rise of the Howe truss shown in Fig. P2.7.17 is 1 in 3. Compute the forces in all the members that meet at joint G.

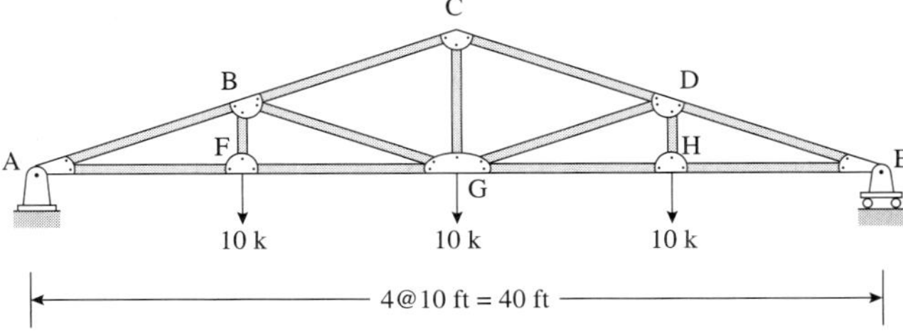

Fig. P2.7.17

2.7.18. The use of truss systems to span long distances is common in buildings such as exhibition halls. We saw an example in Problem 2.7.5. Consider the trussed girder in Fig. P2.7.18. Compute the forces in members KJ, KI, JM, and ST. Assume that the connections between the columns and the truss at A and T are pin connections, and that the truss can be analyzed independently of the frame.

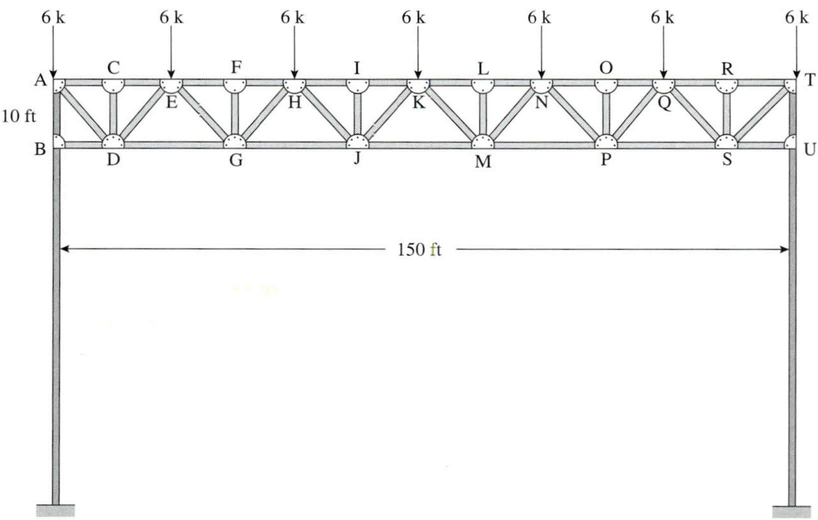

Fig. P2.7.18

Structural Concepts

2.7.19. Without changing the topology and loading of the Warren truss in Example 2.7.3, what is the quickest way to decrease the member forces in members EF and FG?

2.7.20. Could the truss shown in Problem 2.7.13 be used as a cantilevered bridge truss? When is such a bridge desirable?

2.7.5 Zero Force Members

Certain truss configurations—geometry and loading—lead to members that theoretically have zero force in them. A truss model of a crane boom is shown in Fig. 2.7.5.1.

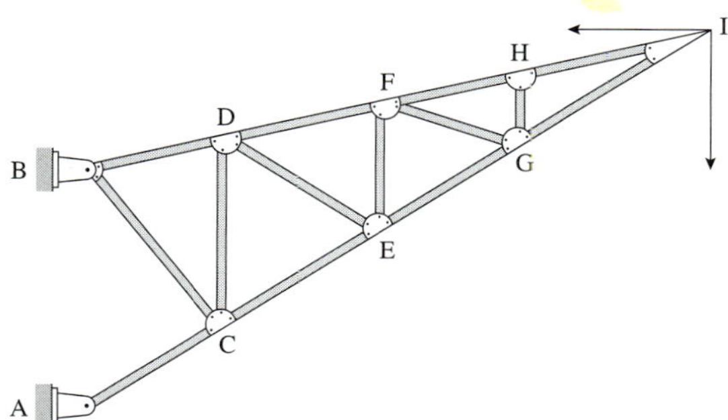

Fig. 2.7.5.1 A crane boom truss model.

The truss is determinate and a joint-by-joint analysis to determine the forces in the bottom and top chord members will show that they are in compression (see Fig. 2.7.5.2). The web member analysis can start at joint H.

Fig. 2.7.5.2
FBD of joints H and G.

From the FBD of joint H, $\sum F_{y'} = 0$ yields the y' component of F_{HG} as zero. Hence F_{HG} is zero. Similarly, from joint G, $\sum F_{y'} = 0$ yields the y' component of F_{GF} as zero. Hence F_{GF} is zero. Proceeding in a similar manner to other joints shows that all web members are zero-force members.

A natural question is, "Why are trusses designed to have zero-force members?" There are primarily three reasons for this. First, some members may be zero-force members under the action of one set of loads but not all. Second, compressive members are prone to buckling. The (Euler) buckling capacity of a member is inversely proportional to the square of the length of the members. In other words, the longer the member, the lower is its resistance to compressive loads. The zero-force web members in the above example reduce the effective length of the top and bottom chords, BI and AI respectively, by splitting them into four members each. These members are much shorter and hence stiffer. Third, in indeterminate trusses, these additional members provide an added level of redundancy (or safety) that may be useful under certain circumstances, for example, if the truss is partially damaged.

Tip: One way to spot zero-force members is to look for joints with three members, two of which are collinear. If there are no external loads applied at the joint, then the non-collinear member is a zero-force member. Another situation is a joint with no external loads and two orthogonal members. In this case both the members are zero-force members.

2.7.6 UNSTABLE TRUSSES

In Section 2.6 the topic of stability was introduced. In this section, we examine the stability of planar truss systems. A truss is clearly unstable if the number of unknowns is less than the number of equations, that is, if $(m + r) < 2j$. Examples illustrating this condition are shown in Fig. 2.7.6.1. In Fig. 2.7.6.1(a), $(m + r) = (5 + 2) < 2j = 2(4)$. Note that there is no restraint in the horizontal direction. In Fig. 2.7.6.1(b), $(m + r) = (4 + 3) < 2j = 2(4)$. Members AD, DC, and CB form an unrestrained set that can move horizontally.

If $(m + r) \geq 2j$, is the truss stable? Unfortunately there is no magical recipe or answer at this stage (the answer is quite straightforward and is discussed in Chapter 6). To answer this question we must look at the truss on a case-by-case basis and ascertain if each joint in the truss satisfies the equations of equilibrium. Consider the truss shown in Fig. 2.7.6.2 and let a joint be introduced at E. The truss appears to be determinate, $(m + r) = (6 + 4) = 2j = 2(5)$.

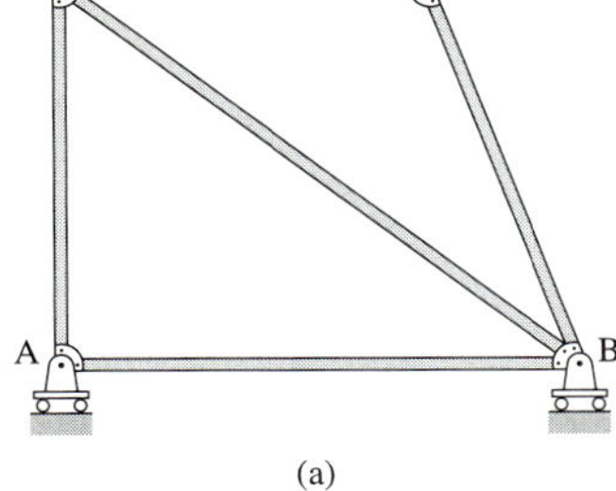

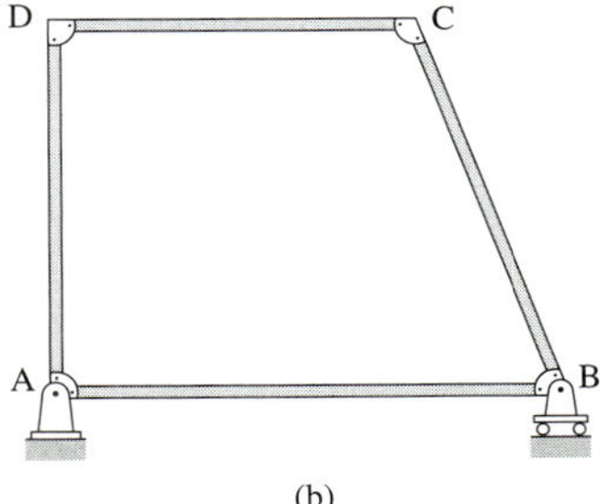

Fig. 2.7.6.1
Truss with (a) five members, two reactions and four joints, and (b) four members, three reactions, and four joints.

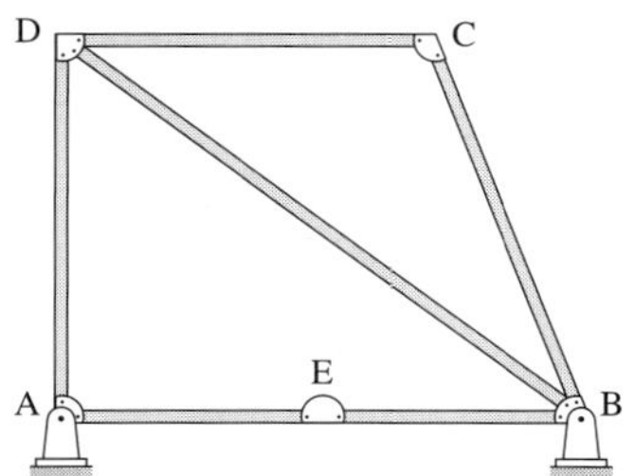

Fig. 2.7.6.2
Unstable truss.

However, considering the FBD of joint E (Fig. 2.7.6.3) shows that $\overset{\uparrow +}{\sum} F_y = 0$ is not satisfied at the joint.

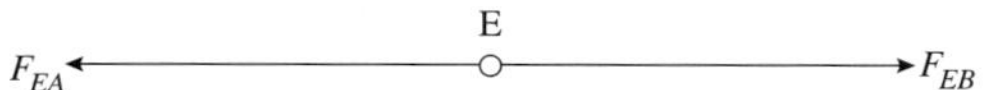

Fig. 2.7.6.3
FBD of joint E.

EXERCISES

Appetizers

2.7.21. Are the trusses in Figs. P2.7.21(a) and (b) stable or unstable?

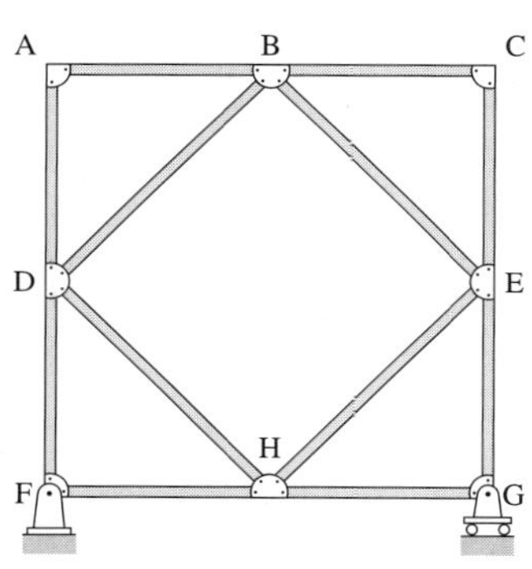

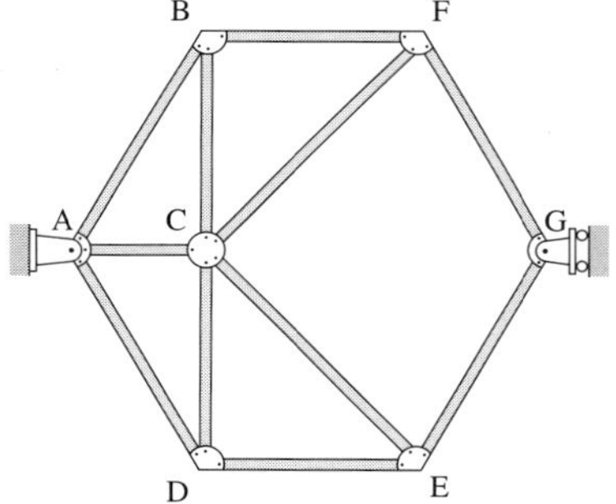

Fig. P2.7.21

Main Course

2.7.22. Identify all the zero-force members in the truss shown in Fig. P2.7.22.

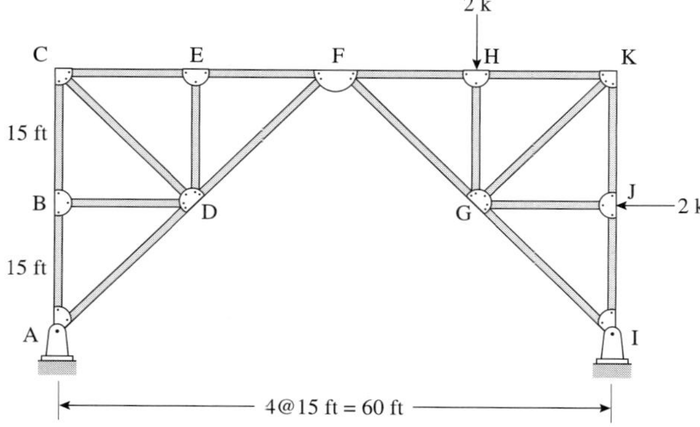

Fig. P2.7.22

2.8 PLANAR FRAME ANALYSIS

It would be quite easy to identify a frame by stating that any skeletal structure (one in which the length of the members is much larger than the member cross-sectional dimensions) that is *not* a truss is a frame. However, we provide a formal definition. A planar frame is a structural system that satisfies the following requirements:

a. The members are slender and prismatic. They can be straight or curved, vertical, horizontal, or inclined. The cross-sectional dimensions are small in comparison to the member lengths. Also, when constructing the frame model, we treat the members as a one-dimensional entity (having length and negligible cross-sectional dimensions).
b. The joints can be assumed to be rigid connection, frictionless pins (or internal hinges), or typical connections.
c. The loads can be concentrated forces, moments that act at joints or on the frame members, or distributed forces acting on the members.

As a consequence of these assumptions, planar frame members have internal forces that can be classified as axial force N acting along the member axis x', shear force V acting normal to the member axis at the cut, or bending moment M acting along the axis that is orthogonal to x' and y' (see Fig. 2.8.1).

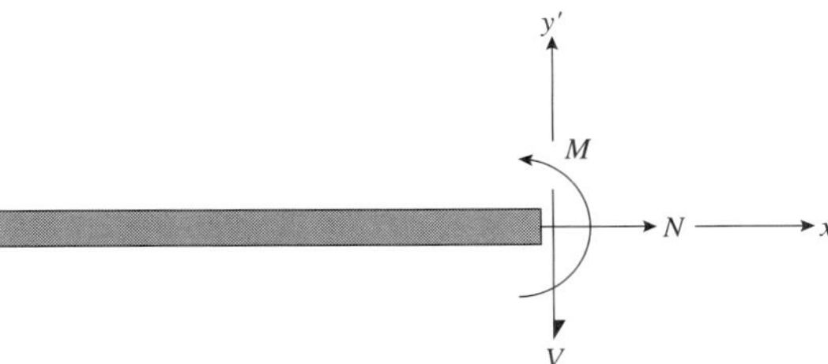

Fig. 2.8.1 Internal forces in a frame member.

We look at the issue of stability in Section 2.8.5. It is sufficient to state now that a frame is determinate (provided it is stable) if the number of unknowns is equal to the number of available equations. Let m be the number of members, r the number of support reactions, j the number of joints, and c the number of equations of condition, appearing in independent FBDs. Then, for a planar frame,

if $(3m + r - 3j - c) < 0$, then the frame is unstable,
if $(3m + r - 3j - c) = 0$, then the frame is determinate (provided it is stable), and
if $(3m + r - 3j - c) > 0$, then the frame is indeterminate unstable (provided it is stable).

Thus in Fig. 2.8.2:

Frame (a): $3m + r - 3j - c = 3(3) + 6 - 3(4) - 0 = 3$. Indeterminate to the third degree.
Frame (b): $3m + r - 3j - c = 3(3) + 5 - 3(4) - 2 = 0$. Determinate.
Frame (c): $3m + r - 3j - c = 3(6) + 6 - 3(6) - 0 = 6$. Indeterminate to the sixth degree.
Frame (d): $3m + r - 3j - c = 3(6) + 6 - 3(6) - 4 = 2$. Indeterminate to the second degree.

2.8.1 Internal Forces

The purpose of analyzing a frame is to determine the magnitude and nature of the internal forces in the members. The axial force acts along the (longitudinal) axis of the member, can be either compressive or tensile in nature, and typically does not vary within a member. The shear force acts tangential to the axis of the member and typically varies along the member. The bending moment acts along the axis that is normal to the plane containing the member and typically varies along the member.

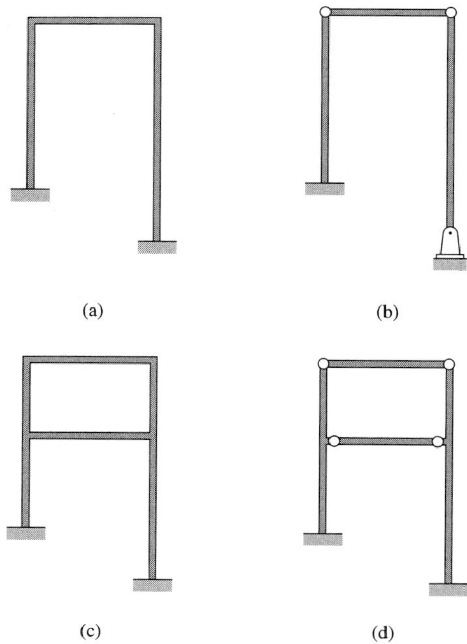

Fig. 2.8.2 Determinate and indeterminate frames.

From a design perspective, these internal forces determine whether a particular member is deemed adequate as per the design requirements. Note that the most common structural materials, steel and concrete, behave quite differently. Concrete is much weaker in tension than in compression. Identifying which part of a member is in tension and the level of that tension determines where and how much steel reinforcement needs to the placed. Similar statements can be made about other behavior modes.

EXAMPLE 2.8.1 ***(Variation of Example 2.5.1)***

Compute the internal forces at (i) the points to the left and right of D, and (ii) the points to the left and right of C, for the beam shown in Fig. E2.8.1(a).

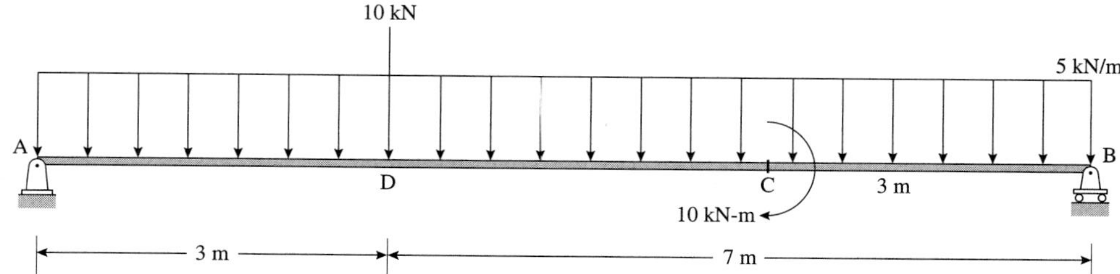

Fig. E2.8.1(a)

SOLUTION

Step 1: Using the structural FBD, we find: $A_y = 31$ kN(↑), $A_x = 0$ and $B_y = 29$ kN(↑).

Step 2: Making a cut to the left of D, we have the FBD in Fig. E2.8.1(b), and we find

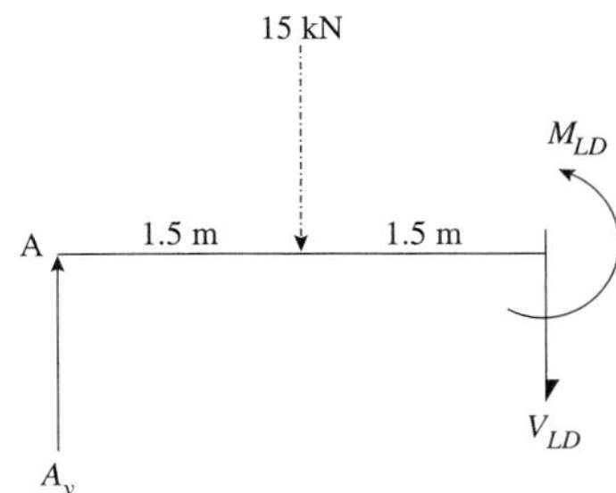

Fig. E2.8.1(b)

$$\overset{\uparrow +}{\sum} F_y = 0 = A_y - 15 - V_{LD} \Rightarrow V_{LD} = 16 \text{ kN}$$

$$\overset{\curvearrowleft +}{\sum} M_{\text{cut}} = 0 = -A_y(3) + 15(1.5) + M_{LD} \Rightarrow M_{LD} = 70.5 \text{ kN-m}$$

Making a cut to the right of D, we have the FBD in Fig. E2.8.1(c).

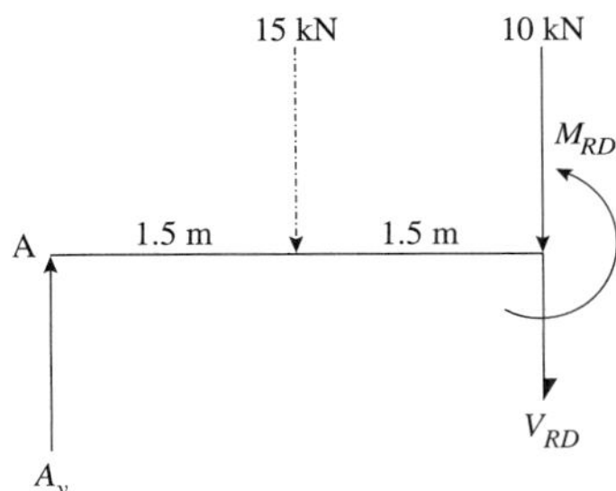

Fig. E2.8.1(c)

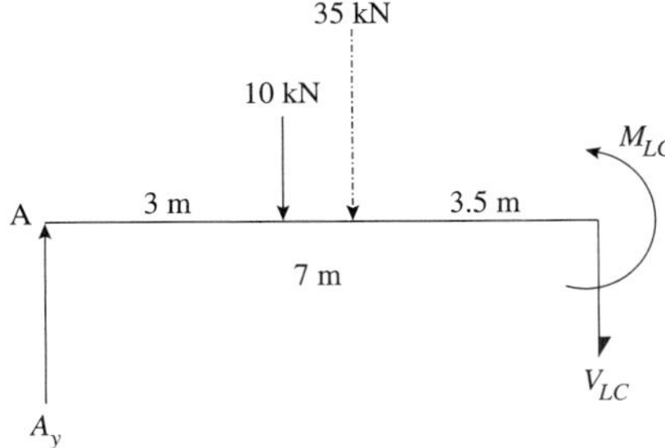

Fig. E2.8.1(d)

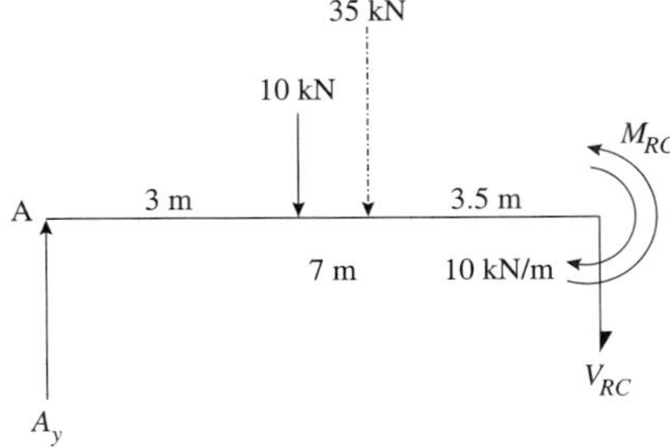

Fig. E2.8.1(e)

$$\overset{\uparrow +}{\sum} F_y = 0 = A_y - 15 - 10 - V_{RD} \Rightarrow V_{RD} = 6 \text{ kN}$$

$$\overset{\curvearrowleft +}{\sum} M_{\text{cut}} = 0 = -A_y(3) + 15(1.5) + M_{RD} \Rightarrow M_{RD} = 70.5 \text{ kN-m}$$

As we can see from the two results, there is an abrupt change of 10 kN (equal to the applied force) in the shear force value (from 16 kN to 6 kN) going from the left of the concentrated force to its right. There is, however, no change in the bending moment values. Similarly, making a cut to the left and right of C yields the results in Fig. E2.8.1(d).

$$\overset{\uparrow +}{\sum} F_y = 0 = A_y - 10 - 35 - V_{LC} \Rightarrow V_{LC} = -14 \text{ kN} \text{ (shear force is actually acting up)}$$

$$\overset{\curvearrowleft +}{\sum} M_{\text{cut}} = 0 = -A_y(7) + 10(4) + 35(3.5) + M_{LC} \Rightarrow M_{LC} = 54.5 \text{ kN-m}$$

$$\overset{\uparrow +}{\sum} F_y = 0 = A_y - 10 - 35 - V_{RC} \Rightarrow V_{RC} = -14 \text{ kN} \text{ (shear force is actually acting upwards)}$$

$$\overset{\curvearrowleft +}{\sum} M_{\text{cut}} = 0 = -A_y(7) + 10(4) + 35(3.5) - 10 + M_{RC} \Rightarrow M_{RC} = 64.5 \text{ kN-m}$$

Again comparing the two results, there is a sudden change of 10 kN-m (equal to the applied moment) from 54.5 kN-m to 64.5 kN-m in the bending moment going from the left of the concentrated moment to its right. There is, however, no change in the shear force values.

Observation: The points of discontinuity in the shear force and bending moment values are important locations and will be identified in the shear force and bending moment diagrams in the next few sections. A discontinuity is defined as an abrupt change in value or an abrupt change in the slope. These points include locations of

- Concentrated forces, e.g., where support reactions occur or where external loads are applied.
- Concentrated moments, e.g., where external loads are applied.
- Starting and ending points of distributed loads, or sudden change in the load intensity of distributed loads.

EXERCISES

Appetizers

2.8.1. Calculate the shear force and bending moment at the center of the beam in Fig. P2.8.1.

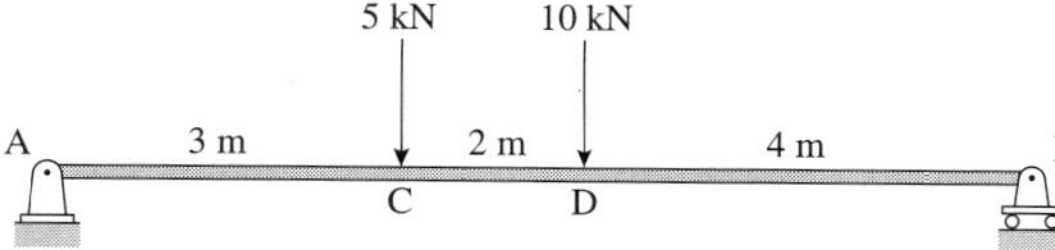

Fig. P2.8.1

2.8.2. Calculate the shear force and bending moment to the left and right of point C in Fig. P2.8.2.

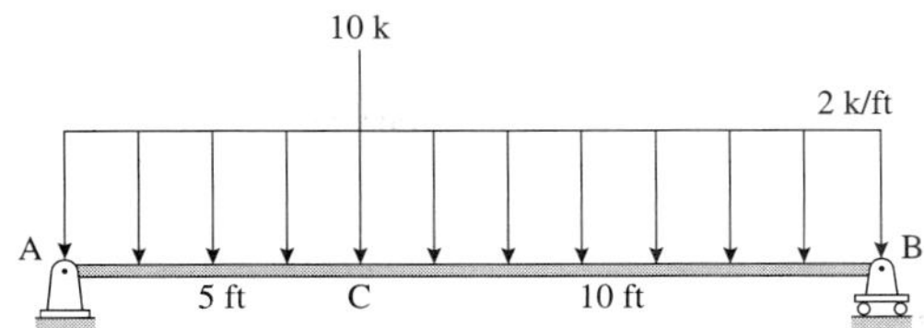

Fig. P2.8.2

2.8.3. Calculate all the internal forces at the beam's interior quarter points in Fig. P2.8.3.

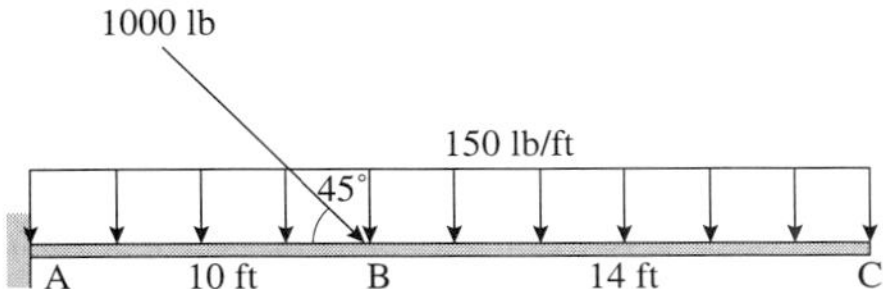

Fig. P2.8.3

2.8.4. Compute the internal forces to the left and right of points B and C in Fig. P2.8.4.

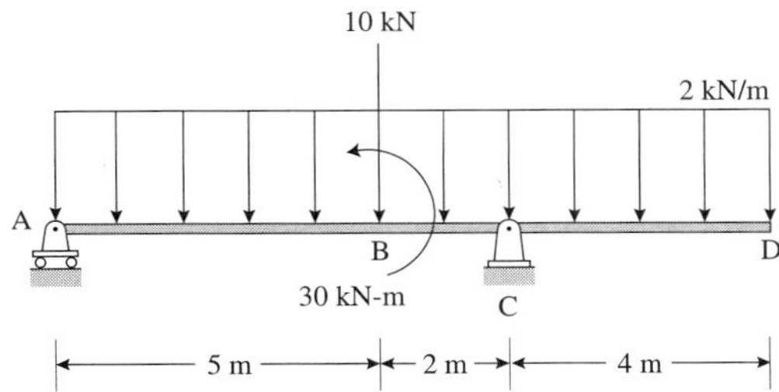

Fig. P2.8.4

Main Course

2.8.5. Calculate the shear force and bending moment to the left and right of point C in Fig. P2.8.5.

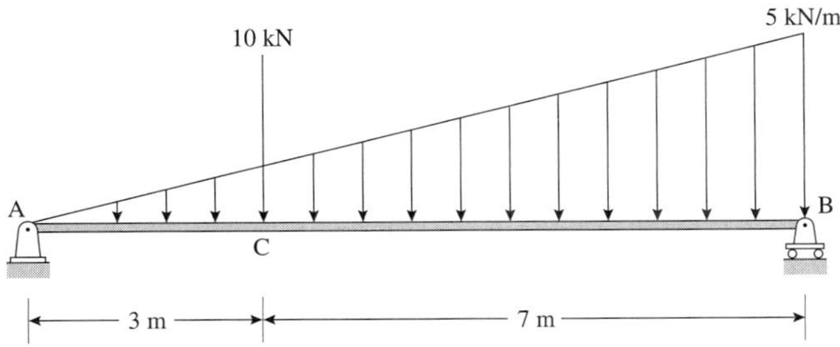

Fig. P2.8.5

2.8.6. Calculate the shear force and bending moment to the left and right of support B in Fig. P2.8.6.

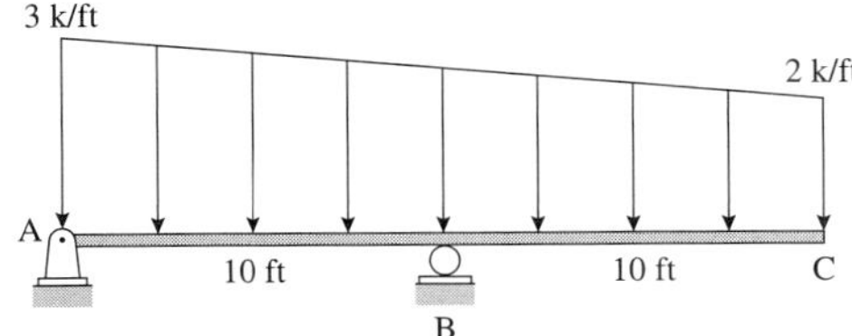

Fig. P2.8.6

2.8.7. Consider the beam shown in Fig. P2.8.7. A is a fixed support, B is an internal hinge, and C is a roller support. Compute the internal forces to the left and right of hinge B.

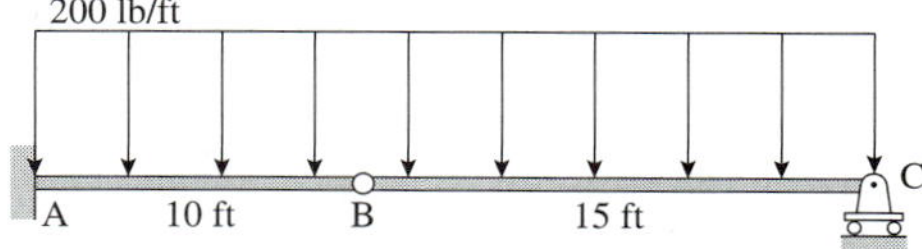

Fig. P2.8.7

2.8.8. Compute the internal forces at the center of the beam in Fig. P2.8.8.

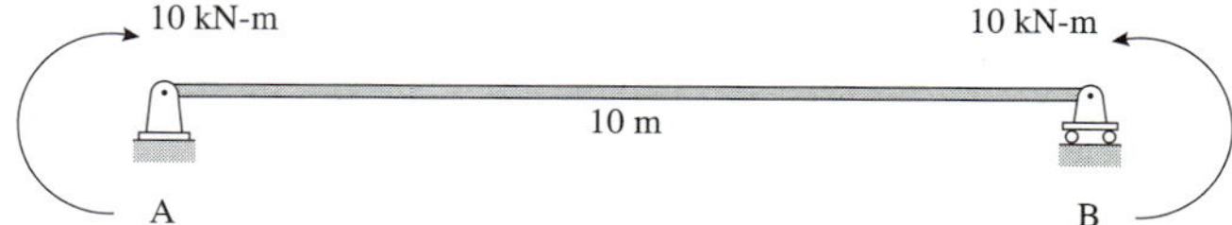

Fig. P2.8.8

Structural Concepts

2.8.9. Why would it be incorrect to state that the largest shear always occurs at the supports?

The following problems deal with graphing polynomials. Graphing details can be found in Appendix E.

Appetizers

2.8.10. (a) Sketch the function $f(x) = x^2 - 5x - 30$ in the range $0 \leq x \leq 15$ using a right-handed coordinate system. (b) Sketch the function $f(x) = x^3 - 2x + 20$ in the range $0 \leq x \leq 10$ using a left-handed coordinate system.

Main Course

2.8.11. (a) Sketch the function $f(x) = x^3 - 10x^2 - 25x + 20$ in the range $0 \leq x \leq 15$. Calculate and locate the extreme values. (b) Sketch the function $f(x) = -x^2 + 5x + 10$ in the range $0 \leq x \leq 10$. Calculate and locate the extreme values.

Structural Concepts

2.8.12. A function is defined as $f(x) = -x^2 + 5x + 10$ in the range $0 \leq x \leq 10$. Transform the function to a new coordinate system x_1 whose origin is located two units to the right of x. Sketch both the functions in their given ranges to verify your result.

2.8.2 Shear Force and Bending Moment Diagrams

In general, the shear force and the bending moments vary along the length of any member. If we are able to construct a graph for the shear force, we will be able to find its properties such as the maximum and minimum values, and the zero value locations. Such as graph is known as the shear force (SF) diagram. Similarly, the graph of the bending moment is called the bending moment (BM) diagram.

Consider a simply supported beam with a distributed load as shown in Fig. 2.8.2.1.

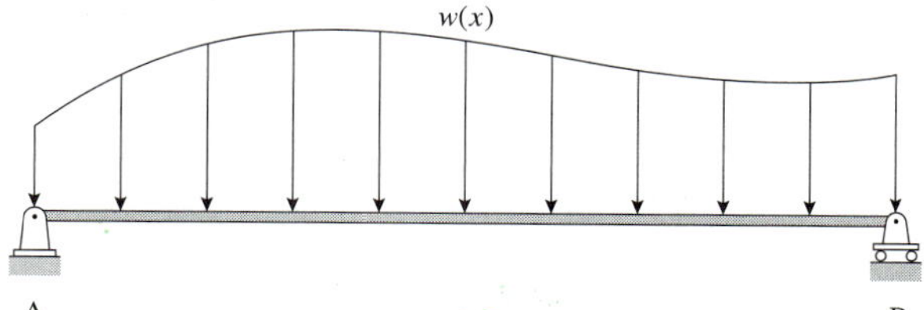

Fig. 2.8.2.1

Make a cut at a distance x from the left support and another cut at a distance $(x + \Delta x)$ from the left support. The resulting FBDs are shown in Fig. 2.8.2.2. The axial forces at the cuts are not shown since they are zero. We can develop the relationship between the distributed load, the shear force, and bending moments by studying the equilibrium of the differential element (Fig. 2.8.2.3).

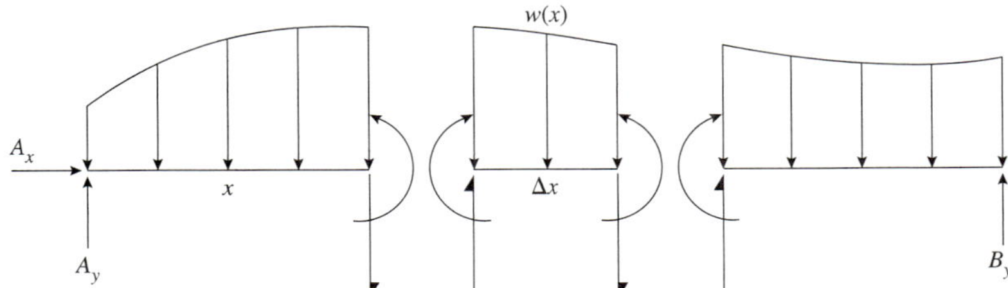

Fig. 2.8.2.2

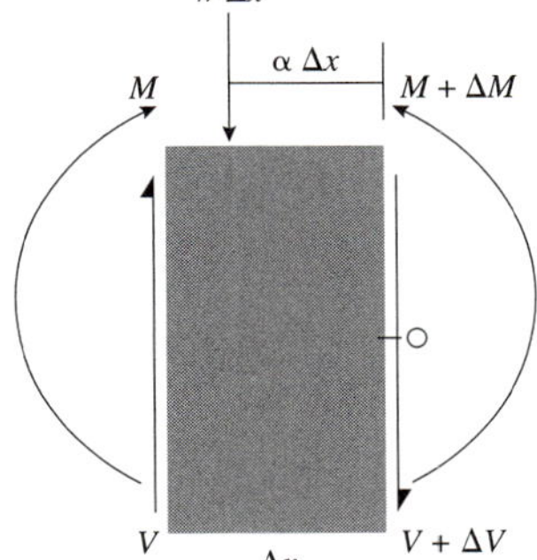

Fig. 2.8.2.3 FBD of the differential element.

The shear force and bending moment on the left face are denoted by V and M respectively. The resultant of the distributed load is $w(x')dx$.[2] On the right face, $(V + \Delta V)$ is the shear force and $(M + \Delta M)$ is the bending moment. ΔV and ΔM denote the changes in the shear force and bending moment between the left and the right faces. Since the entire beam is in equilibrium, the differential element must also be in equilibrium.

$$\overset{\uparrow +}{\sum} F_y = 0 = V - w(x')(\Delta x) - (V + \Delta V) \tag{2.8.2.1}$$

Simplifying the above equation and taking the limit $\Delta x \to 0$, we have

$$\boxed{\frac{dV}{dx} = -w(x)} \tag{2.8.2.2}$$

Relationship 1: The slope of the shear force diagram at any point on the beam is equal to the negative of the intensity of the loading at that point.

Similarly, taking the moment of all the forces about point O yields

$$\overset{\curvearrowleft +}{\sum} M_{Oz} = 0 = -V(\Delta x) - M + w(\Delta x)[\alpha(\Delta x)] + (M + \Delta M) \qquad 0 < \alpha < 1 \tag{2.8.2.3}$$

Simplifying the above equation and taking the limit $\Delta x \to 0$, we have

$$\boxed{\frac{dM}{dx} = V(x)} \tag{2.8.2.4}$$

---

[2] The resultant R is a force that acts on the differential element, i.e., between x and $(x + \Delta x)$. Go back to Section 2.2.1 to verify this observation. In the limit $\Delta x \to 0$, we have $w = w(x)$.

Relationship 2: The slope of the bending moment at any point on the beam is equal to the shear force at the corresponding point.

The two relationships are a powerful tool in drawing and checking the accuracy of the SF and BM diagrams, as summarized in Table 2.8.2.1.

Observation: The two relationships provide the means of locating and computing the extreme values of shear forces and bending moments. The shear force has an extreme value where the load is zero. Similarly, the bending moment has an extreme value where the shear is zero. The extreme value can be a maximum or a minimum value (see Appendix E).

Table 2.8.2.1 Polynomial Relationships Between Loading, Shear, and Bending Moments

Type of Loading Between Two Points, $w(x)$	Nature of the Shear Force Between Those Two Points	Nature of the Bending Moment Between Those Two Points
Zero (no load)	Constant	Linear
Constant (uniformly distributed load)	Linear	Quadratic
Linear (triangular or trapezoidal)	Quadratic	Cubic

EXERCISES

Appetizers

2.8.13.

(a) The expressions for the loading, shear, and bending moments are as follows: $w(x) = -8$, $V(x) = 8x - 3$, $M(x) = 4x^2 - 3x + 10$. Show that they satisfy the relationships between the three quantities.

(b) For the beam shown in Fig. P2.8.13(b), write the expression for the loading $w(x)$.

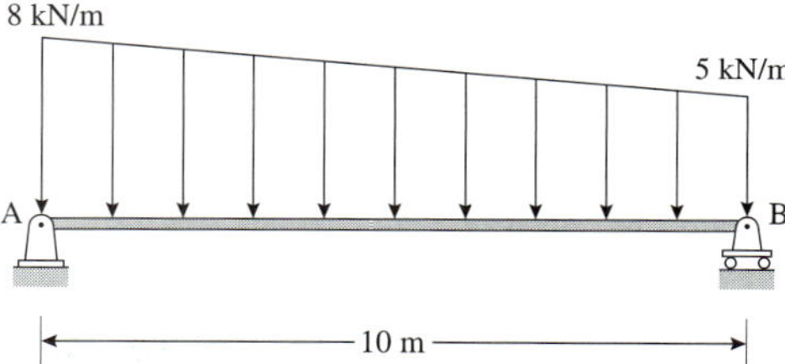

Fig. P2.8.13(b)

Main Course

2.8.14. The loading on a beam is given as $w(x) = 3x + 4$ in the range $0 \le x \le 12$. What are the expressions for the shear force and bending moments given that $V(x = 0) = M(x = 0) = 0$?

Structural Concepts

2.8.15. Are the following true or false? If false, state the reason(s) why (a preferred approach is to give counterexample(s) to show why the statement is false).

(a) The largest shear occurs at the point of zero loading.

(b) The largest moment occurs at the point of zero shear.

(c) There is a change in slope in the bending moment diagram where there is a sudden change in the value of the shear force.

2.8.3 Shear Force and Bending Moment Diagrams for Beams

We now look at the procedure for drawing the shear force and bending moment diagrams. The diagrams can be readily interpreted if we have a sign convention associated with the diagrams.

Sign Convention. When a cut is made in a member it creates two internal surfaces. We designate these two surfaces as one with a right outward normal (RON) and the other with a left outward normal (LON). A shear force is positive on a RON face if it is acting down. Similarly, the bending moment is positive on a RON face if it is acting counterclockwise. One can similarly state the conditions for a LON face. In Fig. 2.8.3.1, all the shear forces and bending moments are positive as per this sign convention.

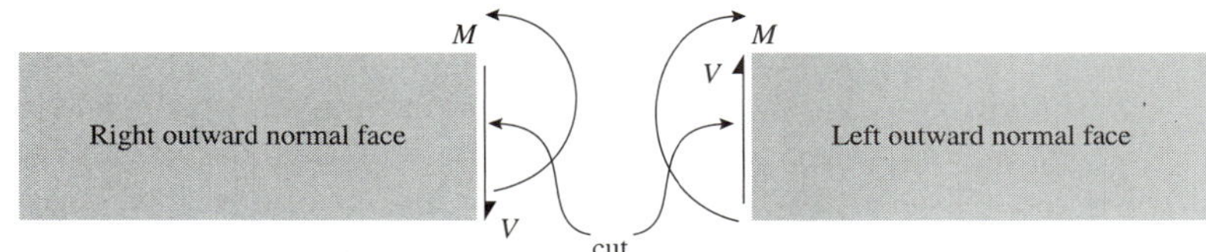

Fig. 2.8.3.1

Sign conventions are arbitrarily chosen. Since they are not universal, one must communicate the sign convention when transmitting the diagrams to a different person.

Note that Eqs. (2.8.2.2) and (2.8.2.4) are derived for a RON face and are valid as such. You should derive similar equations for a left outward normal face.

We now look at the steps in drawing the SF and BM diagrams for beams.

Step 1: Compute all the support reactions using the structural FBD.

Step 2: Identify all the points of discontinuities in the shear force and bending moment diagrams. Since the diagrams must close at the ends of the beams, these two points should be included in the list of points.

Step 3: Between two adjacent points of discontinuities, make a cut in the beam. Use the FBD of either the left or the right half of the beam. Show the shear force V and the bending moment M at the cut, assuming that they are positive as per the sign convention. Now sum the forces in the vertical direction, use the equilibrium condition, and generate the expression $V(x)$. Similarly, sum the moments of all the forces about the cut, use the equilibrium condition, and generate the expression $M(x)$. Locate the zero, maximum and minimum points and values using $V(x)$ and $M(x)$. Repeat the procedure for all the segments (the beam between two adjacent points of discontinuities is defined as a segment).

Use the expressions for $V(x)$ and $M(x)$ to find (a) locations (x values) where the expressions are zero, and (b) maximum and minimum values and the corresponding locations.

It is also extremely useful to carry out checks as you proceed through the solution. Some of the checks are as follows.

- The diagrams must satisfy Eqs. (2.8.2.2) and (2.8.2.4).
- An abrupt change in the shear force diagram takes place where there are concentrated forces (support reactions or applied forces). An abrupt change in the slope of the shear force diagram takes place where there are abrupt changes in the loading.
- An abrupt change in the bending moment diagram takes place where there are concentrated moments. An abrupt change in the slope of the bending moment diagram takes place where there are abrupt changes in the shear force diagram.
- The shear force and bending moments at the end supports are equal to the support reactions. A free end has zero shear and moment.

Step 4: Using the results from step 3, sketch the SF and BM diagrams. There are two options here. Either these diagrams can be drawn for each member separately, or a combined diagram can be drawn for the entire structure. For a structure with a few members, the latter option is preferable.

Observation: There are several approaches to drawing SF and BM diagrams. Some of these procedures integrate Eq. (2.8.2.2) and interpret the result as the difference in shear values between two points as being equal to negative of the area of the loading diagram between those two points. A similar conclusion involving change in moments and area under the shear diagram can be obtained by integrating Eq. (2.8.2.4). The diagrams can usually be constructed intuitively and quickly. However, these procedures are not encouraged here for three reasons. First, they run of out steam when computing areas under curves becomes difficult. Second, with increasing orders of polynomial, they are unable to provide locations of maximum, minimum, and zero values. Third, the expressions for the moments $M(x)$ will be indispensable when computing the deflections of frames and when solving indeterminate structures.

Illustrating the Sign Convention. It is desirable that the sign convention for the shear force and bending moment diagrams be shown along with the diagrams. Figure 2.8.3.2 shows the suggested style for the shear force diagram. The differential element on the right illustrates our sign convention. For example, on a RON surface at A, the shear force acts down.

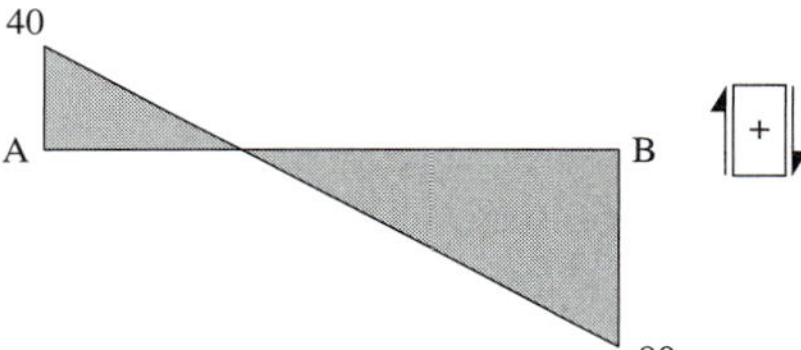

Fig. 2.8.3.2
Shear force diagram (lb).

Figure 2.8.3.3 shows the suggested style for the bending moment. The curvature of the beam is shown in the diagram so that it is clear which side of the beam is under compression or tension.

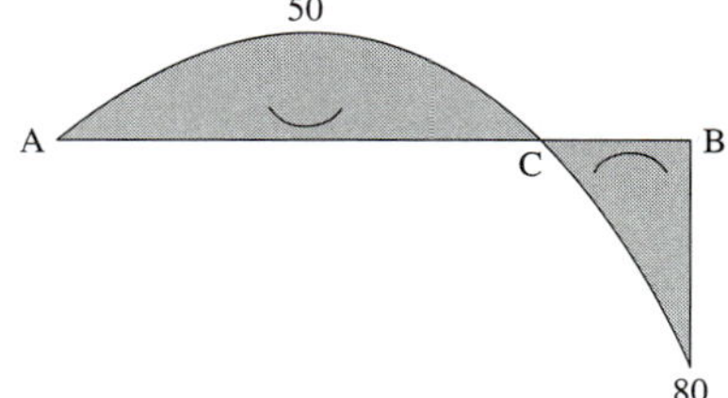

Fig. 2.8.3.3
Bending moment diagram (lb-ft).

When the curvature is shown as ◡, the top fiber is in compression and the bottom fiber is in tension. Similarly, when the curvature is shown as ◠, the bottom fiber is in compression and the top fiber is in tension. Hence in Fig. 2.8.3.3, from A to C the top fiber is in compression, and from C to B, the bottom fiber is in compression.

EXAMPLE 2.8.2 ***Cantilever Beam***

Draw the shear force and bending moment diagrams for the beam shown in Fig. E2.8.2(a).

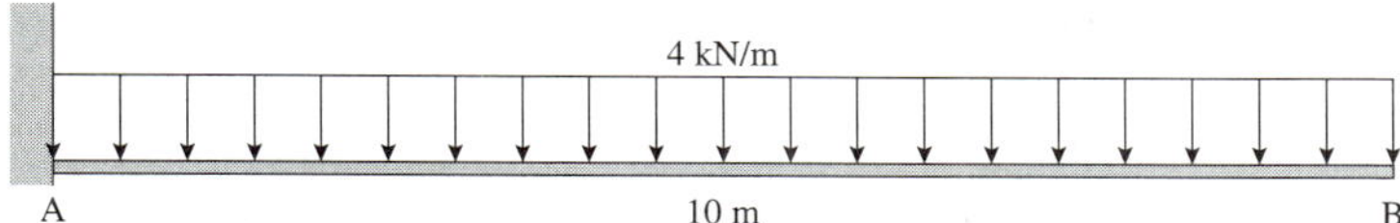

Fig. E2.8.2(a)

SOLUTION

Step 1: R = 4 kN/m × 10 m = 40 kN. Using the structural FBD in Fig. E2.8.2(b), we find

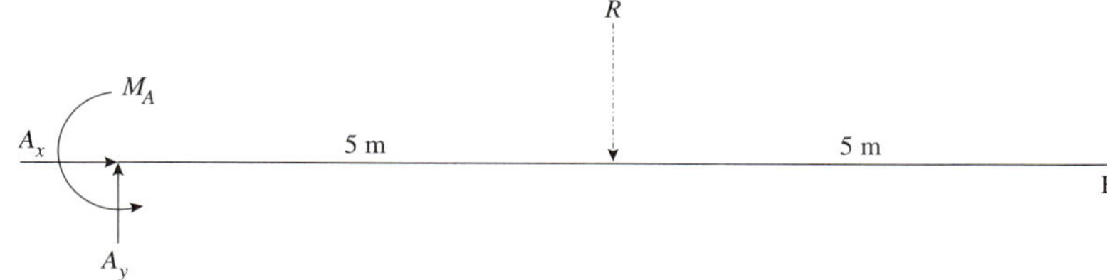

Fig. E2.8.2(b)

(a) $\overset{\curvearrowleft +}{\sum} M_A = 0 = M_A - R(5) \Rightarrow M_A = 200 \text{ kN-m}$

(b) $\overset{\uparrow +}{\sum} F_y = 0 = A_y - R \Rightarrow A_y = 40 \text{ kN}$

(c) $\overset{\rightarrow +}{\sum} F_x = 0 = A_x \Rightarrow A_x = 0 \text{ kN}$

Step 2: There are no points of discontinuities between A and B.

Step 3: Making a cut between A and B (see Fig. E2.8.2(c)), we have (for $0 < x < 10$ m) $\overset{\uparrow +}{\sum} F_y = 0 = A_y - R - V \Rightarrow V(x) = 40 - 4x$. Note that the resultant of the distributed load $R = 4x$. Shear force is zero at $x = 10$, and the linear function indicates that the maximum value is at $x = 0$ and decreases thereafter.

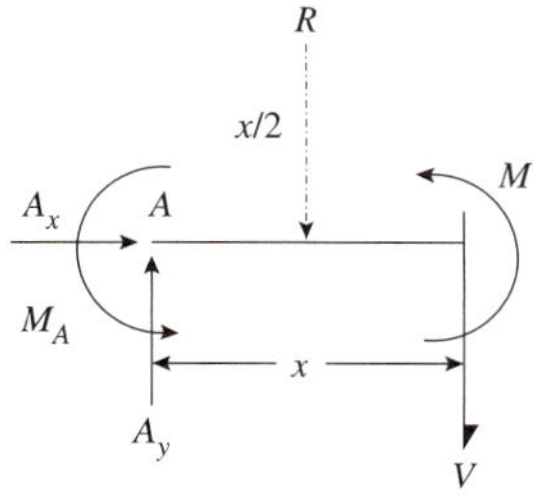

Fig. E2.8.2(c)

$$\overset{\curvearrowleft +}{\sum} M_{\text{cut}} = 0 = M + M_A - A_y(x) + R(x/2) \Rightarrow M(x) = -2x^2 + 40x - 200$$

The quadratic function has its roots at $M = 0 \ @ \ x_{1,2} = \dfrac{-40 \pm \sqrt{40^2 - 4(-2)(-200)}}{-(2)(-2)} = 10, 10$.

Checks:

1. B (x = 10 m) is a free end and the shear and bending moments must be zero. $V = 0 @ x = 40/4 = 10$ m. Also, $M(x = 10) = -2(10)^2 + 40(10) - 200 = 0$.

2. Also, $w(x) = 4x \cdot \dfrac{dV}{dx} = \dfrac{d}{dx}(40 - 4x) = -4x = -w(x)$,

$\dfrac{dM}{dx} = \dfrac{d}{dx}(-2x^2 + 40x - 200) = -4x + 40 = V(x)$ and Eqs. (2.8.2.2) and (2.8.2.4) are satisfied.

Step 4: We now draw the SF and BM diagrams by graphing in the range $0 \le x \le 10$ m for $V(x) = 40 - 4x$ and $M(x) = -2x^2 + 40x - 200$ (see Figs. E2.8.2(d) and (e)).

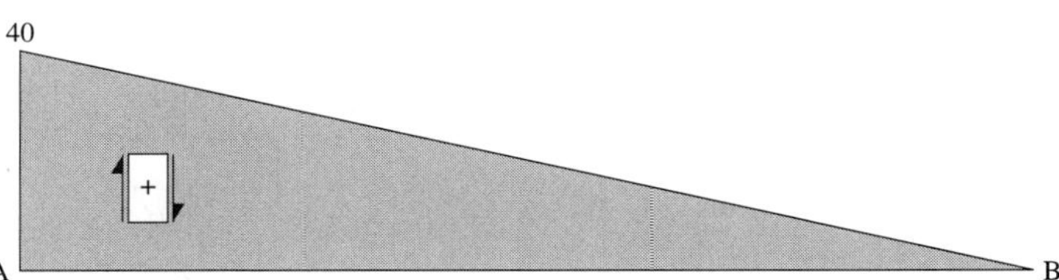

Fig. E2.8.2(d) Shear force diagram (kN).

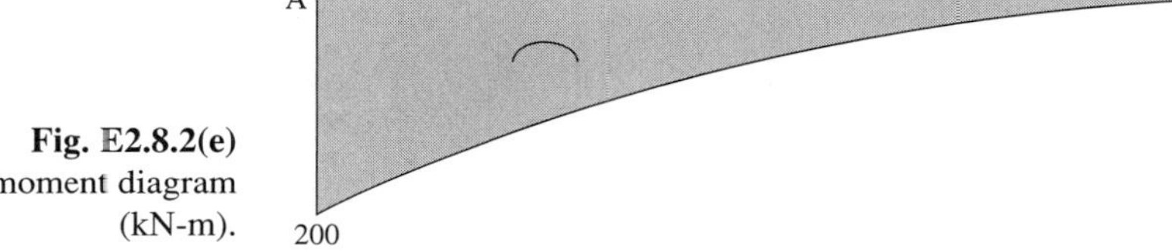

Fig. E2.8.2(e) Bending moment diagram (kN-m).

EXAMPLE 2.8.3 ***Simply Supported Beam (Concentrated Force)***

Draw the shear force and bending moment diagrams for the beam shown in Fig. E2.8.3(a).

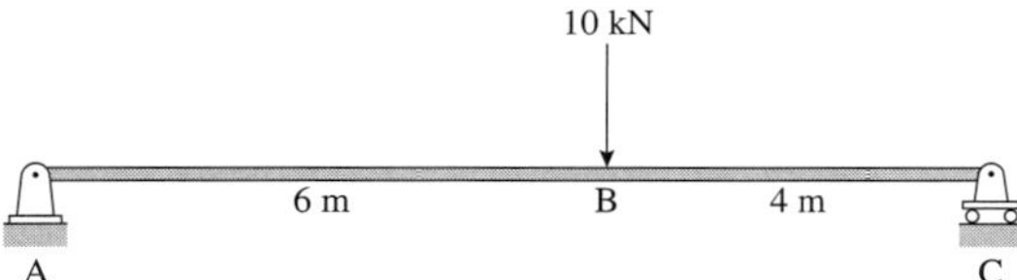

Fig. E2.8.3(a)

SOLUTION

Step 1: Using the structural FBD, we find (see Fig. E2.8.3(b))

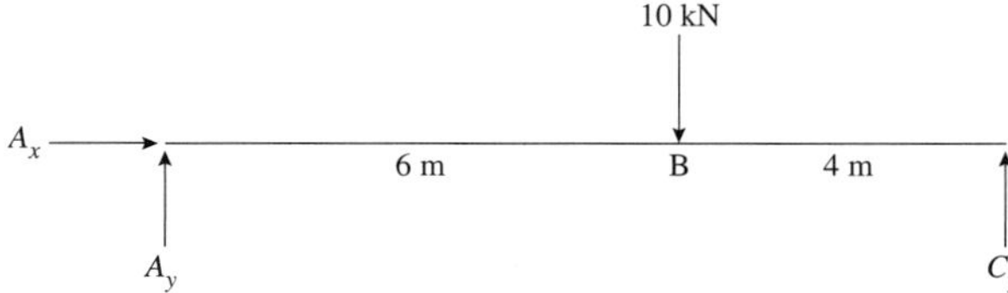

Fig. E2.8.3(b)

(a) $\overset{\curvearrowleft +}{\sum} M_A = 0 = C_y(10) - (10)(6) \Rightarrow C_y = 6$ kN

(b) $\overset{\uparrow +}{\sum} F_y = 0 = A_y - 10 + C_y \Rightarrow A_y = 4$ kN

(c) $\overset{\rightarrow+}{\sum} F_x = 0 = A_x \Rightarrow A_x = 0 \text{ kN}$

Step 2: B is a point of discontinuity. The shear force diagram must show an abrupt change in its value at B (equal to 10 kN) while there will be a change in slope of the bending moment diagram (equal to 10 kN-m/m).

Step 3: We slowly change the process of showing and computing the answers. A four-column table is developed showing the segment, the free-body diagram for the segment, the expression $V(x)$ for the shear force, and the expression $M(x)$ for the bending moment.

Segment (m)	FBD	$V(x)$ (kN)	$M(x)$ (kN-m)
AB $0 < x < 6$	A_x, A, M, x, A_y, V	$\overset{\uparrow+}{\sum} F_y = 0 = A_y - V$ $V(x) = 4$	$\overset{\curvearrowleft+}{\sum} M_{\text{cut}} = 0 = M - A_y(x)$ $M(x) = 4x$
BC $6 < x < 10$	10 kN, A_x, A, 6, $(x-6)$, B, M, x, A_y, V	$\overset{\uparrow+}{\sum} F_y = 0 = A_y - 10 - V$ $V(x) = -6$	$\overset{\curvearrowleft+}{\sum} M_{\text{cut}} = 0$ $M - A_y(x) + 10(x-6) = 0$ $M(x) = -6x + 60$

Checks:

1. Moment zero at A? $M(x) = 4x \ @\ x = 0 \Rightarrow 4(0) = 0$. Yes.
2. Moment zero at C? $M(x) = -6x + 60 \ @\ x = 10 \Rightarrow -6(10) + 60 = 0$. Yes.
3. Shear at A equal to the support reaction? $V(x) = 4 \ @\ x = 0 \Rightarrow 4$. Yes.
4. Shear at C equal to the support reaction? $V(x) = -6 \ @\ x = 10 \Rightarrow -6$. Yes.
5. Jump in shear at B equal to the applied load? $V(x) = 4 \ @\ x = 6 \Rightarrow 4$ and

   $V(x) = -6 \ @\ x = 6 \Rightarrow -6$. Change in shear = –6 – (4) = –10. Yes.
6. Unique value of bending moment at B? $M(x) = 4x \ @\ x = 6 \Rightarrow 4(6) = 24$ and

   $M(x) = -6x + 60 \ @\ x = 6 \Rightarrow -36 + 60 = 24$. Yes.

Step 4: Draw the SF and BM diagrams as in Figs. E2.8.3(c) and (d).

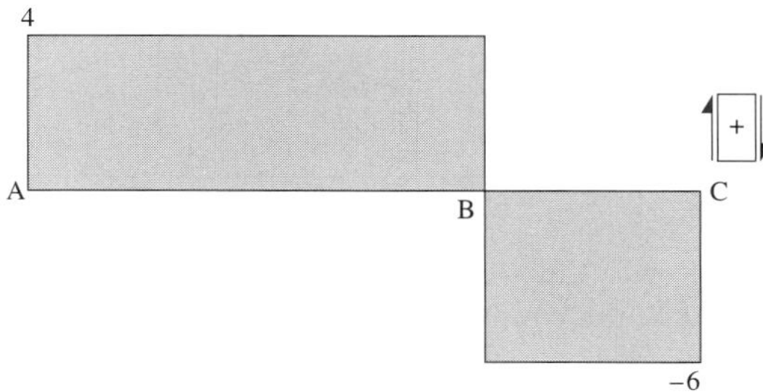

Fig. E2.8.3(c)
Shear force diagram (kN).

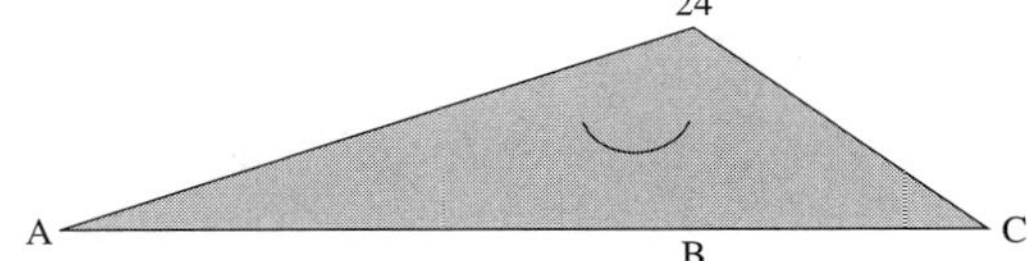

Fig. E2.8.3(d) Bending moment diagram (kN-m).

When considering a RON surface, the shear force diagram shows that the shear force acts down on the surface in the segment AB and then acts up all the way to the right support. The bending moment is such that the top fiber is under compression throughout the beam.

EXAMPLE 2.8.4 ***Determinate Beam with Uniformly Distributed Loading***

Draw the shear force and bending moment diagrams for the beam shown in Fig. E2.8.4(a).

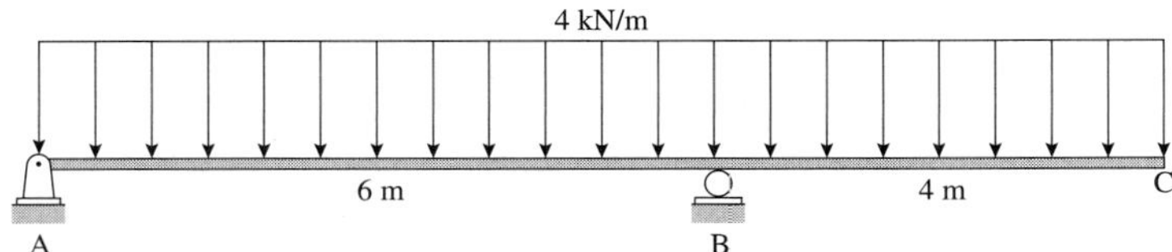

Fig. E2.8.4(a)

SOLUTION

Step 1: Using the structural FBD (not shown), we find

$$\text{(a)}\quad \overset{\curvearrowleft +}{\sum} M_A = 0 = -(4\times 10)(5) + B_y(6) \Rightarrow B_y = 33.33 \text{ kN}$$

$$\text{(b)}\quad \overset{\uparrow +}{\sum} F_y = 0 = A_y - (4\times 10) + B_y \Rightarrow A_y = 6.67 \text{ kN}$$

Step 2: The support reaction at B will cause a sudden change in the shear force value.

Segment (m)	FBD	$V(x)$ (kN)	$M(x)$ (kN-m)
AB $0 < x < 6$	$R = 4x$; A_x; A $(x/2)$; x; A_y; M; V	$\overset{\uparrow +}{\sum} F_y = 0 = A_y - 4x - V$ $V(x) = 6.67 - 4x$ $V = 0 \ @\ x = 1.67$	$\overset{\curvearrowleft +}{\sum} M_{\text{cut}} = 0$ $M - A_y(x) + R(x/2) = 0$ $M(x) = 6.67x - 2x^2$ $M(x = 1.67) = 5.56$ $M = 0 \ @\ x_1 = 0; x_2 = 3.33$
BC $6 < x < 10$	$R = 4x$; A_x; A; $(x/2)$; 6 m; $x-6$; A_y; B_y; M; V	$\overset{\uparrow +}{\sum} F_y = 0$ $A_y - 4x + B_y - V = 0$ $V(x) = 40 - 4x$ $V = 0 \ @\ x = 10$	$\overset{\curvearrowleft +}{\sum} M_{\text{cut}} = 0$ $M - A_y(x) + R(x/2) - B_y(x-6)$ $M(x) = -2x^2 + 40x - 200$ $M(x = 10) = 0$ $M = 0 \ @\ x = 10$

For both the segments, the origin of the coordinate system is at A. While there is a jump in the shear force at B, the bending moment value is unique. From the expression for segment AB, $M_B = M(x = 6) = 6.67(6) - 2(6)^2 = -32$ kN-m. From the expression for segment BC, $M_B = M(x = 6) = -2(6) + 40(6) - 200 = -32$ kN-m.

There are occasions where the FBD associated with the segment with LON is simpler to configure and use. Consider the segment CB with the origin of the coordinate system at C as shown below.

CB $0 < x_1 < 4$	$R = 4x_1$; M; V; $x_1/2$; C; x_1	$\overset{\uparrow +}{\sum} F_y = 0 = V - R$ $V(x_1) = 4x_1$	$\sum M_{cut} = 0$ $-M - 4x_1\left(\frac{x_1}{2}\right) = 0$ $M(x_1) = -2x_1^2$

Using this expression, let us check the bending moment value at B: $M_B = M(x_1 = 4) = -2(4)^2 = -32$ kN-m, the same value as before. For this (LON) segment Eqs. (2.8.2.2) and (2.8.2.4) do not apply and hence cannot be used as a check.[3]

Step 3: Draw the SF and BM diagrams as in Figs. E2.8.4(b) and (c).

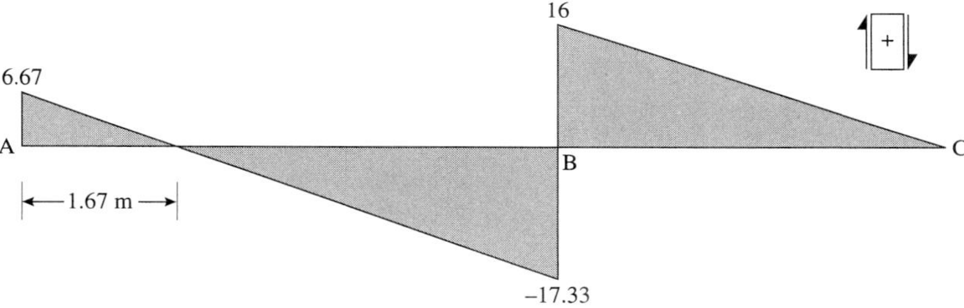

Fig. E2.8.4(b)
Shear force diagram (kN).

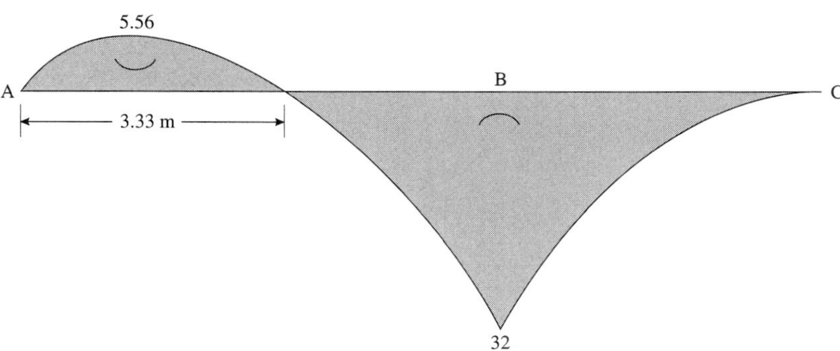

Fig. E2.8.4(c)
Bending moment diagram (kN-m).

When considering a RON surface, the shear force diagram shows that the shear force acts down on the surface from A to 1.67 m from A and then acts up till support B. The direction then changes; the shear force once again acts down in the segment BC. The bending moment is such that the top fiber is under compression from A to 3.33 m from A. Then the bottom fiber is in compression all the way to C.

Checks:

1. The relationships in Eqs. (2.8.3.2) and (2.8.3.4) are satisfied in each segment. The loading is constant, the shear force is piecewise linear and the bending moment is piecewise quadratic.

---

[3] The derivation in Section 2.8.2 was based on a RON segment. Derive Eqs. (2.8.2.2) and (2.8.2.4) for a LON segment.

2. The bending moments at the end pin support (at A) and at the free end C are zero. Note, however, that the bending moment at the internal roller support at B is not zero.
3. The shear force at A is equal to the support reaction. It is zero at the free end at C. The jump at B is equal to the support reaction. The maximum bending moment occurs at the point of zero shear.

EXAMPLE 2.8.5 ***Cantilever Beam with Triangular Loading***

Draw the shear force and bending moment diagrams for the beam shown in Fig. E2.8.5(a).

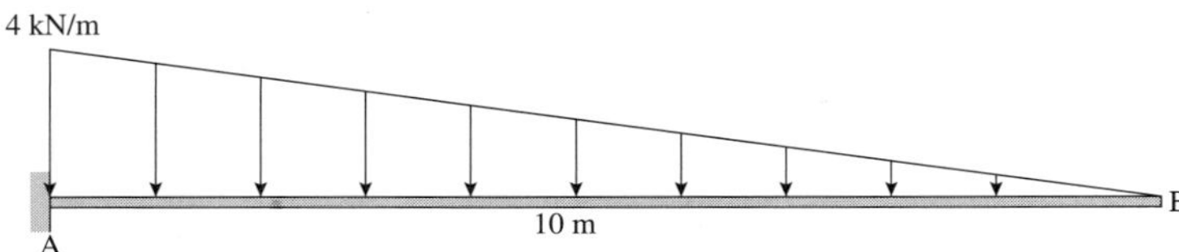

Fig. E2.8.5(a)

SOLUTION

Step 1: The reactions at A are

$$A_y = \frac{1}{2}(10)(4) = 20 \text{ kN and } M_A = 20\left(\frac{10}{3}\right) = 66.67 \text{ kN-m}.$$

Step 2: There are no discontinuities between A and B.

Approach 1: Using a RON segment and the FBD in Fig. E2.8.5(b), for $0 \le x \le 10$ m, we have

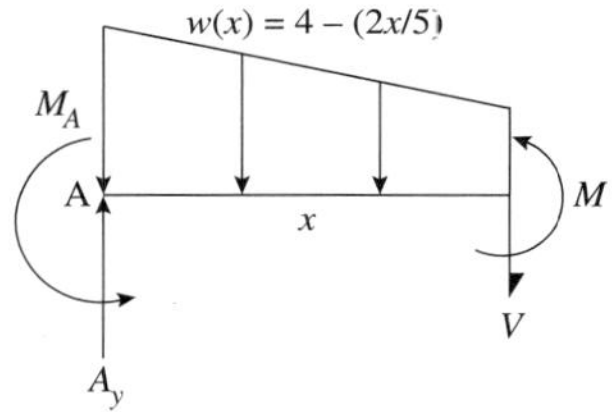

Fig. E2.8.5(b)

$$\overset{\uparrow +}{\sum} F_y = 0 = A_y - \int_0^x w(a)\,da - V \Rightarrow V(x) = 20 - \left[4a - \frac{a^2}{5}\right]_0^x = \frac{x^2}{5} - 4x + 20$$

$$\sum M_{\text{cut}} = 0 = M_A + \int_0^x w(a)(x-a)\,da - A_y x + M$$

$$\text{or } M(x) = -66.67 - \left[4xa - 2a^2 - \frac{a^2 x}{5} + \frac{2a^3}{15}\right]_0^x + 20x = \frac{x^3}{15} - 2x^2 + 20x - 66.67$$

Approach 2: Using a LON segment and the FBD in Fig. E2.8.5(c), we find $0 \le x_1 \le 10$ m

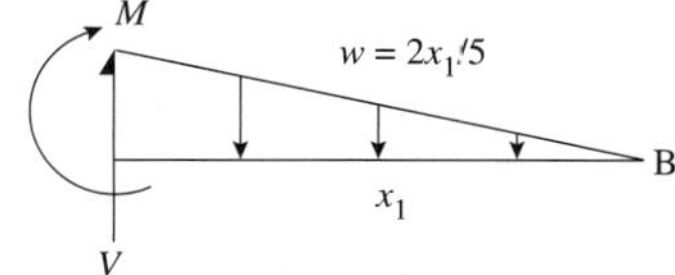

Fig. E2.8.5(c)

$$\overset{\uparrow +}{\sum} F_y = 0 = -\int_0^{x_1} w(a)\,da + V \Rightarrow V(x_1) = \left[\frac{a^2}{5}\right]_0^{x_1} = \frac{x_1^2}{5}$$

$$\sum^{\curvearrowleft +} M_{\text{cut}} = 0 = -M - \int_{0}^{x_1} w(a)(x_1 - a)\,da$$

$$\text{or} \quad M(x_1) = -\left[\frac{a^2 x_1}{5} - \frac{2a^3}{15}\right]_0^{x_1} = \frac{x_1^3}{15}$$

Step 3: Draw the shear force and bending moment diagrams[4] as in Figs. E2.8.5(d) and (e).

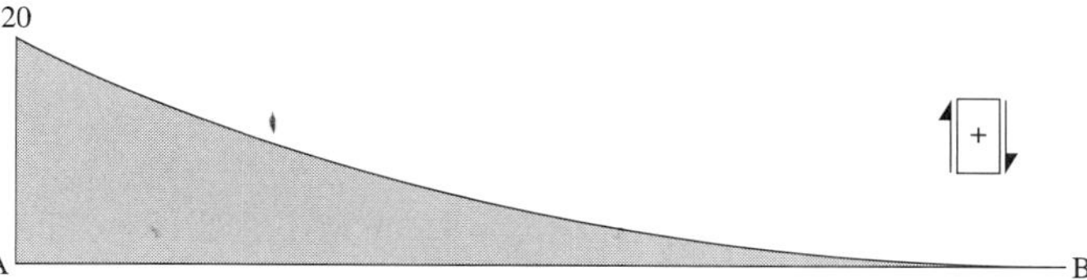

Fig. E2.8.5(d)
Shear force diagram (kN).

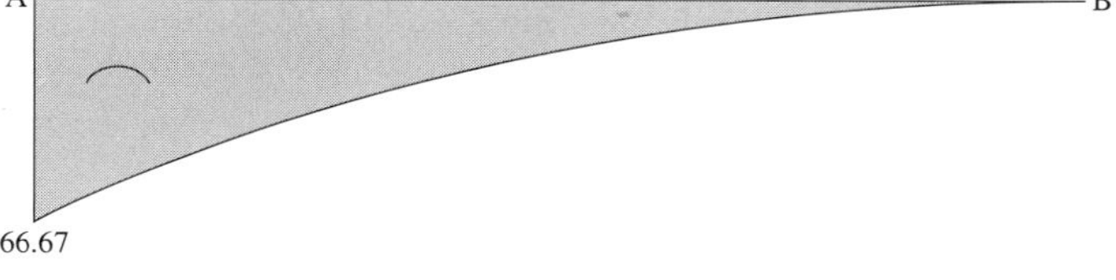

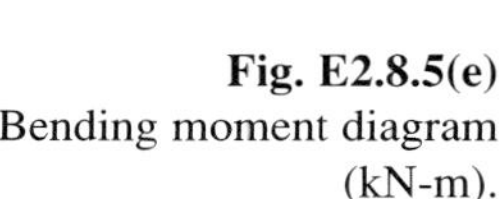

Fig. E2.8.5(e)
Bending moment diagram (kN-m).

When considering a RON surface, the shear force diagram shows that the shear force acts down on the surface from A to B. The bending moment is such that the bottom fiber is under compression from A to B.

Checks:

1. The relationships in Eqs. (2.8.2.2) and (2.8.2.4) are satisfied in each segment.
2. The bending moment and shear force are zero at the free end B.
3. The bending moment and shear force at the fixed end A are equal to the support reactions.

EXAMPLE 2.8.6

Beam with Internal Hinge

Draw the shear force and bending moment diagrams for the beam shown in Fig. 2.8.6(a).

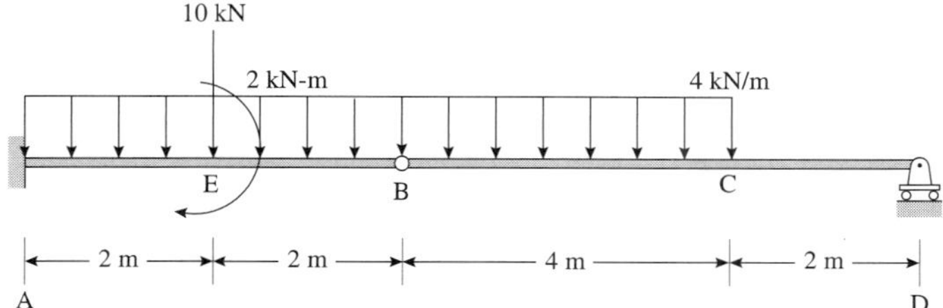

Fig. E2.8.6(a)

SOLUTION

Step 1: Using the FBDs in Fig. E2.8.6(b), we find

---

[4] In approach 2, replace the triangular loading with its resultant R = (1/2) × x_1 × ($2x_1/5$)= x_1^2) acting at $x_1/3$ from B. Rederive the expressions. Similarly, in approach 1 replace the trapezoidal loading with a resultant for the rectangular part and a resultant for the triangular part. This process may be a little more intuitive than the integration approach. See Example 2.8.9.

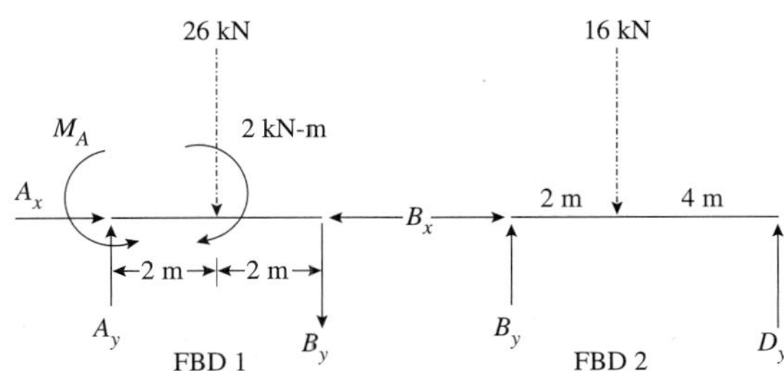

Fig. E2.8.6(b)

FBD 2: $D_y = 5.33\ \text{kN}(\uparrow)$, $B_y = 10.67\ \text{kN}(\uparrow)$, $B_x = 0$

FBD 1: $A_y = 36.67\ \text{kN}(\uparrow)$, $M_A = 96.68\ \text{kN-m}\ (\curvearrowleft)$, $A_x = 0$

Step 2: From the problem data, E and C are the points of discontinuities between A and D. At E there is a sudden change in shear force and bending moment. At C there is a sudden change only in the shear force value. Note that while B is an internal hinge (hence a zero moment location), there is neither a change in value nor a change in slope in the shear force and bending moment diagrams at B. We use the RON segment for AE and LON segments for DC and CE.

Segment (m)	FBD	$V(x)$ (kN)	$M(x)$ (kN-m)
AE $0 < x < 2$	$4x$, M_A, A, $x/2$, M, A_x, x, A_y, V	$\overset{\uparrow+}{\sum} F_y = 0 = A_y - 4x - V$ $V(x) = 36.67 - 4x$	$\overset{\curvearrowleft +}{\sum} M_{\text{cut}} = 0$ $M - A_y(x) + 4x\left(\frac{x}{2}\right) + M_A = 0$ $M(x) = -2x^2 + 36.67x - 96.68$
CE $2 < x_1 < 8$	$4(x_1-2)$, M, $(x_1-2)/2$, C, D, 2 m, x_1, V, D_y	$\overset{\uparrow+}{\sum} F_y = 0$ $D_y - 4(x_1 - 2) + V = 0$ $V(x_1) = 4x_1 - 13.33$ $V = 0\ @\ x_1 = 3.33$	$\overset{\curvearrowleft +}{\sum} M_{\text{cut}} = 0 = -M + D_y(x_1)$ $+ [4(x_1 - 2)]\left[\frac{x_1 - 2}{2}\right]$ $M(x_1) = -2x_1^2 + 13.33x_1 - 8$ $M(x_1 = 3.33) = 14.21$
DC $0 < x_1 < 2$	M, D, x_1, V, D_y	$\overset{\uparrow+}{\sum} F_y = 0$ $D_y + V = 0$ $V(x_1) = -5.33$	$\overset{\curvearrowleft +}{\sum} M_{\text{cut}} = 0$ $-M + D_y(x_1) = 0$ $M(x_1) = 5.33x_1$

Step 3: Draw the shear force and bending moment diagrams as in Figs E2.8.6(c) and (d).

For a RON surface, the shear force diagram shows that the shear force acts down on the surface from A to 6.67 m from A and then acts up all the way to the right support. The bending moment is such that the bottom fiber is under compression from A to B. In segment BD the top fiber is in compression.

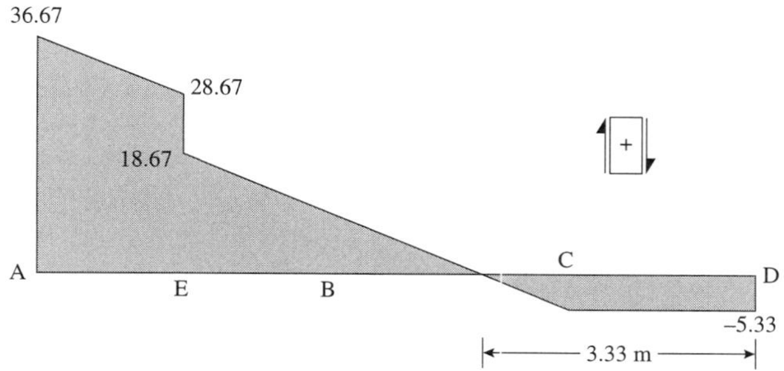

Fig. E2.8.6(c)
Shear force diagram (kN).

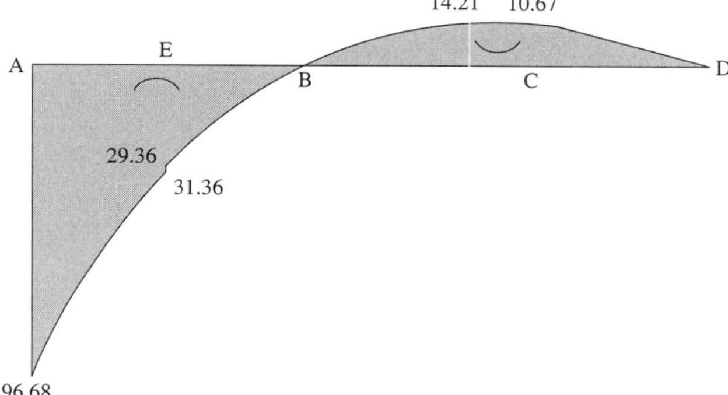

Fig. E2.8.6(d)
Bending moment diagram (kN-m).

Checks:

1. The relationships in Eqs. (2.8.2.2) and (2.8.2.4) are satisfied in each segment.
2. The bending moments at the internal hinge B and the end roller D are zero. The jump in the bending moment at E is 2 kN-m. The bending moment at A is equal to the moment support reaction at A.
3. The shear forces at A and D are equal to the (vertical) support reactions at A and D. The jump in the shear force at E is equal to the concentrated force 10 kN. The maximum bending moment occurs at the point of zero shear.

EXERCISES

Appetizers

Draw the shear force and bending moment diagrams for the beams in Figs. P2.8.16–P2.8.19.

2.8.16.

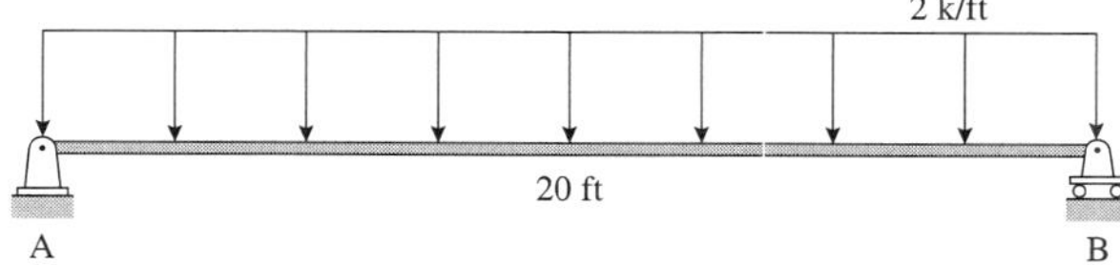

Fig. P2.8.16

2.8.17.

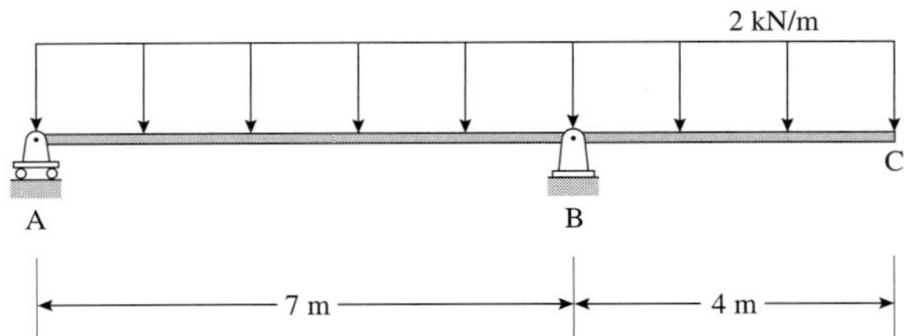

Fig. P2.8.17

2.8.18.

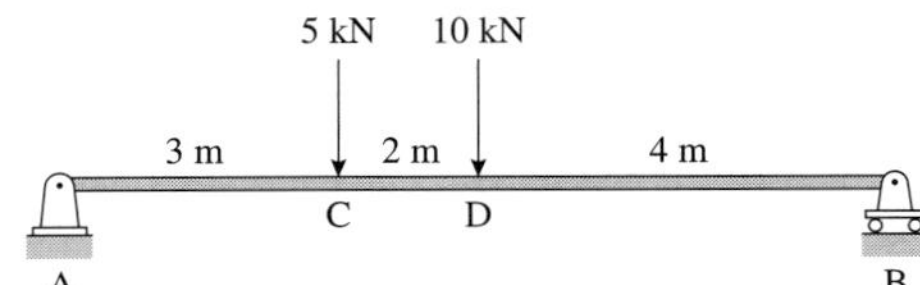

Fig. P2.8.18

2.8.19.

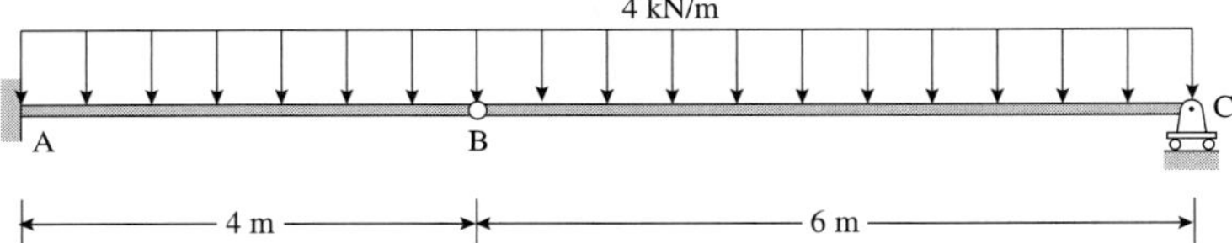

Fig. P2.8.19

Main Course

Draw the shear force and bending moment diagrams for the beams shown in Figs. P2.8.20–P2.8.22.

2.8.20.

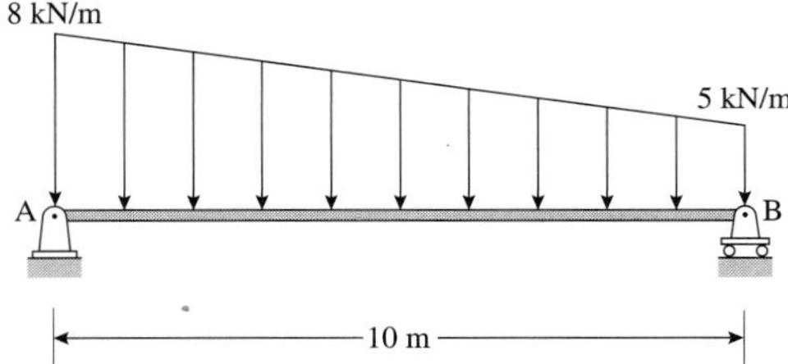

Fig. P2.8.20

2.8.21.

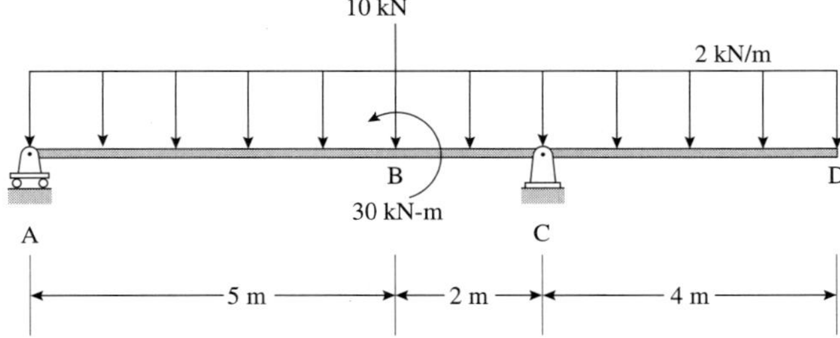

Fig. P2.8.21

2.8.22.

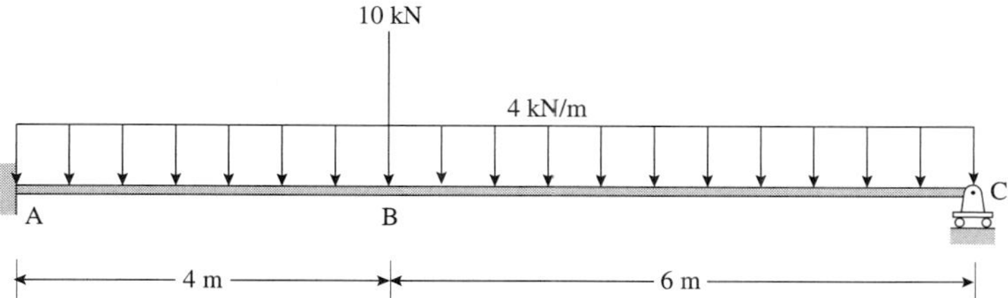

Fig. P2.8.22

2.8.23. What are the largest (magnitudes) shear force and bending moment in the beam in Fig. P2.8.23?

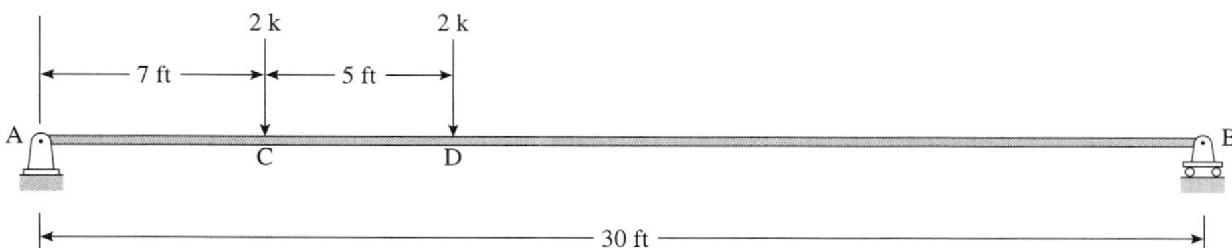

Fig. P2.8.23

Structural Concepts

2.8.24. If you are given the shear force and bending moment diagrams for a beam, can you construct the loading diagram and also show the manner in which the beam is supported?

2.8.25. Figure P2.8.25 shows an idealized model of a continuous beam (known as a balanced cantilever beam), with n equal spans of 10 ft each. In alternate spans, hinges are provided at a distance x from the supports. The idea is to reduce the magnitude of the largest bending moment in the beam. What is x and what is the largest bending moment for this x? Are the results valid for any span?

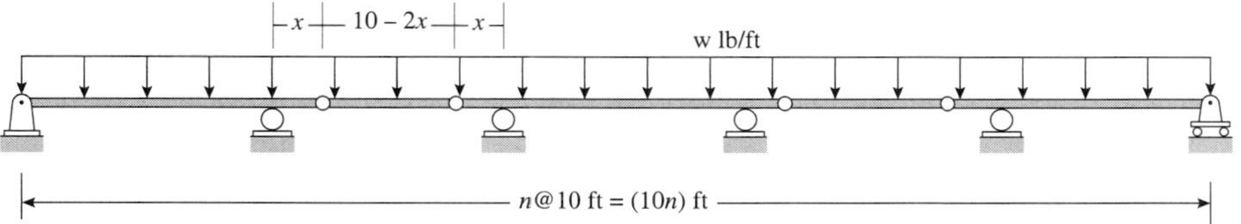

Fig. P2.8.25

2.8.26. The distribution of soil pressure under a footing depends on the soil properties and the relative stiffness of the footing and the soil. Figure P2.8.26 shows a cross-section of a *spread* footing. Assuming that the bearing pressure under the footing is uniform, draw the shear force and bending diagrams for the footing.

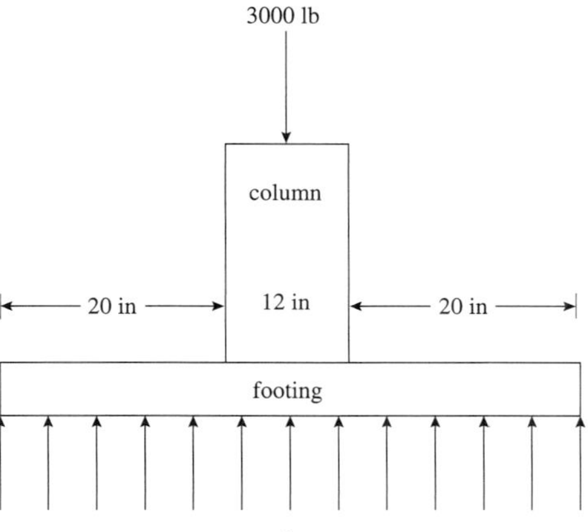

Fig. P2.8.26

2.8.27. Figure P2.8.27 shows a cross-section of a spread footing. The column is subjected to an axial force and a bending moment. As a result, the bearing pressure is trapezoidal, as shown. Draw the shear force and bending diagrams for the footing.

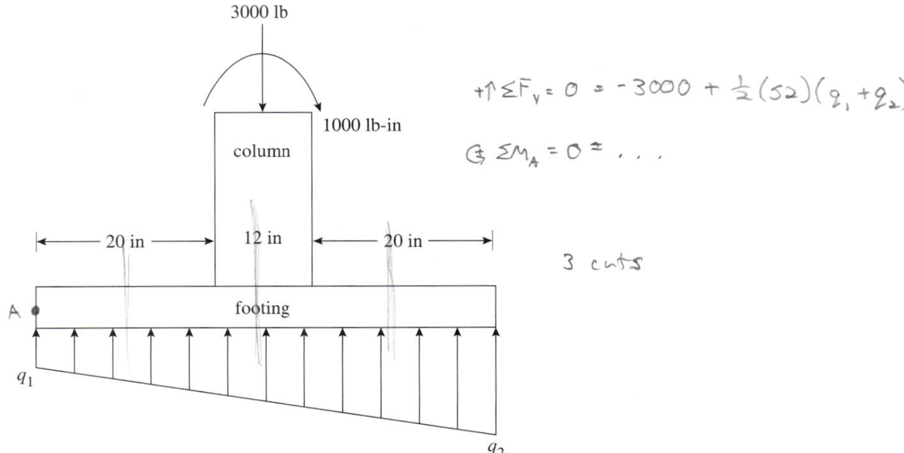

Fig. P2.8.27

2.8.4 Shear Force and Bending Moment Diagrams for Frames

In the preceding section, we looked at the procedure to draw the shear force and bending moment diagrams for beams. It will be necessary to reinvestigate the topic very briefly for two reasons. First, the notion of LON and RON is ambiguous with nonhorizontal members. Second, in a frame, the interaction between different members is crucial to understanding the distribution of shear forces and moments amongst the different members.

Sign Convention: Draw the shear force by assuming a consistent coordinate system for each member. Draw the bending moment on the compressive side of the member. Consider the following FBD (see Example 2.8.6). After obtaining the expression for $M(x)$, if for a certain value of x the moment is positive, then it acts as shown in the FBD. The implication is that the top fiber is in compression. Conversely, if for a certain value of x the moment is negative, then it acts opposite to the direction shown in the FBD in Fig. 2.8.4.1. The implication is that the bottom fiber is in compression. Since a positive value on a graph is placed above the member and a negative value below the member, the graph clearly indicates the compressive side of the member.

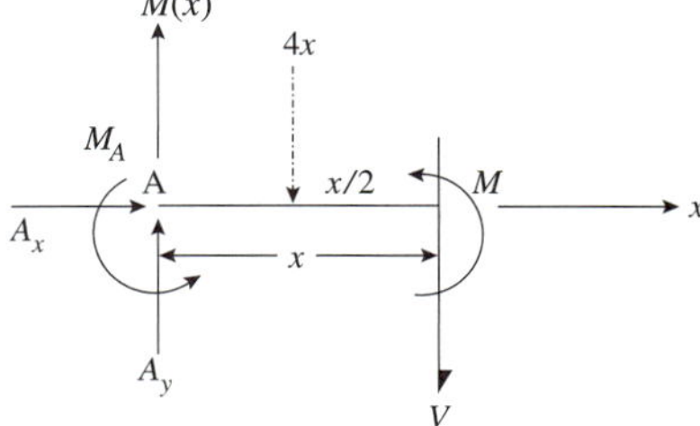

Fig. 2.8.4.1 Interpretation of the sign convention.

A careful examination of the sign convention shows that even though the two statements appear different, the net effect is the same. In fact, there is no need to redraw the BM diagrams from the preceding section—they conform to this new sign convention.

EXAMPLE 2.8.7 ***Simple Determinate Frame***

For the frame in Fig. E2.8.7(a), draw the shear force and bending moment diagrams.

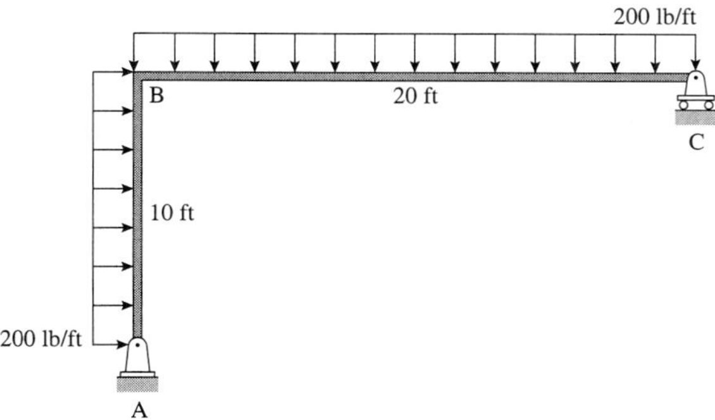

Fig. E2.8.7(a)

SOLUTION

Step 1: Using the structural FBD, we find $A_x = 2000$ lb (←), $A_y = 1500$ lb (↑), and $C_y = 2500$ lb (↑).

Step 2: Column AB has no points of discontinuity and neither does beam BC.

Column AB: As we have done with beams, we make a cut between A and B. The resulting FBD is shown in Fig. E2.8.7(b).

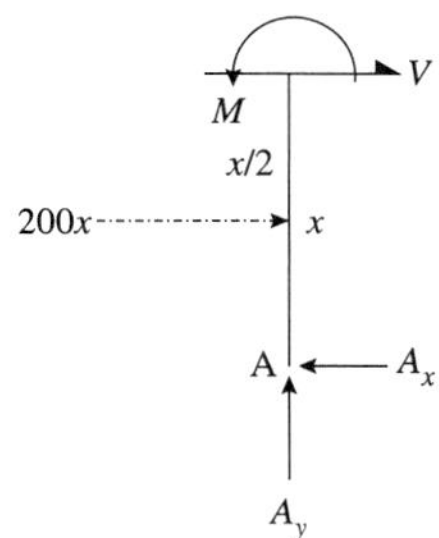

Fig. E2.8.7(b)

For $0 \le x \le 10$ ft,

$$\overset{\rightarrow +}{\sum} F_x = 0 = 200x + V - A_x \Rightarrow V(x) = -200x + 2000$$

$$\overset{\curvearrowleft +}{\sum} M_{\text{cut}} = 0 = M + 200x(x/2) - A_x(x) \Rightarrow M(x) = -100x^2 + 2000x$$

The shear is zero at x = 2000/200 = 10 ft, i.e., at B, and the bending moment at B is –10000 lb-ft.

Beam BC: We first make a cut between B and C, and then obtain the expressions in two different ways—the left FBD and the right FBD.

Approach 1: Left FBD ($0 \le x_1 \le 20$ ft; see Fig. E2.8.7(c)):

$$\overset{\uparrow +}{\sum} F_y = 0 = A_y - 200x_1 - V \Rightarrow V(x_1) = -200x_1 + 1500$$

Also, $V = 0 \text{ @ } x_1 = \dfrac{1500}{200} = 7.5$ ft.

$$\overset{\curvearrowleft +}{\sum} M_{\text{cut}} = 0 = -A_x(10) - A_y(x_1) + 2000(5) + 200x_1(x_1/2) + M$$

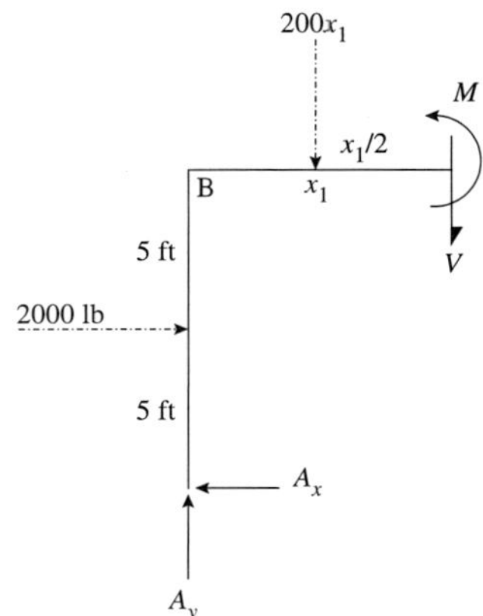

Fig. E2.8.7(c)

or $M(x_1) = -100x_1^2 + 1500x_1 + 10000$

And $M(x_1 = 7.5) = 15625$ lb-ft

Approach 2: Right FBD ($0 \le x_2 \le 20$ ft; see Fig. E2.8.7(d)):

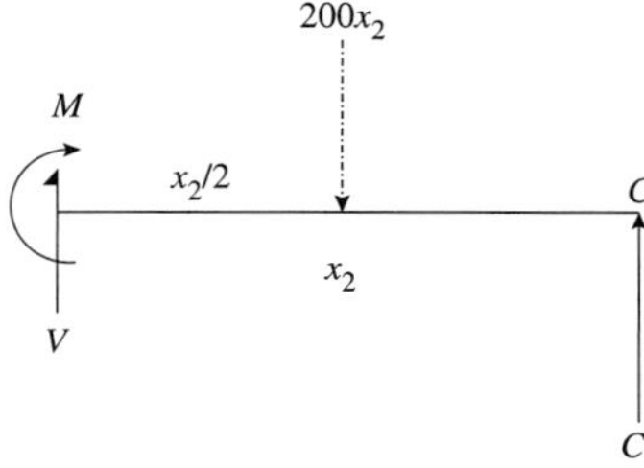

Fig. E2.8.7(d)

$$\overset{\uparrow +}{\sum} F_y = 0 = C_y - 200x_2 + V \Rightarrow V(x_2) = 200x_2 - 2500$$

$V = 0$ @ $x_2 = 12.5'$

$$\overset{\curvearrowleft +}{\sum} M_{\text{cut}} = 0 = -M + C_y(x_2) + 200x_2(x_2/2) \Rightarrow M(x_2) = 2500x_2 - 100x_2^2$$

$M(x_2 = 12.5) = 15625$ lb-ft

Step 3: Draw the shear force and bending moment diagrams, as in Figs. E2.8.7(e) and (f).

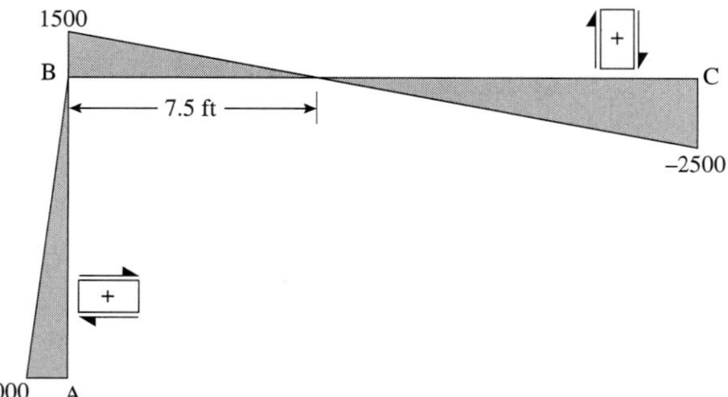

Fig. E2.8.7(e)
Shear force diagram (lb).

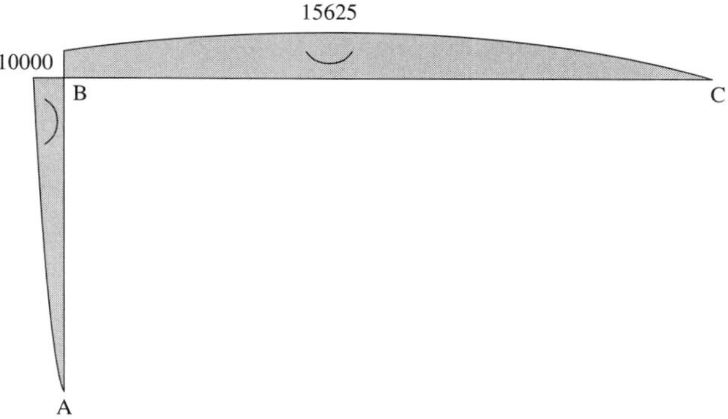

Fig. E2.8.7(f)
Bending moment diagram (lb-ft).

Since the loading on both members is constant, the shear force is expected to be linear and the bending moment a quadratic polynomial. The shear forces at A and C are equal to the appropriate reactions at those points. Similarly, the bending moments are zero at A (pin support) and at C (roller support) and continuous at B. The left side of column AB is in compression. Similarly, the top side of beam BC is in compression. The largest bending moment in BC occurs at the point of zero shear.

EXAMPLE 2.8.8 ***Frame with Inclined Member***

For the frame shown in Fig. E2.8.8(a), draw the shear force and bending moment diagrams.

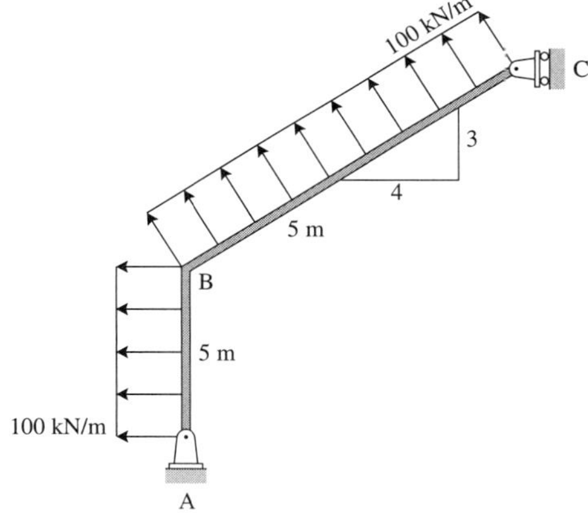

Fig. E2.8.8(a)

SOLUTION

Step 1: Using the structural FBD in Fig. E2.8.8(b), we find

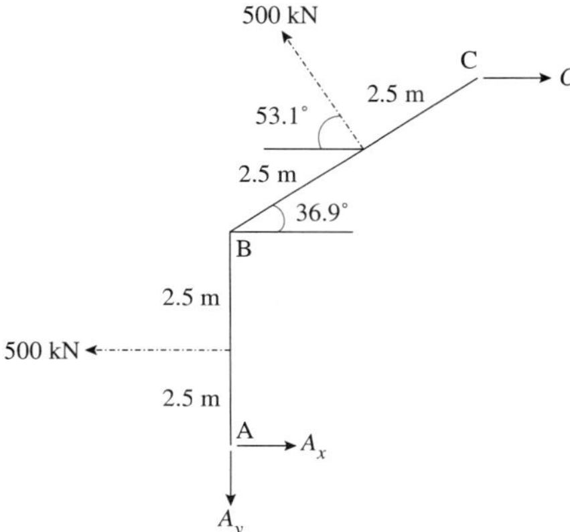

Fig. E2.8.8(b)

$$\overset{\uparrow+}{\sum} F_y = 0 = -A_y + 500\sin(53.1^\circ) \Rightarrow A_y = 400 \text{ kN}$$

$$\overset{\curvearrowleft+}{\sum} M_C = 0 = A_x(5 + 5\sin(36.9^\circ)) + A_y(5\cos(36.9^\circ)) - 500(2.5 + 5\sin(36.9^\circ)) - 500(2.5)$$

$$\text{Or,}\quad A_x = \frac{2750 + 1250 - 1600}{8} = 300 \text{ kN}$$

$$\overset{\rightarrow+}{\sum} F_x = A_x - 500 - 500\cos(53.1^\circ) + C_x \Rightarrow C_x = 500 \text{ kN}$$

Step 2: There are two members, AB and BC, that are subjected to a uniformly distributed loading. Hence we need two segments to construct the shear force and bending moment expressions.

Member AB ($0 \le x \le 5$ m; see Fig. E2.8.8(c)):

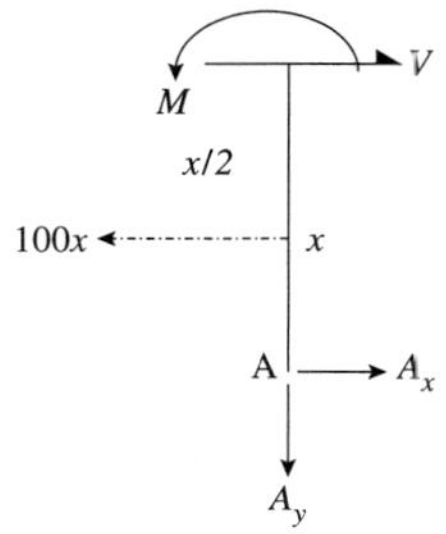

Fig. E2.8.8(c)

$$\overset{\rightarrow+}{\sum} F_x = 0 = -100x + V + A_x \Rightarrow V(x) = 100x - 300$$

And $V = 0$ @ $x = 3$ m.

$$\overset{\curvearrowleft+}{\sum} M_{\text{cut}} = 0 = M - 100x(x/2) + A_x(x) \Rightarrow M(x) = 50x^2 - 300x$$

and $M(x = 3) = 450 - 900 = -450$ kN-m

Member CB ($0 \le x_1 \le 5$ m; see Fig E2.8.8(d)): Note that in this FBD the cut must be such that V is normal to the member axis:

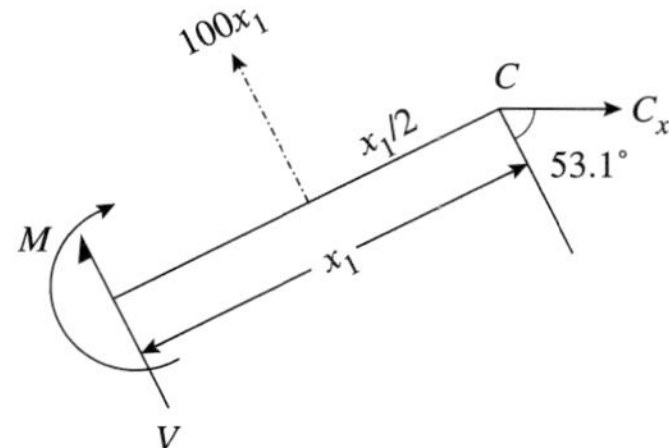

Fig. E2.8.8(d)

$$\overset{\rightarrow+}{\sum} F_T = 0 = 100x_1 + V + -C_x\cos(53.1^\circ) \Rightarrow V(x_1) = -100x_1 + 300$$

And $V = 0$ @ $x_1 = 3$ m

$$\overset{+}{\curvearrowleft}\sum M_{\text{cut}} = 0 = -M + 100x_1(x_1/2) - (C_x \cos(53.1°))(x) \Rightarrow M(x_1) = 50x_1^2 - 300x_1$$

and $M(x_1 = 3) = 450 - 900 = -450$ kN-m

Step 3: Draw the shear force and bending moment diagrams, as in Fig. 2.8.8(e).

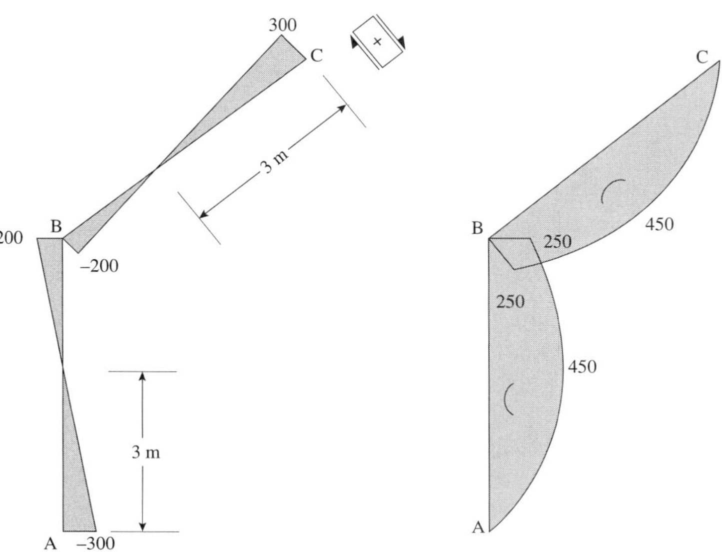

Fig. E2.8.8(e) Shear force diagram (kN) and bending moment diagram (kN-m).

The shear force in members AB and BC varies linearly (since the loading is constant). The bending moments are quadratic polynomials, with the largest magnitudes occurring at points of zero shear. The right side of column AB and the bottom side of member BC are in compression. Note that the bending moments are zero at A (pin support) and C (roller support), and continuous at B.

EXAMPLE 2.8.9

Portal Frame with Triangular Loading

For the frame shown in Fig. E2.8.9(a), draw the shear force and bending moment diagrams.

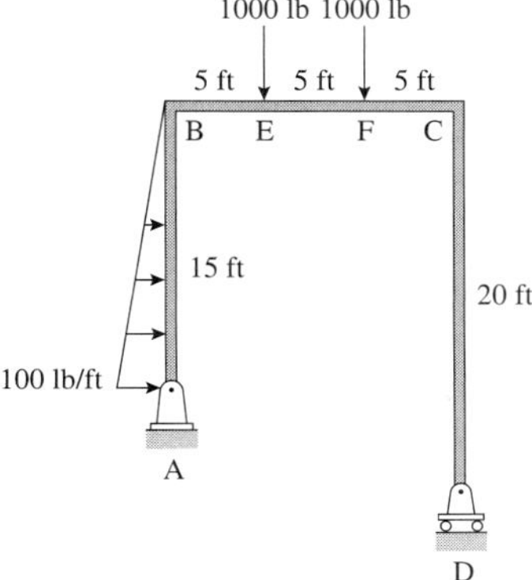

Fig. E2.8.9(a)

SOLUTION

Step 1: Using the structural FBD, we find A_x = 750 lb (←), D_y = 1250 lb (↑), and A_y = 750 lb (↑).

Step 2: The frame is made up of three members—column AB, beam BC, and column CD. We will start with column AB. Since there are no discontinuities between the ends, a cut is made at a distance x from A, as in Fig. E2.8.9(b). With the chosen coordinate system, the trapezoidal loading can be expressed as $w(x) = 100 - (20x/3)$.

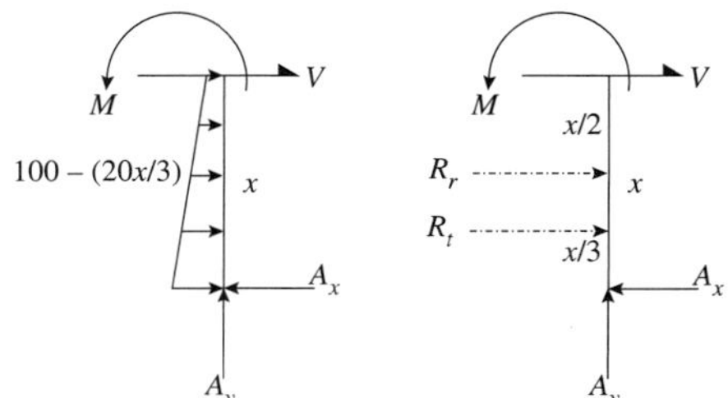

Fig. E2.8.9(b)

The loading has a triangular R_t and a rectangular R_r component. Using $w(x)$, we get

$$R_r = (x)\left(100 - \frac{20x}{3}\right) = -\frac{20x^2}{3} + 100x \text{ acting at distance } x/2 \text{ from A.}$$

$$R_t = \frac{1}{2}(x)\left(\frac{20x}{3}\right) = \frac{10x^2}{3} \text{ acting at distance } x/3 \text{ from A.}$$

$$\text{Hence, } \overset{\rightarrow+}{\sum} F_x = 0 = -A_x + R_r + R_t + V \Rightarrow V(x) = \frac{10x^2}{3} - 100x + 750$$

$$\overset{\curvearrowleft+}{\sum} M_{\text{cut}} = 0 = -A_x(x) + R_r\left(\frac{x}{2}\right) + R_t\left(\frac{2x}{3}\right) + M \Rightarrow M(x) = \frac{10x^3}{9} - 50x^2 + 750x$$

To construct the expressions for beam BC, the internal forces at B must be known. Consider the FBDs in Fig. E2.8.9(c). FBD 1 is constructed by making at cut just below joint B. V_B and M_B can be computed (not by using FBD 1) by substituting $x = 15$ in the expressions for $V(x)$ and $M(x)$. Hence, $V_B = 0$ and $M_B = 3750$ lb-ft. From FBD 1, $N_B = 750$ lb. The FBD of joint B is shown. And finally, in FBD 2, the internal forces at the cut just to the right of joint B are shown. Note that the axial force in column AB becomes the shear force in beam BC, whereas the shear force in column AB becomes the axial force in BC. The bending moment carries through the FBDs to FBD 2.

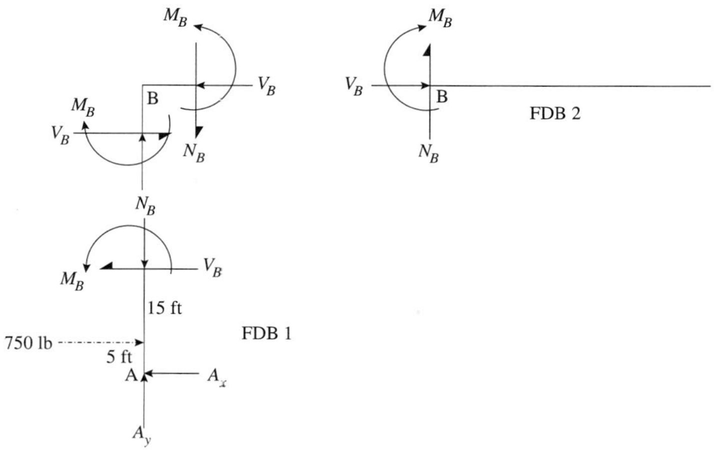

Fig. E2.8.9(c)

Segment (ft)	FBD	$V(x)$ (lb)	$M(x)$ (lb-ft)
BE $0 < x_1 < 5$	3750, B, M, x_1, 750, V	$\overset{\uparrow+}{\sum} F_y = 0 = 750 - V$ $V(x_1) = 750$	$\overset{\curvearrowleft+}{\sum} M_{\text{cut}} = 0$ $-3750 - 750x_1 + M = 0$ $M(x_1) = 750x_1 + 3750$
EF $5 < x_1 < 10$	1000, 3750, M, B, 5 ft, E, x_1, 750, V	$\overset{\uparrow+}{\sum} F_y = 0$ $750 - 1000 - V = 0$ $V(x_1) = -250$	$\overset{\curvearrowleft+}{\sum} M_{\text{cut}} = 0 = -750x_1 - 3750$ $+1000(x_1 - 5) + M$ $M(x_1) = -250x_1 + 8750$
FC $10 < x_1 < 15$	1000, 1000, 3750, M, B, 5 ft, 5 ft, E, F, x_1, 750, V	$\overset{\uparrow+}{\sum} F_y = 0$ $750 - 2000 - V = 0$ $V(x_1) = -1250$	$\overset{\curvearrowleft+}{\sum} M_{\text{cut}} = 0 = -750x_1 - 3750$ $+1000[(x_1 - 5) + (x_1 - 10)] + M$ $M(x_1) = -1250x_1 + 18750$

There are discontinuities in beam BC at E and F.

Finally, to construct the expressions for column CD, the internal forces at C must be known. From the expressions for segment FC, $V(x_1 = 15) = -1250$ lb and $M(x_1 = 15) = 0$. There are no discontinuities in CD. The FBD of the column is given in Fig. 2.8.9(d) ($0 \le x_2 \le 20$).

$$\overset{\rightarrow+}{\sum} F_x = 0 = V \Rightarrow V(x_2) = 0$$

$$\overset{\curvearrowleft+}{\sum} M_{\text{cut}} = 0 = M \Rightarrow M(x_2) = 0$$

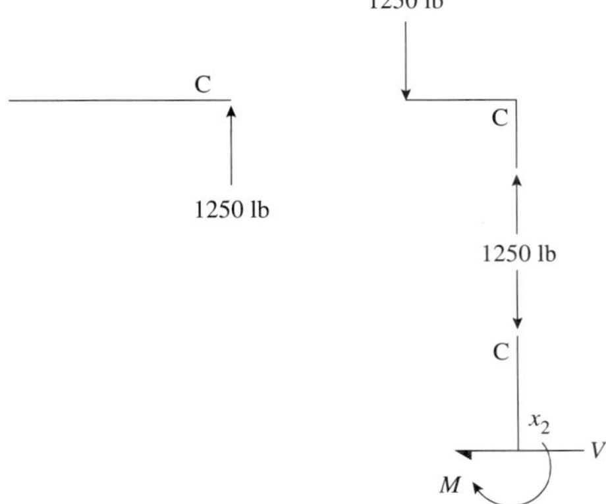

Fig. E2.8.9(d)

Step 3: Draw the shear force and bending moment diagrams, as in Figs. 2.8.9(e) and (f).

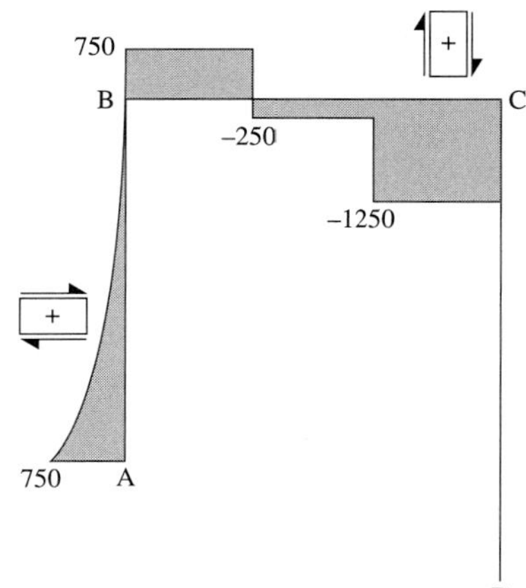

Fig. E2.8.9(e)
Shear force diagram (lb).

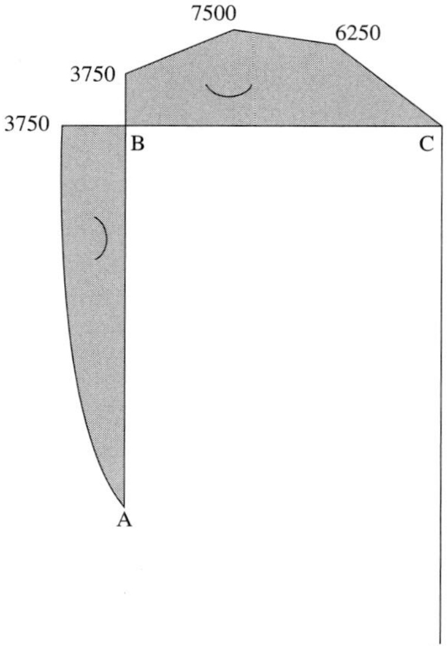

Fig. E2.8.9(f)
Bending moment diagram (lb-ft).

The shear force diagram is quadratic in column AB, shows abrupt changes where the concentrated forces are applied in BC, and is zero in column CD. The bending moment diagram is cubic in AB with the left side of the column in compression. It is piecewise linear in BC with the top side in compression. The column CD has no bending moment (only an axial force equal to the reaction D_y). Note the bending moment continuities at joints B and C.

Finally, Fig. E2.8.9(g) shows alternate FBDs are shown that could have been used. However, as you can see, these are less useful in understanding the structural behavior since they do not draw upon the interaction among the different members in the frame.

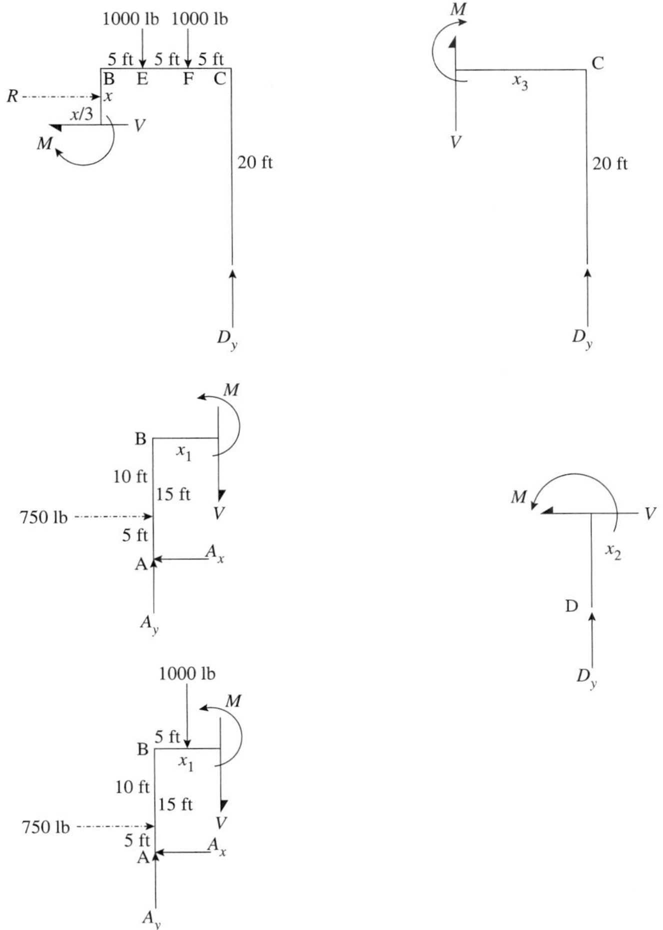

Fig. E2.8.9(g)

EXAMPLE 2.8.10 ***Portal Frame with Internal Hinge and Trapezoidal Loading***

For the frame in Fig. E2.8.10(a), draw the shear force and bending moment diagrams.

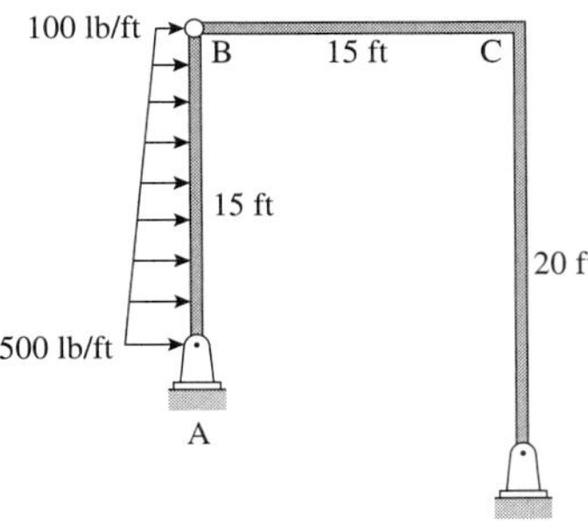

Fig. E2.8.10(a)

SOLUTION

Step 1: Breaking the structure at the hinge (at B), we find FBDs in Fig. E2.8.10(b).

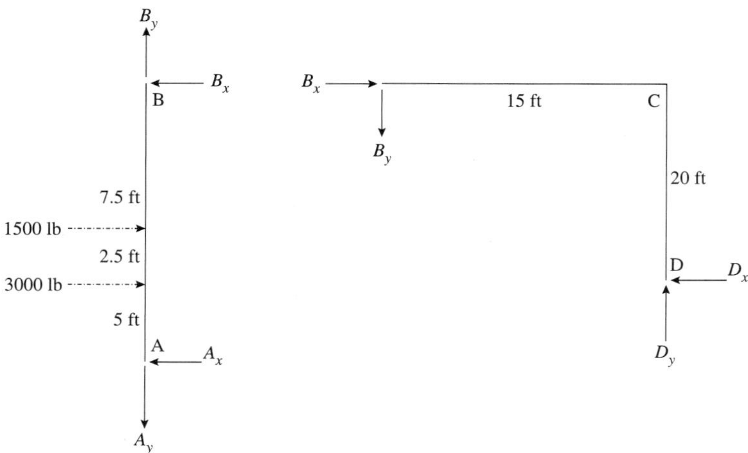

Fig. E2.8.10(b)

$$\text{FBD 1: } \overset{\curvearrowleft +}{\sum} M_B = 0 = -A_x(15) + 3000(10) + 1500(7.5) \Rightarrow A_x = 2750 \text{ lb}$$

$$\text{FBD 1: } \overset{\rightarrow +}{\sum} F_x = 0 = -A_x + 3000 + 1500 - B_x \Rightarrow B_x = 1750 \text{ lb}$$

$$\text{FBD 2: } \overset{\curvearrowleft +}{\sum} M_D = 0 = -B_x(20) + B_y(15) \Rightarrow B_y = 2333 \text{ lb}$$

$$\text{FBD 2: } \overset{\rightarrow +}{\sum} F_x = 0 = B_x - D_x \Rightarrow D_x = 1750 \text{ lb}$$

$$\text{FBD 2: } \overset{\uparrow +}{\sum} F_y = 0 = -B_y + D_y \Rightarrow D_y = 2333 \text{ lb}$$

$$\text{FBD 1: } \overset{\uparrow +}{\sum} F_y = 0 = -B_y + A_y \Rightarrow A_y = 2333 \text{ lb}$$

Step 2: The frame is composed of three members: column AB, beam BC, and column CD. Since there are no discontinuities between the end points, we need to derive one expression for each member.

Column AB: The trapezoidal load can be expressed as $w(x) = 100 + 80x/3$ with the origin of the coordinate system at B. Resultants of the rectangular and the triangular portions are shown in the FBD in Fig. E2.8.10(c).

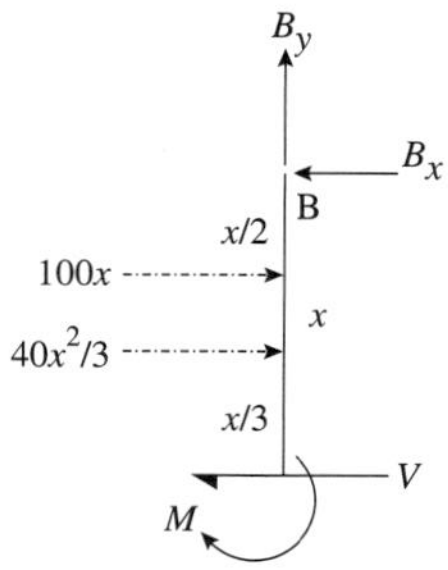

Fig. E2.8.10(c)

For $0 \le x \le 15$ ft,

$$\overset{\rightarrow +}{\sum} F_x = 0 = -B_x + 100x + \frac{40x^2}{3} - V \Rightarrow V(x) = \frac{40x^2}{3} + 100x - 1750$$

And $V = 0 \ @ \ x_{1,2} = \dfrac{-100 \pm \sqrt{100^2 + 4(40/3)(1750)}}{(80/3)} = -15.3, 8.3$

$$\overset{\curvearrowleft +}{\sum} M_{\text{cut}} = 0 = B_x(x) - 100x\left(\frac{x}{2}\right) - \frac{40x^2}{3}\left(\frac{x}{3}\right) - M \Rightarrow M(x) = -\frac{40x^3}{9} - 50x^2 + 1750x$$

and $M(x = 8.3) = 8539$ lb-ft.

Beam BC: It is convenient to draw the FBD anchored at B, as in Fig. E2.8.10(d).

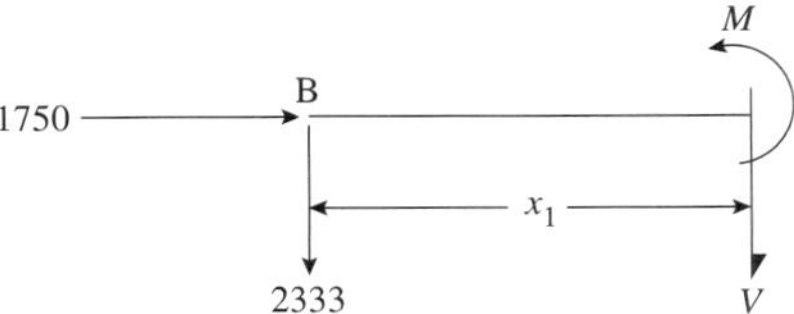

Fig. E2.8.10(d)

For $0 \le x_1 \le 15$ ft,

$$\overset{\uparrow +}{\sum} F_y = 0 = -2333 - V \Rightarrow V(x_1) = -2333$$

$$\overset{\curvearrowleft +}{\sum} M_{\text{cut}} = 0 = 2333(x_1) + M \Rightarrow M(x_1) = -2333x_1$$

Column DC: Similarly, for $0 \le x_2 \le 20$ ft, we have the FBD in Fig. E2.8.10(e), and

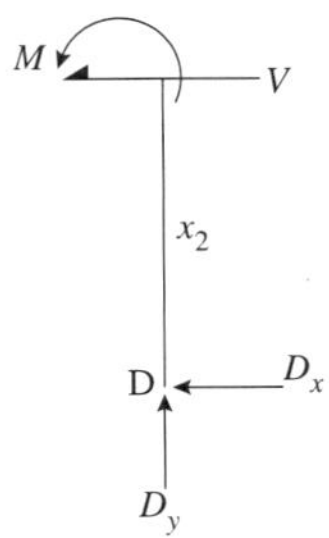

Fig. E2.8.10(e)

$$\overset{\rightarrow +}{\sum} F_x = 0 = -D_x - V \Rightarrow V(x_2) = -1750$$

$$\overset{\curvearrowleft +}{\sum} M_{\text{cut}} = 0 = -D_x(x_2) + M \Rightarrow M(x_2) = 1750x_2$$

Step 3: Draw the shear force and bending moment diagrams, as in Figs. E2.8.10(f) and (g).

The entire left side of column AB is in compression. Similarly, the bottom side of beam BC and the left side of column CD are in compression. Note also that (a) the moments are zero at the pin supports (at A and D); (b) the moment is zero at the internal hinge at B; (c) the moments at end C for members BC and CD are equal to 35000 lb-ft; (d) the maximum moment in AB occurs at the point with zero shear force; (e) shear is quadratic in AB and the bending moment is cubic; (f) shear is constant in BC and CD and the corresponding bending moments are linear; and (g) the shear forces at A and D are equal to the support reactions.

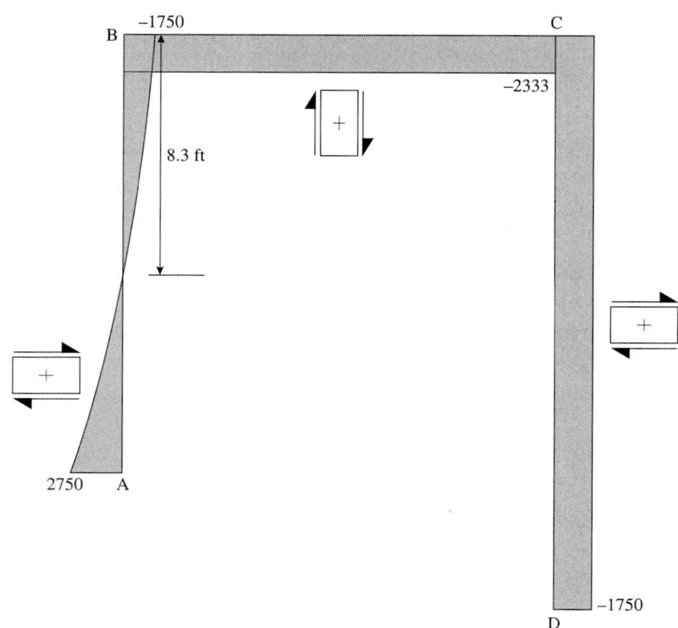

Fig. E2.8.10(f)
Shear force diagram (lb).

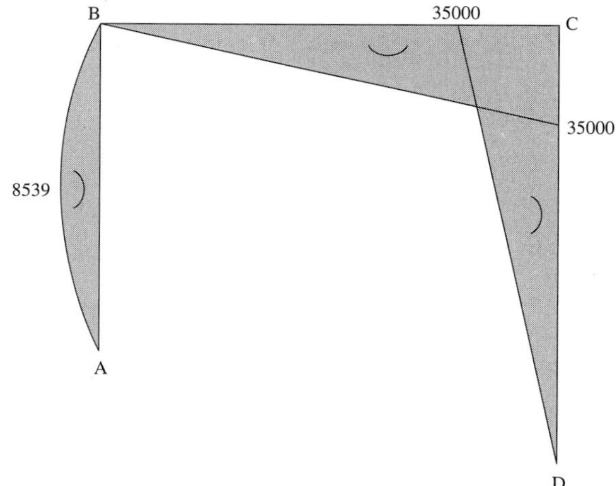

Fig. E2.8.10(g)
Bending moment diagram (lb-ft).

EXAMPLE 2.8.11 ***Portal Frame with Unequal Cantilever Sections***

For the frame in Fig. E2.8.11(a), draw the shear force and bending moment diagrams.

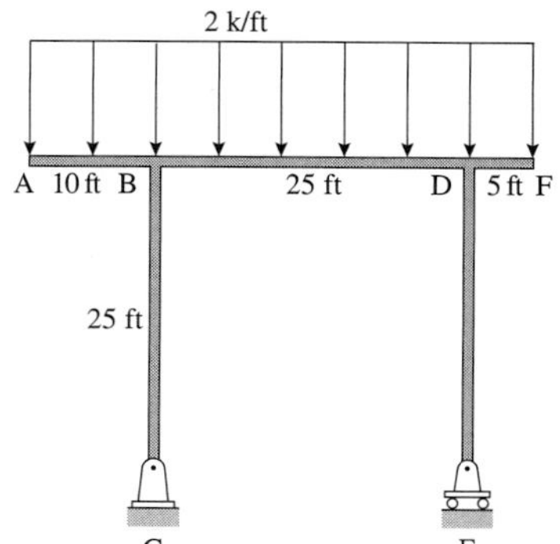

Fig. E2.8.11(a)

SOLUTION

Step 1: Using the structural FBD, we find $C_x = 0$, $C_y = 48$ k (↑) and $Ey = 32$ k (↑).

Step 2: An examination of the columns should make it clear that they are shear- and bending-moment free (only axial forces exist in them). We now look at the three segments of the beam.

Segment (ft)	FBD	$V(x)$ (k)	$M(x)$ (k-ft)
AB $0 < x < 10$	A, 2x, x, M, V	$\overset{\uparrow +}{\sum} F_y = 0 = -2x - V$ $V(x) = -2x$	$\sum M_{\text{cut}} = 0 = 2x\left(\frac{x}{2}\right) + M$ $M(x) = -x^2$
BD $10 < x < 35$	A, x/2, x/2, 2x, B, x–10, 48, M, V	$\overset{\uparrow +}{\sum} F_y = 0 = -2x + 48 - V$ $V(x) = -2x + 48$ $V = 0$ @ $x = 24'$	$\sum M_{\text{cut}} = 0$ $M(x) = -x^2 + 48x - 480$ $M(x = 24) = 96$ $M = 0 @ x = 14.2$ ft, 33.8 ft
FD $0 < x_1 < 5$	M, $2x_1$, $x_1/2$, $x_1/2$, F, V	$\overset{\uparrow +}{\sum} F_y = 0 = -2x_1 + V$ $V(x_1) = 2x_1$	$\sum M_{\text{cut}} = 0 = -x_1^2 - M$ $M(x_1) = -x_1^2$

Step 3: Draw the shear force and bending moment diagrams in Figs. E2.8.11(b) and (c).

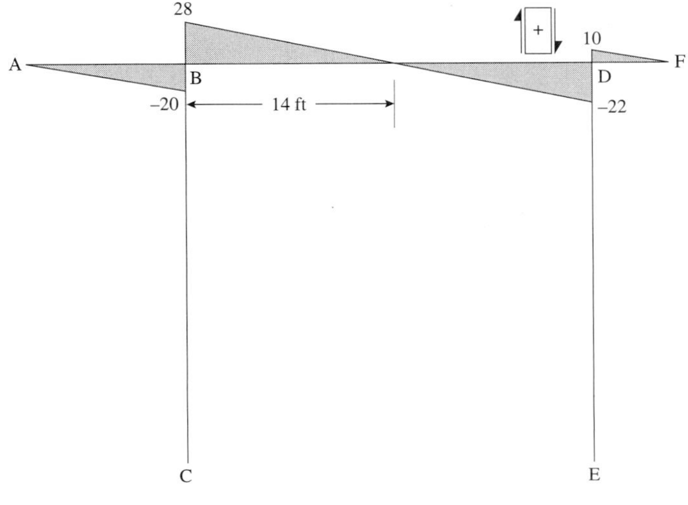

Fig. E2.8.11(b)
Shear force diagram (k).

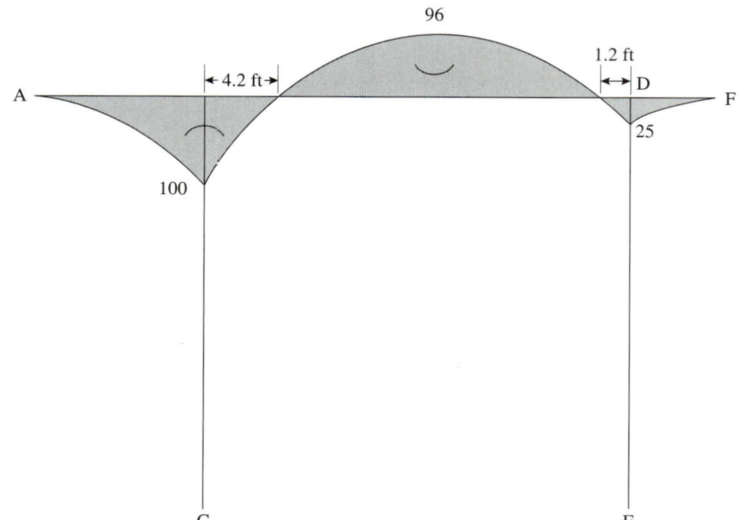

Fig. E2.8.11(c)
Bending moment diagram (k-ft).

Member ABDF is such that the bottom fiber of the beam is under compression from A to 4.2 ft to the right of B and from F to 1.2 ft to the left of D. In the remaining part of the beam, the top fiber is under compression. In the shear force diagrams, notice the jump in the shear values where the beam meets the column. The maximum bending moment occurs at the point of zero shear. However, the largest magnitude is at B.

2.8.5 Unstable Frames

The general ideas governing stability of frames are the same as we have seen before. If the number of unknowns (internal forces and support reactions) is less than the number of equations available from independent FBDs, the frame is unstable. The other types of unstable frames can be detected from a detailed analysis. One or more equations of equilibrium are not satisfied. We look at a few examples of unstable frames.

In Fig. 2.8.5.1, frame (a) is such that $(3m + r - 3j - c) = 3(3) + 4 - 3(4) - 2 = -1 < 0$. Hence it is an unstable frame. For frame (b), $(3m + r - 3j - c) = 3(6) + 4 - 3(6) - 4 = 0$, and the frame appears to be determinate and stable. However, there is no horizontal restraint. Similarly, for frame (c), $(3m + r - 3j - c) = 3(2) + 4 - 3(3) - 1 = 0$ and the frame appears to be determinate and stable. However, there is no restraint to prevent the column from moving horizontally.

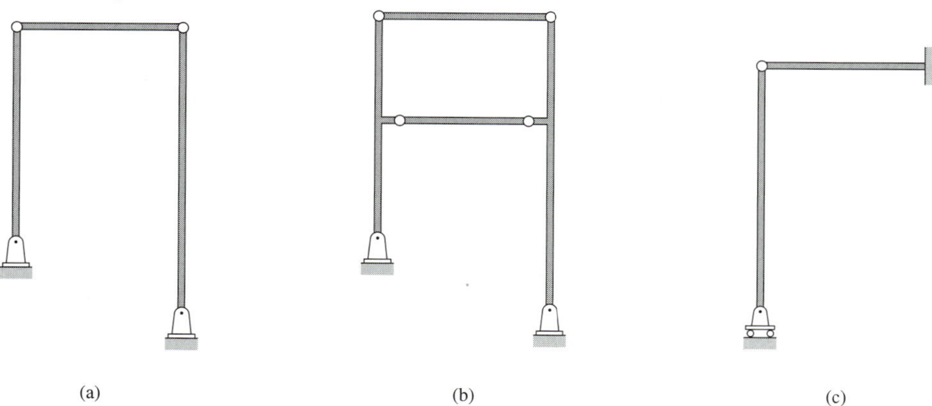

Fig. 2.8.5.1
Examples of unstable frames.

EXERCISES

Appetizers

For the frames shown in Figs. P2.8.28–P2.8.32, draw the shear force and bending moment diagrams.

2.8.28.

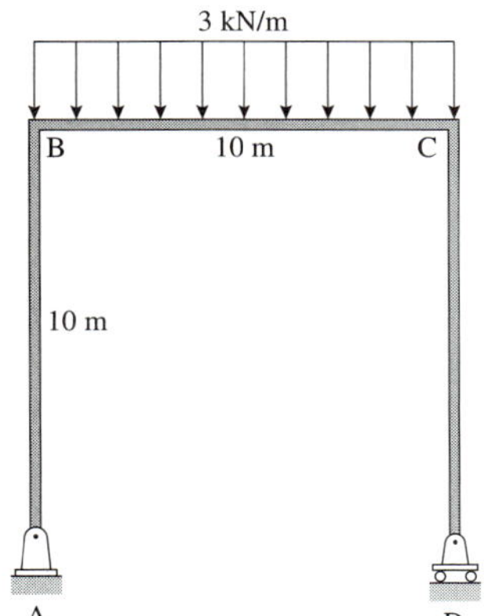

Fig. P2.8.28

2.8.29.

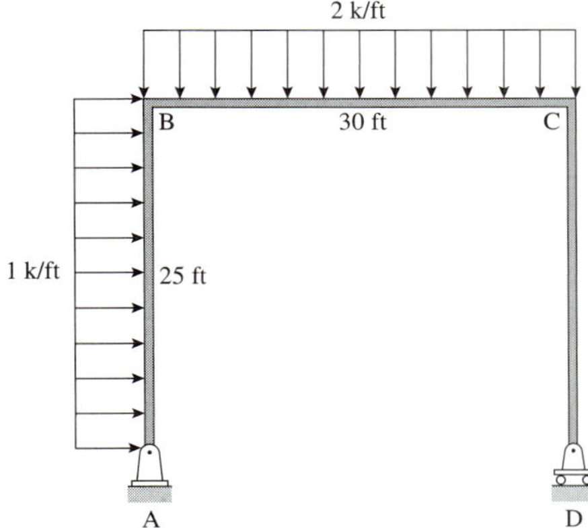

Fig. P2.8.29

2.8.30.

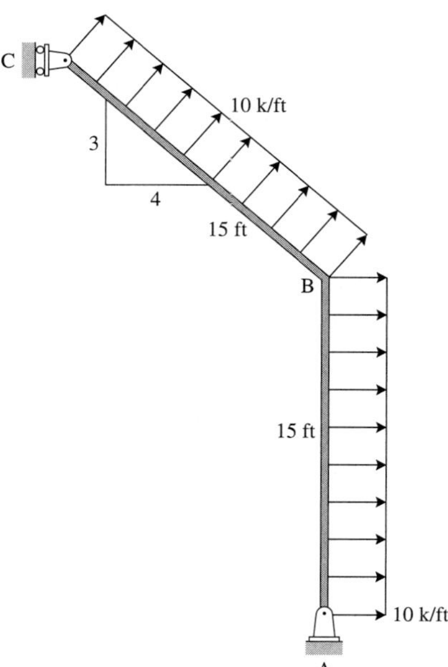

Fig. P2.8.30

2.8.31.

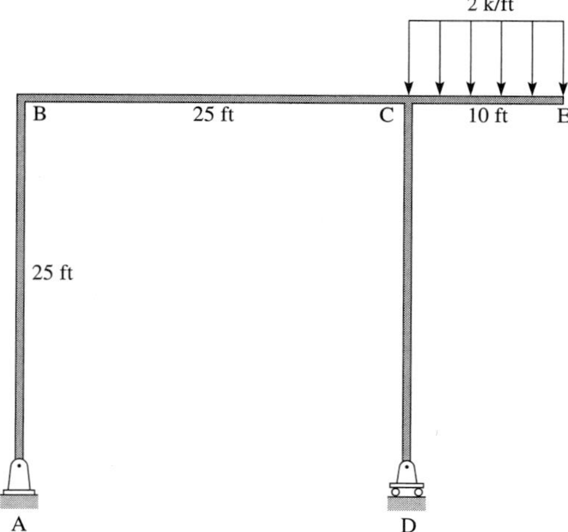

Fig. P2.8.31

2.8.32.

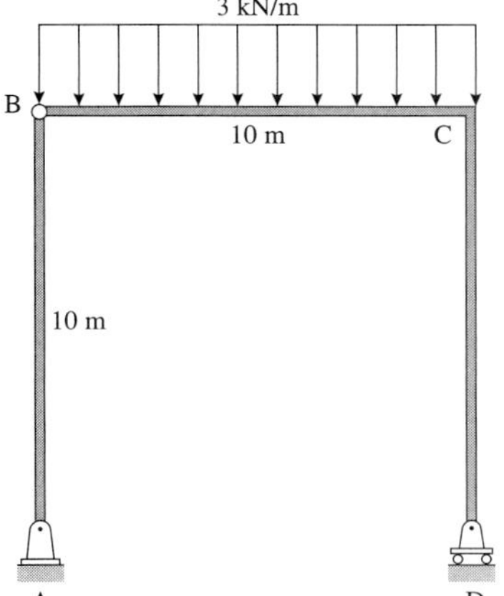

Fig. P2.8.32

Main Course

2.8.33. For the frame in Fig. P2.8.33, draw the shear force and bending moment diagrams.

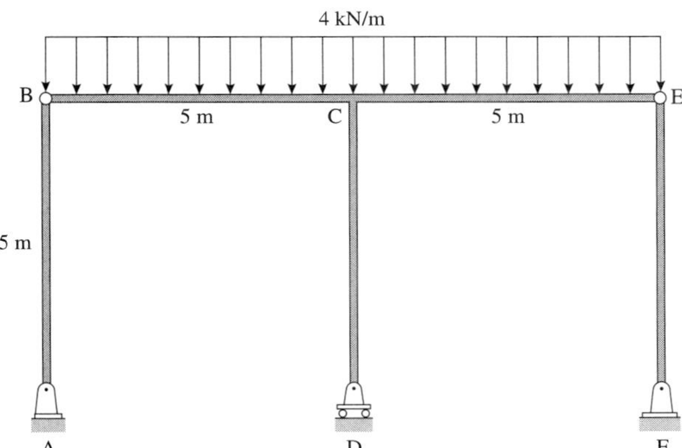

Fig. P2.8.33

2.8.34. The frame in Fig. P2.8.34 is used to hold back water (density 62.5 lb/ft^3) as well as provide an access road on the top. Draw the shear force and bending moment diagrams.

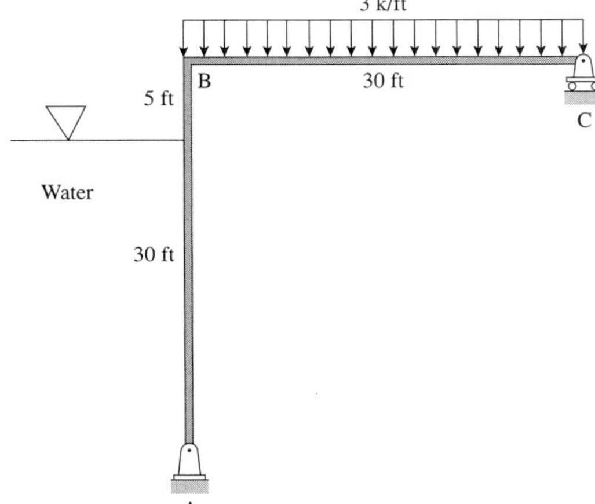

Fig. P2.8.34

2.8.35. For the three-storied frame in Fig. P2.8.35, draw the shear force and bending moment diagrams for the beams BF and CG and the columns AB and BC. Assume that the axial force in beams BF and CG can be neglected.

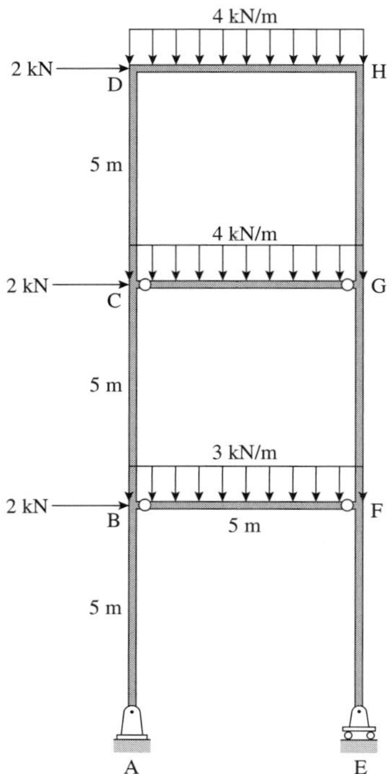

Fig. P2.8.35

Structural Concepts

2.8.36. Is there a need to draw an axial force diagram? Why or why not?

SUMMARY

In this chapter we have taken a second look at determinate structural systems. The analysis of simple and pin-connected structures, trusses, beams, and frames involve the use of (just) two very important concepts—free-body diagrams and static equilibrium. From a design perspective, the motivation is to compute as much of the structural response as possible. This includes support reactions, pin forces, and member internal forces. For framed structures, shear force and bending moment diagrams provide a graphical view of the varying internal forces so that the extreme values and their locations can be tracked easily.

It must be emphasized that while structural analysis may lead to unique solutions, the strategies and steps are usually nonunique. Some of the example problems have been used to drive home this point. Select a style that is clear and logical.

Lastly, one must be aware of the assumptions behind the behavior of all the different structural systems and components. The analysis principles and procedures must satisfy these assumptions; otherwise the results are meaningless. The important step in translating a physical system into the type of mathematical models that we have seen in this chapter is dealt with gradually in the rest of the book.

SUMMARY EXERCISES

Appetizers

2.1. Compute the forces in all the truss members that meet at joint G in Fig. P2.1.

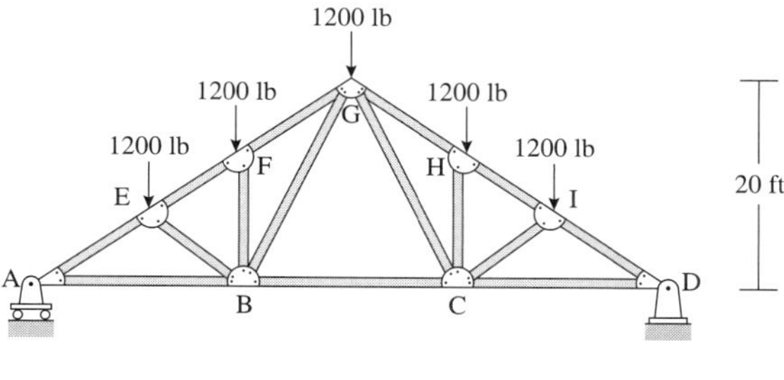

Fig. P2.1

2.2. Draw the shear force and bending moment diagrams for the frame in Fig. P2.2.

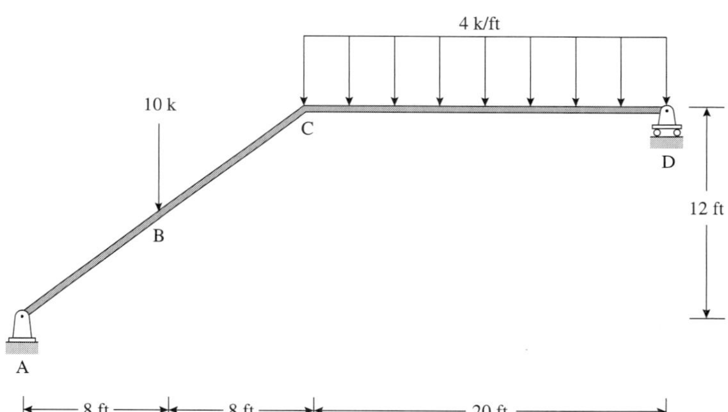

Fig. P2.2

2.3. Figure P2.3 shows a truss. Compute the axial force in all the members.

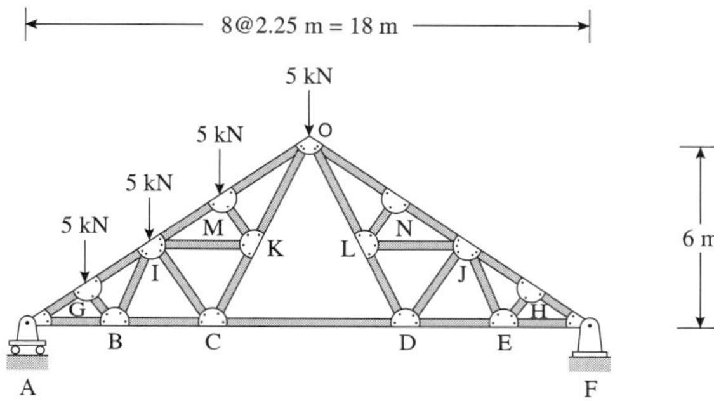

Fig. P2.3

Main Course

2.4. Figure P2.4 shows a composite structure. Member ABCD is a continuous beam. Rods BE and CE are connected to each other and to the beam via pins. Draw the shear force and bending moment diagram of the beam.

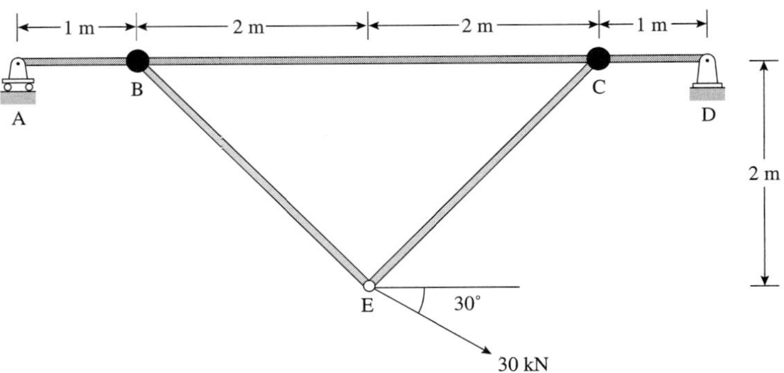

Fig. P2.4

2.5. Draw the shear force and bending moment diagrams for the arch in Fig. P2.5.

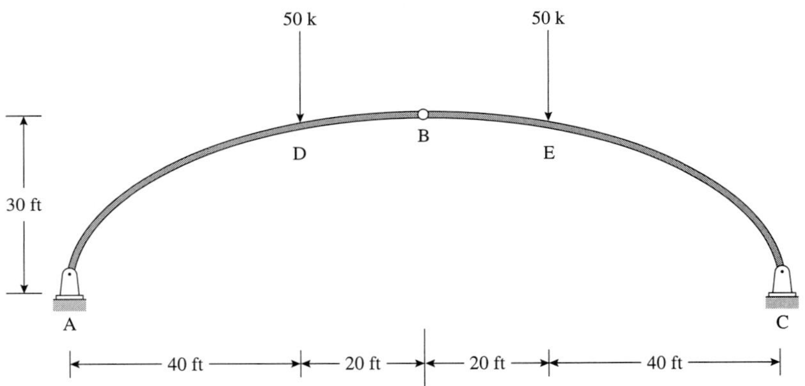

Fig. P2.5

2.6. A parallel chord Pratt truss as shown in Fig. P2.6 is to be hoisted into place by a crane. The hoisting cables are placed at joints 6 and 14. Assuming that each member weighs 200 N/m, compute the member forces.

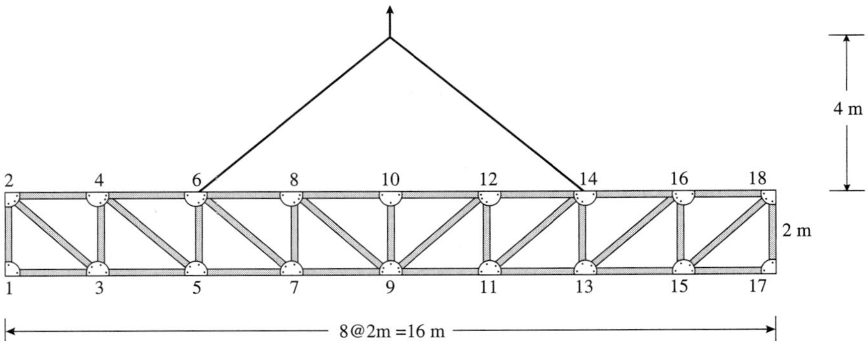

Fig. P2.6

Structural Concepts

2.7. Consider the structure shown in Fig. P2.7. Is this structure a truss or a frame? Is the structure stable?

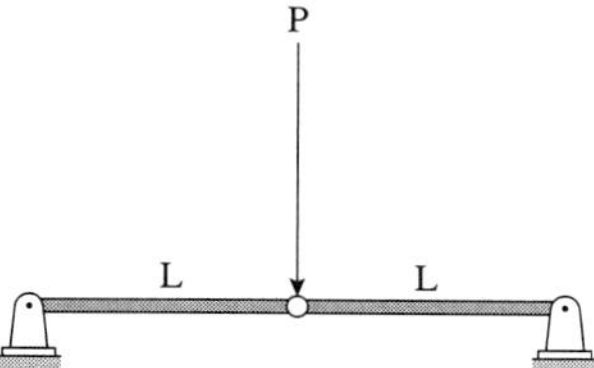

Fig. P2.7

2.8. Consider a two-bar truss subjected to a concentrated force as shown in Fig. P2.8. The results from a linear structural analysis show that the force in member AB is 259 lb (T) and in member BC is 966 lb (C). Joint B moves 2.5 in to the right and 1.5 in down. In the deformed state, check whether the joint is in equilibrium.

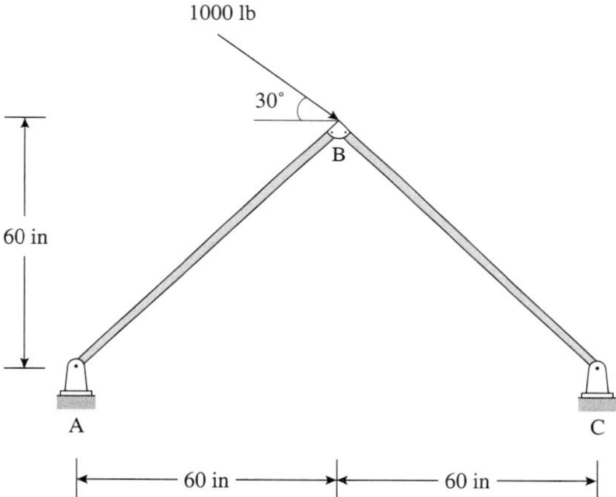

Fig. P2.8

Chapter 3

Structural Design Fundamentals

Successful structural design requires team effort amongst architects, civil engineers and construction specialists, to name a few.

"Most of what you hear you forget; some of what you see, you remember; what you do sticks." *Canoeing Handbook*, ed. R. Rowe, British Canoe Union

"A good scientist is a person with original ideas. A good engineer is a person who makes a design that works with as few original ideas as possible. There are no prima donnas in engineering." Freeman Dyson, British-born U.S. physicist, author of *Disturbing the Universe*

What is structural design? Can design principles be formalized? Can design be taught? These are some of many questions surrounding the topic of engineering design in general and structural design in particular. A look at various definitions will serve to illustrate the different points of view.

> "Thus, when analyzing a structure, *the properties are given and the response is to be determined*. The inverse process is called **design**." (Gere and Timoshenko, 1997)
>
> "Design is generally regarded as a creative process by which new methods, devices, and techniques are developed to solve new or existing problems." (Jaluria, 1998)
>
> "Unlike Athena, who sprang fully grown and fully armed from the brow of Zeus, the designs of mortal engineers must struggle to be born. In imitation of nature, they must evolve as survivors in a hostile environment, called upon to demonstrate their efficacy and survivability in competition with other solutions to the problems they address." (Cooper and Chen, 1985)
>
> "The design process is a sequential and iterative decision-making process." (McGregor, 1988)
>
> "Designing building structures involves the consideration of a wide range of factors. Building structural designers must not only understand structural behavior and how to provide for it adequately, but also be knowledgeable about building construction materials and processes, building codes and standards, and building economics." (Ambrose, 1995)

All these definitions are completely acceptable but present somewhat narrow points of view. An all-encompassing definition might take a complete chapter and is perhaps out of place in the context of the objectives of this text. Instead, we attempt to highlight the different facets of design.

1. Design is a creative process. However, very rarely does it lead to widely used new methods, new techniques, or the like.
2. Structural design requires the knowledge of a wide array of issues and topics—client needs, material behavior, structural analysis, applicable design codes, engineering economics, etc.
3. Design is an iterative process. The process of refinements starts with the preliminary design and goes through several intermediate designs that finally lead to the final design.
4. Design solutions are almost always non-unique. Unlike structural analysis, where there is a unique solution with a defined model, there may exist several designs that are all acceptable solutions to a design problem.
5. Some decision-making processes during design can be quite subjective. Some aspects cannot be quantified. For example, one cannot assign a scale or unit to architectural aspects, aesthetics, or value to society.
6. Structural design requires an understanding of structural analysis. Often analysis is thought to be synonymous with design, much as design is thought to be a process that can take place independent of analysis. This is certainly not true. Structural analysis forms the backbone of the design process.

With the definition out of the way and with very modest design goals in mind, the design process is defined:

1. The first step is to define the functional requirements of the structural system or component to be designed.
2. The functional requirements can be used to define the type of structural system.
3. This is usually followed by establishing the structural requirements that must be met by the system and the computation of the loads that must be resisted or carried by the system.
4. The information from steps 2 and 3 is used to create the mathematical model.
5. A structural analysis is carried out to compute the appropriate structural response. The response values are checked against the structural requirements set in step 3.
6. If the requirements are not met, changes must be made to the structural system. If the requirements are met, a decision must be made whether or not to look for further improvements to the structural system. Otherwise if no changes are to be made, the final design can be identified. If changes are to be made, the designer must decide what to change and by how much; the design process is reinitiated, usually starting from step 2.

While in Section 3.6 we look very briefly at the design codes that are so often used in structural design, the treatment of design in this chapter is more fundamental, more mechanics-based. The treatment of code-based design using steel, concrete, masonry, timber, and other structural materials traditionally follows an introductory course on structural analysis.

OBJECTIVES

- To understand the basic ideas associated with structural design.
- To understand the role of material behavior and performance requirements in the context of structural design.
- To understand the relationship between structural analysis and structural design.
- To understand the role of computer-based tools in the design process.
- To learn how to model some of the different types of structural loads and systems.
- To design simple structural systems using the trial-and-error approach.

ASSUMPTIONS

- Unless otherwise stated, the material properties are assumed to be linear, elastic, and homogeneous.
- The displacements and strains will be assumed to be small.
- Beam and column behavior will be based on a "plane sections remaining plane" assumption.

MATHEMATICAL BACKGROUND

The reader is urged to review the mathematical background necessary for this chapter, as presented in Appendix E.

3.1 MATERIAL BEHAVIOR

The motivation for studying material behavior is quite simple. Structural members, connections, and supports are made of some material and understanding the material behavior is useful in designing these components. In this section, an abridged treatment is presented. A more comprehensive treatment can be found in mechanics of materials textbooks.

3.1.1 Stress and Strain

The internal forces acting on the cut are vectorial in nature. In Chapter 2 we looked at the average values at the cut. In this section we investigate the effects of the internal forces in greater detail.

Normal Stress and Strain

In general, the internal forces acting on the infinitesimal areas at the cut vary in magnitude and direction. The stress is a measure of the intensity of these internal forces and is expressed as force per unit area, e.g. lb/in^2. It cannot be overemphasized that stress (and strain) *at a point* are a function of the plane and direction on which they are assumed to act.[1] Consider a prismatic, homogeneous bar subjected to axial forces that act through the centroid of the cross-section as shown in Fig. 3.1.1.1. The cross-sectional dimensions are small compared to the length of the bar. Furthermore, assume that the bar is in equilibrium.

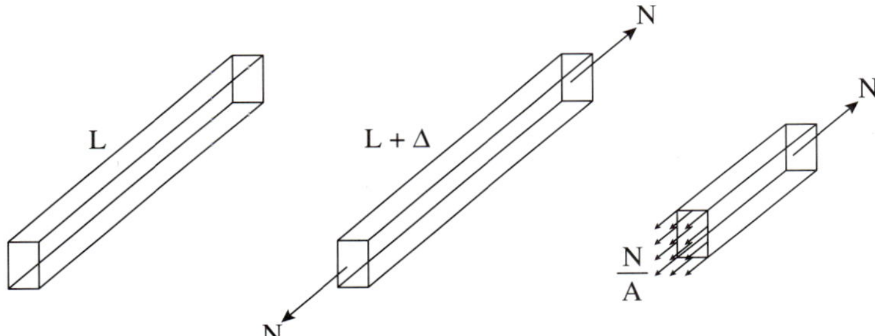

Fig. 3.1.1.1
Normal force and stress.

Let a cut be made normal to the axis of the bar. The internal stresses are exposed at the cut. Since they act normal to the cut, they are known as normal stresses. The normal stress at the cut is uniform throughout the length of the bar provided the cut is sufficiently far from the applied loads (ends of the bar in this case).

Under the action of the tensile forces the bar will increase in length (elongate). Let the elongation of the entire bar be Δ. By assuming that the material is homogeneous, it can be shown that the elongation is directly proportional to the length of the original bar. The normal tensile strain is a measure of the elongation per unit length. Strain as such has no units but is commonly expressed as length/length, e.g. m/m.

While this example deals with tensile forces, stresses, and strains, the similar comments apply to compressive forces, stresses, and strains.

[1] Stress is a second-order tensor and as such requires two subscripts.

Shear Stress and Strain

Now consider the same bar from Fig. 3.1.1.1 but let the cut be inclined to the member axis, as shown in Fig. 3.1.1.2. The resultant force on the cut can be resolved into a component that is normal to the surface of the cut, N_N, and a component that is tangential to the surface of the cut, N_T.

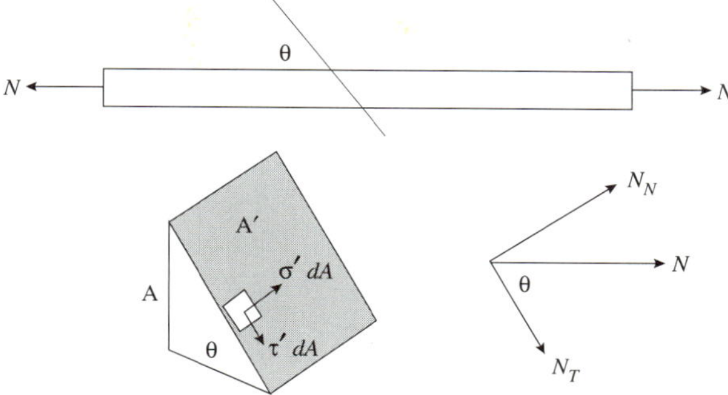

Fig. 3.1.1.2 Normal and shear stress.

While the normal component gives rise to normal stress, the tangential component is associated with the shear stress. Assuming that the normal stress σ′ and the shear stress τ′ are constant over the inclined area A', we have

$$N_N = \int_{A'} \sigma' \, dA = \sigma'_{\text{avg}} A' \qquad N_T = \int_{A'} \tau' dA = \tau'_{\text{avg}} A' \tag{3.1.1.1}$$

However,

$$N_N = N \sin\theta \qquad N_T = N\cos\theta \qquad \text{A} = \text{A}'\sin\theta \tag{3.1.1.2}$$

Using the two equations, we find

$$\sigma'_{\text{avg}} = \frac{N_N}{\text{A}'} = \frac{N}{\text{A}}(\sin\theta)^2 \qquad \tau'_{\text{avg}} = \frac{N_T}{\text{A}'} = \frac{N}{\text{A}}\cos\theta\sin\theta \tag{3.1.1.3}$$

The results show that (a) the average normal and shear stress vary with the orientation of surface on which they act, (b) there are infinite combinations of the normal and shear stress since there are infinite planes that pass through the centroid of the cross-section, and (c) there are special planes where the normal and shear stress have their extreme (largest, smallest, and zero) values.

Let us now develop a more comprehensive derivation of the strain expression so as to get away from the average values. Consider a one-dimensional problem in which every point in a bar can be described by a single displacement $u = u(x)$. On the unloaded bar, consider a segment AB that deforms to A′B′ when the bar is loaded, as shown in Fig. 3.1.1.3. Point A has moved u to the right and B has moved $(u + \Delta u)$ to the right, so that the change in length in AB is Δu.

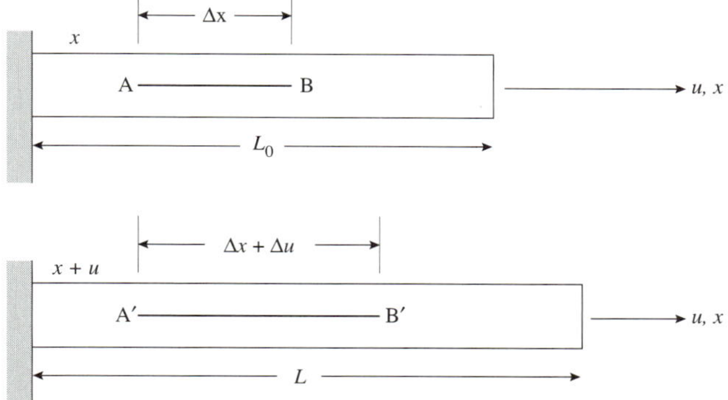

Fig. 3.1.1.3 Uniaxial state of stress and strain.

Using the definition of normal strain, we find

$$\varepsilon_x = \underset{\Delta x \to 0}{Lt} \frac{\Delta u}{\Delta x} = \frac{du}{dx} \tag{3.1.1.4}$$

The expression represents the strain at a point to which Δx shrinks. If the loading on the bar is such that the deformation is uniformly distributed over the length, then

$$\varepsilon_x = \frac{L - L_0}{L_0} = \frac{\Delta}{L_0} \tag{3.1.1.5}$$

where Δ is the change in length of the bar.

Now consider the state of deformation at a point A in a two-dimensional body as shown in Fig. 3.1.1.4(a).

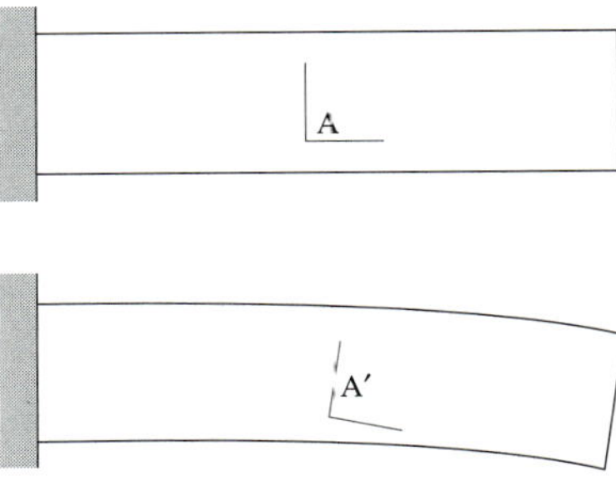

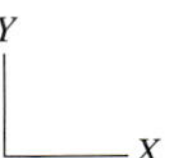

Fig. 3.1.1.4(a) Two-dimensional state of stress and strain.

In this case we need two displacements to describe the displacement field at any point, $u = u(x, y)$ and $v = v(x, y)$. Point A moves to A′ in the deformed state. If we consider a differential element of sides dx and dy, then the total deformation at A can be divided into two components—a change in length of the sides of the differential element with no angular change at A, and a change in the angle at A with no change in the length of the sides. From the first-order Taylor series expansion, the first component is shown in Fig. 3.1.1.4(b).

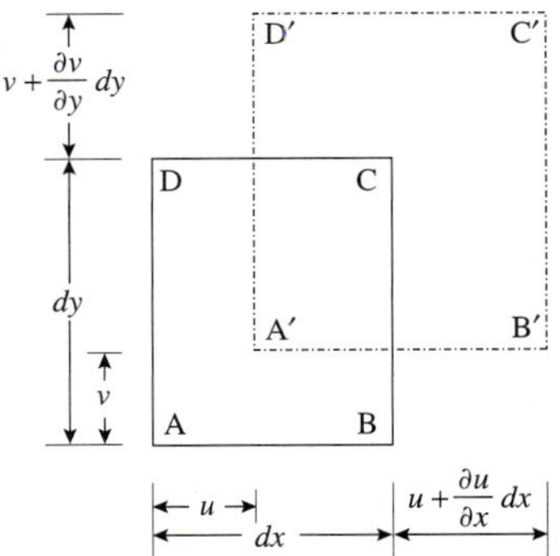

Fig. 3.1.1.4(b) Normal strain.

Using the definition of normal stress, we have

$$\varepsilon_x = \frac{u + \frac{\partial u}{\partial x}dx - u}{dx} = \frac{\partial u}{\partial x} \qquad \varepsilon_y = \frac{v + \frac{\partial v}{\partial y}dy - v}{dy} = \frac{\partial v}{\partial y} \tag{3.1.1.6}$$

Now consider the second component, as shown in Fig. 3.1.1.4(c).

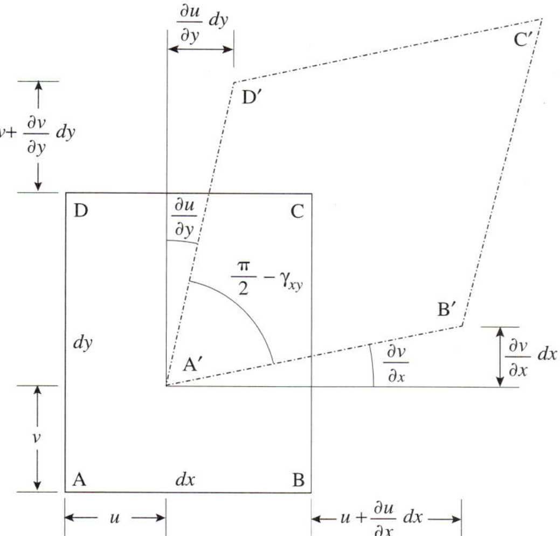

Fig. 3.1.1.4(c)
Shear strain.

If we define shear strain as the change in the angle at A, then

$$\gamma_{xy} = \gamma_{yx} = \frac{\partial u}{\partial y} + \frac{\partial v}{\partial x} \tag{3.1.1.7}$$

The equation is valid as long as the change in angle is small or $\theta \approx \tan\theta$. Note that if the normal or shear strain is small (<< 1), the second-order and higher-order terms in the Taylor series expansion will be small and can be neglected. As a sign convention, let us assume that tensile normal strain is positive and that a decrease in angle indicates a positive shear strain.

To summarize, note that the normal strain is associated with the change in length whereas shear strain is associated with change in angle. In a similar manner we can develop the expressions for strains in a three-dimensional situation. In general, at any point in a body, there are six components of stress and strains:

$$\overline{\sigma} = \{\sigma_x, \sigma_y, \sigma_z, \tau_{xy}, \tau_{yz}, \tau_{zx}\} \tag{3.1.1.8}$$

$$\overline{\varepsilon} = \{\varepsilon_x, \varepsilon_y, \varepsilon_z, \gamma_{xy}, \gamma_{yz}, \gamma_{zx}\} \tag{3.1.1.9}$$

As we see later, for one or two-dimensional problems, some of these components are zero. Finally, it should be noted that strain, not stress, is more fundamental. When a body or structure is loaded, strains can be measured. With these strain values the corresponding stress values can be obtained from the stress-strain relationship.

3.1.2 Material Properties

As we saw in Chapter 2, the fundamental material properties affect the manner in which the member or the structure can resist the loads. Of the tens of fundamental properties, only a few important ones are defined here. Fig. 3.1.2.1 shows a typical stress-strain diagram for a ductile material.

Point 1 is at the end of the linear regime and is called the proportional limit. Point 2 is the beginning of the perfectly plastic regime and is called the yield stress. There is a large deformation in this regime with little or no change in the stress. From point 3 onwards, the material begins to harden, with point 4 representing the ultimate stress. Beyond this point the stress actually decreases till the material (specimen) breaks or fractures at point 5.

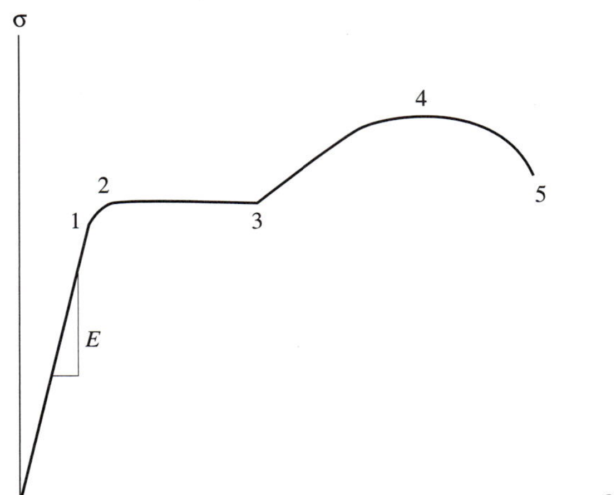

Fig. 3.1.2.1
Typical stress-strain diagram for a ductile metal.

Modulus of Elasticity (E). Also known as elastic modulus or Young's modulus and usually denoted E. In a one-dimensional state of stress it is the constant of proportionality between the normal stress and the normal strain, i.e., $E = \sigma_x / \varepsilon_x$, and has units of stress.

Poisson's Ratio (ν). Again using a one-dimensional state of stress, the Poisson's ratio is the ratio of the lateral strain to the axial strain. It is usually denoted as ν. ν = *lateral strain/normal strain* = $\varepsilon_y / \varepsilon_x = -(\varepsilon_z / \varepsilon_x)$ and hence is unitless.

Shear Modulus (G). Also called modulus of rigidity. For isotropic materials, the shear modulus is a function of the modulus of elasticity and the Poisson's ratio, i.e., $G = E/2(1 + \nu)$.

Yield Stress (σ_y). Also known as yield strength. This represents the stress level at which plastic deformations are initiated.

Ultimate Stress (σ_u). Also known as ultimate strength. This represents the state of stress corresponding to the largest load.

3.1.3 Stress–Strain Relationship

As we saw in the previous section, different materials exhibit different behavior when a material specimen is loaded. In fact, a relationship can be established between the state of stress and strain. For homogenous isotropic materials, the relationship between stress and strain is given by

$$\begin{aligned}
\sigma_x &= 2G\varepsilon_x + \lambda e \\
\sigma_y &= 2G\varepsilon_y + \lambda e \\
\sigma_z &= 2G\varepsilon_z + \lambda e \\
\tau_{xy} &= G\gamma_{xy} \\
\tau_{yz} &= G\gamma_{yz} \\
\tau_{zx} &= G\gamma_{zx}
\end{aligned} \qquad (3.1.3.1)$$

where

$$\lambda = \frac{\nu E}{(1+\nu)(1-2\nu)} \qquad e = \varepsilon_x + \varepsilon_y + \varepsilon_z \qquad (3.1.3.2)$$

In the rest of the book, the above equations will be tailored to handle one- and two-dimensional states of stress and strain.

3.1.4 Principal Stress and Strain

In Section 3.1.1 we briefly saw that an infinite number of stress values describe the state of stress at a point. Two questions naturally arise. First, are all these states of stress (or strain) important? Second, if we know the state of stress on a particular plane that passes through the point, how do we compute the state of stress on another plane?

Let the state of stress be known at a point O with respect to two orthogonal planes, as shown in Fig. 3.1.4.1. In other words, we know $\{\sigma_x, \sigma_y, \tau_{xy}\}$. Let us assume that the state of stress is to be computed on the plane whose normal is in the x' direction, as shown in Fig. 3.1.4.1. The element with the resulting state of stress is shown in Fig. 3.1.4.2.

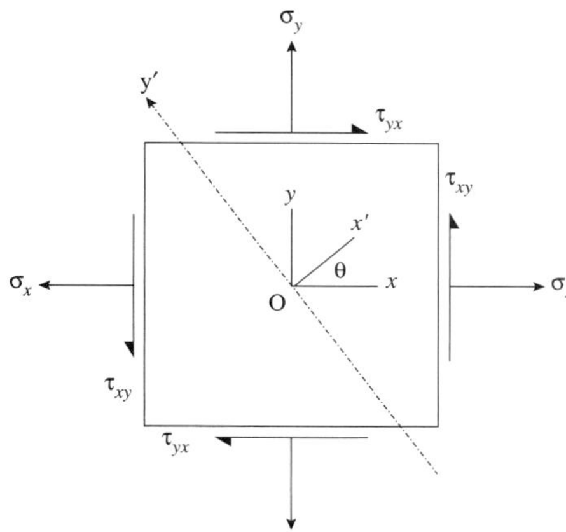

Fig. 3.1.4.1
Stress transformation.

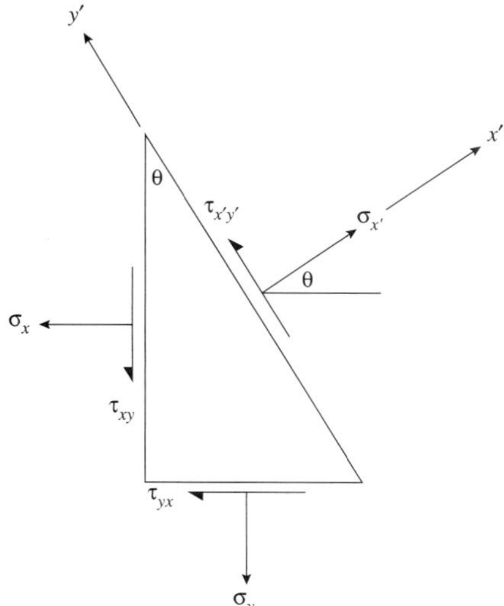

Fig. 3.1.4.2
Stress transformation.

Using the concept of static equilibrium, the following equations can be derived:

$$\sigma_{x'} = \frac{\sigma_x + \sigma_y}{2} + \frac{\sigma_x - \sigma_y}{2}\cos 2\theta + \tau_{xy}\sin 2\theta \tag{3.1.4.1}$$

$$\tau_{x'y'} = -\frac{\sigma_x - \sigma_y}{2}\sin 2\theta + \tau_{xy}\cos 2\theta \tag{3.1.4.2}$$

In addition, $\sigma_{y'}$ can be obtained by substituting $(\theta + \pi/2)$ for θ to yield

$$\sigma_{y'} = \frac{\sigma_x + \sigma_y}{2} - \frac{\sigma_x - \sigma_y}{2}\cos 2\theta - \tau_{xy}\sin 2\theta \tag{3.1.4.3}$$

To find the plane containing the maximum or minimum normal stress, it is necessary to differentiate Eq. (3.1.4.1) and solve for θ as

$$\tan 2\theta = \frac{2\tau_{xy}}{(\sigma_x - \sigma_y)} \tag{3.1.4.4}$$

The solution of the above equation yields two roots, θ and $(\theta + 180^\circ)$. The two planes corresponding to these two roots are shown in Fig. 3.1.4.3. These planes are called the principal planes. The normal stresses are called the principal stresses and the planes are shear-stress-free. Substituting Eq. (3.1.4.4) in (3.1.4.1) yields

$$(\sigma_{x'})_{\text{max/min}} = \frac{\sigma_x + \sigma_y}{2} \pm \sqrt{\left(\frac{\sigma_x - \sigma_y}{2}\right)^2 + \tau_{xy}^2} \tag{3.1.4.5}$$

There are three principal stresses, denoted $\sigma_1 \le \sigma_2 \le \sigma_3$. Similarly, the maximum shear stress can be found by differentiating Eq. (3.1.4.2) and solving for θ as

$$\tan 2\theta = -\frac{(\sigma_x - \sigma_y)}{2\tau_{xy}} \tag{3.1.4.6}$$

Substituting Eq. (3.1.4.6) into (3.1.4.2) yields

$$(\tau)_{\text{max/min}} = \pm\sqrt{\left(\frac{\sigma_x - \sigma_y}{2}\right)^2 + \tau_{xy}^2} \tag{3.1.4.7}$$

The planes containing the extreme values of shear stress are orthogonal to the planes containing the principal stresses.

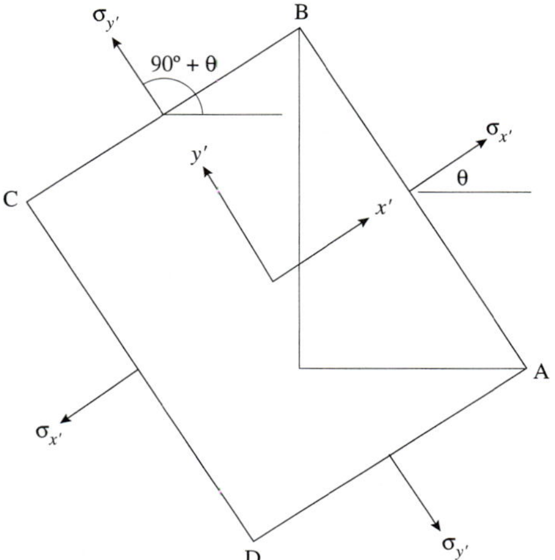

Fig. 3.1.4.3
Principal planes.

There is immense value to computing and locating the extreme values of normal and shear stresses. Some materials, e.g., concrete, are weaker in tension, while other are equally strong in tension and compression, e.g., steel. Similarly, other materials, e.g., wood, are weaker in shear. Computing the principal stresses gives an idea of the magnitude and direction of the largest normal stresses and the largest shear stresses in a member, and can indicate a member's susceptibility to failure.

Mohr's Circle

Mohr's circle provides a graphical technique to view the state of stress at a point and carry out stress transformations. We will discuss the steps in constructing the circle. The sign convention for constructing and interpreting the values from Mohr's circle is as follows. For normal stresses, tensile stresses are positive and compressive stresses are negative. Shear stresses are positive if they produce clockwise couples. Note that this sign convention for shear stresses is opposite to the one shown in Fig. 3.1.4.1. In Fig. 3.1.4.4(a), the normal stresses are positive, and the shear stress τ_{xy} is negative, whereas τ_{yx} is positive.

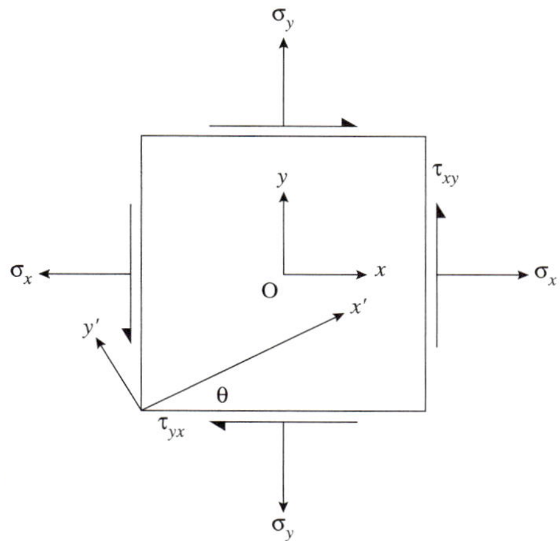

Fig. 3.1.4.4(a) Stress state.

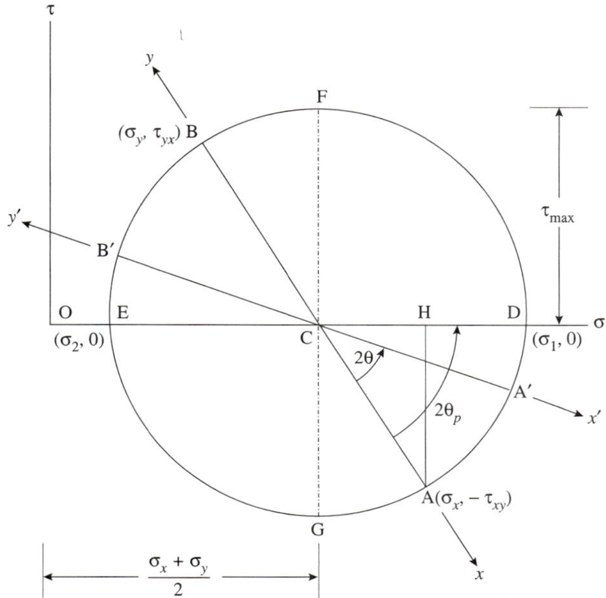

Fig. 3.1.4.4 (b) Mohr's circle.

The steps in constructing the circle are as follows.

1. Using the problem data and the sign convention for the Mohr's circle, draw the stress differential element.
2. Draw the coordinate system showing the σ and the τ axes.
3. Locate the center of the circle C on the σ axis at distance $(\sigma_x + \sigma_y)/2$ from the origin.
4. Locate the point corresponding to the positive x-face of the stress element. With reference to Fig. 3.1.4.4(a), the point A on the Mohr's circle corresponding to the positive x-face would be at $(\sigma_x, -\tau_{xy})$. Locate point B that represents the state of stress on the positive y-face. With reference to Fig. 3.1.4.4(a), the point B would be at (σ_y, τ_{yx}). Note that A and B are diametrically opposite to each other.
5. Now draw the circle with the center at C and radius from C to A or diameter from A to B passing through C. Note that Eq. (3.1.4.7) can be used to compute the radius of the circle.

Observations: The following observations can be made about the Mohr's circle. First, the circle represents the state of stress at a point. Second, the infinite points on the circle represent the state of stress on the infinite planes that pass through the point. Third, an angle 2θ on the circle represents an angle θ on the stress element. Fourth, the Mohr's circle is a two-dimensional or planar representation of the state of stress. In other words, for three-dimensional situations one must draw a circle for each projection of the three-dimensional stress element or use a more sophisticated analytical technique. Lastly, these concepts are applicable to other tensor quantities.

We illustrate the use of Mohr's circle through an example.

EXAMPLE 3.1.1 ***Stress Transformations***

The state of stress at a point is shown in Fig. E3.1.1(a). (a) Compute the principal stresses and planes. (b) Compute the maximum shear stress and the corresponding normal stress. (c) Determine the normal and shear stress on the plane inclined at an angle of 30° with respect to the horizontal.

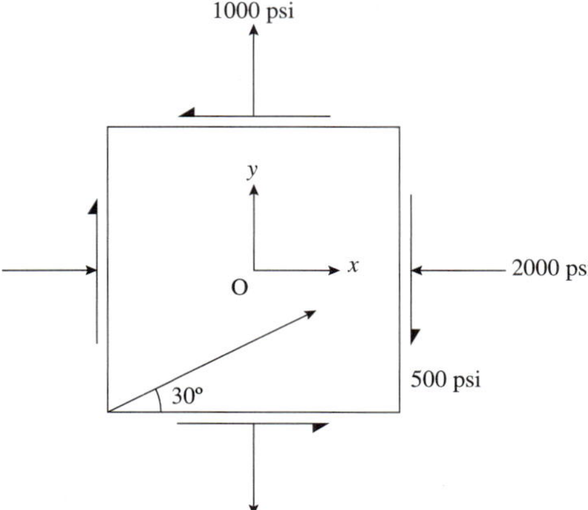

Fig. E3.1.1(a)

SOLUTION

Step 1: Employing the Mohr's circle approach, we have $\sigma_x = -2000$ psi, $\sigma_y = 1000$ psi, and $\tau_{xy} = 500$ psi. The center of the circle C is at $((\sigma_x + \sigma_y)/2, 0) = (-500, 0)$. The positive x-face corresponding point A is at (–2000, 500). The radius of the circle is the distance between C and A, and is given as $R = \sqrt{\left(\frac{\sigma_x - \sigma_y}{2}\right)^2 + \tau_{xy}^2} = \sqrt{\left(\frac{-2000 - 1000}{2}\right)^2 + (500)^2} = 1581.1$. We can now draw the circle in Fig. E3.1.1(b).

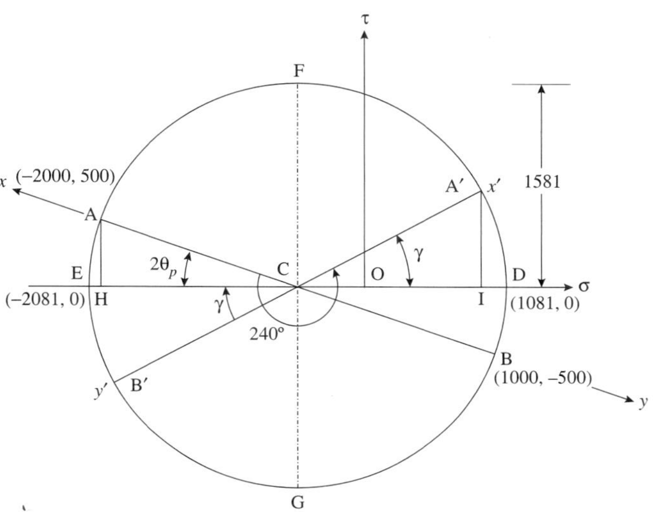

Fig. E3.1.1(b)

Step 2: The principal stresses are given by points D and E.

$$\sigma_1 = \overline{DC} - \overline{OC} = R - \left(\frac{\sigma_x + \sigma_y}{2}\right) = 1581.1 - 500 = 1081.1\,\text{psi}$$

$$\sigma_2 = -(\overline{OC} + \overline{CE}) = -\left(\frac{\sigma_x + \sigma_y}{2}\right) - R = -500 - 1581.1 = -2081.1\,\text{psi}$$

$$2\theta_p = \sin^{-1}\left(\frac{AH}{AC}\right) = \sin^{-1}\left(\frac{500}{1581.1}\right) = 18.4^\circ$$

Hence the principal stresses are –2081.1 psi (and the plane containing this stress is at an angle θ_p, i.e., 9.2° with respect to the x axis), and 1081.1 psi (and the plane containing this stress is at an angle of 9.2° + 90° = 99.2° with respect to the x axis). The maximum shear stress is the radius of the circle and is equal to 1581.1 psi. The normal stress on the plane containing the maximum shear stress is given by the normal stress at the center of the circle, $\sigma = -500$ psi.

Step 3: To locate the point A′ on the circle corresponding to the 30° plane (see Fig. E3.1.1(c)), we first note that the outward normal on that plane is at an angle of (30° + 90°) = 120° (in the counterclockwise sense) with respect to the positive x-face. This corresponds to 240° (in the counterclockwise sense) on the Mohr's circle. Using the triangle A′CI, we can compute the coordinates of A′ as (γ = 240° – 180° – 18.4° = 41.6°).

$$\sigma_{x'} = R\cos(\gamma) - \overline{OC} = 1581.1\cos(41.6^\circ) - 500 = 682.3\,\text{psi}$$

$$\tau_{x'y'} = R\sin(\gamma) = 1581.1\sin(41.6^\circ) = 1049.7\,\text{psi}$$

In a similar manner, we find

$$\sigma_{y'} = -(\overline{OC} + R\cos(\gamma)) = -500 - 1581.1\cos(41.6^\circ) = -1682.3\,\text{psi}$$

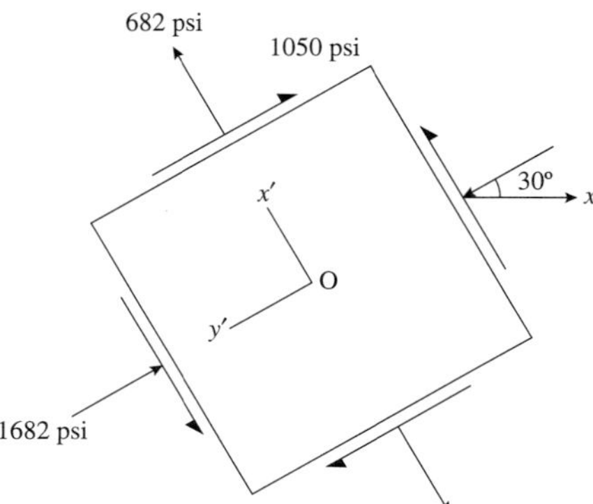

Fig. E3.1.1(c)

Redo the problem using the equations from Section 3.1.4.

EXERCISES

Appetizers

3.1.1. A steel cable (also known as *wire rope*) when unloaded is exactly 10 m long. When loaded it is observed that the new length of the cable is 10.4 m. What is the average normal strain in the cable?

3.1.2. In Problem 3.1.1 the cross-section is made of steel as shown in Fig. P3.1.2. There are seven wires of 3 mm diameter per strand, and seven strands make up the entire cross-section. What is the tensile force in the cable?

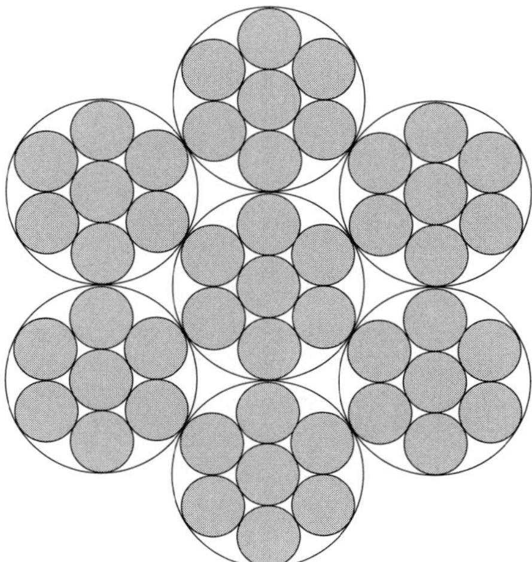

Fig. P3.1.2

3.1.3. A commonly used steel has the properties $E = 200$ GPa and $\nu = 0.3$. What is the value of the shear modulus?

3.1.4. Figure P3.1.4 shows a stress differential element. (a) Compute the principal stresses and planes. (b) Compute the maximum shear stress and the corresponding normal stress.

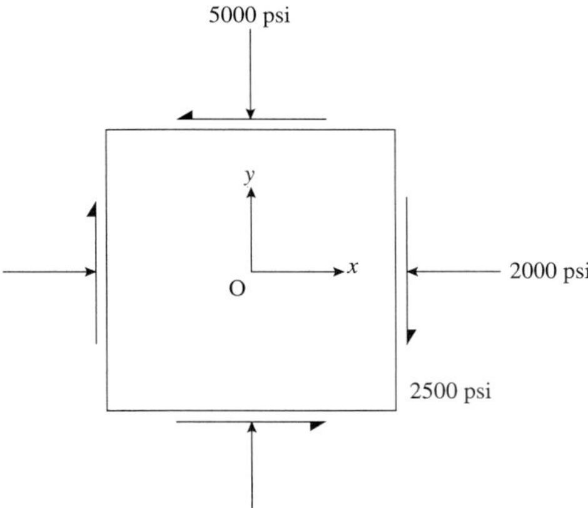

Fig. P3.1.4

Main Course

3.1.5. State whether the following are true or false. If false, state the reasons why.

(a) The SI unit for shear strain is m/m.

(b) The terms *isotropic* and *homogenous* have the same meaning.

(c) A steel specimen is loaded axially so that the state of stress is beyond the yield stress point. The load is removed. The final strain in the specimen is zero.

3.1.6. The strains at a point in a planar body are given as $\{\varepsilon_x, \varepsilon_y, \gamma_{xy}\} = \{100, -200, 30\}10^{-6}$. Compute the state of stress at that point, assuming that the material is 0.2% C hot-rolled steel.

3.1.7. The state of stress at different points are given below. For each point compute the principal stresses, principal planes, and the largest shear stress. For normal stresses, tensile stresses are positive and compressive stresses are negative. Shear stresses are positive if they produce clockwise couples.

(a) $\sigma_x = 25\,\text{MPa},\ \sigma_y = -10\,\text{MPa},\ \text{and}\ \tau_{xy} = 12\,\text{MPa}.$

(b) $\sigma_x = -2\,\text{ksi},\ \sigma_y = -5\,\text{ksi},\ \text{and}\ \tau_{xy} = -3\,\text{ksi}.$

(c) $\sigma_x = 0,\ \sigma_y = -0.5\,\text{MPa},\ \text{and}\ \tau_{xy} = 1.2\,\text{MPa}.$

(d) In part (b), compute the state of stress on the plane that is located 30° counterclockwise with respect to the x-plane.

3.2 STRESS AND STRAIN COMPUTATIONS

In Chapter 2 we saw how to compute the internal forces in planar truss, beam, and frame members. Now we will see how to use the internal forces to compute the stress and strain distribution within these members. As sign convention, tensile stresses are assumed to be positive.

3.2.1 Cross-sectional Properties

Several cross-sectional shapes are used as member cross-sections, e.g., rectangular solid, symmetric I-section, channel section, angle section, etc. These different shapes have different advantages with respect to the manner in which they can be manufactured, connected to other members and components, and basic cross-sectional properties that they provide.

Centroid. The location of the centroid of the cross-section can be easily obtained. In Fig. 3.2.1.1, point C represents the centroid of the plane area and O is the origin of an arbitrarily selected *X-Y* coordinate system.

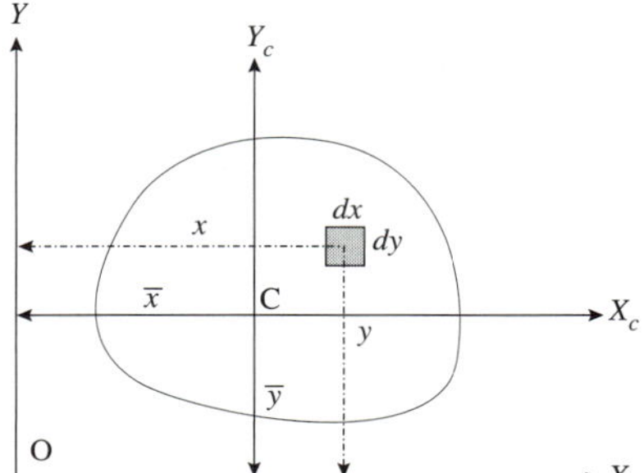

Fig. 3.2.1.1
Centroid.

$$\bar{x} = \frac{\int x\,dA}{\int dA} \qquad \bar{y} = \frac{\int y\,dA}{\int dA} \tag{3.2.1.1}$$

The numerator represents the first moment of the area while the denominator is the area of the cross-section. When the given cross-section is a composite section made up of more than one basic shape (e.g., rectangle, triangle), the centroid can be located as

$$\bar{x} = \frac{\sum_{i=1}^{n} \bar{x}_i A_i}{\sum_{i=1}^{n} A_i} \qquad \bar{y} = \frac{\sum_{i=1}^{n} \bar{y}_i A_i}{\sum_{i=1}^{n} A_i} \tag{3.2.1.2}$$

where $(\bar{x}_i, \bar{y}_i)$ are the (x, y) coordinates of the centroid of the ith basic shape and A_i is its area.

Moments of Inertia. The moments of inertia of planar areas with respect to the (*X-Y*) coordinate axes are given by (see Fig. 3.2.1.1)

$$I_x = \int_A y^2 dA \qquad I_y = \int_A x^2 dA \tag{3.2.1.3}$$

If the moments of inertia are known about the cross-section's centroidal axes ($X_c - Y_c$), then we can use the parallel-axis theorem to find the moment of inertia about any other axis, using

$$I_x = I_{x_c} + A(\bar{y})^2 \qquad I_y = I_{y_c} + A(\bar{x})^2 \tag{3.2.1.4}$$

Cross-sectional properties of commonly used sections and shapes are shown in Appendix B.

EXAMPLE 3.2.1 ***Centroid and Moment of Inertia***

Find the (a) centroid and (b) moment of inertia about the centroidal axes of the W36 × 300 AISC cross-section in Fig. E3.2.1.

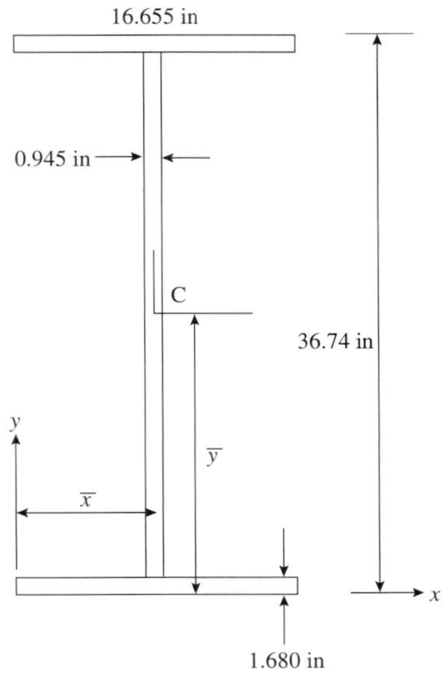

Fig. E3.2.1

SOLUTION

Step 1: A rectangle of width b and height h has the following properties:

$$A = bh \qquad I = \frac{bh^3}{12}$$

Step 2: Breaking the I-section into three rectangles and using the bottom left corner of the section as the origin, we can construct the following table.

Rectangle	$A(\text{in}^2)$	$\bar{x}$ (in)	$\bar{y}$ (in)
Bottom flange	(16.655)(1.680) = 27.9804	8.3275	0.84
Web	(0.945)(33.38) = 31.5441	8.3275	18.37
Top flange	(16.655)(1.680) = 27.9804	8.3275	35.9

Using Eq. (3.2.1.2), we find

$$\bar{x} = \frac{(8.3275)(27.9804) + (8.3275)(31.5441) + (8.3275)(27.9804)}{(27.9804 + 31.5441 + 27.9804)} = 8.3275 \text{ in}$$

$$\bar{y} = \frac{(0.84)(27.9804) + (18.37)(31.5441) + (35.9)(27.9804)}{(27.9804 + 31.5441 + 27.9804)} = 18.37 \text{ in}$$

The results show that the centroid is at the intersection of the cross-section's two axes of symmetry.

Step 3: Now to compute the moment of inertia we construct the following table. The last two columns locate the centroid of each rectangle with respect to the centroid of the entire cross-section.

Rectangle	A (in^2)	$I_{\bar{x}}$ (in^4)	$\bar{x}$ (in)	$\bar{y}$ (in)
Bottom flange	27.9804	$\dfrac{(16.655)(1.68)^3}{12} = 6.581$	0	17.53
Web	31.5441	$\dfrac{(0.945)(33.38)^3}{12} = 2928.93$	0	0
Top flange	27.9804	$\dfrac{(16.655)(1.68)^3}{12} = 6.581$	0	17.53

Using Eq. (3.2.1.4) yields

$$I_{x_c} = 6.581 + 27.9804(17.53)^2 + 2928.93 + 6.581 + 27.9804(17.53)^2 = 20139 \text{ in}^4$$

Note that the contribution to the moment of inertia by the flanges dominates the web contribution.

In a similar fashion we can compute the moment of inertia about the y_c axis. The computed values for area (87.505 in^2) and the moment of inertia (20139 in^4) are slightly smaller than the AISC values of 88.3 in^2 and 20300 in^4. This is because the three rectangles approximate the actual cross-section (which is slightly larger than the three rectangles).

3.2.2 Axial Force

Consider a slender, straight, long member with the coordinate system shown in Fig. 3.2.2.1.

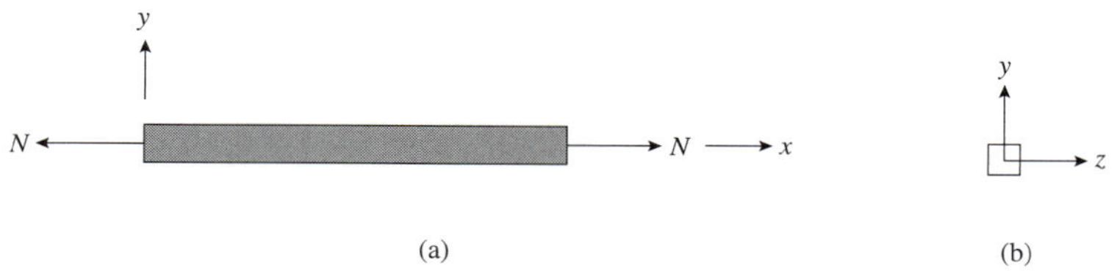

Fig. 3.2.2.1 Orientation of (a) member (b) cross-section.

Assume that the member is subjected only to an axial force. The normal stress σ due to an axial force N passing through the centroid of the cross-section is given by (using the coordinate system shown)

$$\sigma_x = \frac{N_x}{A} \tag{3.2.2.1}$$

The stress distribution is assumed to be constant on the exposed surface. The other stress components can be assumed to be zero. Clearly this situation is valid for truss members that are subjected only to an axial force.

3.2.3 Bending Moment

Consider a segment of slender, straight, long beam that is subjected to pure bending as shown in Fig. 3.2.3.1. Let the material be such that it has the same modulus of elasticity in tension and compression. The beam axis is oriented along x while the beam transverse displacement v is oriented along y.

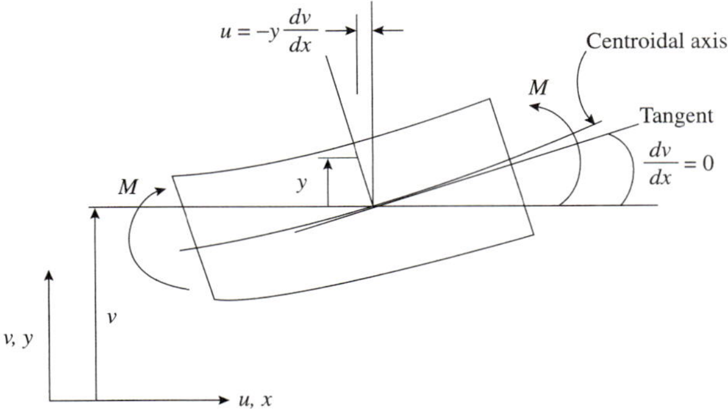

Fig. 3.2.3.1 Beam under pure bending.

In the simple beam theory the following assumptions are made:

$$\varepsilon_y = \varepsilon_z = 0 \qquad \gamma_{xy} = \gamma_{yz} = \gamma_{zx} = 0 \tag{3.2.3.1}$$

From Eq. (3.1.1.6), $\varepsilon_y = 0$ is possible only if $v = v(x)$. In other words, the transverse displacement is the same for all points on a given cross-section (having the same x). Since $\gamma_{xy} = 0$, using Eq. (3.1.1.7), we find

$$\frac{\partial u}{\partial y} = -\frac{dv}{dx} \tag{3.2.3.2}$$

Integrating, yields

$$u = -y\frac{dv}{dx} + g(x) \tag{3.2.3.3}$$

Since $u(y = 0) = 0$ represents the state of axial deformation on the centroidal axis, $g(x)$ is zero. For small deformations $dv/dx \approx \theta$ (as shown in Fig. 3.2.3.1) and

$$u = -y\theta \tag{3.2.3.4}$$

Since the axial displacement at a section is now a linear function of y, this equation states that plane sections remain plane.

Now consider the beam section loaded by pure moment, as shown in Fig. 3.2.3.2. Consistent with the simple beam assumptions, the state of normal stress at an arbitrary cut is shown in the figure. The only nonzero component of stress is given as a linear function as

$$\sigma_x = cy \tag{3.2.3.5}$$

Clearly $y = 0$ is the neutral axis and contains the neutral plane. Since the beam is in equilibrium,

$$\overset{\rightarrow +}{\sum} F_x = 0 = \int_A \sigma_x \, dA \tag{3.2.3.6}$$

$$\overset{\curvearrowleft +}{\sum} M_{\text{cut}} = 0 = M + \int_A (y)(\sigma_x dA) \tag{3.2.3.7}$$

Substituting Eq. (3.2.3.5) into Eqs. (3.2.3.6) and (3.2.3.7) yields

$$c\int_A y \, dA = 0 \tag{3.2.3.8}$$

$$-c\int_A y^2 \, dA = M \tag{3.2.3.9}$$

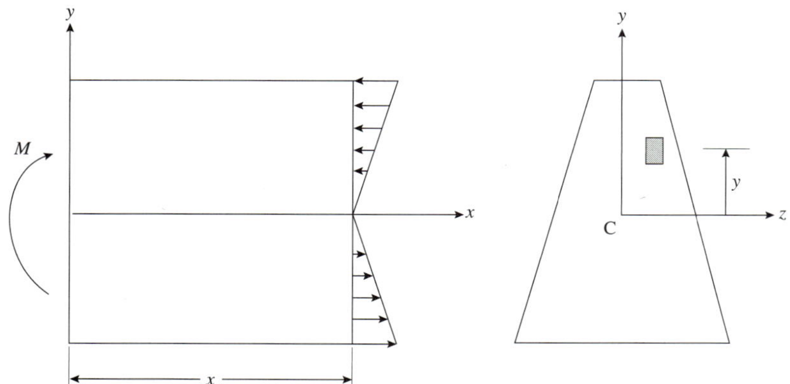

Fig. 3.2.3.2 Stress distribution under pure bending.

From Eq. (3.2.3.8), $\int y\,dA = 0$ since $c \neq 0$. The interpretation is that the neutral and the centroidal axes are the same. From Eq. (3.2.3.9), note that $\int_A y^2\,dA = I_z$. Hence $c = -(M/I_z)$. Substituting this result in Eq. (3.2.3.5) yields

$$\sigma_x = -\frac{M_z y}{I_z} \tag{3.2.3.10}$$

The equation states that a positive moment causes a compressive stress for positive values of y. We can rewrite the above equation for a specified value of the bending moment as

$$(\sigma_x)_{\max} = \frac{M_z}{S} \tag{3.2.3.11}$$

where $S = I_z / y_{\max}$ is called the *section modulus.* Since the distance to the fiber varies from point to point, finding the largest y value leads to the largest normal stress. The normal stress on the exposed surface varies linearly with the zero value on the neutral axis and the maximum compressive or tensile stress values at the outer fibers.

3.2.4 Shear Force

Usually a bending moment at a section also indicates the presence of shear force. The simple beam theory assumes that the shear strains and stresses can be neglected (Eq. (3.2.3.1)). However, from a viewpoint of satisfying equilibrium, they exist and in certain cases, their effects cannot be neglected.

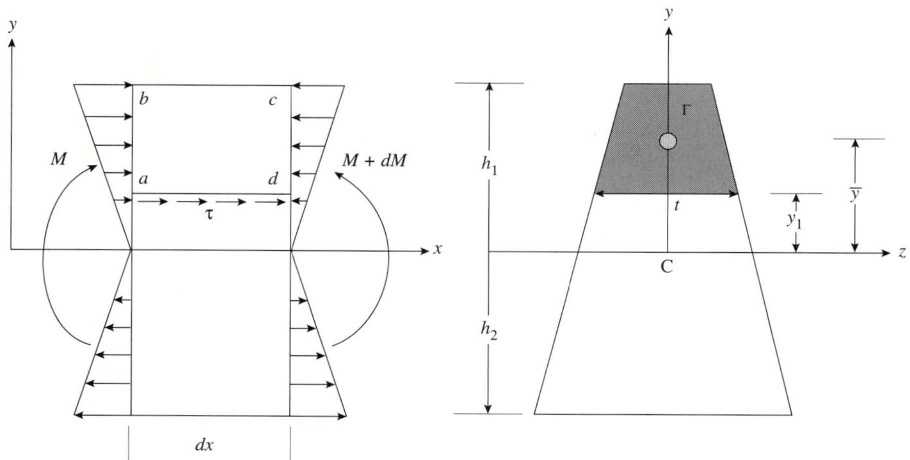

Fig. 3.2.4.1 Shear stresses in a beam.

Consider Fig. 3.2.4.1. The beam segment of length dx is shown. The top and the bottom surfaces are shear-stress-free. However, the shear stress does exist on other longitudinal surfaces, such as *a-d* (width t). On a given surface (given y value) it can be assumed that the shear stress is a constant. Consider the block a-b-c-d (shaded area Γ). On the left face (*a-b*) using Eq. (3.2.3.10), we find

$$\int_{-t/2}^{t/2}\int_{y_1}^{h_1}(\sigma_x)_{a-b}\,dy\,dz = \int_{\Gamma} -\frac{My}{I}\,d\Gamma \tag{3.2.4.1}$$

Similarly, on the right face (c-d)

$$\int_{-t/2}^{t/2}\int_{y_1}^{h_1}(\sigma_x)_{c-d}\,dy\,dz = \int_{\Gamma} -\frac{(M+dM)y}{I}\,d\Gamma \tag{3.2.4.2}$$

Since the block is in equilibrium, summing the forces on the block horizontally yields

$$\tau\, t dx - \int_{\Gamma}\frac{(M+dM)y}{I}\,dA + \int_{\Gamma}\frac{My}{I}\,dA = 0 \tag{3.2.4.3}$$

Solving, we find

$$\tau = \frac{1}{It}\int_{\Gamma}\frac{dM}{dx}\,y\,dA \tag{3.2.4.4}$$

We use Eq. (2.8.3.4) to get

$$\tau = \frac{V}{It}\int_{\Gamma} y\,dA = \frac{VQ}{It} \tag{3.2.4.5}$$

where

$$Q = \int_{\Gamma} y\,dA = \Gamma\bar{y} \tag{3.2.4.6}$$

represents the first moment of the shaded area about the neutral axis.

The shear stress τ due to a shear force V passing through the shear center[2] of the cross-section and acting along the y axis is given by

$$\tau_{xy} = \frac{V_y Q}{I_z t} \tag{3.2.4.7}$$

We can rewrite the above equation for a specified value of the shear force and cross-section as

$$\left(\tau_{xy}\right)_{\max} = \frac{V_y}{SF} \tag{3.2.4.8}$$

where $SF = I_z(t/Q)_{\min}$ is referred to as the *shear factor* in this text. The ratio t/Q can vary over the cross-section, and finding the location with the *smallest* ratio leads to the largest shear stress.

Finally it should be noted that Eq. (3.2.4.7) has its limitations. It is applicable to cases where the shear stresses act parallel to the y axis and are uniform across the width t.

3.2.5 Combined Stresses

When a member in a planar frame (beam included) is subjected to external loads, the internal forces include axial force, shear force, and bending moments. This situation requires that the separate effects of these internal forces be combined to find the state of stress.

---

[2] In this text, the focus is on cross-sections that have two symmetric planes. When sections have one or no plane of symmetry, the resultant of the shear force must pass through a special point called the shear center. If it does not, the beam will bend *and twist*.

Assume that the internal forces at a specified cross-section are as shown in Fig. 3.2.5.1. The normal stress at the section due to normal force and bending moment is given by

$$\sigma_x = \frac{N_x}{A} \mp \frac{M_z y}{I_z} \tag{3.2.5.1}$$

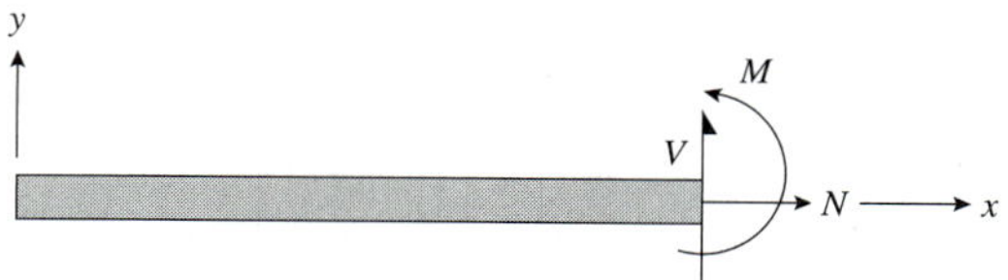

Fig. 3.2.5.1 All the internal forces at a cut.

The shear stress at the section due to the shear force is given by

$$\tau_{xy} = \frac{V_y Q}{I_z t} \tag{3.2.5.2}$$

In the following examples, the intent is to find the locations of the extreme values of σ_x and τ_{xy} for three commonly used cross-sectional shapes. Based on the discussions in the previous sections, for each cross-section we will select a few sample points called *critical points* that are candidate points for the extreme values.

Rectangular Cross-section

Consider the section to be rectangular, as shown in Fig. 3.2.5.2.

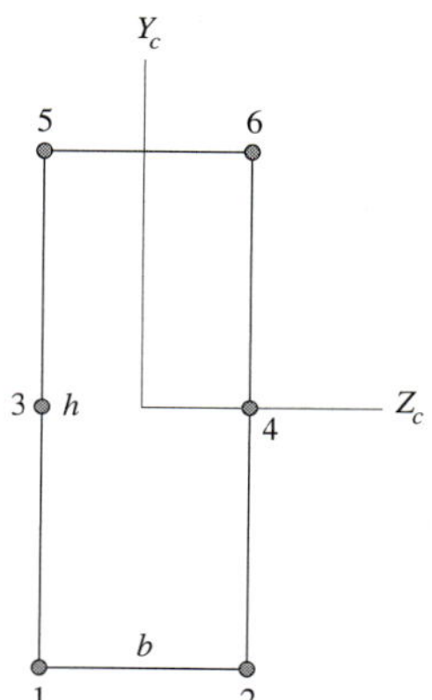

Fig. 3.2.5.2 Critical points.

Point	σ_x	τ_{xy}
1	$\frac{N_x}{bh} + \frac{6M_z}{bh^2}$	0
2	same as 1	0
3	$\frac{N_x}{bh}$	$\frac{3V_y}{2bh}$
4	same as 3	same as 3
5	$\frac{N_x}{bh} - \frac{6M_z}{bh^2}$	0
6	same as 5	0

Circular Cross-section

Consider the circular cross-section shown in Fig. 3.2.5.3.

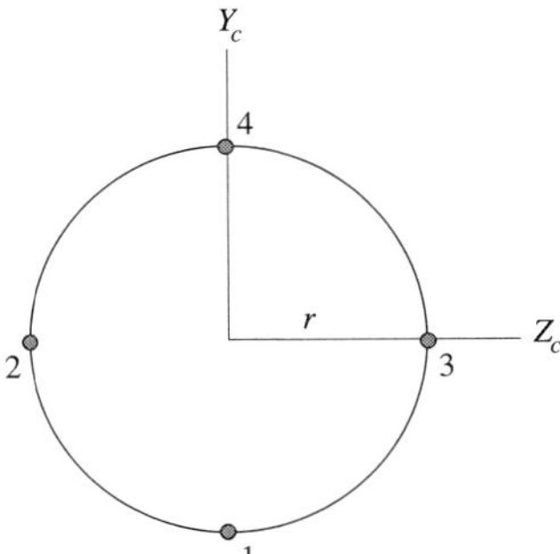

Fig. 3.2.5.3
Critical points.

Point	σ_x	τ_{xy}
1	$\frac{N_x}{\pi r^2}+\frac{4M_z}{\pi r^3}$	0
2	$\frac{N_x}{\pi r^2}$	$\frac{4V_y}{3\pi r^2}$
3	same as 2	same as 2
4	$\frac{N_x}{\pi r^2}-\frac{4M_z}{\pi r^3}$	0

Symmetric I-Section[3]

Consider the symmetric I-section shown in Fig. 3.2.5.4.

$$A = 2w_f t_f + t_w d_w \qquad I = \frac{w_f t_f^3}{6} + \frac{w_f t_f d_w^2}{4} + \frac{t_w d_w^3}{12}$$

$$S = \frac{2I}{d_w + 2t_f} \qquad SF = \frac{8It_w}{4w_f t_f (d_w + t_f) + t_w d_w^2}$$

The same idea can be applied to other cross-sectional shapes. Note that the above procedure does not find the largest normal and the largest shear stress in the cross-section. The largest values can be found by searching for the point with the largest principal stresses and the largest shear stresses, as discussed in Section 3.1.5. However, the values found by the above procedure are acceptably close to the actual extreme values.

[3] AISC sections having this shape are known as S (standard) sections, W (wide flange) sections, etc.

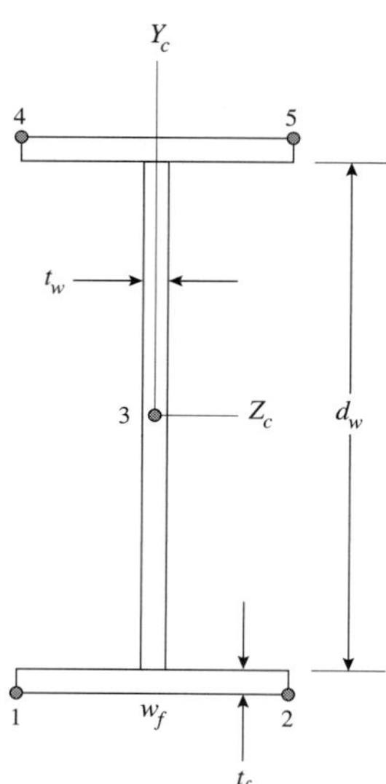

Fig. 3.2.5.4
Critical points.

Point	σ_x	τ_{xy}
1	$\frac{N_x}{A} + \frac{M_z}{S}$	0
2	same as 1	0
3	$\frac{N_x}{A}$	$\frac{V_y}{SF}$
4	$\frac{N_x}{A} - \frac{M_z}{S}$	0
5	same as 4	0

EXAMPLE 3.2.2

Stress Distribution in a Beam

Fig. E3.2.2 shows a simply supported timber beam subjected to a uniformly distributed loading. Compute (a) the largest normal stress in the beam, and (b) largest shear stress at the glued joint.

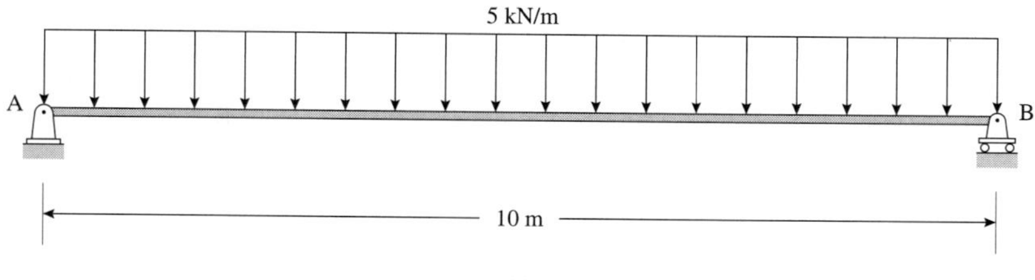

Fig. E3.2.2(a)
Simply supported beam.

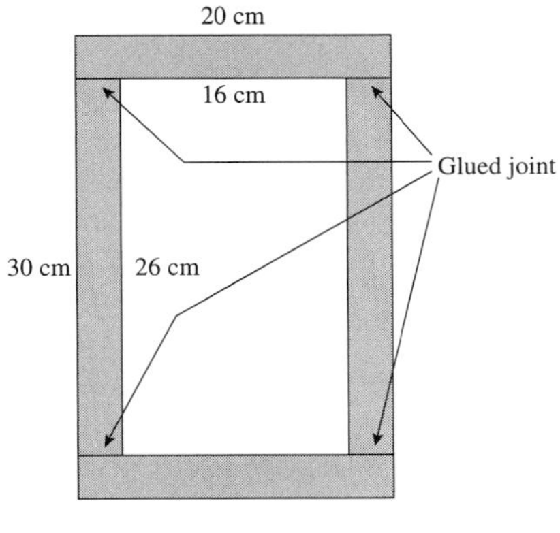

Fig. E3.2.2(b)
Beam cross-section.

SOLUTION

Step 1: Eqs. (3.2.5.1) and (3.2.5.2) can be used to compute the normal and the shear stress at any point in the beam. Since the beam cross-section is uniform along the length of the beam, we need to answer two questions to solve this problem. First, where are the maximum bending moment and maximum shear force along the beam? Note that there are no axial forces in this problem. This will indicate the location along the length of the beam of the two maximum stresses. Second, where is the critical point(s) in the cross-section?

Step 2: Using the bending moment diagram and the shear force diagram, we obtain the following for a simply supported beam of length L loaded by a uniformly distributed load w:

$$M_{\max} = \frac{wL^2}{8} \quad \text{occurring at the center of the beam}$$

$$V_{\max} = \frac{wL}{2} \quad \text{along the entire length of the beam}$$

We compute the maximum normal stress first. Since the normal stress is directly proportional to the distance of the fiber from the centroidal axis, the largest normal stresses occur at the top and the bottom of the beam. Hence

$$(\sigma_x)_{\max} = \mp\frac{(M_z)_{\max}(h/2)}{I_z}$$

Using the problem data, we find

$$I_z = \frac{20(30)^3}{12} - \frac{16(26)^3}{12} = 21565.3\,\text{cm}^4 \quad \text{and } h/2 = 30/2 = 15 \text{ cm}$$

$$(M_z)_{\max} = \frac{(5000)(10)^2}{8} = 62500\,\text{N-m}$$

Substituting into the expression for the normal stress (converting to m)

$$(\sigma_x)_{\max} = \mp\frac{(62500)(0.15)}{(21565\times10^{-8})} = 43.5(10^6)\,\text{Pa} = 43.5\,\text{MPa}$$

Step 3: The maximum shear stress at the glued joint is given by

$$(\tau_{xy})_{\max} = \frac{(V_y)_{\max}}{I_z}\left(\frac{Q}{t}\right)_{\text{joint}}$$

Using the problem data, we find

$$(V_y)_{\max} = \frac{(5000)(10)}{2} = 25000\,\text{N}$$

$$(Q)_{\text{joint}} = A\bar{y} = (20)(2)(15-1) = 560\,\text{cm}^3 \text{ and } (t)_{\text{joint}} = (2+2) = 4\,\text{cm}$$

Substituting into the expression for the shear stress (converting to m) yields

$$\left(\tau_{xy}\right)_{\text{joint}} = \frac{(25000)}{(21565\times10^{-8})}\frac{560\times10^{-6}}{4\times10^{-2}} = 1.62(10^6)\,\text{Pa} = 1.62\,\text{MPa}$$

Observations: Why did we not compute the principal stresses using, say, the Mohr's circle approach to determine the largest normal stress? This is a perfectly valid question. A detailed answer can be obtained by solving Problem 3.2.7. A short answer is that (a) most structural components are subjected to bending moments and shear forces such that the normal stress due to the bending moment dominates the shear stress due to shear force, and (b) at a cross-section, the location of the largest normal stress due to bending is also the location of zero stress due to the shear force and vice-versa. Hence the assumption is that the largest normal stress due to bending moment alone is the largest normal stress in the structural component. Obviously there are exceptions to these observations that we should be aware of.

EXAMPLE 3.2.3 ***Stress Distribution in a Frame (Example 2.8.7)***

Fig. E3.2.3 shows a planar frame. The column AB is W24 × 84 and beam BC is $W18 \times 50$. Compute the largest normal stress and the largest shear stress in the frame.

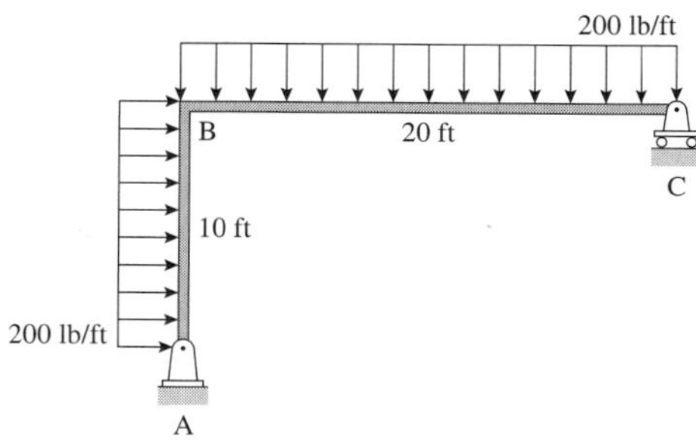

Fig. E3.2.3

SOLUTION

Step 1: The properties of the wide flange sections (W sections) are as follows. The Q_z value is computed for the top half of the cross-section (since the largest shear stress occurs at the centroidal axis) as

$$\text{W18}\times50{:}\, Q_z = (7.495)(0.570)\left(\frac{17.99-0.57}{2}\right) + \frac{1}{2}\left(\frac{17.99}{2}-0.57\right)^2(0.355) = 49.81\,\text{in}^3$$

$$\text{W24}\times84{:}\, Q_z = (9.02)(0.77)\left(\frac{24.1-0.77}{2}\right) + \frac{1}{2}\left(\frac{24.1}{2}-0.77\right)^2(0.47) = 110.92\,\text{in}^3$$

Cross-section	Area, A (in^2)	Moment of inertia, I_z (in^4)	Section modulus, S_z (in^3)	Second moment of area, Q_z(in^3)	Web thickness, t(in)
W18 × 50	14.7	800	88.9	49.81	0.355
W24 × 84	24.7	2370	196	110.92	0.470

As in the previous problem, we need to locate the maximum bending moment and the maximum shear force, but for each member of the frame. Also, in this problem, while the beam has no axial force, the column is subjected to an axial force that will affect the normal stress in the column.

Step 2: We first analyze the column. From the bending moment and shear force diagrams (see Example 2.8.7), $V_{max} = 2000$ lb (at A), $N_{max} = 2500$ lb (compression throughout the column) and $M_{max} = 10000$ lb-ft (at B). Using these values (converting to in), we find

$$(\sigma_x)_{max} = \frac{(N_x)_{max}}{A} \pm \frac{(M_z)_{max}}{(S_z)_{min}} = -\frac{2500}{24.7} - \frac{(10000)(12)}{196} = -713.5\,\text{psi}$$

This compressive stress occurs at B on the outside (or, left) face. And

$$(\tau_{xy})_{max} = \frac{(V_y)_{max}}{I_z}\left(\frac{Q}{t}\right)_{max} = \frac{2000}{2370}\frac{110.92}{0.470} = 199.2\,\text{psi}$$

This shear stress occurs at A on the centroidal axis. Both these stresses are small.

Step 3: Now we analyze the beam. From the bending moment and shear force diagrams, $V_{max} = 2500$ lb (at B), $N_{max} = 0$, and $M_{max} = 13125$ lb-ft (at 12.5 ft from B). Hence,

$$(\sigma_x)_{max} = \frac{(N_x)_{max}}{A} \pm \frac{(M_z)_{max}}{(S_z)_{min}} = \pm\frac{(13125)(12)}{88.9} = \pm 1772\,\text{psi}$$

This compressive stress occurs at 12.5 ft from B on the top (compression) and bottom (tension) fibers. And

$$(\tau_{xy})_{max} = \frac{(V_y)_{max}}{I_z}\left(\frac{Q}{t}\right)_{max} = \frac{2500}{800}\frac{49.8}{0.355} = 438.4\,\text{psi}$$

This shear stress occurs at B on the centroidal axis. Again, both these stresses are small.

EXERCISES

Appetizers

3.2.1. Find the moment of inertia of the W10 × 19 AISC cross-section about its centroidal axes.

3.2.2. Find the cross-sectional properties of the built-up section shown in Fig. P3.2.2.

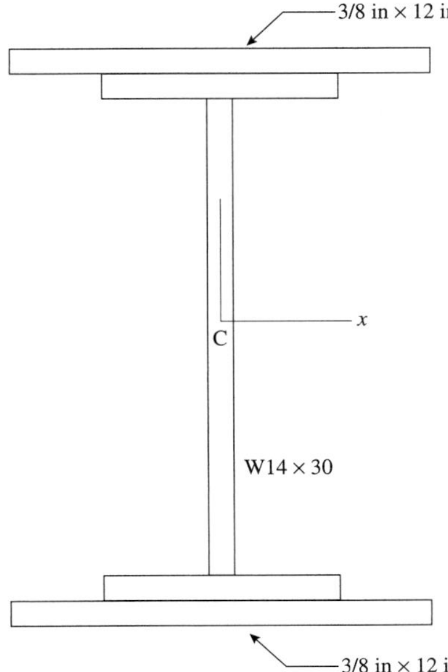

Fig. P3.2.2

3.2.3. Figure P3.2.3 shows the free-body diagram of a member in a frame. The cross-section is rectangular with the height 20 cm and the width 10 cm. Compute the (a) largest compressive and tensile stress in the member due to axial force and bending moment, and (b) the largest shear stress due to shear force.

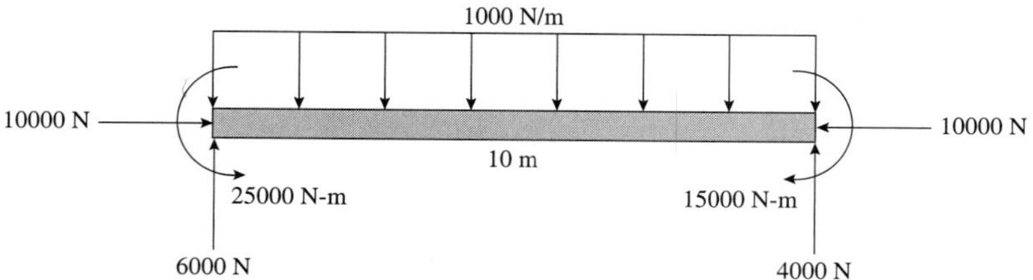

Fig. P3.2.3

Main Course

3.2.4. Figure shows a cantilever beam whose cross-section is W10 × 19 AISC cross-section. Compute the largest tensile and the largest compressive stress in the beam. Include the self-weight of the beam.

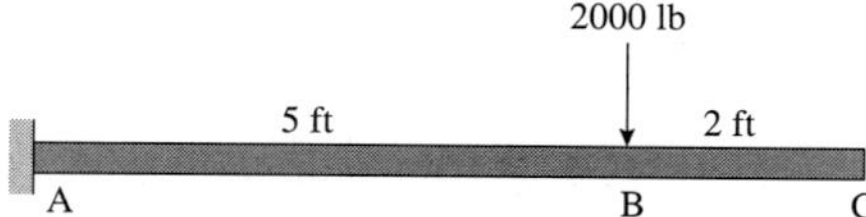

Fig. P3.2.4

3.2.5. For the frame shown in Fig. P3.2.5, compute the largest normal stress and shear stress.

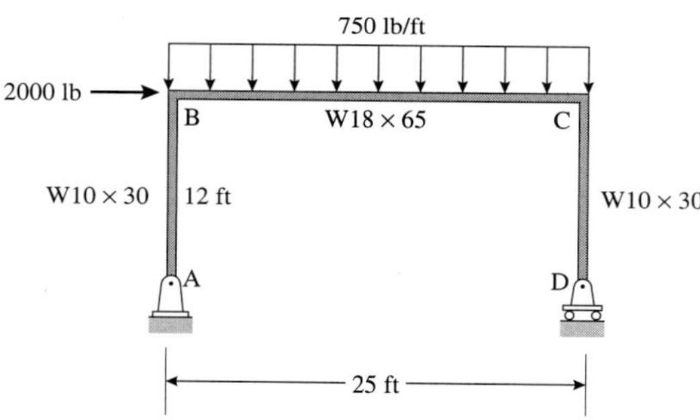

Fig. P3.2.5

3.2.6. Fig. P3.2.6(a) shows the cross-section of a wooden beam made of four pieces—two flanges and two webs. The webs are connected to the flanges by plastic screws whose allowable load in shear is 250 lb. Determine the spacing of the screws p if the beam is loaded as shown in Fig. P3.2.6(b).

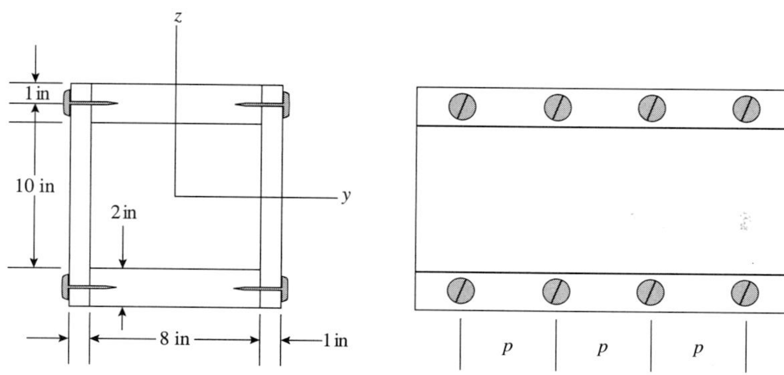

Fig. P3.2.6(a)

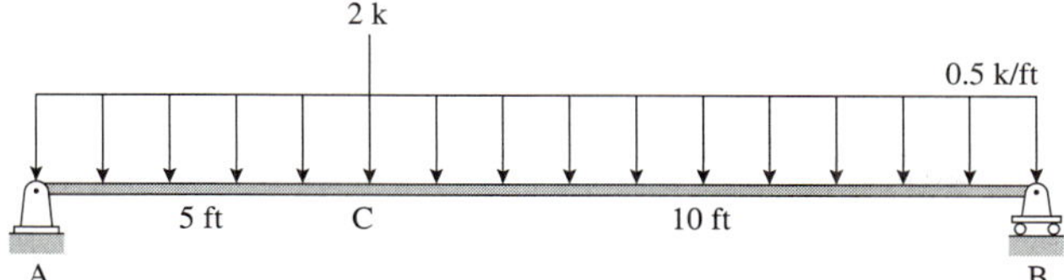

Fig. P3.2.6(b)

Structural Concepts

3.2.7. Consider a beam made of solid rectangular cross-section of height h and width b. The beam is subjected to a loading such that at a given section the bending moment is M and the shear force is V.

(a) Customize Eqs. (3.2.5.1) and (3.2.5.2) for this cross-section by writing the expressions for a point located at a distance y from the centroidal axis.

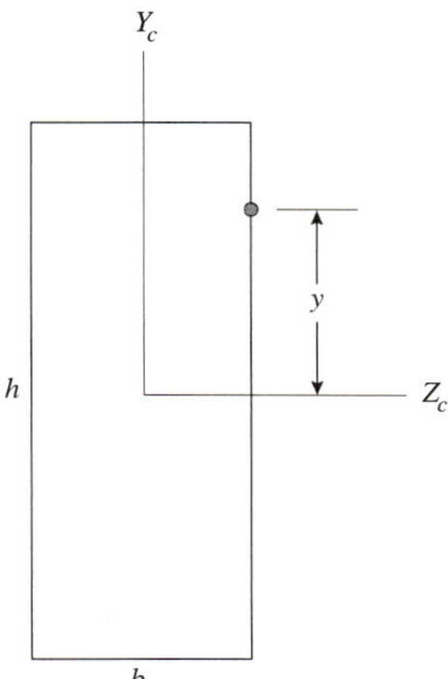

Fig. P3.2.7(a)

(b) Now assume that $h = c_1 b$ and $M = c_2 V$ where the constants c_1 and c_2 have the appropriate units. Write the expression for the principal stress in terms of y, c_1, c_2, h, and M. Assume values for h and M. Now vary c_1 and c_2, and plot σ_1 and σ_2 as a function of y. Write down your conclusions about the location of the point with the largest normal stress as a function of the problem parameters.

3.3 THEORIES OF FAILURE

There are at least two major aspects to establishing the design requirements. The first is purely functional: structural systems are conceived to meet certain practical and useful needs. The second is requirement is that they should not fail when being built and later when they are operational. In this text failure will imply structural failure. There are nonstructural failures that a designer must be concerned with, but discussion of those is outside the scope of this text.

3.3.1 Some Causes of Structural Failures

Yielding. Yielding occurs when there are plastic (or permanent) deformations in ductile material. Members, connections, and supports should not yield excessively. Excessive yielding may cause the structure to deform to unacceptable levels or may even cause catastrophic failure. Some yielding may be localized, e.g., in connections around bolt holes.

Low Stiffness. This is especially a concern with regard to loads that can cause the structure to resonate, e.g., wind loads, earthquake loads. The structure must be designed to be adequately stiff.

Buckling. Compressive forces may produce sudden major changes in the geometry of the structure (global or overall buckling) or members (local buckling). It should be noted that buckling can occur even when the stress levels in the structure are well within the elastic region. We look at certain aspects of buckling in the next section.

Crushing. Compressive forces may also cause material failure through crushing, where a brittle material may split or a ductile material may deform excessively (crush).

Fracture. Fracture is a phenomenon in which surfaces are created in the form of cracks, either new or extensions of old cracks. Sometimes repeated application of loads or cyclic loading may cause fatigue fracture. A brittle fracture failure may occur with very little plastic deformation.

3.3.2 Failure Criteria

Failure criteria are associated with material failure that is said to occur when the material reaches a certain limiting value. This value can be a measure of stress, strain, energy, or other suitable quantity.

Von Mises Criterion. Also known as octahedral shearing stress criterion or strain energy density of distortion criterion. The yielding of an isotropic material takes place when

$$\tau_{\text{oct}} = \frac{1}{3}\sqrt{(\sigma_x - \sigma_y)^2 + (\sigma_y - \sigma_z)^2 + (\sigma_z - \sigma_x)^2 + 6(\tau_{xy}^2 + \tau_{yz}^2 + \tau_{zx}^2)^2} \geq \frac{\sqrt{2}}{3}\bar{\sigma} \qquad (3.3.2.1)$$

where τ_{oct} is known as the octahedral shear stress and $\bar{\sigma}$ is some limiting value, usually taken as the yield stress from a uniaxial test. This failure criterion is typically applied to ductile material.

Maximum Principal Stress Criterion. Tensile fracture surfaces form in an isotropic material when the largest principal tensile stress exceeds some limiting value σ_1:

$$\sigma_1 \geq \bar{\sigma} \qquad (3.3.2.2)$$

$\bar{\sigma}$ is usually taken as the yield stress from a uniaxial test. This failure criterion is typically applied to brittle material.

Mohr's Criterion. The failure of an isotropic material through fracture or yielding takes place when

$$\frac{\sigma_1}{(\sigma_t)_f} - \frac{\sigma_3}{(\sigma_c)_f} \geq 1 \qquad (3.3.2.3)$$

where $(\sigma_t)_f$ and $(\sigma_c)_f$ are the magnitudes of the stress at failure in uniaxial tensile and compressive tests, respectively. The criterion is usually applied to brittle materials that are much stronger in compression than tension.

Norris Criterion. Failure in an anisotropic material occurs when at least one of the following conditions is true:

$$\begin{aligned}
&\frac{\sigma_{11}^2}{(\sigma_{11})_f^2} - \frac{\sigma_{11}\sigma_{22}}{(\sigma_{11})_f(\sigma_{22})_f} + \frac{\sigma_{22}^2}{(\sigma_{22})_f^2} + \frac{\tau_{12}^2}{(\tau_{12})_f^2} \geq 1 \\
&\frac{\sigma_{22}^2}{(\sigma_{22})_f^2} - \frac{\sigma_{22}\sigma_{33}}{(\sigma_{22})_f(\sigma_{33})_f} + \frac{\sigma_{33}^2}{(\sigma_{33})_f^2} + \frac{\tau_{23}^2}{(\tau_{23})_f^2} \geq 1 \\
&\frac{\sigma_{33}^2}{(\sigma_{33})_f^2} - \frac{\sigma_{33}\sigma_{11}}{(\sigma_{33})_f(\sigma_{11})_f} + \frac{\sigma_{11}^2}{(\sigma_{11})_f^2} + \frac{\tau_{31}^2}{(\tau_{31})_f^2} \geq 1
\end{aligned} \tag{3.3.2.4}$$

where (1,2,3) are the principal material directions[4] and $(\sigma_{ii})_f$ and $(\tau_{ij})_f$ are experimentally determined failure stress values. This criterion can be applied to orthotropic material such as wood where the axial (along the grain), radial, and tangential directions can be taken as the principal material directions.

Buckling. The buckling phenomenon merits special treatment, as it is one of the most common failure mode. Significant portions of most structural systems are subjected to compressive forces sometime during their life and the buckling phenomenon is associated with such forces. We now look at a specific problem—the elastic buckling of columns (slender members subjected to compressive forces). Consider an initially straight, slender column (see Fig. 3.3.2.1) subjected to an axial force that passes through the centroid of the cross-section.

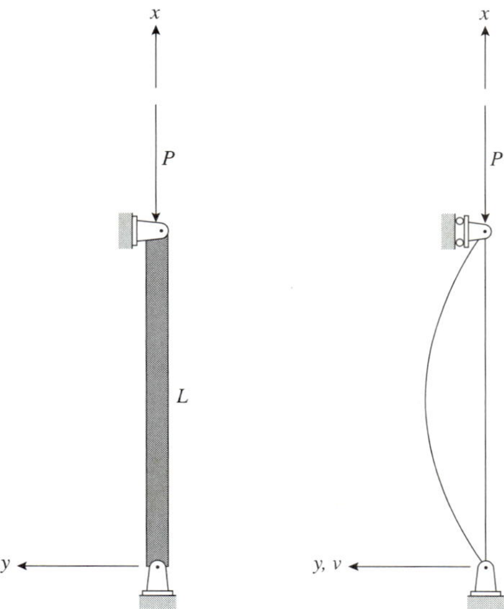

Fig. 3.3.2.1 Euler buckling.

As the axial force is gradually increased from zero, the column is subjected to a uniform compressive stress. If a small lateral load is applied, the column bends. This deflection disappears if the lateral load is removed. This state is known as stable equilibrium. At some load value, neutral equilibrium is reached. The load value is called the critical load P_{cr}. At this state, the column may be bent. A small lateral load may produce a lateral deflection that does not disappear if the lateral load is removed. In other words, the column is in equilibrium and the shape can be one of many possibilities. With any further increase in the load, the column becomes unstable and is likely to collapse. Mathematically, the column is stable if $P < P_{cr}$, is in neutral equilibrium if $P = P_{cr}$, and is unstable if $P > P_{cr}$.

[4] Three mutually perpendicular planes of elastic symmetry exist in orthotropic materials. The principal material axes are normal to the planes of symmetry.

We can write the differential equation for column buckling (Euler buckling[5]) as

$$EI\frac{d^2v}{dx^2} + Pv = 0 \tag{3.3.2.5}$$

The solution of this differential equation gives the critical load and the corresponding deflected shape of the buckled column. Let

$$k = \sqrt{\frac{P}{EI}} \tag{3.3.2.6}$$

Substituting in the differential equation, we have

$$\frac{d^2v}{dx^2} + k^2v = 0 \tag{3.3.2.7}$$

The solution to this differential equation is

$$v = a\sin kx + b\cos kx \tag{3.3.2.8}$$

where a and b are constants of integration that can be determined from the boundary (or end) conditions. For the column in Fig. 3.3.2.1, $v(x = 0) = 0$ and $v(x = L) = 0$. Substituting these conditions in Eq. (3.3.2.8), we obtain

$$a\sin kL = 0 \tag{3.3.2.9}$$

from which we can conclude that either $a = 0$ or $\sin kL = 0$. The former represents a trivial solution. The latter can be interpreted as

$$\sin kL = 0 \Rightarrow kL = n\pi \quad n = 1, 2, \ldots \tag{3.3.2.10}$$

Using Eq. (3.3.2.6) yields

$$P = \frac{n^2\pi^2EI}{L^2} \qquad n = 1, 2, \ldots \tag{3.3.2.11}$$

and the corresponding deflection is

$$v = a\sin\frac{n\pi x}{L} \qquad n = 1, 2, \ldots \tag{3.3.2.12}$$

The lowest critical load and the deflected shape are (with $n = 1$)

$$P_{cr} = \frac{\pi^2EI}{L^2} \qquad v = a\sin\frac{\pi x}{L} \tag{3.3.2.13}$$

A few observations can be made about the results. First, the critical load is directly proportional to the flexural rigidity of the column, EI. I is the moment of inertia about the buckling axis. Hence, everything else being the same, the column will buckle about the weaker (principal) axis. Second, the critical load is inversely proportional to the square of the length, L. Third, the only material property that influences the critical load is the modulus of elasticity. Fourth, as we see next, the end conditions also affect the critical load value.

The results given by Eq. (3.3.2.13) are valid for the case when the ends of the column are pinned. We can analyze the column for different support conditions and obtain a general result as

$$P_{cr} = \frac{\pi^2EI}{L_e^2} = \frac{\pi^2EI}{(KL)^2} \tag{3.3.2.14}$$

where L_e is the effective length and K is the effective-length factor. The results are summarized in Table 3.3.2.1 for the commonly encountered end conditions.

[5] The credit for this buckling analysis goes to the mathematician Leonhard Euler, who first published the results in 1744. The derivation of the differential equation can be found in Appendix E.

Table 3.3.2.1

Pinned-pinned	Fixed-free	Fixed-fixed	Fixed-pinned
$P_{cr} = \frac{\pi^2 EI}{L^2}$	$P_{cr} = \frac{\pi^2 EI}{4L^2}$	$P_{cr} = \frac{4\pi^2 EI}{L^2}$	$P_{cr} = \frac{2.046\pi^2 EI}{L^2}$
P_{cr}, L	P_{cr}, L	P_{cr}, L	P_{cr}, L
$L_e = L$ $K = 1$	$L_e = 2L$ $K = 2$	$L_e = 0.5L$ $K = 0.5$	$L_e = 0.7\text{L}$ $K = 0.7$

The state of stress at the critical load can be computed as

$$\sigma_{cr} = \frac{P_{cr}}{A} = \frac{\pi^2 EI}{AL^2} \tag{3.3.2.15}$$

It is common to relate the state of stress to two parameters: the *radius of gyration r*, given as

$$r = \sqrt{I/A} \tag{3.3.2.16}$$

and the *slenderness ratio k*, given as

$$k = L/r \tag{3.3.2.17}$$

so that

$$\sigma_{cr} = \frac{\pi^2 E}{(L/r)^2} = \frac{\pi^2 E}{(k)^2} \tag{3.3.2.18}$$

What we have seen so far is called the overall elastic buckling of a column. There are other types of buckling of structural components. While long compression members fail by elastic buckling, short (or stub) columns may fail through material yielding (or crushing). However, failure through *inelastic buckling* is quite common. Here a portion of the cross-section yields before buckling occurs. All commonly used cross-sections, whether rolled shapes or built-up sections, are made up of plate sections. When the member is under compressive loads, the elastic or inelastic buckling may not take place. Instead, *local buckling* may occur in the one of the plate sections and the member's capacity to take on any additional loads is drastically reduced. Plate theory can be generally used to investigate the local buckling phenomena and we briefly discuss this topic in Chapter 9.

3.4 COMMONLY USED STRUCTURAL MATERIALS

There are tens if not hundreds of different materials used in structural systems. Of these, the most common ones are briefly discussed below.

Steel. Almost all structural systems use steel in one manner or another—beams and columns in high-rise buildings, members in roofs and bridges, connection components such as bolts and gusset plates, connector plates and nails in wood structures, reinforcements in concrete and masonry structures, tendons in prestressed concrete, and so on. Structural steel members are available in two categories: hot-rolled steel (heat-treated alloy steel, carbon steel, and high-strength low-alloy steel) and cold-rolled steel. The manufacture of hot-rolled steel is an involved process and requires sophisticated manufacturing techniques. However, the end product is a material that is for all practical purposes homogeneous, isotropic, and ductile, with very desirable structural properties. The manufacture of cold-rolled steel starts with steel sheet or strip steel. The members are formed through press-braking or roll forming. The American Institute of Steel Construction (AISC) code governs the usage of hot-rolled steel, and the American Iron and Steel Institute (AISI) code governs the usage of cold-rolled steel. While the load and resistance factor design (LRFD) code has been in existence for quite some time now, the traditional allowable stress design (ASD) code is still quite popular today.

Concrete. In very many ways concrete is very different from steel. It is not homogeneous, isotropic or ductile. Concrete has four ingredients—cement, water, sand, and aggregates. Chemical reaction between the cement and the water form a paste that holds the sand and the aggregates together. Concrete's strength in tension is about a tenth of its strength in compression. It is brittle, easily susceptible to cracking and fracture. To overcome this susceptibility to tensile and shear stresses, steel reinforcements or bars are used. Shrinkage and creep cause short-term and long-term problems. Yet concrete is one of the most widely used structural material. Not only can drawbacks be overcome with care, but concrete structures are extremely economical, require very little maintenance, and can be formed into a variety of structural shapes. Concrete is used primarily in three forms: reinforced, composite, and prestressed concrete structures. The American Concrete Institute (ACI) design code was among the first to embrace the idea of strength design methodology as an advancement over the traditional working stress design.

Masonry. Masonry structures can be made of a variety of products. However, the two most common products are burned-clay bricks (including building bricks) and hollow concrete blocks. *Mortar* bonds these "building blocks" to one another. The properties of concrete and brick masonry are similar to concrete: they are strong in compression and weak in tension. The mortar is composed of cementitious material, aggregate, additives, and water. Just like concrete, it is becoming increasingly common to see steel reinforced masonry structures such as walls. *Grout* is used to fill the void and bond the reinforcements to the concrete blocks. Masonry units such as walls provide effective barriers to noise and fire. They are certainly more durable and permanent than wood.

Wood. The widespread use of wood makes it the most common material for typically low-rise structures—single-family homes, apartments, small industrial buildings, etc. While some of the material characteristics are the same as the other structural materials, wood has its own unique properties. First, it is a composite material. The orientation of the grains gives wood direction-dependent properties. Typical wood specimens are comparatively strong in tension and compression parallel to the grain and weak perpendicular to the grain. Wood can also split along the grain laminations, and hence is considered weak in shear. Second, it is found naturally in several different forms. The implication is that the material properties can have a wide range of values. Third, it requires special treatment to make it less susceptible to environmental effects (due to, for example, moisture, heat, and termites). The types of failure occurring in timber are many. Some of these can be attributed to naturally occurring defects—knots, checks, shakes, splits, etc. Most softwoods (fir, pine, etc.) are fairly ductile and do not undergo sudden failure. In spite of the complexity in the material characteristics of wood, its structural use is sustained by the advances and innovations in the wood industry. Introduction of wooden I-joists and glulam beams make it possible to design members with longer spans and heavier load-carrying capacities. Glulam (glued laminated lumber) members are fabricated as layered material made from thin laminates. The cross-sectional properties (area, moments of inertia) and material properties (bending stress, tensile, compressive, and shear stresses parallel to the grain, and tensile and compressive stress perpendicular to the grain) make it possible to compute the allowable stress values. Recent wood design codes using the ASD approach and the impending use of the LRFD methodology put wood design in the same league as other structural materials.

The material properties of these and other materials are shown in Appendix A. We will study the design of steel and concrete structures in Chapter 9.

3.5 MODELING THE STRUCTURE AND THE LOADS

In Chapter 1 we saw the basic elements of a mathematical model. Almost all the models considered in Chapter 2 and to be considered in this and following chapters are two-dimensional or planar models. While structural systems are almost always placed in three-dimensional space, the question that needs to be answered is how do we construct an approximate but sufficiently accurate planar model or models? An equally important question is whether these approximations are always accurate? We provide basic answers to some of these questions using several examples.

EXAMPLE 3.5.1

Footbridge

Consider the model of a footbridge shown in Fig. E3.5.1.1(a). The deck of the bridge (not shown) rests on a system of longitudinal and transverse beams. The beams are then supported by the main supporting system—three planar frames.

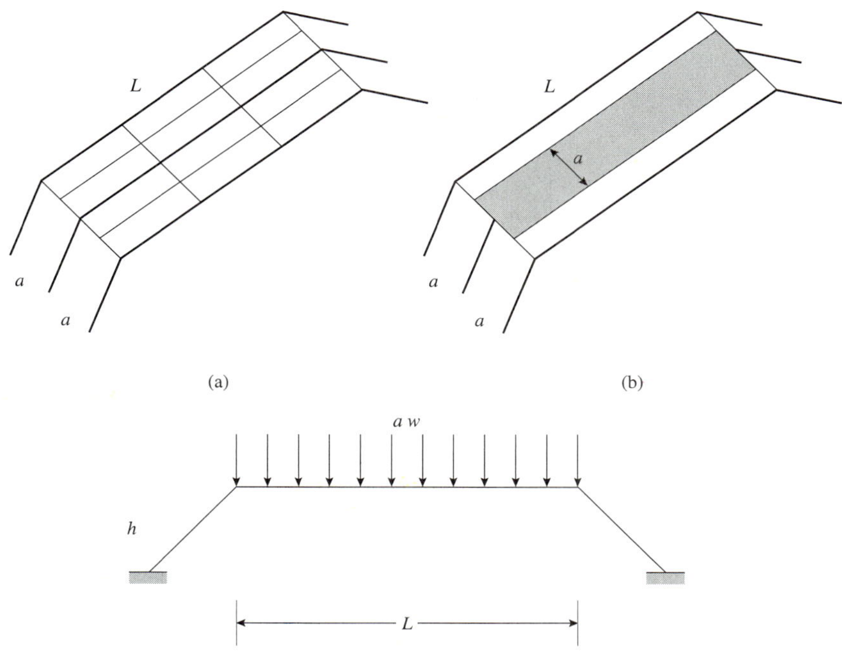

Fig. E3.5.1 (a) Model of the footbridge. (b) Tributary area for the center frame. (c) Equivalent loading on a single frame.

While the model as shown is three-dimensional, can we make appropriate assumptions to reduce the analysis problem to an analysis of a planar frame? If the loading on the bridge deck is uniform in intensity w force/unit area, then it can be assumed that the center frame supports half the load while the end frames support the other half of the load. The *tributary area* (loaded area supported by a frame) for the center frame is shown in Fig. E3.5.1(b). The load on the tributary area can be converted to a uniformly distributed load aw force/unit length acting on the center frame as shown in Fig. E3.5.1(c). This two-dimensional approximation is fairly accurate as long as $a << L$, the loading is symmetric about the center frame, and the three frames are identical.

While we have addressed the model of the primary load-bearing system, we have not looked at the modeling of the secondary system—the longitudinal and transverse beams.

EXAMPLE 3.5.2

Residential Roof

Consider the system of trusses that forms a roof system for a residential home. Fig. E3.5.2(a) shows a part of the system. The trusses of span L are placed at a constant spacing a, with $a << L$.

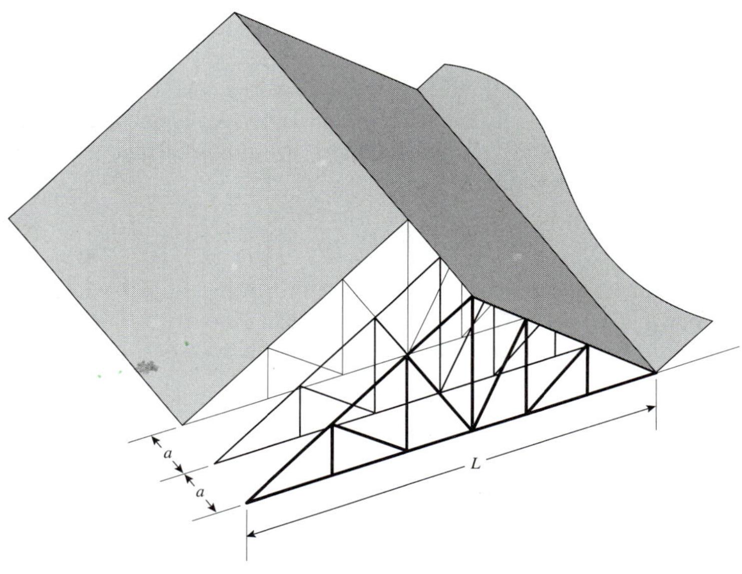

Fig. E3.5.2(a) Residential roof truss system.

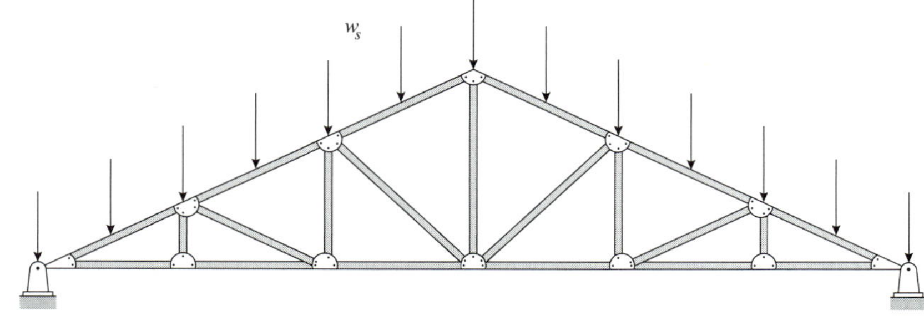

Fig. E3.5.2(b) Equivalent gravity loads on a single truss.

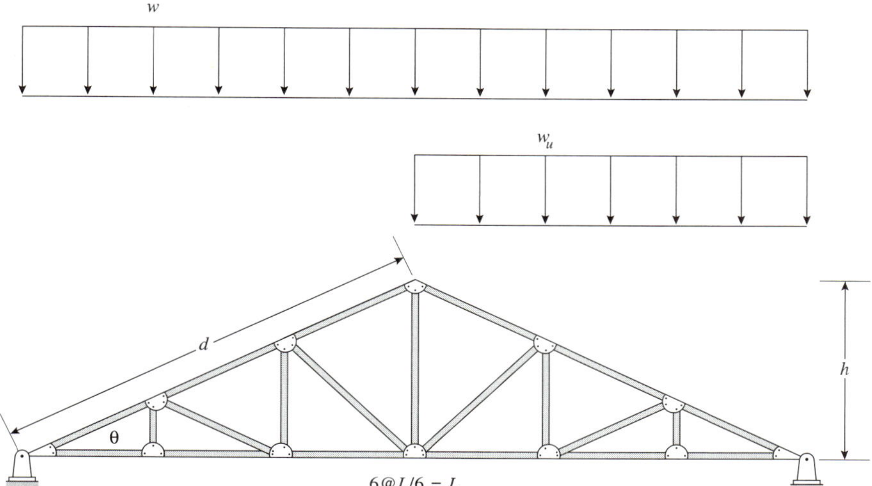

Fig. E3.5.2(c) Equivalent *projected* loads on a single truss.

Typically the roof system consists of a deck, rafters and purlins that support the deck, and trusses on which the purlins rest. The deck is primarily made of plywood sheets. The purlins are the horizontal beams between the trusses. The trusses are finally supported on walls, piers, or columns.

As an example, we evaluate the snow loads acting on a single truss. Let the load be calculated from a design code as q force/unit area applied over the entire roof surface. To compute the equivalent load (see Fig. E3.5.2(b)), w_s force/unit length acting on a single truss, the truss spacing is used first, as we did in the previous example. In other words, the width of the tributary area for single truss is a. Hence

$$w_s = qa \tag{3.5.1}$$

Design drawings and calculations typically represent the loads acting on the *projected* length or area. (The practice is based on the idea of separating gravity and lateral loads; the former act "vertically" and the latter "horizontally.") The top of Fig. E3.5.2(c) shows the equivalent snow load acting on the projected length of the truss. In other words,

$$w\frac{L}{2} = w_s d \Rightarrow w = w_s\frac{d}{L/2} \tag{3.5.2a}$$

Or $w = w_s/\cos\theta = w_s\sqrt{1+p^2}$ (3.5.2b)

where p is the pitch (rise h over run $L/2$) of the roof. The bottom of Fig. E3.5.2(c) shows the equivalent snow load acting on the projected length of the truss for the unbalanced condition (wind from the left causes the snow to accumulate on the leeward roof surface).

If we now further assume that the single roof truss does behave like a truss, we need to compute the equivalent forces acting on the joints of the top chord of the truss (Fig. E3.5.2(d)). Noting that the joints on the top chord (or, panel points) are equally spaced and using the tributary area concept, we can compute the joint loads as

$$(n-1)P = wL \Rightarrow P = \frac{wL}{n-1} \tag{3.5.3}$$

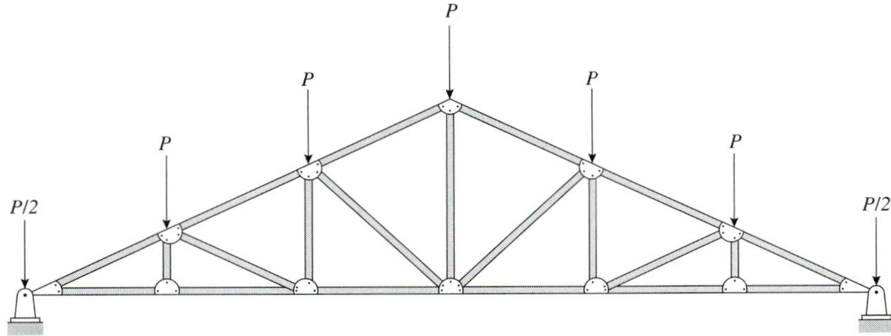

Fig. E3.5.2(d) Joint loads for uniformly distributed loading.

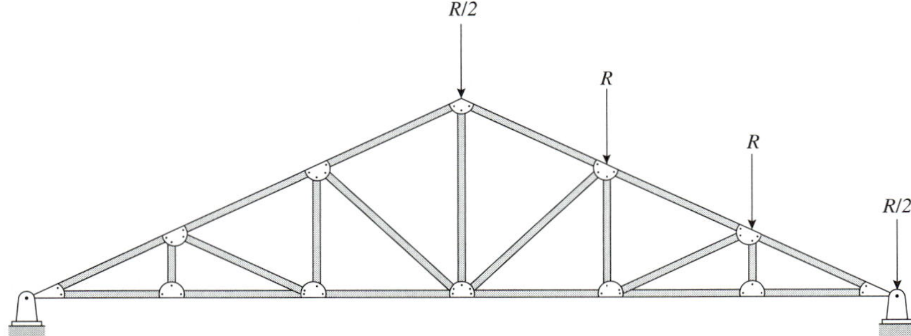

Fig. E3.5.3(e) Joint loads for unbalanced loading.

where n is the number of loaded joints and supports. The tributary area for an interior joint is twice as much as the end ones; hence the loads (P) on the interior joints are twice as much as at the ends ($P/2$).

Similarly, for the unbalanced loading (Fig. E3.5.2(e))

$$R = \frac{w_u L}{2(n-1)} \tag{3.5.4}$$

where n is once again the number of loaded joints and supports.

How good is this planar truss model? The most obvious question is whether the structure is really a truss. The top and bottom chords are continuous pieces. The connections between the web members and the chord are not pins. With the roof deck connected continuously to the top chord, loading on the truss, such as snow, is distributed. Clearly the basic assumptions of a truss are not satisfied.

How does one model the supports? Does the flexibility of the walls, piers, and columns affect the modeling of the support and the behavior of the truss?

The overall loads acting on a truss are modeled fairly accurately as long as the basic assumptions outlined earlier are followed.

EXAMPLE 3.5.3 ***Multi-Story Office Building***

Figures E3.5.3(a)–(d) show several views of a low-rise multi-story office building. The locations on the horizontal grid in Fig. E3.5.3(b) are numbered much like a chessboard: the four columns at the south end of the building are located at a1, b1, c1, and d1. The four girders span the entire N-S length and are supported at the four columns located at 1-5-9-13. The 13 beams spanning in the E-W direction support the floor and are connected to the girders. The frame at the south end of the building is shown in Fig. E3.5.3(c) and that at the east end of the building is shown in Fig. E3.5.3(d). The grid numbering in the vertical direction starts at 0 (ground level) and extends to 3 (roof level): e.g., the roof at the SW corner of the building is located at 1a-3.

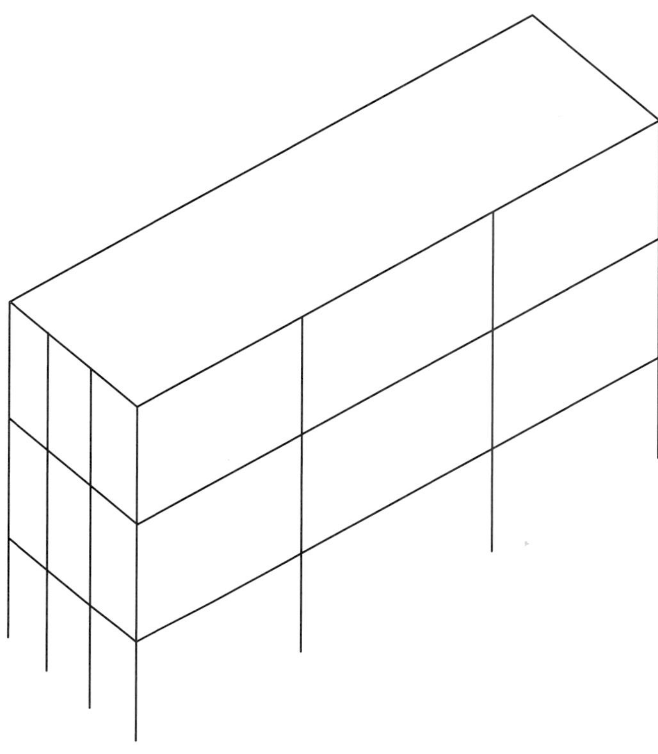

Fig. E3.5.3(a)
Multi-storied office building.

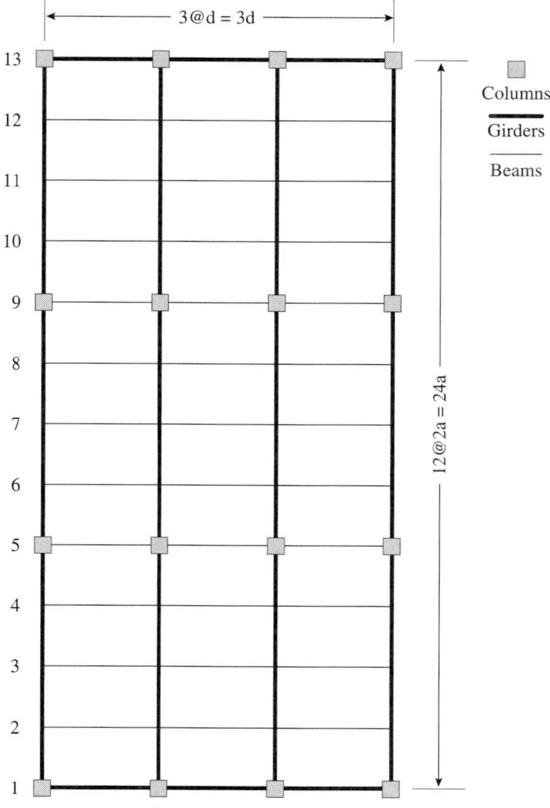

Fig. E3.5.3(b)
Plan view of one floor.

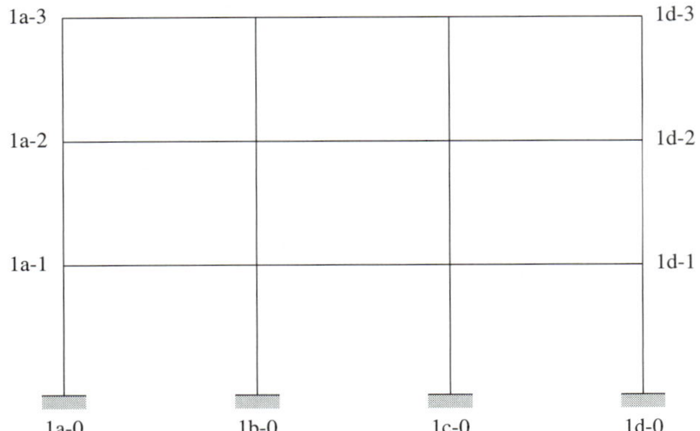

Fig. E3.5.3(c) South end frame.

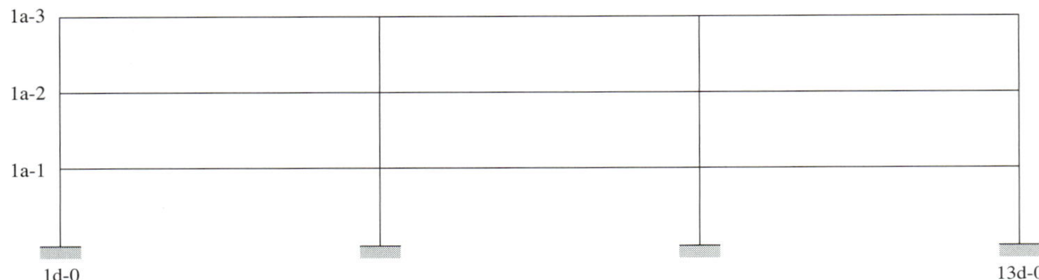

Fig. E3.5.3(d) East end frame.

The modeling of a moderately complex structure such as this multi-story office building is much more of a challenge than the previous two examples. First, we will look at the dimensions of the structure. The beam spacing, 2a, is smaller than the beam span d. We assume that the tops of the beams and the girders are at the same level. The floor is partly supported by the beams and partly by the girders. Second, the nature of the connection between the beam and the girder and the girder the column determines the type of approximation that can be made. Let us assume that, in this example, all the connections are pin connections.[6] This enables us to analyze the beams and girders independently (since these structural components are now statically determinate).

For those loads acting as uniformly distributed loads on the floor, the basic load-distribution pattern is shown in Fig. E3.5.3(e). The tributary areas for the two beams and two girders that form the rectangular pattern are shown in the top right. The oblique lines make 45° angles with the sides. This pattern is repeated for all the rectangular domains. Hence the tributary area for a typical segment on the girder is shown for b9-b10 and b10-b11—two equilateral triangles. Similarly, the tributary area for a typical beam is shown for c11-d11—two trapezoids.

---

[6] While this assumption simplifies the analysis, it renders the structure unstable unless additional members such as cross-bracing members are used.

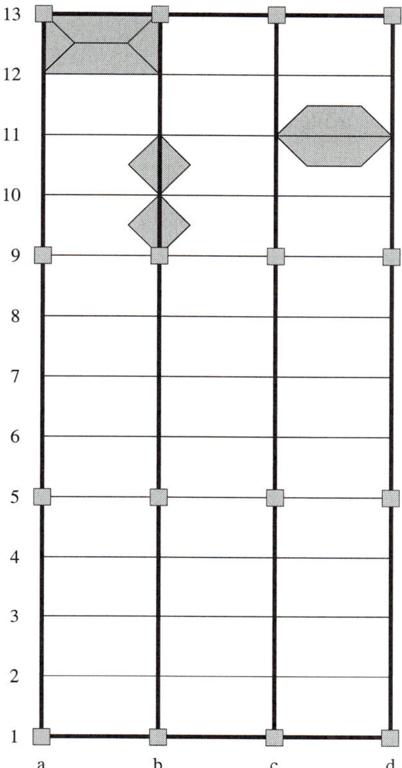

Fig. E3.5.3(e)
Tributary loading on the girders and beams.

Now we are in a position to draw the FBD of a typical beam and girder. The tributary area and the FBD of a typical beam are shown in Fig. E3.5.3(f). If the intensity of the distributed floor load is w force/unit area, then the distributed load on the beam is $2aw$ force/unit length. Since the beam is simply supported, the reactions at the ends of the beam are

$$R_{c11} = R_{d11} = \frac{1}{2}\left[(d-2a)2aw + 2\left(\frac{1}{2} \times a \times 2aw\right)\right] = aw(d-a) \tag{3.5.5}$$

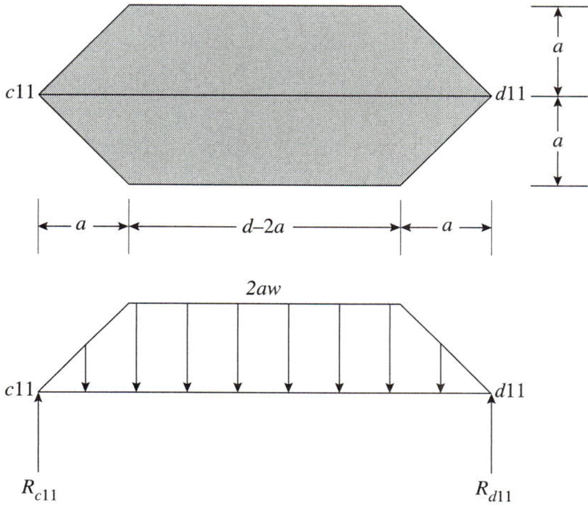

Fig. E3.5.3(f)
Loads on a typical beam.

We can draw the tributary area and FBD of a typical girder in a similar form; Fig. E3.5.3(g) shows the details.

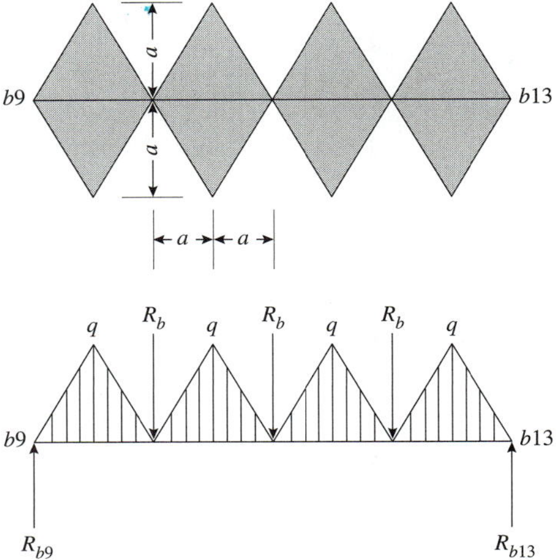

Fig. E3.5.3(g)
Loads on a typical girder.

If the intensity of the distributed floor load is w force/unit area, then

$$q = 2aw \tag{3.5.6}$$

and $R_b = R_{c11} + R_{d11}$ is the reaction at the end of a typical beam (see Eq. (3.5.5)). Using the girder FBD, we find

$$R_{b9} = R_{b13} = \frac{3}{2}R_b + 2aq \tag{3.5.7}$$

We can also compute the force transmitted to a typical column. Consider column b9. Forces are transmitted to it from the beams a9-b9 and b9-c9, girders b5-b9 and b9-b13. Hence the applied force N on column b9 is

$$N = 2aw(d - a) + 3R_b + 4aq \tag{3.5.8}$$

The above process makes it possible to handle gravity loads. The issue of modeling the structure with the lateral loads, on the other hand, is more challenging. The building as shown has 16 planar frames. Two of those frames are shown in Figs. E3.5.3 (c)–(d). Once the lateral loads are computed for a frame, the analysis of the planar frames can take place.

3.6 DESIGN SPECIFICATIONS

From a structural practitioner's viewpoint, arriving at the right loads that act on a structural system is half the work. While this viewpoint is slightly exaggerated, computing the right loads is crucial to the survival and performance of the structural system. There is sufficient concern in published literature about teaching and learning the issue of loads on structural systems.[7] The different aspects of structural loading are addressed in this section so that we can begin to build some of the structural models for analysis and design. The intent is to present the ideas here (planting the seeds, so to speak). Formal structural design courses are perhaps the right place to investigate this topic in sufficient detail.

---

[7] R. J. Schimdt, "When Are Loads Taught? A Case Study and National Trends," *Proc. Structural Congress XII*, Atlanta, 1994, pp. 1358–1363; D. S. Ellifritt, "Where Do the Arrows Come From? Evaluating Structural Loads," *Proc. Structural Congress XII*, Atlanta, 1994, pp. 1364–1368.

3.6.1 Design Codes

In the introductory section we looked at the design steps. Having defined the structural system (Step 2) to meet the functional requirements (Step 1), the designer must establish the structural requirements to be met by the system and then compute the loads to be carried by the system (Step 3). There are several publications called standards or codes that help the designer establish the service loads, safe stress levels, and acceptable deflections. They also provide a list of commonly available members and materials and their properties. The applicable codes for the commonly used structural materials are listed below.

Steel

AISC LRFD Manual of Steel Construction, American Institute of Steel Construction, 1994.
AISI LRFD Manual, American Iron and Steel Institute, 1991.

Concrete

Building Code Requirements for Structural Concrete, ACI 318-99, American Concrete Institute, 1999.
Commentary on Building Code Requirements for Structural Concrete, American Concrete Institute, 1999.

Masonry

Masonry Design Manual, Masonry Institute of America, 1989.
Building Code Requirements for Masonry Structures, ACI530-92/ASCE5-92, American Concrete Institute and American Society of Civil Engineers, 1992.
Specifications for Masonry Structures, ACI530.1-92/ASCE6-92, American Concrete Institute and American Society of Civil Engineers, 1992.

Wood

National Design Specification for Wood Construction and Supplement, National Forest Products Association, 1991.

Loads

American National Standard Minimum Design Loads for Buildings and Other Structures, ANSI A58.1-1982, American National Standards Institution, 1982.
Minimum Design Loads for Buildings and Other Structures, ASCE 7-95, American Society of Civil Engineers, 1995.
Uniform Building Code (UBC), International Conference of Building Officials, 1997.
AASHTO LRFD Bridge Design Specifications, American Association of State Highway and Transportation Officials, 1997.

Loads that act on a structural system can be classified as (a) those caused by gravity and (b) lateral loads. Dead, live, and snow loads are examples of gravity-induced loads whereas wind and earthquake loads are lateral loads. We look at more detailed explanations of these loads next.

3.6.2 Dead Loads

Dead loads include the weight of all the stationary and permanent structural and non-structural components that constitute the structural system. Examples include the weight of beams, columns, walls, floors, roofs, piping, conduits, lighting fixtures, etc. Commonly used dead-load values are shown in Appendix C and can be estimated with good accuracy.

3.6.3 Live Loads

Live loads are contributed by nonstructural components that are not permanently attached to the structural system. Examples include the weight of movable partitions, furniture, human occupants, equipment, vehicles, etc. Commonly used live-load values are shown in Appendix C. Unlike dead loads, live loads can only be estimated because of the transient nature of the loads. As we see later, even with the load magnitudes, the designer must configure the locations of the loads so that the worst effect on the structure can be simulated.

The roof live loads are handled differently from other live loads. Section 4.9 of the ASCE 7-95 Design Code gives guidance on computing the minimum roof live loads. The live load L_r (in psf of *horizontal projection*) on ordinary flat, pitched, and curved roofs is defined as

$$L_r = 20R_1R_2 \tag{3.6.3.1}$$

where R_1 and R_2 are reduction factors. These factors are determined as

$$\begin{aligned} R_1 &= 1 \text{ for } A_t \le 200\,\text{sq ft} \\ R_1 &= 1.2 - 0.001A_t \text{ for } 200\,\text{sq ft} \le A_t \le 600\,\text{sq ft} \\ R_1 &= 0.6 \text{ for } A_t \ge 600\,\text{sq ft} \end{aligned} \tag{3.6.3.2}$$

where A_t is the tributary area in square foot supported by any structural member and

$$\begin{aligned} R_2 &= 1 \text{ for } F \le 4 \\ R_2 &= 1.2 - 0.05F \text{ for } 4 < F < 12 \\ R_2 &= 0.6 \text{ for } F \ge 12 \end{aligned} \tag{3.6.3.3}$$

where, for a pitched roof, F is number of inches of rise per foot. Note that the purpose of defining the reduction factors is to provide some relief, since it is unlikely that, the larger the roof, the entire roof will be subjected to the live load at the same time. The value of the live load, however, must lie in the range 12 psf $\le L_r \le$ 20 psf.

EXAMPLE 3.6.1 ***Computing Dead and Live Loads***

A typical residential roof truss is shown in Fig. E3.6.1(a). The trusses are spaced 2 ft on centers. The roof deck is made of 5/8-in wood sheathing (4 psf per inch) and covered with insulation and shingles that weigh 2 psf. The ceiling is 1/8-in gypsum board weighing 0.55 psf and is covered with 2-in loose insulation (0.5 psf per inch). Compute the dead and the live loads acting on the roof truss.

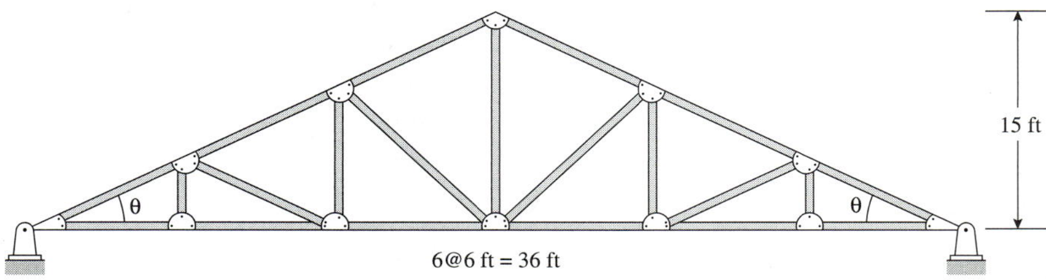

Fig. E3.6.1(a)

SOLUTION Since the truss spacing is 2 ft on centers, the computed distributed loads must finally be multiplied by 2 to obtain the actual loading on the truss. The pitch p of the roof is p = 15 ft/18 ft = 0.833, or F = 10 in/ft.

Dead Loads: We first compute the dead loads acting on the top chord.

Item	Unit load	Value	Distributed load (psf)
Deck	4 psf per inch	5/8"	2.5
Insulation and shingles	2 psf		2.0
	Total		4.5

The dead loads on the bottom chord are as follows.

Item	Unit load	Value	Distributed load (psf)
Gypsum board	0.55 psf		0.55
Insulation	0.5 psf per inch	2"	1.0
	Total		1.55

Live Loads: These loads, acting on the top chord, will be computed using the ASCE 7-95 Design Code. The tributary area for either the left or the right top chord is $A_t = \left(\sqrt{18^2 + 15^2}\right)(2) = 47\,\text{sqf}$. Hence, $R_1 = 1.0$. Since $F = 10$ in/ft, $R_2 = 1.2 - (0.05)(10) = 0.7$. Substituting in Eq. (3.6.3.1), we get

$$L_r = 20R_1R_2 = (20)(1.0)(0.7) = 14\,\text{psf} > 12\,\text{psf}$$

Now, computing the total load on a single truss, we have the following.

Item	Dead load (lb/ft)	Live load (lb/ft)
Top chord	$4.5 \times 2 = 9$	$14 \times 2 = 28$
Bottom chord	$1.55 \times 2 = 3.1$	

To represent the dead load on the top chord as load on the projected area, using Eq. (3.5.2b), we have $w_{DL} = (9)\sqrt{1 + 0.833^2} = 11.7\,\text{lb/ft}$. The computed dead and live loads are shown in Fig. E3.6.1(b).

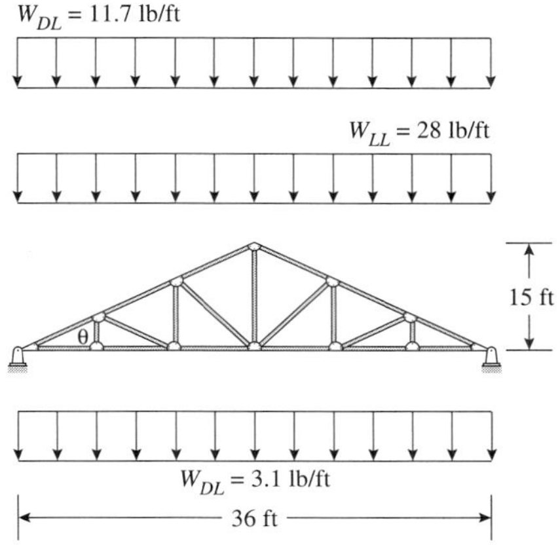

Fig. E3.6.1(b)

3.6.4 Wind Loads

The computation and consideration of wind loads for the design of structural systems has attracted renewed interest with the enormous damage caused by hurricanes, tornadoes, and other natural phenomena. Design codes set provisions to mitigate property losses and prevent loss of lives. With the advances in wind research, a more rational approach can now be used to determine the effects due to wind loads. We first discuss the background, then look at selected tables from the ASCE 7-95 Design Code, and finally discuss an abbreviated procedure to compute the wind design loads.

Background. A moving mass of air has kinetic energy. The amount of this energy is directly proportional to the square of the wind velocity:

$$V = \frac{1}{2}mv^2 \tag{3.6.4.1}$$

where V is the kinetic energy, m is the wind mass, and v its velocity. This kinetic energy translates into primarily strain energy when it encounters a stationary object, through deformations induced in that object. The dynamic nature of wind makes this interaction complex, so that simplified analyses are not adequate for all structural systems. For example, models of tall buildings or structures are subjected to wind loads in a wind tunnel in order to better understand the interactions. Nevertheless, enough knowledge has been gained over the years so that a rational procedure can now be used to find the load intensities on most common structural systems. The starting point in the computation of wind design loads is the wind velocity v, which is systematically converted to an equivalent static pressure q. In this section we see how this is achieved.

Use of ASCE 7-95 Design Code. The basic equation to determine the static pressure of wind loads is drawn from Eq. (3.6.4.1) as

$$q_z = 0.00256K_zK_{zt}v^2I \tag{3.6.4.2}$$

where

- q: effective velocity pressure (psf); q_z is based on K_z at any height z above ground; q_h is based on K_h at mean roof height h
- K_z: exposure velocity pressure coefficient, which reflects change in wind speed with height and terrain roughness (see ASCE Table 6.1)
- K_{zt}: topographic factor (ASCE Fig. 6.2)
- I: importance factor (see Table 3.6.4.1)
- v: basic wind speed in miles per hour

It should be noted that the effective velocity pressure is used to compute the design pressure using the equations of ASCE Table 6-1. The design pressure p is the load that then acts on various parts of the structure.

In order to understand the parameters in the different equations that appear in this section, we discuss some commonly used terms and definitions.

Positive pressure. This is the pressure exerted by the wind on a surface that is perpendicular to its direction, i.e., on the windward side, and that acts towards the surface. For quantities that affect the pressure, a positive sign (or value) indicates a positive pressure situation.

Negative pressure. The moving air causes a "suction effect" on the leeward side. The direction of this pressure is outwards, i.e., it acts away from the surface. For quantities that affect the pressure, a negative sign (or value) indicates a negative pressure situation.

Drag. Surfaces that are parallel to the wind direction can be subjected to positive and negative pressures and aerodynamic drag forces acting parallel to the wind direction.

While wind blows in different directions, for the purposes of design we assume that its direction is parallel to the ground, or horizontal. This movement of the air over structures, however, can cause pressures on the internal and external surfaces that are then assumed to act normal to the surface (see Fig. 3.6.4.1(a)).

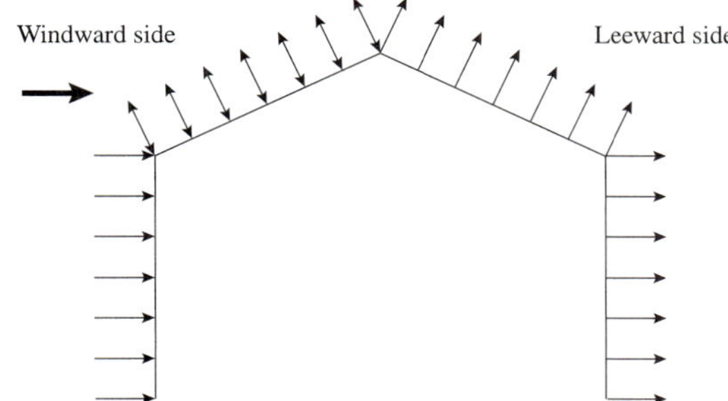

Fig. 3.6.4.1(a)
Wind pressure on external surfaces.

Fig. 3.6.4.1(a) shows the pressure on the external surfaces. The pressure on the sloping windward side can act towards as well as away from the external surface. On the leeward side there is a suction effect, which must also be considered with flat roofs. Fig. 3.6.4.1(b) shows one of the corresponding situations for the interior surfaces. There is pressure on the interior surfaces since most buildings allow for wind movements through them—they are partially open. Fig. 3.6.4.1(c) shows the other possibility. In Fig. 3.6.4.1(b) one can assume that the openings in the structure are primarily on the windward side, whereas in Fig. 3.6.4.1(c), they are assumed to be primarily on the leeward side. The critical design wind loads is a combination of all of these effects. An explanation of some of the terms in the ASCE standard follows.

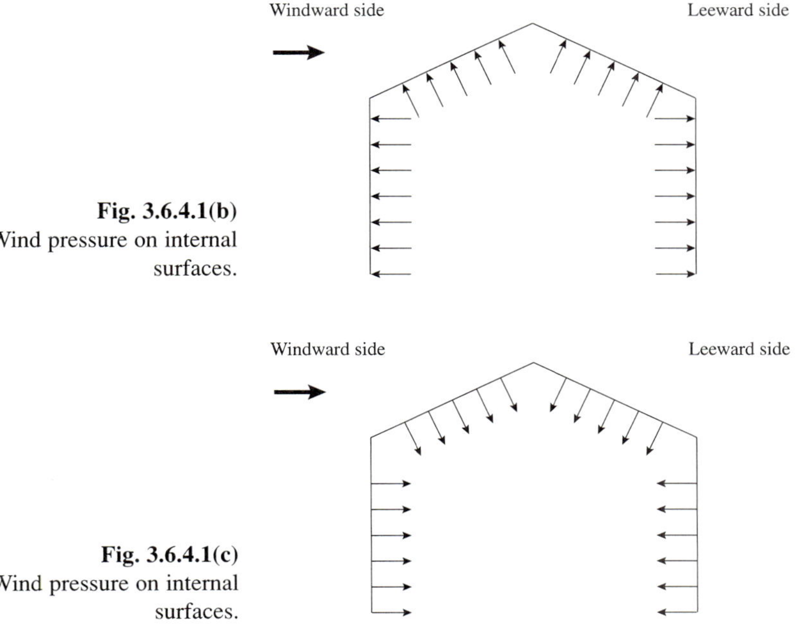

Fig. 3.6.4.1(b)
Wind pressure on internal surfaces.

Fig. 3.6.4.1(c)
Wind pressure on internal surfaces.

Basic wind speed v. A 3-second gust speed at 33 ft above the ground for a certain exposure and mean recurrence interval. A wind gust is a localized phenomenon within the general moving mass of air and is associated with a brief increase in wind velocity (hence pressure).

Enclosed building. A building that does not comply with the requirements for open or partially enclosed buildings.

Open building. A structure having all walls at least 80% open.

Partially enclosed building. A building that complies with both of the following conditions:

1. The total area of openings in a wall that receives positive external pressure exceeds the sum of the area of openings in the balance of the building envelopes (walls and roof) by more than 10%; and
2. The total area of openings in a wall that receives positive external pressure exceeds 4 sq ft or 1% of the area of that wall, whichever is smaller, and the percentage of openings in the balance of the building envelopes does not exceed 20%.

Low-rise building. Enclosed or partially enclosed buildings that comply with the following conditions:

1. Mean roof height h less than or equal to 60 ft;
2. Mean roof height h does not exceed least horizontal dimension.

Main wind-force resisting system (MWFRS). An assemblage of structural elements assigned to provide support and stability for the overall structure. The system generally receives wind loading from more than one surface.

Components and cladding (C&C). Elements that do not qualify as part of the main wind-force resisting system.

Flexible buildings and other structures. Slender buildings and other structures that have a fundamental natural frequency less than 1 Hz. Included are buildings and other structures that have a height h exceeding four times the least horizontal dimension.

Importance factor I. A factor that accounts for the degree of hazard to human life and damage to property.

Design force F. Equivalent static force to be used in the determination of wind loads for open buildings and other structures.

Design pressure p. Equivalent static pressure to be used in the determination of wind loads for buildings. The pressure is denoted as:

p_z = pressure that varies with height in accordance with the velocity pressure q_z evaluated at height z, or

p_h = pressure that is uniform with respect to height as determined by the velocity pressure q_h evaluated at mean roof height h.

Exposure categories. An exposure category that adequately reflects the characteristics of ground surface irregularities. The different categories are as follows.

1. *Exposure A*. Large city centers with at least 50% of the buildings having a height in excess of 70 ft (21.3 m). Use of this exposure category shall be limited to those areas for which terrain representation of Exposure A prevails in the upwind direction for a distance of at least 0.5 mi (0.8 km) or 10 times the height of the building or other structure, whichever is greater. Possible channeling effects or increased velocity pressures due to the building or structure being located in the wake of adjacent buildings shall be taken into account.
2. *Exposure B*. Urban and suburban areas, wooded areas, or other terrain with numerous closely spaced obstructions having the size of single-family dwellings or larger. Use of this exposure category shall be limited to those areas for which terrain representative of Exposure B prevails in the upwind direction for a distance of at least 1500 ft (460 m) or 10 times the height of the building or other structure, whichever is greater.
3. *Exposure C*. Open terrain with scattered obstructions having heights generally less than 30 ft (9.1 m). This category includes flat open country and grasslands.
4. *Exposure D*. Flat, unobstructed areas exposed to wind flowing over open water for a distance of at least 1 mi (1.61 km). This exposure shall apply only to those buildings and other structures exposed to the wind coming from over the water. Exposure D extends inland from the shoreline a distance of 1500 ft (460 m) or 10 times the height of the building or other structure, whichever is greater.

The parameters that appear in ASCE Table 6-1 have the following meanings.

A_f: area of open buildings and other structures either normal to the surface or projected on a plane normal to the wind direction, in square feet (square meters) except where C_f is given for surface area.

C_f: force coefficient (for other structures).

C_p: external pressure coefficient (for buildings).

G: gust effect factor. For MWFRS of buildings and other structures, and for components and cladding of open buildings and other structures, the value shall be 0.8 for exposure A and B, and 0.85 for exposure C and D.

G_f: gust effect factor for MWFRS of flexible buildings and structures.

GC_p: product of external pressure coefficient and gust effect factor (for buildings).

GC_{pf}: product of equivalent external pressure coefficient and gust effect factor (for MWFRS of low-rise buildings).

GC_{pi}: product of equivalent internal pressure coefficient and gust effect factor for buildings.

Table 3.6.4.1 Classification of Buildings and Other Structures for Wind, Snow, and Earthquake Loads (courtesy ASCE).

Nature of occupancy	Category
Buildings and other structures that represent a low hazard to human life in the event of failure including, but not limited to: • Agricultural facilities • Certain temporary facilities • Minor storage facilities	I
All buildings and other structures except those listed in Categories I, III, and IV.	II

continued

Table 3.6.4.1 Continued

Nature of occupancy	Category
Buildings and other structures that represent a substantial hazard to human life in the event of failure including, but not limited to: • Buildings and other structures where more than 300 people congregate in one area • Buildings and other structures with elementary school, secondary school, or day-care facilities with capacity greater than 250 • Buildings and other structures with a capacity greater than 500 for colleges or adult education facilities • Health-care facilities with a capacity of 50 or more resident patients but not having surgery or emergency treatment facilities • Jails and detention facilities • Power generation stations and other public utility facilities not included in category IV • Buildings and other structures containing sufficient quantities of toxic or explosive substances to be dangerous to the public if released	III
Buildings and other structures designated as essential facilities including, but not limited to: • Hospitals and other health-care facilities having surgery or emergency treatment facilities • Fire, rescue, and police stations and emergency vehicle garages • Designated earthquake, hurricane, or other emergency shelters • Communications centers and other facilities required for emergency response • Power generating stations and other public utility facilities required in an emergency • Buildings and other structures having critical national defense functions	IV

The importance factor values are 0.87, 1.00, 1.15, and 1.15 for structures that belong to Categories I, II, III, and IV respectively.

Wind analysis procedure. Now we examine in some detail the analytical process (versus wind tunnel tests) of computing the wind loads using the ASCE 7-95 Design Code. The following steps are applicable (with certain exceptions) to the main load-bearing frames.

1. Use Fig. 3.6.4.2 to find the applicable basic wind speed.
2. Topographic factor K_{zt} has a minimum value of 1.0. However, ASCE Fig. 6.2 should be used if a structure is on isolated hills and escarpments located in Exposure B, C, or D with

$$K_{zt} = (1 + K_1K_2K_3)^2$$

3. Use Table 3.6.4.1 to determine the importance coefficient I.
4. Now compute the effective velocity pressure as $p_z = 0.00256K_z\mathrm{K}_{zt}\mathrm{v}2I$ (see Eq. (3.6.4.2)) in terms of K_z.
5. For different heights z (based on the structural characteristics), determine K_z from ASCE Table 6.3 . Compute the mean roof height h. For this value, determine K_h.

Now use ASCE Table 6-1 to compute the design wind pressure. Once the design pressure has been computed, the structure can be modeled and analyzed. The following example illustrates this general procedure.

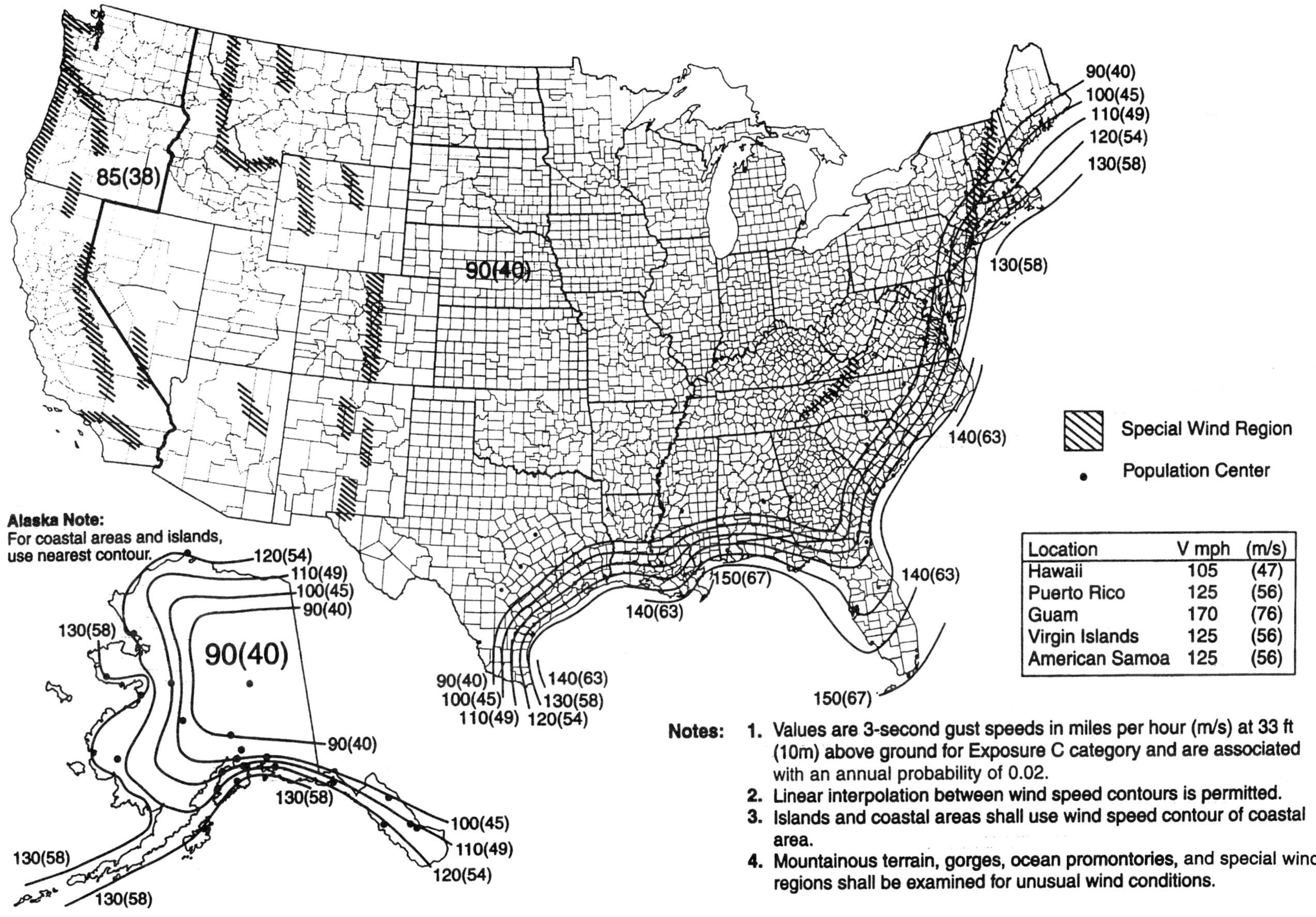

Location	V mph	(m/s)
Hawaii	105	(47)
Puerto Rico	125	(56)
Guam	170	(76)
Virgin Islands	125	(56)
American Samoa	125	(56)

Notes:

1. Values are 3-second gust speeds in miles per hour (m/s) at 33 ft (10m) above ground for Exposure C category and are associated with an annual probability of 0.02.
2. Linear interpolation between wind speed contours is permitted.
3. Islands and coastal areas shall use wind speed contour of coastal area.
4. Mountainous terrain, gorges, ocean promontories, and special wind regions shall be examined for unusual wind conditions.

Fig. 3.6.4.2
Basic wind speeds (courtesy ASCE).

EXAMPLE 3.6.2 ***Computing Wind Loads***

Figure E3.6.2(a) shows a commercial warehouse building located in a suburb of Phoenix. The building is 150 ft × 300 ft with the eaves located at a height of 15 ft. The frames are located 20 ft apart. Compute the wind forces acting on a single frame.

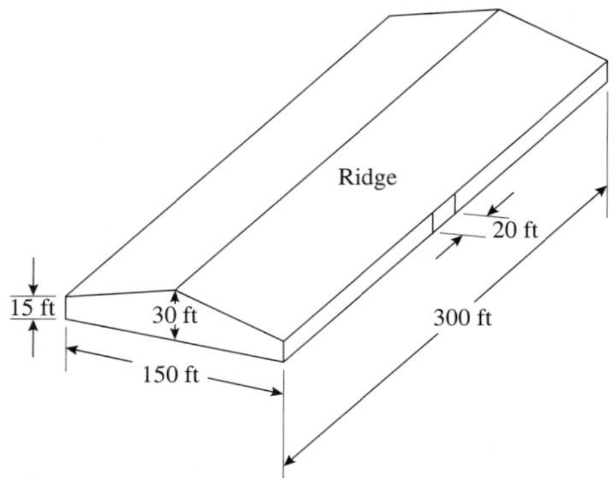

Fig. E3.6.2(a)

SOLUTION

Step 1: Phoenix is not located in a special wind region. Hence, the basic wind speed is 90 mph.

Step 2: The building is located on a flat terrain, with exposure category C being applicable. Hence, the topographic factor K_{zt} has the minimum value of 1.0.

Step 3: The building is not considered to be an essential facility or likely to be occupied by 300 people at one time. Hence it belongs to Category II with importance coefficient $I = 1.0$.

Step 4: Substituting in Eq. (3.6.4.2), we have

$$q_z = 0.00256K_z(1.0)(90)^2(1.0) = 20.7K_z \text{ psf}$$

Step 5: The mean roof height $h = 0.5(15 + 30) = 22.5$ ft

Computation of Velocity Pressures q_z, psf

Height, ft	K_z	q_z
0–15	0.85	17.6
20	0.90	18.7
22.5	0.92	$q_h = 19.0$
25	0.94	19.5
30	0.98	20.3

Step 6: Design wind pressure is computed per ASCE Table 6-1. The basic formula is

$$p = qGC_p - q_h(GC_{pi})$$

where

$q = q_z$ for the windward wall at height z, $q = q_h$ for the leeward wall, side walls, and roof

$G = 0.85$ for exposure C (see definition of G, gust effect factor)

C_p is obtained from ASCE Fig. 6-3

(GC_{pi}) is obtained from ASCE Table 6-4

The two cases of wind direction to be considered for this example are when (a) the wind is normal to the ridge, and (b) the wind is parallel to the ridge.

Wall Pressure Coefficients, C_p (Using ASCE Fig. 6-3)			
Surface	Wind direction	L/B	C_p
Windward wall	All	All	0.80
Leeward wall	Perpendicular to ridge	0.5	–0.5
	Parallel to ridge	2.0	–0.3
Side wall	All	All	–0.70

Wind Normal to the Ridge

Roof C_p. The roof angle is $\tan^{-1}(15/75) = 11.3°$ and $h/L = 22.5/150 = 0.15$. Using ASCE Fig. 6-3 and interpolating, we have the following. Note that since the roof angle is less than 15°, the windward sloping roof is subjected only to suction forces.

Surface	10°	11.3°	15°
Windward	–0.7	–0.65	–0.5
Leeward	–0.3	–0.35	–0.5

Internal GC_{pi}. Assuming that the openings are evenly distributed in the walls, $GC_{pi} = \pm 0.18$ using ASCE Table 6-4.

Now we can compute the design wind pressure for the frames by substituting in $p = qGC_p - q_h\,(GC_{pi})$.

Design Pressures					
				Net pressure, psf, with $p = qGC_p - qh(GC_{pi})$	
Surface	z (ft)	q (psf)	C_p	$+(GC)_{pi}$	$-(GC)_{pi}$
Windward wall	0–15	17.6	0.8	[17.6(0.85)(0.8) – 19.0(0.18)] 8.6	[17.6(0.85)(0.8) – 19.0(0.18)] 15.4
Leeward wall	All	19.0	–0.5	[19.0(0.85)(–0.5) – 19.0(0.18)] –11.5	[19.0(0.85)(–0.5) – 19.0(0.18)] –4.7
Side walls	All	19.0	–0.7	[19.0(0.85)(–0.7) – 19.0(0.18)] –14.7	[19.0(0.85)(–0.7) – 19.0(0.18)] –7.9
Windward roof	–	19.0	–0.65	[19.0(0.85)(–0.65) – 19.0(0.18)] –13.9	[19.0(0.85)(–0.65) – 19.0(0.18)] –7.1
Leeward Roof	–	19.0	–0.35	[19.0(0.85)(–0.35) – 19.0(0.18)] –9.1	[19.0(0.85)(–0.35) – 19.0(0.18)] –2.2

Two load cases are generated by the two internal pressures, as shown in Figs. E3.6.2(b), (c).

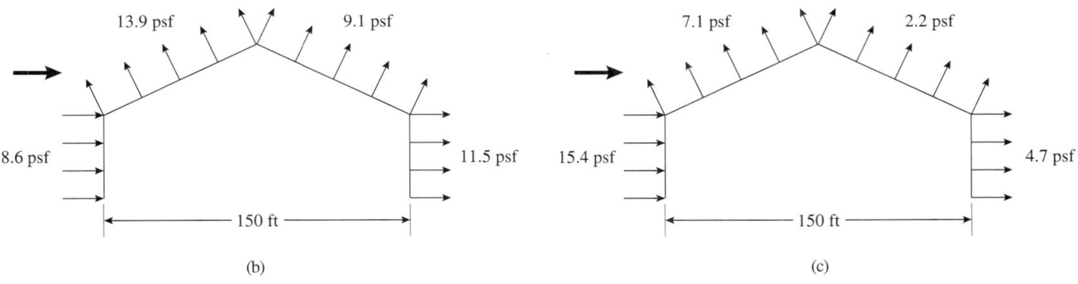

Fig. E3.6.2
(b) Wind loads with positive internal pressure coefficient. (c) Wind loads with negative internal pressure coefficient.

Wind Parallel to the Ridge

Roof C_p. The roof angle is 11.3° and $h/L = 22.5/300 = 0.075$. Using ASCE Fig. 6-3 we have the following.

Surface	h/L	Distance from windward edge	C_p
Roof	≤ 0.5	0 to h	–0.9
		h to $2h$	–0.5
		> $2h$	–0.3

Internal GC_{pi}. Assuming that the openings are evenly distributed in the walls, $GC_{pi} = \pm 0.18$ using ASCE Fig. 6-3.

Now we can compute the design wind pressure for the frames by substituting in $p = qGC_p - q_h(GC_{pi})$.

Design Pressures					
				Net pressure, psf, with $p = qGC_p - q_h(GC_{pi})$	
Surface	z (ft)	q (psf)	C_p	$+(GC)_{pi}$	$-(GC)_{pi}$
Windward wall	0–15	17.6	0.8	8.6	15.4
	20	18.7	0.8	9.3	16.1
	25	19.5	0.8	9.8	16.7
	30	20.3	0.8	10.4	17.2
Leeward wall	All	19.0	–0.3	–8.3	–1.4
Side walls	All	19.0	–0.7	–14.7	–7.9
Roof	0-h	19.0	–0.9	–18.0	–11.1
	h–$2h$	19.0	–0.5	–11.5	–4.7
	>$2h$	19.0	–0.3	–8.3	–1.4

Two load cases are generated by the two internal pressures, as shown in Figs. E3.6.2(d), (e).

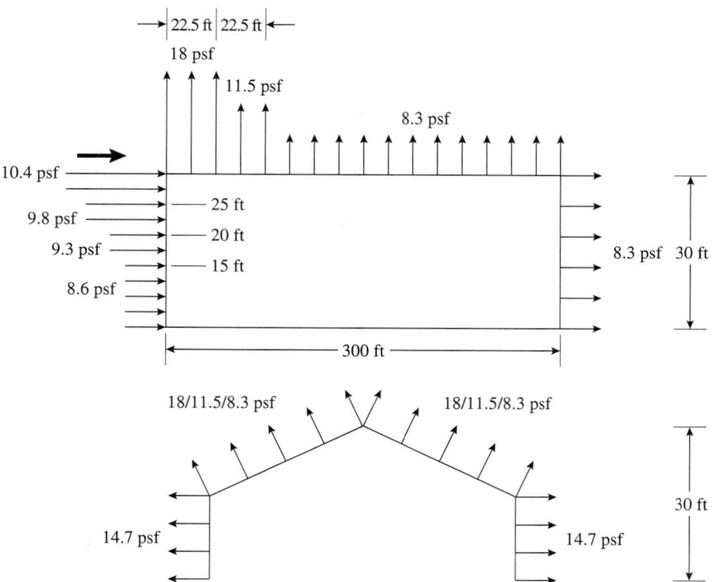

Fig. E3.6.2(d) Wind loads with positive internal pressure coefficient.

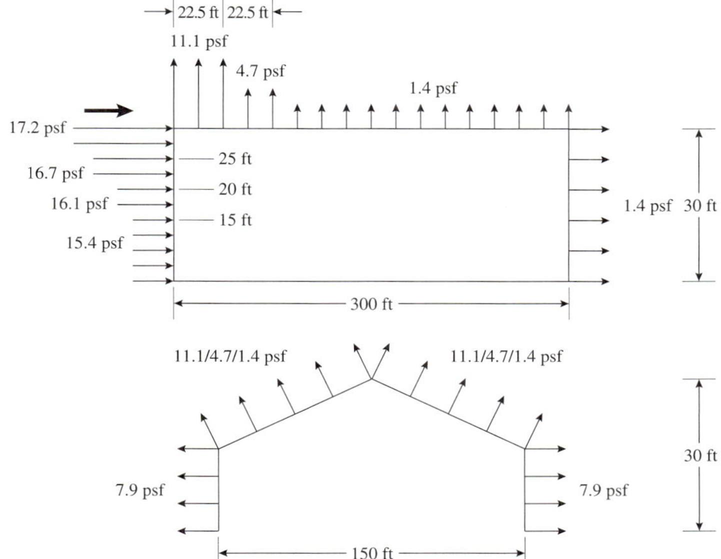

Fig. E3.6.2(e) Wind loads with negative internal pressure coefficient.

To compute the wind forces acting on single frame, the wind pressures must be multiplied by the appropriate tributary area.

3.6.5 Snow and Rain Loads

Snow Loads

Snow loads acting on roofs depend on a variety of factors. Apart from the obvious, the location of the structure, snow loads are a function of several other factors such as the pitch and exposure of the roof. There is less accumulation of snow on steep and smooth roofs. Wind also plays a role in moving snow off the roof, or building greater depths of snow due to drifting. If the structure is not heated there is greater likelihood of snow accumulating on the roof. Larger loads can act on a roof due to the nature of the snow and the deflection of the roof due to the accumulated snow. For example, fresh snow is much lighter than wet snow.

Snow loads acting on roofs can be estimated using the ASCE 7-95 Design Code. The first step is to find the ground snow load for the site (see Fig. 3.6.5.1). The next step is to classify whether the roof is flat or sloped. For flat roofs where the slope is equal to or less than 5°, the snow load p_f (in psf) is given by

$$p_f = 0.7C_e C_t I p_g \qquad (3.6.5.1)$$

where

p_g = ground snow load (see Fig. 3.6.5.1)
C_e = exposure factor (see ASCE Table 7-2)
C_t = thermal factor (see ASCE Table 7-3)
I = importance factor

The 0.7 factor is the basic exposure factor that accounts for wind blowing some of the snow off the roof. The exposure factor C_e accounts for local wind effects that are a function of the immediate terrain. The thermal factor C_t incorporates the effect of heating within the structure. As mentioned earlier, more snow tends to accumulate on unheated buildings. The importance factor was discussed in the preceding section with respect to wind loads. The basic idea of the importance factor is the same when considered with snow loads. The importance factor is 0.8, 1.0, 1.1, and 1.2 for structures that belong to categories I, II, III, and IV respectively.

Structures have different exposure classifications as follows:

Partially exposed. All roofs except as indicated below.

Fully exposed. Roofs exposed on all sides with no shelter afforded by terrain, higher structures or trees. Roofs that contain several large pieces of mechanical equipment or other obstructions are not in this category.

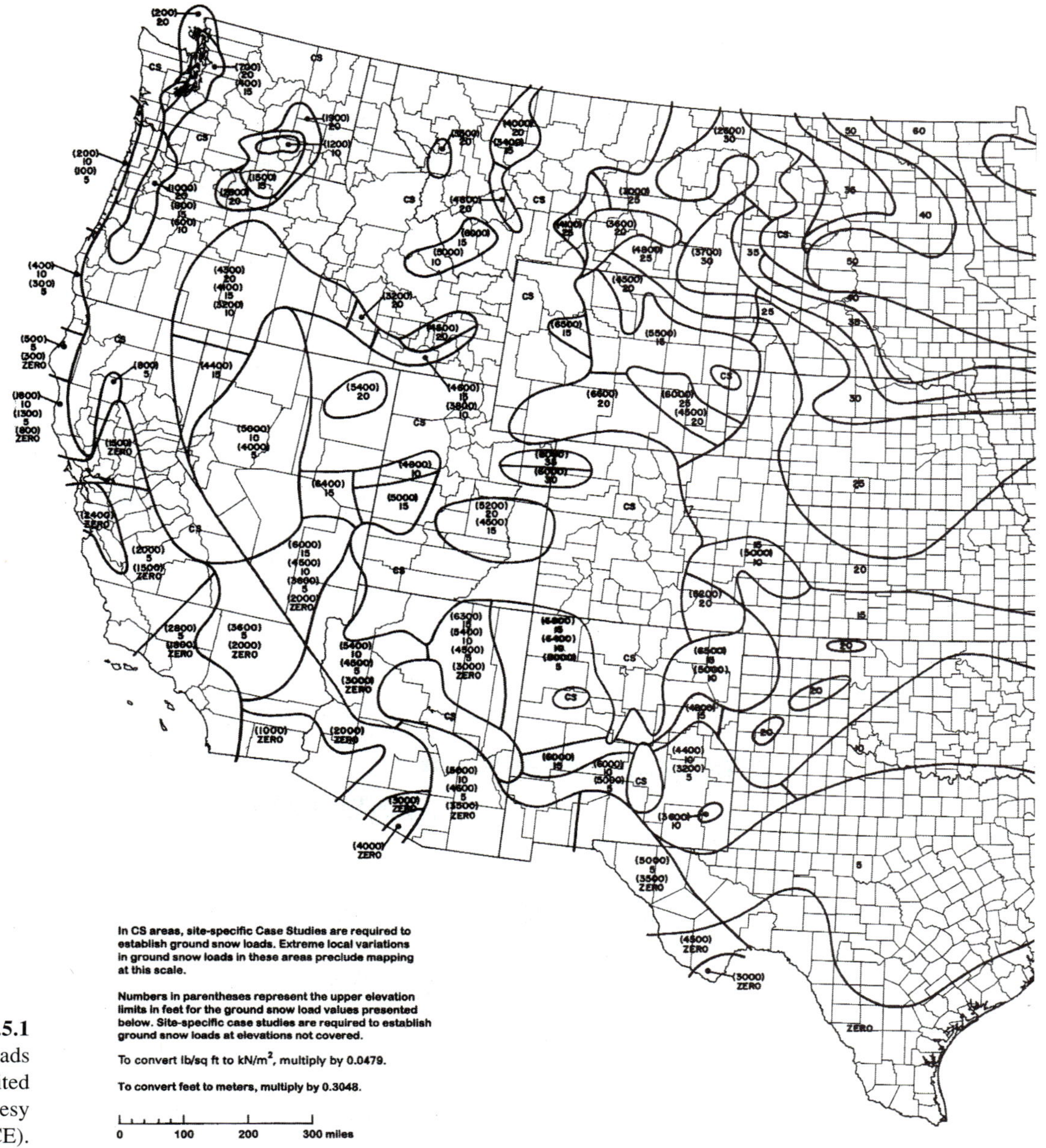

Fig. 3.6.5.1 Ground snow loads p_g for the United States (psf) (courtesy ASCE).

Sheltered. Roofs located tight in among conifers that qualify as obstructions.

For sloped roofs where the slope is greater than 5°, the snow load p_s (in psf) is given by

$$p_s = C_s p_f \tag{3.6.5.2}$$

where C_s = slope factor (see ASCE Fig. 7-2). Snow loads acting on sloped surfaces are assumed to act on the horizontal projection of that surface.

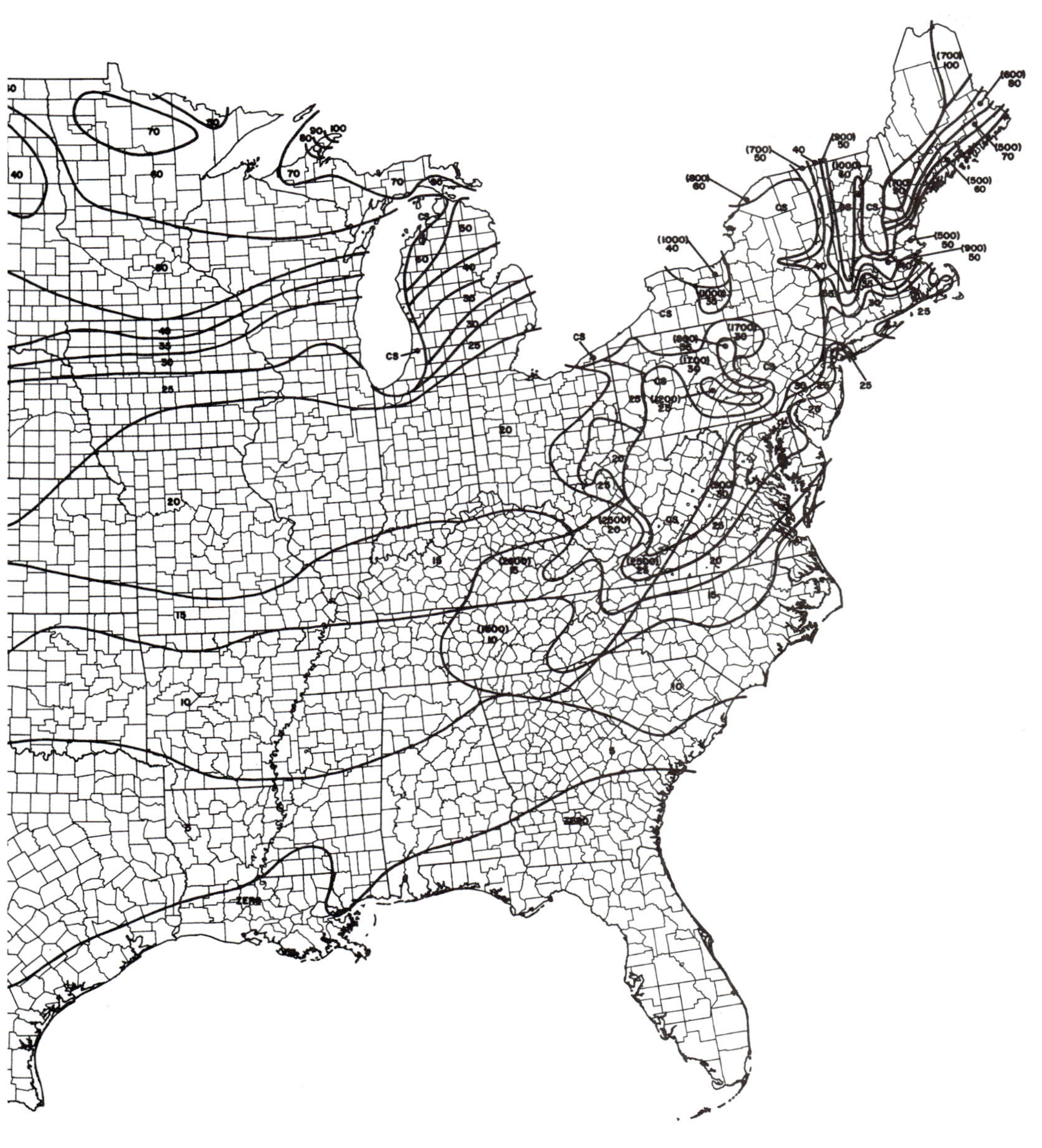

While applications of Eqs. (3.6.5.1) and (3.6.5.2) are quite straightforward, additional considerations are important. For example, for flat roofs, there are minimum allowable values of p_f, live load reductions do not apply, the shape of the roof determines the value for C_s, the roof can be subjected to balanced and unbalanced loads, and roofs need to be designed to sustain localized loads from snow drifts, rain-on-snow surcharge loads, sliding snow loads, etc. A detailed procedure is shown in the ASCE 7-95 standard.

Step 1: Find the ground snow load for the site using Fig. 3.6.5.1.

Step 2: Classify whether the roof is flat or sloped and compute the snow load using Eq. (3.6.5.1) or (3.6.5.2).

Step 3: Determine unbalanced loads for hip and gable roofs. The unbalanced snow loads is considered for roofs with $15^\circ \leq \theta \leq 70^\circ$ with a uniform load on the leeward side equal to $1.3p_s/C_e$ and the windward side free of snow.

Step 4: Consider additional effects whenever applicable, such as snow drifts, wet snow, and ponding.

Rain Loads

Roof drainage systems are designed to handle all the flow associated with intense, short-duration rainfall events (ASCE 7-95). A typical design includes a primary drainage system dimensioned to handle a specified rainfall rate. The primary system collects the rainwater and directs it to storm drains. However, to take into account those situations that can temporarily overload the primary system, secondary drainage systems are also required. Secondary systems are also recommended in roof areas where features such as expansion joints, and parapet walls can create the likelihood of additional water accumulation. To alleviate this situation, large holes or tubes in walls (called scuppers) are sometimes used to drain this excess water.

The flow rate through a single drainage system is given as

$$Q = 0.0104Ai \tag{3.6.5.3}$$

where

Q = flow rate in gallons per minute
A = roof area serviced in square feet
i = design rainfall intensity in inches per hour

The flow rate can be related to the hydraulic head d_h as shown in ASCE Table C8-1. With the data provided in that table, the rain loads R (in psf) on undeflected roofs are computed as

$$R = 5.2(d_s + d_h) \tag{3.6.5.4}$$

where

d_s = depth of water (in inches) on the undeflected roof up to the inlet of the secondary drainage system when the primary drainage system is blocked,
d_h = additional depth of water (in inches) on the undeflected roof above the inlet of the secondary drainage system at its design flow (zero if water overflows over the roof edge).

The undeflected roof terminology is used to separate the effect due to *ponding* from other factors that influence the depth of water. With flat roofs, if the water is not drained rapidly enough, the accumulated water can cause the roof to deflect (this is called ponding). The deflection causes additional water to accumulate on the roof and the cumulative effects can cause failure of the roof.

EXAMPLE 3.6.3

Computing Snow Loads

The residential roof truss in Fig. E3.6.3(a) is used in a residential building in Iowa City, Iowa. The truss eaves are located well above the ground. The building is exposed and is heated from the inside. Compute the snow loads on the roof.

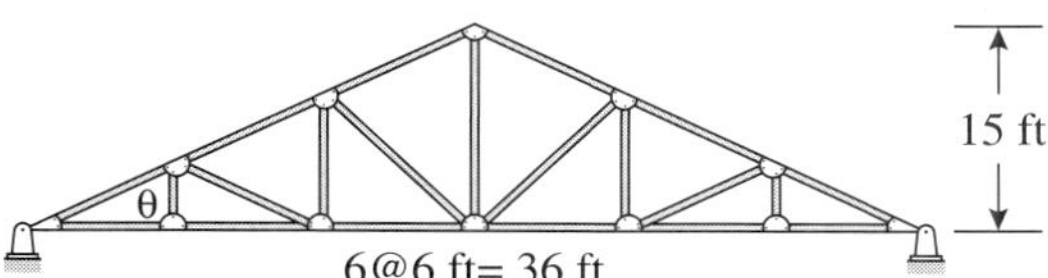

Fig. E3.6.3(a)

SOLUTION

Step 1: From Fig. 3.6.5.1, the ground snow load for Iowa City is 30 psf.

Step 2: The roof pitch is $\tan^{-1}(15/18) = 40^{\circ}$. Hence it is a sloped roof. Exposure C applies. The building is fully exposed. Using the flat roof formula $p_f = 0.7C_eC_tIp_g$ with $C_e = 0.9$, $C_t = 1.0$, $I = 1.0$ (assuming Category II structure), we have $p_f = 18.9$ psf. Now for sloped roofs, we can compute the slope factor C_s as (instead of using ASCE Fig. 7-2)

$$\text{For warm roofs}\begin{cases} C_s = 1 & \text{for } 0 \le \theta \le 30^{\circ} \\ C_s = 1 - \dfrac{\theta - 30^{\circ}}{40^{\circ}} & \text{for } 30^{\circ} \le \theta \le 70^{\circ} \\ C_s = 0 & \text{for } 0 \ge 70^{\circ} \end{cases}$$

$$\text{For cold roofs}\begin{cases} C_s = 1 & \text{for } 0 \le \theta \le 45^{\circ} \\ C_s = 1 - \dfrac{\theta - 45^{\circ}}{25^{\circ}} & \text{for } 45^{\circ} \le \theta \le 70^{\circ} \\ C_s = 0 & \text{for } 0 \ge 70^{\circ} \end{cases}$$

Hence, $C_s = 0.75$ and $p_s = 0.75(18.9) = 14.2$ psf.

Step 3: The unbalanced snow load p_u is given as

$$p_u = 1.3\frac{p_s}{C_e} = 1.3\left(\frac{14.2}{0.9}\right) = 20.5\,\text{psf}$$

The two load conditions, balanced and unbalanced, are shown in Fig. E3.6.3(b).

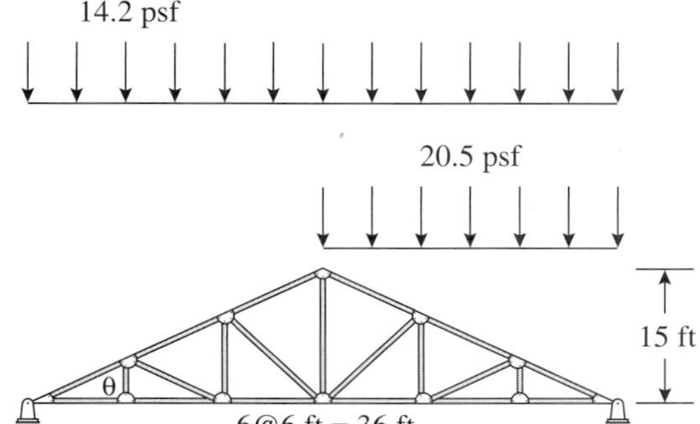

Fig. E3.6.3(b) Balanced and unbalanced snow loads.

As mentioned in earlier examples, the load must be converted using the tributary area into line loads acting on a single truss.

3.6.6 Earthquake Loads

Most codes give the designer two options when designing for earthquake-induced loads: (a) carry out a dynamic structural analysis, or (b) for certain classes of structural systems, estimate an equivalent static load. The treatment of dynamic analysis is outside the scope of this text.

The ground motion during an earthquake can cause significant forces to develop in a structure. The inertial effects are due to the mass of the structure and the induced acceleration. Typically the deformations in the structure cause shear forces to develop. A simplified model for analysis can be generated using the UBC code. UBC permits the use of the static method under clearly specified conditions. When those provisions are not satisfied, a dynamic method of analysis is required. The base shear or the total lateral inertial force V on the structure is given by

$$V = \frac{C_v I}{RT} W \tag{3.6.6.1}$$

C_v is the *seismic coefficient* that is a function of the *seismic zone factor Z* and the *soil profile type S;* i.e., it accounts for the amount of seismic risk present in the site where the structure is located. There are six zone coefficients—0, 1, 2A, 2B, 3, and 4. For example, most of California is in Zone 4 (highest risk), while most of Texas is in Zone 0 (lowest risk). The zone factors represent the effective peak ground acceleration that have only a 10% chance of being exceeded in 50 years. There are six soil profile types: S_A, S_B, S_C, S_D, S_E, and S_F. For example, hard rock is categorized as S_A whereas S_E is soft soil. I is the *importance factor* (either 1.0 or 1.25), with an intent similar to the one we saw before with wind loads. T is the fundamental period of the structure. In the absence of a dynamic analysis, the fundamental period can be determined from the equation

$$T = C_t (h_n)^{3/4} \tag{3.6.6.2}$$

where h_n is the height of the building from the base to level n and C_t is 0.035 for steel moment-resisting frames, 0.030 for reinforced concrete moment-resisting frames and eccentrically braced frames and 0.020 for all other buildings. The weight W of the structure is the total dead load, including the weight of pipes, ducts, equipment, etc. R is a "numerical coefficient representative of the inherent overstrength and global ductility capacity of lateral force-resisting systems." This value varies between 2.2 and 8.5, and is a partly empirical/judgmental factor that is used to adjust the base shear value.

The topic of computing earthquake loads and carrying out seismic-resistant design is a continuously evolving area that will improve as the understanding of the earthquake phenomenon improves.

3.6.7 Other Design Issues

There are basically two design philosophies used in the design with most materials—the *allowable stress* (or *working stress* or *service load* method) design (ASD) and the *limit states design* (LSD) (referred to as *strength design* in concrete and *load and resistance factor design* (LRFD) in the steel industry). We examine these design approaches in Chapter 9. Structural systems are designed to have adequate strength (do not collapse), stiffness (do not unduly deflect), and durability (last the design service life of the system). Irrespective of the design principles, assumptions are made about all the parameters and factors that constitute the design model. For example, assumptions are made about the loads acting on the structure.

Should the structure fail if it is overloaded? The overloading can arise through underestimating the actual loads, or if the use of the structural system changes over the life of the structure, or if the level of sophistication in the structural analysis is inadequate to predict the structural response. There may be other uncertainties with respect to the members used (do the actual dimensions correspond to the design data?), the material properties (have the members been constructed or fabricated to yield the assumed material behavior?), and the construction procedures. Both the ASD and the LSD approaches guard against overload and understrength but attempt to achieve them in quite different ways.

Load Combinations. To take into account for the worst effect on the structure under the action of the different types of loads, *factored* loads are used in the LSD procedure. The ASCE 7-95 standard prescribes the following.

> Structures, components, and foundations shall be designed so that their design strength equals or exceeds the effects of the factored loads in the following combinations:
>
> 1. $1.4D$
> 2. $1.2(D + F + T) + 1.6(L + H) + 0.5(L_r \text{ or } S \text{ or } R)$
> 3. $1.2D + 1.6(L_r \text{ or S or } R) + (0.5L \text{ or } 0.8W)$ (3.6.7.1)
> 4. $1.2D + 1.3W + 0.5L + 0.5(L_r \text{ or } S \text{ or } R)$
> 5. $1.2D \pm 1.0E + 0.5L + 0.2S$
> 6. $0.9D \pm (1.3W \text{ or } 1.0E)$
>
> where
>
> D = dead load
> E = earthquake load
> F = load due to fluids with well-defined pressures and maximum heights
> F_a = flood load
> H = load due to the weight and lateral pressure of soil and water in soil
> L = live load
> L_r = roof live load
> R = rain load
> S = snow load
> T = self-straining force
> W = wind load

Exception: The load factor L in the combinations 3–5 shall equal 1.0 for garages, areas occupied as public assembly, and all areas where the live load is greater than 100 psf (4.79 kN/m^2).

For the ASD procedure, the ASCE 7-95 standard prescribes the following.

> Loads listed herein shall be considered to act in the following combinations, whichever produces the most unfavorable effect in the building, foundation, or structural member being considered. Effects of one or more loads not listed shall be investigated.
>
> 1. D
> 2. $D + L + F + H + T + (L_r \text{ or } S \text{ or } R)$
> 3. $D + (W \text{ or } E)$ (3.6.7.2)
> 4. $D + L + (L_r \text{ or } S \text{ or } R) + (W \text{ or } E)$

The most unfavorable effects from both wind and earthquake loads shall be investigated, where appropriate, but they need not be assumed to act simultaneously.

Safety Factor. The factor of safety or safety factor is defined as the ratio of the load that would cause failure to the load that the structure or member is designed for (service load):

$$SF = \frac{F_f}{F_s} \tag{3.6.7.3}$$

Most design codes tie the concept of safety factor to allowable values. For example, if the allowable stress σ_a of a material is given as $0.6\sigma_y$, then the implied safety factor is $SF = \sigma_y/0.6\sigma_y = 1.67$.

Serviceability Considerations. The last major design issue deals with serviceability—issues related to the occupancy of structures, e.g., cracking, deflections, vibrations, etc. We will learn more about computing deflections in Chapter 4. The other considerations are covered in Chapter 9.

EXAMPLE 3.6.4 ***Computing Factored Loads***

Using the truss shown in Examples 3.6.1 and 3.6.3, compute the following factored loads:

(a) $1.2D + 1.6L + 0.5S$
(b) $1.2D + 1.6S + 0.5L$

SOLUTION

Step 1: For the roof truss we take the live loads L to be the same as the roof live load L_r. Summarizing the previously derived results and noting that the trusses are spaced 2 ft apart, we have for the projected area

Dead load $D = 11.7$ lb/ft
Live load $L = 28$ lb/ft
Balanced snow load $S = (14.2)(2) = 28.4$ lb/ft
Unbalanced snow load $S = (20.5)(2) = 41$ lb/ft

Step 2: Now we compute the factored loads.

(a) For the balanced snow loads,

$$1.2D + 1.6L + 0.5S = 1.2(11.7) + 1.6(28) + 0.5(28.4) = 73 \text{ lb/ft}$$

For the unbalanced snow loads, on the windward side

$$1.2D + 1.6L = 1.2(11.7) + 1.6(28) = 58.8 \text{ lb/ft}$$

and on the leeward side

$$1.2D + 1.6L + 0.5S = 1.2(11.7) + 1.6(28) + 0.5(41) = 79.3 \text{ lb/ft}$$

(b) For the balanced snow loads,

$$1.2D + 1.6S + 0.5L = 1.2(11.7) + 1.6(28.4) + 0.5(28) = 73.5 \text{ lb/ft}$$

For the unbalanced snow loads, on the windward side

$$1.2D + 0.5L = 1.2(11.7) + 0.5(28) = 28 \text{ lb/ft}$$

and on the leeward side

$$1.2D + 1.6S + 0.5L = 1.2(11.7) + 1.6(41) + 0.5(28) = 93.6 \text{ lb/ft}$$

EXERCISES

Appetizers

3.6.1. A cross-section of a flooring system is shown in Fig. P3.6.1. Compute the dead load as lb/ft^2. Take the spacing of the floor beams as 6 ft o.c. (on-center) and the girders as 24 ft o.c. The suspended ceiling system can be assumed to be 1 psf.

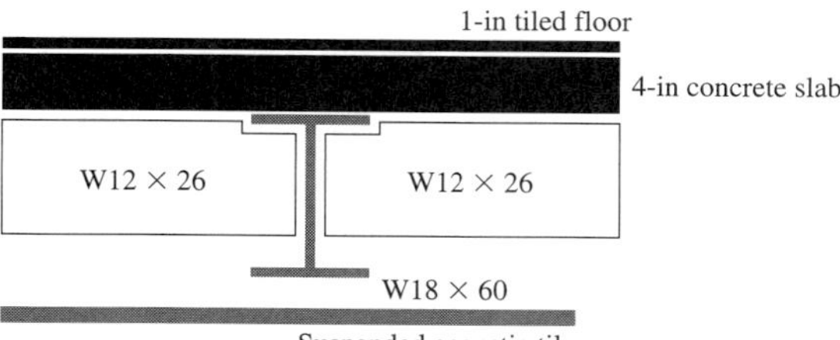

Fig. P3.6.1

3.6.2. Fig. P3.6.2 shows the plan view of a typical floor in a building frame.

Assuming that the load on the floor is 2 kN/m^2, compute

(a) the loading on the beam A-B,

(b) the loading on beam D-C, and

(c) the loading on beam E-F.

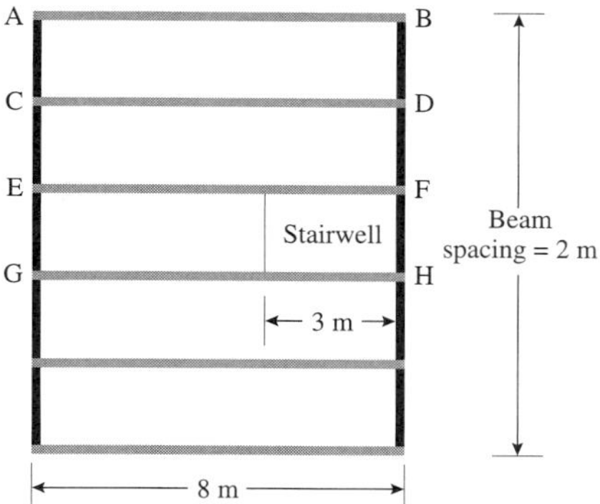

Fig. P3.6.2

3.6.3. Figure P3.6.3 shows the plan for an interior bay of an office building. The floor is 8-in reinforced concrete. The beams can be assumed to be 30 lb/ft and the girders are 60 lb/ft. Assume that the connections between the beams, girders, and columns can be approximated as internal hinges. Compute the dead loads acting on the beams, girders, and columns.

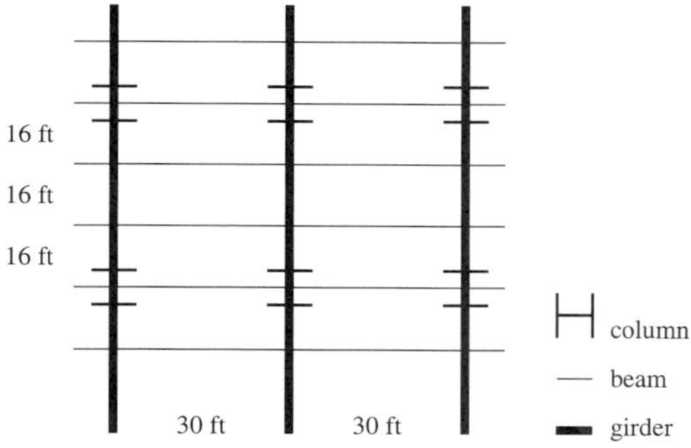

Fig. P3.6.3

3.6.4. For the office building shown in Fig. P3.6.3,

(a) Compute the live loads acting on the beams, girders, and columns.

(b) Compute the factored load $1.2D + 1.6L$.

(c) Assume that all the floors have the same floor plan, the building is 5 stories tall, each floor is 15 ft and a typical column weighs 100 lb/ft. Compute the dead load acting on the interior columns in the first floor. Ignore the loads contributed by the roof of the building.

Main Course

3.6.5. For the truss shown in Fig. P3.6.5, compute the wind loading. Assume that the house is in the city where your school is located. The eaves are located 15 ft off the ground and the plan dimensions of the house are 80 ft in length by 48 ft in width. The trusses are spaced 2 ft o.c.

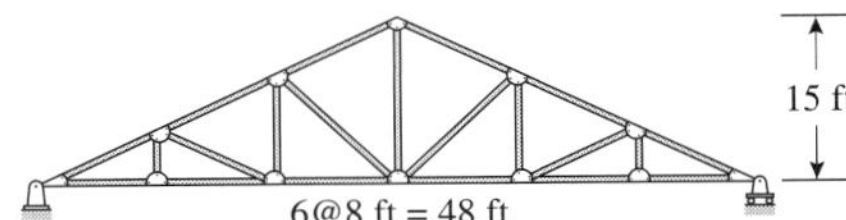

Fig. P3.6.5

3.6.6. Assume that the house in Problem 3.6.5 is located in suburban St. Louis, Missouri. Compute the snow loads.

3.6.7. For the house shown in Problem 3.6.5, compute the reactions for a typical interior truss due to

(a) Wind load

(b) Wind load plus snow load combination

Structural Concepts

3.6.8. Figure P3.6.8 shows the model of a building beam containing three spans. The beam is made of reinforced concrete. The cross-section is rectangular with a depth of 12 in and a width of 8 in. The live load is computed to be 1000 lb/ft. Place the live load (covering the entire span) so as to generate the critical loads for the dead-load-plus-live-load combination.

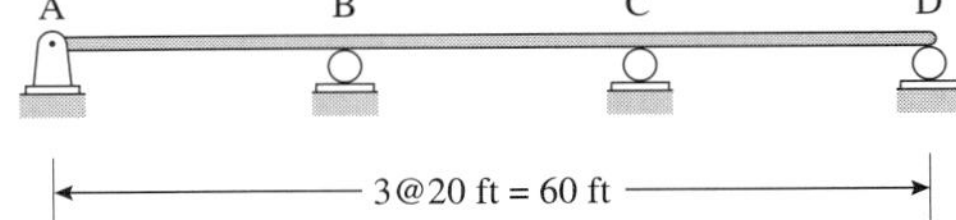

Fig. P3.6.8

3.6.9. Derive the value 0.00256 in Eq. (3.6.4.2).

3.7 SIMPLE DESIGN EXAMPLES

In this section, we look at several different aspects of design using several examples. Recall that at the beginning of this chapter, it was stated that design is an iterative process in which progressively designs are modified to meet the design objectives.

3.7.1 Mathematical Background

Interpretation of Derivatives (Design Sensitivity Analysis). Consider a smooth, differentiable function $f(x)$ of one variable. From the first-order approximation

$$f(x_{i+1}) \cong f(x_i) + f'(x_i)(x_{i+1} - x_i) \tag{3.7.1.1}$$

we can draw the following conclusion: if $x_{i+1} > x_i$, then $f(x_{i+1}) > f(x_i)$ only if $f'(x_i) > 0$. The implication is that if we are searching for the *largest* value of $f(x)$, then we should look at the derivative (or, gradient) of the function at the current point, and move in the direction of *pos-*

itive gradient. Conversely, the move in the direction of negative gradient will move in the direction of the smallest value of $f(x)$.

Example: Given $f(x) = x^3 - 9x^2 + 24x - 17$. If the current value of the variable is $x = 1$, should x be increased or decreased to increase $f(x)$?

Using $f(x)$, we have $f'(x) = 3x^3 - 18x + 24$. Substituting the current value of x, we have $f(x = 1) = -1$ and $f'(x = 1) = 9$. Hence increasing x increases the value of $f(x)$. Check: $f(x = 1.1) = -0.159 > -1$.

Example: Given $f(x) = x^3 - 9x^2 + 24x - 17$. If the current value of the variable is $x = 3$, should x be increased or decreased to increase $f(x)$?

Substituting the current value of x, we have $f(x = 3) = 1$ and $f'(x = 3) = -3$. Hence decreasing x increases the value of $f(x)$. Check: $f(x = 2.9) = 1.3 > 1$.

Now consider a smooth, differentiable function $f(x, y)$ of two variables. The first-order approximation is given by

$$f(x_{i+1}, y_{i+1}) \cong f(x_i, y_i) + \frac{\partial f}{\partial x}(x_{i+1} - x_i) + \frac{\partial f}{\partial y}(y_{i+1} - y_i) \tag{3.7.1.2}$$

If we hold y constant then if $x_{i+1} > x_i$, then $f(x_{i+1}, y_i) > f(x_i, y_i)$ only if $\partial f/\partial x > 0$. Similarly, if we hold x constant then if $y_{i+1} > y_i$, then $f(x_i, y_{i+1}) > f(x_i, y_i)$ only if $\partial f/\partial y > 0$.

Example: Consider $f(x, y) = 2x^2 - 4y^2 - 3x + 7y - 10$. If the current values of the variables are $x = 1$, $y = 2$, (a) should x be increased or decreased to increase $f(x, y)$ and (b) should y be increased or decreased to increase $f(x, y)$?

Use the function $\partial f/\partial x = 4x - 3$ and $\partial f/\partial y = -8y + 7$ and $f(x, y) = -13$. Substituting the current values yields $\partial f/\partial x = 1$ and $\partial f/\partial y = -9$. Hence increasing x increases $f(x, y)$ and decreasing y increases $f(x, y)$. Check: let $(x, y) = (1.1, 2)$, then $f(x, y) = -12.88 > -13$; let $(x, y) = (1, 1.9)$, then $f(x, y) = -12.14 > -13$.

3.7.2 Design Problems and Issues

Most structural systems can be viewed as "economic engines." They are built to serve one or more purposes. For example, an elementary school is designed and built as a facility for fulfilling the educational needs of young children. A strip mall, on the other hand, can have several uses. It may house small shops and restaurants, or provide office space for small businesses, or (usually) both.

Design of structural systems typically starts with a client's needs. For example, a developer may set aside a certain piece of land for several residential homes or for an office building. When these needs have been described in terms of tangible specifications, they are transformed from the specification to the final product—residential homes or the office building with all the accompanying facilities. One can quickly see that even the simplest design situation involves several distinct disciplines that must mesh together as seamlessly as possible. The residential homes example involves the supply and distribution of utilities such as electricity, gas, telephone, cable, and water, handling solid and liquid wastes, access with respect to roads, perhaps a neighborhood park, and finally the layout and design of homes themselves. A market analysis may be necessary to ascertain what types of homes provide the best opportunity for success. The layout of the roads, utilities, parks, and homes is closely tied to the market analysis. Architectural plans are then necessary to describe the functional layout of individual homes. The layout is then translated to geotechnical, structural, electrical, and mechanical design specifications. From the specifications spring forth the actual design details. And, finally, the construction process transforms the design drawings to the physical homes.

3.7.2.1 Design Scenarios

In this text, the "structural" design scenarios discussed earlier are split into four categories, as discussed next.

Establishing design requirements. Structural design is usually carried out to design a structural component or system so as to meet the design specifications in design codes. Some of the codes were discussed in Section 3.6. In this text, most of the design requirements do *not* involve meeting the different code requirements.[8] Instead, in some examples and exercises, the loads are computed using the applicable design code. However, the requirements are mechanics based, meeting stress/strain and deflection requirements, etc.

Design checks. The simplest design example is one in which, given a structural component (e.g., a member) or system (a collection of joints, members, and supports forming a self-standing structure), the design requirements are checked to see whether they are satisfied or not. Look at Examples 3.7.1 and 3.7.2 as simple examples to illustrate the concepts.

Design of components. This is the most common design example in the published literature. A typical instance is the design of a single member in a truss or frame to satisfy a set of design requirements. The starting point is the member data (forces acting on the member, member end conditions, and material used). The end product of the design exercise is to specify a cross-section that satisfies the requirements. More often than not, the component is determinate and there is no goal driving the design. Example 3.7.3 shows the design of a system with a single component—a beam.

System design. This is the most difficult and most ignored topic. The starting point is a set of specifications and requirements. The end product is a set of data that defines the structural system as completely as possible. The starting point may define some broad guidelines and limitations. The designer must use his or her design experience, intuition, etc., to formalize the structural definition—choice of the structural system (truss, frame, etc.), material(s) to be used (concrete, steel, etc.), loading to be considered, applicable design codes, and finally, a preliminary design (or an initial guess). The preliminary design then must be carefully evaluated, changed, and moved towards a final design. Examples 3.7.4 and 3.7.5 illustrate some of the system design concepts associated with the design of planar trusses and frames.

3.7.2.2 Solution Techniques

The formal solution techniques to solve general design problems will be discussed in Chapter 8. In this section we introduce the terminology and the techniques to solve *simple* design problems involving one or two design parameters and several constraints. Consider the following design problem:

Find $\mathbf{x} = \{x_1, x_2\}$ (3.7.2.1)

Minimize $f(\mathbf{x}) = (x_1 - 2)^2 + (x_2 - 3)^2$ (3.7.2.2)

Subject to $g_1(\mathbf{x}) \equiv 2x_2 - x_1^2 \le 0$ (3.7.2.3)

$$1 \le x_1 \le 5; \; 1 \le x_2 \le 5 \tag{3.7.2.4}$$

The objective of the design problem is to find the values of the *design variables* $\mathbf{x} = \{x_1, x_2\}$ so that the value of the *objective function* $f(\mathbf{x})$ is minimized while simultaneously satisfying the constraints $g(\mathbf{x})$ and the *bounds on the design variables* expressed as $x_i^L \le x_i \le x_i^U$. Most engineering design problems can be posed in the above form. For structural design problems involving truss and frame structures, the design variables are typically cross-sectional dimen-

[8] Every engineering curriculum has one or more courses that discuss specific design code(s). These courses follow the first course on structural analysis, the major topic of this text. However, see Chapter 9.

sions or properties. Examples include the width and height of rectangular cross-sections, radius of circular cross-sections, etc; examples of the objective function include cost, weight, volume, etc. The constraints are used to obtain the required structural response. Examples include limiting stresses and displacements to allowable values typically obtained from design codes. The bounds on the design variables ensure that the values of the design variables are acceptable. For example, if the design variables are cross-sectional dimensions, the values can neither be small/negative nor large.

Several questions immediately come to mind.

1. What is the nature of the design variables? Are they continuously varying? Do we have to select the values from a set of predefined values?
2. What is the nature of the objective function—is it linear or nonlinear? Do we always have a single objective function, or do we try to minimize (or maximize) several functions simultaneously? Is the function an explicit function of the design variables?
3. What is the nature of the constraints—are they linear or nonlinear functions, equality or inequality? Are the functions explicit functions of the design variables?
4. Is there a unique solution? Are there multiple solutions? Do we in fact always have a solution?
5. How do we find the solution(s) in a systematic manner?

We defer the answers to some of these questions. With reference to Eqs. (3.7.2.1)–(3.7.2.4), we can find infinite number of $\mathbf{x} = \{x_1, x_2\}$ that satisfy the constraints. This set is called the *feasible set* or domain. A particular combination $\{x_1, x_2\}$ is called a feasible solution. Similarly, there exist infinite $\mathbf{x} = \{x_1, x_2\}$ that do not satisfy the constraints. This set is called the *infeasible set* or domain. Unless the problem is properly posed, there may be no feasible solution to the problem. From the feasible domain, the best solution is called the *global minimum or optimal solution*.

The basic fact remains that if, mathematically speaking, the problem is a nonlinear programming problem (NLP), then there is no unique manner in which the design problem can be solved. There are tens of methods of solving the NLP problem. In this book and especially in this section, we look at some simple techniques for obtaining reasonable solution to these problems.

Exhaustive search. The most straightforward and also the most computationally expensive technique is exhaustive search. The basic idea is to create a grid pattern by varying the values of the design variables between the lower and upper bounds of the design variables. The objective function and the constraints are then evaluated at these grid points. When this method is used with the problem described by Eqs. (3.7.2.1)–(3.7.2.4), the results are as shown below. The resolution of the grid is taken as 0.5 for both x_1 and x_2. Only the feasible solutions are shown and the best answer is shaded.

X1	X2	Obj Func	Constr
1.5	1	4.25	-0.25
2	1	4	-2
2	1.5	2.25	-1
2.5	1	4.25	-4.25
2.5	1.5	2.5	-3.25
2.5	2	1.25	-2.25
2.5	2.5	0.5	-1.25

continued

X1	X2	Obj Func	Constr
2.5	3	0.25	-0.25
3	1	5	-7
3	1.5	3.25	-6
3	2	2	-5
3	2.5	1.25	-4
3	3	1	-3
3	3.5	1.25	-2
3	4	2	-1
3.5	1	6.25	-10.25
3.5	1.5	4.5	-9.25
3.5	2	3.25	-8.25
3.5	2.5	2.5	-7.25
3.5	3	2.25	-6.25
3.5	3.5	2.5	-5.25
3.5	4	3.25	-4.25
3.5	4.5	4.5	-3.25
3.5	5	6.25	-2.25
4	1	8	-14
4	1.5	6.25	-13
4	2	5	-12
4	2.5	4.25	-11
4	3	4	-10
4	3.5	4.25	-9
4	4	5	-8
4	4.5	6.25	-7
4	5	8	-6
4.5	1	10.25	-18.25
4.5	1.5	8.5	-17.25
4.5	2	7.25	-16.25
4.5	2.5	6.5	-15.25
4.5	3	6.25	-14.25
4.5	3.5	6.5	-13.25
4.5	4	7.25	-12.25
4.5	4.5	8.5	-11.25
4.5	5	10.25	-10.25
5	1	13	-23
5	1.5	11.25	-22
5	2	10	-21
5	2.5	9.25	-20
5	3	9	-19
5	3.5	9.25	-18
5	4	10	-17
5	4.5	11.25	-16
5	5	13	-15

Clearly, this method is computationally expensive. To find the solution, the objective function and constraints were evaluated a total of 81 times each—there were 9 points on the x_1 axis and 9 points on the x_2 axis. If the number of design variables is increased, the number of evaluations will increase exponentially. If there are n design variables and m points on each axis, then we have a total of m^n points!

Graphical. The graphical method is preferable since it gives a graphical view of the design space. However, it cannot be used effectively for more than two design variables. We illustrate the process in Fig. 3.7.2.2.1 using the above problem for problems involving two design variables. The first step is to draw the grid that bounds the design space. For the problem at hand this means $1 \le x_1 \le 5$, $1 \le x_2 \le 5$. The next step is to sketch the constraints. This is necessary to identify the feasible and the infeasible domains.

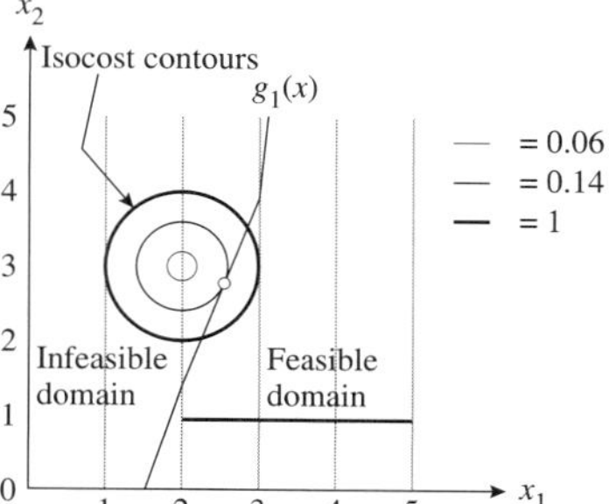

Fig. 3.7.2.2.1 Design space showing the objective function and constraint.

The only constraint in this problem is $g_1(\mathbf{x}) \equiv 2x_2 - x_1^2 \le 0$. To sketch the constraint function that is a nonlinear function, we will set $g_1(x) = 0$ so as to identify the boundary between the feasible ($g_1 \le 0$) and infeasible ($g_1 > 0$) domain. Setting $g_1(x) = 0$ yields $x_2 = x_1^2/2$.

x_1	x_2
1	0.5
2	2.0
3	4.5
4	8
5	12.5

Now we can set up the table to show the (x_1, x_2) values by computing x_2 for appropriate values of x_1. The area to the left of the curve (or above the curve) is such that $g_1 > 0$, or is in other words the infeasible domain. To sketch the objective function one must note that the function is nonlinear in both x_1 and x_2, and that we are looking for the minimum value of the function (not a specific value). Hence we must draw contours. These curves are often called isocost contours. Typically, we start by first picking values c of the cost or objective function. Then we set $f(x) = c$ and solve one design variable in terms of another. Finally, we create a table of (x_1, x_2) values as we have for the constraint function. However, we recognize that the objective function in this example is the equation of a circle with its center at (2, 3). Selecting three different values of the radius as (0.25, 0.375, 1), we can sketch the three isocost contours as shown in Fig. 3.7.2.2.1. Since the feasible domain is to the right of g_1, then intent is to find the smallest circle that just reaches the feasible domain. This explains the selection of the 0.375 value. The graphical optimal solution is shown in Fig. 3.7.2.2.1 at approximately (2.4, 2.9).

Trial and error. One can also use a trial-and-error procedure that is guided by (mathematical) intuition and experience. To facilitate the process, we first compute the gradients of the objective function and the constraints:

$$\nabla f(x) = \left\{ \frac{\partial f(\mathbf{x})}{\partial x_1}, \frac{\partial f(\mathbf{x})}{\partial x_2} \right\} = \{2(x_1 - 2), 2(x_2 - 3)\} \tag{3.7.2.5}$$

$$\nabla g_1(\mathbf{x}) = \left\{ \frac{\partial g_1(\mathbf{x})}{\partial x_1}, \frac{\partial g_1(\mathbf{x})}{\partial x_2} \right\} = \{-2x_1, 2\} \tag{3.7.2.6}$$

1. We start the iterative process at an arbitrarily chosen point (1, 1). At this point $f = 5$ but $g_1 = 1$—an infeasible point. We need to decrease g_1. Looking at $\nabla g_1(\mathbf{x})$, we can conclude that g_1 decreases if x_1 increases. We increase the value of x_1.
2. Let us choose (3, 1) as the next point. At this point $f = 5$ and $g_1 = -7$, a feasible point. We now can decrease the objective function by moving in the $-\nabla f(\mathbf{x})$ direction. At the current point $-\nabla f(\mathbf{x}) = (-2, 4)$, implying that x_1 needs to decrease and x_2 needs to increase to decrease f. Let us now decrease x_1 and increase x_2.
3. Let us choose (2.5, 2) as the next point. At this point $f = 1.25$ and $g_1 = -2.25$—a feasible point. As before, we can decrease the objective function by moving in the $-\nabla f(\mathbf{x})$ direction. At the current point $-\nabla f(\mathbf{x}) = (-1, 2)$, implying as before that x_1 needs to decrease and x_2 needs to increase to decrease f.
4. Let us choose (2.4, 2.25) as the next point. At this point $f = 0.72$ and $g_1 = -1.26$, a feasible point. We can continue this process until we are satisfied with the results.

Constraint controlled optimal design.[9] More often than not, engineering designs are controlled by constraints. In other words, the optimal solution is one in which one or more inequality constraints are just satisfied (or are satisfied as an equality). If there are several constraints, then one must use trial-and-error to ascertain which constraint or constraints control the optimal design. We illustrate the use of this technique through examples.

We have looked at four different techniques to solve a simple engineering design problem formulated and posed as an NLP problem. The graphical technique is clearly preferable if one is to gain insight into a design problem. However, it is useful only if the number of design variable is restricted to two. The trial-and-error process is guided by the designer, and its effort and success is a function of the experience of the user. If one has access to a computer program and function evaluations are inexpensive, then the exhaustive search will do very well with problems involving a few design variables. In Chapter 8 we will look at the topic of engineering design problem formulation and solution methodologies in greater detail.

EXAMPLE 3.7.1 ***Design Checks for a Wooden Beam***

Figure E3.7.1(a) shows a simply supported beam. The beam is made of Douglas fir (modulus of elasticity = 1800 ksi, mass density = 1.0 slugs/ft^3). The cross-section is rectangular, with width b = 6 in and height h = 8 in. Is the maximum normal stress less than 2 ksi and is the shear stress less than 0.1 ksi?

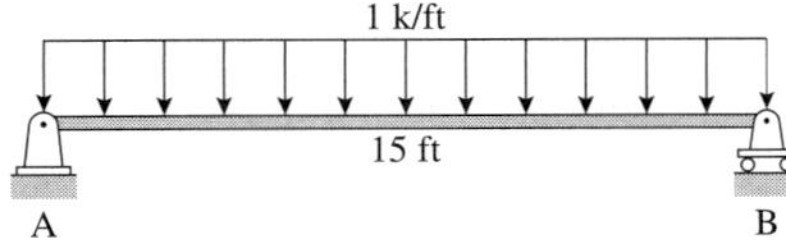

Fig. E3.7.1(a)

SOLUTION

Step 1: A strategy to solve the problem is first to analyze the structure and draw the shear force and bending moment diagrams. This gives us the largest shear force and bending moment values. These values can then be used to compute the largest normal and shear stress and compare them to the given values.

[9] This approach is presented formally in Chapter 8, Section 8.3.1, under the heading Kuhn-Tucker conditions.

Instead of using numerical values directly, we solve the problem symbolically. The intent is to establish the relationship between different problem parameters. Let the loading on a simply supported beam of length L be w. Structural analysis shows that the largest bending moment occurs at the center and $M_{max} = wL^2/8$, and the largest shear force occurs at the supports and $V_{max} = wL/2$.

Step 2: Using the largest bending moment, we find that the largest normal stress due to bending is

$$\sigma_{max} = \frac{M_{max}c_{max}}{I} = \frac{M_{max}(h/2)}{(bh^3)/12} = \frac{(wL^2/8)(6)}{bh^2} = \frac{3wL^2}{4bh^2}$$

This stress occurs on the outer fibers. Similarly, we find that the largest shear stress due to shear force is

$$\tau_{max} = \frac{V_{max}Q_{max}}{It} = \frac{(wL/2)(bh^2/8)}{(bh^3/12)(b)} = \frac{3wL}{4bh}$$

and occurs at the points on the centroidal axis.

Step 3: Substituting the numerical values $L = 180$ in, $b = 6$ in, $h = 8$ in, $w = 1000$ lb/ft = 83 lb/in, we have

$$\sigma_{max} = \frac{3(83)(180)^2}{4(6)(8)^2} = 5250\text{psi} > 2000\text{psi}$$

$$\tau_{max} = \frac{3(83)(180)}{4(6)(8)} = 233\text{psi} > 100\text{psi}$$

Hence the beam does not satisfy the specified requirements.

EXAMPLE 3.7.2 ***Design Checks and Redimensioning a Wooden Beam***

Redo Example 3.7.1 but now assume that the beam cross-section is rectangular, with width $b = 10$ in and height $h = 12$ in.

SOLUTION

Step 1: Substituting the numerical values $L = 180$ in, $b = 10$ in, $h = 12$ in, $w = 83$ lb/in, we have

$$\sigma_{max} = \frac{3(83)(180)^2}{4(10)(12)^2} = 1400\text{psi} < 2000\text{psi}$$

$$\tau_{max} = \frac{3(83)(180)}{4(10)(12)} = 93\text{psi} < 100\text{psi}$$

Hence the beam satisfies the specified requirements. However, we did not consider the weight of the beam. Can this be ignored?

Step 2: The cross-sectional area of the beam is bh. The equivalent uniformly distributed load w_w on the beam is

$w_w = (bh)(\text{weight density}) = bh(32.2 \text{ lb/ft}^3) = bh(0.00058 \text{ lb/in}^3) = (10)(12)(0.00058) = 0.07$ lb/in

which is less than 1% of the applied load. Hence the self-weight of the beam can be ignored for this problem, and the beam satisfies the specified requirements.

An examination of the expressions for the stress shows why these dimensions are superior to the previous beam:

$$\sigma_{max} = \frac{3wL^2}{4bh^2} \qquad \tau_{max} = \frac{3wL}{4bh}$$

In both the equations, the stresses are inversely proportional to the width of the beam and square of the height of the beam. In other words, the stresses *decrease* if the width of the beam or the height of the beam is *increased*. However, the decrease is *more by changing the height of the beam* versus the width of the beam. The process of comparing the effect of changes in the different design parameters (in this case the height and width of the beam) to certain responses (in this case the stresses) is known as design sensitivity analysis. Taking the derivative of the normal stress with respect to the height and the width, we obtain

$$\frac{\partial \sigma_{\max}}{\partial b} = -\frac{3wL^2}{4b^2h^2} \qquad \frac{\partial \sigma_{\max}}{\partial h} = -\frac{3wL^2}{2bh^3}$$

Substituting the numerical value: $L = 180$ in, $b = 6$ in, $h = 8$ in, $w = 83$ lb/in, we find

$$\frac{\partial \sigma_{\max}}{\partial b} = -\frac{3(83)(180)^2}{4(6)^2(8)^2} = -875 \text{ psi/in}$$

$$\frac{\partial \sigma_{\max}}{\partial h} = -\frac{3(83)(180)^2}{2(6)(8)^3} = -1310 \text{ psi/in}$$

The two numbers indicate that the normal stress will *decrease* (due to the negative value) by 875 psi to 4375 psi if the width is *increased* by unity (from 6 in to 7 in holding the height at 8 in), and by 1310 psi to 3940 psi if the height is *increased* by unity (from 8 in to 9 in holding the width at 6 in). In reality the stress values will decrease in both cases but not by the computed amounts (4500 psi with 7 × 8-in beam and 4150 *psi* with a 6 × 9-in beam), since the expressions are derivatives (∂), not the change (Δ). From a design perspective, it is more effective to increase the height of the beam than to increase the width in order to decrease the normal stress. The numerical results bear out the interpretation.

Similarly,

$$\frac{\partial \tau_{\max}}{\partial b} = -\frac{3wL}{4b^2h} \quad \text{and} \quad \frac{\partial \tau_{\max}}{\partial h} = -\frac{3wL}{4bh^2}$$

Substituting the numerical value: $L = 180$ in, $b = 6$ in, $h = 8$ in, $w = 83$ lb/in, we find

$$\frac{\partial \tau_{\max}}{\partial b} = -\frac{3(83)(180)}{4(6)^2(8)} = -39 \text{ psi/in}$$

$$\frac{\partial \tau_{\max}}{\partial h} = -\frac{3(83)(180)}{4(6)(8)^2} = -28 \text{ psi/in}$$

While the values of b and h are not equal, the derivatives indicate that the maximum shear stress will decrease by increasing the width and the height, though not by the same amount.

EXAMPLE 3.7.3 ***Design of a Wooden Beam***

Consider the beam in Example 3.7.1. Find the width b and height h so that the maximum normal stress is less than 2 ksi and the shear stress is less than 0.1 ksi. Neglect self-weight of the beam.

SOLUTION

Step 1: Let us examine the problem statement carefully. With the same problem we looked at two solutions—one that did not satisfy the stress requirements with the beam cross-section as 6 × 8 in and another that did satisfy the stress requirements with the cross-section as 10 × 12 in. With this information, one obvious question is whether a beam of size 9.5 × 12 in or even 10 × 12.5 in will satisfy the requirements. An examination of these beam dimensions shows that both those dimensions satisfy the requirements and that there really is no unique answer to the problem. In fact, there are infinite combinations of b and h that meet the prescribed design requirements.

Step 2: The design problem statement plays a pivotal role in determining the design solution. Let us modify the problem statement in four different ways.

(a) Find the smallest square beam so that the maximum normal stress less than 2 ksi and the shear stress is less than 0.1 ksi.

Analytical solution. This problem statement reduces the design problem to a *single-parameter design problem* since $b = h$. Using this condition,[10] we now have

$$\sigma_{max} = \frac{3wL^2}{4b^2} \qquad \tau_{max} = \frac{3wL}{4b^2} \tag{E3.7.3a}$$

It should be recognized that with the smallest permissible beam cross-section, the resulting stress will be just at the stress limit. Hence, substituting the numerical values for the normal stress, we have

$$\sigma_{max} = 2000 = \frac{3wL^2}{4b^3} \text{ from which } b^3 = \frac{3wL^2}{8000} = \frac{3(83)(180)^2}{8000} \Rightarrow b = 10 \text{ in}$$

In other words, a *minimum* width and height of 10 in is necessary to hold the maximum normal stress to 2000 psi.

Now, substituting the numerical values for the shear stress gives

$$\tau_{max} = 100 = \frac{3wL}{4b^2} \text{ from which } b^2 = \frac{3wL}{400} = \frac{3(83)(180)}{400} \Rightarrow b = 10.6 \text{ in}$$

In other words, a *minimum* width and height of 10.6 in is necessary to hold the maximum shear stress to 100 psi. Hence, the solution to the problem is: a beam of width and height of 10.6 in is the smallest square beam to meet the specified requirements.

Graphical solution. We will solve the problem graphically. The first step is to graph the design space. In other words, we draw the graphs of the normal stress and the shear stress requirements as a function of the design parameter b. Hence we have

$$\text{Normal stress requirement: } \sigma_{max} = \frac{3wL^2}{4b^3} \le 2000 \Rightarrow \frac{1008.45}{b^3} - 1 \le 0 \tag{E3.7.3b}$$

$$\text{Shear stress requirement: } \tau_{max} = \frac{3wL}{4b^2} \le 100 \Rightarrow \frac{112.05}{b^2} - 1 \le 0 \tag{E3.7.3c}$$

As shown in Fig. E3.7.3(a), values of b that yield normal and shear stress requirements less than zero are acceptable. Clearly the smallest value of b that we can find yields the smallest beam. From the graph, the smallest value of b that satisfies the normal stress requirement is 10 in, and the smallest value of b that satisfies the shear stress requirement is about 10.5 in (10.6 in, to be exact). The answers are the same as the analytical solution.

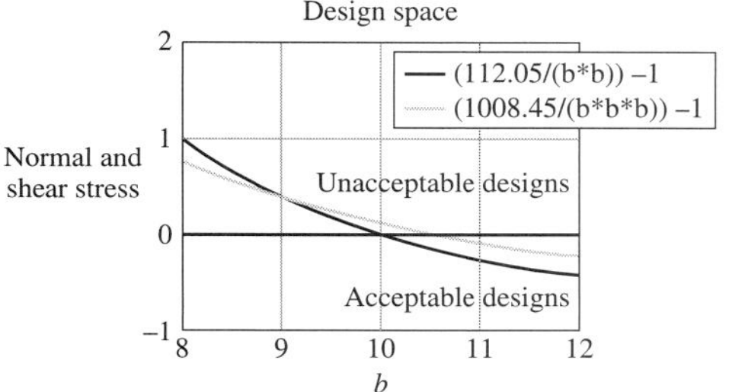

Fig. E3.7.3(a) One-parameter design space.

[10] This equation is in fact an additional (equality) constraint that is then used to reduce the number of independent design parameters.

(b) Find the width b and height h of the lightest beam so that the maximum normal stress less than 2 ksi and is the shear stress less than 0.1 ksi.

Solution. This is a *two-parameter design problem*. Since the beam is assumed to be made of one material, the lightest beam is also the one with the least volume. The relevant equations to be used are given below. In these equations, $L = 180$ in, $w = 83$ lb/in.

Beam volume: $V(b, h) = bhL = 180bh$ (E3.7.3d)

Maximum normal stress requirement: $\sigma_{max}(b,h) = \dfrac{3wL^2}{4bh^2} = \dfrac{2.0169(10^6)}{bh^2} - 2000 \le 0$ (E3.7.3e)

Maximum shear stress requirement: $\tau_{max}(b,h) = \dfrac{3wL}{4bh} = \dfrac{11205}{bh} - 100 \le 0$ (E3.7.3f)

Analytical solution. As we observed in the single-parameter design example, typically the best solution is the design point where one or more constraints are active. What this means is that the inequality constraint turns into an equality constraint at the design point.

Let us assume that the normal stress constraint governs this design. Hence, $2.0169(10^6)/bh^2 - 2000 = 0$, from which $b = 1008.45/h^2$. Substituting this into the volume of the beam we have $V = 180bh = 181521/h$. To minimize the volume, we need $dV/dh = 0 = -181521/h^2$. This condition yields no finite value for h. In other words, we have an unbounded solution. A very large value of h will yield the lightest beam since the corresponding value of b obtained from $b = 1008.45/h^2$ will be extremely small.

Let us assume that the shear stress constraint governs this design. Hence $11205/bh - 100 = 0$, from which $b = 112.05/h$. Substituting this into the volume of the beam yields $V = 180bh = 20169$. To minimize the volume, we need, $dV/dh = 0$. However, this condition yields no value for h or for b. In other words, any appropriate combination will yield a volume of 20169 in^3.

Let us assume that the shear stress *and* normal stress constraints govern the design. With both the constraints as equality constraints we have a set of nonlinear equations, $b = 1008.45/h^2$ and $b = 112.05/h$. Solving, we find $h = 9$ in, $b = 12.45$ in, and $V = 200169$ in^3. The design space is shown in Fig. E3.7.3(b). Clearly, this is the best solution obtained.

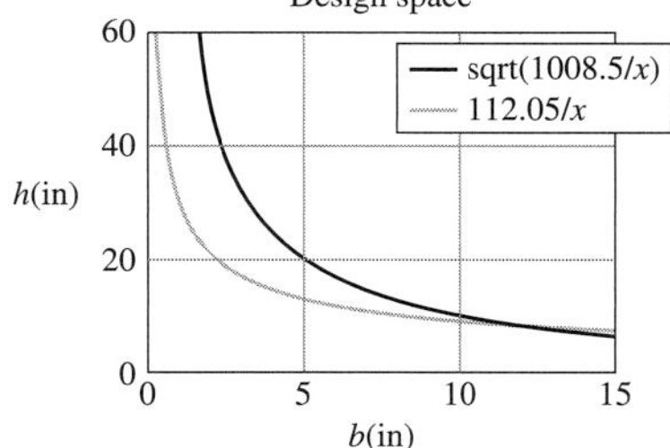

Fig E3.7.3(b) Two-parameter design space.

(c) Find the width b and height h so that the maximum normal stress less than 2 ksi and the shear stress less than 0.1 ksi so that the resulting beam is the lightest beam. The height of the beam should not exceed twice its width.

The problem is the same as in part (b) except we now have an additional constraint:

$h \le 2b$ (E3.7.3g)

If we assume that the normal stress constraint governs this design, then $b = 1008.45/h^2$. We also require Eq. (E3.7.3g) to be satisfied. In the previous part, we noted that larger heights yielded better solutions. Hence let $h = 2b$. Substituting, we have $b = 1008.45/h^2 = 1008.45/4h^2$. Solving yields $b = \sqrt[3]{252.1} = 6.3$ in and $h = 12.6$ in, $V = 14288$ in^3.

If we assume that the shear stress constraint governs this design, then $b = 112.05/h$. We also require Eq. (E3.7.3g) to be satisfied. Also, as before, let $h = 2b$. Substituting, we have $b = 112.05/h = 56.025/b$. Solving yields $b = \sqrt{56.025} = 7.5$ in and $h = 15$ in, $V = 20250$ in^3.

We have two sets of answers. Which one is correct? Even though we are trying to find the lightest beam, we need to pick the larger values (shear stress governs) since the lower values (normal stress governs) do *not* satisfy the shear stress constraint. The correct solution is $b = 7.5$ in and $h = 15$ in!

In this case we have a meaningful result. By imposing the additional constraint that the height be no more than twice the width, we obtained the beam dimensions as $(b, h) = (7.5$ in, 15 in$)$.

What solution would we obtain if we assumed that both the shear stress and normal stress constraints govern the design? With both the constraints as equality constraints we have a set of nonlinear equations, $b = 1008.45/h^2$ and $b = 112.05/h$. Solving yields $h = 9$ in, $b = 12.45$ in, and $V = 20088$ in^3. This also satisfies the requirement that the height be no more than twice the width.[11] *This is the best solution* and is the same as the previous problem formulation.

Observation: A design problem with n parameters can have *at most n* constraints that govern the design. In the absence of a rigorous theory (we look at this aspect in Chapter 8), one must check all possible combinations to draw the correct conclusion.

In the above problem formulation we had two design parameters and three inequality constraints. How would the solution change is we imposed the smallest acceptable dimension for both the height and the width as 1 in instead of constraint (E3.7.3g)?

(d) In the previous parts we have assumed that both the height and the width can change continuously. Find the lightest beam from Table B-1 in Appendix B so that the maximum normal stress is less than 2 ksi and the shear stress is less than 0.1 ksi.

In this part we are no longer accepting continuous values for (b, h). Instead we are using predetermined or predefined values. What should be the approach to solve this problem? One of the easiest ways is to rearrange the table by reordering the cross-sections going from the lightest to the heaviest section. Then we can systematically select the cross-sections, check the requirements, and pick the *first* cross-section that meets the requirements. The table below presents the calculations and results. The 8 × 16-in beam is the lightest section that meets the requirements.

Actual dimensions		Area	Shear stress	Normal stress
1.5	2.5	3.75	2888	213146.6667
2.5	1.5	3.75	2888	356577.7778
1.5	3.5	5.25	2034.285714	107768.7075
3.5	1.5	5.25	2034.285714	254126.9841
1.5	5.5	8.25	1258.181818	42451.79063
5.5	1.5	8.25	1258.181818	160989.899
2.5	3.5	8.75	1180.571429	63861.22449
3.5	2.5	8.75	1180.571429	90205.71429
1.5	7.25	10.875	930.3448276	23582.24336
7.25	1.5	10.875	930.3448276	121647.5096

continued

[11] Normally the height is larger than the width. The reason why the width is larger than the height in this problem is due to the severe shear stress requirement.

Actual dimensions		Area	Shear stress	Normal stress
3.5	3.5	12.25	814.6938776	45043.73178
2.5	5.5	13.75	714.9090909	24671.07438
5.5	2.5	13.75	714.9090909	56676.36364
1.5	9.25	13.875	707.5675676	13715.6075
1.5	11.25	16.875	564	8624.526749
2.5	7.25	18.125	518.2068966	13349.34602
7.25	2.5	18.125	518.2068966	42513.10345
3.5	5.5	19.25	482.0779221	17050.76741
5.5	3.5	19.25	482.0779221	27936.92022
1.5	13.25	19.875	463.7735849	5659.190697
2.5	9.25	23.125	384.5405405	7429.3645
3.5	7.25	25.375	341.5763547	8963.818583
7.25	3.5	25.375	341.5763547	20710.76707
2.5	11.25	28.125	298.4	4374.716049
5.5	5.5	30.25	270.4132231	10123.21563
3.5	9.25	32.375	246.1003861	4735.260357
2.5	13.25	33.125	238.2641509	2595.514418
2.5	15.25	38.125	193.9016393	1469.174953
3.5	11.25	39.375	184.5714286	2553.368607
5.5	7.5	41.25	171.6363636	4519.59596
7.5	5.5	41.25	171.6363636	6890.358127
3.5	13.25	46.375	141.6172507	1282.510299
5.5	9.5	52.25	114.4497608	2063.460086
3.5	15.25	53.375	109.9297424	477.9821093
7.5	7.5	56.25	99.2	2781.037037
5.5	11.5	63.25	77.1541502	772.985049
7.5	9.5	71.25	57.26315789	979.8707295
5.5	13.5	74.25	50.90909091	12.22097518
5.5	15.5	85.25	31.43695015	−473.5597389
7.5	11.5	86.25	29.91304348	33.52236925
9.5	9.5	90.25	24.15512465	352.5295233
7.5	13.5	101.25	10.66666667	−524.3712849
9.5	11.5	109.25	2.562929062	−394.5876032
7.5	15.5	116.25	−3.612903226	−880.6104752
9.5	13.5	128.25	−12.63157895	−835.0299617
7.5	17.5	131.25	−14.62857143	−1121.85034
11.5	11.5	132.25	−15.27410208	−673.7897592
7.5	19.5	146.25	−23.38461538	−1292.746
9.5	15.5	147.25	−23.9049236	−1116.271428
11.5	13.5	155.25	−27.82608696	−1037.633447
9.5	17.5	166.25	−32.60150376	−1306.723953
11.5	15.5	178.25	−37.13884993	−1269.963353
13.5	13.5	182.25	−38.51851852	−1180.206269
9.5	19.5	185.25	−39.51417004	−1441.641579
11.5	17.5	201.25	−44.32298137	−1427.2937
13.5	15.5	209.25	−46.4516129	−1378.116931
11.5	19.5	224.25	−50.03344482	−1538.747392
11.5	21.5	247.25	−54.68149646	−1620.570461
11.5	23.5	270.25	−58.53839038	−1682.405968

EXAMPLE 3.7.4 ***Design of a Planar Truss***

Consider the cantilever truss in Example 2.7.1 (see Fig. E3.7.4). The truss members are made of steel and have a circular hollow cross-section. Select three cross-sections for the top chord, the bottom chord, and the web members respectively. Design the lightest truss so that the normal stress in the members is less than 24 ksi and Euler buckling criterion is satisfied.

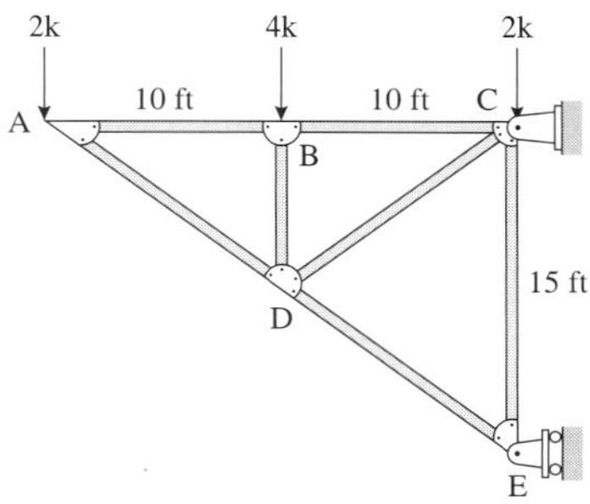

Fig E3.7.4

SOLUTION

Using our experience with previously solved problems, we will set the overall strategy to solve the problem. To compute the normal stress in the member we need to know the axial force in the member. A structural analysis is required. The buckling capacity of a truss member is not a function of the member force but depends merely on the length, moment of inertia, and modulus of elasticity. In the prior examples we were able to write the equations for these requirements explicitly in terms of the design variables. Is this possible with this problem? The answer is yes.

Observation: A determinate structure can be analyzed to determine the internal forces without knowledge of the material and cross-sectional properties of the members. This is not the situation with indeterminate structures, where the internal forces are dependent on the choice of material and cross-sectional dimensions.

As in the previous problem, two design variables define the cross-section of a member. To obtain meaningful values for these two dimensions we could establish lower and upper bounds on them. Instead, we require as previously that the wall thickness be at least 20% of the inner radius; in other words,

$$\frac{t}{r_i} \geq 0.2 \tag{E3.7.4a}$$

is required for all the three cross-sections. The lightest section is typically achieved when this constraint is active, i.e., $t = 0.2r_i$.

Step 1: The axial forces have been computed in Example 2.7.1. We use *k* and in as the problem units.

Step 2: There are two possibilities with each member. First, the lightest structure is one in which the stress in each member is exactly at the specified allowable stress limit, i.e., 24 ksi (this process is known as fully stressed design in published literature). We can then use this condition to compute the cross-sectional radius as follows. With *P* as the axial force in the member, we have

$$\frac{P}{A} = 24 \text{ or, } \frac{P}{\pi\left((1.2r_i)^2 - r_i^2\right)} = 24 \text{ or, } r_i = \sqrt{\frac{P}{15.36\pi}} \quad \text{(E3.7.4b)}$$

Second, the buckling constraint governs the design of compressive members. Hence,

$$P = P_{cr} = \frac{\pi^2 EI}{L^2} \text{ or, } P = \frac{\pi^2 E \frac{\pi}{4}\left((1.2r_i)^4 - r_i^4\right)}{L^2} \text{ or, } r_i = \sqrt[4]{4.1435(10^{-6})PL^2} \quad \text{(E3.7.4c)}$$

As in the previous problem, we will find the larger of the two cross-sections thus obtained, so as to ensure that both the axial stress and the buckling constraints are satisfied.

Step 3: The calculations are shown below. The first part of the table shows the case when the stress constraint is assumed to govern the design. The second part of the table shows the calculations only for the compressive members and it is assumed that buckling governs the design of the member.

Member	P (k)	r_i (in)	t (in)	A (in^2)	I (in^4)	P/A (ksi)	L (in)	P_{cr} (k)
AB	2.67	0.283693078	0.056738616	0.11125	0.005461694	24	120	
BC	2.67	0.283693078	0.056738616	0.11125	0.005461694	24	120	
BD	4	0.347234688	0.069446938	0.166666667	0.012258146	24	90	0.433149185
CD	3.33	0.316821923	0.063364385	0.13875	0.008495585	24	150	
CE	4	0.347234688	0.069446938	0.166666667	0.012258146	24	180	
AD	3.33	0.316821923	0.063364385	0.13875	0.008495585	24	150	0.10807083
DE	6.67	0.44839011	0.089678022	0.277916667	0.034084465	24	180	0.301098919
BD	4	0.605311211	0.121062242	0.506477289	0.113200213	7.897688769	90	3.999999668
AD	3.33	0.746447302	0.14928946	0.77019527	0.261775494	4.323578879	150	3.329999724
DE	6.67	0.972770016	0.194554003	1.308045627	0.755045424	5.099210504	180	6.669999446

Step 4: The final answers are

Top chord (AB and BC): $(r_i, t) = (0.28 \text{ in}, 0.06 \text{ in})$. Stress governs the design.

Web members (BD, DC, CE): $(r_i, t) = (0.61 \text{ in}, 0.12 \text{ in})$. Buckling of member BD governs the design.

Bottom chord (AD, DE): $(r_i, t) = (0.97 \text{ in}, 0.19 \text{ in})$. Buckling of member DE governs the design.

Note that, as in the previous problem, we have neglected to include the self-weight of the truss. Can this assumption be justified?

EXAMPLE 3.7.5 ***Design of a Planar Frame***

Consider the planar frame in Example 2.8.9 (see Fig. E3.7.5). Design the lightest steel frame such that the largest normal stress is less than 1000 psi. Use the same AISC W section for all the members, selecting from the list below.

Sections to consider: W4, W5, W6, W8, W10.

SOLUTION

Conceptually the design of this frame will be similar to the design of the truss in the previous example. The frame is determinate; the internal forces are not functions of the member properties. The problem units are lb and in.

Step 1: We use the previously computed structural response (see Example 2.8.9) to find largest internal forces in the three members.

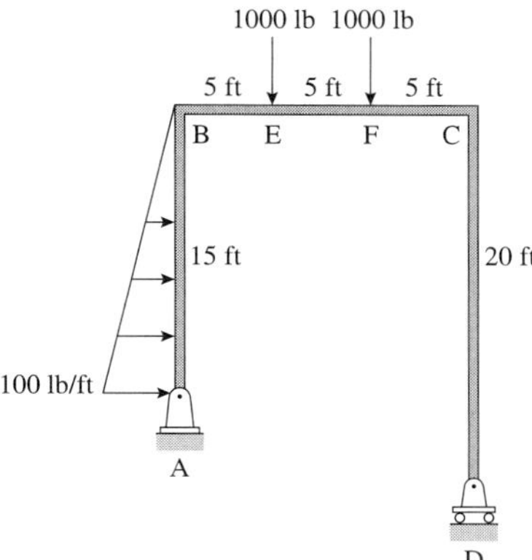

Fig E3.7.5

Member AB: Largest axial force $N = 750$ lb(C), largest shear force $V = 750$ lb, and largest bending moment $M = 3750$ lb-ft = 45000 lb-in.

Member BC: Largest axial force $N = 15$ lb (T), largest shear force $V = 750$ lb, and largest bending moment $M = 7500$ lb-ft = 90000 lb-in.

Member CD: Largest axial force $N = 1250$ lb (C), largest shear force $V = 0$, and largest bending moment $M = 0$.

Step 2: Due to axial force and bending moment, the normal stress is given by

$$\sigma = \frac{N}{A} \pm \frac{Mc}{I} = \frac{N}{A} \pm \frac{M}{S} \tag{E3.7.5a}$$

where S is the section modulus. The computed values (tension or compression) must be less than 1000 psi.

Step 3: We first sort the sections in order of ascending cross-sectional area values. Then we apply Eq. (E3.7.5a) to each of the members to compute the largest normal stress due to axial force and bending moment. The results are shown in the table below. The maximum normal stress in members AB and CD is compressive while the normal stress in member BC is tensile.

Step 4: Scanning the table shows that the lightest section that satisfies the requirement is $W10 \times 88$.

Once again, justify whether the self-weight of the members can be neglected.

Section	Area (in^2)	Section modulus (in^3)	Normal stress in AB (psi)	Normal stress in BC (psi)	Normal stress in CD (psi)
W6X9	2.68E+00	5.56E+00	8373	16193	466
W8X10	2.96E+00	7.81E+00	6015	11529	422
W10X12	3.54E+00	1.09E+01	4340	8261	353
W6X12	3.55E+00	7.31E+00	6367	12316	352
W4X13	3.83E+00	5.46E+00	8438	16487	326
W8X13	3.84E+00	9.91E+00	4736	9086	326
W10X15	4.41E+00	1.38E+01	3431	6525	283
W6X15	4.43E+00	9.72E+00	4799	9263	282
W8X15	4.44E+00	1.18E+01	3982	7630	282
W5X16	4.68E+00	8.51E+00	5448	10579	267
W6X16	4.74E+00	1.02E+01	4570	8827	264
W10X17	4.99E+00	1.62E+01	2928	5559	251

continues

Section	Area (in^2)	Section modulus (in^3)	Normal stress in AB (psi)	Normal stress in BC (psi)	Normal stress in CD (psi)
W8X18	5.26E+00	1.52E+01	3103	5924	238
W5X19	5.54E+00	1.02E+01	4547	8826	226
W10X19	5.62E+00	1.88E+01	2527	4790	222
W6X20	5.87E+00	1.34E+01	3486	6719	213
W8X21	6.16E+00	1.82E+01	2594	4947	203
W10X22	6.49E+00	2.32E+01	2055	3882	193
W8X24	7.08E+00	2.09E+01	2259	4308	177
W6X25	7.34E+00	1.67E+01	2797	5391	170
W10X26	7.61E+00	2.79E+01	1711	3228	164
W8X28	8.25E+00	2.43E+01	1943	3706	152
W10X30	8.84E+00	3.24E+01	1474	2779	141
W8X31	9.13E+00	2.75E+01	1719	3274	137
W10X33	9.71E+00	3.50E+01	1363	2573	129
W8X35	1.03E+01	3.12E+01	1515	2886	121
W10X39	1.15E+01	4.21E+01	1134	2139	109
W8X40	1.17E+01	3.55E+01	1332	2536	107
W10X45	1.33E+01	4.91E+01	973	1834	94
W8X48	1.41E+01	4.33E+01	1092	2080	89
W10X49	1.44E+01	5.46E+01	876	1649	87
W10X54	1.58E+01	6.00E+01	797	1501	79
W8X58	1.71E+01	5.20E+01	909	1732	73
W10X60	1.76E+01	6.67E+01	717	1350	71
W8X67	1.97E+01	6.04E+01	783	1491	63
W10X68	2.00E+01	7.57E+01	632	1190	63
W10X77	2.26E+01	8.59E+01	557	1048	55
W10X88	2.59E+01	9.85E+01	486	914	48
W10X100	2.94E+01	1.12E+02	427	804	43
W10X112	3.29E+01	1.26E+02	380	715	38

SUMMARY

In this rather long chapter we looked at a number of issues dealing with structural design. While several tools are used to design a structural system, the basic tool is structural analysis. An understanding of the relationship between the structural parameters and the response of the structure is crucial in designing efficient structural components and systems. A number of problems are presented at the end of this chapter to investigate areas and topics in order better to understand the design of a structure.

Lastly, it should be carefully noted that the design codes are recommendations to a very large extent. The designer must still use "common sense" in interpreting code provisions, recognizing when they are *not* applicable as much as when they are applicable to the structural system being designed. In this regard there is some justification in criticism of the manner in which the codes, and the textbooks dealing with the code, present the code provisions. Too often equations are presented in an if-then-else manner without due attention to the concepts and assumptions behind them. Structural engineering has always been a healthy mix of (mechanics) fundamentals and engineering judgment. There is simply no substitute for sound engineering principles.

SUMMARY EXERCISES

Appetizers

3.1. Compute the largest load P that can be applied to the beam in Fig. P3.1 without violating the following requirements: the normal stress should be less than 15 ksi and the shear stress should be less than 5 ksi. The beam cross-section is rectangular with a height of 12 in and a width of 8 in.

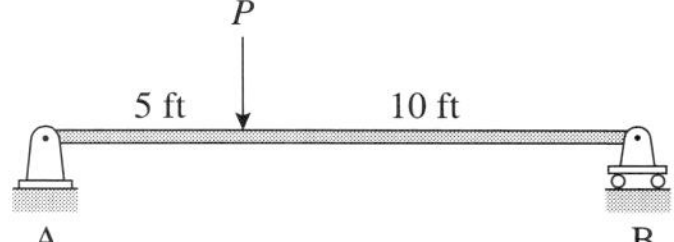

Fig. P3.1

3.2. Check the beam in Example 7.4.1 for the following requirements: (a) the maximum shear stress due to shear force is less than 1 MPa, and (b) the maximum normal stress due to bending moments is less than 2 MPa.

3.3. Redo Problem 3.1 but assume that the cross-section is $W16 \times 31$.

Main Course

3.4. Design the lightest steel frame using AISC W-sections for the frame shown in Problem 2.8.29. Use the same cross-section for all the three members. The normal stress should be less than 20 ksi and the shear stress less than 10 ksi. Start with the list of sections considered in Example 3.7.5 and use heavier sections if required.

3.5. Design the lightest planar frame by finding the cross-sections for members ABC and CD in Fig. P3.5. The cross-sections are AISC W-sections. The normal stress should be less than 20 ksi and the shear stress less than 10 ksi. Prevent Euler buckling in member ABC.

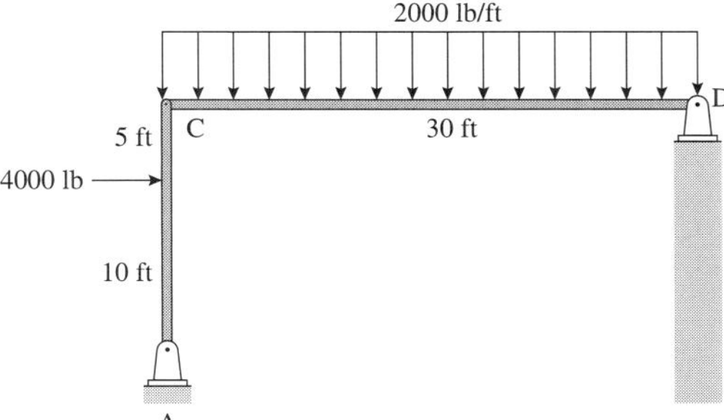

Fig. P3.5

3.6. It is required to design a support bracket as shown in Fig. P3.6. Member ADC is W16 × 31. Member BD has a circular hollow cross-section. Supports A and B are pin supports and connection at D is a pin connection. Design the lightest steel member BD so that the normal stress in the member is less than 10 ksi and Euler buckling is prevented with a safety factor of 2. The wall thickness of the pipe cannot be less than 15% of the inner radius.

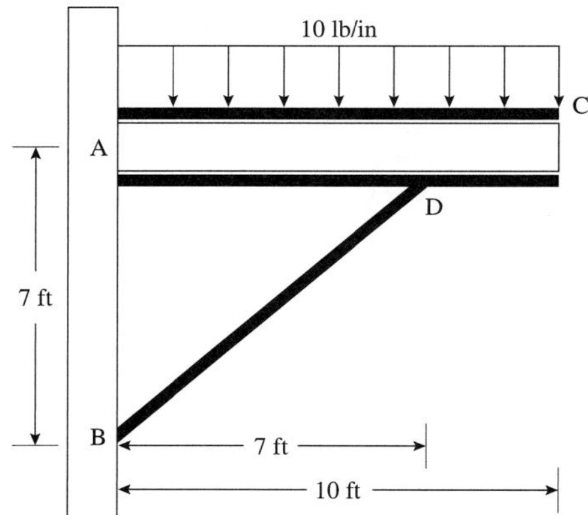

Fig. P3.6

Structural Concepts

3.7. A project usually starts with the client (business venture, city, state etc.) describing its needs in a project document called a Request for Proposal (RFP). You are a design engineer working for a city that wishes to build a pedestrian bridge. Write the specifications for the project in the RFP so that potential firms can bid on the project.

Chapter 4

Computation of Deflections

Construction of the Sun Devil Stadium (circa 1958), host to several events including the super bowl.

"It is customary to think of engineering as a part of trilogy, pure science, applied science and engineering. It needs emphasis that this trilogy is only one of a triad of trilogies into which engineering fits. . . . Many engineering problems are as closely allied to social problems as they are to pure science." Hardy Cross

Computation of deflections in structural systems serves two purposes. First, one of the serviceability requirements (see Section 3.6.7) from a design viewpoint deals with limiting deflections. Second, in earlier courses in statics and deformable solids (or strength of materials), most of the structural systems were statically determinate. The analyses of these structures were carried out using the concept of static equilibrium and free-body diagrams. In Chapter 5 we will look at two quite different techniques to solve for the response of indeterminate systems. Both these solution methodologies require the use of structural deflections in different ways.

One could categorize the classical methods for computing deflections of structural systems as either geometry based[1] or energy-based. In this chapter we look at two geometric methods to compute the deflection of beams. Later in Section 4.5 we will study a very powerful method, the virtual work method, that can used be used to compute the deflection of the usual truss, beam, and frame systems.

OBJECTIVES

- To investigate several different techniques for computing deflections of truss, beam, and frame systems.
- To understand the concepts associated with strain energy and work done, and compute these quantities in truss, beam, and frame systems.

ASSUMPTIONS

- Small deformation theory will apply.
- Axial and shear deformations of frame members will be assumed small in comparison to the bending or flexural deformations.

MATHEMATICAL BACKGROUND

The reader is urged to review the mathematical background necessary for this chapter, as presented in Appendix E.

4.1 BEAM DEFLECTION DIFFERENTIAL EQUATION

Consider a simply supported beam as in Fig. 4.1.1, loaded as shown. The deflection (or vertical displacement) v in the y direction varies along the beam. The deflected shape of the beam is called the elastic curve. In fact, in addition to the displacement there is a rotation of the beam. The rotation θ at any point is the angle between the x axis and the tangent to the elastic curve.

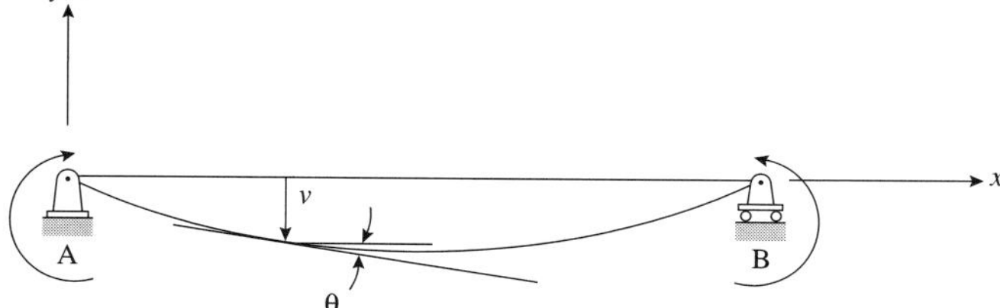

Fig. 4.1.1 Deformation in a simply-supported beam

[1] The solution of the differential equation(s) describing the deflection of a beam falls in this category.

We extend the discussions involving beams subjected to bending moments from Section 3.2. Figure 4.1.2(a) shows the deflected axis of the beam. The radius of the axis is ρ and the center of curvature is O. Consider a typical block PQRS as shown in Fig. 4.1.2(b) (DE represents the neutral axis). The angle subtended at O by two beam sections PR and QS is $\Delta\theta$. From Fig. 4.1.2(a), we see that $\Delta s = \rho\Delta\theta$. The curvature κ can now be defined as

$$\kappa = \frac{1}{\rho} = \underset{\Delta s \to 0}{Lt} \frac{\Delta\theta}{\Delta s} = \frac{d\theta}{ds} \tag{4.1.1}$$

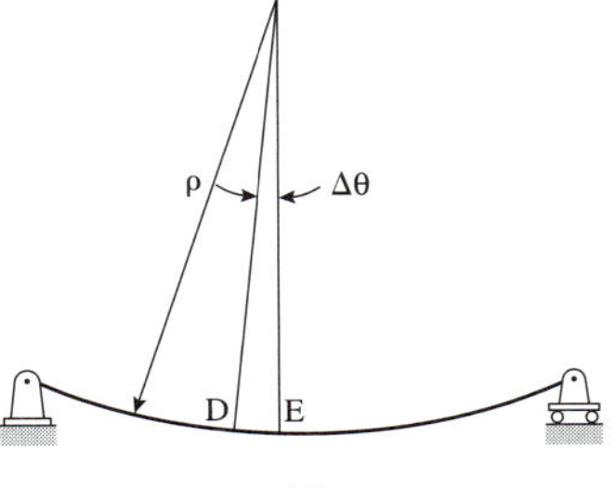

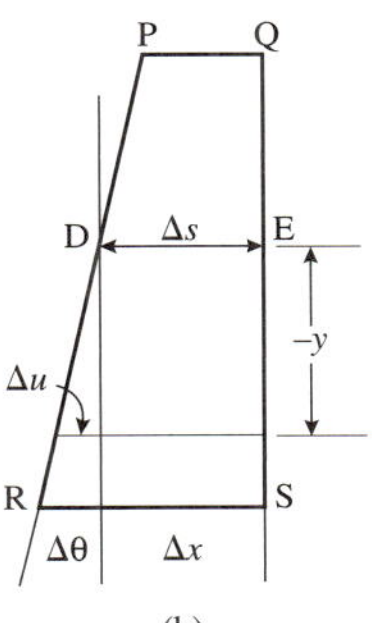

Fig. 4.1.2 Deformations in the beam. (a) The elastic curve. (b) Deformations in a typical block.

From the geometry of the deformation, $\Delta u = -y\Delta\theta$. The negative sign is due to the fact that an elongation occurs for a negative y. Dividing both sides of the equation by Δs we have

$$\underset{\Delta s \to 0}{Lt} \frac{\Delta u}{\Delta s} = -y \underset{\Delta s \to 0}{Lt} \frac{\Delta\theta}{\Delta s} \Rightarrow \frac{du}{ds} = -y\frac{d\theta}{ds} \tag{4.1.2a}$$

Also, since du/ds is the axial strain in the fiber at distance y from the neutral axis, we find

$$\frac{du}{ds} = \varepsilon \tag{4.1.2b}$$

Using Eqs. (4.1.1) and (4.1.2b) in (4.1.2a), we have

$$\frac{1}{\rho} = \kappa = -\frac{\varepsilon}{y} \tag{4.1.3}$$

But since $\varepsilon = \sigma/E$ and $\sigma = -My/I$, substituting above we have

$$\frac{1}{\rho} = \frac{M}{EI} \tag{4.1.4a}$$

or

$$d\theta = \frac{M}{EI}dx \tag{4.1.4b}$$

Analytic geometry gives us another definition of curvature as

$$\frac{1}{\rho} = \frac{d^2v/dx^2}{\sqrt{(1+dv/dx)^3}} \tag{4.1.5}$$

Using Eq. (4.1.4a) gives us

$$\frac{M}{EI} = \frac{d^2v/dx^2}{\sqrt{(1+dv/dx)^3}} \tag{4.1.6}$$

For small deflections, $dv/dx << 1$. Hence the denominator on the right-hand side is approximately equal to 1, leading to

$$\frac{d^2v}{dx^2} = \frac{M}{EI} \tag{4.1.7}$$

This is the governing (ordinary) differential equation for the transverse deflection of a beam. The differential equation can be solved if the boundary conditions are known.

Tips: Here are some helpful hints (and a brief view of material from Chapter 7) for planar beams and frames.

End roller support. The displacement normal to the support surface is zero. The displacement along the support and the rotation are nonzero. The bending moment at the support is zero.

End pin support. The two orthogonal displacements are zero but the rotation is nonzero. The bending moment at the support is zero.

Interior roller support. The displacement normal to the support surface is zero.

Interior pin support. The two orthogonal displacements are zero.

Fixed support. The two orthogonal displacements and the rotation are zero.

4.2 MOMENT-AREA METHOD

The differential equation (4.1.7) can be solved using the integration technique where the constants of integration can be evaluated by using the boundary conditions. The method is easy to use when the moment expression is continuous over the entire length of the beam. However, it is very easy to create a problem when a single $M(x)$ is not sufficient to describe the moment diagram. For example, a concentrated force causes such a discontinuity. We looked at several other sources and examples in Chapter 2. To tackle these problems, additional constraints in the form of continuity equations are applied at the points of moment discontinuities. The process can become cumbersome. In this section and the next, we look at two powerful techniques applicable for problems with general loading.

Consider a simply supported beam of length L as shown in Fig. 4.2.1. The elastic curve for the beam is shown in Fig. 4.2.2.

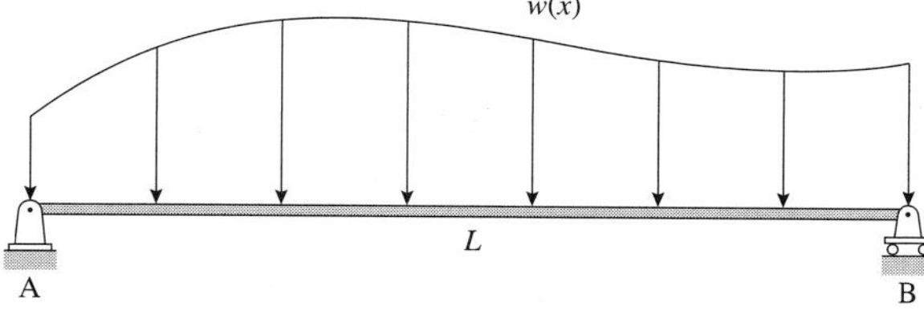

Fig. 4.2.1
Simply-supported beam.

There are four straight lines that represent the tangents to the elastic curve—tan A and tan B are the tangents to the elastic curves at A and B respectively, and the two tangents at the ends of the element dx. We will measure angles in the counterclockwise sense. The M/EI diagram for the beam is shown in Fig. 4.2.3.

Assuming that the displacements and rotations are small, the change in angle between the ends of the element dx is $d\theta$. Restating Eq. (4.1.4b) gives us

$$d\theta = \frac{M}{EI}dx \tag{4.2.1}$$

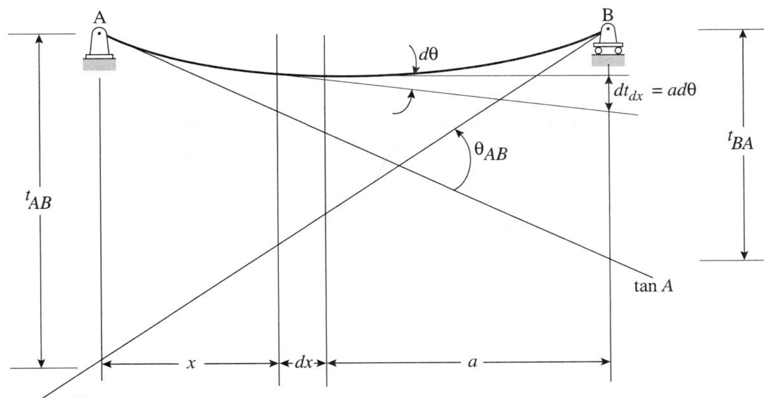

Fig. 4.2.2
Elastic curve.

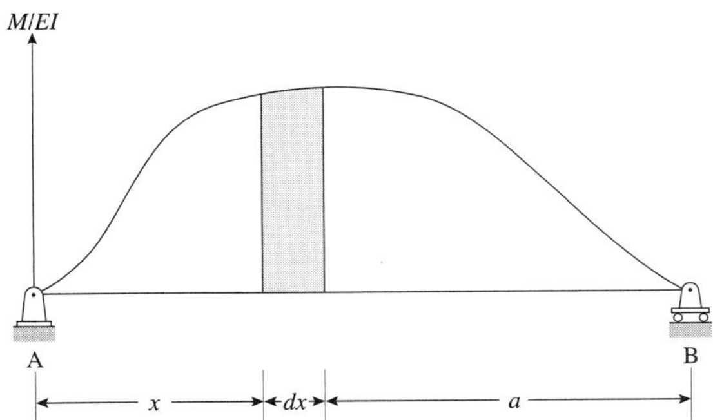

Fig. 4.2.3
M/EI diagram.

The quantity on the right-hand side is the shaded area under the *M/EI* diagram in Fig. 4.2.3. We can represent the change in slope between two points A and B as

$$\theta_{AB} = \int_A^B d\theta \tag{4.2.2}$$

Substituting Eq. (4.2.1) in the above equation, we have

$$\theta_{AB} = \int_A^B \frac{M}{EI} dx \tag{4.2.3}$$

This leads us to the first theorem.

Theorem 1. The change in angle between two arbitrary points A and B on the elastic curve is equal to the area under the *M/EI* diagram between those two points.

Note that the theorem does not give the value of the slope at a point. It gives the slope at a point relative to another. For example, in Fig. 4.2.2, θ_{AB} represents the change in angle going from tangent at A to the tangent at B. If the elastic curve is correct, θ_{AB} should be positive since the slope increases going from A to B—the slope at A is negative and the slope at B is positive.

The intersection of the tangents at the ends of the element *dx* with the vertical line through B creates an intercept dt_{dx} such that

$$dt_{dx} = ad\theta \tag{4.2.4}$$

Substituting Eq. (4.2.1) in the above equation, we have

$$dt_{dx} = a\frac{M}{EI}dx \tag{4.2.5}$$

The right-hand side represents the moment of the shaded area $(M/EI)dx$ about B since a is the lever arm. Going back to Fig. 4.2.2, extending the tangents to intersect the vertical lines through different points gives us the deviations of the tangents from these points. For example, t_{AB} represents the tangential deviation of point A on the elastic curve with respect to tangent through B. Similarly, t_{BA} represents the tangential deviation of point B on the elastic curve with respect to tangent through A. It should be noted that in general, $t_{ij} \neq t_{ji}$. Equation (4.2.5) can used to find the tangential deviation between two arbitrary points A and B as

$$t_{BA} = \int_A^B dt_{dx} = \int_A^B a\frac{M}{EI}\,dx = \int_A^B (L-x)\frac{M}{EI}\,dx \tag{4.2.6}$$

In a similar fashion it can be shown that

$$t_{AB} = \int_A^B x\frac{M}{EI}\,dx \tag{4.2.7}$$

We can now state the second moment-area theorem.

Theorem 2. The deflection of point B on the elastic curve with respect to the tangent through point A on the elastic curve is equal to the moment of the area under the M/EI diagram between those two points taken about B.

Once again, the theorem does not give the value of the displacement at a point. We obtain the relative displacement of a point on the elastic curve with respect to the tangent through another point on the elastic curve. As a matter of interpretation, t_{ij} is positive if i on the elastic curve is *above* the tangent through j. From Fig. 4.2.2, if the elastic curve is correct, t_{AB} should be positive since A on the elastic curve is above tangent B. The same comments apply for t_{BA}.

A few points about the moment-area method. First, as with most methods in this chapter, the flexural (or, bending) action is the only action considered. The axial and shear deformations are not taken into account since they are usually assumed to be small. Second, the method assumes that there are no discontinuities in the structure such as internal hinges. Internal hinges cause discontinuities in the elastic curve that are not reflected in the relationship between the curvature of the beam and the M/EI values, as we saw in the previous section. Third, the application of Eqs. (4.2.3) and (4.2.7) requires care. The integrals on the right-hand side can be either negative or positive. This fact must be correlated with the geometry of the elastic curve.

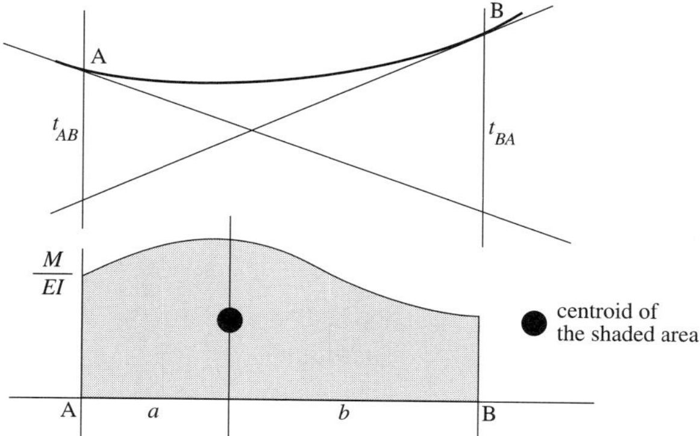

Fig. 4.2.4 Valid elastic curve and M/EI diagram.

For example, consider the elastic curve and the corresponding *M/EI* diagram between two points A and B in Fig. 4.2.4. From the geometry of the elastic curve, both t_{AB} and t_{BA} are positive since the corresponding points on the elastic curve are above the tangents. From the second theorem, we find

$$t_{AB} = \left(\text{Area}_{M/EI}\right)_{AB}(a) \tag{4.2.8}$$

Distance a is always positive. Hence the area of the *M/EI* diagram between A and B must be positive for the right-hand side to be a positive quantity. As is evident from Fig. 4.2.4, the *M/EI* diagram is positive between A and B.

Let us consider another case shown in Fig. 4.2.5. From the elastic curve it is evident that t_{BA} is negative. From the second theorem, we get

$$t_{BA} = \left(\text{Area}_{M/EI}\right)_{AB}(b) \tag{4.2.9}$$

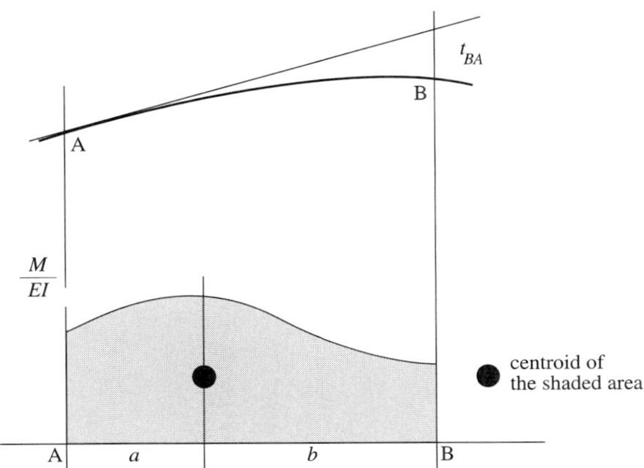

Fig. 4.2.5
Inconsistent elastic curve and M/EI diagram.

Distance b is positive. Hence the area of the *M/EI* diagram between A and B must be negative for the right-hand side to be a negative quantity. As can be seen in Fig. 4.2.5, the *M/EI* diagram is positive between A and B. Hence, either the elastic curve is incorrect or the *M/EI* diagram is incorrect.

General Procedure

Step 1: Construct the moment diagram for the structure. Now construct the *M/EI* diagram by dividing the moment value *M* by the product of the modulus of elasticity *E* and the moment of inertia *I*.

Step 2: Draw an exaggerated elastic curve. Note that convex elastic curve as shown in Fig. 4.2.4 requires that the top fiber be in compression. Similarly, a concave elastic curve as shown in Fig. 4.2.5 requires that the bottom fiber be in compression. There is a change of curvature at points of zero bending moment when the *M/EI* diagram changes sign.

Step 3: Draw the appropriate tangents to the elastic curve and label the diagram with the angular changes θ_{ij} and the tangential deviations t_{ij}. Use the first theorem to compute the change in angle between two points. If the slope at one of these points is known, the slope at the other point can be computed. Similarly, use the second theorem to compute the relative displacement between two points. If the displacement at one of these points is known, the displacement at the other point can be computed. As the examples illustrate later, *the deflection computations should be carried out using the geometry of the elastic curve*. If the results from the assumed elastic curve do not match the values obtained from the moment-area theorems, the elastic curve must be revised.

Step 4: Finally, the sign or sense of the displacement and/or rotation can be assigned once it is verified that the elastic curve is correct.

EXAMPLE 4.2.1

Deflection of a Cantilever Beam

Compute the vertical displacement and rotation at B of the beam shown in Fig. E4.2.1(a). *EI* is a constant.

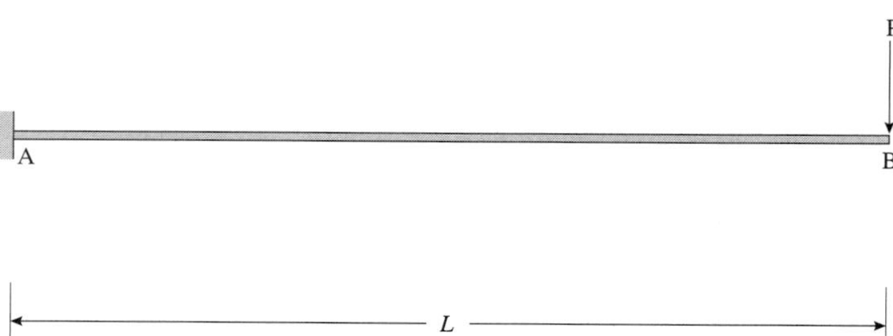

Fig. E4.2.1(a)

SOLUTION

Step 1: We first construct the *M/EI* diagram. This will help us sketch the elastic curve. Using the FBD in Fig. E4.2.1(b), we find

$$M(x) = P(x - L)$$

The *M/EI* diagram is shown in Fig. E4.2.1(c).

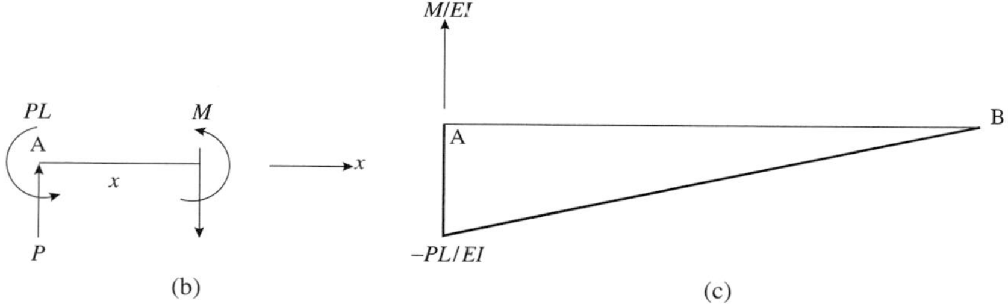

Fig. E4.2.1(b) and (c)

Step 2: Sketch the elastic curve. Using the moment diagram as a guide, we can sketch the elastic curve of the beam. Since the entire bottom of the beam is in compression the beam must deflect concave down. At fixed support A both the displacement and rotation are zero.

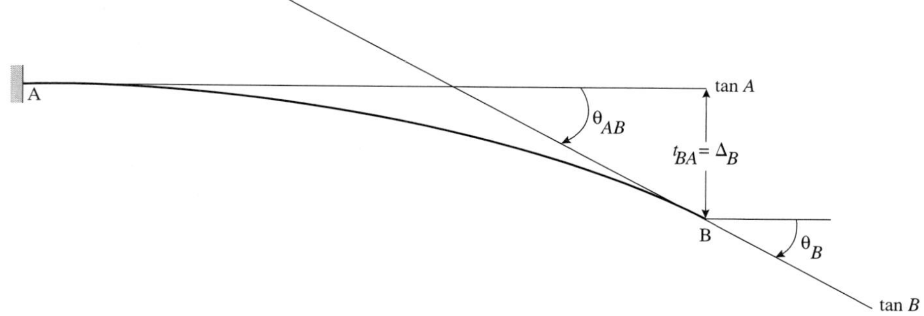

Fig. E4.2.1(d)

A few things should be noted about the elastic curve. We have drawn two tangents to the elastic curve, tan *A*, which is tangent to the elastic curve at A (hence horizontal) and tan *B*, that is tangent to the elastic curve at B. The slopes or rotations are measured with reference to the undeformed state. If the elastic curve is correct, the slope at B is and should be negative. θ_{AB} is the change in slope going from point A to point B. t_{BA} is the tangential deviation of point B on the elastic curve with respect to tan *A*. The relationships that we intend to use are.

$$\theta_B = \theta_{AB} \tag{1}$$

$$\Delta_B = t_{BA} \tag{2}$$

Step 3: Use the moment-area theorems. From Theorem 1 and noting that $\theta_A = 0$, we have

$$\theta_{AB} = \theta_B - \theta_A = \theta_B = \int_A^B \frac{M}{EI}\,dx = \int_0^L \frac{P(x-L)}{EI}\,dx = -\frac{PL^2}{2EI} \tag{3}$$

The negative sign indicates that the slope at B is less than the slope at A. This is corroborated by the assumed elastic curve. Hence,

$$\theta_B = \frac{PL^2}{2EI}(\curvearrowright)$$

While we strongly encourage the use of integration as in (3), note that from Theorem 1, θ_{AB} is the area under the *M*/*EI* diagram between A and B. In this problem, this is the area of the triangle between A and B: $\frac{1}{2}(L)\left(-\frac{PL}{EI}\right) = -\frac{PL^2}{2EI}$. The negative sign indicates that the net area is below the *x*-axis and that the slope at B should be less than the slope at A.

From Theorem 2, we find

$$t_{BA} = \int_A^B \frac{M(x)}{EI}(L-x)\,dx = \int_0^L \frac{P(x-L)}{EI}(L-x)\,dx = -\frac{PL^3}{3EI} \tag{4}$$

The negative sign indicates that B on the elastic curve should be below tan *A*. Once again, this is corroborated by the assumed elastic curve. Hence,

$$\Delta_B = \frac{PL^3}{3EI}(\downarrow)$$

The nonintegral approach to computing (2) is to take the moment of the area of the *M*/*EI* diagram between A and B about B:

$$t_{BA} = \left[\frac{1}{2}(L)\left(-\frac{PL}{EI}\right)\right]\left[\frac{2L}{3}\right] = -\frac{PL^3}{3EI}$$

where the first term represents the area of the *M*/*EI* diagram between A and B (a triangle) and the second term is the distance from the centroid of the triangle to B.

EXAMPLE 4.2.2 ***Deflection of a Simply Supported Beam***

Compute the rotations at A and C of the beam in Fig. E4.2.2(a). Take E = 30000 ksi and I = 700 in^4.

SOLUTION

Step 1: The problem units are k, ft. Without using the numerical *EI* value, the *M*/*EI* diagram consists of two right triangles. The centroids of the triangles are marked with the crosshairs.

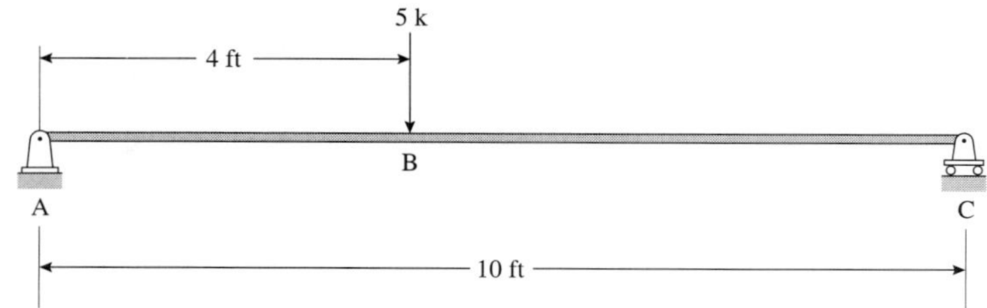

Fig. E4.2.2(a)

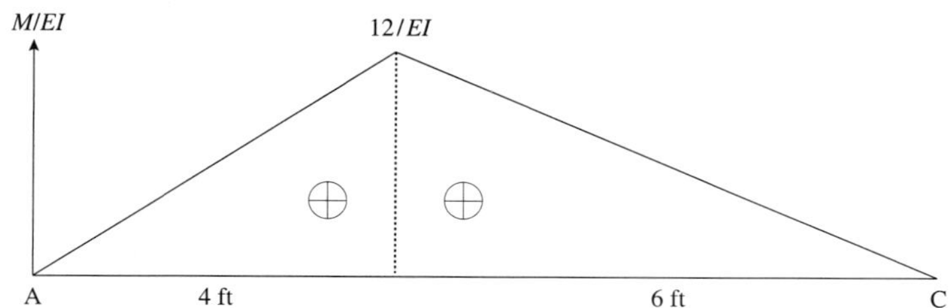

Fig. E4.2.2(b)

Step 2: Elastic curve. Noting that the displacements at A and C in Fig. E4.2.2(c) are zero and that the entire top fiber of the beam is in compression, we can sketch the elastic curve. From the elastic curve and triangle ACC′, we see that

$$\theta_A \approx \tan\theta_A = t_{CA}/L \tag{1}$$

Similarly, from triangle CAA′ we see that

$$\theta_C = t_{AC}/L \tag{2}$$

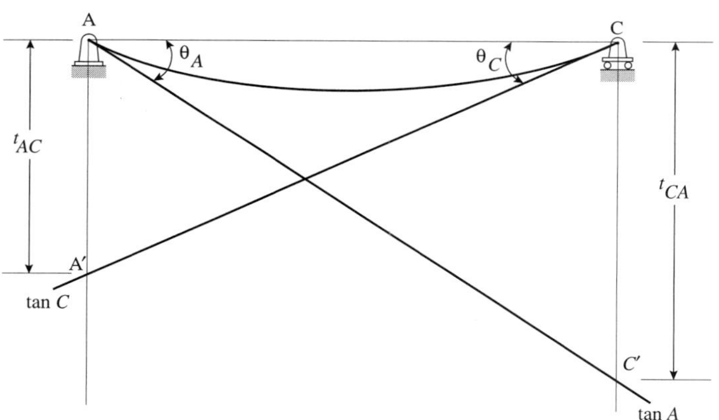

Fig. E4.2.2(c)

Step 3: Use the moment-area theorems. From Theorem 2, we have

$$t_{CA} = \left[\frac{1}{2}(4)\left(\frac{12}{EI}\right)\right]\left[\frac{4}{3}+6\right]+\left[\frac{1}{2}(6)\left(\frac{12}{EI}\right)\right]\left[\frac{2}{3}(6)\right]=\frac{320}{EI}$$

where the first term is the moment of the left triangle about C and the second term is the moment of the right triangle about C. The positive sign demands that C on the elastic curve be above tan *A*. The assumed elastic curve meets that requirement.

Similarly, we have

$$t_{AC} = \left[\frac{1}{2}(4)\left(\frac{12}{EI}\right)\right]\left[\frac{2}{3}(4)\right]+\left[\frac{1}{2}(6)\left(\frac{12}{EI}\right)\right]\left[4+\frac{1}{3}(6)\right]=\frac{280}{EI}$$

Once again, the positive sign demands that A on the elastic curve be above tan C, and the assumed elastic curve meets that requirement.

We have

$$EI = 21(10^6)\ \text{k-in}^2\left(\frac{1\ \text{ft}^2}{144\ \text{in}^2}\right) = 145833\ \text{k-ft}^2$$

Hence, we find

$$\theta_A = \frac{t_{CA}}{10} = \frac{16}{EI} = 0.22(10^{-3})\ \text{rad}\ (\circlearrowleft)$$

$$\theta_C = \frac{t_{AC}}{10} = \frac{14}{EI} = 0.19(10^{-3})\ \text{rad}\ (\curvearrowleft)$$

As an exercise, solve this problem using the integration approach.

EXAMPLE 4.2.3 ***Deflection of a Simply Supported Beam***

For the beam in Example 4.2.2, compute the rotation and vertical displacement at B.

SOLUTION

Step 1: The M/EI diagram from the previous problem is repeated in Fig. E4.2.3(a).

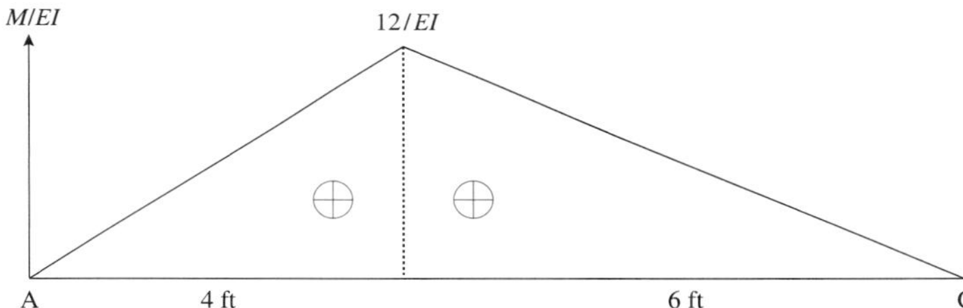

Fig. E4.2.3(a)

Step 2: Elastic curve. Figure E4.3.2(b) shows a modified diagram with the same elastic curve so that we may compute the displacement and rotation at B. The distance DB is the required displacement at B, Δ_B, and θ_B is shown as the slope at B. From Fig. E4.2.3(b), we have

$$\theta_A = \theta_{AB} + \theta_B \tag{1}$$

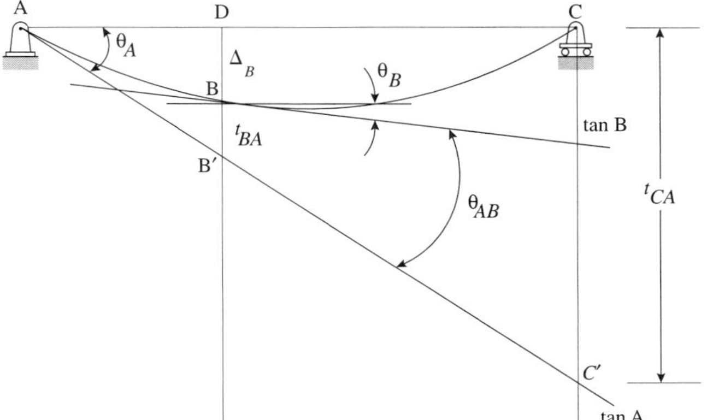

Fig. E4.2.3(b)

where all the quantities are their respective magnitudes (signs excluded). From similar triangles, we find

$$\theta_A = \frac{\Delta_B + t_{BA}}{4} = \frac{t_{CA}}{10} \tag{2}$$

where once again all the quantities are their respective magnitudes (signs excluded).

Step 3: Use the moment-area theorems. From Theorem 1, we find

$$\theta_{BA} = \frac{1}{2}(4)\left(\frac{12}{EI}\right) = \frac{24}{EI}$$

The positive sign indicates that the slope at B is greater than the slope at A. From Theorem 2 we find

$$t_{BA} = \left[\frac{1}{2}(4)\left(\frac{12}{EI}\right)\right]\left[\frac{4}{3}\right] = \frac{32}{EI}$$

The positive sign indicates that B on the elastic curve is above tan A. Using these results in (1) yields

$$\theta_B = \frac{32}{EI} - \frac{24}{EI} = \frac{8}{EI}(\circlearrowleft) = 5.49(10^{-5})\text{ rad}$$

$$\Delta_B = 4\theta_A - t_{BA} = 4\left(\frac{32}{EI}\right) - \frac{32}{EI} = \frac{96}{EI}(\downarrow) = 6.58(10^{-4})\text{ ft} = 0.008\text{ in}$$

EXAMPLE 4.2.4

Deflection of a Beam

Compute the vertical displacement at C and rotation at B of the beam in Fig. E4.2.4(a). EI is a constant.

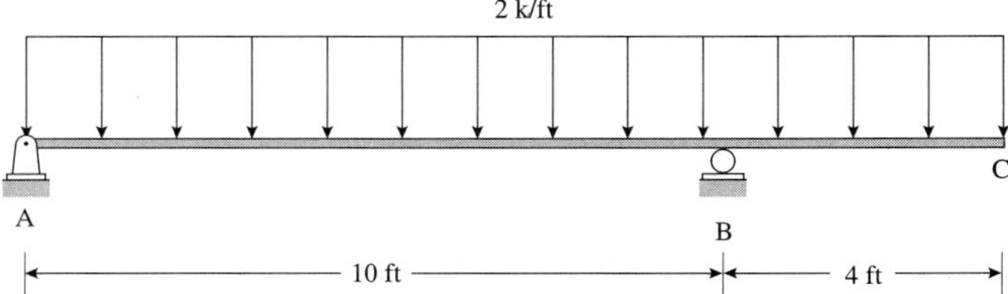

Fig. E4.2.4(a)

SOLUTION

Step 1: The problem units are k, ft. The M/EI diagram is shown in Fig. E4.2.4(b). For segment AB, $0 < x < 10$ ft, $M(x) = -x^2 + 8.4x$. For segment CB, $0 < x_1 < 4$ ft, $M(x_1) = -x_1^2$.

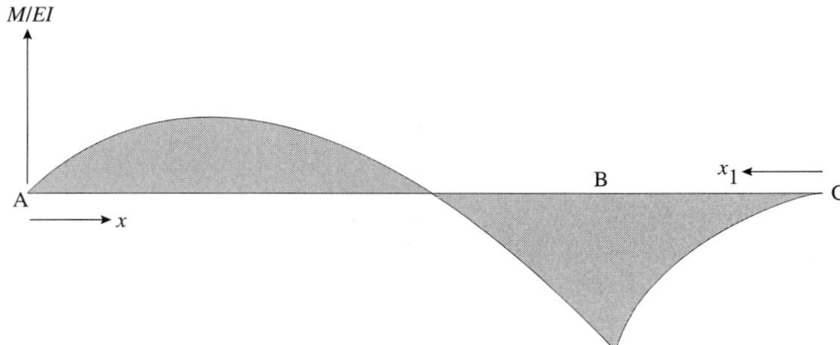

Fig. E4.2.4(b)

Step 2: Elastic curve. The elastic curve in Fig. E4.2.4(c) is constructed noting that (a) the displacements at A and B are zero, (b) the top of the beam is in compression from A to the zero moment point and the bottom of the beam is in compression from that point to C.

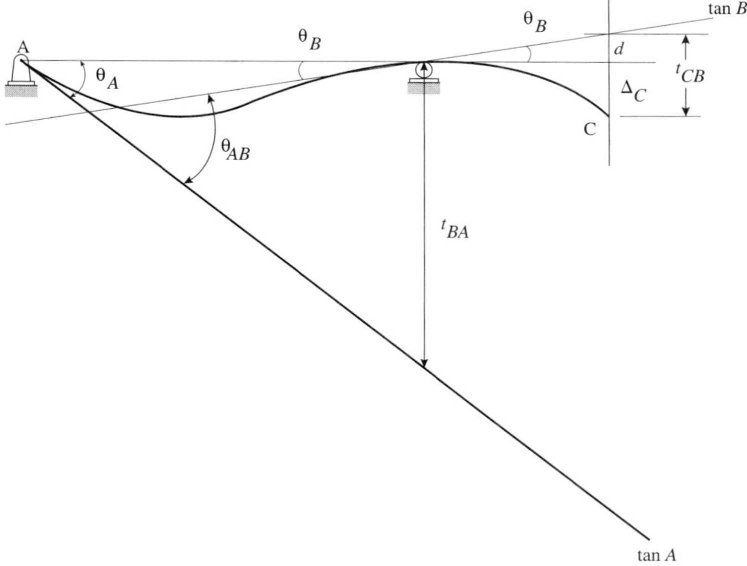

Fig. E4.2.4(c)

From the elastic curve, we have

$$\theta_A = t_{BA}/10 \tag{1}$$

$$\theta_{AB} = \theta_A + \theta_B \tag{2}$$

From (1) and (2) we can solve for θ_B. Also we find

$$t_{CB} = \Delta_C + d \tag{3}$$

$$d = \theta_B(5) \tag{4}$$

From (3) and (4) we can solve for Δ_C. In the above equations all the quantities are their respective magnitudes (signs excluded).

Step 3: Use the moment-area theorems. Theorem 1 implies

$$\theta_{AB} = \frac{1}{EI}\int_0^{10}\left(-x^2 + 8.4x\right)dx = \frac{86.67}{EI}$$

The positive sign indicates that the slope at B is greater than the slope at A. This is borne out by the assumed elastic curve. From Theorem 2, we find

$$t_{BA} = \frac{1}{EI}\int_0^{10}(-x^2 + 8.4x)(10 - x)\,dx = \frac{566.67}{EI}$$

The positive sign indicates that B on the elastic curve is above tan A. This is borne out by the assumed elastic curve. Using (1) and (2) yields

$$\theta_A = \frac{t_{BA}}{10} = \frac{56.67}{EI}$$

$$\theta_B = \theta_{AB} - \theta_A = \frac{86.67}{EI} - \frac{56.67}{EI} = \frac{30}{EI}\ (\curvearrowleft)$$

Now to compute the vertical displacement at B, we first compute t_{CB}:

$$t_{CB} = \frac{1}{EI}\int_0^4(-x_1^2)(x_1)\,dx_1 = -\frac{64}{EI}$$

The negative sign indicates that C on the elastic curve is below tan *B* and the elastic curve supports that view. Now, from the right triangle, we have

$$d = 4\theta_B = 4\left(\frac{30}{EI}\right) = \frac{120}{EI}$$

From the elastic curve (where all the quantities are their respective magnitudes, signs excluded), we find

$$t_{CB} = d + \Delta_C$$

In other words, t_{CB} is greater than both d and Δ_C. However, our results show that $d > t_{CB}$. Hence the elastic curve is not correct! The corrected elastic curve is shown in Fig. E4.2.4(d). The corrections are in the segment BC.

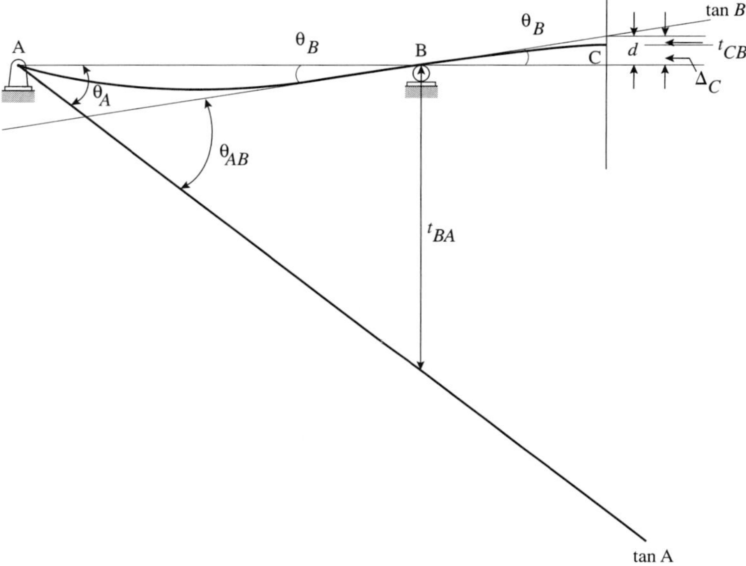

Fig. E4.2.4(d)

From the corrected elastic curve, we find

$$d = t_{CB} + \Delta_C$$

Using the computed values gives us

$$\Delta_C = d - t_{CB} = \frac{120}{EI} - \frac{64}{EI} = \frac{56}{EI}(\uparrow)$$

In other words, the net displacement at C is upwards!

EXERCISES

Appetizers

4.2.1. The beam has a constant *EI* value. Using the moment-area method,

(a) compute the vertical displacement and rotation at B, and

(b) compute the vertical displacement and rotation at C.

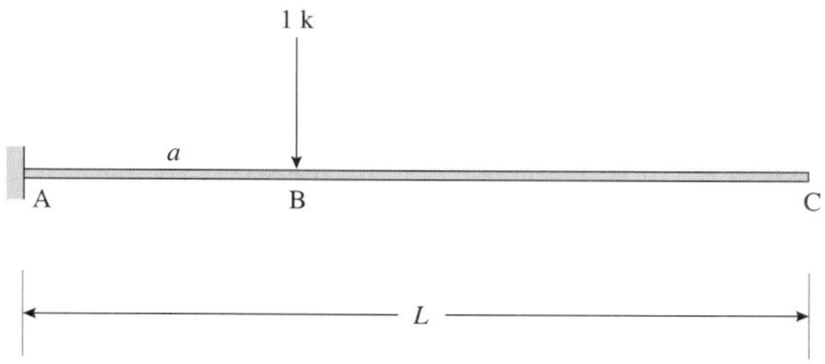

Fig. P4.2.1

(c) What is the rotation at C if the load at B is replaced with a 500 lb force?

4.2.2. The beam in Fig. P4.2.2 has a constant EI value. Using the moment-area method, compute the vertical displacement and rotation at B.

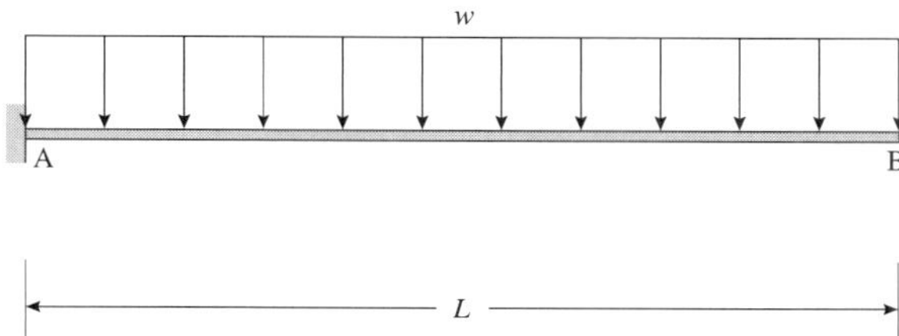

Fig. P4.2.2

4.2.3. The beam in Fig. P4.2.3 has a constant EI value. Compute (a) the rotation at A, and (b) the vertical displacement at the center of the beam, using the moment-area method.

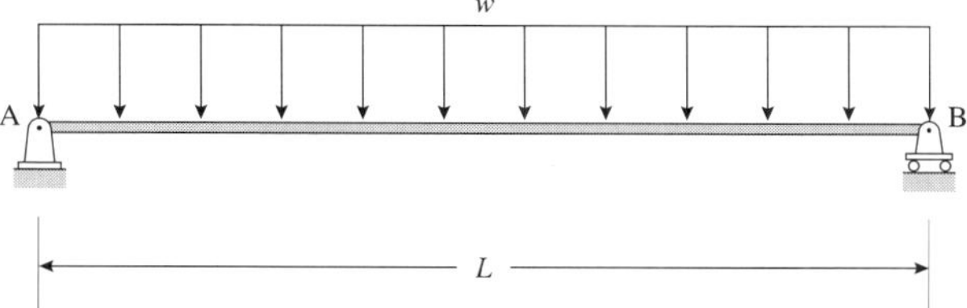

Fig. P4.2.3

4.2.4. The beam in Fig. P4.2.4 has a constant EI value. Take a = 10 m and b = 5 m.

(a) Compute the vertical displacement and rotation at C using the moment-area method.

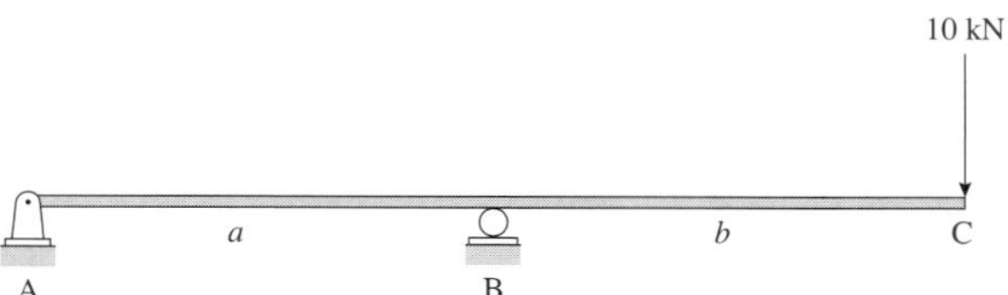

Fig. P4.2.4

(b) Assume that the beam is made of steel. What should be the moment of inertia so that the vertical displacement at C is less than 0.5 cm?

Main Course

4.2.5. The beam in Fig. P4.2.5 has a constant *EI* value. Using the moment-area method, (a) compute the vertical displacement and rotation at C, and (b) compute the vertical displacement at D.

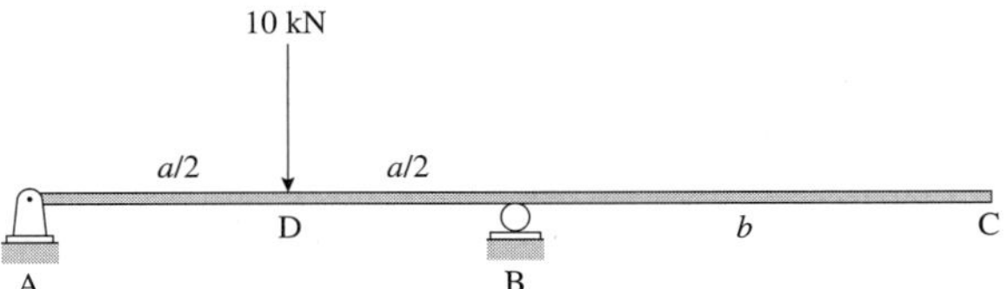

Fig. P4.2.5

4.2.6. The beam in Fig. P4.2.6 has a constant *EI* value. Using the moment-area method, (a) compute the vertical displacement at the center of the beam, and (b) compute the rotation at A.

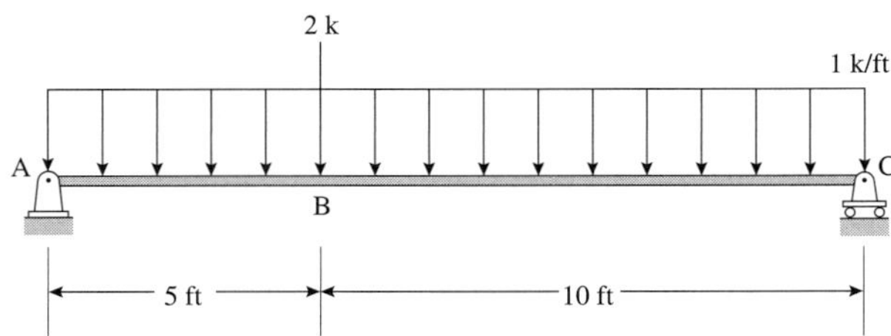

Fig. P4.2.6

Structural Concepts

4.2.7. Sketch a qualitative elastic curve for the beam in Fig. P4.2.7. Justify your answer.

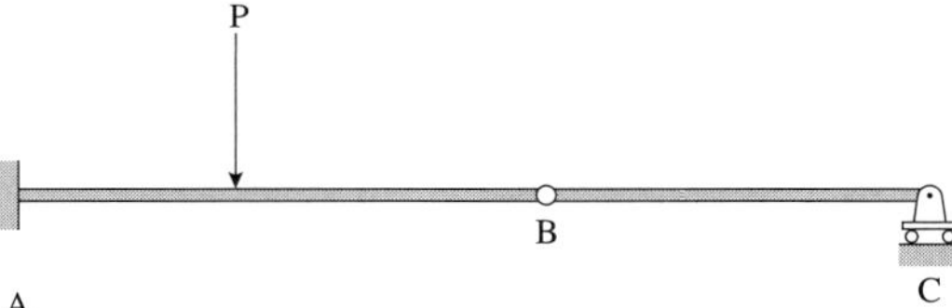

Fig. P4.2.7

4.3 CONJUGATE BEAM METHOD

The conjugate beam method is an extremely versatile method. From Chapter 2, the relationships between the loading, shear, and bending moments are given by

$$\frac{d^2M}{dx^2} = \frac{dV}{dx} = -w(x) \tag{4.3.1}$$

Similarly, from Section 4.1 we have the following:

$$\frac{d^2v}{dx^2} = \frac{d\theta}{dx} = \frac{M}{EI} \tag{4.3.2}$$

A comparison of the two set of equations indicates that if M/EI is the loading on an *imaginary* beam, the resulting shear and moment in the beam are the slope and displacement of the *real* beam (see Fig. 4.3.1).

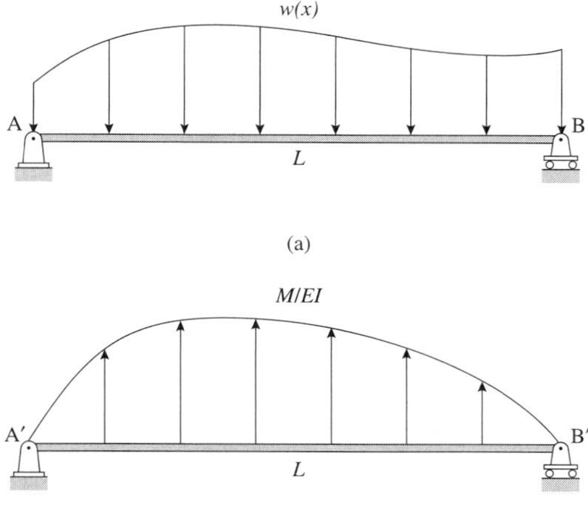

Fig. 4.3.1 (a) Real beam. (b) Corresponding conjugate beam.

We now summarize the results from the conjugate beam method.

Theorem 1. The slope θ at a point in the real beam is given by the shear V at the corresponding point in the conjugate beam.

Theorem 2. The displacement Δ of a point in the real beam is given by the bending moment M at the corresponding point in the conjugate beam.

We use these two theorems to help construct the conjugate beam by establishing the support and interior continuity conditions between the real and the conjugate beams, as in Table 4.3.1, where L and R denote the left and right sides of the support or hinge.

General Procedure

Step 1: For the real beam, draw the M/EI diagram.

Step 2: Construct the conjugate beam. The conjugate beam has the same length as the real beam. Its internal and external continuity and support conditions match the corresponding internal and external continuity and support conditions in the real beam as per the details in Table 4.3.1. The loading on the conjugate beam is the M/EI diagram constructed in Step 1. The direction of this loading is outward normal to the compressive fiber (see Fig. 4.3.1(b)).

Step 3: Analyze the conjugate beam. If necessary, draw the shear force and bending moment diagrams of the conjugate beam. Use Theorems 1 and 2 to obtain the slope and displacement of the real beam. We continue to use the same sign convention for the shear force and bending moment as in Chapter 2. A positive shear in the conjugate beam translates to a positive slope (counterclockwise) at the corresponding point in the real beam. Similarly, a positive bending moment in the conjugate beam translates to a positive displacement (upwards) at the corresponding point in the real beam.

Observation: The moment-area method is extremely powerful in quickly arriving at a qualitative elastic curve. On the other hand, the conjugate beam method is more amenable to problem solving. The steps are straightforward, involving no guesswork. In the next few sections we see more powerful techniques that can be used to solve for the deflections of a variety of structural systems, not merely beams.

Table 4.3.1 Relationship Between the Real Beam and Corresponding Conjugate Beam

Support or connection real beam	Support or connection conjugate beam
End roller support (θ = ?, $\Delta = 0$)	End roller support (V = ?, $M = 0$)
End pin support (θ = ?, $\Delta = 0$)	End pin support (V = ?, $M = 0$)
Fixed support ($\theta = 0$, $\Delta = 0$)	Free end ($V = 0$, $M = 0$)
Free end (θ = ?, Δ = ?)	Fixed support (V = ?, M = ?)
Interior roller/pin support (θ = ?, $\Delta = 0$)	Internal hinge (V = ?, $M = 0$)
Internal hinge (θ_L = ?, θ_R = ?, Δ = ?)	Interior roller support (V_L = ?, V_R = ?, M = ?)

EXAMPLE 4.3.1

Deflection of a Cantilever Beam (Example 4.2.1)

Compute the vertical displacement and rotation at B of the beam in Fig. E4.3.1(a). EI is a constant.

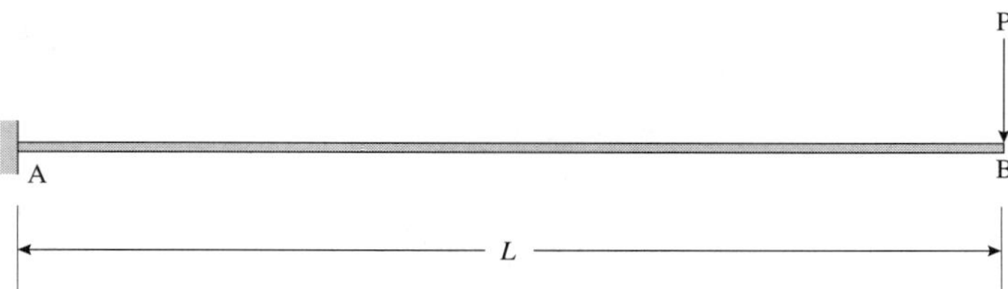

Fig. E4.3.1(a)

SOLUTION

Step 1: We first construct the M/EI diagram in Fig. E4.3.1(b) (see Example 4.2.1).

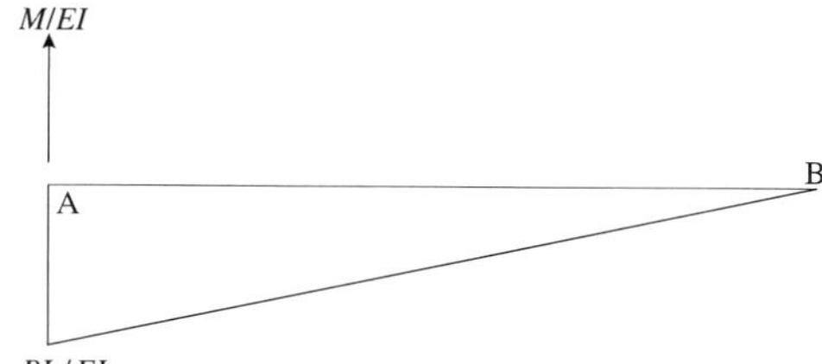

Fig. E4.3.1(b)

Step 2: Conjugate beam. The next step is to construct the conjugate beam. We use Table 4.3.1. The fixed end A maps into a free end A′. Similarly, the free end B maps into a fixed end B′. Since the bottom of the beam is under compression all the way from A to B, the load acting on the conjugate beam is downwards. The conjugate beam is shown in Fig. E4.3.1(c).

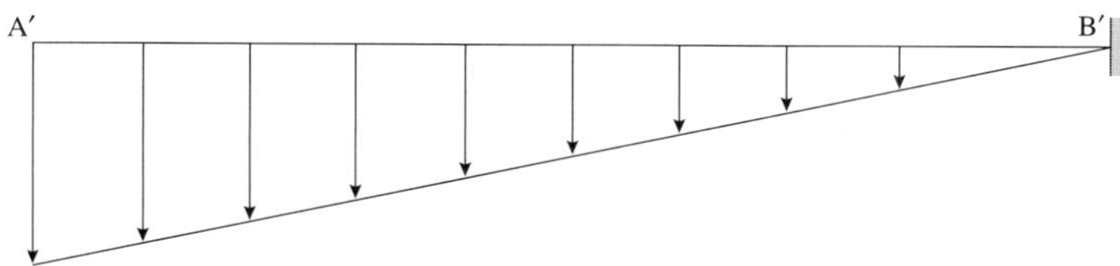

Fig. E4.3.1(c)

Step 3: Solution of the conjugate beam. Since it is required to compute the slope and displacement at B, we need to compute the shear force and bending moment in the conjugate beam at B′. The FBD is shown in Fig. E4.3.1(d). $PL^2/2EI$ represents the resultant of the triangular loading.

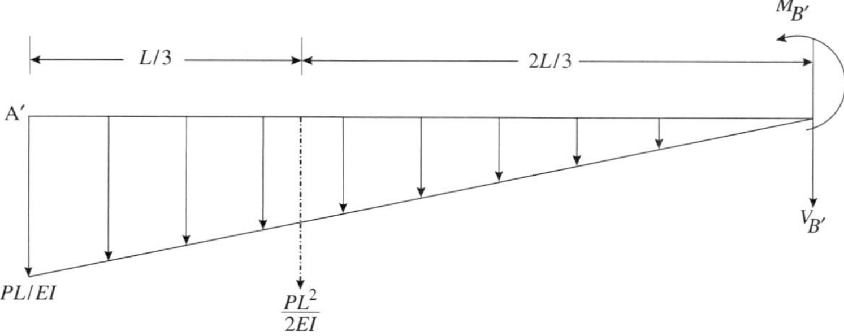

Fig. E4.3.1(d)
Conjugate beam.

Using the equilibrium equation, we have

$$\overset{\uparrow +}{\sum} F_y = -\frac{PL^2}{2EI} - V_{B'} \quad \text{or} \quad V_{B'} = -\frac{PL^2}{2EI} \Rightarrow \theta_B = \frac{PL^2}{2EI} \ (\circlearrowright)$$

Note that a negative shear in the conjugate beam translates to a clockwise slope in the real beam. Similarly, we find

$$\overset{\circlearrowleft +}{\sum} M_B = 0 = \frac{PL^2}{2EI}\left(\frac{2L}{3}\right) + M_{B'} \quad \text{or} \quad M_{B'} = -\frac{PL^2}{2EI} \Rightarrow \Delta_B = \frac{PL^2}{2EI}(\downarrow)$$

A negative moment in the conjugate beam translates into a downward displacement in the real beam. The answers obtained are the same as those obtained using the moment-area method.

EXAMPLE 4.3.2 ***Deflection of a Beam (Example 4.2.4)***

Compute the slope at B and the vertical displacement at C of the beam in Fig. E4.3.2(a). *EI* is a constant.

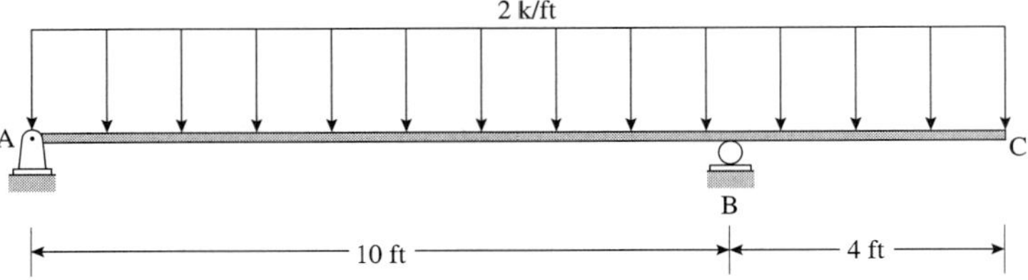

Fig. E4.3.2(a)

SOLUTION

Step 1: We first construct the *M*/*EI* diagram in Fig. E4.3.2(b) (see Example 4.2.4).

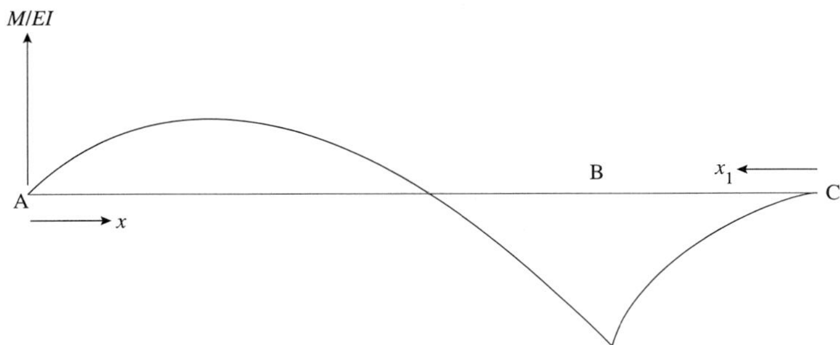

Fig. E4.3.2(b)

Step 2: Conjugate beam. We construct the support conditions of the conjugate beam. The end pin support at A (in the real beam) maps into an end pin support at A′ in the conjugate beam. The internal roller at B maps into an internal hinge at B′. And finally, the free end at C maps into a fixed support at C′. The load on the conjugate beam is simply the *M*/*EI* from the real beam with the direction of the loading being directed outwards (or away from the beam surface) from the compressive fiber (see Fig. E4.3.2(c)).

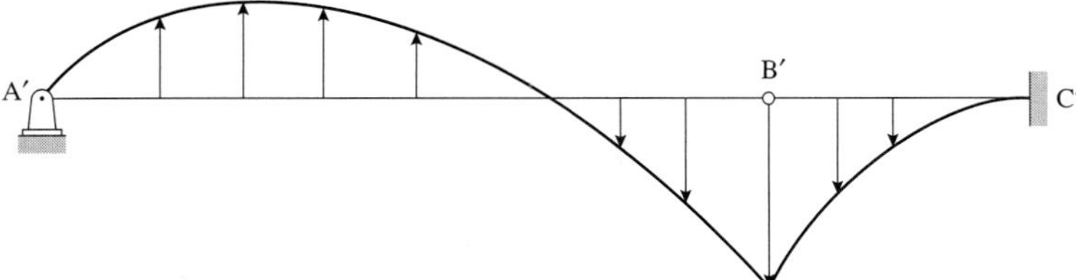

Fig. E4.3.2(c)

At first glance, it appears that the conjugate beam is statically indeterminate. However, if we ignore the axial effects, there are four unknown reactions and two FBDs (see Fig. E4.3.2(d)). Each FBD yields two equations; hence the beam is determinate.

Step 3: Solution of the conjugate beam. It will be helpful to review the mathematical background (Appendix E), since the concepts discussed there are used extensively here. Note that x is a right-handed coordinate system and x_1 is a left-handed system. To compute the slope at B we need to compute the shear at B′. First we compute B_y' from FBD I:

$$\overset{+}{\curvearrowleft}\sum M_{A'} = 0 = \int_0^{10} [w(x)dx][x] - B_y' \ (10)$$

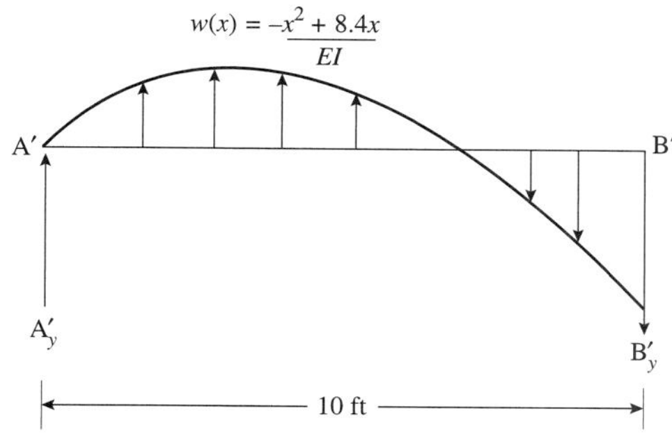

Fig. E4.3.2(d)

FBD I

FBD II

$$\text{or} \quad \frac{1}{EI}\int_0^{10}\left(-x^2+8.4x\right)x\,dx - 10B'_y = 0 \Rightarrow B'_y = \frac{30}{EI}$$

Using our sign convention for shear, we should recognize that $V_{B'} = B'_y = 30/EI \Rightarrow \theta_B = 30/EI$ (↶). The slope is counterclockwise (or, positive) since the shear at B′ is positive.

To compute the displacement at C we need to compute the moment at C′. Using FBD II yields

$$\overset{+}{\curvearrowleft}\sum M_{C'} = 0 - \left(B'_y\right)(4) - \int_0^4 \left[w(x_1)dx_1\right]\left[x_1\right] + C'_M$$

$$\text{or} \quad C'_M = 4\left(\frac{30}{EI}\right) + \frac{1}{EI}\int_0^4 \left(-x_1^2\right)\left(x_1\right)\,dx_1 = \frac{56}{EI}$$

Using the sign convention for the bending moment yields $M_{C'} = C'_M = 56/EI \Rightarrow \Delta_C = 56/EI\,(\uparrow)$. The displacement is upwards since the bending moment at C′ is positive.

Observation: A direct comparison of the moment-area method and the conjugate beam method using this example problem as a guide shows that the latter method is more straightforward (recipe-oriented). However, the moment-area method requires us to understand how the beam would deform. Only then can we use the geometry of the deformed shape to compute the deformations.

One can also compute the largest vertical displacement in the beam by plotting the bending moment diagram of the conjugate beam.

EXAMPLE 4.3.3 *Deflection of a Beam with Internal Hinge*

Compute the vertical displacement and rotation at B of the beam in Fig. E4.3.3(a). Take E = 30000 ksi and I = 700 in^4.

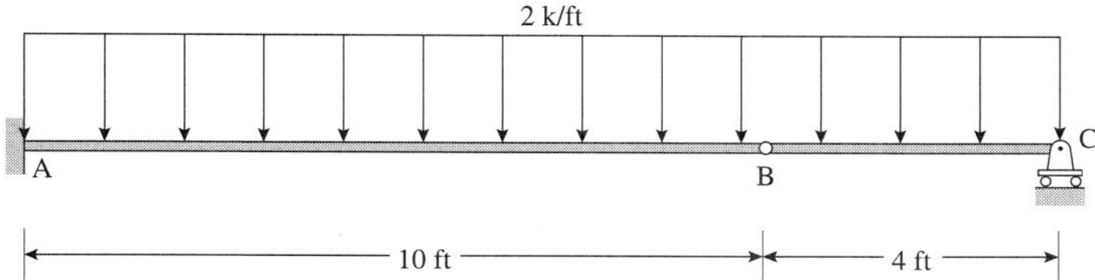

Fig. E4.3.3(a)

SOLUTION

Step 1: The M/EI diagram for the beam is shown in Fig. E4.3.3(b). For the segment CA, $0 < x < 14$ ft, $M(x) = 4x - x^2$. Note that x is a left-handed coordinate system.

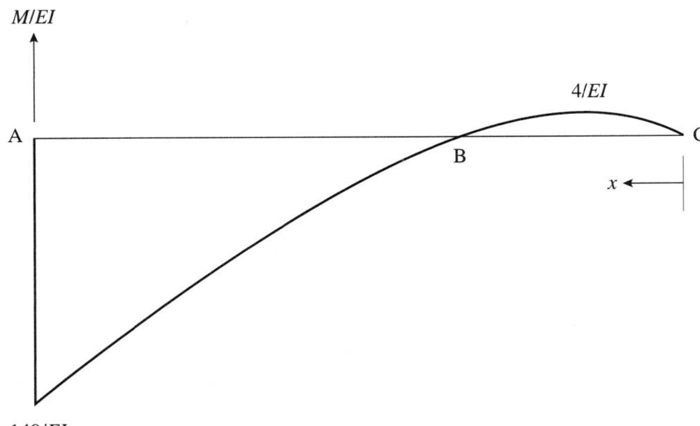

Fig. E4.3.3(b)

Step 2: Conjugate beam. To construct the conjugate beam we note that (a) A, a fixed support, maps into a free end A′, (b) B, an internal hinge, maps into an internal roller B′, and (c) C, an end roller, remains an end roller C′ (see Fig. E4.3.3(c)). The conjugate beam is shown below. The beam appears unstable but is computationally stable if we ignore the axial effects!

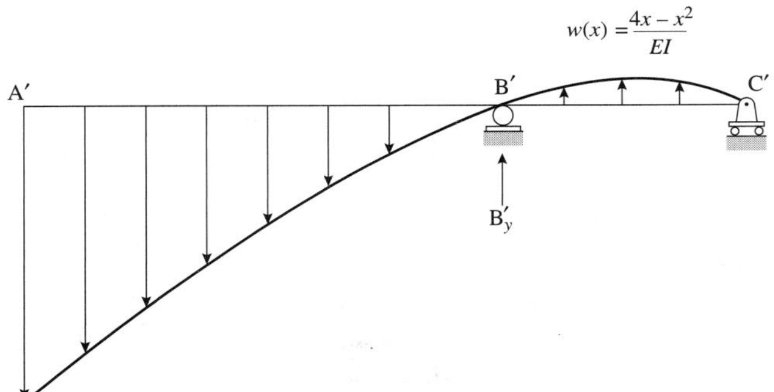

Fig. E4.3.3(c)

Step 3: Solution of the conjugate beam. To obtain the slope at B we should note that B is an internal hinge. Hence, there is a slope to the left and to the right of B. In other words, we need to compute the shear to the left and right of B′. To obtain the displacement at B, we need to compute the moment at B′. First we compute the support reaction at B′:

$$\overset{+}{\curvearrowleft}\sum M_C = 0 = -\int_0^{14} [w(x)dx]x - B'_y(4)$$

$$\text{or } 4B'_y = -\frac{1}{EI}\int_0^{14} \left(4x - x^2\right) x\, dx \quad \text{or} \quad B'_y = \frac{1486.33}{EI}(\uparrow)$$

To compute the slope to the left of B and the displacement at B, we use the FBD in Fig. E4.3.3(d).

Fig. E4.3.3(d)

The resultant of the distributed load is $R = \frac{1}{EI}\int_4^{14} w(x)\,dx = \frac{1}{EI}\int_4^{14}\left(4x - x^2\right)dx = -\frac{533.3}{EI}$.

The negative sign indicates that the load is acting in the negative y-direction (or downwards). Hence,

$$\overset{\uparrow +}{\sum} F_y = 0 = R - V_{B'L} \Rightarrow V_{B'L} = -\frac{533.3}{EI} \Rightarrow \theta_{BL} = \frac{533.3}{EI}\ (\curvearrowright)$$

Similarly, we find

$$\overset{\curvearrowleft +}{\sum} M_B = 0 = M_{B'} - \int_0^{10}\left[w(x_1)\,dx_1\right](x_1)$$

or $M_{B'} = \frac{1}{EI}\int_4^{14}\left(4x - x^2\right)(x-4)\,dx = -\frac{3833.33}{EI} \Rightarrow \Delta_{B'} = \frac{3833.33}{EI}(\downarrow)$

To compute the slope to the right of B we use the FBD in Fig. E4.3.3(e):

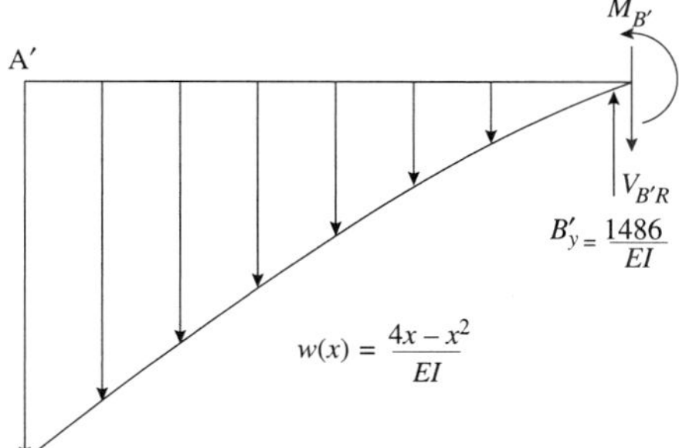

Fig. E4.3.3(e)

$$\overset{\uparrow +}{\sum} F_y = 0 = -R + B'_y - V_{B'R} \Rightarrow V_{B'R} = 953/EI$$

Hence, $\theta_{BR} = \frac{953}{EI}\ (\curvearrowleft)$. Since $EI = 30(10^3)(700)(1/144) = 145833$ k-ft^2, we have

$\theta_{BL} = 0.0037$ rad $(\circlearrowright)$ $\theta_{BR} = 0.0065$ rad $(\curvearrowleft)$ $\Delta_B = 0.026$ ft $= 0.32''(\downarrow)$

EXERCISES

Appetizers

4.3.1. Resolve Problem 4.2.1 using the conjugate beam method.

4.3.2. Resolve Problem 4.2.2 using the conjugate beam method.

4.3.3. Resolve Problem 4.2.3 using the conjugate beam method.

4.3.4. Resolve Problem 4.2.4 using the conjugate beam method.

Main Course

4.3.5. Resolve Problem 4.2.5 using the conjugate beam method.

4.3.6. Resolve Problem 4.2.6 using the conjugate beam method.

4.3.7. The beam in Fig. P4.3.7 has a constant *EI* value. Using the conjugate beam method, compute the rotation to the left and right of internal hinge B.

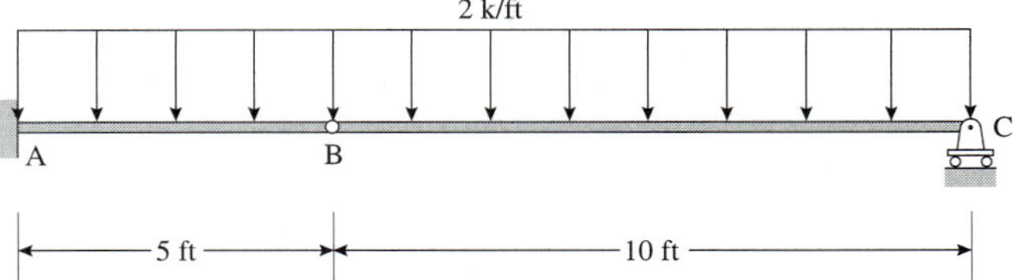

Fig. P4.3.7

4.3.8. The beam in Fig. P4.3.8 has a constant *EI* value. Using the conjugate beam method, compute (a) the rotation at A and (b) the vertical displacement and rotation at B.

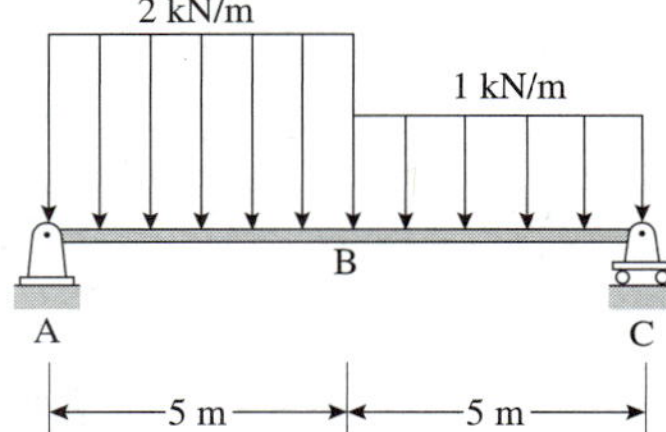

Fig. P4.3.8

4.3.9. The beam in Fig P4.3.9 has a nonuniform cross-section: $I_{AB} = I_{CD} = 2I$, and $I_{BC} = I$. Given that $E = 200$ GPa, find I so that the largest vertical displacement is less than 1 cm. Use the conjugate beam method.

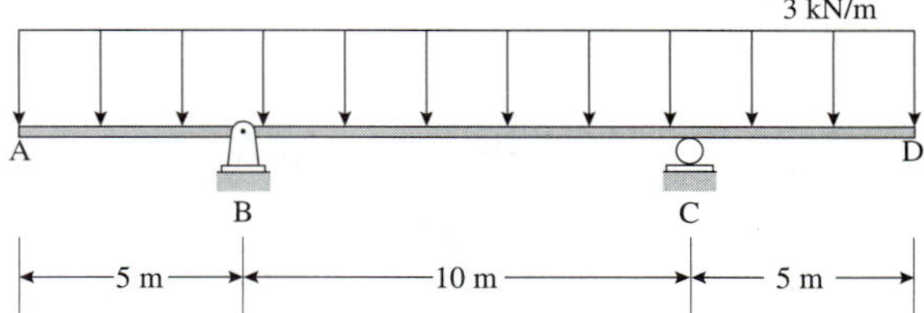

Fig. P4.3.9

4.3.10. The beam in Fig. P4.3.10 has a constant *EI* value. Using the conjugate beam method, compute (a) the rotation at A and (b) the horizontal displacement at the mid-height of the beam.

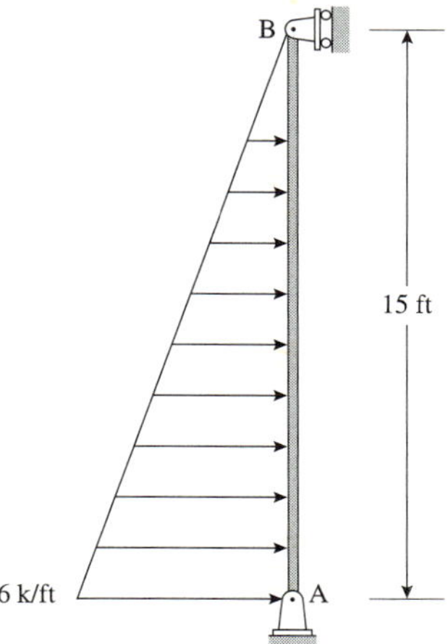

Fig. P4.3.10

4.4 ENERGY PRINCIPLES

In Chapter 3 we saw the computation of stresses and strain, strain-displacement relationships, and characterization of material behavior. These concepts are very useful in computing energy-related quantities such as work and strain energy. These can then be used in the computation of deflections.

Conservative Systems

A system is conservative if the work done around a closed path is equal to zero. A conservative system obeys the conservation of energy principle. Most systems are conservative. A system subjected to frictional forces is an example of a nonconservative system. Consider two systems shown in Fig. 4.4.1. In the first, the work W necessary to raise the block of mass m through height h is $-mgh$. The change in potential energy is $+mgh$. The work necessary to return the block to its original state is $+mgh$. The change in potential energy is $-mgh$. Similarly, with respect to the *linear elastic*[2] spring system, the work done by the force P that is gradually applied is $P\Delta$. The increase in the strain energy of the system is $\left(\frac{1}{2}\right)k\Delta^2$. When the force P is gradually removed, the spring retains its original shape and state. The external work done by the force P is stored in the spring in the form of strain energy.

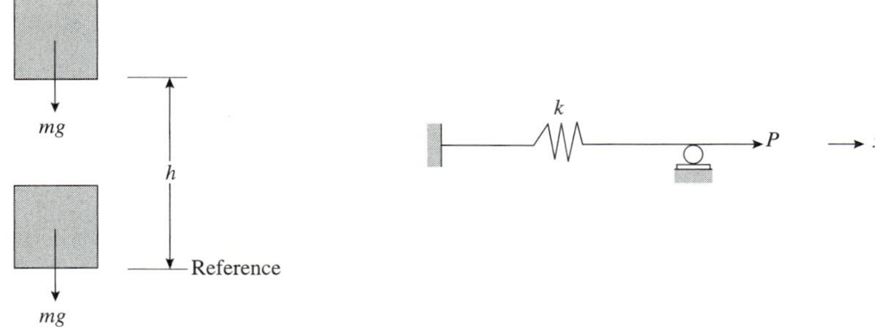

Fig. 4.4.1
Examples of conservative systems.

[2] A force system is said to be conservative if the virtual work δW vanishes for a virtual displacement that carries the system around a closed path. We will see the concept of virtual work in the next section. A structural system is said to be elastic if the internal forces are conservative, in which case the function $U(x)$ is called strain energy.

Work

The differential work dW done by force F undergoing displacement dx is given as

$$dW = Fdx \tag{4.4.1}$$

provided the displacement is in the same direction as the applied force. If the final displacement is x, then the total work can be computed as

$$W = \int_0^x Fdx \tag{4.4.2}$$

Similarly, the differential work done by a moment M undergoing rotation $d\theta$ is given as

$$dW = Md\theta \tag{4.4.3}$$

provided the rotation is in the same direction as the applied moment. If the final rotation is θ, then the total work can be computed as

$$W = \int_0^\theta Md\theta \tag{4.4.4}$$

We now customize these equations for specific situations.

Axial Force. If an axial force F is gradually applied (increasing from 0 to a final value N) to an elastic bar of cross-sectional area A and modulus of elasticity E, then the bar will elongate. Let the final elongation of the bar be Δ. The load-deflection graph is shown in Fig. 4.4.2. From the graph it is clear that $F(x) = (N/\Delta)x$. Using Eq. (4.4.2), we find

$$W = \int_0^\Delta \frac{N}{\Delta} x\, dx = \frac{N}{\Delta} \frac{x^2}{2}\bigg|_0^\Delta = \frac{1}{2} N\Delta \tag{4.4.5}$$

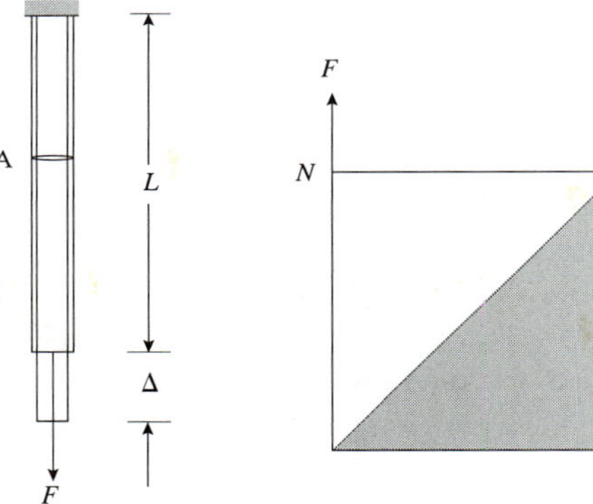

Fig. 4.4.2
Load-deflection graph (uniaxial bar and loading)

The right-hand side expression represents the area under the load-deflection graph.

Bending Moment. If bending moment M is gradually applied to a system and the corresponding rotation has a final value of θ then the work done is expressed (using Eq. 4.4.4) as

$$W = \frac{1}{2} M\theta \tag{4.4.6}$$

Strain Energy

When external loads are applied on an elastic body they deform. The work done is transformed into elastic strain energy U that is stored in the body. We will develop expressions for the (complementary) strain energy for different types of loads.

Axial Force. Figure 4.4.3 shows a bar made of linear elastic material. The bar is subjected to an axial force that is gradually applied. The external work done by the axial force is stored in the bar in the form of linear strain energy.

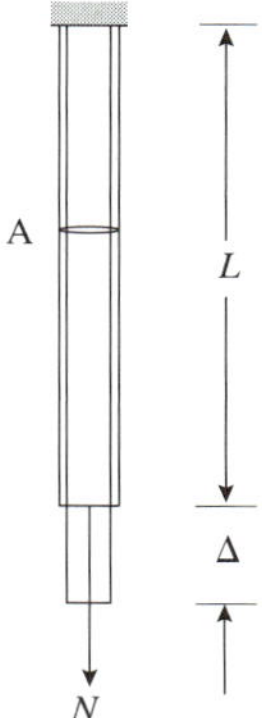

Fig. 4.4.3
Axially loaded bar.

From Hooke's law, $\sigma = E\varepsilon$. Since $\sigma = N/A$ and $\varepsilon = \Delta/L$, substituting these in the stress-strain relationship yields

$$\Delta = \frac{NL}{AE} \tag{4.4.7}$$

Substituting Eq. (4.4.7) into (4.4.5), we have

$$U = \frac{N^2 L}{2AE} \tag{4.4.8}$$

Bending Moment. The strain energy in a structural system subjected to bending moments can be computed using Eq. (4.4.6). Consider a beam subjected to a general loading as shown in Fig. 4.4.4. The rotation of the differential element can be expressed as

$$d\theta = \frac{M}{EI} dx \tag{4.4.9}$$

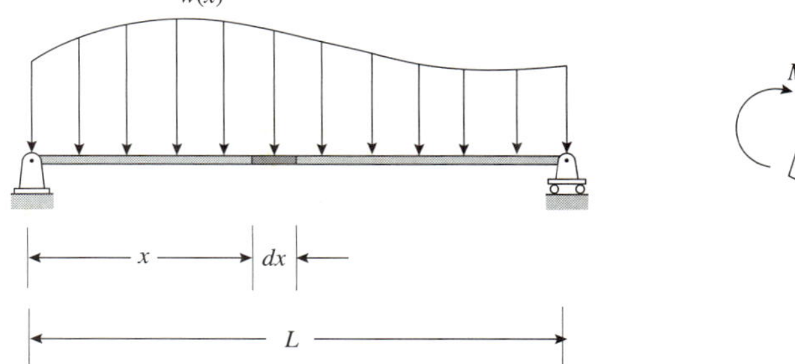

Fig. 4.4.4
Bending of a simply supported beam.

Hence, the strain energy in the differential element can be computed as

$$dU = \frac{1}{2} M d\theta = \frac{1}{2} M \frac{M}{EI} dx = \frac{M^2}{2EI} dx \tag{4.4.10}$$

To compute the strain energy in the entire beam, we need to integrate the above equation as

$$U = \int_0^L \frac{M^2}{2EI} dx \tag{4.4.11}$$

where it is clear that $M = M(x)$ (and, if necessary, $I = I(x)$, $E = E(x)$) for the integration to be possible.

Shear Force. Continuing with the beam in Fig. 4.4.4, the strain energy in the beam is a function of both the bending moment and the shear force. For most beam (and frames), the strain energy due to the shear force can be ignored. However, for some situations, such as a deep beam (one whose depth is comparable to its length), the shear strain energy can be appreciable. The shear force at a section causes a shear strain that contributes to the beam's strain energy (see Fig. 4.4.5).

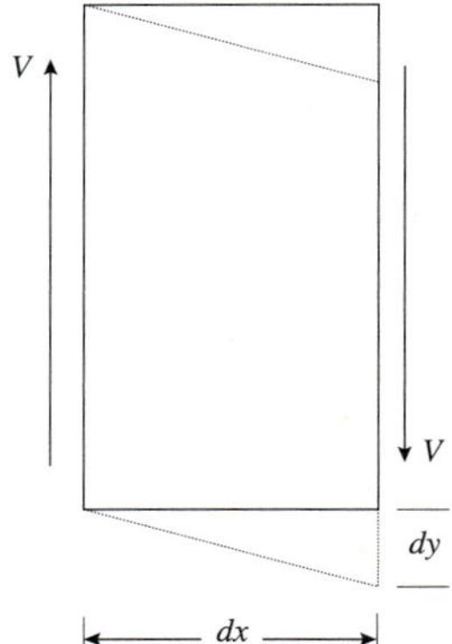

Fig. 4.4.5 Deformation due to shear force.

The shearing strain γ from Fig. 4.4.5, is given as $\gamma = dy/dx$. From Hooke's Law, the shear stress τ is related to the shear strain as $G = \tau/\gamma$, where G is the shear modulus. Combining the two equations, we have

$$dy = \frac{\tau}{G} dx \tag{4.4.12}$$

The evaluation of the above equation is complicated by the fact that the shear stress at a section varies from location to location and hence is a function of the shape of the section. A reasonable approximation is to express the shear stress as

$$\tau = K\frac{V}{A} \tag{4.4.13}$$

where K is called the form factor[3] of the cross-sectional shape and A is the cross-sectional area. Hence we find

$$dU = \frac{1}{2}V\,dy = \frac{1}{2}K\frac{V^2}{GA} \tag{4.4.14}$$

The strain energy in the beam can be computed as before by integrating over the length of the beam:

$$U = \int_0^L \frac{KV^2}{2GA} dx \tag{4.4.15}$$

[3] The form factors for commonly used shapes are 6/5 for a rectangle, and 10/9 for a circle.

Stress-Strain Components. There is a more general and powerful approach to computing the strain energy using the stress and strain distribution in a body. The strain energy density (strain energy per unit volume) U_0 is defined as

$$U_0 = \frac{1}{2}\boldsymbol{\sigma}^T \boldsymbol{\epsilon} \tag{4.4.16}$$

Hence the strain energy in a body can be computed by integrating the strain energy density over the entire volume of the body as

$$U = \int_V U_0 \, dV = \int_V \frac{1}{2}\boldsymbol{\sigma}^T \boldsymbol{\epsilon} \, dV \tag{4.4.17}$$

We will customize this equation for previously derived expressions. For the axially loaded bar (Fig. 4.4.3), at a point there is only one nonzero component of stress. Hence $\varepsilon = \sigma/E$ and $\sigma = N/A$. Substituting in Eq. (4.4.16) yields

$$U_0 = \frac{1}{2}\frac{N}{A}\frac{N}{AE} = \frac{1}{2}\frac{N^2}{A^2 E} \tag{4.4.18}$$

Hence, noting that none of the parameters are functions of x, we have

$$U = \int_0^L \frac{1}{2}\frac{N^2}{A^2 E} A \, dx = \frac{1}{2}\frac{N^2}{A^2 E}(AL) = \frac{N^2 L}{2AE} \tag{4.4.19}$$

which is the same as Eq. (4.4.8). In a similar manner, for the case with the bending moment (Fig. 4.4.4), $\varepsilon = \sigma/E$ and $\sigma = [M(x)y]/I$. Substituting in Eq. (4.4.16) yields

$$U_0 = \frac{1}{2}\frac{M(x)y}{I}\frac{M(x)y}{IE} = \frac{1}{2}\frac{[M(x)]^2 y^2}{I^2 E} \tag{4.4.20}$$

Hence, assuming that the moment of inertia and modulus of elasticity are not functions of x, we get

$$U = \int_0^L \int_A \frac{1}{2}\frac{[M(x)]^2 y^2}{I^2 E} \, dA \, dx = \int_0^L \frac{[M(x)]^2}{2EI} \, dx \tag{4.4.21}$$

which is the same as Eq. (4.4.11).

Potential Energy

The potential energy Π of an elastic system is defined as

$$\Pi = U + WP \tag{4.4.22}$$

where WP is defined as the work potential or the potential energy of the applied loads. In the case of concentrated force N and moment M, we find

$$WP = -\frac{1}{2}N\Delta \tag{4.4.23}$$

$$WP = -\frac{1}{2}M\theta \tag{4.4.24}$$

With distributed forces and moments, the work potential can be obtained through integration.

Compatibility

The concept of compatibility, though used directly less often, is as important as that of equilibrium. A system should deform so that the members mesh with each other without creating voids or overlaps while satisfying the deformation shape afforded by the connections and support conditions. Otherwise the system does not meet the compatibility requirements. The beam at the top of Fig. 4.4.6 shows a beam deformation that is compatible with the manner in which the beam is supported. However, the beam deformation at the bottom of the figure does not satisfy compatibility. The transverse displacement at the roller support should be zero. Figure 4.4.7 shows another problem with the deformed shape: The displacement field is such that a void develops in the beam. Clearly if the original beam did not have a void, then there is no reason why a void should develop when the beam is loaded (assuming that the loading does not initiate material failure).

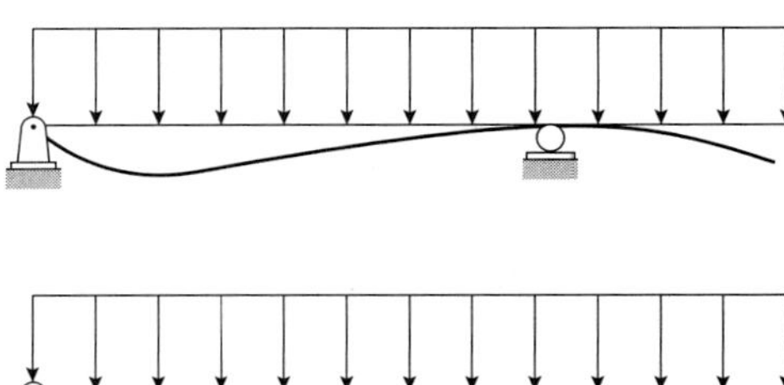

Fig. 4.4.6
Top: Beam deformation satisfies compatibility. Bottom: Beam deformation does not satisfy compatibility.

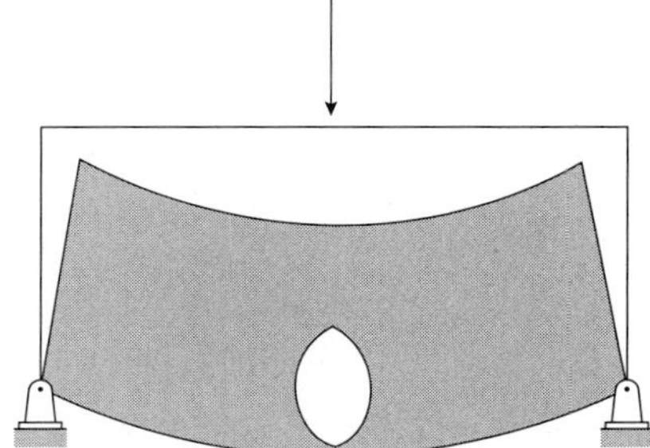

Fig. 4.4.7
Structural incompatibility.

Conservation of Energy

The first law of thermodynamics captures the principle of conservation of energy (energy cannot be created nor destroyed) as

$$\text{Work done} = \text{Change in energy} \tag{4.4.25}$$

For a system in which no heat is generated, added to, or removed from the system and where the loads are applied slowly, we have

$$W_e = U \tag{4.4.26}$$

where the term on the left is the total external work done by the applied loads. We can rewrite this equation in a different form as

$$W_e + W_i = 0 \tag{4.4.27}$$

expressing the fact, due to Eq. (4.4.25), that the sum of the external work and the internal work must be zero.

EXAMPLE 4.4.1 ***Energy in a Cantilever Beam***

Compute the strain energy in the beam in Fig. E4.4.1. EI is a constant.

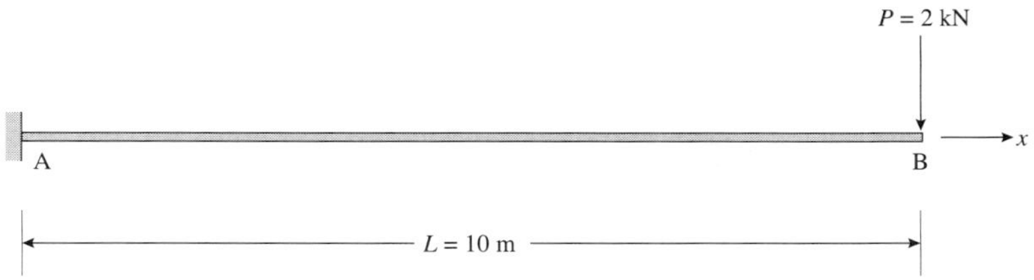

Fig. E4.4.1

SOLUTION

Step 1: Compute the internal moment. We need to compute a single moment expression covering the region from A to B. Using the coordinate system shown in Fig. E4.4.1, and with kN, m as the problem units, we have

For $0 < x < 10$ m, $M(x) = 2x - 20$

Step 2: Use Eq. (4.4.11) to compute the strain energy:

$$U = \int_0^L \frac{M^2}{2EI}\,dx = \int_0^{10} \frac{(2x-20)^2}{2EI}\,dx = \frac{666.7}{EI}\ \text{kN-m}$$

with EI expressed in consistent units. Let $E = 200$ GPa and $I = 10^{-5}$ m^4. Then

$$U = \frac{666.7}{2(10^8)(10^{-5})} = \frac{1}{3}\ \text{kN-m}$$

As you can see, the strain energy is inversely proportional to the modulus of elasticity and the moment of inertia of the beam. In other words, stiffer beams store less strain energy than their less stiff counterparts when loaded and supported in a similar manner.

We cannot compute the total potential energy at this stage since we have not developed the methodology to compute the deflection and hence the work potential term in Eq. (4.4.22).

EXERCISES

Appetizers

4.4.1. Consider the loaded spring in Fig. P4.4.1. Compute the (a) strain energy in the spring, and (b) the total potential energy in the system.

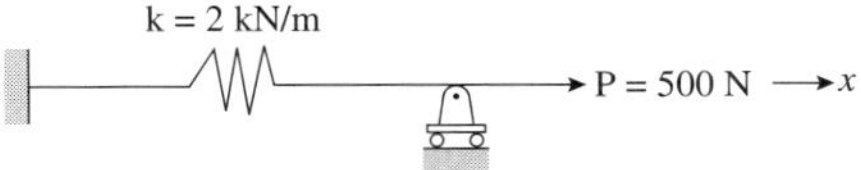

Fig. P4.4.1

4.4.2. Consider the loaded beam in Fig. P4.4.2. Compute the strain energy in the beam. Use $E = 29(10^6)$ psi, $I = 500$ in^4, $a = 5$ ft, and $L = 15$ ft.

L

Fig. P4.4.2

Main Course

4.4.3. The planar truss in Fig. P4.4.3 is made of steel, $E = 200$ GPa. The cross-sectional area of both the members is 0.01 m^2. Compute the strain energy in the system.

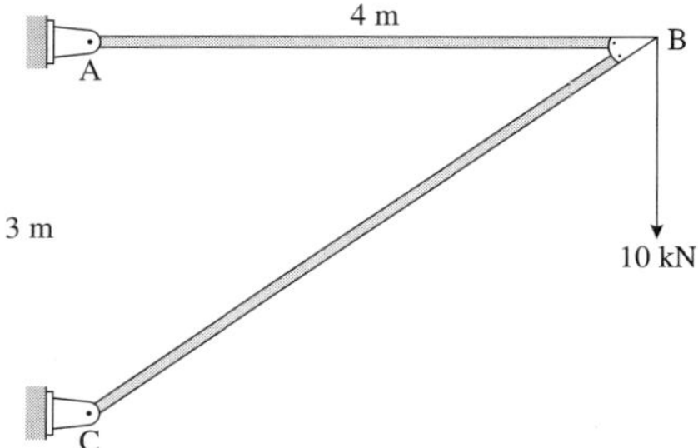

Fig. P4.4.3

4.4.4. Consider the loaded beam in Fig. P4.4.4. Compute the strain energy in the beam.

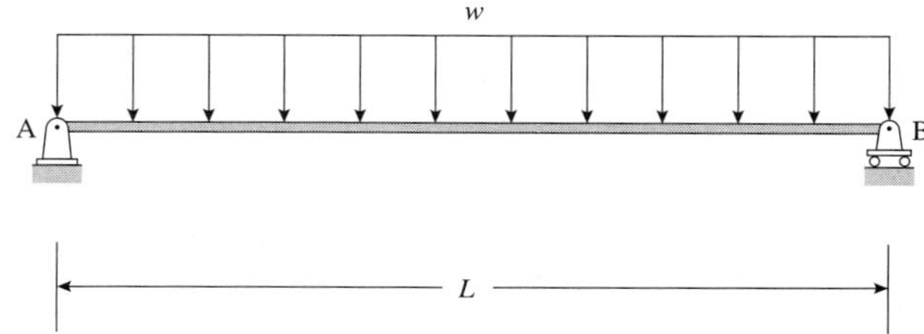

Fig. P4.4.4

Structural Concepts

4.4.5. Repeat Problem 4.4.4 but include the shear strain energy. Let $L = 10$ ft, $w = 2$ k/ft, $E = 6(10^6)$ psi and $G = E/2$.
(a) Assume that the cross-section is rectangular of dimensions 8 in (w) × 12 in (h).
(b) Assume that the cross-section is rectangular of dimensions 6 in (w) × 24 in (h).

4.5 PRINCIPLE OF VIRTUAL WORK

The development of the principle of virtual work is credited to John Bernoulli in 1717. Consider a deformable body as shown in Fig. 4.5.1(a). The body is subjected to a number of forces P_1, P_2,... and is adequately supported to be in stable equilibrium. We can image the squares inside the body as being its constituent parts or elements. Stresses σ develop within these elements due to the action of the external loads.

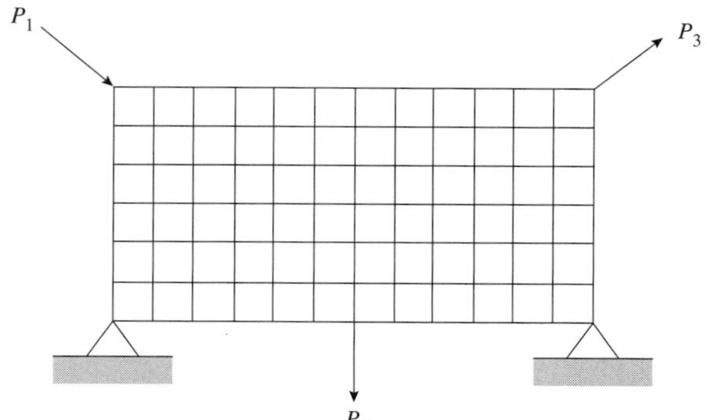

Fig. 4.5.1(a) Deformable body subjected to external loads.

Now imagine the body is subjected to a set of compatible *virtual* displacements δD. These displacements are fictitious or imaginary displacements as shown in Fig. 4.5.1(b). The symbol δ is used to indicate that the displacements are virtual, not real. The resulting virtual strains in the body are denoted $\delta\varepsilon$.

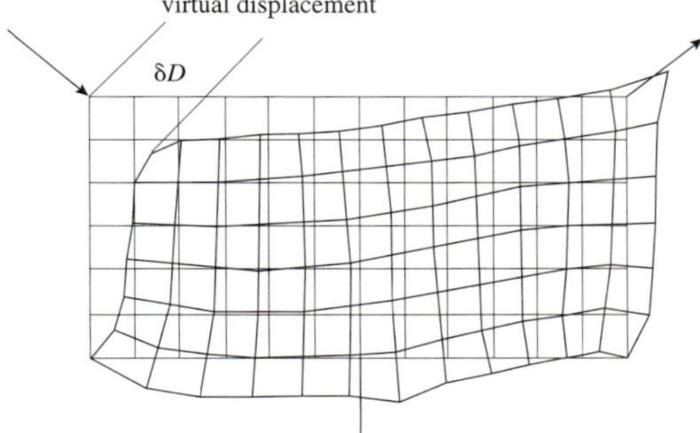

Fig. 4.5.1(b) Deformable body subjected to virtual displacements.

While the body is being displaced, the real forces acting on the body move through these virtual displacements. These forces and virtual displacements must satisfy the principle of conservation of energy. Using Eq. (4.4.27) yields

$$\delta W_e + \delta W_i = 0 \tag{4.5.1}$$

or

$$\sum_{i=1}^{n} P_i(\delta D)_i = \int_V \sigma(\delta\varepsilon)\, dV \tag{4.5.2}$$

This is the principle of virtual work: A deformable body that is in equilibrium under the action of external loads P and is subjected to compatible virtual deformations is such that the external virtual work done by the external loads P is equal to internal virtual work done by the stress field σ.

To make the principle a practical tool, we need to interchange the role of the forces and displacements. In the above discussion, the structure acted on by real forces was subjected to virtual displacements. What if the structure acted on by virtual forces is subjected to real displacements? Then we could write Eq. (4.5.2) as

$$\sum_{i=1}^{n} D_i(\delta P)_i = \int_V \varepsilon(\delta\sigma)\, dV \tag{4.5.3}$$

This is the principle of complementary virtual work, and we will use this to compute the displacements. Note that since the virtual stresses/forces attain their full values before the real deformations are applied, the 1/2 factor (see Eq. (4.4.5)) does not appear on both sides of the equation. To ease the computation of the term on the right-hand side, we will customize the expression for beams/frames and trusses in the next few sections. The above derivation is quite general. We can imagine the body in Fig. 4.5.1 to be a beam, truss, frame, or any other structural system.

To determine the deflection at a point on a body, consider the body in Fig. 4.5.2. It is required to compute the displacement at point A along the direction A-B (recall that displacements are vectors—they have a magnitude and a direction). We first apply a virtual force δF at A along A-B (Fig. 4.5.3(a)). This would result in a virtual force δf in a typical element in the body. This can be computed using the concepts that we saw in Chapter 2 in truss, beam, and frame structures.

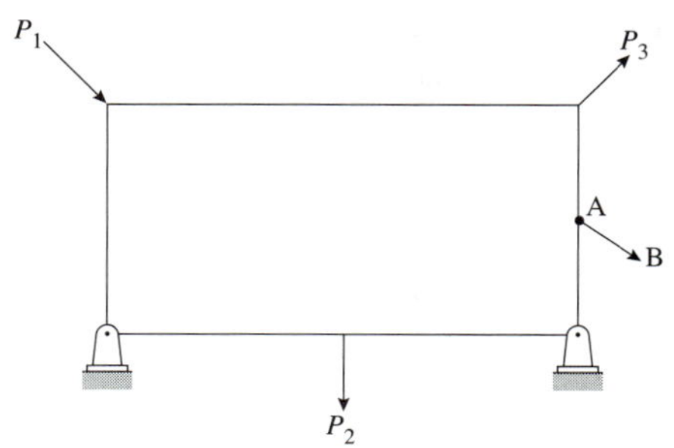

Fig. 4.5.2
Computing displacement at A along A-B

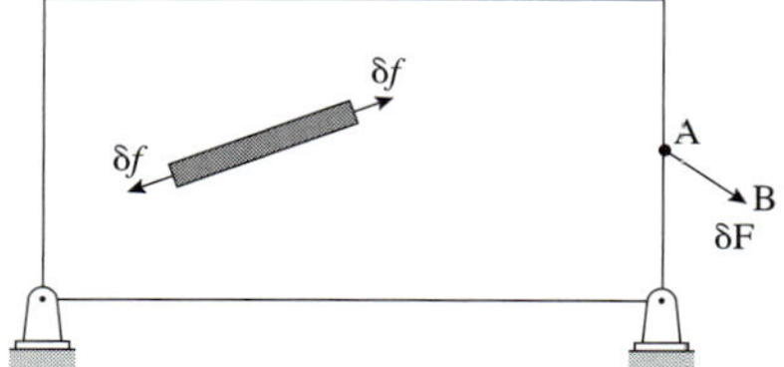

Fig. 4.5.3(a)
Body subjected to a virtual force δF.

Next, while the virtual force remains on the body, we apply the external loads (or real forces) acting on the body as shown in Fig. 4.5.3(b). Point A moves or displaces to A′. The displacement along A-B is Δ. The (internal) elements deform ΔL and the deformations can be computed (see for example Eq. 4.4.7). Hence using Eq. (4.5.3), we have

$$\delta F \cdot \Delta = \sum \delta f \cdot \Delta L \tag{4.5.4}$$

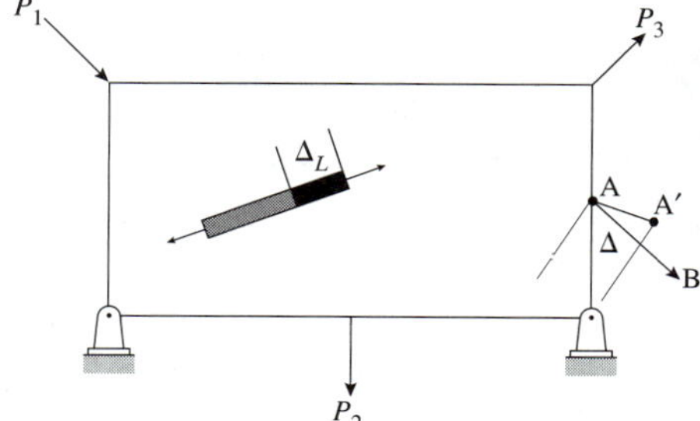

Fig. 4.5.3(b)
Body subjected to real forces P_i.

where the left-hand side represents the external work done by the virtual force δF moving through real displacement Δ, and the right-hand side represents the internal work done by the virtual element forces δf moving through a deformation ΔL. In the above equation no limitations are placed on the magnitudes of δF or δf. Therefore it is most convenient to choose δF as unity from which comes another name for the method, the unit load method. Hence,

$$1 \cdot \Delta = \sum f \cdot \Delta L \tag{4.5.5}$$

Stated simply, we have

- 1 virtual unit force applied at a point where the displacement is to be computed along a specified direction
- Δ (real) displacement at the point along the direction of the unit virtual force
- f internal forces caused by the virtual forces
- ΔL (real) internal deformations

In a similar fashion, we can compute the rotation or slope at a point as

$$1 \cdot \theta = \sum f_\theta \cdot \Delta L \tag{4.5.6}$$

Stated simply, we have

- 1 virtual unit moment applied at a point where the rotation is to be computed about a specified axis
- θ (real) rotation at the point along the direction of the unit virtual moment
- f_θ internal forces caused by the virtual forces
- ΔL (real) internal deformations

The computation of rotations or slopes is valid only for beams and frames.

It is necessary to customize the expression on the right-hand side for the different types of structural systems (truss, beam, and frame). We derive these expressions in the next few sections.

4.5.1 Unit Load Method for Beams and Frames

The internal work done in beams and frames can be due to several factors—bending moment, shear force, axial force, temperature change, etc. Among these, the strain energy due to bending moment dominates the other factors.

Consider the beam in Fig. 4.5.1.1(a). The beam has a constant EI value and length L. It is required to compute the vertical displacement at B, Δ_B. Hence a unit load is applied at B in the vertical direction (see Fig. 4.5.1.1(b)) As we will see later, it does not matter whether the unit load acts up or down.

Fig. 4.5.1.1(a) Cantilever beam with real loads.

Fig. 4.5.1.1(b) Cantilever beam with the virtual unit force.

To compute the displacement we need to compute the internal virtual work (right-hand side of Eq. (4.5.5)). Since the virtual work done is primarily due to bending, we will compute the virtual work done due to the internal moment $m(x)$ rotating through $d\theta$ (see Fig. 4.5.1.2). From Fig. 4.5.1.2 and Eq. (4.4.9), we see that $d\theta = (M/EI)dx$. Hence the internal virtual work done is $md\theta = (mM/EI)dx$ at the cut and $\int_0^L \frac{mM}{EI} dx$ over the length of the beam. The external virtual work done is $(1)(\Delta_B)$. Equating the two expressions (which can then be solved for Δ_B), we have

$$(1)(\Delta_B) = \int_0^L \frac{mM}{EI} dx \tag{4.5.1.1}$$

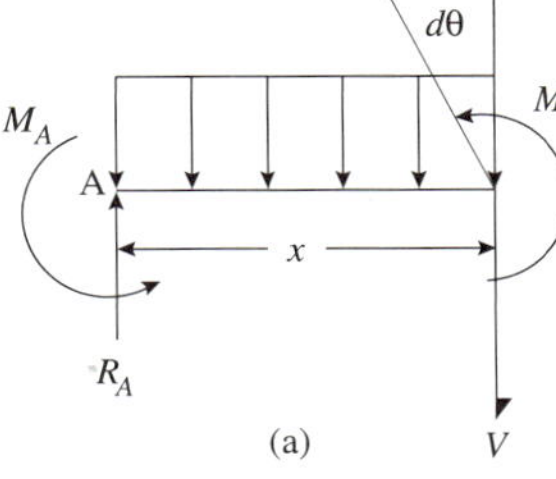

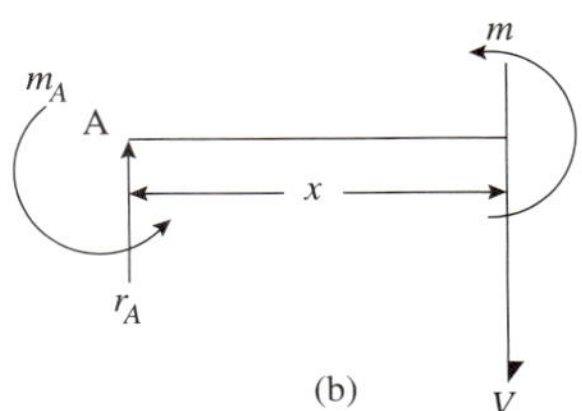

Fig. 4.5.1.2 FBDs showing the internal moments in (a) beam with real loads and (b) beam with virtual load.

where $m(x)$ is the internal moment in the beam due to the unit force, and $M(x)$ is the internal moment in the beam due to the (real) external loads.

The rotation at B, θ_B, can be computed in a similar manner. We first apply a unit virtual moment at B (Fig. 4.5.1.3). Again the direction of the moment is not important; i.e., it does not matter whether the moment is clockwise or counterclockwise.

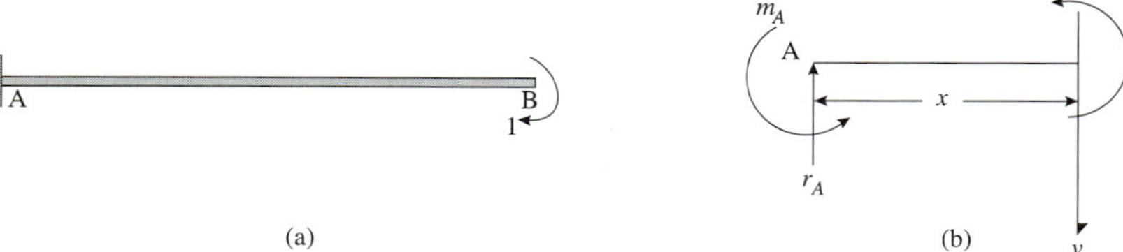

Fig. 4.5.1.3 Cantilever beam (a) with virtual unit moment (b) FBD showing internal moment.

To compute the internal virtual work (right-hand side of Eq. (4.5.6)) due to bending, we need to compute the work done due to the internal moment $m_\theta(x)$ rotating through $d\theta$. The final equation obtained by equating the external virtual work done to the internal virtual work done is

$$(1)(\theta_B) = \int_0^L \frac{m_\theta M}{EI}\, dx \tag{4.5.1.2}$$

where $m_\theta(x)$ is the internal moment in the beam due to the unit moment, and $M(x)$ is the internal moment in the beam due to the (real) external loads.

The computation of the right-hand side of the above two equations must be done with care. Let us assume that in the above example, the vertical displacement at C is to be computed. The virtual unit load is applied at A. This load causes point C to be a point of discontinuity in the moment diagram for the beam with the virtual load. Hence we need two expressions for the moments—one valid from A to C, and the other from C to B. The final form of the displacement equation is

$$(1)(\Delta_C) = \int_A^C \frac{mM}{EI}\, dx + \int_C^B \frac{mM}{EI}\, dx \tag{4.5.1.3}$$

from which Δ_C can be computed, as in Fig. 4.5.1.4.

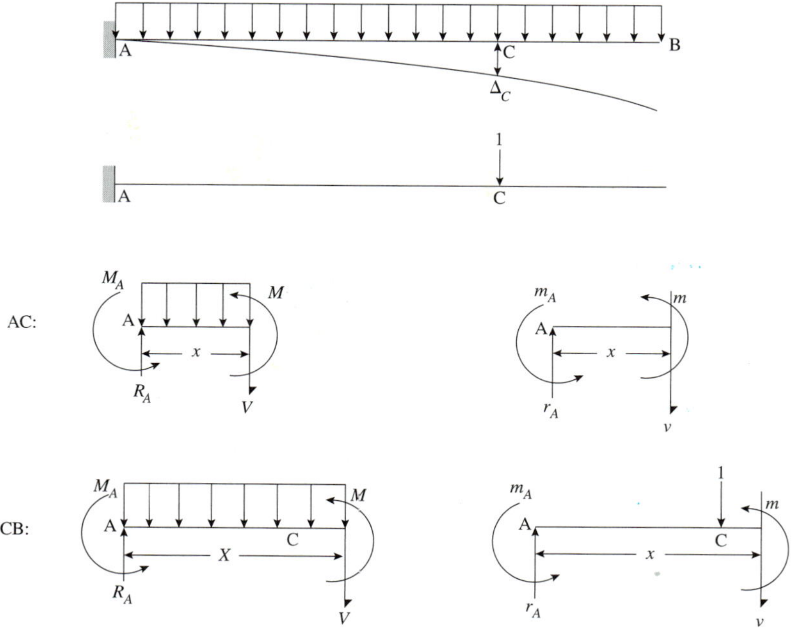

Fig. 4.5.1.4 Steps in computing the vertical displacement at C.

Now consider a further extension of the previous example. Let us add a concentrated force acting at D on the real beam. It is again required to compute the vertical displacement at C. Using the same approach as before to identify the points of discontinuity in the two moment diagrams, we conclude that three segments are needed to compute the internal virtual work done. The final form of the displacement equation is

$$(1)(\Delta_C) = \int_A^D \frac{mM}{EI}\,dx + \int_D^C \frac{mM}{EI}\,dx + \int_C^B \frac{mM}{EI}\,dx \tag{4.5.1.4}$$

from which Δ_C can be computed, as in Fig. 4.5.1.5.

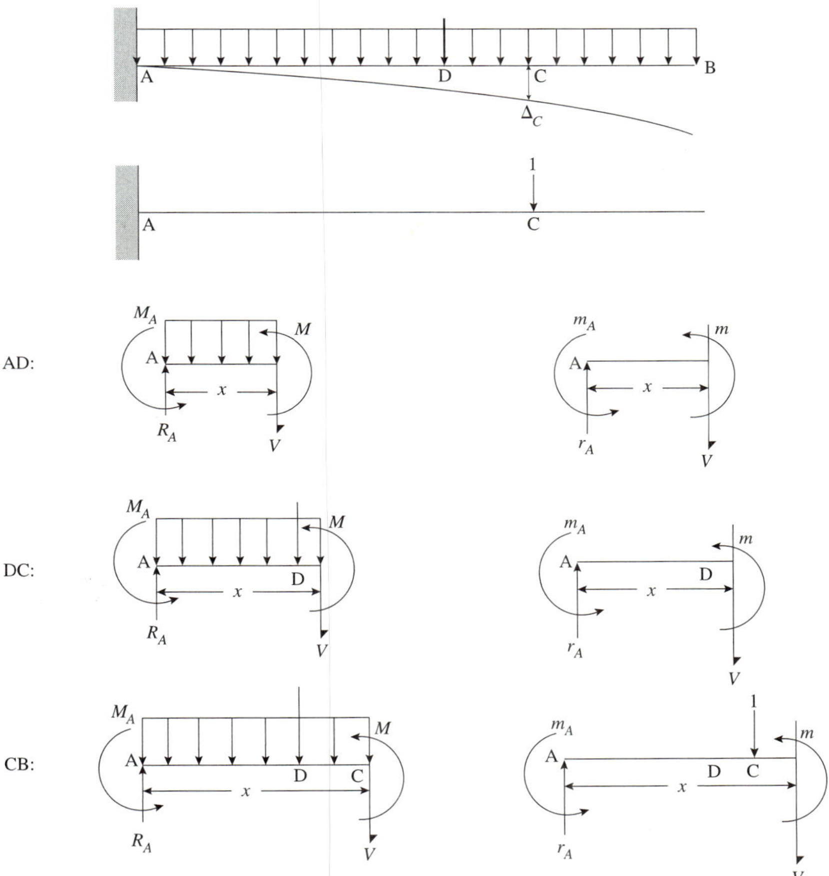

Fig. 4.5.1.5
Steps in computing the vertical displacement at C.

General Procedure

Note that there are two structures involved in the steps below. The structure with only the real loads (no virtual load) will be called SRL. The structure with only the virtual unit load acting on it (no real loads) will be called SVL.

Step 1: To set up the SVL, establish the point where the displacement (rotation) is to be computed and the desired direction. Apply a unit force (unit moment) at the point along the desired direction.

Step 2: Now scan *both* the structures and determine the *maximum* number of segments necessary to obtain the moment expressions valid for both the structures. Review Section 2.8 in Chapter 2 to refresh the concepts associated with moment diagrams. It may be necessary to compute the support reactions.

Step 3: Make a table as shown below.

Segment	FBD (SRL)	$M(x)$	FBD (SVL)	$m(x)$ or, $m_\theta(x)$	$\int \frac{M(x)m(x)}{EI}dx$ or, $\int \frac{M(x)m_\theta(x)}{EI}dx$

In both the FBDs, the internal moment $M(x)$ and $m(x)$ (or $m_\theta(x)$) must be shown exactly the same way (same coordinate system and same direction). Otherwise the virtual work expression in the last column will not be consistent.

Step 4: Apply Eq. (4.5.1.1) or (4.5.1.2) to obtain the deflection as

$$\Delta = \sum\left[\int \frac{M(x)m(x)}{EI}\right]dx \quad \text{or} \quad \theta = \sum\left[\int \frac{M(x)m_\theta(x)}{EI}\right]dx$$

Note that the term in the parentheses is the sum of the last column in the table. If the right-hand side is positive, then the displacement (rotation) is in the same direction as the unit force (moment); otherwise the direction is opposite to the direction of the unit force (moment).

EXAMPLE 4.5.1

Deflection of a Cantilever Beam

Compute the vertical displacement and rotation at B of the beam in Fig. E4.5.1. EI is a constant.

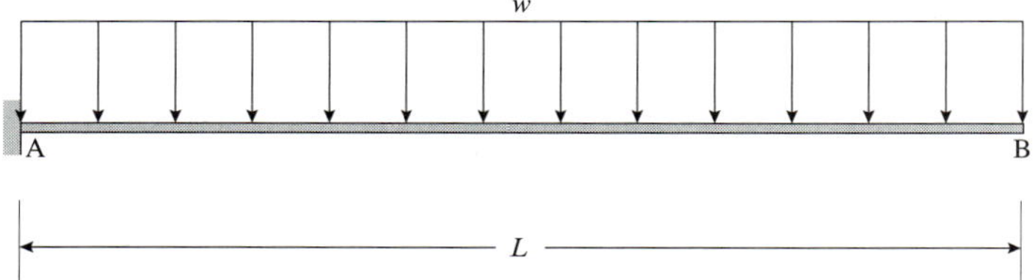

Fig. E4.5.1

SOLUTION

Vertical displacement at B

Step 1: Apply the unit force at B as shown in Fig. 4.5.1.1(b).

Step 2: We need only one segment to write the moment expressions. Making a cut between A and B, we use a right-hand segment to develop the moment expressions.

Segment	FBD (SRL)	$M(x)$	FBD (SVL)	$m(x)$	$\int \frac{M(x)m(x)}{EI}dx$
BA ($0 < x < L$)	M, V, x, B	$-\frac{wx^2}{2}$	1, m, V, x, B	$-x$	$\int_0^L \frac{\left[-\frac{wx^2}{2}\right][-x]}{EI}dx$
				Sum	$\frac{wL^4}{8EI}$

Step 3: Hence the vertical displacement at B can be computed as

$$(1)(\Delta_B) = \frac{wL^4}{8EI} \Rightarrow \Delta_B = \frac{wL^4}{8EI}(\downarrow)$$

Note that since the displacement value is positive, the displacement is in the same direction as the unit force.

Rotation or slope at B

Step 1: Apply the unit moment at B as shown in Fig. 4.5.1.3(a).

Step 2: Again, we need only one segment to write the moment expressions. Making a cut between A and B, we use a right-hand segment to develop the moment expressions.

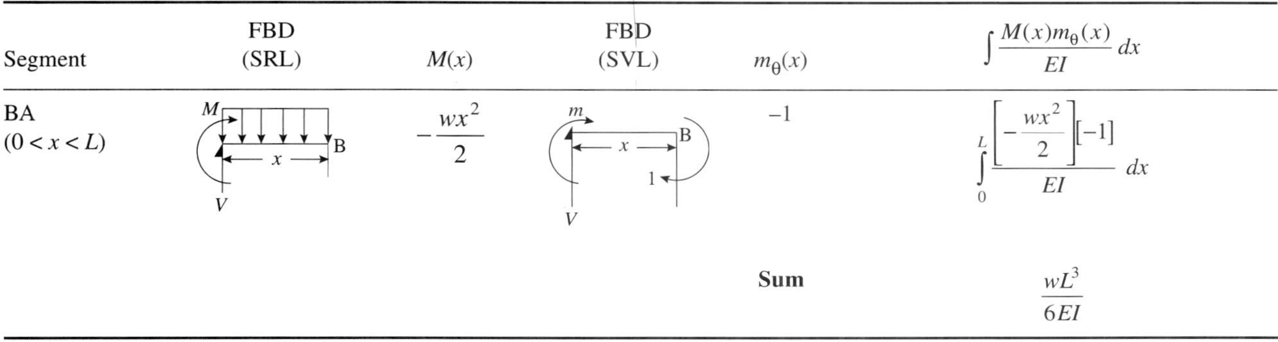

Segment	FBD (SRL)	$M(x)$	FBD (SVL)	$m_\theta(x)$	$\int \frac{M(x)m_\theta(x)}{EI}dx$
BA $(0 < x < L)$		$-\frac{wx^2}{2}$		-1	$\int_0^L \frac{\left[-\frac{wx^2}{2}\right][-1]}{EI}dx$
				Sum	$\frac{wL^3}{6EI}$

Step 3: Hence the rotation at B can be computed as

$$(1)(\theta_B) = \frac{wL^3}{6EI} \Rightarrow \theta_B = \frac{wL^3}{6EI}(\curvearrowright)$$

Note that since the rotation value is positive, the rotation is in the same direction as the unit moment, i.e., clockwise.

EXAMPLE 4.5.2 ***Deflection of a Beam (Examples 4.2.4 and 4.3.2)***

Compute the vertical displacement and rotation at C of the beam in Fig. E4.5.2(a). EI is a constant.

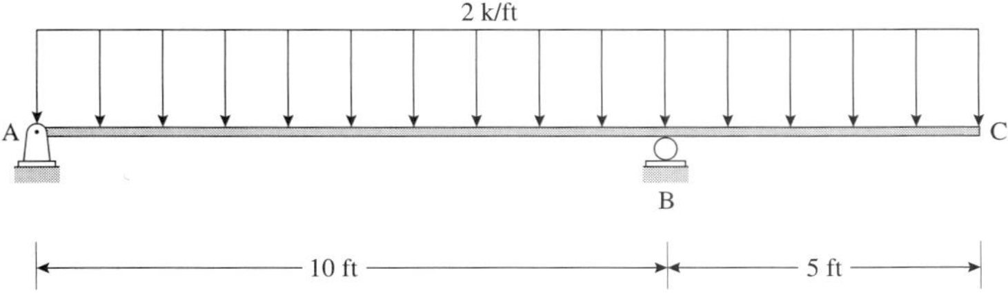

Fig. E4.5.2(a)

SOLUTION We will use k, ft as the problem units.

Vertical displacement at C

Step 1: Since the vertical displacement at C is to be computed, we apply a unit force at C.

Step 2: Noting the structure with the real loads and the structure with the unit load, we need two segments to develop the moment equations—AB and CB (see Fig. E4.5.2(b)). We need not compute all the support reactions and the relevant equations are not shown. However, for the structure with the real loads, $A_y = (2 \times 15)(2.5)/10 = 7.5$ k. Similarly for the structure with the virtual load, $A_y = (1)(5)/10 = .5$ k.

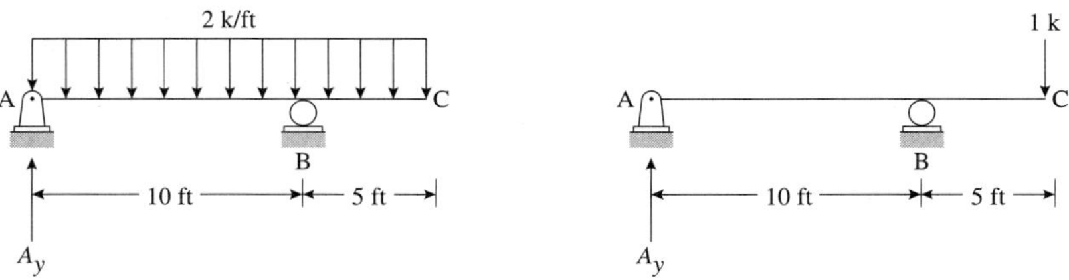

Fig. E4.5.2(b)

Segment	FBD (SRL)	$M(x)$	FBD (SVL)	$m(x)$	$\int \frac{M(x)m(x)}{EI} dx$
AB ($0 < x < 10'$)	2 k/ft, M, A, x, 7.5, V	$-x^2 + 7.5x$	A, m, x, 0.5, V	$-0.5x$	$\int_0^{10} \frac{(-x^2 + 7.5x)(-0.5x)}{EI} dx = 0$
CB ($0 < x_1 < 5'$)	M, 2 k/ft, x_1, C, V	$-x_1^2$	m, 1 k, x_1, C, V	$-x_1$	$\int_0^{5} \frac{(-x_1^2)(-x_1)}{EI} dx_1 = \frac{625}{4EI}$
				Sum	$\frac{156.25}{EI}$

Step 3: Hence the vertical displacement at C can be computed as

$$(1)(\Delta_C) = \frac{156.25}{EI} \Rightarrow \Delta_C = \frac{156.25}{EI} \text{ ft}(\downarrow)$$

Note that since the displacement value is positive, the displacement is in the same direction as the unit force, i.e., the displacement at C is downwards.

Rotation at C

Step 1: Since the rotation at C is to be computed, we apply a unit moment at C (see Fig. E4.5.2(c)).

Step 2: We reuse as much of the previously developed moment expressions as possible. For the structure with the virtual load, $A_y = 0.1$ k.

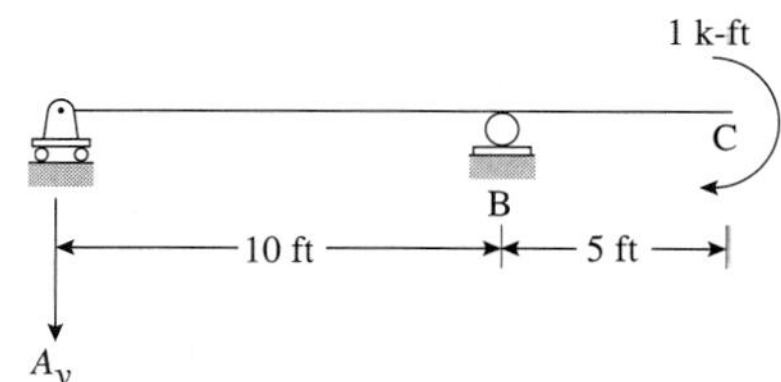

Fig. E4.5.2(c)

Segment	FBD (SRL)	$M(x)$	FBD (SVL)	$m_\theta(x)$	$\int \frac{M(x)m_\theta(x)}{EI}dx$
AB ($0 < x < 10'$)		$-x^2 + 7.5x$		$-0.1x$	$\int_0^{10} \frac{(-x^2 + 7.5x)(-0.1x)}{EI} dx = 0$
CB ($0 < x_1 < 5'$)		$-x_1^2$		-1	$\int_0^5 \frac{((-x_1^2)(-1)}{EI} dx_1 = \frac{125}{3EI}$
				Sum	$\frac{41.67}{EI}$

Step 3: Hence the rotation at C can be computed as

$$(1)(\theta_C) = \frac{41.67}{EI} \Rightarrow \theta_C = \frac{41.67}{EI} \text{ rad}(\curvearrowright)$$

Note that since the computed rotation value is positive, the rotation is in the same direction as the unit moment, i.e., the rotation at C is clockwise.

Observation: It is beneficial to compute the right-hand side terms separately so as to find the sensitivity of the problem parameters to the computed deflected values. For example, in this problem, the vertical deflection and rotation at C are both unaffected by the *EI* values of segment AB.

EXAMPLE 4.5.3

Deflection of a Frame

Compute the horizontal displacement and rotation at B of the frame in Fig. E4.5.3(a). Take E = 200 GPa and $I = 50(10^{-5})m^4$ for the two columns and $I = 80(10^{-5})m^4$ for the beam.

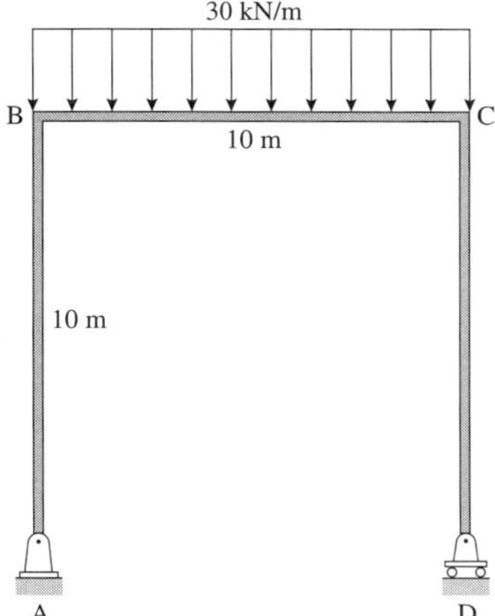

Fig. E4.5.3(a)

SOLUTION

Step 1: The problem units are kN, m. We first compute the support reactions for the frame as shown above: $A_y = D_y = 150$ kN(↑) and $A_x = 0$.

Horizontal displacement at B

Step 2: We apply a unit force at B, as in Fig. E4.5.3(b).

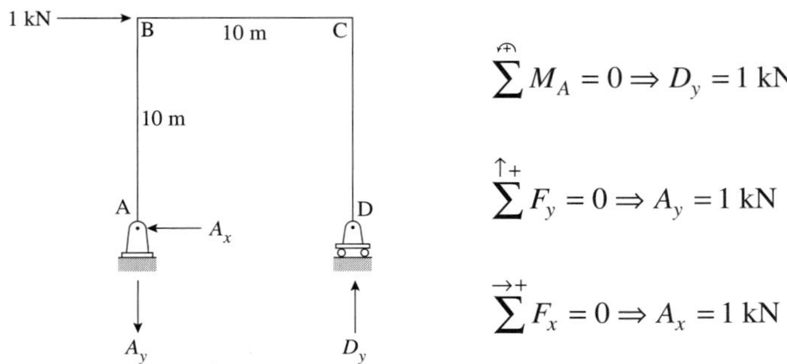

$$\overset{\curvearrowleft+}{\sum} M_A = 0 \Rightarrow D_y = 1 \text{ kN}$$

$$\overset{\uparrow+}{\sum} F_y = 0 \Rightarrow A_y = 1 \text{ kN}$$

$$\overset{\rightarrow+}{\sum} F_x = 0 \Rightarrow A_x = 1 \text{ kN}$$

Fig. E4.5.3(b)

Step 3: Compute the internal moments:

Segment	FBD (SRL)	$M(x)$	FBD (SVL)	$m(x)$	$\int \frac{M(x)m(x)}{EI}dx$
AB ($0 < x_1 < 10$)		0	—	—	0
DC ($0 < x_2 < 10$)		0	—	—	0
BC ($0 < x < 10$)		$-15x^2 + 150x$		$-x + 10$	$\int_0^{10} \left(-15x^2 + 150x\right)(-x + 10)\,dx = \frac{12500}{EI_{BC}}$
				Sum	$\frac{12500}{EI_{BC}}$

$$EI_{BC} = \frac{200(10^9)}{10^3}\frac{\text{kN}}{\text{m}^2} \times 80(10^{-5})\ \text{m}^4 = 160\,000\ \text{kN - m}^2$$

Hence, $(\Delta_B)_x = 12500/160000 = 0.078$ m(→). Since the computed value is positive, the displacement is in the same direction as the unit load—to the right.

Rotation of joint B

Step 2: We apply a unit moment at B, as in Fig. E4.5.3(c). The direction or sense of the moment is arbitrarily selected.

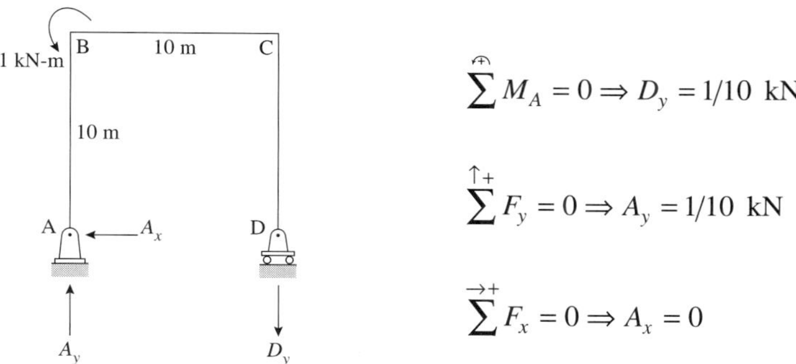

Fig. E4.5.3(c)

Step 3: Compute the internal moments:

Segment	FBD (SRL)	$M(x)$	FBD (SVL)	$m_\theta(x)$	$\int \frac{M(x)m_\theta(x)}{EI}dx$
AB $(0 < x_1 < 10)$	same as before	0	—	—	0
DC $(0 < x_2 < 10)$	same as before	0	—	—	0
BC $(0 < x < 10)$	same as before	$-15x^2 + 150x$	m_θ; B; x; 1; 10 m; A; 1/10	$\frac{x}{10} - 1$	$\int_0^{10}\left(-15x^2 + 150x\right)\left(\frac{x}{10} - 1\right)dx = -\frac{1250}{EI_{BC}}$
				Sum	$-\frac{1250}{EI_{BC}}$

Hence, $\theta_B = -1250/160000 = -0.0078$ rad $\Rightarrow \theta_B = 0.0078$ rad (↷).

Observation: The horizontal displacement and rotation at B can be reduced by increasing the moment of inertia of the beam BC alone. The columns do not affect the deflections at B.

4.5.2 Unit Load Method for Trusses

The methodology for computing the displacement of a truss joint is very similar to the manner in which we computed the displacements in beams and frames. However unlike beams and frames, the internal work is entirely due to the axial force.

External Loads. Consider the truss in Fig. 4.5.2.1. The truss is subjected to external forces and P_1 and P_2. Let us assume that is required to compute the vertical displacement at joint A, Δ_A. Our task now is to customize the general expression on the right-hand side of Eq. (4.5.5). Under the action of the (real) external loads, let the force in a typical member by N. The deformation in the member via Eq. (4.4.7) is $\Delta = NL/AE$.

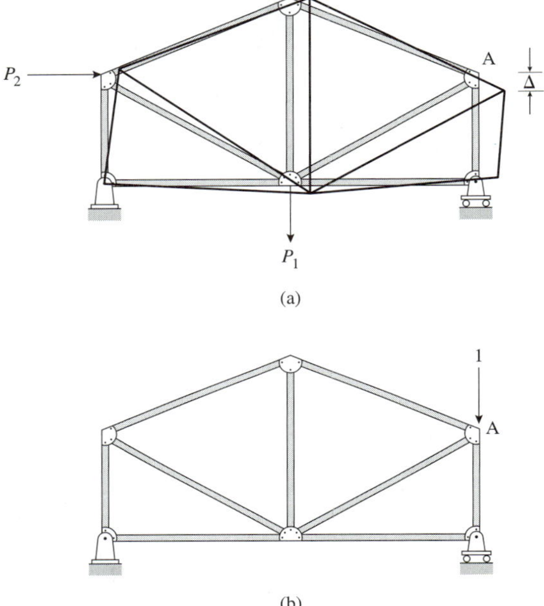

Fig. 4.5.2.1
(a) Truss subjected to real external loads. (b) Truss subjected to unit vertical load at A.

Substituting in Eq. (4.5.5), we have

$$(1)(\Delta) = \sum \frac{nNL}{AE} \tag{4.5.2.1}$$

where n is the axial force in a truss member due to the unit load, N is the axial force in a truss member due to the externally applied (real) loads, and the summation is over all the members in the truss. The above equation can be solved for Δ. Once again, note that the left-hand side represents the virtual work by the external unit load and the right-hand side represents the internal virtual work.

Fabrication Errors. The same concept used with the external loads can be used to compute the displacement when fabrication errors exist in members. Customizing the general expression on the right-hand side of Eq. (4.5.5) for the case when one or more members is either ΔL too long or short yields

$$(1)(\Delta) = \sum (n)(\Delta L) \tag{4.5.2.2}$$

where n is the axial force in a truss member due to the unit load, ΔL is the fabrication error (positive if the member is ΔL too long, negative if the member is ΔL too short), and the summation is over all the members with fabrication errors.

Thermal Loads. Temperature changes can also cause deflections in trusses. When a truss member of length L and coefficient of thermal expansion α is subjected to a temperature change ΔT, the change in length of the member $\Delta L = \alpha L(\Delta T)$. Hence

$$(1)(\Delta) = \sum (n)(\Delta L) = \sum (n)(\alpha L)(\Delta T) \tag{4.5.2.3}$$

where n is the axial force in a truss member due to the unit load, ΔT is the change in temperature of the member (positive for an increase in temperature in the element, negative for a decrease in temperature in the element), and the summation is over all the members with temperature changes.

Sign Convention: We will use the following sign convention: tensile forces are positive and compressive forces are negative.

General Procedure

Note that two structures are involved in the steps below. The structure with only the real loads (no virtual load) will be called SRL. The structure with only the virtual unit load acting on it (no real loads) will be called SVL.

Step 1: To set up the SVL, establish the joint where the displacement is to be computed and the desired direction. Apply a unit force at the joint along the desired direction. Compute the force n in each member. The method of joints is preferable.

Step 2: For the truss with the real external loads (SRL), compute the force N in each member. The method of joints is preferable.

Step 3: Now compute the displacement Δ using (a tabular form of computations is recommended)

$$\Delta = \sum \frac{nNL}{AE} + \sum (n)(\Delta L) + \sum (n)(\alpha L)(\Delta T) \tag{4.5.2.4}$$

The idea of linear superposition is used in combining the effects due to external loads, fabrication errors and thermal loads. If Δ is positive then the displacement is in the same direction as the unit force.

Tip: To compute the member forces in the two trusses SRL and SVL, one can assume that all members are in tension when drawing the FBD. Then the computed values can be entered into the table as is. In other words, members in tension will have a positive value and members in compression will have a negative value. This conforms to the assumed sign convention!

EXAMPLE 4.5.4 ***Deflection of a Planar Truss***

Compute the vertical displacement at A of the truss shown below. Take $E = 29(10^3)$ ksi. The member cross-sectional areas are shown in Fig. E4.5.4(a)

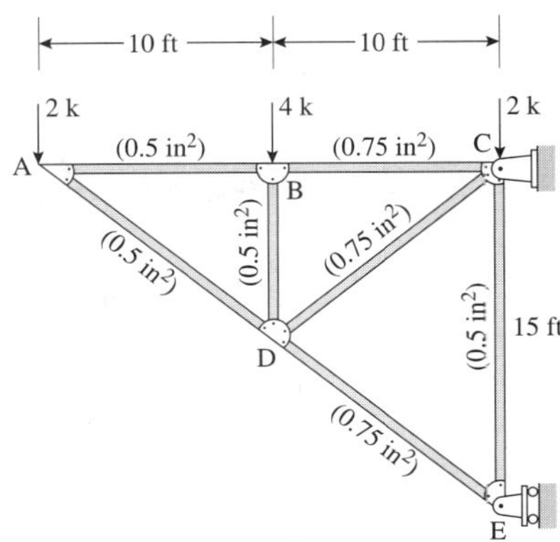

Fig. E4.5.4(a)

SOLUTION

Step 1: We use k, in as the problem units. We can analyze the truss shown above using the method of joints. The details are not shown (see Example 2.7.1). The analysis yields the values of the N force in each member (Eq. (4.5.2.4)).

Step 2: To compute the vertical displacement at A, we apply a unit force at A as shown in Fig. E4.5.4(b).

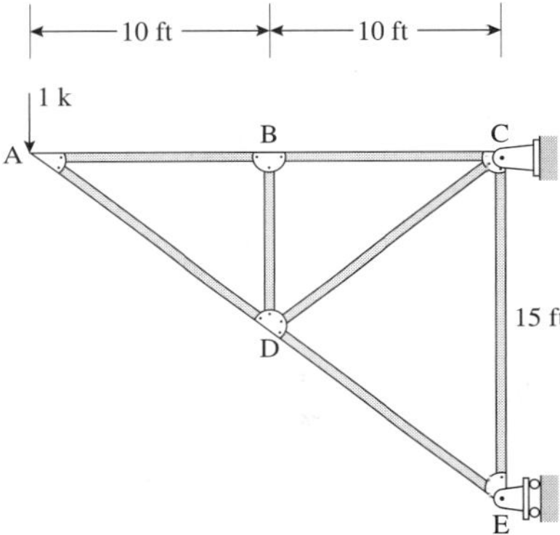

Fig. E4.5.4(b)

This truss can be analyzed using the method of joints. The details are not shown here. The analysis yields the values of the n force in each member (Eq. (4.5.2.4)).

Step 3: Now we are ready to compute the vertical displacement at A. Note the sign convection used in columns N and n. A negative sign indicates a member in compression.

Member	$N(k)$	$n(k)$	L/A(1/in)	$NnL/A(k^2$/in)
AB	2.67	1.33	240	852
BC	2.67	1.33	160	568
AD	–3.33	–1.67	300	1668
BD	–4	0	—	0
DC	3.33	0	—	0
DE	–6.67	–1.67	200	2228
CE	4	1	360	1440
			Sum	6756

Finally, substituting the value of the modulus of elasticity and using Eq. (4.5.2.4) gives

$$(1)(\Delta_A)_V = \frac{6756}{29000} = 0.23 \Rightarrow (\Delta_A)_V = 0.23''(\downarrow)$$

Since the displacement value is positive, the displacement is in the same direction as the unit force–downwards.

EXAMPLE 4.5.5

Deflection of a Planar Truss

For the truss in Example 4.5.4, introduce a fabrication error in member CE so that the vertical displacement at A is zero under the action of the applied loads.

SOLUTION

Step 1: Under the combined action of external loads and fabrication error(s) we have

$$(1)(\Delta_A) = \sum \frac{NnL}{AE} + \sum n(\Delta L)$$

where ΔL is the fabrication in a specific member.

Step 2: From the problem data, we have

$$\sum \frac{NnL}{AE} + \sum n(\Delta L) = 0$$

Using the results from Example 4.5.4, we find

$$\sum \frac{NnL}{AE} + \sum n(\Delta L) = 0.23 + (1)(\Delta L)_{CE} = 0$$

where $(\Delta L)_{CE}$ is the fabrication error in member CE. Solving yields $(\Delta L)_{CE} = -0.23$ in. In other words, member CE needs to be 0.23 in shorter for the net vertical displacement at A to be zero. On can imagine a turnbuckle in member CE that can be used to shorten the member. The force needed in the turnbuckle to achieve this would be $k(\Delta) = AE/L(\Delta) = [(0.5)(29000)/(15 \times 12)](0.23) = 18.5$ k.

Observation: Ideas such as the one illustrated in this example are used routinely. For example, camber is deliberately provided in roof trusses so that the net deflection in the truss (due to dead and other loads) is less than if an initial camber were not provided.

EXERCISES

Solve the following problems using the method of virtual work (unit load method).

Appetizers

4.5.1. Resolve Problem 4.2.1.

4.5.2. Resolve Problem 4.2.2.

4.5.3. Resolve Problem 4.2.3.

4.5.4. Resolve Problem 4.2.4.

Main Course

4.5.5. Resolve Problem 4.2.5.

4.5.6. Resolve Problem 4.2.6.

4.5.7. Resolve Problem 4.3.7.

4.5.8. Resolve Problem 4.3.8.

4.5.9. For the planar frame in Fig. P4.5.9, compute the horizontal displacement at B. *EI* is a constant for the entire frame.

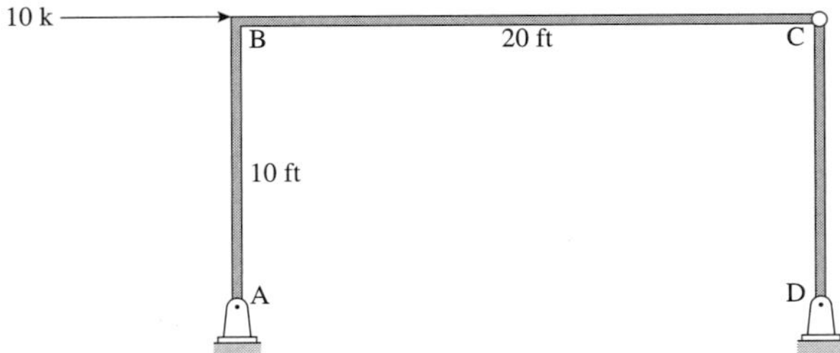

Fig. P4.5.9

4.5.10. For the planar frame in Fig. P4.5.10, compute (a) the rotation at A and (b) the horizontal displacement at C. *EI* is a constant for the entire frame.

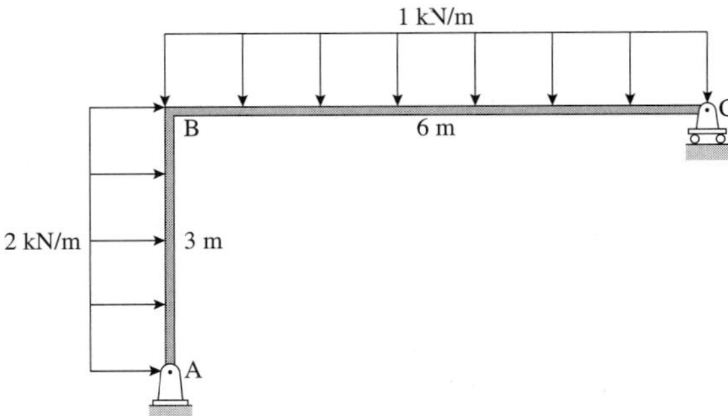

Fig. P4.5.10

4.5.11. For the planar frame in Fig. P4.5.11, compute the horizontal displacement and vertical displacement at C. *EI* is a constant for the entire frame.

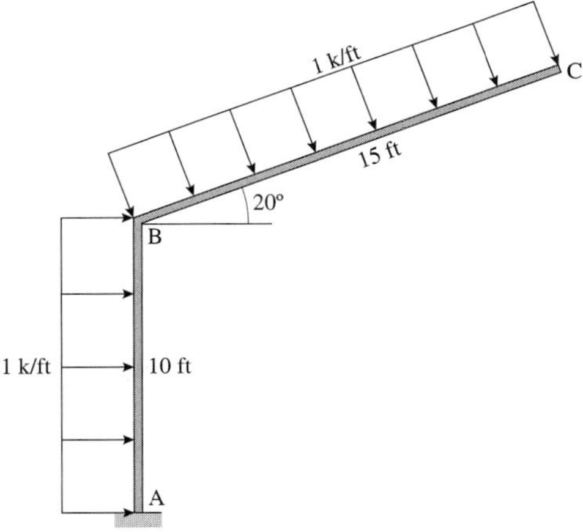

Fig. P4.5.11

4.5.12. The planar frame in Fig. P4.5.12 has a constant EI value. Compute (a) the horizontal displacement at C and (b) the rotation at A.

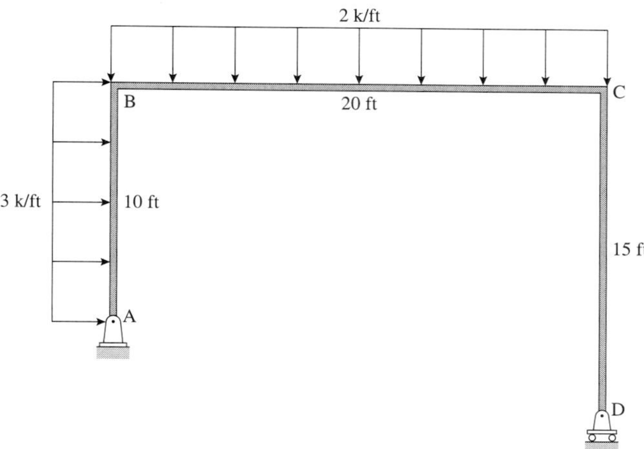

Fig. P4.5.12

4.5.13. The planar frame in Fig. P4.5.13 is such that $I_{AB} = I_{DE} = 2I$, and $I_{BC} = I_{CD} = I$. Compute the horizontal displacement at C.

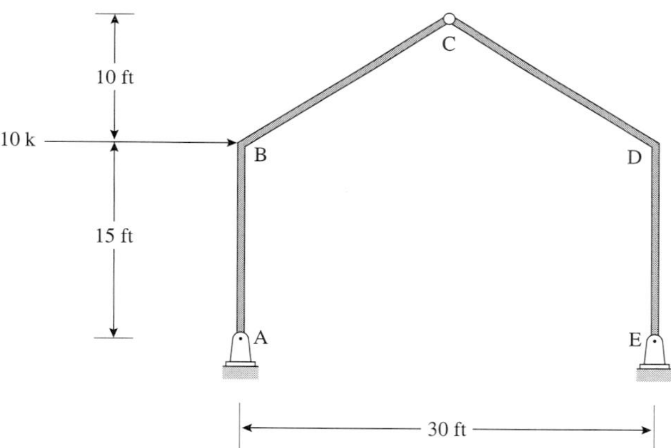

Fig. P4.5.13

4.5.14. The planar frame in Fig. P4.5.14 is such that $I_{BC} = I_{AB} = 2I$. Compute the vertical displacement at C.

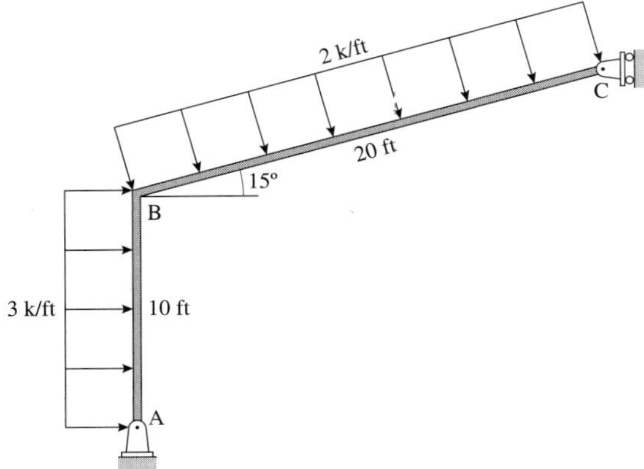

Fig. P4.5.14

4.5.15. The planar truss in Fig. P4.5.15 is made of steel, $E = 200$ GPa. The cross-sectional area of both the members is 0.01 m^2. Compute the vertical and the horizontal displacements at B.

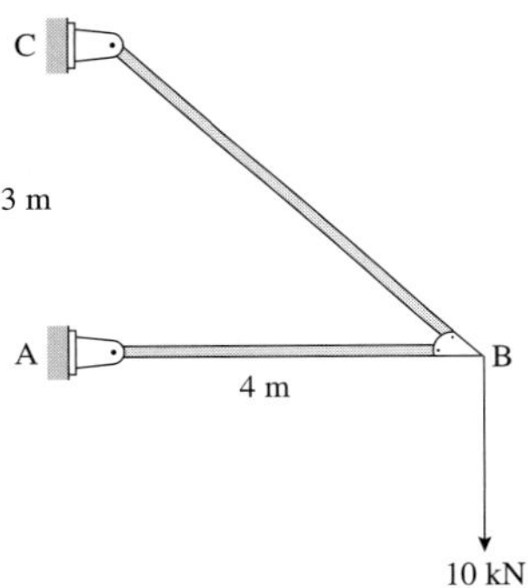

Fig. P4.5.15

4.5.16. The planar truss in Fig. P4.5.16 is made of steel: $E = 200$ GPa. The cross-sectional area of both the members is 0.01 m^2. Compute the vertical and the horizontal displacements at B.

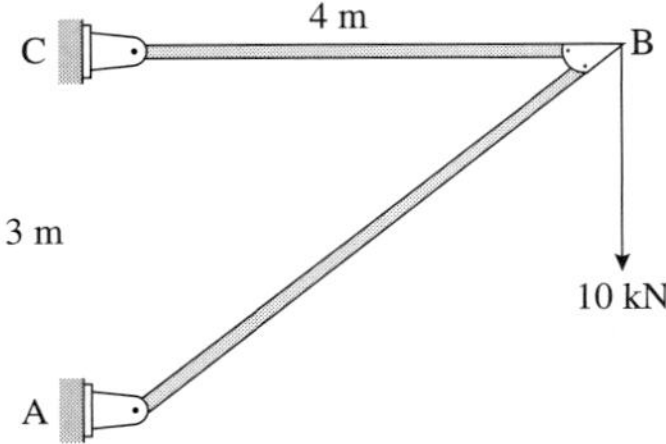

Fig. P4.5.16

4.5.17. The planar truss in Fig. P4.5.17 is made of steel: $E = 30(10^6)$ psi. The cross-sectional area of all the members is 2 in^2. Compute (a) the vertical displacement at C and (b) the horizontal displacement at D.

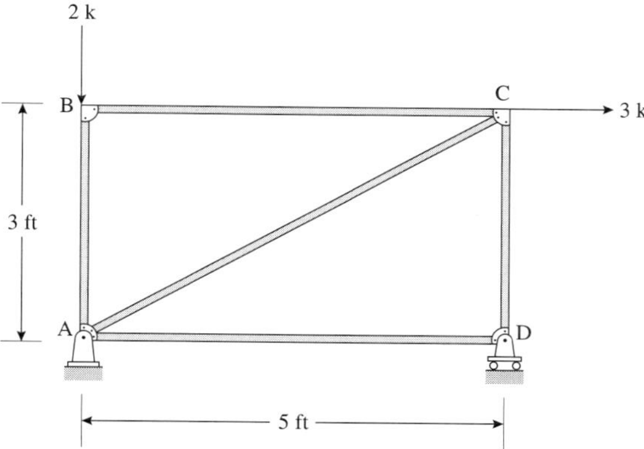

Fig. P4.5.17

4.5.18. The planar truss in Fig. P4.5.18 is made of steel: $E = 30(10^6)$ psi. The cross-sectional area of all the front chord members is 2 in^2, of the back chord members is 3 in^2, and of the web members is 2.5 in^2. Compute the horizontal displacement at A.

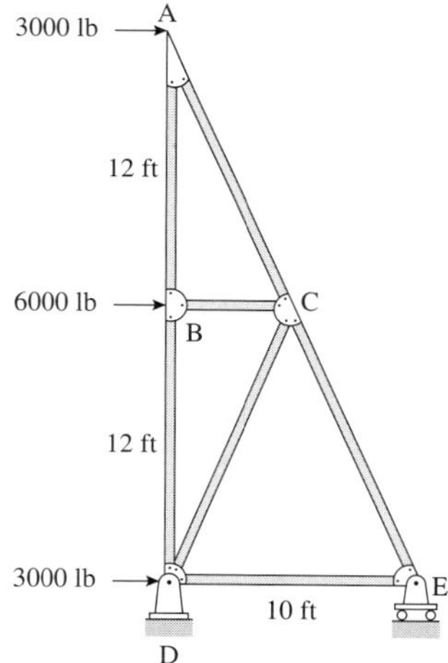

Fig. P4.5.18

4.5.19. The cross-sectional area of all the members in Fig. P4.5.19 is 0.01 m^2. Compute the vertical displacements at D and E. The planar truss is made of steel: $E = 200$ GPa.

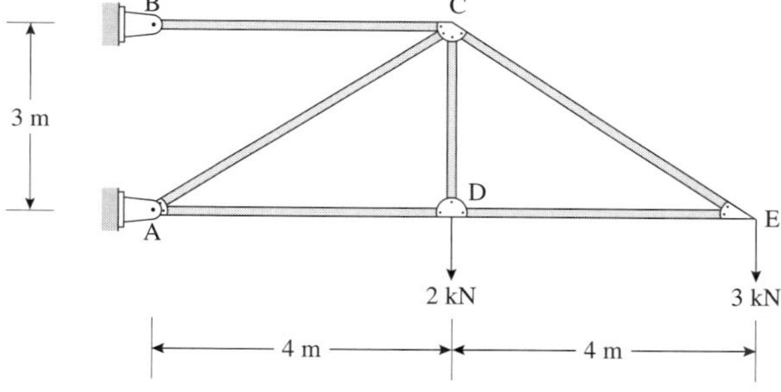

Fig. P4.5.19

Structural Concepts

4.5.20. Is the principle of virtual work applicable only to linear, elastic structural systems?

4.5.21. Another powerful method to compute deflections is called the Castigliano's (First) Theorem. Alberto Castigliano (1847–1884) was a Italian railroad engineer who developed two theorems as a part of his dissertation for an engineering degree in Turin.

Castigliano's first theorem. If the strain energy U for a conservative, linear system subjected to conservative loads is expressed in terms of independent displacements D_1, D_2,..., then the load P_i that corresponds to D_i is given by

$$P_i = \frac{\partial U}{\partial D_i}$$

Consider the spring-roller system in Fig. P4.5.21(a).

Fig. P4.5.21(a)

Consider the spring system in Fig. P4.5.21(b). There is only one independent displacement D needed to express the displaced state of the system.

Fig. P4.5.21(b)

The strain energy $U = \frac{1}{2}kD^2$. Hence using the first theorem, we have

$$P = \frac{\partial U}{\partial D} = kD \Rightarrow D = \frac{P}{k}$$

The spring system shown in Fig. P4.5.21(c) has two independent displacements D_1 and D_2. The strain energy

$$U = \frac{1}{2}k_1 D_1^2 + \frac{1}{2}k_2\left(D_2 - D_1\right)^2$$

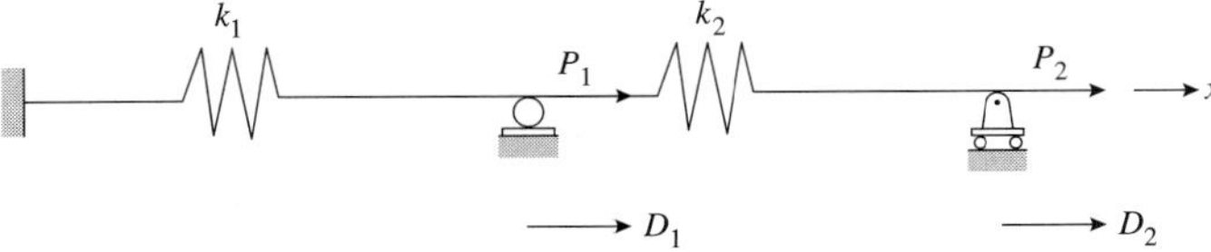

Fig. P4.5.21(c)

Using the first theorem, we have

$$\frac{\partial U}{\partial D_1} = k_1 D_1 - k_2\left(D_2 - D_1\right) = P_1, \qquad \frac{\partial U}{\partial D_2} = k_2\left(D_2 - D_1\right) = P_2$$

These two equations can be solved to obtain the displacements D_1 and D_2. Extend the ideas presented here to beams, frames, and trusses and solve a few of the exercise problems.

4.5.22. The work of two pioneers, E. Betti and Lord Rayleigh, is captured in an important theorem. The theorem by itself is not as useful as the application in other more practical methods.

Reciprocal theorem. If an elastic body is subjected to two systems of forces F_1 and F_2, the work done by the first system F_1 going through the displacements D_1 due to a second system of forces is equal to the work done by the second set of forces F_2 going through the displacements D_2 due to the first system of forces.

Consider the cantilever beam in Fig. P4.5.22. Let P represent the first force system and M represent the second force system. Let the corresponding displacements D_1 be Δ, and D_2 be θ.

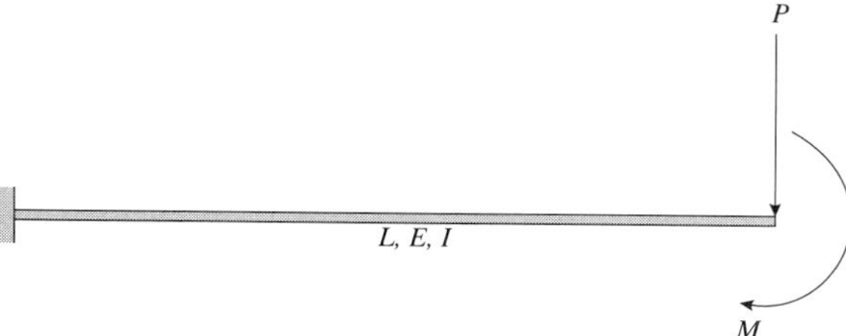

Fig. P4.5.22

The following result can be easily derived: $\Delta = ML^2/2EI$, i.e., the tip displacement Δ due to a tip moment M alone. Now how do we use the reciprocal theorem to compute the rotation θ due to a tip force P? From the reciprocal theorem, we have

$$(P)(\Delta) = (M)(\theta)$$

Substituting yields $(P)(ML^2/2EI) = (M)(\theta)$. Hence, $\theta = PL^2/2EI$.

Using this idea and the results from Example 4.2.2, compute the vertical displacement at B due to a 100 k-in counterclockwise moment at C.

This theorem will be useful with the force method discussed in the next chapter.

Chapter 5

Indeterminate Structural Systems

Indeterminate steel frames are commonly used in the design of multi-storied buildings.

"A good theory is worth a thousand computer runs."

"A precedent embalms a principle." Benjamin Disraeli

"#3 pencils and quadrille pads." Seymour Cray, when asked what CAD tools he used to design Cray I; he also recommended using the back side of the pages so that the lines were not so dominant

Several classical methods have been developed over the years to solve for the forces and the displacements of statically determinate and indeterminate systems. The list includes the force method (also called the method of consistent displacements or deformations), the moment distribution method, three moment equation, Castigliano's theorems, the slope-deflection method, and a host of approximate methods especially for building frames. Two factors have played an important role in paring this list. First, time constraints in any undergraduate curriculum permit coverage of only a limited number of topics. Second, with the ready availability of sophisticated computer programs, it is perhaps necessary to look at the important fundamental concepts so that one can use these computer programs efficiently and correctly. We cover two important methods in this chapter—the force method and the slope-deflection method—that help us understand the fundamentals as well as lay the groundwork for the numerical methods covered in the next chapter.

In this chapter we sow the seeds for more powerful techniques that can be generally classified as displacement-based energy methods. The basic idea is to use the concepts associated with equilibrium, compatibility, and material behavior (or constitutive law) in designing the solution process.

Finally, a few words about indeterminate versus determinate structural systems. There are several advantages in designing indeterminate systems. These include the design of lighter and more rigid structures. With the added redundancy in the system there is an increase in the overall factor of safety. The primary disadvantage of indeterminate systems is the effect of support settlements, temperature changes, fabrication errors, etc., on the performance of the members. Large stresses or stress reversals may result from these actions.

OBJECTIVES

- To understand and apply the force method in solving statically indeterminate structural systems. We will reinforce the ideas associated with linear superposition and compatibility.
- To understand and apply the slope-deflection method in solving statically determinate and indeterminate structural systems.
- To understand the strengths and limitations of the classical solution techniques.

ASSUMPTIONS

- Small deformation theory applies.
- Axial and shear deformations of frame members are assumed small in comparison to the bending or flexural deformations.

5.1 FORCE METHOD

The force method is one of the two classical methods that we will investigate to solve for the response of statically indeterminate systems. The method owes its origin to the contributions made by James Clerk Maxwell, Heinrich Muller-Breslau, and Otto Mohr in the 1860s.

In Chapter 2 we discussed determinate systems. Recall that a determinate system is one where the unknowns (support reactions, pin forces, member internal forces) are obtained by using the concept of equilibrium only. However, to solve indeterminate systems, we must combine the concept of equilibrium with compatibility. This is the basic idea behind the force method. We introduce the concepts in stages, starting with beams that are statically indeterminate to degree one.

5.1.1 Beams

Consider the beam shown in Fig. 5.1.1.1. The beam is statically indeterminate to degree one—four support reactions and three equations of equilibrium.

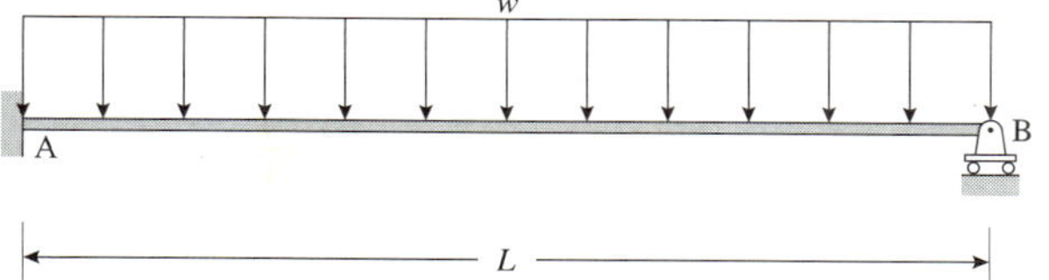

Fig. 5.1.1.1 Statically indeterminate beam.

If we remove an appropriate reaction, R, from the original indeterminate beam, the resulting beam is stable and determinate. Let us label this *beam A*. Using the concepts from the previous sections, we should be able to compute the deflection of this determinate beam. Now consider the same determinate beam without the external loads. Let us label this *beam B*. If we now apply the reaction R (which was removed from the original beam) as an external load on this determinate beam, we again should be able to compute the deflection of the beam. However, the expression for the deflection will be in terms of R. As we noted in Chapter 3, the deflection at a support along the support reaction is zero. With the concept of superposition, the superposition of beams A and B should yield the original indeterminate beam. This situation is shown in Fig. 5.1.1.2. We have used the vertical reaction at B, B_y, as the reaction or *redundant*. This extra support reaction is called a redundant since we can remove and still have a stable structure.

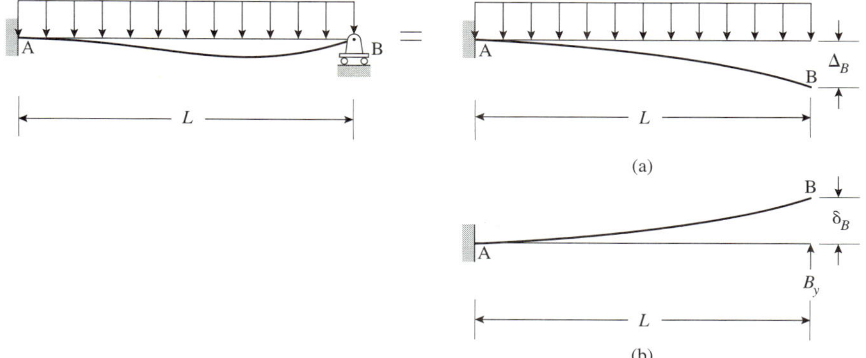

Fig. 5.1.1.2 Superposition of two determinate beams A and B to yield the original indeterminate beam.

It is clear that beams A and B yield the original beam provided the net deflection at B is zero. In other words, with an appropriate sign convention, we have

$$\Delta_B + \delta_B = 0 \tag{5.1.1.1}$$

This is the compatibility condition that ensures compatibility between the original indeterminate beam and the two determinate beams. Hence the method is also known as method of consistent displacements.

How do we use the above equation? Consider the beam shown in Fig. 5.1.1.3. The beam is identical to beam B except that a unit force replaces the redundant B_y. We call this beam *beam C*.

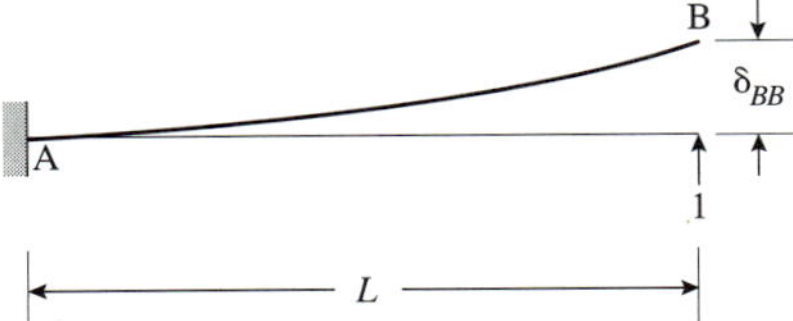

Fig. 5.1.1.3 Beam C.

For linear systems, it is clear that $\delta_B = B_y\delta_{BB}$. In other words, the displacements in the two beams are proportional to each other. A note about the nomenclature: δ_{ij} is the displacement at i due to a unit force applied at j. We can now utilize this fact to modify Eq. (5.1.1.1) as

$$(\uparrow+)\ \Delta_B + B_y\delta_{BB} = 0 \tag{5.1.1.2a}$$

or even

$$(\downarrow+)\ \Delta_B + B_y\delta_{BB} = 0 \tag{5.1.1.2b}$$

from which B_y can be solved for. The sign convention for the displacement is arbitrarily chosen. It should be emphasized that (a) the above equation is first written in a symbolic form as shown, before it is actually used to solve for the redundant, (b) while a sign convention is used with the above equation, we do not assign signs to the two displacements when we write it in a symbolic form, since that would mean that we know the correct directions of the displacements, and (c) the beam with the unit force was used so that both Δ_B and δ_{BB} could be computed using the results from that beam. As we will see later, *after* we have computed the two displacements, we can assign the appropriate signs to the two displacements as per the sign convention.

If we use the unit load method[1] to compute Δ_B, we can use beams A (to obtain $M(x)$) and C (to obtain $m(x)$) as

$$\Delta_B = \int \frac{M(x)m(x)}{EI}\,dx \tag{5.1.1.3}$$

To compute δ_{BB} we can use beam C (to obtain $M(x)$) and beam C again (to obtain $m(x)$) as

$$\delta_{BB} = \int \frac{m(x)m(x)}{EI}\,dx \tag{5.1.1.4}$$

Since δ_{BB} is the displacement at B due to a unit force at B, the structure with the real load is beam C and the structure with the virtual load is also beam C! An examination of the above two equations shows that we need to compute only two moment expressions—$M(x)$ from beam A and $m(x)$ from beam C.

Why did we select B_y as the redundant? Is the choice of the redundant unique? The answer to the first question is that selection was done arbitrarily. The answer to the second question is that, usually, the choice of the redundant is not unique. Note that we can use only those reactions as the redundant (a) that cannot be computed using the structural FBD and equilibrium equations, and (b) whose removal renders the structure determinate *and* stable. Going back to the previous example, we could have chosen M_A as the redundant, since M_A cannot be found from the structural FBD and removing M_A leaves the beam determinate and stable (see Fig. 5.1.1.4).

It is clear that beams A and B yield the original beam provided the net rotation at A is zero. In other words, with an appropriate sign convention

$$\theta_A + \alpha_A = 0 \tag{5.1.1.5}$$

[1] It is not necessary to use the unit load method to compute the displacements. One could use any method. The difference is that the unit load method is usually more efficient. With most other methods one would have to compute, for example, δB in terms of the redundant B_y.

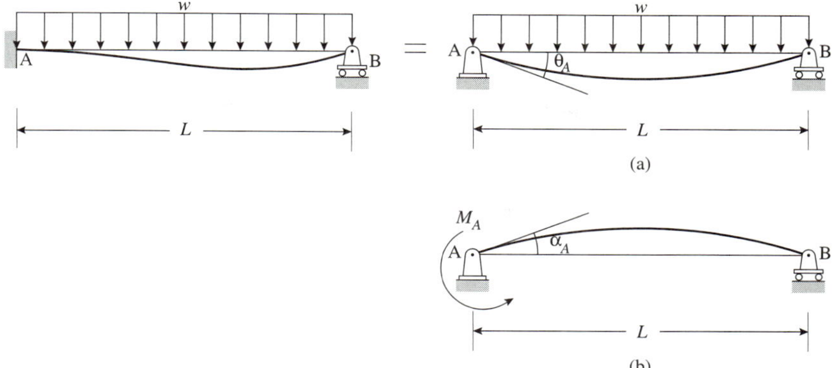

Fig. 5.1.1.4
M_A as the redundant.

As before, to enable effective computation of the rotations we can use the beam with a unit moment as in Fig. 5.1.1.5.

(5.1.1.6a)

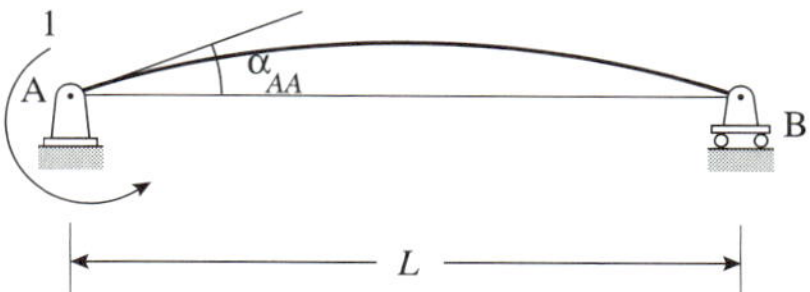

Fig. 5.1.1.5

$$(\curvearrowright +)\theta_A + M_A\alpha_{AA} = 0$$

or even

$$(\curvearrowleft +)\theta_A + M_A\alpha_{AA} = 0 \tag{5.1.1.6b}$$

from which M_A can be solved.

General Procedure for Beams with a Single Redundant

Step 1: Identify the redundant. If the redundant is removed from the original structure, the resulting beam must be stable and determinate. Now create the two beams whose superposition results in the original indeterminate beam.

Remove the redundant from the original beam but leave the external loads. This is beam DSRL (Determinate Structure with Real Loads) that we previously called beam A.

Remove the redundant and all loads from the original beam. Assume a direction for the redundant. Now apply a unit force (if the redundant is a force reaction) or unit moment (if the redundant is a moment reaction) along the assumed direction of the redundant. This is beam DSUL (Determinate Structure with Unit Load) that we previously called beam C.

Write the single compatibility equation in the symbolic form. Select a sign convention for the associated displacements appearing in the equation. This equation should contain the redundant.

Step 2: Compute the deflection from beam DSRL. We saw the procedure in Chapter 4.

Step 3: Compute the deflection from beam DSUL. We saw the procedure in Chapter 4.

Step 4: Now substitute the deflections from Steps 2 and 3 into the compatibility equation. Use the sign convention to assign the correct sign to the two displacements. Solve the compatibility equation for the redundant. If the answer is positive, the assumed direction for the redundant is correct. Otherwise, flip the direction.

Step 5: The other support reactions can now be computed using the free-body diagram of the original beam (or through superposition of the two determinate beams).

We now look at examples dealing with beams that are statically indeterminate to degree one.

EXAMPLE 5.1.1 ***Statically Indeterminate Beam***

Compute the support reactions of the beam in Fig. E5.1.1(a). *EI* is a constant.

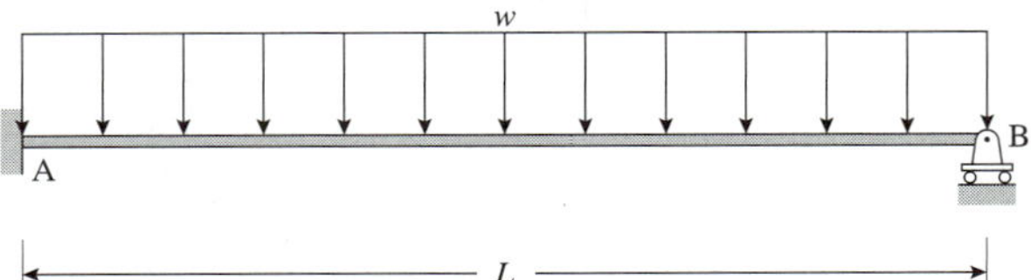

Fig. E5.1.1(a)

SOLUTION

Step 1: There are four support reactions and three equations of static equilibrium. The beam is statically indeterminate to degree one. One must ask what support reactions can be used as the redundant. There are several options with this beam: A_y, M_A, or B_y. A_x cannot be used since removing A_x will render the resulting beam unstable. We select B_y as the redundant (see Fig. E5.1.1(b)).

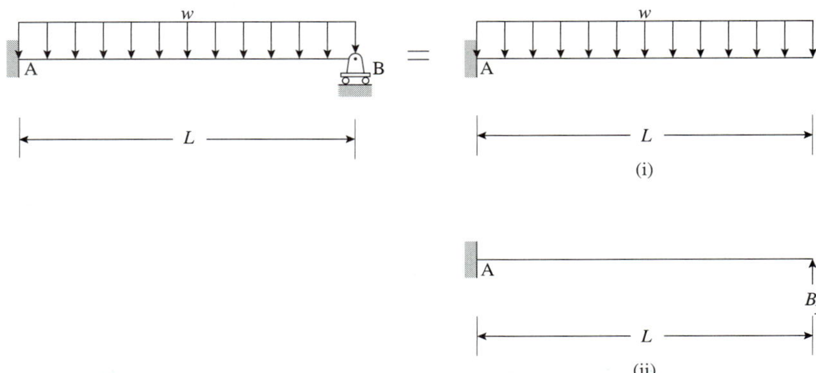

Fig. E5.1.1(b) Superposition of the determinate beams.

The compatibility equation can be written as

$$(\uparrow+)\,\Delta_B + \delta_B = \Delta_B + B_y\delta_{BB} = 0$$

The quantities in the compatibility equation are shown in Fig. E5.1.1(c).

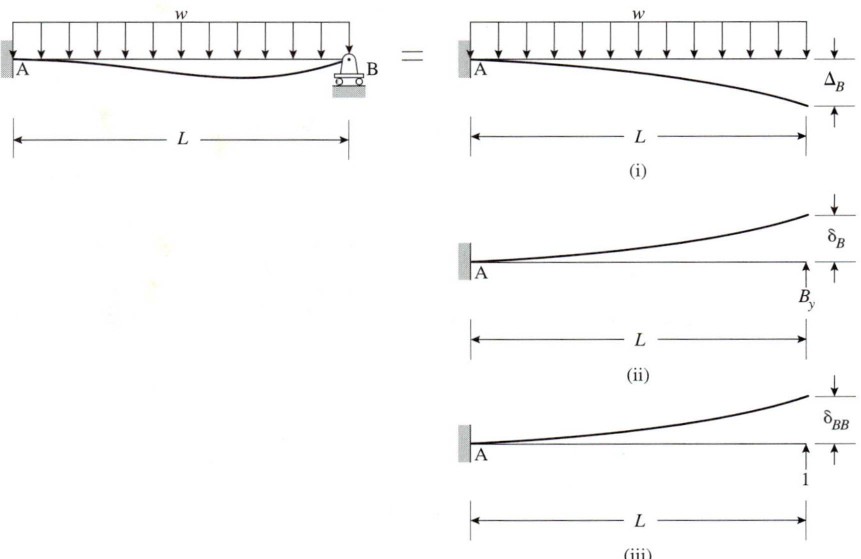

Fig. E5.1.1(c)

We compute the deflections as follows:

$$\Delta_B = \int_0^L \frac{M(x)m(x)}{EI}dx \quad \text{(transverse displacement at B due to the external loads)}$$

$$\delta_{BB} = \int_0^L \frac{m(x)m(x)}{EI}dx \quad \text{(transverse displacement at B due to a unit load at B)}$$

where $M(x)$ is obtained from the determinate beam (i) in Fig. E5.1.1(c), and $m(x)$ is obtained from the determinate beam (iii) in the same figure.

Using Beam (i) in Fig. E5.1.1(c), we find

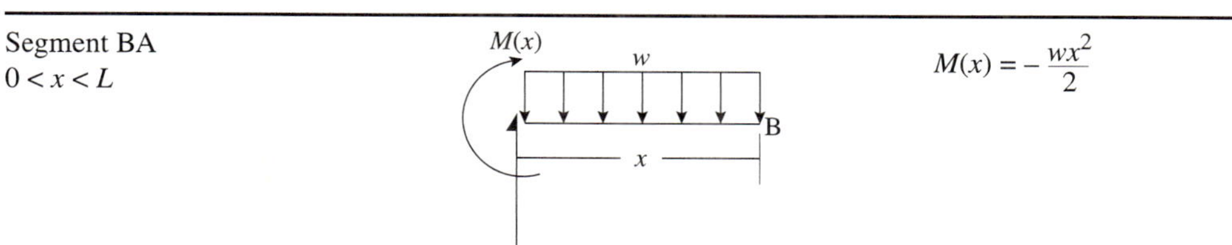

Using beam (iii) in Fig. E5.1.1(c), we find

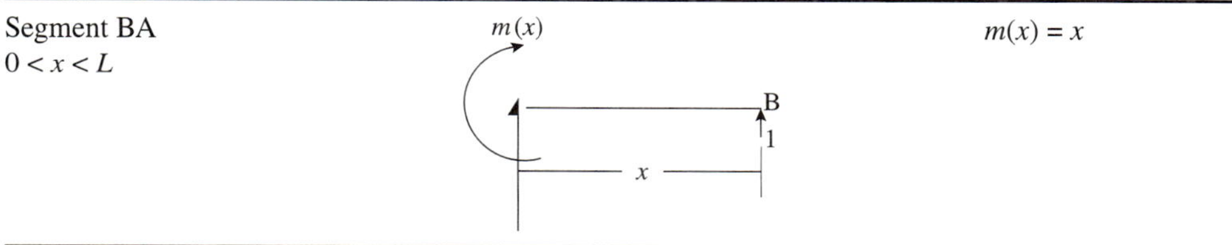

Hence,

$$EI\Delta_B = \int_0^L M(x)m(x)dx = \int_0^L \left(-\frac{wx^2}{2}\right)(x)dx = -\frac{wL^4}{8}.$$

The negative sign indicates that the displacement is opposite in direction to the direction of the unit load. Hence, $\Delta_B = (wL^4/8EI)(\downarrow)$.

$$EI\delta_{BB} = \int_0^L m(x)m(x)dx = \int_0^L (x)(x)dx = \frac{L^3}{3}.$$

The positive sign indicates that the displacement is in the same direction of the unit load. Hence, $\delta_{BB} = (L^3/3EI)(\uparrow)$.

Step 2: Construct the compatibility equation and solution:

$$(\uparrow +)\,\Delta_B + B_y\delta_{BB} = 0$$

Using the sign convention and the computed quantities, we get

$$-\frac{wL^4}{8EI} + B_y\left(\frac{L^3}{3EI}\right) = 0 \qquad \text{or} \qquad B_y = \frac{3wL}{8}$$

The positive sign indicates that the assumed direction (see Fig. E5.1.1(b)) is correct. Now we can use the FBD of the beam to compute the other support reactions, as in Fig. E5.1.1(d)):

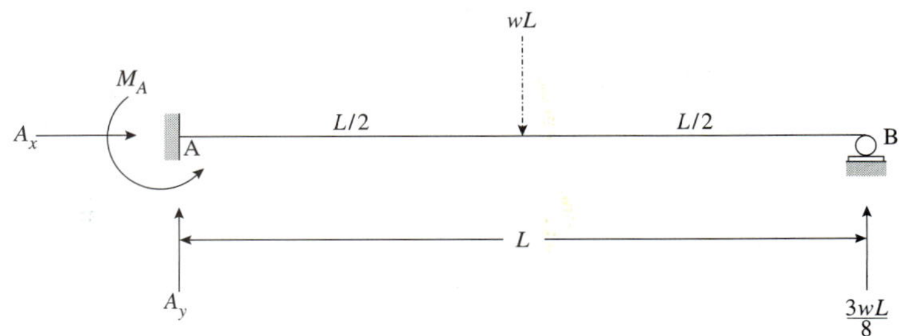

Fig. E5.1.1(d)

$$\overset{\rightarrow+}{\sum} F_x = 0 = A_x \Rightarrow A_x = 0$$

$$\overset{\uparrow+}{\sum} F_y = 0 = A_y - wL + \frac{3wL}{8} \Rightarrow A_y = \frac{5wL}{8}$$

$$\sum M_A = 0 = M_A + -(wL)(L/2) + \left(\frac{3wL}{8}\right)(L) \Rightarrow M_A = \frac{wL^2}{8}$$

Resolve this problem using M_A as the redundant.

EXAMPLE 5.1.2 ***Statically Indeterminate Beam***

Compute the support reactions of the beam in Fig. E5.1.2(a). *EI* is a constant.

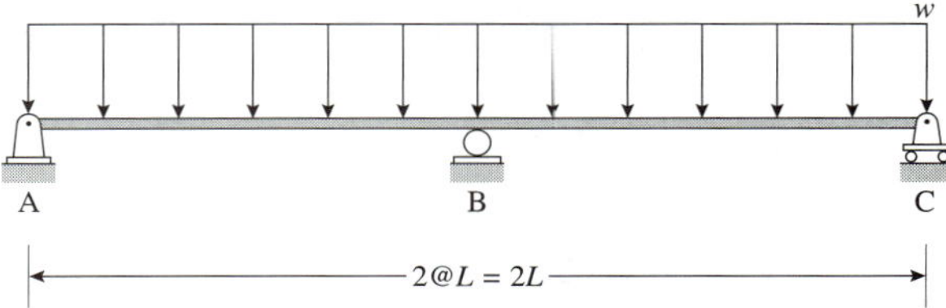

Fig. E5.1.2(a)

SOLUTION

Step 1: The beam is statically indeterminate to degree one. Let the reaction at B, B_y, be the redundant. The two determinate beams are shown in Fig. E5.1.2(b). Hence the compatibility equation can be written as

$$(\uparrow+)\,\Delta_B + B_y\delta_{BB} = 0$$

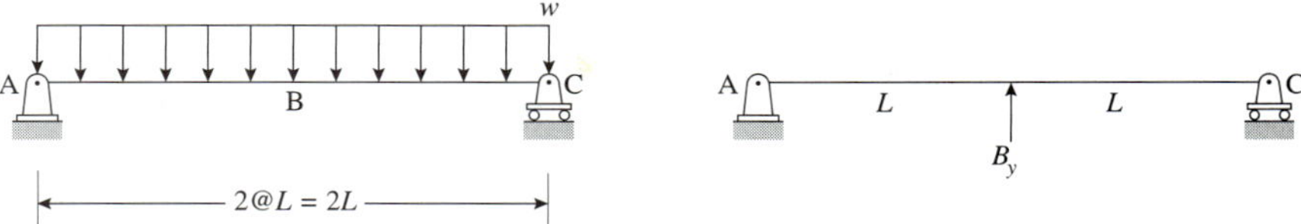

Fig. E5.1.2(b)

Step 2: Computation of internal moments in the determinate beams. From examining both the beams in Fig. E5.1.2(b), it should be clear that we need two segments to compute the moment expressions—AB and CB. Using the beam on the left in Fig. E5.1.2(b), we find

Segment	FBD	$M(x)$
AB $0 < x < L$	A, wx, M, x, wL	$wLx - \frac{wx^2}{2}$
CB $0 < x_1 < L$	M, wx_1, C, x_1, wL	$wLx_1 - \frac{wx_1^2}{2}$

Using the beam on the right in Fig. E5.1.2(b), we find

Segment	FBD	$m(x)$
AB $0 < x < L$	A, m, x, 1/2	$-\frac{x}{2}$
CB $0 < x_1 < L$	m, C, x_1, 1/2	$-\frac{x_1}{2}$

Step 3: Computation of the deflections:

$$\Delta_B = \frac{1}{EI}\int_0^L \left(wLx - \frac{wx^2}{2}\right)\left(-\frac{x}{2}\right)dx + \frac{1}{EI}\int_0^L \left(wLx_1 - \frac{wx_1^2}{2}\right)\left(-\frac{x_1}{2}\right)dx_1$$

Or $\Delta_B = -\frac{5wL^4}{48EI} - \frac{5wL^4}{48EI} = -\frac{5wL^4}{24EI} \Rightarrow \Delta_B = \frac{5wL^4}{24EI}(\downarrow)$

$$\delta_{BB} = \frac{1}{EI}\int_0^L \left(-\frac{x}{2}\right)\left(-\frac{x}{2}\right)dx + \frac{1}{EI}\int_0^L \left(-\frac{x_1}{2}\right)\left(-\frac{x_1}{2}\right)dx_1$$

Or $\delta_{BB} = \frac{L^3}{12EI} + \frac{L^3}{12EI} = \frac{L^3}{6EI}(\uparrow)$

Step 4: Generating and solving the compatibility equation

$$-\frac{5wL^4}{24EI} + B_y\left(\frac{L^3}{6EI}\right) = 0$$

we have, $B_y = 5wL/4$. Using this value and the FBD of the beam, we can compute the support reactions at A. The final beam FBD is given in Fig. E5.1.2(c):

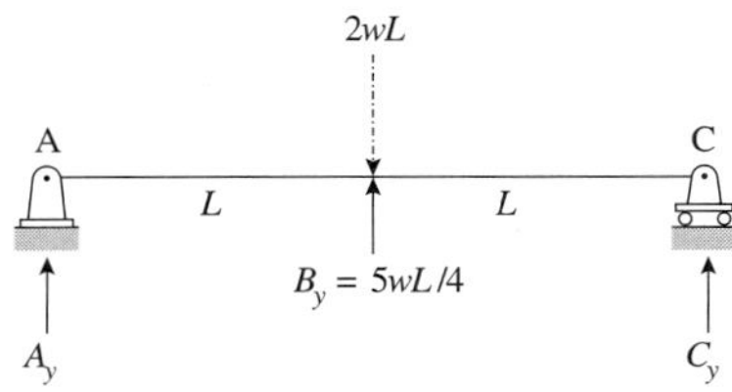

Fig. E5.1.2(c)

$$\overset{\curvearrowleft+}{\sum} M_A = 0 = -\left(\frac{3wL}{4}\right)(L) + (C_y)(2L) \Rightarrow C_y = \frac{3wL}{8}(\uparrow)$$

$$\overset{\uparrow+}{\sum} F_y = 0 = A_y - 2wL + \frac{5wL}{4} + C_y \Rightarrow A_y = \frac{3wL}{8}(\uparrow)$$

Finally, a word about the results. In Chapter 3 we saw some simplifying assumptions made in modeling loads acting on indeterminate systems. Fig. E5.1.2(d) shows the ratio of the reactions when the simplifying assumption is made to distribute the loads evenly in each span to the two supports. However, as we have seen in this example, the ratio of the reactions is different if analyzed as an indeterminate beam (Fig. E5.1.2(e)). Note, however, that the increase in the vertical "reaction" at B is from wL (determinate beams) to $5wL/4$ (indeterminate beam).

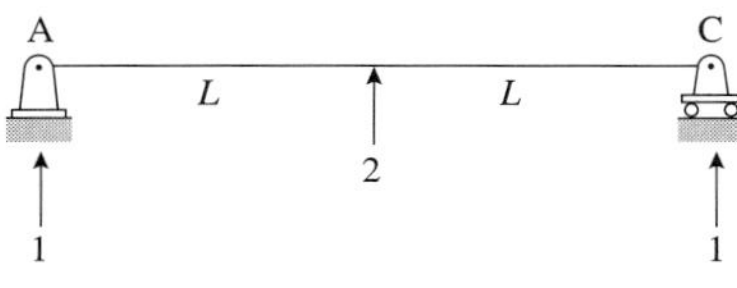

Fig. E5.1.2(d) Ratio of the reactions (assuming two determinate beams).

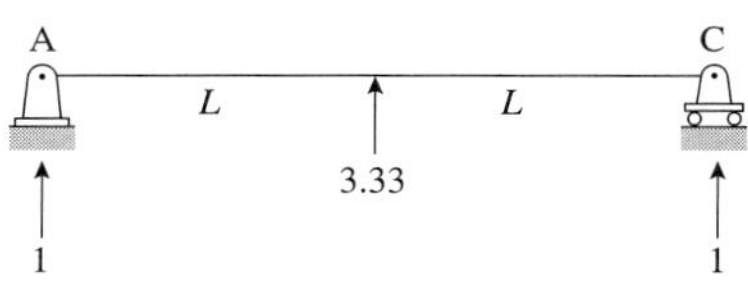

Fig. E5.1.2(e) Ratio of the reactions (indeterminate continuous beam).

EXAMPLE 5.1.3 ***Statically Indeterminate Beam***

Compute the support reactions of the beam in Fig. E5.1.3(a). Take E = 200 GPa and I = 600(10^6) mm^4.

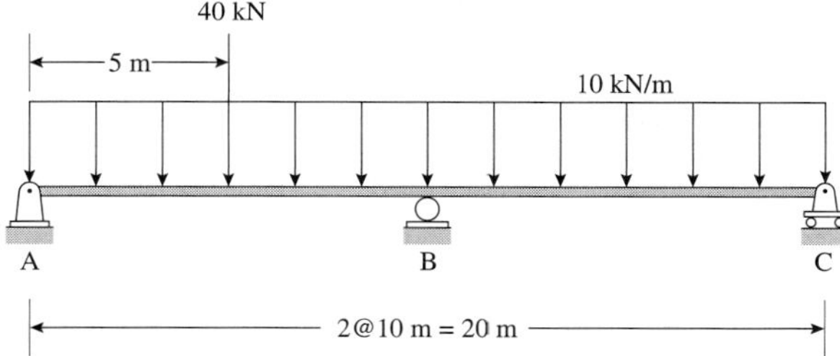

Fig. E5.1.3(a)

SOLUTION

Step 1: The beam is statically indeterminate to degree one. Let us select B_y as the redundant. The two determinate beams are shown in Fig. E5.1.3(b). The compatibility equation is

$$(\uparrow+)\ \Delta_B + B_y\delta_{BB} = 0$$

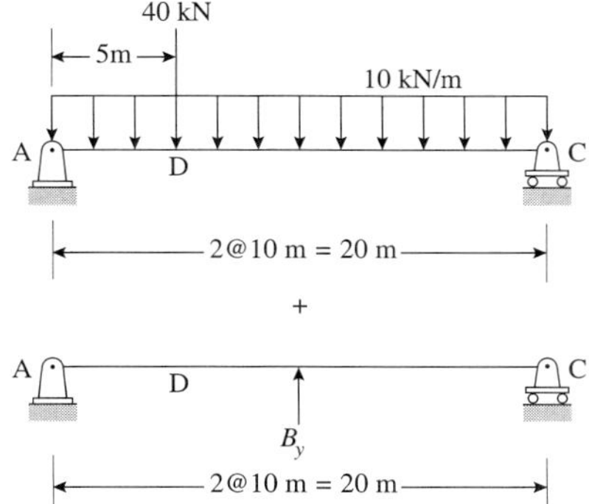

Fig. E5.1.3(b)

Step 2: Computation of the internal moments. An examination of the point of discontinuities in the bending moment diagram of the two beams shows that we need *three* segments—AD, DB, and BC. The calculations of the support reactions for the two structures are not shown—the FBDs should make it clear what their values and directions are.

Segment	FBD	$M(x)$	FBD	$m(x)$
AD $(0 < x < 5)$	$10x$, M, A, x, 130	$-5x^2 + 130x$	A, m, x, $1/2$	$-(x/2)$
DB $(5 < x < 10)$	40, $10x$, M, A, $x - 5$, D, x, 130	$-5x^2 + 90x + 200$	A, m, x, $1/2$	$-(x/2)$
CB $(0 < x_1 < 10)$	M, $10x_1$, C, x_1, 110	$-5x_1^2 + 110x_1$	m, C, x_1, $1/2$	$-(x_1/2)$

Step 3: Computation of the deflections:

$$EI\Delta_B = \int_0^5 \left(-5x^2 + 130x\right)\left(-\frac{x}{2}\right)dx + \int_5^{10} \left(-5x^2 + 90x + 200\right)\left(-\frac{x}{2}\right)dx$$

$$+\int_0^{10} \left(-5x_1^2 + 110x_1\right)\left(-\frac{x_1}{2}\right)dx_1 = -2317.7 - 11015.6 - 12083.3 = -25416.66$$

Hence, $\Delta_B = (25416.66/EI)(\downarrow)$.

$$EI\delta_{BB} = \int_0^{10}\left(\frac{x}{2}\right)^2 dx + \int_0^{10}\left(\frac{x_1}{2}\right)^2 dx_1 = \frac{250}{3} + \frac{250}{3} = \frac{500}{3}$$

Hence, $\delta_{BB} = (500/3EI)(\uparrow)$.

Step 4: Compatibility equation. Substituting in the compatibility equation, we have

$$-\frac{25416.66}{EI} + B_y\left(\frac{500}{3EI}\right) = 0$$

Solving, we find $B_y = 152.5$ kN($\uparrow$).

Step 5: Support reactions. Using the FBD of the entire beam in Fig. E5.1.3(c), we can compute the rest of the support reactions.[2]

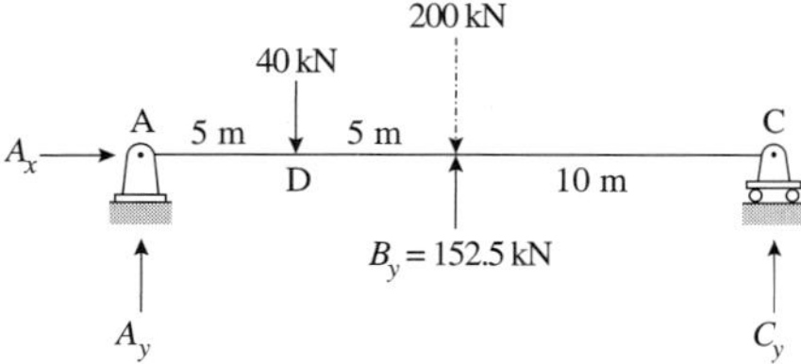

Fig. E5.1.3(c)

$$\overset{\curvearrowleft +}{\sum} M_A = 0 = -40(5) - 200(10) + 152.5(10) + C_y(20) \Rightarrow C_y = 33.75\,\text{kN}(\uparrow)$$

$$\overset{\uparrow +}{\sum} F_y = 0 = A_y - 40 - 200 + 152.5 + C_y \Rightarrow A_y = 53.75\,kN(\uparrow)$$

$$\overset{\rightarrow +}{\sum} F_x = 0 \Rightarrow A_x = 0$$

Since EI is a constant, the actual values are not needed to compute the support reactions. The shear force and the bending moment diagrams are shown in Fig. E5.1.3(d).

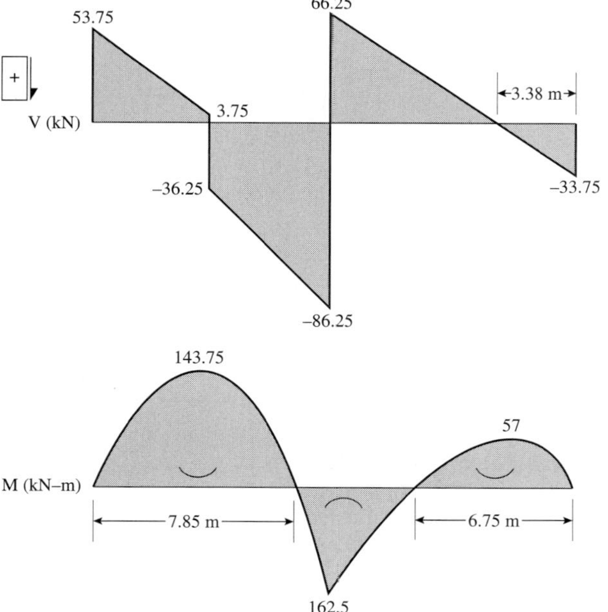

Fig. E5.1.3(d) Shear force and bending moment diagrams.

[2] The concept of linear superposition can be used to compute the support reactions. In other words, the results from the two determinate structures can be added algebraically. For beams and frames, this is left as an exercise. A later example (Example 5.1.8) illustrates the process for a truss structure.

EXAMPLE 5.1.4 ***Statically Indeterminate Beam with Support Settlement***

Compute the support reactions of the beam in Fig. E5.1.4(a). Support B settles 0.1 m. Take E = 200 GPa and $I = 600(10^6)$ mm^4.

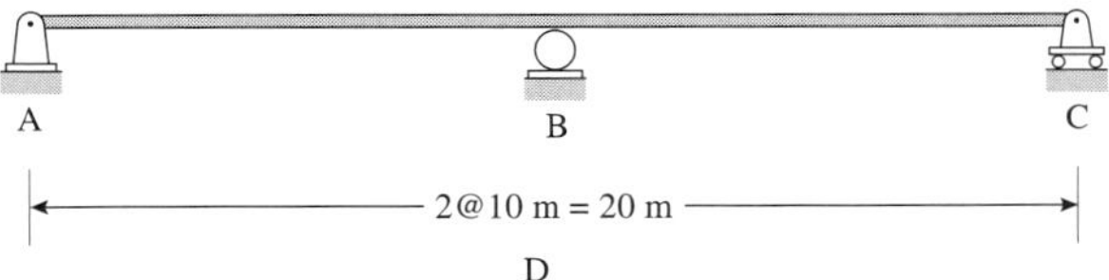

Fig. E5.1.4(a)

SOLUTION

Step 1: Selection of redundant. In the previous problems, the right-hand side of the compatibility equation was zero since the (known) displacement at the support is zero. In this problem the known displacement at the support is at B (see Fig. E5.1.4(b)). Hence it is natural to select B_y as the redundant, as in the previous problem.

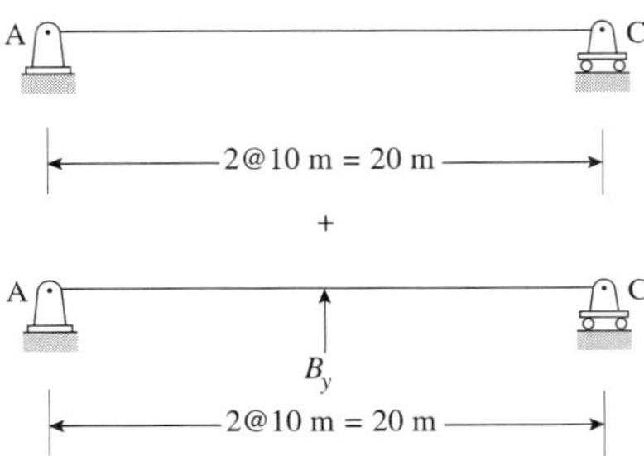

Fig. E5.1.4(b)

We use N, m as the problem units. The compatibility equation can be written as

$$(\downarrow +) B_y \delta_{BB} = 0.1$$

Note the sign convection dictates the sign associated with the known displacement.

Step 2: Computation of the deflections. Using the problem units, $EI = (200 \times 10^9)(600 \times 10^{-6}) = 120(10^6)$ N-m^2. Using the result from the previous problem, we find

$$\delta_{BB} = \frac{500}{3EI} = \frac{500}{360(10^6)} = 1.3889(10^{-6})\,\text{m}(\uparrow)$$

Step 3: Compatibility equation. Substituting in the compatibility equation, we get

$$B_y(-1.3889 \times 10^{-6}) = 0.1$$

Solving yields $B_y = -72000\,\text{N} \Rightarrow B_y = 72\,\text{kN}(\downarrow)$

Step 4: Support reactions. The structural FBD is shown in Fig E5.1.4(c).

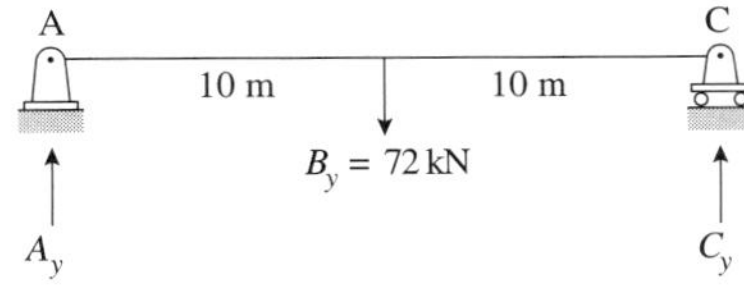

Fig. E5.1.4(c)

Observation: Using the structural FBD, we have $A_y = C_y = 36$ kN(↑). The support settlement appears to be small—about 0.5% of the span of the beam. Yet the support reactions (and hence the internal forces in the beam) are quite large. The support reactions are a function of the beam stiffness, EI. The stiffer the beam, the larger the reactions due to support settlements. Uncontrolled support settlements can be quite detrimental and the structural engineer must be aware of their effects on the structure.

EXAMPLE 5.1.5 ***Deflection of Statically Indeterminate Beam***

Compute the moment M_{BA} needed to cause rotation θ_B at support B of the beam in Fig. E5.1.5(a). EI is a constant.

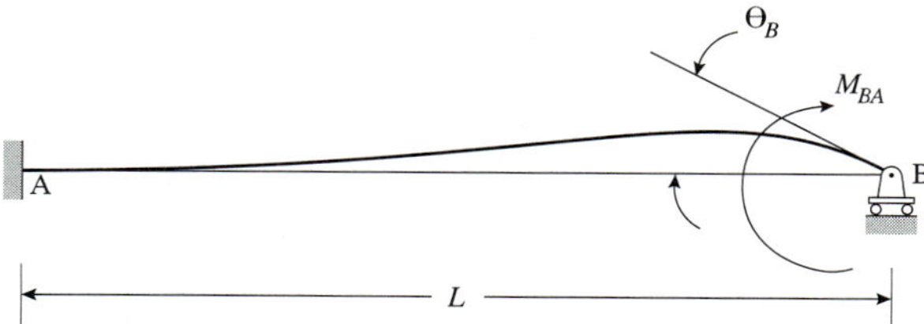

Fig. E5.1.5(a)

SOLUTION

Step 1: This is an unusual (but not uncommon) problem. We need to compute the deflection of an indeterminate beam. The beam is statically indeterminate to degree one. We select B_y as the redundant (see Fig. E5.1.5(b)). Hence the compatibility equation can be written as follows:

$$(\uparrow +)\,\Delta_B + B_y\delta_{BB} = 0$$

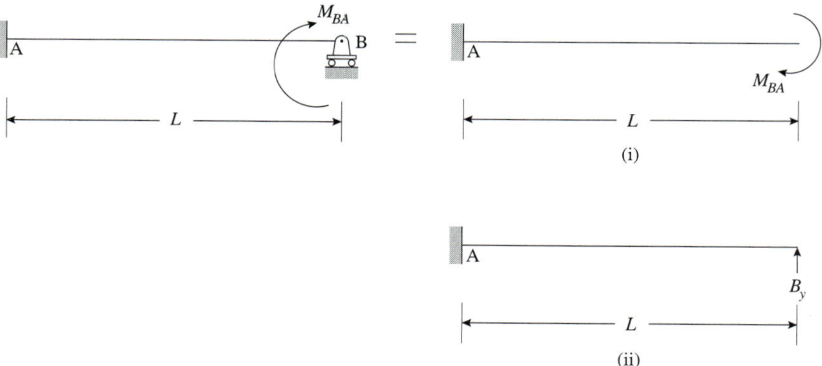

Fig. E5.1.5(b) Superposition of the determinate beams.

Step 2: Computation of internal moments in the determinate beams. Using beam (i) in Fig. E5.1.5(b), we find

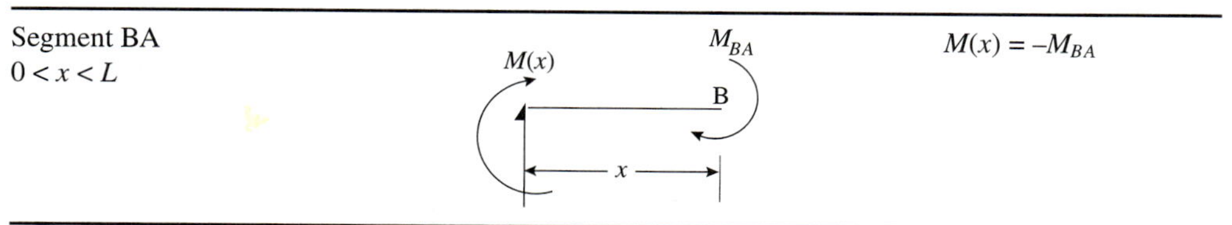

Using beam (ii) in Fig. E5.1.5(b) with unit load for B_y, we find

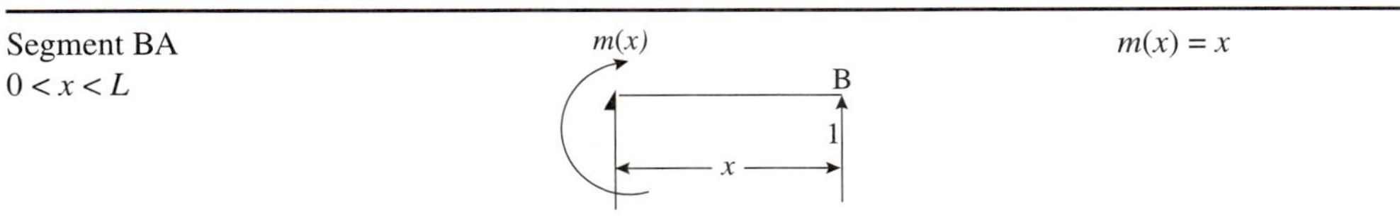

Step 3: Computation of the deflections:

$$\Delta_B = \frac{1}{EI}\int_0^L M(x)m(x)\,dx = \frac{1}{EI}\int_0^L (-M_{BA})(x)\,dx = -\frac{M_{BA}L^2}{2EI} \Rightarrow \Delta_B = \frac{M_{BA}L^2}{2EI}(\downarrow)$$

$$\delta_{BB} = \frac{1}{EI}\int_0^L m(x)m(x)\,dx = \frac{1}{EI}\int_0^L (x)(x)\,dx = \frac{L^3}{3EI} \Rightarrow \delta_{BB} = \frac{L^3}{3EI}(\uparrow)$$

Step 4: Generating and solving the compatibility equation:

$$-\frac{M_{BA}L^2}{2EI} + B_y\left(\frac{L^3}{3EI}\right) = 0$$

Solving, we find $B_y = 3M_{BA}/2L$. Using this value and the FBD of the beam, we can compute the support reactions at A. The final beam FBD is given in Fig. E5.1.5(c).

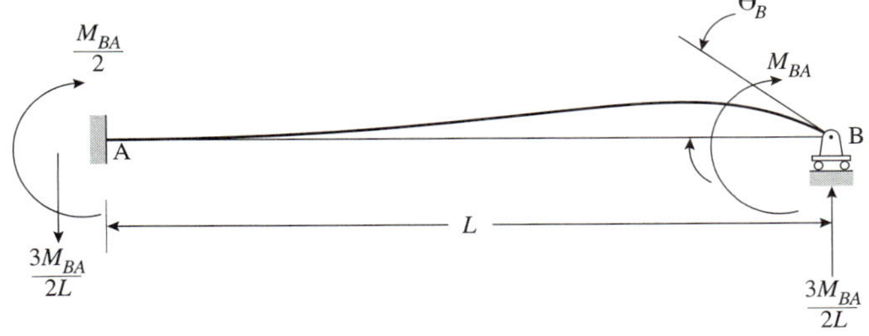

Fig. E5.1.5(c) Support reactions with M_{BA} as the applied load.

Now to find the relationship between the applied moment at B and the resulting rotation at B, we need to use the unit load method. A unit moment is applied at B as shown in Fig. E5.1.5(d). There is no need to solve this indeterminate beam since the reactions for this beam are proportional to the original beam.

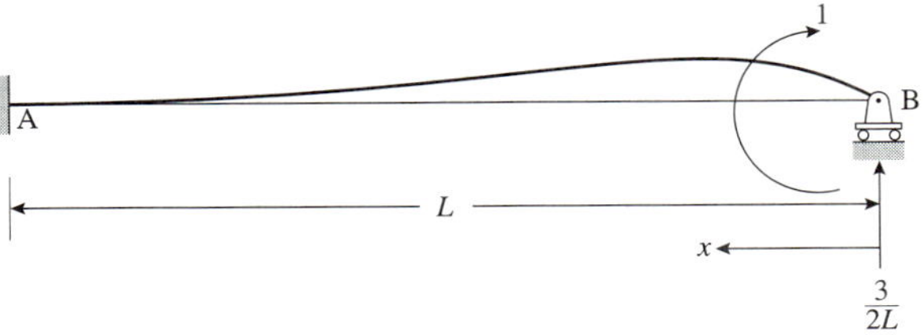

Fig. E5.1.5(d) Support reaction at B with a unit moment as the applied load.

Hence, $EI\theta_B = \int_0^L M(x)m_\theta(x)\,dx$ where from Fig. E5.1.5(c) $M(x) = (3M_{BA}x/2L) - M_{BA}$, and from Fig. E5.1.5(d) $m(x) = (3x/2L) - 1$. Substituting yields

$$EI\theta_B = \int_0^L M(x)m_\theta(x)\,dx = \int_0^L \left(\frac{3M_{BA}x}{2L} - M_{BA}\right)\left(\frac{3x}{2L} - 1\right)dx = \frac{M_{BA}L}{4}$$

Hence the moment M_{BA} required to cause a rotation θ_B at B is such that

$$M_{BA} = \frac{4EI}{L}\theta_B$$

$$M_{AB} = \frac{M_{BA}}{2} = \frac{2EI}{L}\theta_B$$

EXERCISES

Appetizers

Some of the following problems are variations of the problems from Sections 4.2 and 4.3. The results from the solution to those problems can be used here.

5.1.1.

(a) Find all the support reactions that can be used as a redundant in Fig. P5.1.1. For each possibility, draw the determinate beams whose superposition gives the original beam, and write the equation of compatibility.

(b) Use the reaction at C as the redundant and solve for all the support reactions. EI is a constant.

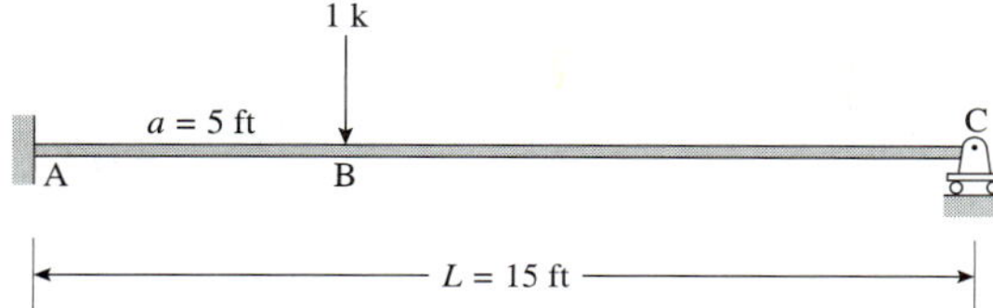

Fig. P5.1.1

5.1.2. Use the moment reaction at A in Fig. P5.1.2 as the redundant and solve for all the support reactions. EI is a constant.

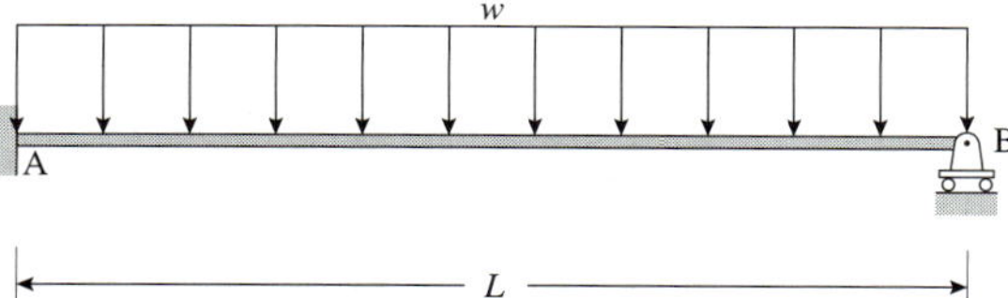

Fig. P5.1.2

5.1.3.

(a) Find all the support reactions in Fig. P5.1.3 that can be used as a redundant. For each possibility, draw the determinate beams whose superposition gives the original beam, and write the equation of compatibility.

(b) First solve for all the support reactions and then draw the shear force and bending moment diagrams. EI is a constant. Use the reaction at C as the redundant. Take $a = 10$ m and $b = 5$ m.

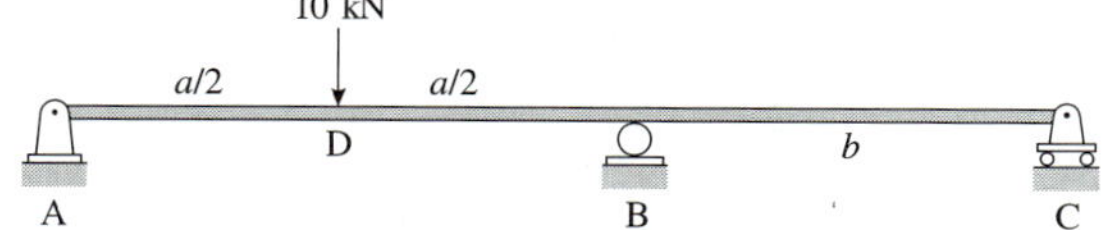

Fig. P5.1.3

5.1.4. First solve for all the support reactions in Fig. P5.1.4 and then draw the shear force and bending moment diagrams. EI is a constant. Use the moment reaction at A as the redundant. Verify your answer by using the reaction at C as the redundant.

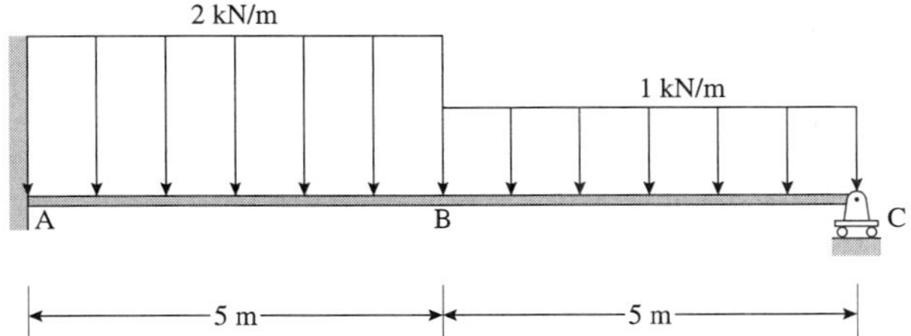

Fig. P5.1.4

5.1.5. Compute the support reactions for the beam in Fig. P5.1.5. Take EI as constant.

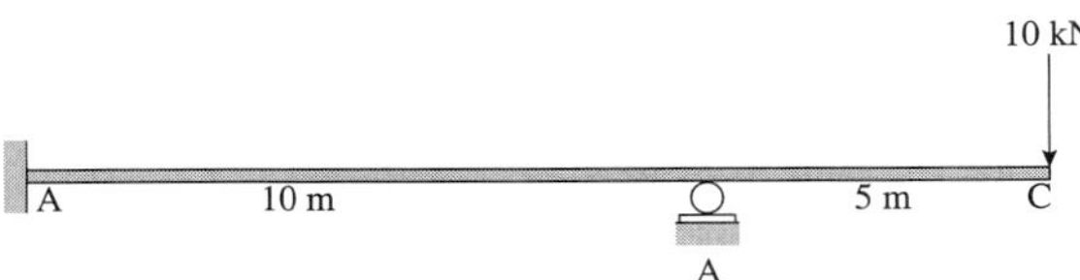

Fig. P5.1.5

Main Course

5.1.6. For Fig. P5.1.6, take $E = 30000$ ksi, $I = 400$ in^4, $L = 20$ ft, and $w = 0$. Support B settles 1 in. Solve for all the support reactions, and draw the shear force and bending moment diagrams.

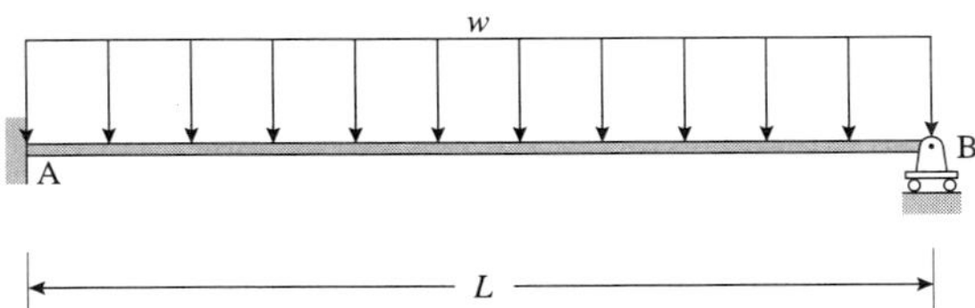

Fig. P5.1.6

5.1.7. For the beam in Problem 5.1.6, take $E = 30000$ ksi, $I = 400$ in^4, $L = 20$ ft, and $w = 1$ k/ft. Support B settles 1 in. Solve for all the support reactions, and draw the shear force and bending moment diagrams. Compare the two solutions.

5.1.8. Compute the support reactions for the beam in Fig. P5.1.8. B is an internal hinge. Take EI as constant.

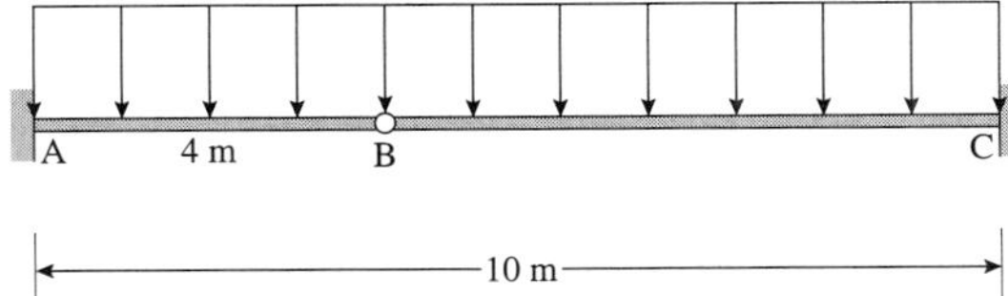

Fig. P5.1.8

5.1.9. First solve for all the support reactions in Fig. P5.1.9 and then draw the shear force and bending moment diagrams. EI is a constant.

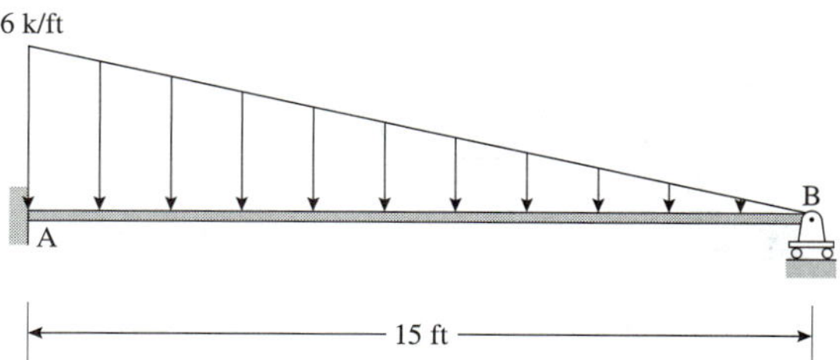

Fig. P5.1.9

5.1.10. Compute the rotation at C for the beam in Problem 5.1.4.

5.1.2 Frames

Indeterminate frames can be solved in the same manner as indeterminate beams. If the frame is statically indeterminate to degree one, then one of the support reactions must be selected as the redundant. Consider the indeterminate frame in Fig. 5.1.2.1. Using C_y as the redundant, the compatibility equation is

$$(\uparrow +)\ \Delta_C + \delta_C = \Delta_C + C_y \delta_{CC} = 0 \tag{5.1.2.1}$$

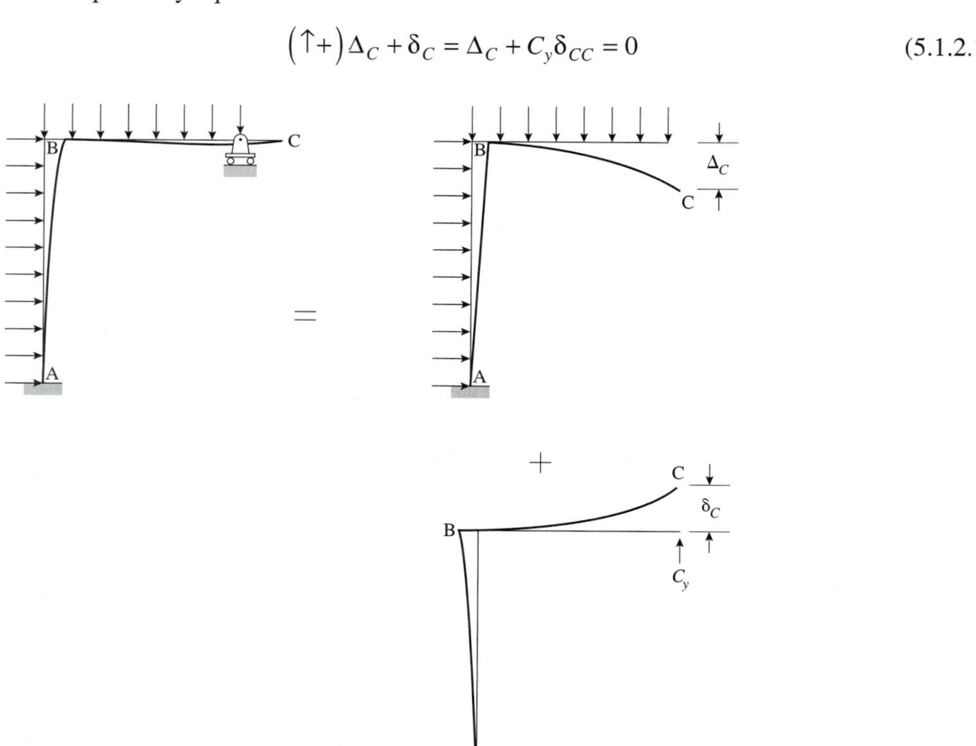

Fig. 5.1.2.1 Superposition of two determinate frames to yield the original indeterminate frame.

where the deflection in the structure with the unit force is shown in Fig. 5.1.2.2. The same general procedure as outlined for an indeterminate beam applies here.

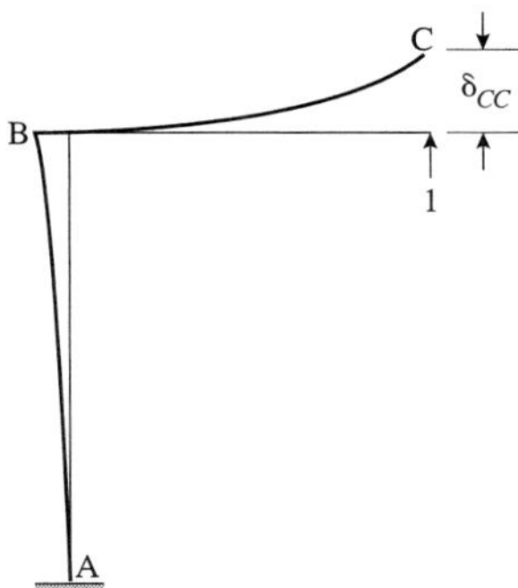

Fig. 5.1.2.2 Displacement due to the unit load.

EXAMPLE 5.1.6 ***Statically Indeterminate Frame***

Compute the support reactions of the frame in Fig. E5.16(a). Take E = 200 GPa, and $I = 10^6$ mm^4 for the column and $I = 2(10^6)$ mm^4 for the beam.

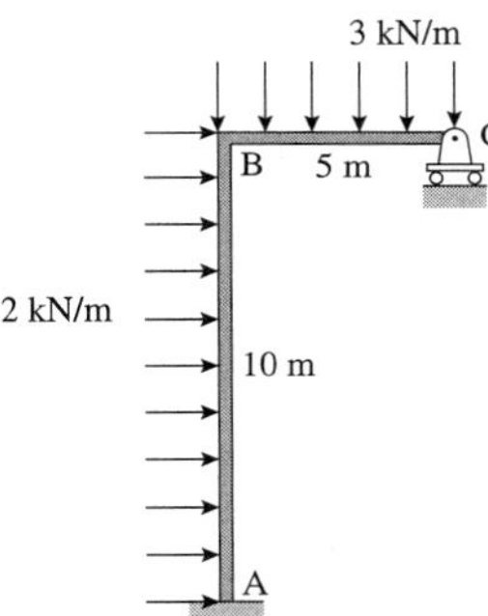

Fig. E5.1.6(a)

SOLUTION

Step 1: The frame is statically indeterminate to degree one. Let us select C_y as the redundant. The two determinate frames are shown in Fig. E5.1.6(b). The compatibility equation is

$$\left(\uparrow +\right)\Delta_C + C_y\delta_{CC} = 0$$

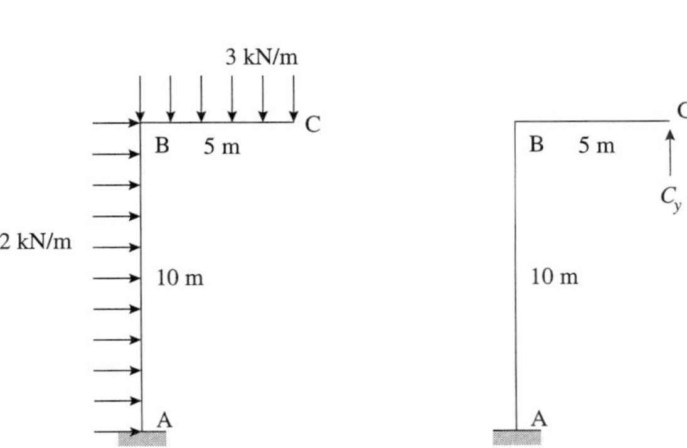

Fig. E5.1.6(b)

Step 2: Computation of the deflections. We need to compute the internal moments in the column and the beam. The calculations of the support reactions for the two structures are not shown—the FBDs should make it clear what their values and directions are.

Segment	FBD	$M(x)$	FBD	$m(x)$
CB $(0 < x < 5)$	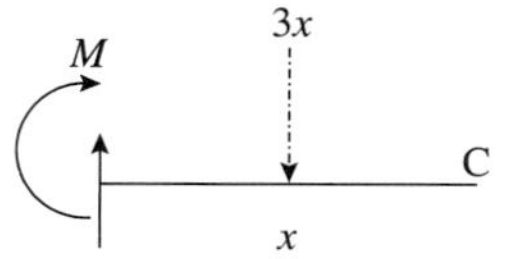	$-(3x^2/2)$	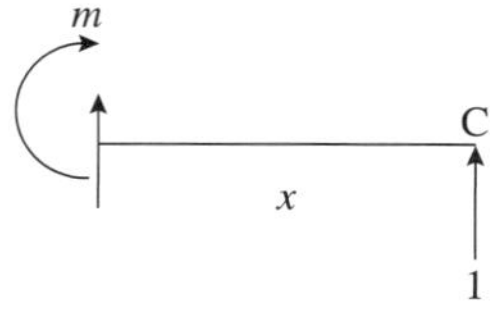	x
AB $(0 < x_1 < 10)$	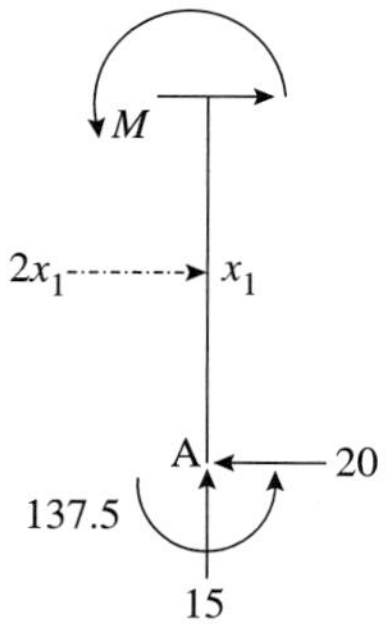	$-x_1^2 + 20x_1 - 137.5$	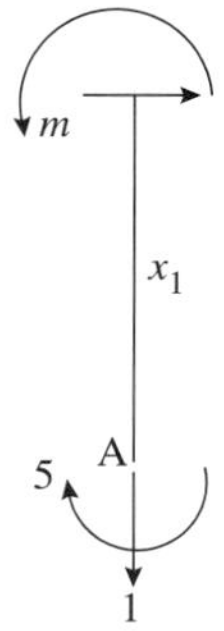	5

The *EI* values are different for the two segments. However, the ratio of the *EI* values is necessary to compute the support reactions not the numerical values. This will become clear in Step 3 when the compatibility equation is solved for the redundant.

$$\Delta_C = \frac{1}{2EI}\int_0^5 \left(-\frac{3x^2}{2}\right)(x)\,dx + \frac{1}{EI}\int_0^{10}\left(-x_1^2 + 20x_1 - 137.5\right)(5)\,dx_1$$

$$\text{or } \Delta_C = -\frac{234.375}{2EI} - \frac{3541.67}{EI} = -\frac{3658.85}{EI} \Rightarrow \Delta_C = \frac{3658.85}{EI}(\downarrow)$$

$$\delta_{CC} = \frac{1}{2EI}\int_0^5 (x)(x)\,dx + \frac{1}{EI}\int_0^{10}(5)(5)\,dx_1$$

$$\text{or } \delta_{CC} = \frac{125}{6EI} + \frac{250}{EI} = \frac{270.833}{EI} \Rightarrow \delta_{CC} = \frac{270.833}{EI}(\uparrow)$$

Step 3: Compatibility equation. Substituting in the compatibility equation, we have

$$-\frac{3658.85}{EI} + C_y\left(\frac{270.833}{EI}\right) = 0$$

Since *EI* appears in the denominator in both the terms on the left, it can be cancelled. Solving, we find

$$C_y = 13.5\,\text{kN}(\uparrow)$$

Step 4: Support reactions. Now using the structural FBD in Fig. E5.1.6(c), we can compute the rest of the support reactions:

$$\overset{\uparrow+}{\sum} F_y = 0 = A_y + C_y + 15 \Rightarrow A_y = 1.5\,\text{kN}(\uparrow)$$

$$\overset{\rightarrow+}{\sum} F_x = 0 = -A_x + 20 \Rightarrow A_x = 20\,\text{kN}(\leftarrow)$$

$$\overset{\curvearrowleft+}{\sum} M_A = 0 = M_A - 20(5) - 15(2.5) + 13.5(5) \Rightarrow M_A = 70\,\text{kN-m}(\circlearrowleft)$$

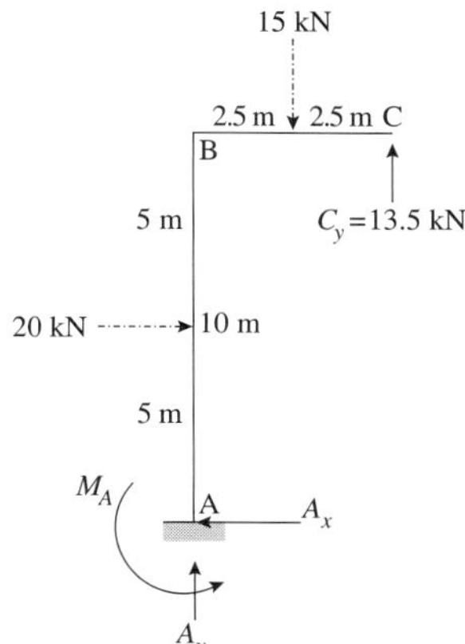

Fig. E5.1.6(c) Structural FBD necessary to compute the other support reactions.

EXAMPLE 5.1.7

Statically Indeterminate Frame

Compute the support reactions of the frame in Fig. E5.1.7(a). Draw the shear force and bending moment diagrams. E is a constant.

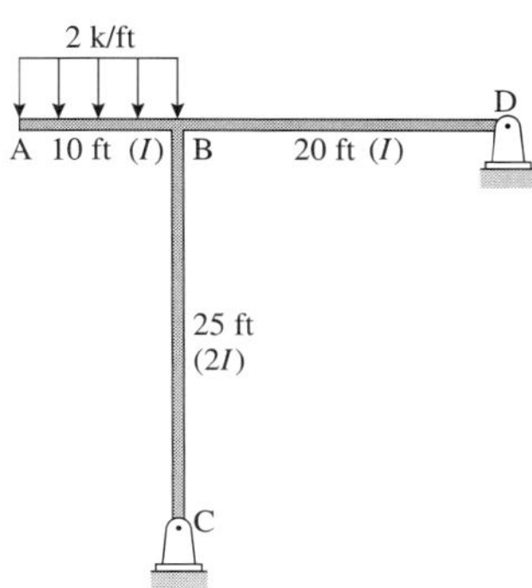

Fig. E5.1.7(a)

SOLUTION

Step 1: The frame is statically indeterminate to degree one. Let us select D_x as the redundant. The two determinate frames are shown in Fig. E5.1.7(b). The compatibility equation is

$$(\overset{+}{\leftarrow})(\Delta_D)_x + D_x(\delta_{D_xD_x}) = 0$$

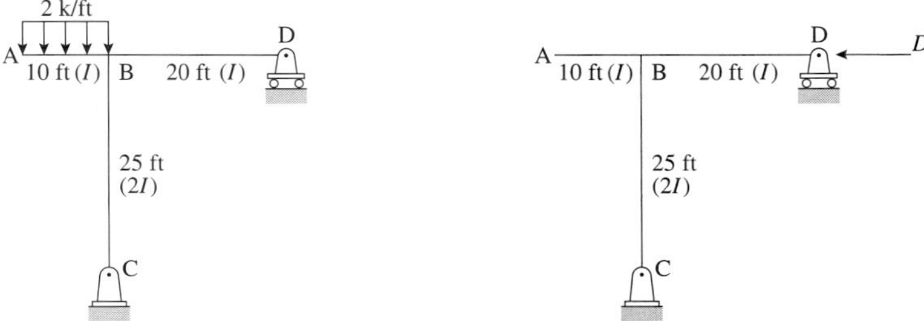

Fig. E5.1.7(b)

Step 2: Computation of the deflections.

We need to compute the internal moments in the three segments—AB, DB, and CB. The calculations of the support reactions for the two structures are not shown—the FBDs should make it clear what their values and directions are. Note that the *EI* values are different for the three segments.

Segment	FBD	$M(x)$	FBD	$m(x)$
AB ($0 < x < 10$ ft)		$-x^2$		0
DB ($0 < x_1 < 20$ ft)		$-5x_1$		$-1.25x_1$
CB ($0 < x_2 < 25$ ft)		0		$-x_2$

$$(\Delta_D)_x = \frac{1}{EI}\int_0^{10}\left(-\frac{x^2}{2}\right)(0)\,dx + \frac{1}{EI}\int_0^{20}(-5x_1)(-1.25x_1)\,dx_1 + \frac{1}{2EI}\int_0^{25}(0)(-x_2)\,dx_2$$

$$\text{or } \Delta_C = 0 + \frac{16666.7}{EI} + 0 = \frac{16666.7}{EI} \Rightarrow (\Delta_D)_x = \frac{16666.7}{EI}(\leftarrow)$$

$$\delta_{D_xD_x} = \frac{1}{EI}\int_0^{10}(0)(0)\,dx + \frac{1}{EI}\int_0^{20}(-1.25x_1)(-1.25x_1)\,dx_1 + \frac{1}{2EI}\int_0^{25}(-x_2)(-x_2)dx_2$$

$$\text{or } \delta_{D_xD_x} = 0 + \frac{4166.67}{EI} + \frac{2604.17}{EI} = \frac{6770.84}{EI} \Rightarrow \delta_{D_xD_x} = \frac{6770.84}{EI}(\leftarrow)$$

Step 3: Compatibility equation. Substituting in the compatibility equation, we have

$$\frac{16666.7}{EI} + D_x\left(\frac{6770.84}{EI}\right) = 0$$

Solving yields $D_x = -2.46\,k \Rightarrow D_x = 2.46\,\text{k}(\rightarrow)$.

Step 4: Support reactions. Now using the structural FBD in Fig. E5.1.7(c), we can compute the rest of the support reactions:

$$\overset{+}{\curvearrowleft}\sum M_A = 0 = 20(5) - 2.46(25) - D_y(20) \Rightarrow D_y = 1.925\,\text{k}(\downarrow)$$

$$\overset{\uparrow+}{\sum} F_y = 0 = -20 + C_y - D_y \Rightarrow C_y = 21.925\,\text{k}(\uparrow)$$

$$\overset{\rightarrow+}{\sum} F_x = 0 = -C_x + 2.46 \Rightarrow C_x = 2.46\,\text{k}(\leftarrow)$$

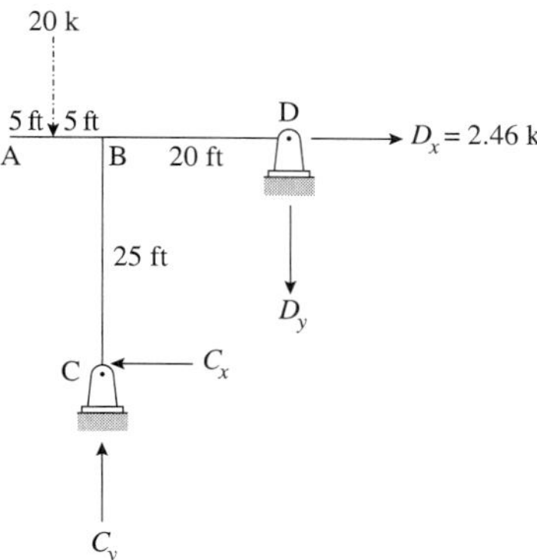

Fig. E5.1.7(c)

Step 5: Shear force and bending moment diagrams. Using the FBD of each member, the shear force and bending moment diagrams can be obtained, as in Figs. E5.1.7(d) and (e). Each member needs a single cut.

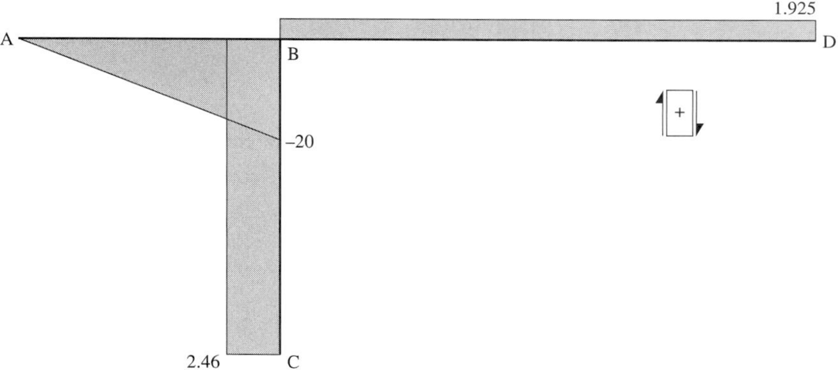

Fig. E5.1.7(d)
Shear force diagram (k).

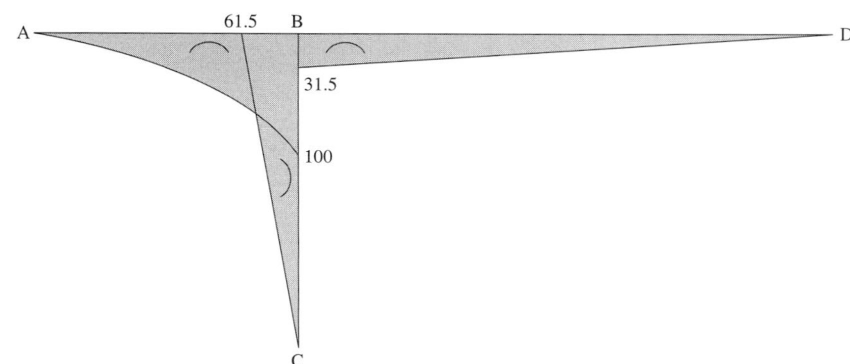

Fig. E5.1.7(e)
Bending moment diagram (k-ft).

EXERCISES

Appetizers

Some of the following problems are variations of the problems from Sections 4.2 and 4.3. The results from the solution to those problems can be used here.

5.1.11. Compute the support reactions for the frame in Fig. P5.1.11. Assume that *EI* is a constant.

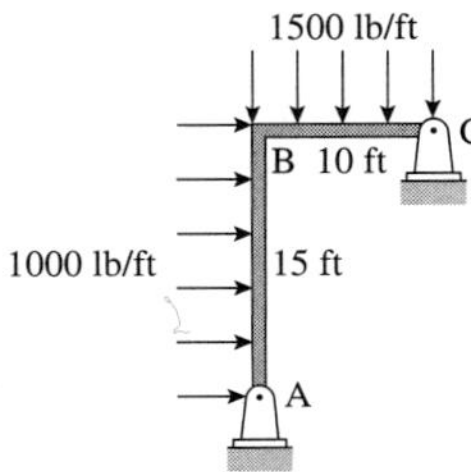

Fig. P5.1.11

5.1.12. Draw the shear force and bending moment diagrams for the frame in Fig. P5.1.12.

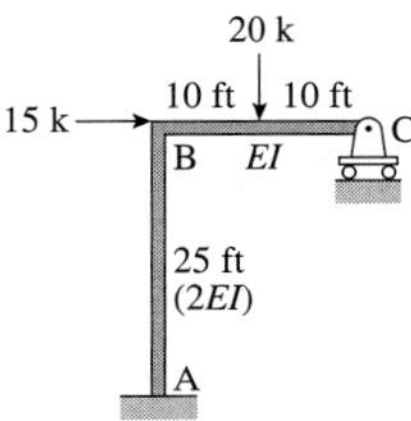

Fig. P5.1.12

5.1.13. Compute the support reactions for the frame in Fig. P5.1.13. Assume that *EI* is a constant.

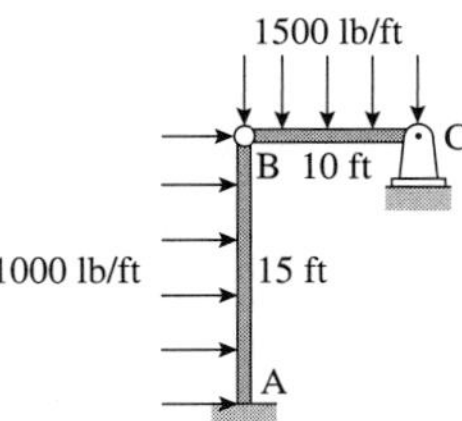

Fig. P5.1.13

5.1.14. Compute the support reactions for the frame in Fig. P5.1.14. Assume that *EI* is a constant.

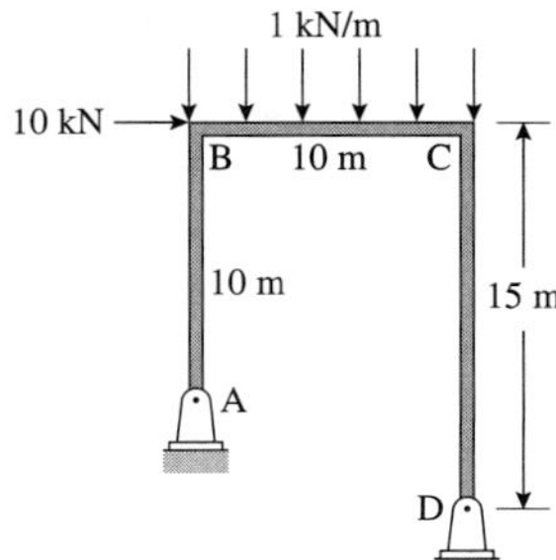

Fig. P5.1.14

5.1.15. Compute the support reactions for the frame in Fig. P5.1.15. Assume that EI is a constant.

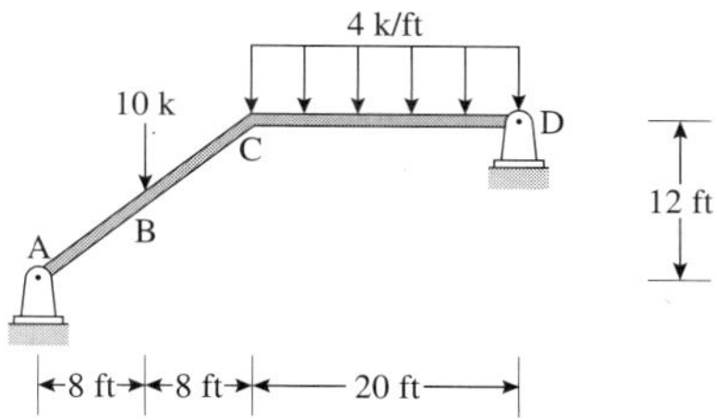

Fig. P5.1.15

Main Course

5.1.16. Draw the shear force and bending moment diagrams for the frame in Fig. P5.1.16. Assume that EI is a constant.

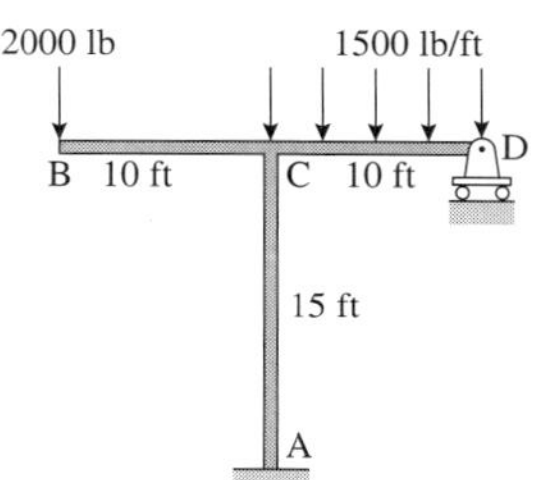

Fig. P5.1.16

5.1.17. Compute all the support reactions for the frame in Fig. P5.1.17.

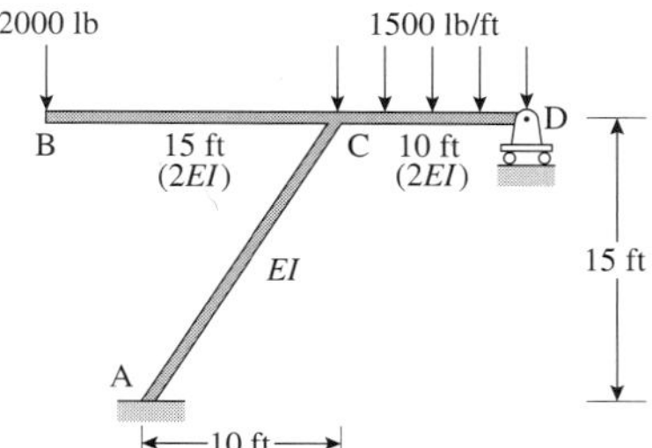

Fig. P5.1.17

5.1.18. Compute all the support reactions for the frame in Fig. P5.1.18.

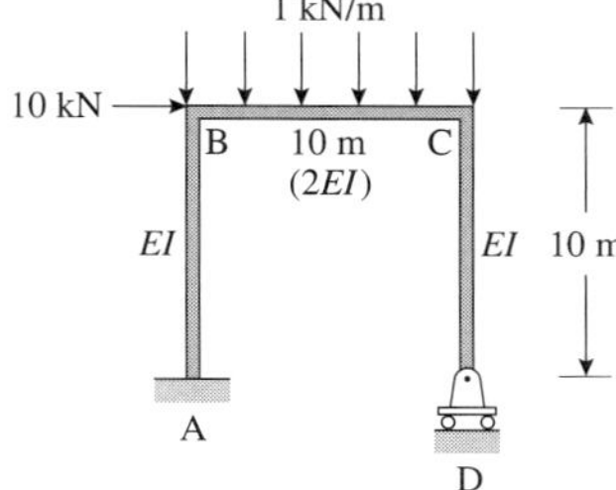

Fig. P5.1.18

5.1.19. Draw the shear force and bending moment diagrams for the frame in Fig. P5.1.19. Assume that *EI* is a constant.

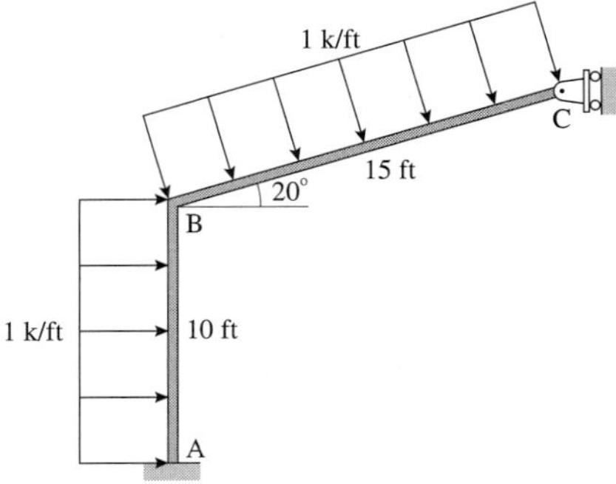

Fig. P5.1.19

5.1.20. Compute all the support reactions for the frame in Fig. P5.1.20. Assume that *EI* is a constant.

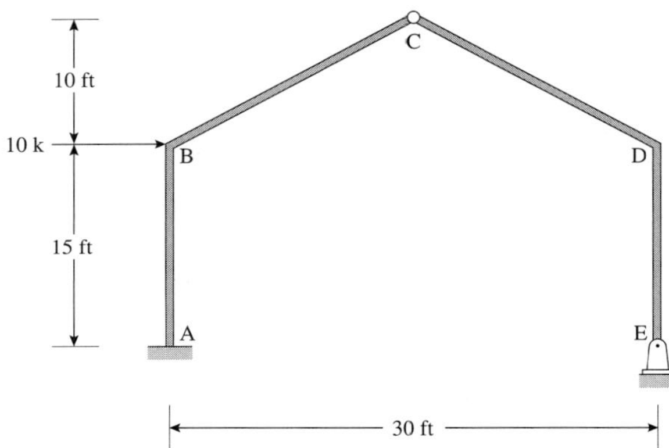

Fig. P5.1.20

5.1.3 Trusses

Trusses can be statically indeterminate due to a variety of reasons—redundant support reactions (externally indeterminate), redundant members (internally indeterminate), or a combination of both. Figure 5.1.3.1(a) shows a truss that is externally indeterminate $((m + r) - 2j = 7 + 4 - 2(5) = 1)$. An internally indeterminate truss is shown in Fig. 5.1.3.1(b) $((m + r) - 2j = 8 + 3 - 2(5) = 1)$.

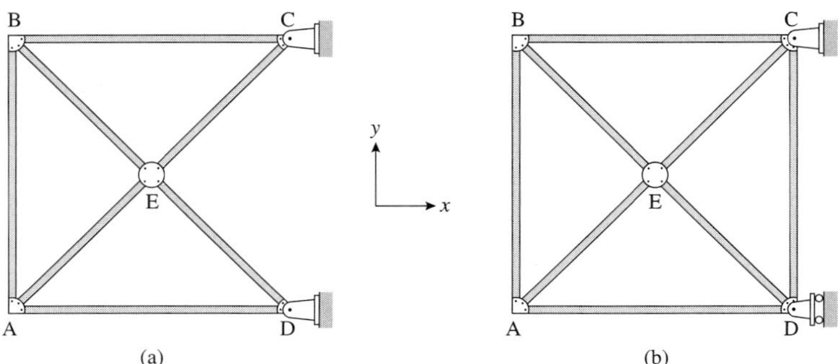

Fig. 5.1.3.1(a) Externally indeterminate truss. (b) Internally indeterminate truss.

For the truss in Fig. 5.1.3.1(a), one can select one of those support reactions as the redundant that cannot be computed using statics. Using the truss FBD and the appropriate equilibrium equations, one can compute C_x and D_x. Hence the possible redundants are C_y and D_y. The procedure for solving such a problem is very similar to the indeterminate beams and frames that we saw in the previous sections. An example of the compatibility equation using the vertical reaction at D as the redundant is

$$(\uparrow +)(\Delta_D)_y + D_y(\delta_{D_yD_y}) = 0 \tag{5.1.3.1}$$

which can then be solved for D_y. Note that

$$(\Delta_D)_y = \sum \frac{NnL}{AE} \tag{5.1.3.2a}$$

$$\left(\delta_{D_yD_y}\right) = \sum \frac{n^2L}{AE} \tag{5.1.3.2b}$$

where the N forces are the member forces in the determinate truss under the action of external forces and the n forces are the member forces in the determinate truss with the unit force applied along the redundant D_y. Using the same nomenclature as before, $\delta_{D_yD_y}$ denotes the vertical displacement at D due to a unit vertical force applied at D.

For the internally indeterminate truss in Fig. 5.1.3.1(b), one can select any of the truss members as the redundant. We now examine such as truss in greater detail. Since the degree of static indeterminacy is one, let us select member AC as the redundant. Removing that member results in a stable, determinate truss.

We now superpose the two determinate trusses. First, the determinate truss is subjected to the external loads as shown in Fig. 5.1.3.2(a). We assume that member AC is in tension.

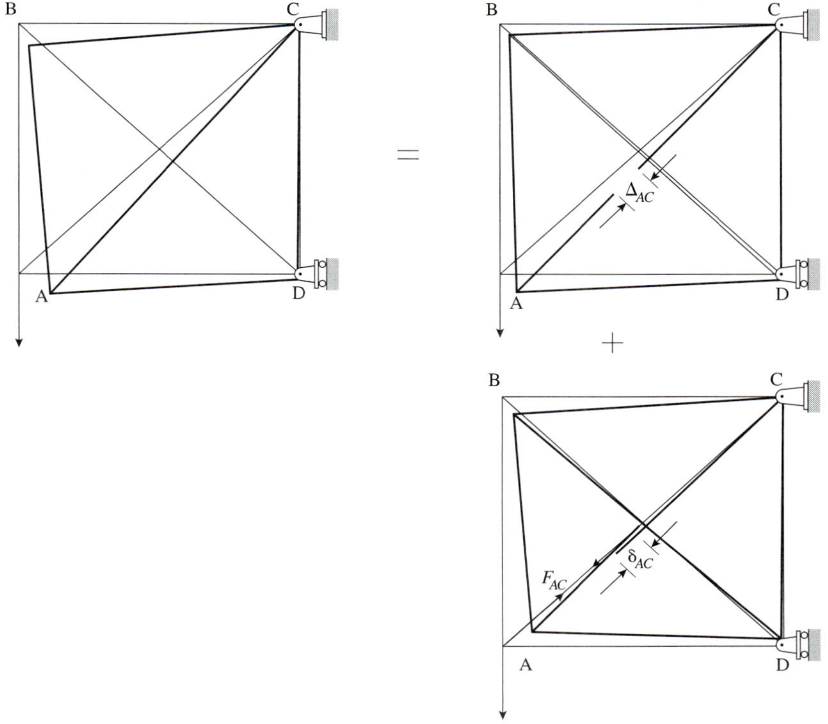

Fig. 5.1.3.2 Internally determinate truss as a linear superposition of (a) determinate truss with AC as the redundant and (b) F_{AC} applied to the determinate truss.

This results in joints AC moving away from each other by an amount Δ_{AC}. We will call this DTRL (Determinate Truss with Real Loads). Second, the determinate truss is subjected to the redundant force F_{AC}. The joints AC now move δ_{AC} closer to each other. We will call this DTRF (Determinate Truss with Redundant Force). Since the original truss has no overlap or a gap in member AC, the compatibility equation is

$$\Delta_{AC} + \delta_{AC} = 0 \tag{5.1.3.4}$$

To compute these two displacements, we can use the unit load method. Let us apply a unit force along AC as shown in Fig. 5.1.3.3. We call this DTUL (Determinate Truss with Unit Load). The two joints now move $(\delta_{AC})_{AC}$ closer to each other, and the following relationship is true:

$$\delta_{AC} = F_{AC}(\delta_{AC})_{AC} \tag{5.1.3.5}$$

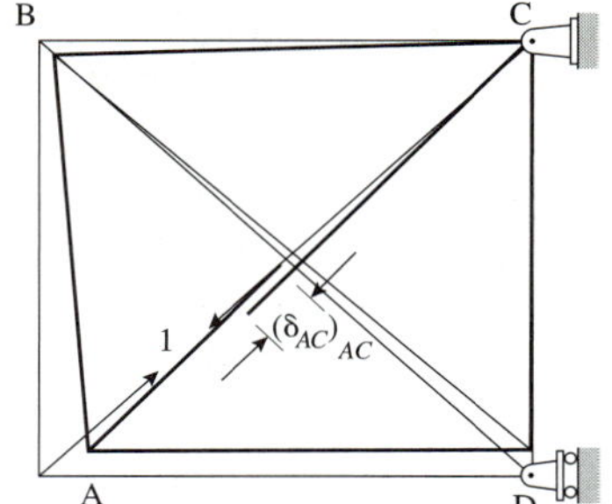

Fig. 5.1.3.3 Unit force applied along AC.

Substituting Eq. (5.1.3.5) into (5.1.3.4), we have

$$\Delta_{AC} + F_{AC}(\delta_{AC})_{AC} = 0 \tag{5.1.3.6}$$

from which F_{AC} can be found. If the answer is positive, then the member is in tension as we originally assumed. Note that

$$\Delta_{AC} = \sum \frac{NnL}{AE} \tag{5.1.3.7a}$$

$$(\delta_{AC})_{AC} = \sum \frac{n^2 L}{AE} \tag{5.1.3.7b}$$

where the N forces are the member forces in the determinate truss under the action of external forces (Fig. 5.1.3.2(a)) and the n forces are the member forces in the determinate truss with the unit force applied along the redundant AC (Fig. 5.1.3.3).

Figure 5.1.3.4 shows a planar truss that is externally and internally indeterminate. Such a truss can be solved by a combination of support reactions and member forces selected as redundants.

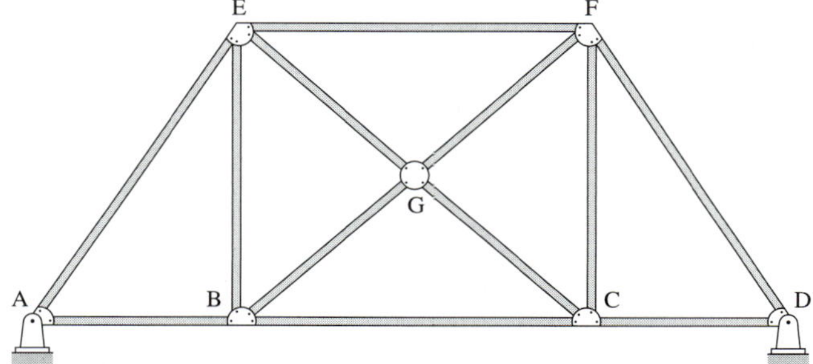

Fig. 5.1.3.4 Externally and internally indeterminate truss.

General Procedure for Internally Indeterminate Truss

Step 1: Identify the redundant member (ij). If the member is removed from the original structure, the resulting truss must be stable and determinate. Now create the two trusses whose superposition results in the original indeterminate truss.

Remove the redundant from the original truss but leave the external loads. This is truss DTRL.

Remove the redundant and all loads from the original truss. Assume that the redundant member is in tension. Now apply unit tensile forces along the redundant member. This is truss DTUL.

Write the single compatibility equation in the symbolic form. This equation should contain the redundant member force F_{ij}.

Step 2: Compute the displacement along ij from truss DTRL.

Step 3: Compute the displacement along ij from truss DTUL.

Step 4: Now substitute the displacement from Steps 2 and 3 into the compatibility equation. Solve the compatibility equation for the redundant. If the answer is positive, the redundant is in tension. Otherwise, the member is in compression.

Step 5: The other member forces can be computed through superposition of the two determinate trusses.

Table 5.1.3.1 can be used to consolidate the calculations. Note a few points about this table:

Table 5.1.3.1

Member	N	n	NnL/AE	n^2L/AE	$F = N + F_{ij}n$
ij	0	1			F_{ij}
		Sum	A	B	

1. *Sign convention for the member forces N and n-tension is positive and compression is negative*. This is the sign convention we used in Chapter 4.
2. Note that $F_{ij} = -(A/B)$.
3. Watch the row corresponding to the redundant ij. The force in this member in truss DTRL is obviously zero. The force in this member in truss DTUL is 1 since we applied a unit force along the member.
4. The last column represents the member force F in the indeterminate truss. The values are calculated after the member force F_{ij} is computed, and represents the superposition of trusses DTRL and DTUL scaled by the factor F_{ij}.

Tip: When solving for the member forces in truss DTRL and DTUL, assume all member forces to be in tension. Then the computed values can be entered into the table as is.

EXAMPLE 5.1.8 ***Externally Indeterminate Planar Truss***

Compute the support reactions and the member forces for the truss in Fig. E5.1.8(a). Take E = 30 ksi and A = 0.5 in^2 for all the members.

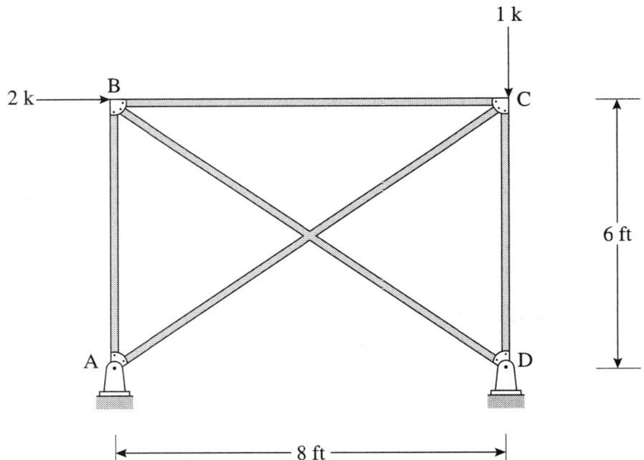

Fig. E5.1.8(a)

SOLUTION

Step 1: The truss is externally indeterminate. We select the horizontal reaction at A as the redundant. The two determinate trusses are shown in Fig. E5.1.8(b) along with the support reactions.

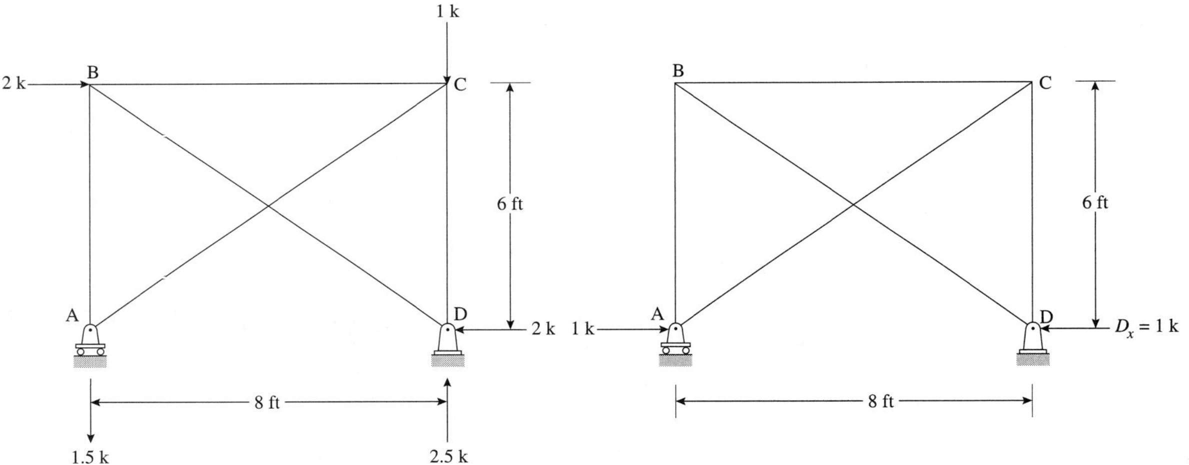

Fig. E5.1.8(b)

The compatibility equation is

$$(\rightarrow +)\ (\Delta A)_x + A_x\left(\delta_{A_xA_x}\right) = 0$$

$$\text{where}\quad (\Delta A)_x = \sum \frac{NnL}{AE} \quad \text{and} \quad \left(\delta_{A_xA_x}\right) = \sum \frac{n^2L}{AE}.$$

Step 2: Solve for the member forces in the two determinate trusses.

The details are not presented here. The method of joints has been used to compute all the member forces. The table summarizes the results:

Member	N (k)	n (k)	L/A (1/in)	NnL/A	n^2L/A
AC	0	−1.25	240	0	375
AB	1.5	0.75	144	162	81
BD	−2.5	−1.25	240	750	375
CD	−1	0.75	144	−108	81
BC	0	1	192	0	192
			Sum	**804**	**1104**

Step 3: Solve the compatibility equation. Substituting the values in the compatibility equation, we have

$$\frac{804}{E} + A_x\left(\frac{1104}{E}\right) = 0 \Rightarrow A_x = -0.73\,\text{k}$$

Hence $A_x = 0.73$ k(←).

Step 4: Support reactions and member forces. The concept of superposition can be used to compute the member forces as well as the support reactions. The force F in a typical member is given by

$$F = N + A_x n$$

where the first term on the right represents truss DTRL and the second term represents truss DTRF.

Member	N (k)	n (k)	$F = N + A_x n$ (k)
AC	0	−1.25	0.91 (T)
AB	1.5	0.75	0.95 (T)
BD	−2.5	−1.25	−1.59 (C)
CD	−1	0.75	−1.55 (C)
BC	0	1	−0.73 (C)

As usual, a positive sign represents tension and a negative sign represents compression.

We use the same strategy for the support reactions. A positive sign indicates an upward or "left to right" reaction. The first term on the right side of the equality represents truss DTRL and the second term represents truss DTRF.

$$A_y = -1.5 + 0(-0.73) = -1.5\,\text{k} \Rightarrow A_y = 1.5\,\text{k}(\downarrow)$$

$$A_x = 0 + (1)(-0.73) = -0.73\,\text{k} \Rightarrow A_x = 0.73\,\text{k}(\leftarrow)$$

$$D_y = 2.5 + 0(-0.73) = 2.5\,\text{k} \Rightarrow D_y = 2.5\,\text{k}(\uparrow)$$

$$D_x = -2 + (-1)(-0.73) = -1.27\,\text{k} \Rightarrow D_x = 1.27\,\text{k}(\leftarrow)$$

EXAMPLE 5.1.9 ***Internally Indeterminate Planar Truss***

Compute the member forces in Fig. E5.1.9(a). Take $E = 29(10^3)$ ksi and $A = 1.5$ in^2.

SOLUTION

Step 1: The truss is internally indeterminate to degree one. We select member CD as the redundant. The two determinate trusses are shown in Fig. E5.1.9(b).

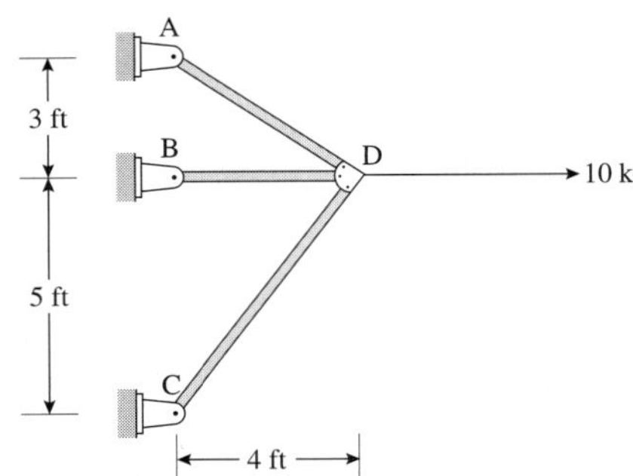

Fig. E5.1.9(a)

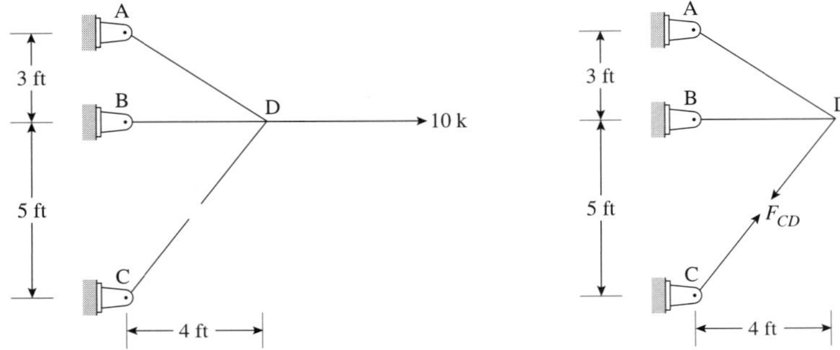

Fig. E5.1.9(b) Truss to yield N forces and n forces.

The compatibility equation is

$$\sum \frac{NnL}{AE} + F_{CD} \sum \frac{n^2 L}{AE} = 0$$

Step 2: Solve for the member forces in the two determinate trusses. The details are not presented here. The method of joints was used to compute all the member forces. The table summarizes the results (sign convention for member forces: tension is positive).

Member	N (k)	n (k)	L/A (1/in)	NnL/A	n^2L/A
AD	0	1.301	40	0	67.70
BD	10	–1.665	32	–532.8	88.71
CD	0	1	51.2	0	51.22
			Sum	**–532.8**	**207.63**

Step 3: Solve the compatibility equation. Substituting the values in the compatibility equation, we have

$$-\frac{532.8}{E} + F_{CD}\left(\frac{207.63}{E}\right) = 0 \Rightarrow F_{CD} = 2.57\,\text{k}$$

Step 4: Member forces. The member forces can be computed using the idea of superposition. The last column in the following table computes the final member forces. As usual, we should check the equilibrium of joint D to verify the results.

Member	N(k)	n(k)	$F = N + F_{CD}n$(k)
AD	0	1.301	3.34 (T)
BD	10	−1.665	5.73 (T)
CD	0	1	2.57 (T)

EXERCISES

Appetizers

5.1.21.

(a) Find all the members in Fig. P5.1.21 that can be used as a redundant. For each case, show the determinate trusses whose superposition leads to the original truss, and write the compatibility equation.

(b) The members in the truss have a cross-sectional area of 0.01 m^2. Compute all the member forces.

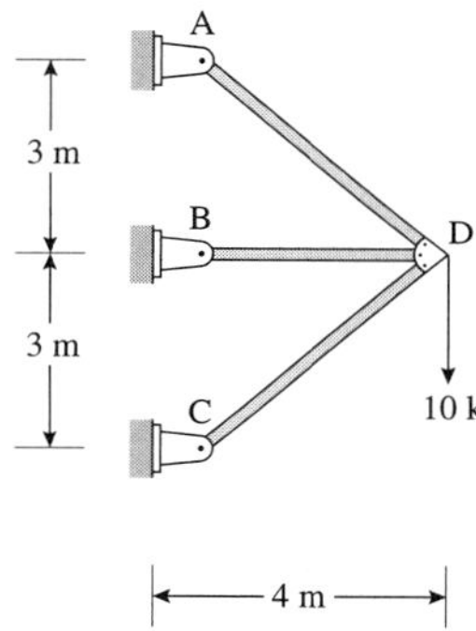

Fig. P5.1.21

5.1.22.

(a) Find all the members in Fig. P5.1.22 that can be used as a redundant. For each case, show the determinate trusses whose superposition leads to the original truss, and write the compatibility equation.

(b) The members in the truss have the following cross-sectional areas: $A_{AC} = A_{CE} = 0.01$ m^2, $A_{BD} = A_{DF} = 0.02$ m^2; the cross-sectional area for the rest of the members is 0.005 m^2. Compute all the member forces.

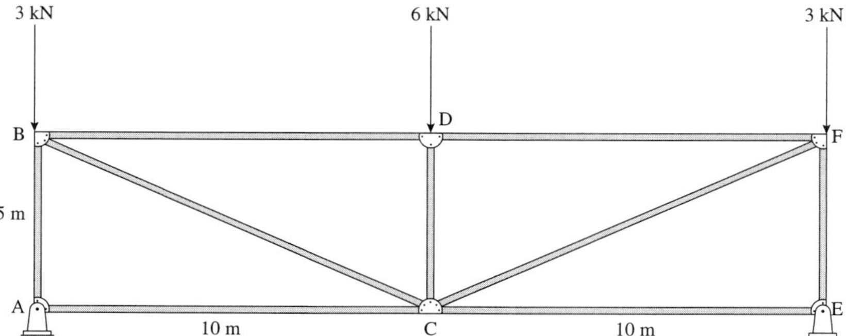

Fig. P5.1.22

Main Course

5.1.23. The members in the truss in Fig. P5.1.23 have the following cross-sectional areas: top chord = 2 in^2, bottom chord = 3 in^2, and web members = 2.5 in^2. Compute all the member forces.

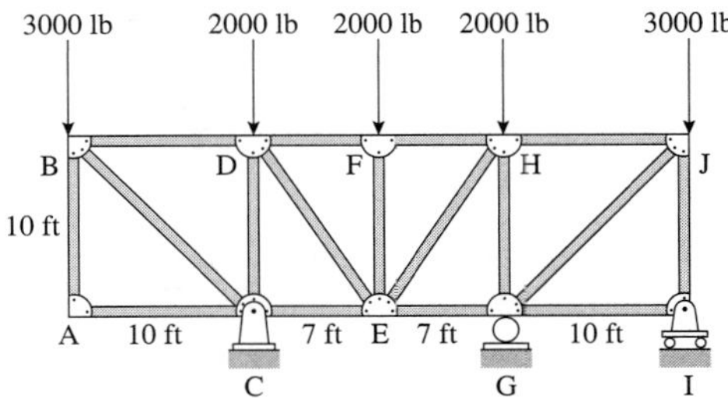

Fig. P5.1.23

5.1.24. Assuming that AE = constant for all the members in Fig. P5.1.24, compute the member forces.

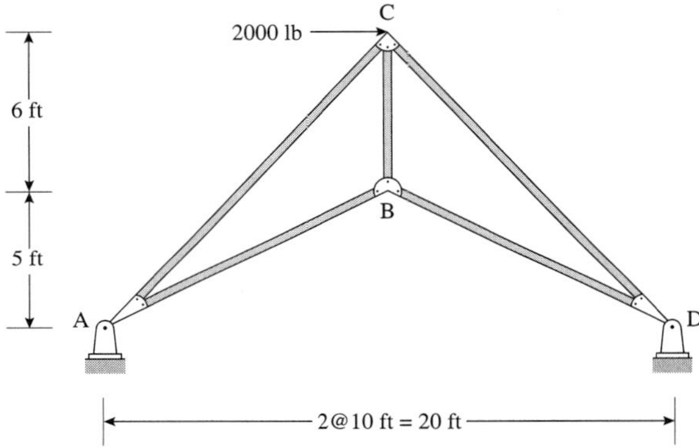

Fig. P5.1.24

5.1.25. Assuming that AE = constant for all the members in Fig. P5.1.25, compute the member forces.

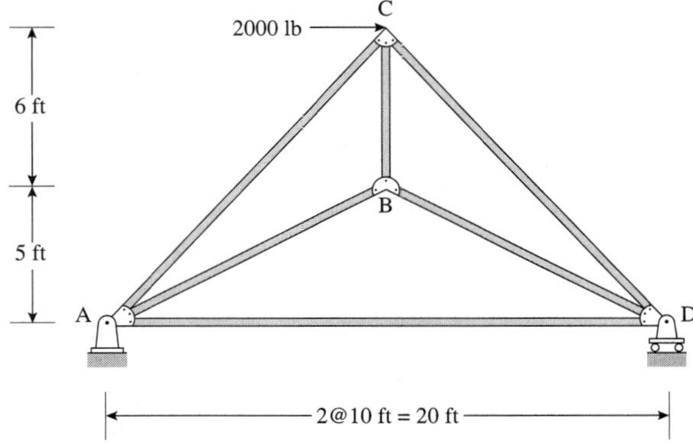

Fig. P5.1.25

5.1.26. Compute the support reactions and the member forces for the truss in Fig. P5.1.26. Take $E = 30$ ksi and $A = 0.5$ in^2.

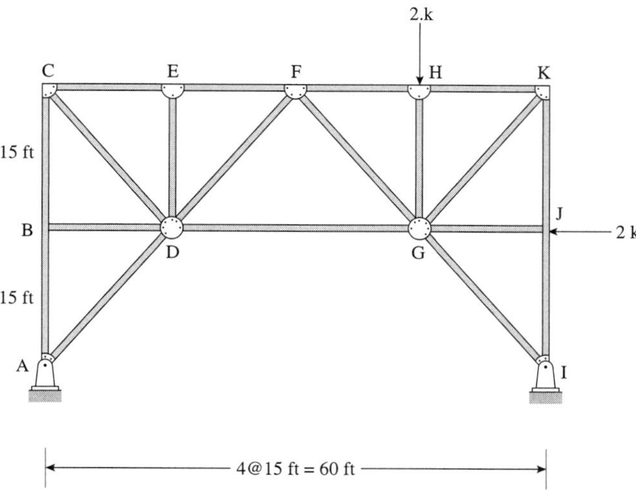

Fig. P5.1.26

Structural Concepts

5.1.27. Here we investigate the differences between the response of determinate and indeterminate trusses. Both the trusses in Figs. P5.1.27(a) and (b) are made of steel and the cross-section of each member is 0.5 in^2.

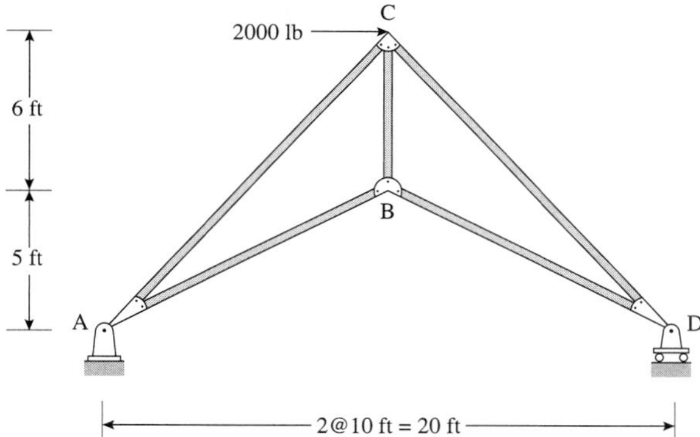

P5.1.27(a)
Determinate truss.

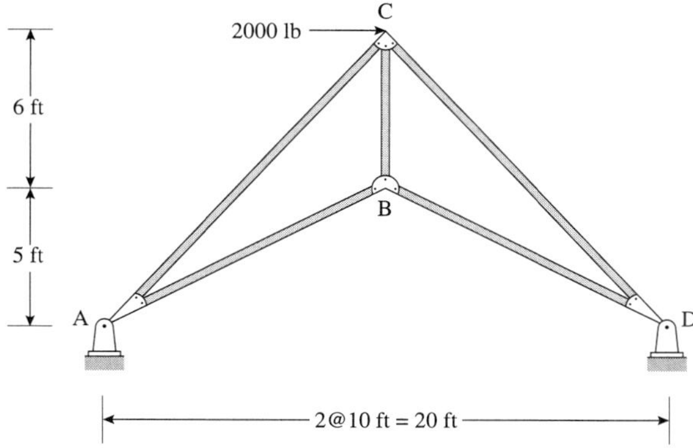

P5.1.27(b)
Indeterminate truss.

Compute the member forces assuming that in addition to the 2000 lb force there is a temperature change of 100°F in each member. What conclusions can you draw?

5.1.4 Higher Degrees of Indeterminacy

Statically indeterminate beams, frames, and truss of higher degree of indeterminacy can be solved using the same concepts and process as we used with structures that were statically indeterminate to degree one. Consider the beam shown in Fig. 5.1.4.1 that is statically indeterminate to degree two. We have selected the redundants as the support reactions B_y and C_y. Once again using the concept of superposition, we can generate the two equations of compatibility that are necessary to give us back the original beam:

$$(\uparrow +)\,\Delta_B + B_y\delta_{BB} + C_y\delta_{BC} = 0 \qquad (5.1.4.1)$$

$$(\uparrow -)\,\Delta_C + B_y\delta_{CB} + C_y\delta_{CC} = 0 \qquad (5.1.4.2)$$

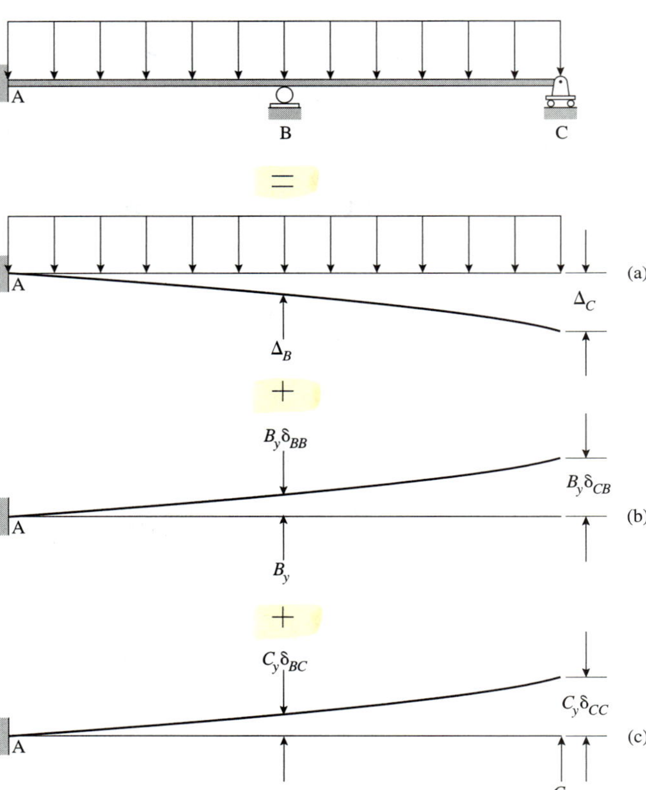

Fig. 5.1.4.1 Statically indeterminate to degree two beam.

The first equation accounts for the net displacement at B being zero. Similarly, the second equation is for the net displacement at C being zero. These two equations can now be solved for the redundants. To compute the displacements in the two equations, we can use the unit load method, as in Fig. 5.1.4.2:

$$\Delta_B = \int \frac{M(x)m_B(x)}{EI}\,dx \qquad (5.1.4.3a)$$

$$\Delta_C = \int \frac{M(x)m_C(x)}{EI}\,dx \qquad (5.1.4.3b)$$

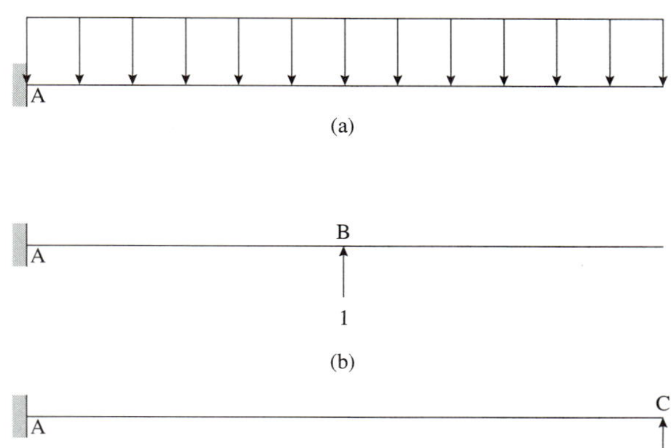

Fig. 5.1.4.2
(a) Beam generating $M(x)$. (b) Beam generating $m_B(x)$. (c) Beam generating $m_C(x)$.

$$\delta_{BB} = \int \frac{m_B(x) m_B(x)}{EI} dx \tag{5.1.4.3c}$$

$$\delta_{CC} = \int \frac{m_C(x) m_C(x)}{EI} dx \tag{5.1.4.3d}$$

$$\delta_{BC} = \int \frac{m_B(x) m_C(x)}{EI} dx = \delta_{CB} \tag{5.1.4.3e}$$

The selection of redundants is not unique for this problem. We present another set of redundants, M_A and B_y as in Fig. 5.1.4.3:

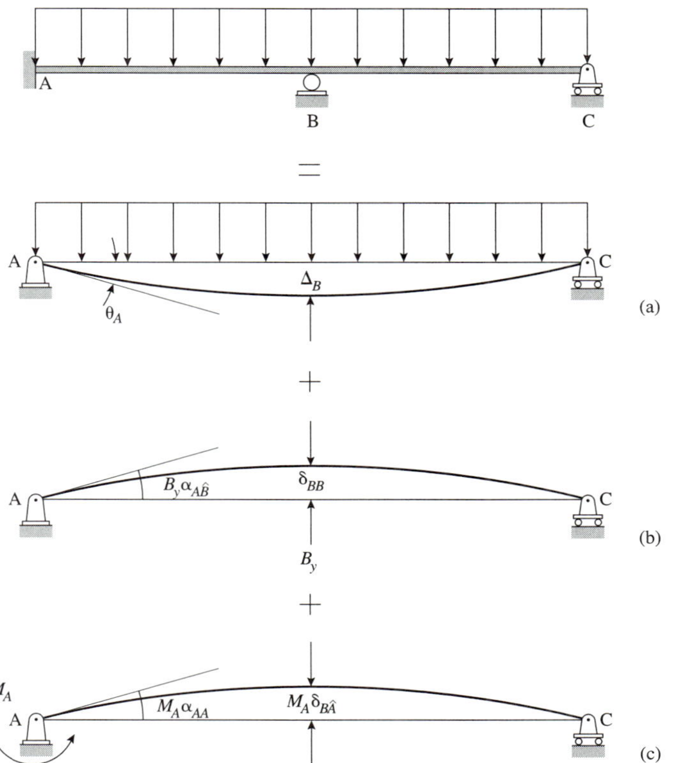

Fig. 5.1.4.3
Beam that is statically indeterminate to degree two: M_A and B_y as redundants.

$$(\curvearrowleft +)\theta_A + M_A\alpha_{AA} + B_y\alpha_{A\hat{B}} = 0 \tag{5.1.4.4}$$

$$(\uparrow +)\Delta_B + M_A\delta_{B\hat{A}} + B_y\delta_{BB} = 0 \tag{5.1.4.5}$$

The first equation accounts for the net rotation at A being zero. Similarly, the second equation is for the net displacement at B being zero. These two equations can now be solved for the redundants (see Fig. 5.1.4.4):

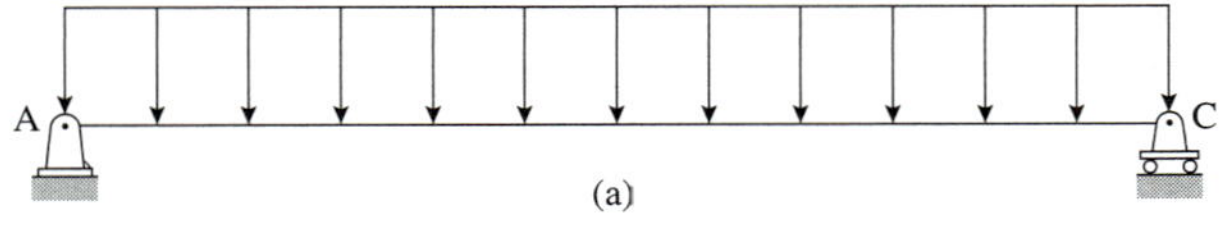

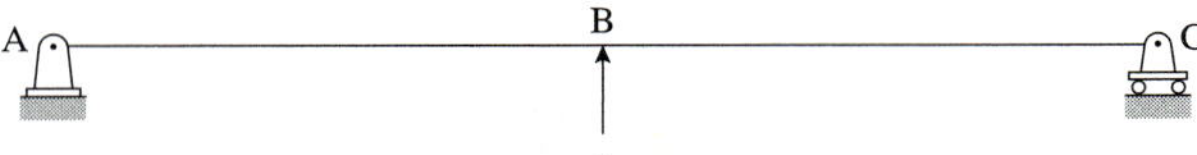

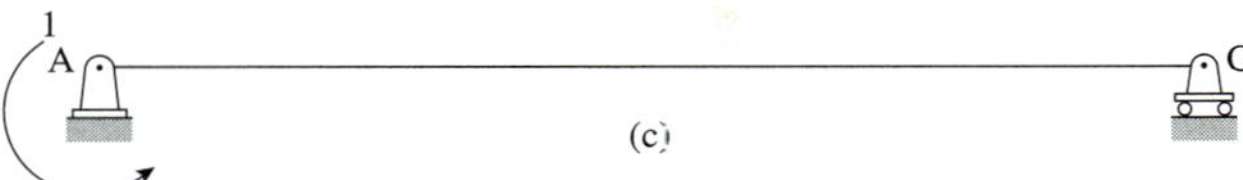

Fig. 5.1.4.4
(a) Beam generating $M(x)$.
(b) Beam generating $m_B(x)$.
(c) Beam generating $m_{\theta A}(x)$.

$$\theta_A = \int \frac{M(x)m_{\theta A}(x)}{EI}dx \tag{5.1.4.6a}$$

$$\Delta_B = \int \frac{M(x)m_B(x)}{EI}dx \tag{5.1.4.6b}$$

$$\alpha_{AA} = \int \frac{m_{\theta A}(x)m_{\theta A}(x)}{EI}dx \tag{5.1.4.6c}$$

$$\delta_{BB} = \int \frac{m_B(x)m_B(x)}{EI}dx \tag{5.1.4.6d}$$

$$\alpha_{A\hat{B}} = \int \frac{m_B(x)m_{\theta A}(x)}{EI}dx = \delta_{B\hat{A}} \tag{5.1.4.6e}$$

The nomenclature is modified here to take into account the combination of displacements and rotations arising from unit force and unit moment. $\alpha_{A\hat{B}}$ is the rotation at A due to unit force at B. Similarly, $\delta_{B\hat{A}}$ is the displacement at B due to a unit moment at A. The last equation illustrates the result derived from Betti reciprocal theorem (see Problem 4.5.22).

Finally, an example dealing with an internally indeterminate truss. Figure 5.1.4.5 shows a degree-two indeterminate truss. To make the truss determinate, we need to select two members as redundants. Let us select members AB and CD as the redundants. Hence the compatibility equations are

$$\Delta_{AB} + F_{AB}(\delta_{AB})_{AB} + F_{CD}(\delta_{AB})_{CD} = 0 \tag{5.1.4.7}$$

$$\Delta_{CD} + F_{AB}(\delta_{CD})_{AB} + F_{CD}(\delta_{CD})_{CD} = 0 \tag{5.1.4.8}$$

where

$$\Delta_{AB} = \sum \frac{Nn_{AB}L}{AE} \tag{5.1.4.9a}$$

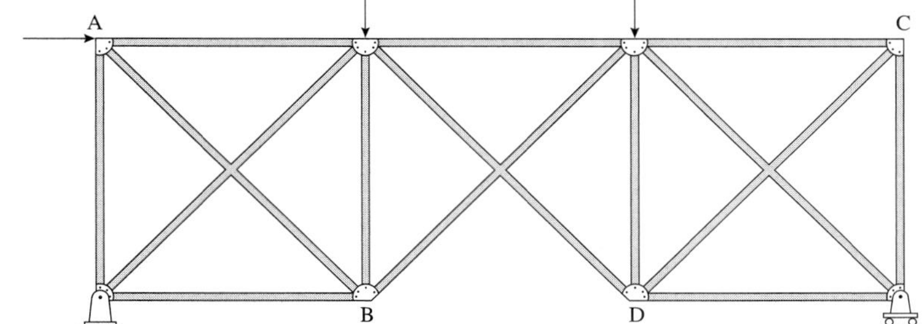

Fig. 5.1.4.5
Statically indeterminate to degree two truss.

$$\Delta_{CD} = \sum \frac{N n_{CD} L}{AE} \tag{5.1.4.9b}$$

$$\left(\delta_{AB}\right)_{AB} = \sum \frac{n_{AB}^2 L}{AE} \tag{5.1.4.9c}$$

$$\left(\delta_{CD}\right)_{CD} = \sum \frac{n_{CD}^2 L}{AE} \tag{5.1.4.9d}$$

$$\left(\delta_{AB}\right)_{CD} = \sum \frac{n_{AB} n_{CD} L}{AE} = \left(\delta_{CD}\right)_{AB} \tag{5.1.4.9e}$$

One can look at the superposition pictorially as in Fig. 5.1.4.6.

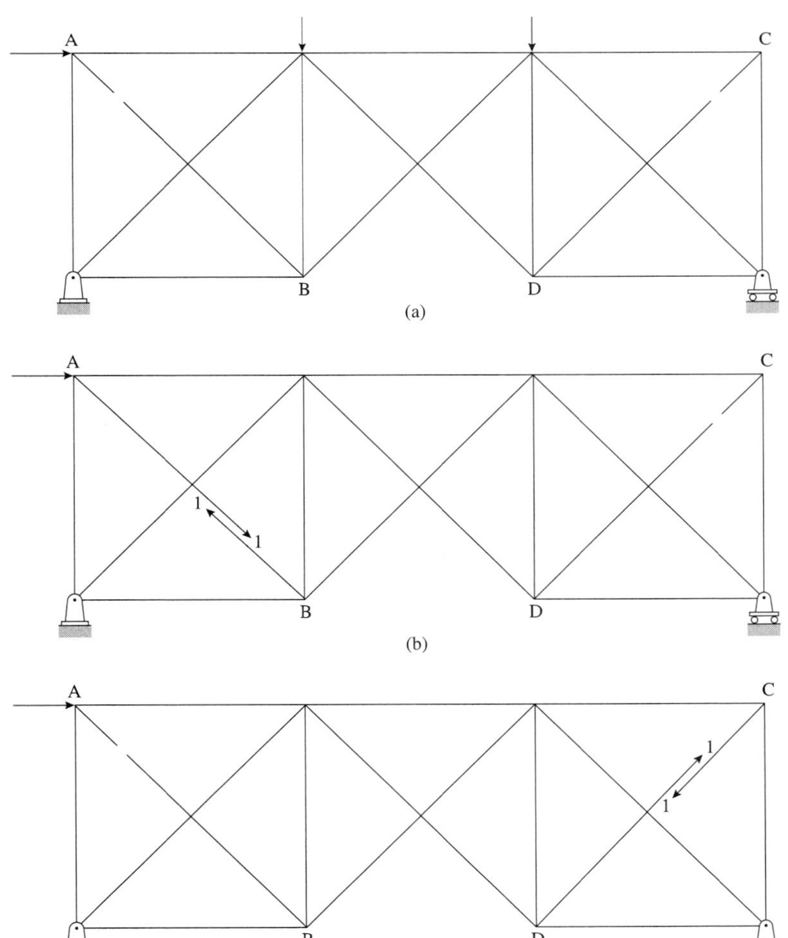

Fig. 5.1.4.6
Trusses for (a) member forces N. (b) Member forces n_{AB}. (c) Member forces n_{CD}.

We present several examples to illustrate the solution of structures with degree two indeterminacy. These results will be extremely useful when we derive the slope-deflection method. Solving even higher-order indeterminate structures using the force method is recommended only for hardy souls.

EXAMPLE 5.1.10 ***Statically Indeterminate to Degree Two Beam***

Compute the support reactions of the beam in Fig. E5.1.10(a). *EI* is a constant.

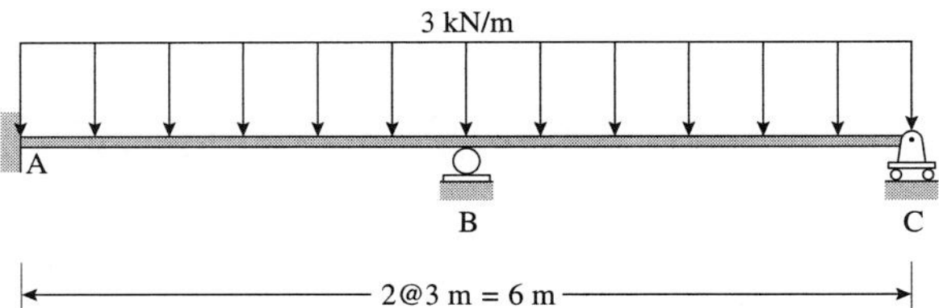

Fig. E5.1.10(a)

SOLUTION

Step 1: The beam is statically indeterminate to degree two. Let the two redundants be M_A and B_y. The superposition of the determinate beams is shown in Fig. E5.1.10(b).

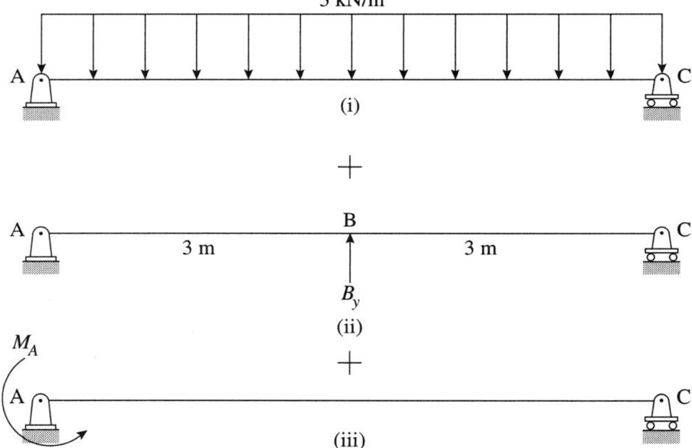

Fig. E5.1.10(b) Beams to obtain (i) $M(x)$; (ii) $m_3(x)$, with 1 kN replacing B_y; (iii) $m_{\theta A}(x)$, with 1 kN-m replacing M_A.

The compatibility equations are

$$(\curvearrowleft +)\ \theta_A + M_A\alpha_{AA} + B_y\alpha_{A\hat{B}} = 0$$

$$(\uparrow +)\ \Delta_B + M_A\delta_{B\hat{A}} + B_y\delta_{BB} = 0$$

where

$$\theta_A = \int \frac{M(x)m_{\theta A}(x)}{EI}dx \qquad \Delta_B = \int \frac{M(x)m_B(x)}{EI}dx$$

$$\alpha_{AA} = \int \frac{m_{\theta A}(x)m_{\theta A}(x)}{EI}dx \qquad \delta_{BB} = \int \frac{m_B(x)m_B(x)}{EI}dx$$

$$\alpha_{A\hat{B}} = \int \frac{m_B(x)m_{\theta A}(x)}{EI}dx = \delta_{B\hat{A}}$$

Step 2: Computation of the internal moments.

Segment	FBD	$M(x)$
AB $0 < x < 3m$	$3x$, M, A, x, 9	$-\frac{3x^2}{2} + 9x$
CB $0 < x_1 < 3m$	M, $3x_1$, C, x_1, 9	$-\frac{3x_1^2}{2} + 9x_1$

Segment	FBD	$m_B(x)$
AB $0 < x < 3m$	1/2, m_B, A, x	$-\frac{x}{2}$
CB $0 < x_1 < 3m$	m_B, 1/2, C, x_1	$-\frac{x_1}{2}$

Segment	FBD	$m_{\theta A}(x)$
AB $0 < x < 3m$	$m_{\theta A}$, A, 1, x, 1/6	$\frac{x}{6} - 1$
CB $0 < x_1 < 3m$	$m_{\theta A}$, 1/6, C, x_1	$-\frac{x_1}{6}$

Step 3: Computation of the deflections.

$$\theta_A = \int_0^3 \frac{\left(-\frac{3x^2}{2} + 9x\right)\left(\frac{x}{6} - 1\right)}{EI} dx + \int_0^3 \frac{\left(-\frac{3x_1^2}{2} + 9x_1\right)\left(-\frac{x_1}{6}\right)}{EI} dx_1 = -\frac{27}{EI} \Rightarrow \theta_A = \frac{27}{EI} \ (\circlearrowleft)$$

$$\Delta_B = \int_0^3 \frac{\left(-\frac{3x^2}{2} + 9x\right)\left(-\frac{x}{2}\right)}{EI} dx + \int_0^3 \frac{\left(-\frac{3x_1^2}{2} + 9x_1\right)\left(-\frac{x_1}{2}\right)}{EI} dx_1 = -\frac{405}{8EI} \Rightarrow \Delta_B = \frac{405}{8EI} (\downarrow)$$

$$\alpha_{AA} = \int_0^3 \frac{\left(\frac{x}{6}-1\right)^2}{EI}dx + \int_0^3 \frac{\left(-\frac{x_1}{6}\right)^2}{EI}dx_1 = \frac{2}{EI}\ (\curvearrowleft)$$

$$\delta_{BB} = \int_0^3 \frac{\left(-\frac{x}{2}\right)^2}{EI}dx + \int_0^3 \frac{\left(-\frac{x_1}{2}\right)^2}{EI}dx_1 = \frac{9}{2EI}(\uparrow)$$

$$\alpha_{A\hat{B}} = \int_0^3 \frac{\left(\frac{x}{6}-1\right)\left(-\frac{x}{2}\right)}{EI}dx + \int_0^3 \frac{\left(\frac{x_1}{6}\right)\left(-\frac{x_1}{2}\right)}{EI}dx_1 = \frac{9}{4EI}\ (\curvearrowleft)$$

$$\delta_{B\hat{A}} = \frac{9}{4EI}(\uparrow)$$

Step 4: Compatibility equations. Substituting in the two compatibility equations, we get

$$-\frac{27}{EI} + M_A\left(\frac{2}{EI}\right) + B_y\left(\frac{9}{4EI}\right) = 0$$

$$-\frac{405}{8EI} + M_A\left(\frac{9}{EI}\right) + B_y\left(\frac{9}{2EI}\right) = 0$$

Solving, we find $M_A = 1.93$ kN-m($\curvearrowleft$) and $B_y = 10.29$ kN($\uparrow$). The final structural FBD appears in Fig. E5.1.10(c).

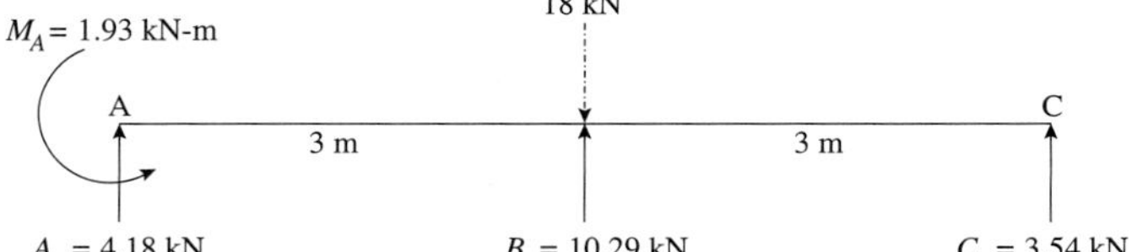

Fig. E5.1.10(c)

EXAMPLE 5.1.11 ***Statically Indeterminate to Degree Two Beam***

Compute the support reactions of the beam in Fig. E5.1.11(a). EI is a constant.

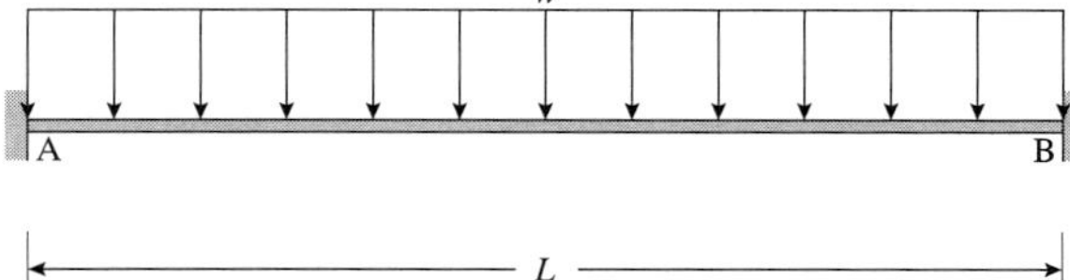

Fig. E5.1.11(a)

SOLUTION

Step 1: The beam is statically indeterminate to degree three. Ignoring the axial effects, the two horizontal reactions can be discarded so that the beam is now statically indeterminate to degree two. We select the two reactions at B as the redundants, B_y and M_B (see Fig. E5.1.11(b)).

Fig. E5.1.11(b) Superposition of the determinate beams.

Hence the compatibility equations are

$$(\uparrow+)\,\Delta_B + B_y\delta_{BB} + M_B\delta_{B\hat{B}} = 0$$

$$(\curvearrowleft+)\,\theta_B + B_y\alpha_{B\hat{B}} + M_B\alpha_{BB} = 0$$

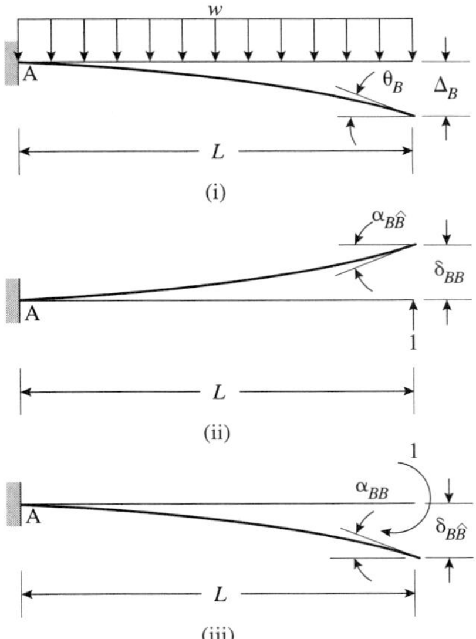

Fig. E5.1.11(c) Deflections used in the determinate beams.

Step 2: Computation of internal moments in the determinate beams. Using beam (i) in Fig. E5.1.11(c), we find

Segment BA $0 < x < L$	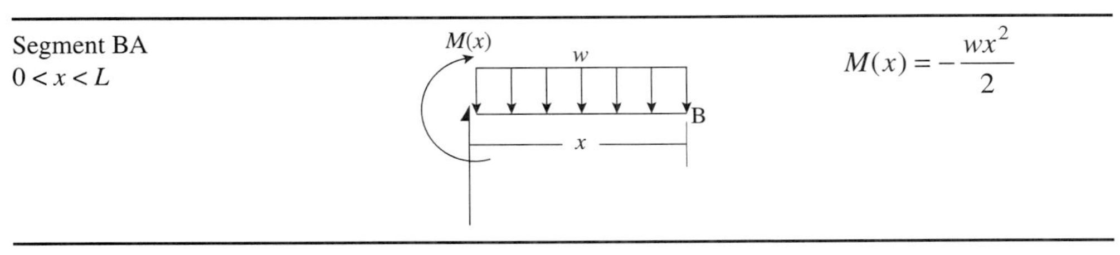 	$M(x) = -\dfrac{wx^2}{2}$

Using beam (ii) in Fig. E5.1.11(c), we find

Segment BA $0 < x < L$	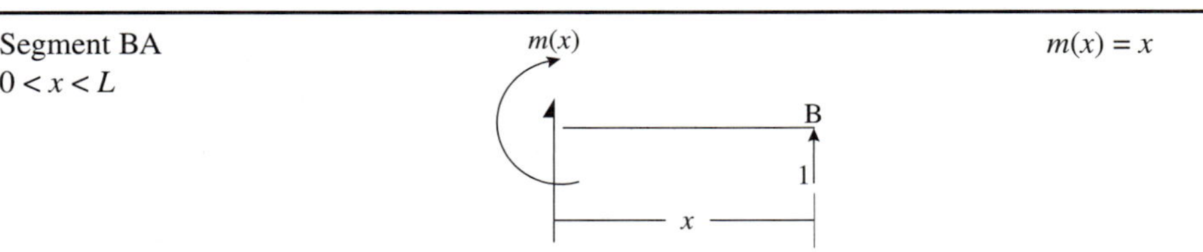	$m(x) = x$

Using beam (iii) in Fig. E5.1.11(c), we find

Segment BA $0 < x < L$	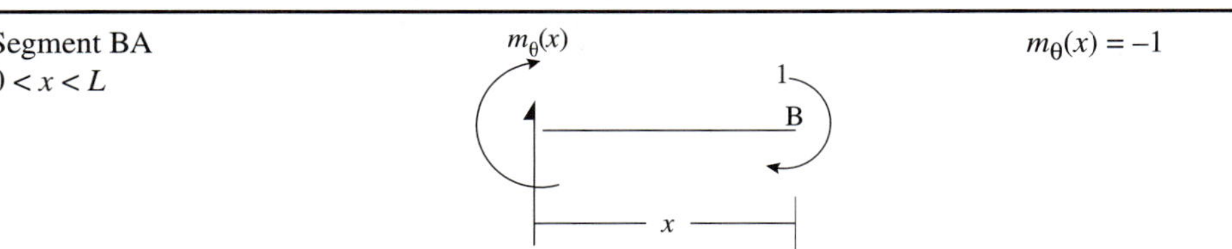	$m_\theta(x) = -1$

Step 3: Computation of the deflections.

$$EI\Delta_B = \int_0^L M(x)m(x)\,dx = \int_0^L \left(-\frac{wx^2}{2}\right)(x)\,dx = -\frac{wL^4}{8} \Rightarrow \Delta_B = \frac{wL^4}{8EI}(\downarrow)$$

$$EI\theta_B = \int_0^L M(x)m_\theta(x)\,dx = \int_0^L \left(-\frac{wx^2}{2}\right)(-1)\,dx = \frac{wL^3}{6} \Rightarrow \theta_B = \frac{wL^3}{6EI}(\circlearrowright)$$

$$EI\delta_{BB} = \int_0^L m(x)m(x)\,dx = \int_0^L (x)(x)\,dx = \frac{L^3}{3} \Rightarrow \delta_{BB} = \frac{L^3}{3EI}(\uparrow)$$

$$EI\alpha_{BB} = \int_0^L m_\theta(x)m_\theta(x)\,dx = \int_0^L (-1)(-1)\,dx = L \Rightarrow \alpha_{BB} = \frac{L}{EI}(\circlearrowright)$$

$$EI\delta_{B\hat{B}} = \int_0^L m_\theta(x)m(x)\,dx = \int_0^L (-1)(x)\,dx = -\frac{L^2}{2} \Rightarrow \delta_{B\hat{B}} = \frac{L^2}{2EI}(\downarrow)$$

$$EI\alpha_{B\hat{B}} = \int_0^L m(x)m_\theta(x)\,dx = \int_0^L (x)(-1)\,dx = -\frac{L^2}{2} \Rightarrow \alpha_{B\hat{B}} = \frac{L^2}{2EI}(\curvearrowleft)$$

Step 4: Generating and solving the compatibility equations:

$$-\frac{wL^4}{8EI} + B_y\left(\frac{L^3}{3EI}\right) + M_B\left(-\frac{L^2}{2EI}\right) = 0$$

$$-\frac{wL^3}{6EI} + B_y\left(\frac{L^2}{2EI}\right) + M_B\left(-\frac{L}{EI}\right) = 0$$

Multiplying the first equation by 24 EI and the second equation by $-12EIL$ gives us

$$-3wL^4 + B_y\left(8L^3\right) + M_B\left(-12L^2\right) = 0 \qquad 2wL^4 + B_y\left(-6L^3\right) + M_B\left(12L^2\right) = 0$$

Adding the two equations yields

$$-wL^4 + B_y\left(2L^3\right) = 0$$

Solving, we have $B_y = wL/2$, and substituting in any one of the two equations yields $M_B = wL^2/12$. Using these values and the FBD of the beam we can compute the two support reactions at A. The final beam FBD is given in Fig. E5.1.11(d).

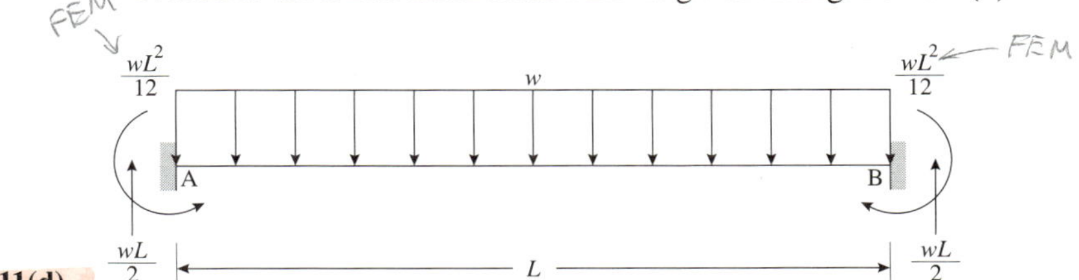

Fig. E5.1.11(d)

EXAMPLE 5.1.12 ***Statically Indeterminate to Degree Two Beam***

Compute the support reactions of the beam in Fig. E5.1.12(a). *EI* is a constant.

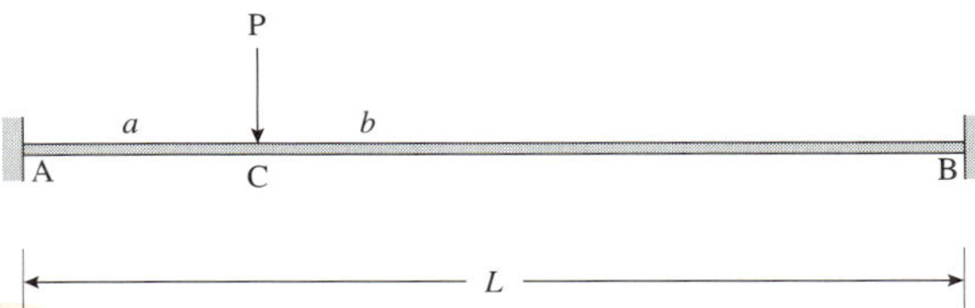

Fig. E5.1.12(a)

SOLUTION

Step 1: The beam is statically indeterminate to degree three. Ignoring the axial effects, the two horizontal reactions can be discarded so that the beam is now statically indeterminate to degree two. We select the two reactions at B as the redundants, B_y and M_B (see Fig. E5.1.12(b)). This problem is similar to the previous problem and we use some of the previous results in solving it. Hence the compatibility equations are as follows:

$$(\uparrow +)\,\Delta_B + B_y \delta_{BB} + M_B \delta_{B\hat{B}} = 0$$

$$(\curvearrowleft +)\,\theta_B + B_y \alpha_{B\hat{B}} + M_B \alpha_{BB} = 0$$

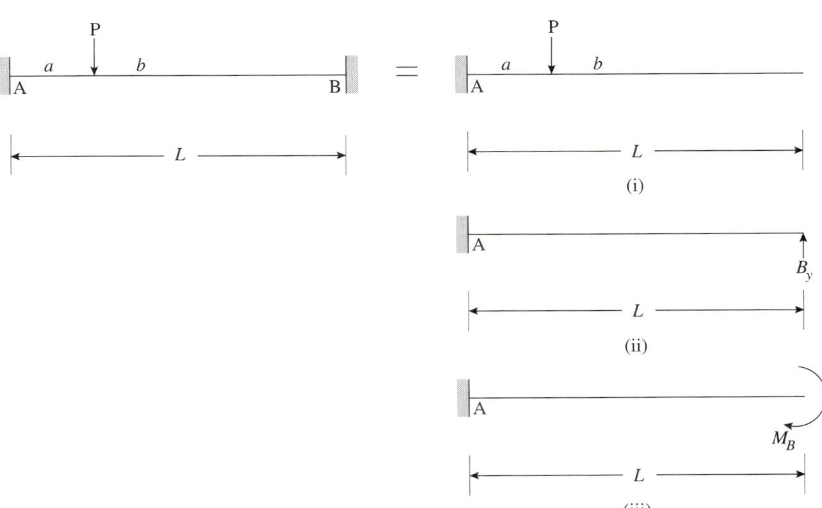

Fig. E5.1.12(b) Superposition of the determinate beams.

Step 2: Computation of internal moments in the determinate beams. Using beam (i) in Fig. E5.1.12(b), we find

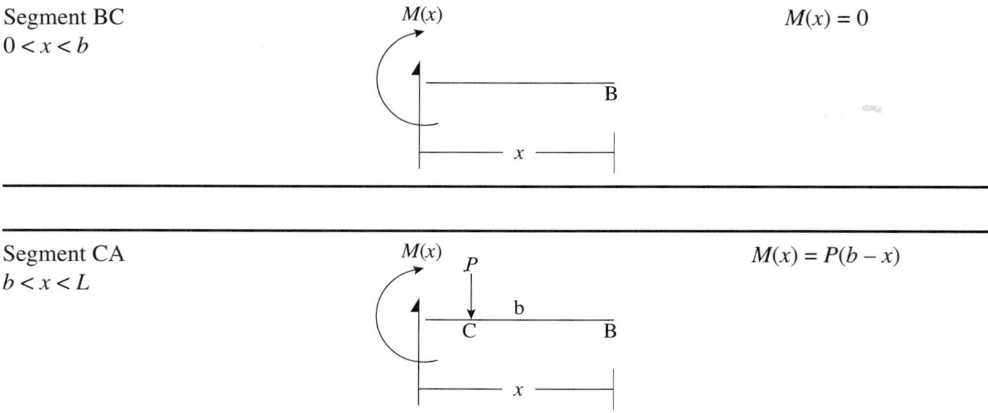

Using beam (ii) in Fig. E5.1.12(b), we have $m(x) = x$, and beam (iii) in Fig. E5.1.10(b) yields $m_\theta(x) = -1$ for $0 < x < b$ and $b < x < L$.

Step 3: Computation of the deflections.

$$\Delta_B = \frac{1}{EI}\int_0^L M(x)\,m(x)\,dx = \frac{1}{EI}\int_b^L P(b-x)(x)\,dx = \frac{P}{6EI}\left(3bL^2 - 2L^3 - b^3\right)(\uparrow)$$

$$\theta_B = \frac{1}{EI}\int_0^L M(x)\,m_\theta(x)\,dx = \frac{1}{EI}\int_0^L P(b-x)(-1)\,dx = \frac{P}{2EI}\left(L^2 - 2bL + b^2\right)(\circlearrowleft)$$

The rest of the deflections are the same as in Example 5.1.11.

Step 4: Generating and solving the compatibility equations (multiplying throughout by EI).

$$\frac{P}{6}\left(3bL^2 - 2L^3 - b^3\right) + B_y\left(\frac{L^3}{3}\right) + M_B\left(-\frac{L^2}{2}\right) = 0$$

$$-\frac{P}{2}(L-b)^2 + B_y\left(\frac{L^2}{2}\right) + M_B(-L) = 0$$

Multiplying the first equation by 6 and second equation by $-2L$ yields

$$P\left(3bL^2 - 2L^3 - b^3\right) + B_y\left(2L^3\right) + M_B\left(-3L^2\right) = 0$$

$$P(L-b)^2(2L) + B_y\left(-2L^3\right) + M_B\left(4L^2\right) = 0$$

Adding the two equations gives us

$$P(-bL^2 + b^3 - 2Lb^2) + M_B\left(L^2\right) = 0$$

Solving, we have $M_B = Pa^2b/L^2$ and substituting in one of the two equations yields $B_y = Pa^2(\mathrm{L} + 2b)/L^3$. Using these values and the FBD of the beam we can compute the two support reactions at A. The final beam FBD is given in Fig. E5.1.12(c).

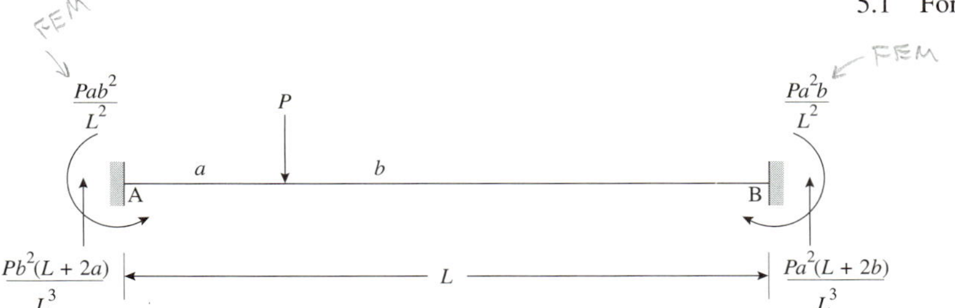

Fig. E5.1.12(c)

EXAMPLE 5.1.13 ***Statically Indeterminate to Degree Two Beam***

Compute the support reactions of the beam in Fig. E5.1.13(a) due to support settlement Δ at B. EI is a constant.

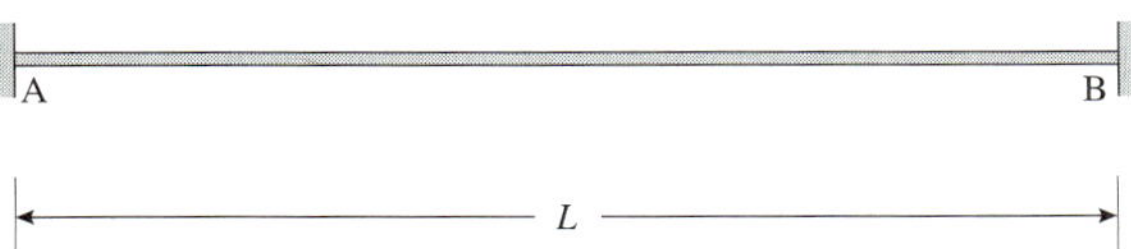

Fig. E5.1.13(a)

SOLUTION

Step 1: We select the two reactions at B as the redundants, B_y and M_B (see Fig. E5.1.13(b)). This problem is similar to the previous two problems and we use some of the previous results in solving it. Hence the compatibility equations are

$$(\uparrow +)\,\Delta_B + B_y\delta_{BB} + M_B\delta_{B\hat{B}} = -\Delta$$

$$(\curvearrowleft +)\,\theta_B + B_y\alpha_{B\hat{B}} + M_B\alpha_{BB} = 0$$

Fig. E5.1.13(b) Superposition of the determinate beams.

Note that the sign associated with the displacement Δ is a function of the (assumed) sign convention governing the equation.

Step 2: Computation of internal moments in the determinate beams. Using beam (i) in Fig. E5.1.13(b), $M(x) = 0$, $0 < x < L$.

Step 3: Computation of the deflections.

$$\Delta_B = \frac{1}{EI}\int_0^L M(x)\,m(x)\,dx = 0$$

$$\theta_B = \frac{1}{EI}\int_0^L M(x)\,m_\theta(x)\,dx = 0$$

The rest of the deflections are the same as in Example 5.1.12.

Step 4: Generating and solving the compatibility equations (multiplying throughout by EI):

$$B_y\left(\frac{L^3}{3}\right) + M_B\left(-\frac{L^2}{2}\right) = -EI\Delta$$

$$B_y\left(\frac{L^2}{2}\right) + M_B(-L) = 0$$

Multiplying the first equation by 6 and second equation by $-2L$ yields

$$B_y\left(2L^3\right) + M_B\left(-3L^2\right) = -6EI\Delta$$

$$B_y\left(-2L^3\right) + M_B\left(4L^2\right) = 0$$

Adding the two equations gives us

$$M_B\left(L^2\right) = -6EI\Delta$$

Solving, we have $M_B = -\,(6EI\Delta/L^2)$ and substituting in one the two equations yields $B_y = -\,(12EI\Delta/L^3)$. Using these values and the FBD of the beam, we can compute the two support reactions at A. The final beam FBD is given in Fig E5.1.13(c).

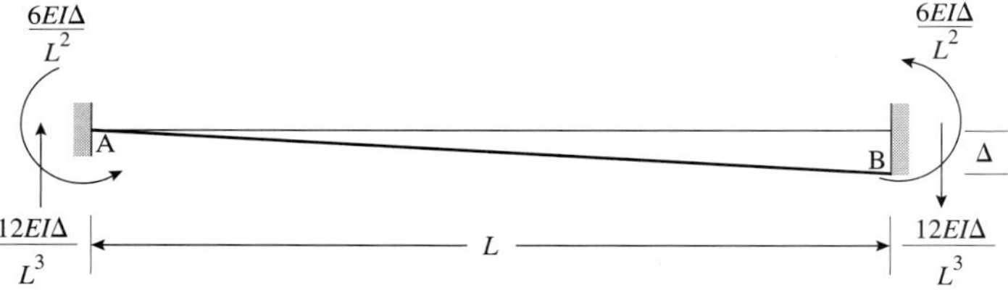

Fig. E5.1.13(c)

EXAMPLE 5.1.14 ***Statically Internally Indeterminate to Degree Two Truss***

Compute the member forces of the truss beam in Fig E5.1.14(a). Take E = 29000 ksi and A = 2 in^2 for all the members.

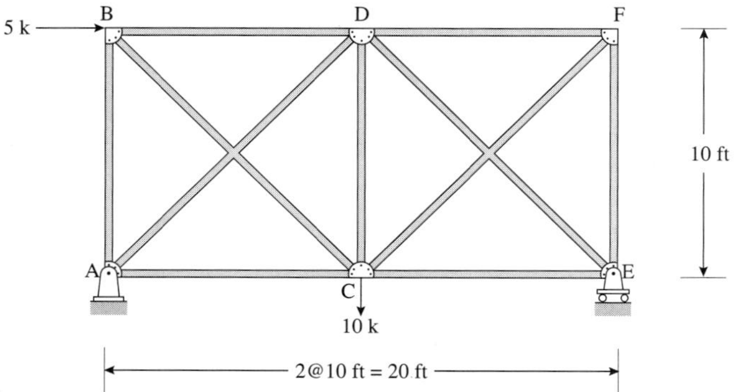

Fig. E5.1.14(a)

SOLUTION

Step 1: From the given truss, $j = 6$, $r = 3$, and $m = 11$. Hence $(m + r) - 2j = 2$. The truss is statically indeterminate to degree two. The truss is externally determinate but internally indeterminate. We select BC and CF as the redundant members and (k, in) as the problem units. The determinate trusses are shown in Fig. E5.1.14(b)–(d).

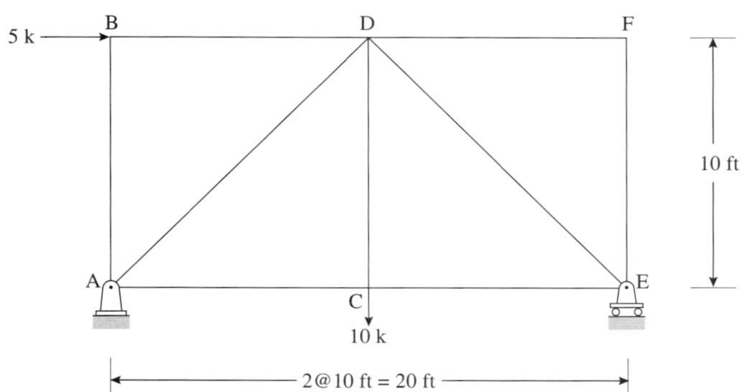

Fig. E5.1.14(b) Determinate truss to compute N forces.

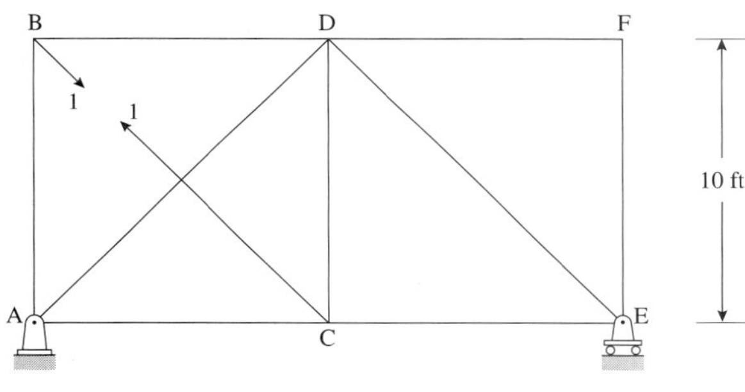

Fig. E5.1.14(c) Determinate truss to compute n_{BC} forces.

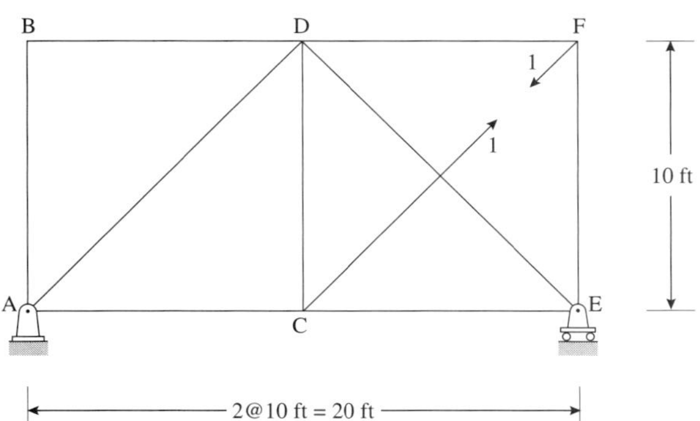

Fig. E5.1.14(d) Determinate truss to compute n_{CF} forces.

The compatibility equations are

$$\Delta_{BC} + F_{BC}\left(\delta_{BC}\right)_{BC} + F_{CF}\left(\delta_{BC}\right)_{CF} = 0 \tag{1}$$

$$\Delta_{CF} + F_{BC}\left(\delta_{CF}\right)_{BC} + F_{CF}\left(\delta_{CF}\right)_{CF} = 0 \tag{2}$$

where

$$\Delta_{BC} = \sum \frac{N n_{BC} L}{AE} \tag{3}$$

$$\Delta_{CF} = \sum \frac{Nn_{CF}L}{AE} \tag{4}$$

$$(\delta_{BC})_{BC} = \sum \frac{n_{BC}^2 L}{AE} \tag{5}$$

$$(\delta_{CF})_{CF} = \sum \frac{n_{CF}^2 L}{AE} \tag{6}$$

$$(\delta_{BC})_{CF} = \sum \frac{n_{BC}n_{CF}L}{AE} = (\delta_{CF})_{BC} \tag{7}$$

Step 2: Computation of the member forces.

Member	L/A(1/in)	N(k)	n_{BC}(k)	n_{CF}(k)	$Nn_{BC}L/A$	$Nn_{CF}L/A$
AB	60	0	–0.707	0	0	0
CD	60	10	–0.707	–0.707	–424.2	–424.4
EF	60	0	0	–0.707	0	0
BD	60	–5	–0.707	0	212.1	0
DF	60	0	0	–0.707	0	0
AC	60	7.5	–0.707	0	–318.15	0
CE	60	7.5	0	–0.707	0	–318.15
BC	84.85	0	1	0	0	0
AD	84.85	–3.54	1	0	–300.37	0
CF	84.85	0	0	1	0	0
DE	84.85	–10.61	0	1	0	–900.26

Step 3: Computation of the displacements. Using the above results, we have

$$\Delta_{BC} = \sum \frac{Nn_{BC}L}{AE} = \frac{-830.62}{29000} = -0.0286$$

$$\Delta_{CF} = \sum \frac{Nn_{CF}L}{AE} = \frac{-1642.81}{29000} = -0.0566$$

$$(\delta_{BC})_{BC} = \sum \frac{n_{BC}^2 L}{AE} = \frac{289.7}{29000} = 0.01$$

$$(\delta_{CF})_{CF} = \sum \frac{n_{CF}^2 L}{AE} = \frac{289.7}{29000} = 0.01$$

$$(\delta_{BC})_{CF} = \sum \frac{n_{BC}n_{CF}L}{AE} = (\delta_{CF})_{BC} = \frac{30}{29000} = 0.0010$$

Step 4: Compatibility equations. Substituting the above in the compatibility equations, we have

$$-0.0286 + 0.01F_{BC} + 0.0010F_{CF} = 0$$
$$-0.0566 + 0.0010F_{BC} + 0.01F_{CF} = 0$$

Solving gives us

$F_{BC} = 2.32$ k $\qquad F_{CF} = 5.43$ k

Step 5: Final member forces. As done before with indeterminate trusses, we use the concept of superposition to compute the final member forces. We should check the equilibrium of a few joints just to ensure that our computations are correct.

Member	N(k)	n_{BC}(k)	n_{CF}(k)	$F = N + F_{BC}n_{BC} + F_{CF}n_{CF}$ (k)
AB	0	−0.707	0	−1.64 (C)
CD	10	−0.707	−0.707	4.52 (T)
EF	0	0	−0.707	−3.84 (C)
BD	−5	−0.707	0	−6.64 (C)
DF	0	0	−0.707	−3.84 (C)
AC	7.5	−0.707	0	5.86 (T)
CE	7.5	0	−0.707	3.66 (T)
BC	0	1	0	2.32 (T)
AD	−3.54	1	0	−1.22 (C)
CF	0	0	1	5.43 (T)
DE	−10.61	0	1	−5.18 (C)

EXERCISES

5.1.28. Compute all the support reactions for the beam in Fig. P5.1.28. EI is a constant.

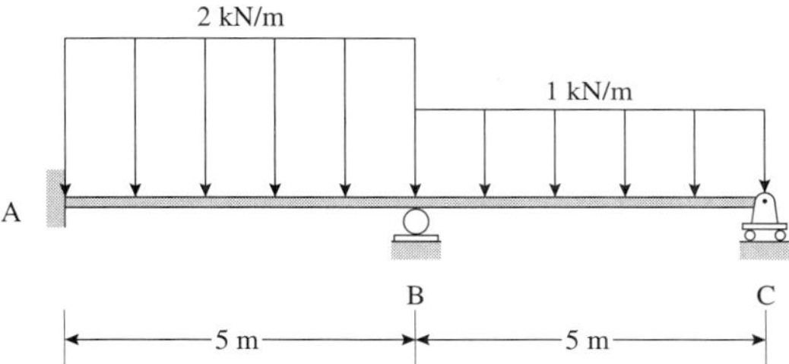

Fig. P5.1.28

5.1.29. Compute all the support reactions for the frame in Fig. P5.1.29. EI is a constant. Draw the shear force and bending moment diagrams.

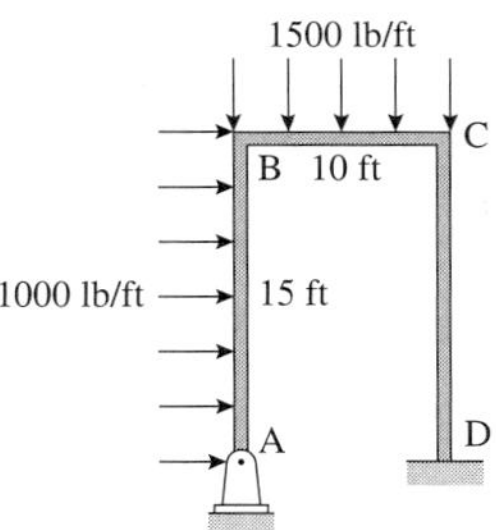

Fig. P5.1.29

5.1.30. Compute the support reactions and the member forces for the truss in Fig. P5.1.30. Take E = 30 ksi and $A = 0.5$ in^2 for all the members.

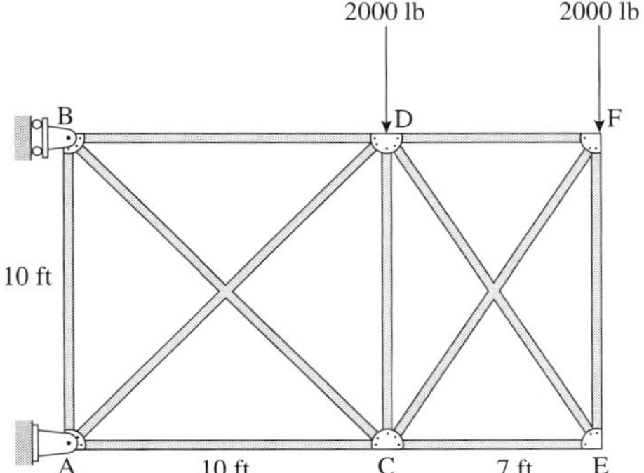

Fig. P5.1.30

5.2 SLOPE-DEFLECTION METHOD

The slope-deflection method is the second of the two classical methods presented in this chapter. While redundant forces were used in the force method to solve indeterminate systems, in the slope-deflection method the primary unknowns are deflections. George Maney is credited with the major development of this methodology. In 1915, while working at the University of Minnesota, he published the details of the method, which provided a very convenient means of analyzing primarily beams and frames. However, in the absence of a computing tool such as the computer, the practical usefulness of the method was restricted to solving problems with a small number of unknowns.

We now describe the major distinction between the force method and the slope-deflection method. The former is used to solve statically indeterminate structures. The latter is used to solve kinematically indeterminate structures—first the deflections are solved for and then the internal member forces are computed. The term *kinematic* refers to deflections. Every point in a planar beam or frame has potentially three deflection components—two orthogonal displacements and one rotation. These deflections are also known as degrees of freedom. We will see this term more often in the next chapter. In the slope-deflection method, the relationship is established between moments at the ends of the members and the corresponding rotations and displacements. The basic assumption used in the derivation is that a typical member can flex but the shear and axial deformations are negligible. These assumptions are no different from those used with the force method.

Consider the beam in Fig. 5.2.1. The beam is statically indeterminate to degree one. Also, there is one kinematic unknown[3], θ_B.

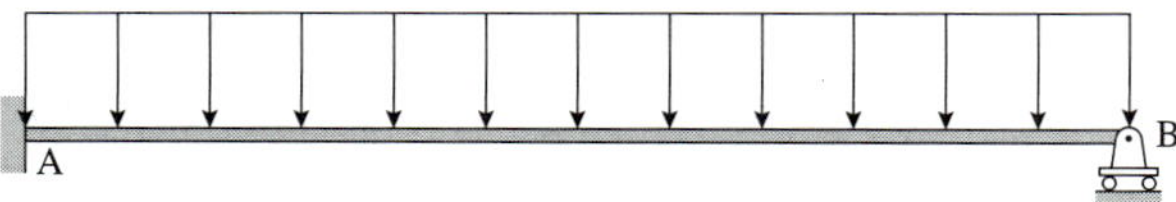

Fig. 5.2.1
Propped cantilever.

Now consider the beam in Fig. 5.2.2. The beam is statically indeterminate to degree one. However, there are three kinematic unknowns, θ_A, θ_B, and θ_C.

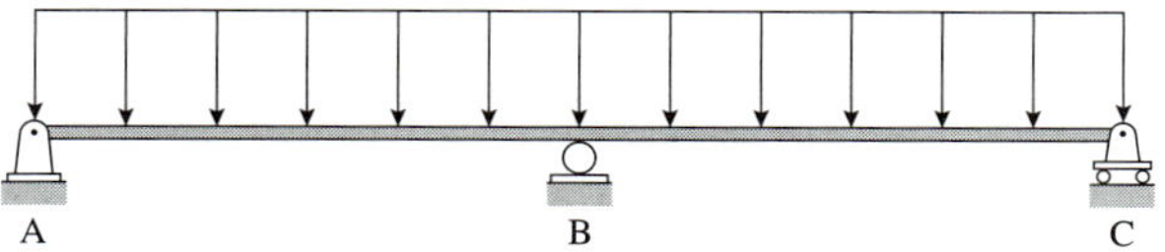

Fig. 5.2.2
Continuous beam.

Consider the beam in Fig. 5.2.3. The beam is statically indeterminate to degree two. Also, there are two kinematic unknowns, θ_B and θ_C.

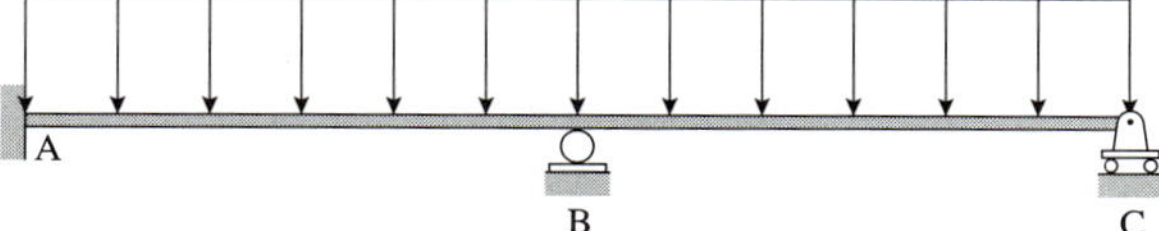

Fig. 5.2.3
Continuous beam.

[3] We ignore axial and shear deformations in the rest of this chapter. Also, we do not use this method to solve truss systems since the derivation is similar to the approach taken in Chapter 6.

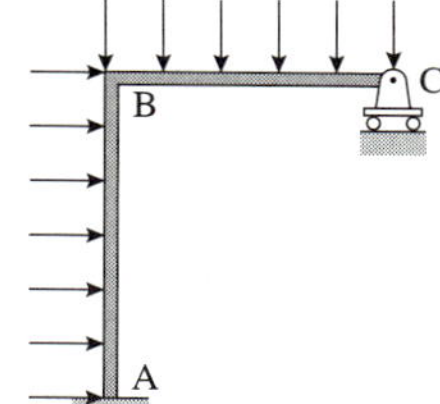

Fig. 5.2.4 Simple frame.

Fig. 5.2.4 shows a planar frame. The frame is statically indeterminate to degree one. However, there are three kinematic unknowns, θ_B, θ_C, and Δ, the horizontal displacement of points B and C. Since we do not assume axial deformations to take place in the beams or columns, (a) points B and C must have the same horizontal displacement and (b) the vertical displacement at B is zero.

Finally, consider the frame in Fig. 5.2.5. The frame is statically indeterminate to degree six! However, there is only one kinematic unknown, θ_B. Since members cannot elongate or contract, point B cannot move up or down, right or left.

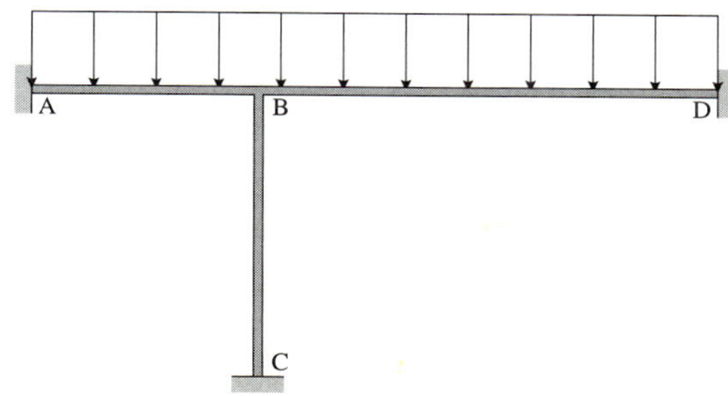

Fig. 5.2.5 Highly indeterminate frame.

Sign convention: We use the following sign convention: All clockwise internal moments and end rotations are positive.

We now derive the fundamental slope-deflection equation. Consider a typical segment of the continuous beam in Fig. 5.2.6. The load on the segment is a known load $w(x)$ and EI is a constant. The primary objective is to use the concept of equilibrium to relate the *internal* moments M_{AB} and M_{BA} at the ends of the segment to the rotations at the ends θ_A and θ_B and to the relative displacement between the ends Δ. All the quantities in the figure are positive as per the sign convention.

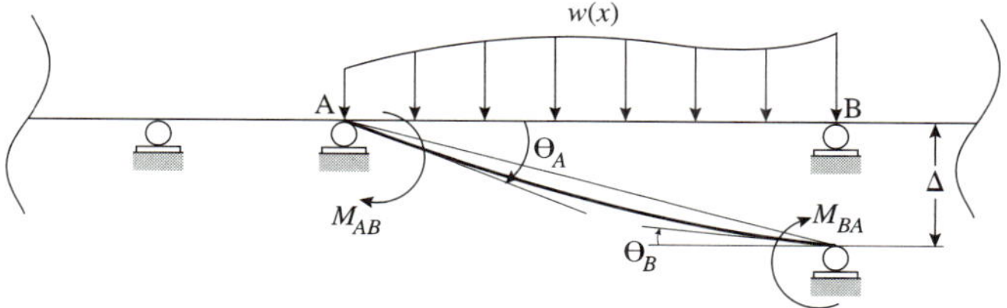

Fig. 5.2.6 Beam segment AB.

To find this relationship we can use the concept of linear superposition as we did with the force method. The components of the superposition are as follows.

a. Find the moments at the ends of the segment that are equivalent to the loading $w(x)$, as shown in Fig. 5.2.7.
b. Find the moments at the ends of the segment when end A is allowed to rotate θ_A, as shown in Fig. 5.2.8.
c. Find the moments at the ends of the segment when end B is allowed to rotate θ_B, as shown in Fig. 5.2.9.
d. Find the moments at the ends of the segment when end B displaces Δ relative to the other end, as shown in Fig. 5.2.10.

Case (a): Fixed-End Moments. The basic idea in the slope-deflection method is to impose equilibrium conditions of the structure at specific locations, the joints. Hence it is necessary to replace the loading on the segment with an equivalent moment system acting at the ends of the segment. These moments are called the fixed-end moments (FEMs) and can be computed a variety of ways.

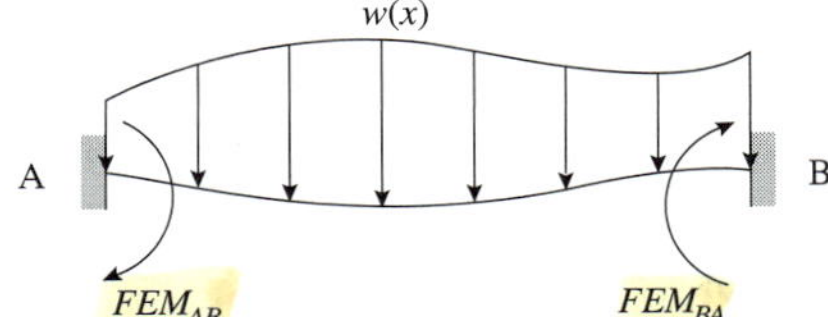

Fig. 5.2.7 Fixed-end moments.

$$M_{AB} = FEM_{AB} \qquad M_{BA} = FEM_{BA} \tag{5.2.1}$$

We have already computed the FEMs for two specific cases in Examples 5.1.11 and 5.1.12 (see Figs. E5.1.11(d) and E5.1.12(c)).

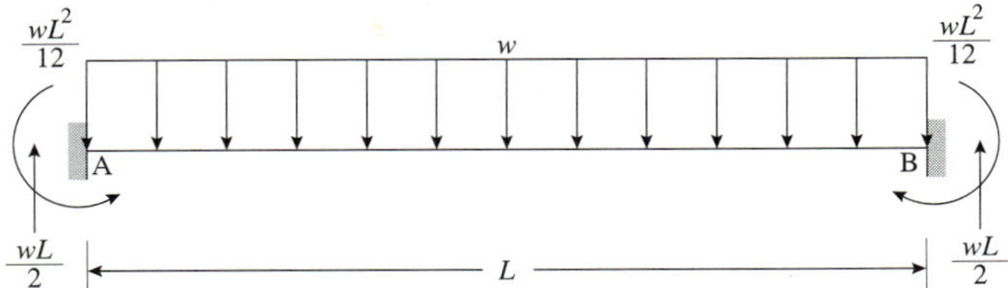

Fig. E5.1.11(d)

$$FEM_{AB} = -\frac{wL^2}{12} \qquad FEM_{BA} = +\frac{wL^2}{12} \tag{5.2.2}$$

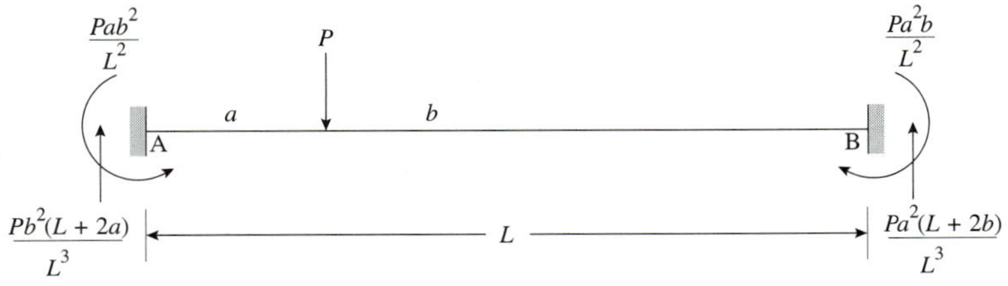

Fig. E5.1.12(c)

$$FEM_{AB} = -\frac{Pab^2}{L^2} \qquad FEM_{BA} = +\frac{Pa^2b}{L^2} \tag{5.2.3}$$

Case (b): Rotation at A, θ_A. We now release end A so that it is free to rotate while the other end is held fixed. The clockwise rotation at A is θ_A. We need to compute the moment M_{AB} necessary to cause this rotation at A.

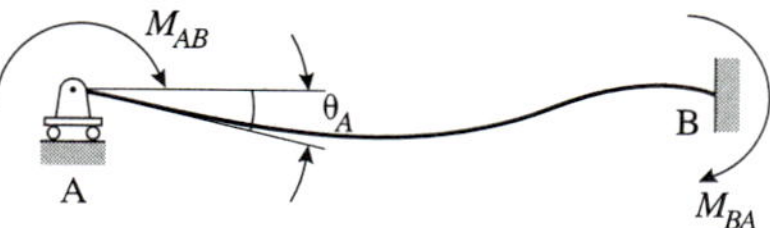

Fig. 5.2.8 End moments due to rotation at A.

We answered this question in Example 5.1.5 and computed

$$M_{AB} = \frac{4EI}{L}\theta_A \qquad M_{BA} = \frac{2EI}{L}\theta_A \tag{5.2.4}$$

Case (c): Rotation at B, θ_B. This case is similar to case (b). We need to compute the moment M_{BA} necessary to cause rotation θ_B at B:

$$M_{AB} = \frac{2EI}{L}\theta_B \qquad M_{BA} = \frac{4EI}{L}\theta_B \tag{5.2.5}$$

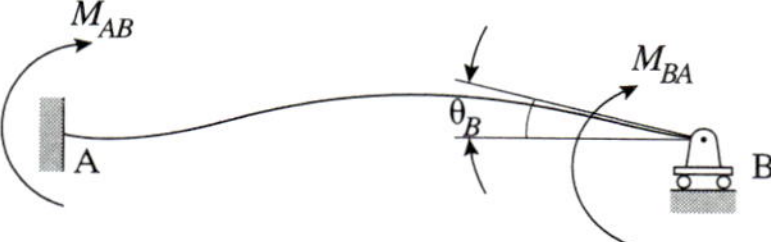

Fig. 5.2.9 End moments due to rotation at B.

Case (d): Displacement of End B Relative to End A, Δ. Finally, we compute the moment (reactions) when end B displaces Δ relative to end A so that the (chord) rotation[4] of the member is clockwise. We have answered this question in Example 5.1.13 (see Fig. E5.1.13(c)):

$$M_{AB} = -\frac{6EI}{L^2}\Delta \qquad M_{BA} = -\frac{6EI}{L^2}\Delta \tag{5.2.6}$$

from example 5.1.13 (p. 291)

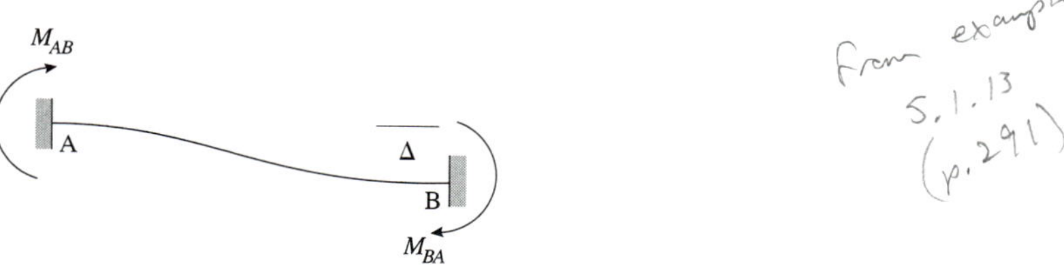

Fig. 5.2.10 End moments due to relative displacement.

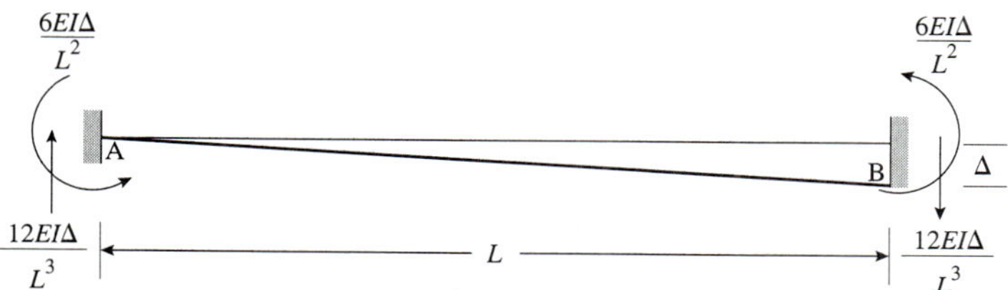

Fig. E5.1.13(c)

[4] A chord can be defined as the straight line connecting A and B.

Note that, in each case, the sign of the moment is positive if it is clockwise, otherwise negative. Combining the four cases, we have

$$M_{AB} = \frac{2EI}{L}\left[2\theta_A + \theta_B - 3\left(\frac{\Delta}{L}\right)\right] + FEM_{AB} \tag{5.2.7}$$

$$M_{BA} = \frac{2EI}{L}\left[2\theta_B + \theta_A - 3\left(\frac{\Delta}{L}\right)\right] + FEM_{BA} \tag{5.2.8}$$

These two equations are to be applied to each segment of the structure. Sometimes it more convenient to remember the formula as

$$M_{nf} = \frac{2EI}{L}\left[2\theta_n + \theta_f - 3\psi\right] + FEM_{nf} \tag{5.2.9}$$

where the symbol n represents the near end of the segment and f the far end, and $\psi = \Delta/L$ is the chord rotation of the segment.

5.2.1 Beams

The analysis of beams via the slope-deflection method can be carried out systematically by applying Eqs. (5.2.7) and (5.2.8). The general procedure is outlined below.

General Procedure

Step 1: Scan the beam and identify the number of (a) segments and (b) kinematic unknowns. A segment is the portion of the beam between two nodes. Kinematic unknowns are those rotations and displacements that are not zero and must be computed. The support or end conditions of the beam will help answer the question.

Step 2: For each segment, generate the two equations. Check the end conditions to see whether one of the end rotations is zero or not (it is not possible for both the end rotations and other deflection components to be zero). If there are no element loads, the *FEM* term is zero. If there are one or more element loads, use the appropriate formula to compute the *FEM* for each element load and then sum all the *FEM*s. If one end of the segment displaces relative to the other, compute the chord rotation; otherwise it is zero.

Step 3: The total number of unknowns—kinematic and the internal moments—will be greater than the number of equations generated in step 2. The additional equations will deal with an equilibrium condition for *each* kinematic unknown. In other words, there will be as many additional equilibrium equations as there are kinematic unknowns. A typical equation involves one or more internal moments. For example, in a continuous beam ABCD, at B the FBD of the joint is as in Fig. 5.2.11.

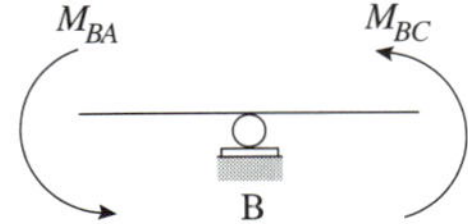

Fig. 5.2.1.1

$$\overset{+}{\sum} M_B = 0 = M_{BA} + M_{BC}$$

Substitute the appropriate expressions for these internal moments from step 2. Now the equations are entirely in terms of the kinematic unknowns. Solve for these unknowns. If the answer is positive, the rotation is clockwise, or the displacement causes clockwise chord rotation.

Step 4: Substitute these kinematic unknowns in the slope-deflection equations (Step 2). If the moments are positive, they act clockwise. Draw the FBD for each segment and show the internal moments and element loads, if any. The shear forces acting at the ends of the segments can be found using the equations of equilibrium for the FBD. Appropriate FBDs can then be used to compute the support reactions.

EXAMPLE 5.2.1

Statically Indeterminate Beam (Example 5.1.1)

For the beam shown in Fig. E5.2.1(a), compute the support reactions. *EI* is a constant.

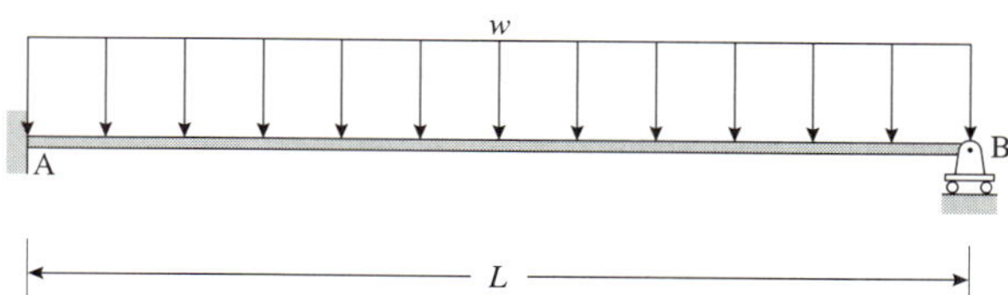

Fig. E5.2.1(a)

SOLUTION

Step 1: The boundary conditions are such that $\theta_A = 0$, $\theta_B = ?$ and $\Delta = 0$. The problem has one kinematic unknown. Using the given loading, we get

$$FEM_{AB} = -\frac{wL^2}{12} \qquad FEM_{BA} = \frac{wL^2}{12}$$

Step 2: Slope-deflection equations. There is only one segment, AB, and we can develop the two equations as follows:

$$M_{AB} = \frac{2EI}{L}(0 + \theta_B - 0) - \frac{wL^2}{12} = \frac{2EI}{L}\theta_B - \frac{wL^2}{12} \tag{1}$$

$$M_{BA} = \frac{2EI}{L}(2\theta_B + 0 - 0) + \frac{wL^2}{12} = \frac{4EI}{L}\theta_B + \frac{wL^2}{12} \tag{2}$$

Step 3: Additional equation. The additional equation will be connected to the only kinematic unknown θ_B. Since B is an end roller support, $M_{BA} = 0$. Using (2) gives us

$(4EI/L)\,\theta_B + (wL^2/12) = 0$

Solving, we find

$$\theta_B = -\frac{wL^3}{48EI} \Rightarrow \theta_B = \frac{wL^3}{48EI}(\curvearrowleft) \tag{3}$$

Recall the sign convention—internal moments and rotations are clockwise if positive.

Step 4: Computation of the internal moments. Substituting the result (3) in (1) and (2), we have

$$M_{AB} = \frac{2EI}{L}\left(-\frac{wL^3}{48EI}\right) - \frac{wL^2}{12} = -\frac{wL^2}{8}$$

$$M_{BA} = \frac{4EI}{L}\left(-\frac{wL^3}{48EI}\right) + \frac{wL^2}{12} = 0$$

The FBD of the beam at this stage is shown in Fig. E5.2.1(b). We can use the equilibrium conditions for the beam to compute the support reactions:

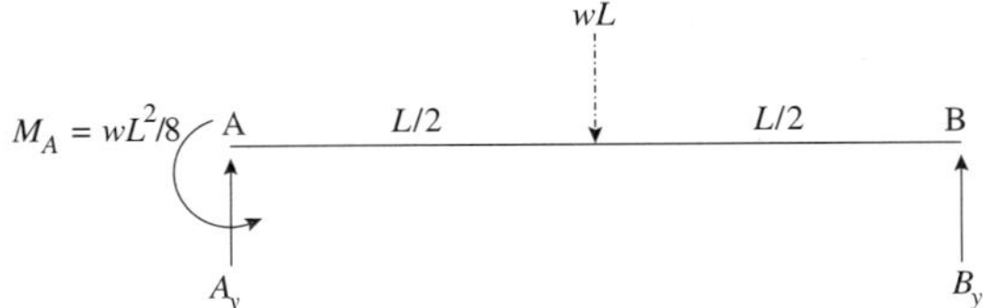

Fig. E5.2.1(b)

$$\sum M_A = 0 \Rightarrow B_y = \frac{3wL}{8}(\uparrow) \qquad \sum F_y = 0 \Rightarrow A_y = \frac{5wL}{8}(\uparrow)$$

We can now develop the shear force, bending moment diagrams and the elastic curve, as shown in Fig. E5.2.1(c). The zero point in the bending moment diagram is the location of zero curvature in the elastic curve.

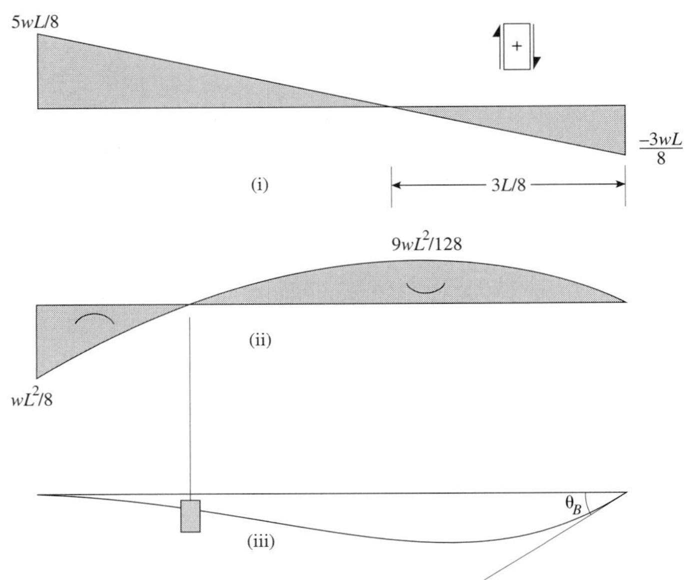

Fig. E5.2.1(c)
(i) Shear force diagram.
(ii) Bending moment diagram. (iii) Elastic curve.

EXAMPLE 5.2.2 ***Statically Indeterminate Continuous Beam***

For the continuous beam in Fig. E5.2.2(a), compute the support reactions. EI is a constant.

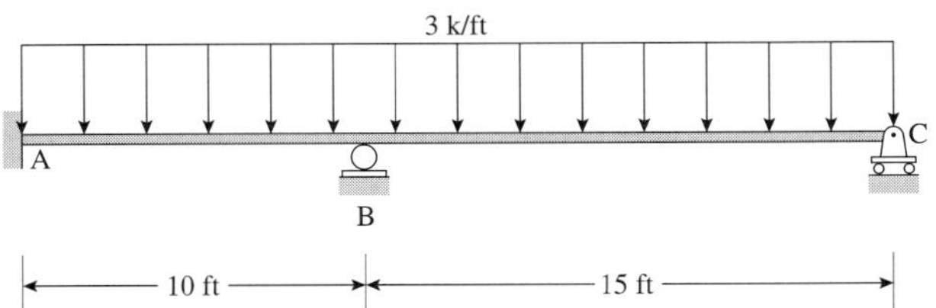

Fig. E5.2.2(a)

SOLUTION

Step 1: The boundary conditions are such that $\theta_A = 0$, $\theta_B = ?$, $\theta_C = ?$ and $\Delta = 0$. The problem has two kinematic unknowns. The problem units are k, ft. Using the given loading, we have

$$FEM_{AB} = -\frac{wL^2}{12} = -\frac{(3)(10)^2}{12} = -25\,\text{k-ft} = -FEM_{BA}$$

$$FEM_{BC} = -\frac{wL^2}{12} = -\frac{(3)(15)^2}{12} = -56.25\,\text{k-ft} = -FEM_{CB}$$

Step 2: Slope-deflection equations. We can now write the four equations for the two segments AB and BC:

$$M_{AB} = \frac{2EI}{10}(0 + \theta_B - 0) - 25 = \frac{EI}{5}\theta_B - 25 \tag{1}$$

$$M_{BA} = \frac{2EI}{10}(2\theta_B + 0 - 0) + 25 = \frac{2EI}{5}\theta_B + 25 \tag{2}$$

$$M_{BC} = \frac{2EI}{15}(2\theta_B + \theta_C - 0) - 56.25 = \frac{4EI}{15}\theta_B + \frac{2EI}{15}\theta_C - 56.25 \tag{3}$$

$$M_{CB} = \frac{2EI}{15}(2\theta_C + \theta_B - 0) + 56.25 = \frac{2EI}{15}\theta_B + \frac{4EI}{15}\theta_C + 56.25 \tag{4}$$

Step 3: Additional equations. The additional equations will be connected to the two kinematic unknowns θ_B and θ_C. The FBDs are crucial to solving the problem.

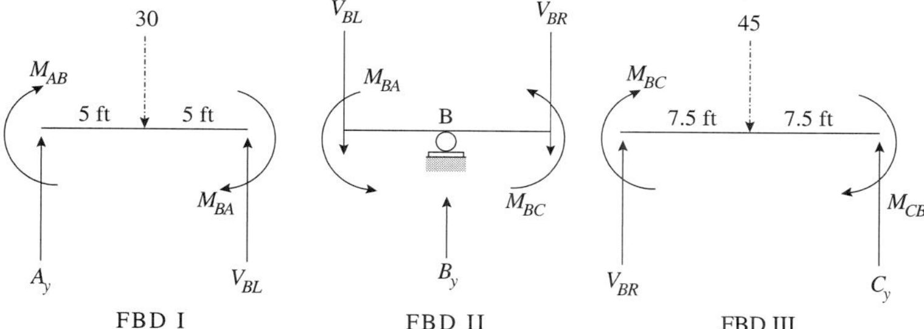

Fig. E5.2.2(b)

A couple of points to note about the FBDs of members AB and BC, and joint B (see Fig. E5.2.2(b)) are:

- The internal moments shown in the FBD of the beam segments (FBD I and FBD III) *must* be shown clockwise since clockwise moments are assumed to be positive. The directions of the end shears, A_y, V_{BL}, etc., are assumed.
- In the FBD of the joint (FBD II), the moments and the shear force are shown equal and opposite to the corresponding internal forces in the beam segments.

$$\text{At joint B:} \quad \overset{\curvearrowright +}{\sum} M_B = 0 = M_{BA} + M_{BC} \tag{5}$$

$$\text{At joint C:} \quad \overset{\curvearrowright +}{\sum} M_C = 0 = M_{CB} \quad \text{(see Fig. E5.2.2(c))} \tag{6}$$

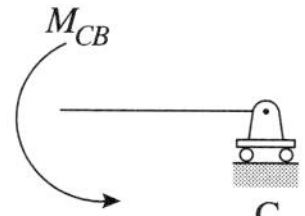

Fig. E5.2.2(c)

Substituting (2)–(4) in (5) and (6) and simplifying, we have

$$\frac{10EI}{15}\theta_B + \frac{2EI}{15}\theta_C = 31.25 \qquad \frac{2EI}{15}\theta_B + \frac{4EI}{15}\theta_C = -56.25$$

Solving the two equations gives us

$$\theta_B = \frac{98.96}{EI}(\circlearrowright) \qquad \theta_C = -\frac{260.42}{EI} \Rightarrow \theta_C = \frac{260.42}{EI}(\circlearrowleft)$$

Step 4: Computation of the internal moments. Substituting the results in (1)–(4), we have

$$M_{AB} = -5.21\text{ k-ft} \qquad M_{BA} = 64.58\text{ k-ft}$$

$$M_{BC} = -64.58\text{ k-ft} \qquad M_{CB} = 0$$

We should check the results by going back to Eqs. (5) and (6). Our results satisfy both those equations!

The FBDs of the beam at this stage are shown in Fig. E5.2.2(d). We can use the equilibrium conditions to compute the end shears and support reactions.

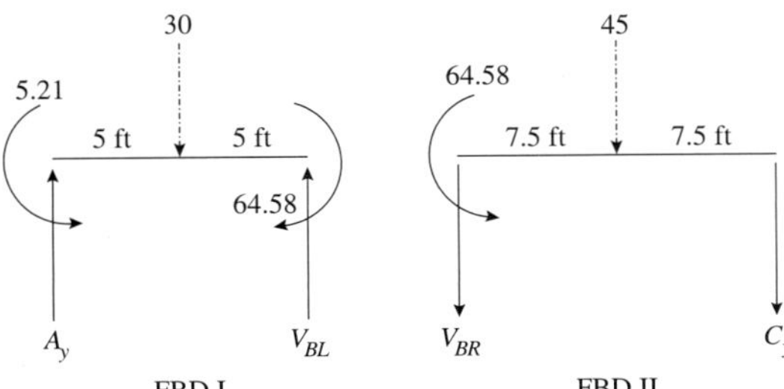

Fig. E5.2.2(d)

$$\text{FBD I:} \quad \overset{\curvearrowleft+}{\sum} M_B = 0 = 5.21 + (30)5 - 64.58 - A_y(10) \Rightarrow A_y = 9.06\,\text{k}(\uparrow)$$

$$\overset{\uparrow+}{\sum} F_y = 0 = A_y - 30 + V_{BL} \Rightarrow V_{BL} = 20.94\,\text{k}(\uparrow)$$

$$\text{FBD II:} \quad \overset{\curvearrowleft+}{\sum} M_B = 0 = 64.58 - (45)7.5 + C_y(15) \Rightarrow C_y = 18.19\,\text{k}(\uparrow)$$

$$\overset{\uparrow+}{\sum} F_y = 0 = V_{BR} - 45 + C_y \Rightarrow V_{BR} = 26.81\,\text{k}(\uparrow)$$

Using FBD II in Fig. 5.2.2(b) gives (see Figs. E5.2.2(e) and (f)):

$$\overset{\uparrow+}{\sum} F_y = 0 = -V_{BL} + B_y - V_{BR} \Rightarrow B_y = 47.75\,\text{k}(\uparrow)$$

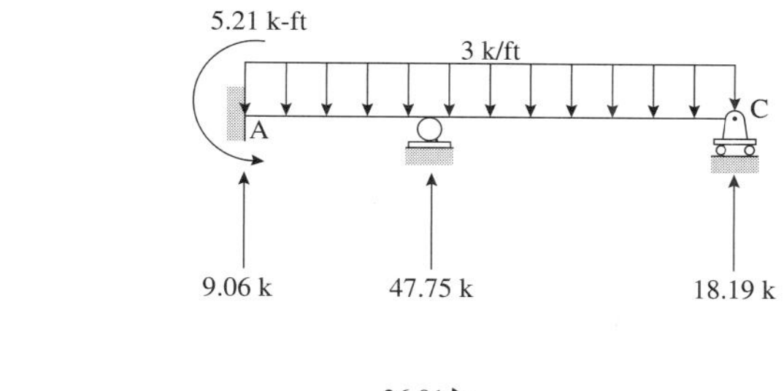

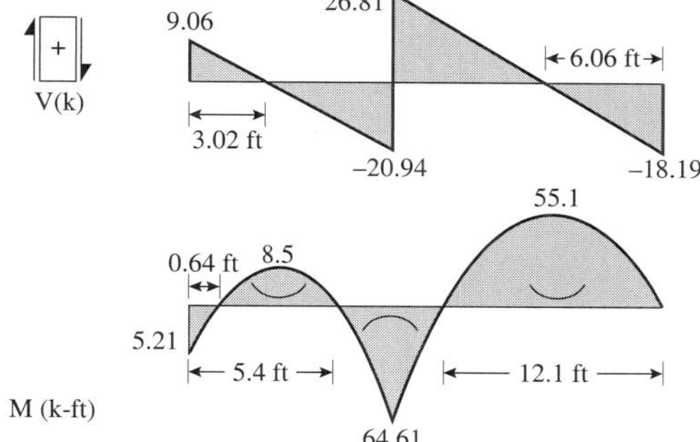

Fig. E5.2.2(e) Support reactions, shear force and bending moment diagrams.

Fig. E5.2.2(f) Elastic curve.

EXAMPLE 5.2.3

Statically Indeterminate Continuous Beam with Support Settlement

For the beam in Fig. E5.2.3(a) compute the support reactions. Support B settles 0.5". Take E = 30000 ksi and I = 1500 in^4.

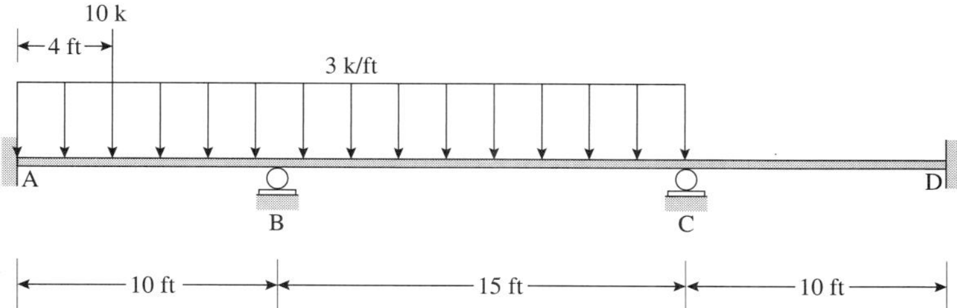

Fig. E5.2.3(a)

SOLUTION

Step 1: The problem units are k, ft. The boundary conditions are such that $\theta_A = 0$, $\theta_B = ?$, $\theta_C = ?$ and $\Delta_B = (0.5/12) = 0.041667$ ft. Due to the support settlement, the chord rotations are

$$\psi_{AB} = \frac{\Delta_B}{10} = 0.0041667, \quad \psi_{BC} = -\frac{\Delta_B}{15} = -2.777(10^{-3}), \quad \psi_{CD} = 0$$

As we have seen before, clockwise rotations are positive and counterclockwise rotations are negative. This is the reason that the chord rotation in AB is positive and in BC is negative (see Fig. E5.2.3(b)).

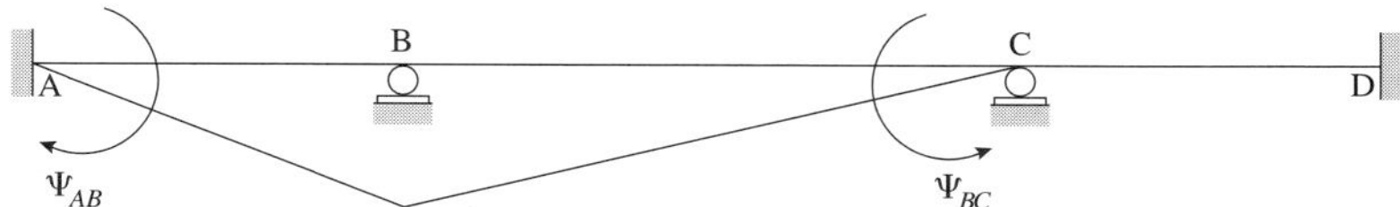

Fig. E5.2.3(b)

The problem has two kinematic unknowns. Using the given loading, we have

$$FEM_{AB} = -\frac{wL^2}{12} - \frac{Pab^2}{L^2} = -\frac{(3)(10)^2}{12} - \frac{10(4)(6)^2}{10^2} = -39.4 \text{ k-ft}$$

$$FEM_{BA} = \frac{wL^2}{12} + \frac{Pa^2b}{L^2} = \frac{(3)(10)^2}{12} + \frac{10(4)^2(6)}{10^2} = 34.6 \text{ k-ft}$$

$$FEM_{BC} = -\frac{wL^2}{12} = -\frac{(3)(15)^2}{12} = -56.25 \text{ k-ft} = -FEM_{CB}$$

In this problem, even though EI is a constant, the beam stiffness will influence the support reactions:

$$EI = (30000)\left(\frac{1500}{144}\right) = 312500 \text{ k-ft}^2$$

Step 2: Slope-deflection equations:

$$M_{AB} = \frac{2EI}{10}(0 + \theta_B - 3\psi_{AB}) - 39.4 = 62500\,\theta_B - 820.65 \quad (1)$$

$$M_{BA} = \frac{2EI}{10}(2\theta_B + 0 - 3\psi_{AB}) + 34.6 = 125000\,\theta_B - 746.65 \quad (2)$$

$$M_{BC} = \frac{2EI}{15}(2\theta_B + \theta_C - 3\psi_{BC}) - 56.25 = 83333.3\,\theta_B + 41666.7\,\theta_C + 290.97 \quad (3)$$

$$M_{CB} = \frac{2EI}{15}(2\theta_C + \theta_B - 3\psi_{BC}) + 56.25 = 41666.7\,\theta_B + 83333.3\,\theta_C + 403.47 \quad (4)$$

$$M_{CD} = \frac{2EI}{10}(2\theta_C + 0 - 0) = 125000\,\theta_C \quad (5)$$

$$M_{DC} = \frac{2EI}{10}(0 + \theta_C - 0) = 62500\,\theta_C \quad (6)$$

Step 3: Additional equations. The additional equations are generated in a manner similar to the previous problem (see Fig. E5.2.3(c)):

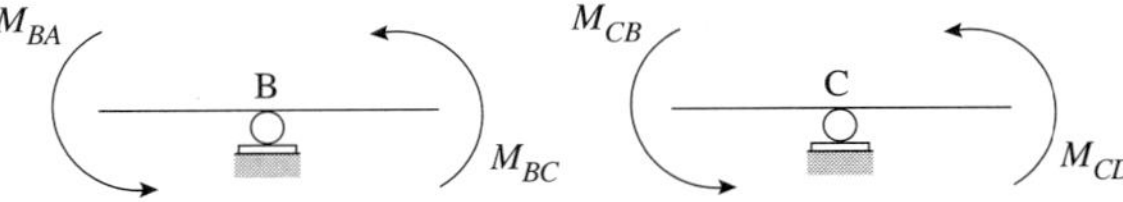

Fig. E5.2.3(c)

$$\text{At joint B: } \sum M_B = 0 = M_{BA} + M_{BC} \quad (7)$$

$$\text{At joint C: } \sum M_C = 0 = M_{CB} + M_{CD} \quad (8)$$

Substituting (2)–(5) in the above two equations, we have

$$208333.3\theta_B + 41666.7\theta_C = 455.68 \quad (A)$$

$$41666.7\theta_B + 208333.3\theta_C = -403.47 \quad (B)$$

Solving, we get

$$\theta_B = 2.682(10^{-3})\,\text{rad} \qquad \theta_C = -2.473(10^{-3})\,\text{rad}$$

Step 4: Computation of the internal moments. Substituting the results in (1)–(6), we have

$$M_{AB} = -653.0\,\text{k - ft} \qquad M_{BA} = -411.4\,\text{k - ft}$$

$$M_{BC} = 411.4\,\text{k - ft} \qquad M_{CB} = 309.1\,\text{k - ft}$$

$$M_{CD} = -309.1\,\text{k - ft} \qquad M_{DC} = -154.6\,\text{k - ft}$$

We should check the results by going back to Eqs. (7) and (8). Our results satisfy both those equations!

The FBDs of the beam at this stage are shown in Fig. E5.2.3(d). We can use the equilibrium conditions to compute the end shears and support reactions:

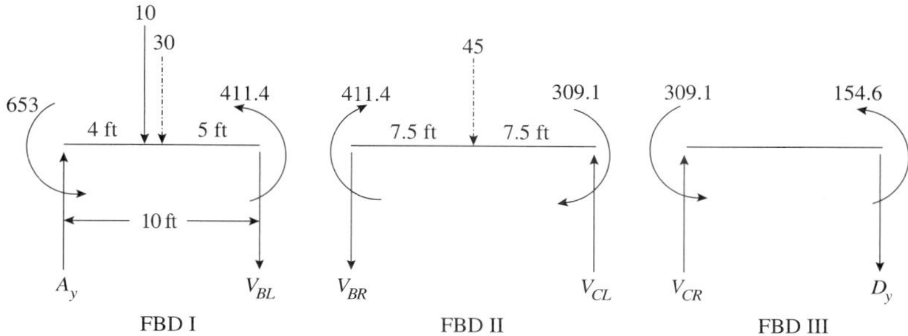

Fig. E5.2.3(d)

FBD I: $\overset{\curvearrowleft +}{\sum} M_A = 0 = 653 + 411.4 - (10)4 - 30(5) - V_{BL}(10) \Rightarrow V_{BL} = 87.4\,\text{k}(\downarrow)$

$\overset{\uparrow +}{\sum} F_y = 0 = A_y - 40 - 87.4 \Rightarrow A_y = 127.4\,\text{k}(\uparrow)$

FBD II: $\overset{\curvearrowleft +}{\sum} M_C = 0 = -411.4 - 309.1 + 45(7.5) + V_{BR}(15) \Rightarrow V_{BR} = 25.5\,\text{k}(\downarrow)$

$\overset{\uparrow +}{\sum} F_y = 0 = -25.5 - 45 + V_{CL} \Rightarrow V_{CL} = 70.5\,\text{k}(\uparrow)$

FBD III: $\overset{\curvearrowleft +}{\sum} M_C = 0 = 309.1 + 154.6 - D_y(10) \Rightarrow D_y = 46.4\,\text{k}(\downarrow)$

$\overset{\uparrow +}{\sum} F_y = 0 = V_{CR} - D_y \Rightarrow V_{CR} = 46.4\,\text{k}(\uparrow)$

Finally, using the FBDs of the supports B and C, we have B_y = 112.9 k(↓) and C_y = 116.9 k(↑).

5.2.2 Frames Without Sidesway

The analysis of frames via the slope-deflection method can also be carried out systematically by applying Eqs. (5.2.7) and (5.2.8). In this section we look at frames that do not displace sideways. A sidesway will not occur if (a) the frame geometry and loading are symmetric, and (b) sidesway is prevented due to supports. Figures 5.2.2.1 through 5.2.2.3 show examples of frames that do not have sidesway. In examining these frames it is important to keep in mind that the slope-deflection method does not take into account or allow axial and shear deformations.

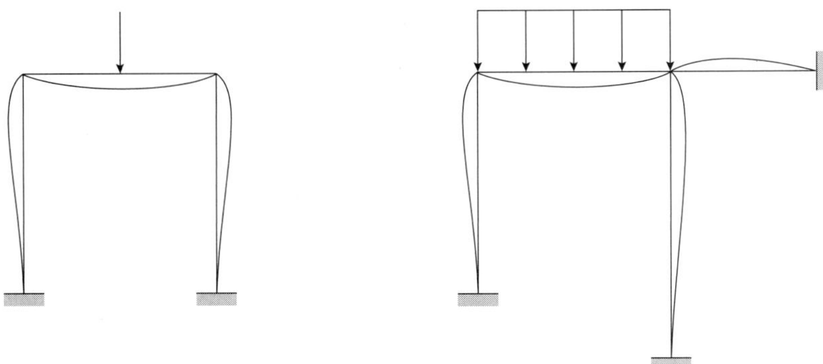

Fig. 5.2.2.1
No sidesway due to symmetry.

Fig. 5.2.2.2
No sidesway due to supports.

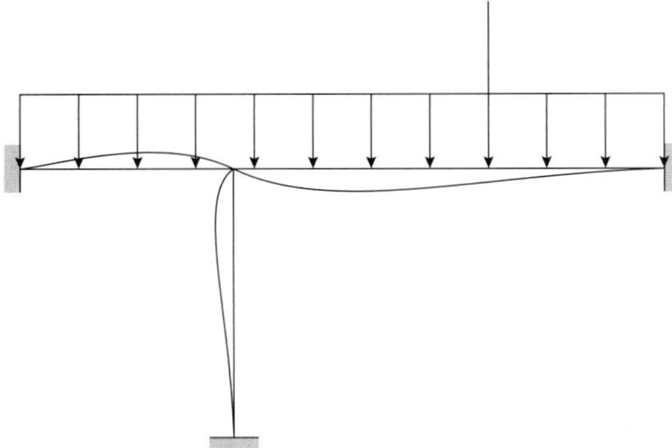

Fig. 5.2.2.3
No sidesway due to supports.

The general procedure for analysis is the same as that used with beams.

EXAMPLE 5.2.4 ***Statically Indeterminate Frame***

Consider the frame in Fig. E5.2.4(a). The modulus of elasticity is $2(10^{11})$ Pa and the moment of inertia is 0.001 m^4 for both the members. Compute the support reactions.

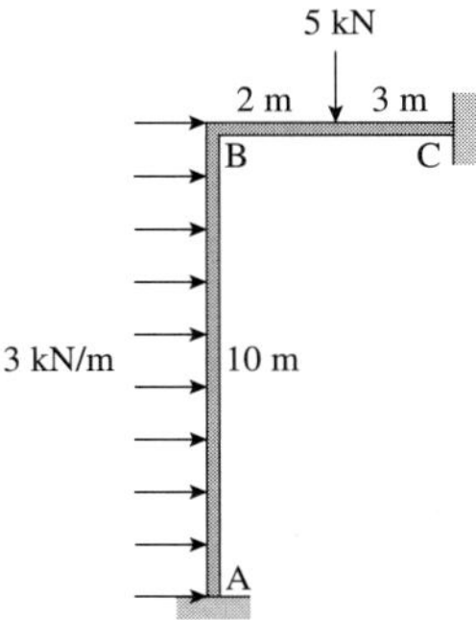

Fig. E5.2.4(a)

SOLUTION

Step 1: The problem units are kN, m. The boundary conditions are such that $\theta_A = \theta_C = 0$ and θ_B = ?. Hence there is only one kinematic unknown. We first compute the fixed-end moments:

$$FEM_{AB} = -\frac{wL^2}{12} = -\frac{(3)(10^2)}{12} = -25\,\text{kN - m} = -FEM_{BA}$$

$$FEM_{BC} = -\frac{Pab^2}{L^2} = -\frac{5(2)(3^2)}{25} = -3.6\,\text{kN - m}$$

$$FEM_{CB} = \frac{Pa^2b}{L^2} = \frac{5(2^2)(3)}{25} = 2.4\,\text{kN - m}$$

$$EI = 2(10^8)(0.001) = 2(10^5)\,\text{kN - m}^2$$

Step 2: Develop the slope-deflection equations:

$$M_{AB} = \frac{2EI}{10}(0 + \theta_B - 0) - 25 = \frac{2EI\theta_B}{10} - 25 \quad (1)$$

$$M_{BA} = \frac{2EI}{10}(2\theta_B + 0 - 0) + 25 = \frac{4EI\theta_B}{10} + 25 \quad (2)$$

$$M_{BC} = \frac{2EI}{5}(2\theta_B + 0 - 0) - 3.6 = \frac{8EI\theta_B}{10} - 3.6 \quad (3)$$

$$M_{CB} = \frac{2EI}{5}(0 + \theta_B - 0) + 2.4 = \frac{4EI\theta_B}{10} + 2.4 \quad (4)$$

Step 3: Generate the additional equation and solve for the kinematic unknowns. From the FBD of joint B in Fig. E5.2.4(b), we have

$$\overset{\curvearrowleft +}{\sum} M_B = 0 = M_{BA} + M_{BC} \quad (5)$$

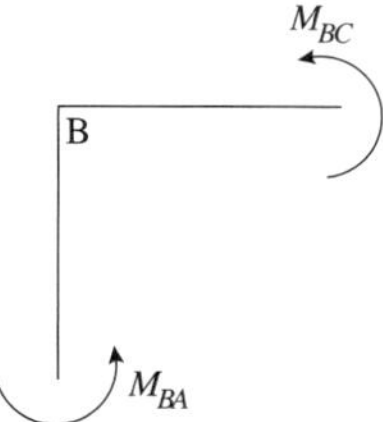

Fig. E5.2.4(b)

Substituting (2) and (3) in (5), we have $(12EI/10)\theta_B + 21.4 = 0 \Rightarrow \theta_B = -17.833/EI$. Hence $\theta_B = 8.9165(10^{-5})$ rad ($\curvearrowright$).

Step 4: Computation of internal moments. Substituting the result into (1)–(4) yields

$$M_{AB} = -28.6\,\text{kN - m} \qquad M_{BA} = 17.9\,\text{kN - m}$$

$$M_{BC} = -17.9\,\text{kN - m} \qquad M_{CB} = -4.73\,\text{kN - m}$$

Using the FBD of column AB in Fig. E5.2.4(c) gives us

$$\overset{\curvearrowleft +}{\sum} M_B = 0 \Rightarrow A_x = 16.1\,\text{kN}(\leftarrow) \qquad \overset{\uparrow +}{\sum} F_x = 0 \Rightarrow B_x = 13.9\,\text{kN}(\leftarrow)$$

Using the FBD of beam BC in Fig. E5.2.4(c) gives us

$$\overset{\curvearrowleft +}{\sum} M_B = 0 \Rightarrow C_y = 2.53\,\text{kN}(\downarrow) \qquad \overset{\uparrow +}{\sum} F_y = 0 \Rightarrow B_y = 7.53\,\text{kN}(\uparrow)$$

And finally, $C_x = 13.9$ kN($\leftarrow$) and $A_y = 7.53$ kN($\uparrow$).

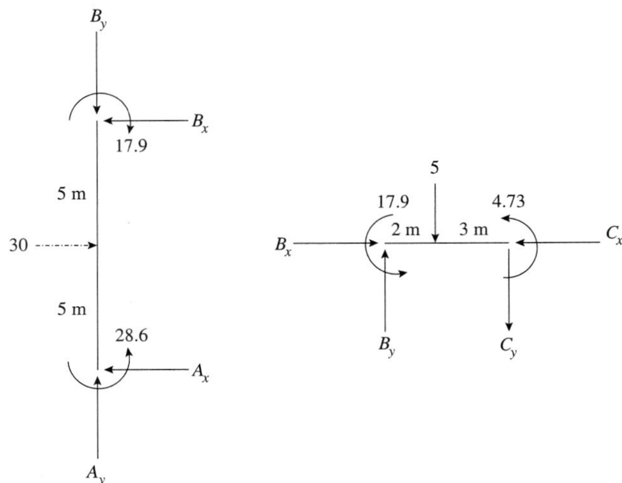

Fig. E5.2.4(c)

The shear force and bending moment diagrams and the elastic curve are shown in Figs. E5.2.4(d) and (e).

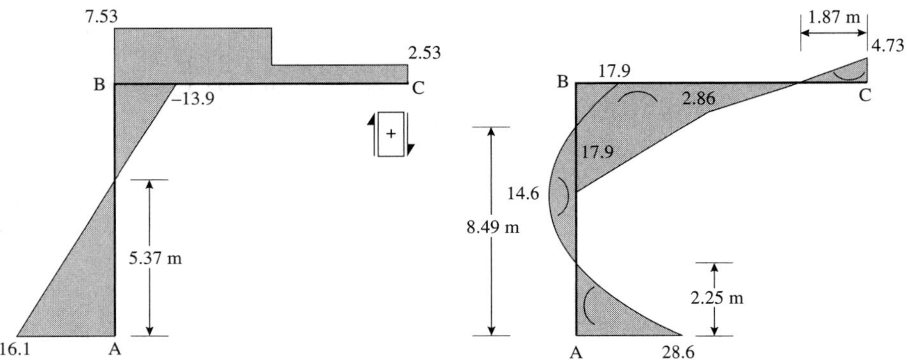

Fig. E5.2.4(d) Shear force (kN) and bending moment (kN-m) diagrams.

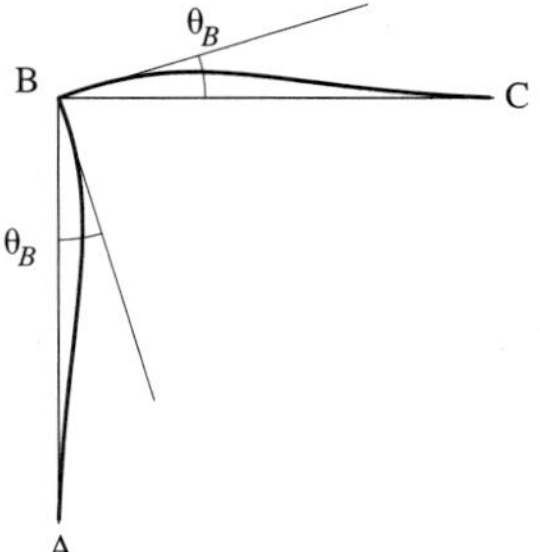

Fig. E5.2.4(e) Deformed shape.

EXAMPLE 5.2.5 ***Statically Indeterminate Frame with Internal Hinge***

Consider the frame in Fig. E5.2.5(a). The modulus of elasticity is $2(10^{11})$ Pa and the moment of inertia is 0.001 m^4 for both the members. Compute the support reactions.

SOLUTION

Step 1: The problem is the same as the previous problem except for one major difference: the boundary conditions are such that $\theta_A = \theta_C = 0$. Since B is an internal hinge, the members that meet at B are free to rotate independent of each other. Hence the kinematic unknown are θ_{BA} and θ_{BC}. We first compute the fixed-end moments.

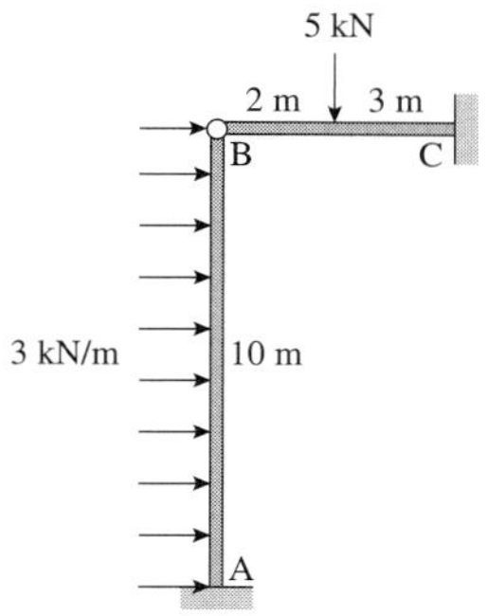

Fig. E5.2.5(a)

$$FEM_{AB} = -\frac{wL^2}{12} = -\frac{(3)(10^2)}{12} = -25\,\text{kN - m} = -FEM_{BA}$$

$$FEM_{BC} = -\frac{Pab^2}{L^2} = -\frac{5(2)(3^2)}{25} = -3.6\,\text{kN - m}$$

$$FEM_{CB} = \frac{Pa^2b}{L^2} = \frac{5(2^2)(3)}{25} = 2.4\,\text{kN - m}$$

$$EI = 2(10^8)(0.001) = 2(10^5)\,\text{kN - m}^2$$

Step 2: Develop the slope-deflection equations:

$$M_{AB} = \frac{2EI}{10}(0 + \theta_{BA} - 0) - 25 = \frac{2EI\theta_{BA}}{10} - 25 \quad (1)$$

$$M_{BA} = \frac{2EI}{10}(2\theta_{BA} + 0 - 0) + 25 = \frac{4EI\theta_{BA}}{10} + 25 \quad (2)$$

$$M_{BC} = \frac{2EI}{5}(2\theta_{BC} + 0 - 0) - 3.6 = \frac{8EI\theta_{BC}}{10} - 3.6 \quad (3)$$

$$M_{CB} = \frac{2EI}{5}(0 + \theta_{BC} - 0) + 2.4 = \frac{4EI\theta_{BC}}{10} + 2.4 \quad (4)$$

Step 3: Generate the additional equations and solve for the kinematic unknowns. Since B is an internal hinge, we have

$$M_{BA} = 0 = \frac{4EI\theta_{BA}}{10} + 25 \quad (5)$$

$$M_{BC} = 0 = \frac{8EI\theta_{BC}}{10} - 3.6 \quad (6)$$

Solving the two equations, gives us

$$\theta_{BA} = -\frac{62.5}{EI} \Rightarrow \theta_{BA} = 3.125(10^{-4})\text{ rad}(\curvearrowleft)$$

$$\theta_{BC} = \frac{4.5}{EI} \Rightarrow \theta_{BC} = 2.25(10^{-5})\text{ rad}(\circlearrowright)$$

Step 4: Computation of internal moments. Substituting the results into (1)–(4) yields

$$M_{AB} = -37.5\,\text{kN - m} \qquad M_{BA} = 0$$
$$M_{BC} = 0 \qquad M_{CB} = 4.2\,\text{kN - m}$$

Using the FBD of column AB in Fig. E5.2.5(b), we have

$$\overset{\curvearrowleft +}{\sum} M_B = 0 \Rightarrow A_x = 18.75\,\text{kN}(\leftarrow) \qquad \overset{\uparrow +}{\sum} F_x = 0 \Rightarrow B_x = 11.25\,\text{kN}(\leftarrow)$$

Using the FBD of beam BC (Fig. E5.2.5(b)), we have

$$\overset{\curvearrowleft +}{\sum} M_B = 0 \Rightarrow C_y = 2.84\,\text{kN}(\uparrow) \qquad \overset{\uparrow +}{\sum} F_y = 0 \Rightarrow B_y = 2.16\,\text{kN}(\uparrow)$$

And finally, C_x = 11.25 kN(←) and A_y = 2.16 kN(↑)

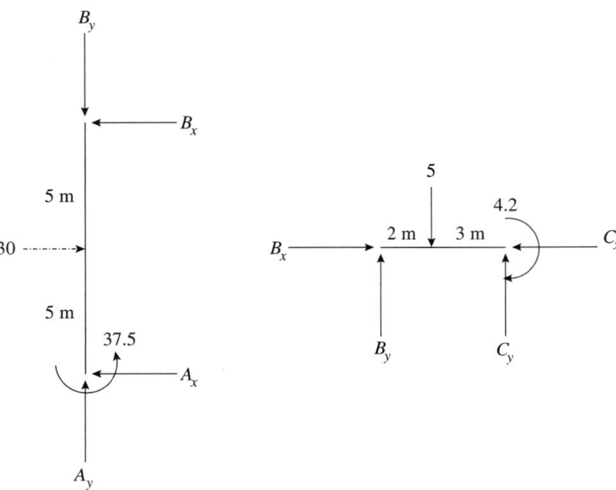

Fig. E5.2.5(b)

The shear force and bending moment diagrams and the elastic curve are shown in Figs. E5.2.5(c) and (d).

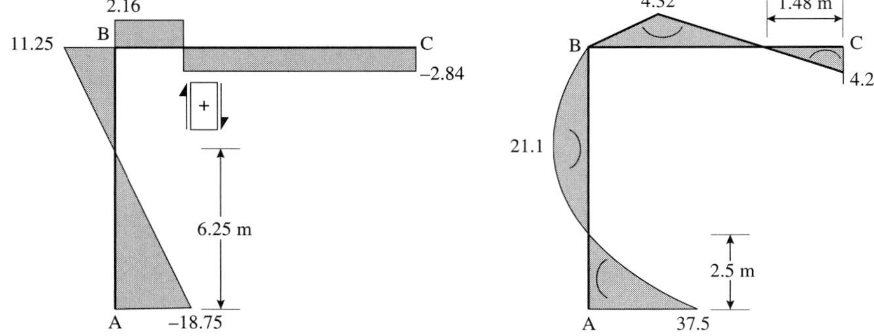

Fig. E5.2.5(c) Shear force (kN) and bending moment (kN-m) diagrams.

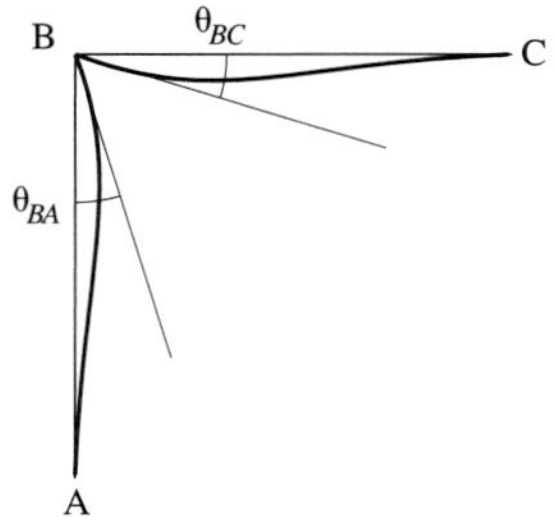

Fig. E5.2.5(d) Deformed shape.

EXAMPLE 5.2.6

Statically Indeterminate Frame

Figure E5.2.6(a) shows a planar frame. The material is steel, $E = 2(10)^{11}$ Pa and the cross-sectional properties are such that $A = 0.01$ m^2 and $I = 0.0001$ m^4. Solve for the member nodal forces and the support reactions.

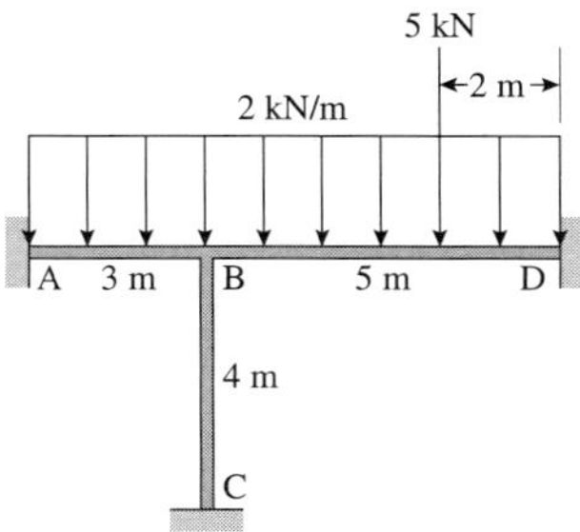

Fig. E5.2.6(a)

SOLUTION

Step 1: The problem units are N, m. The boundary conditions are such that $\theta_A = \theta_C = \theta_D = 0$. The only kinematic unknown is θ_B. We first compute the fixed-end moments.

$$FEM_{AB} = -\frac{wL^2}{12} = -\frac{(2000)(3^2)}{12} = -1500\text{ N - m} \quad \text{and} \quad FEM_{BA} = 1500\text{ N - m}$$

$$FEM_{BD} = -\frac{wL^2}{12} - \frac{Pab^2}{L^2} = -\frac{(2000)(5^2)}{12} - \frac{5000(3)(2^2)}{5^2} = -6566.7\text{ N - m}$$

$$FEM_{DB} = \frac{wL^2}{12} + \frac{Pa^2b}{L^2} = \frac{(2000)(5^2)}{12} + \frac{5000(3^2)(2)}{5^2} = 7766.7\text{ N - m}$$

Step 2: Develop the slope-deflection equations:

$$M_{AB} = \frac{2EI}{L}\left(2\theta_A + \theta_B - 3\left(\frac{\Delta}{L}\right)\right) + FEM_{AB} = \frac{2EI\theta_B}{3} - 1500 \tag{1}$$

$$M_{BA} = \frac{2EI}{L}\left(2\theta_B + \theta_A - 3\left(\frac{\Delta}{L}\right)\right) + FEM_{BA} = \frac{4EI\theta_B}{3} + 1500 \tag{2}$$

$$M_{BD} = \frac{2EI}{L}\left(2\theta_B + \theta_D - 3\left(\frac{\Delta}{L}\right)\right) + FEM_{BD} = \frac{4EI\theta_B}{5} - 6566.7 \tag{3}$$

$$M_{DB} = \frac{2EI}{L}\left(2\theta_D + \theta_B - 3\left(\frac{\Delta}{L}\right)\right) + FEM_{DB} = \frac{2EI\theta_B}{5} + 7766.7 \tag{4}$$

$$M_{BC} = \frac{2EI}{L}\left(2\theta_B + \theta_C - 3\left(\frac{\Delta}{L}\right)\right) + FEM_{BC} = EI\theta_B \tag{5}$$

$$M_{CB} = \frac{2EI}{L}\left(2\theta_C + \theta_B - 3\left(\frac{\Delta}{L}\right)\right) + FEM_{CB} = \frac{EI\theta_B}{2} \tag{6}$$

Step 3: Generate the additional equation and solve for the kinematic unknown. Eqs. (1–6) deal with seven unknowns and hence we need to generate the additional equation needed to solve for the unknowns. The last equation deals with equilibrium of joint B whose FBD is shown in Fig. E5.2.6(b):

$$\overset{\curvearrowleft +}{\sum} M_B = 0 = M_{BA} + M_{BC} + M_{BD} \tag{7}$$

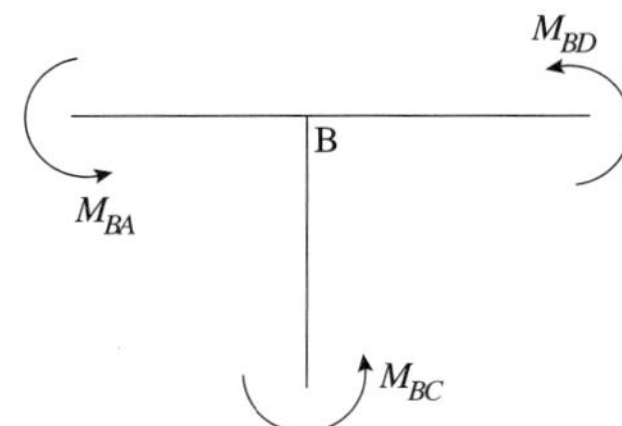

Fig. E5.2.6(b)

Substituting Eqs. (2), (3), and (5) into (7) yields

$$\frac{4EI\theta_B}{3} + 1500 + \frac{4EI\theta_B}{5} - 6566.7 + EI\theta_B = 0$$

$$\text{Or } \frac{47EI\theta_B}{15} = 5066.7 \Rightarrow EI\theta_B = 1617.03 \Rightarrow \theta_B = \frac{1617.03}{EI}(\circlearrowright) \qquad (8)$$

Step 4: Compute the member end moments. Substituting Eq. (8) into Eqs. (1)–(6), we have

$M_{AB} = \text{-}422\ \text{N - m}$ $\qquad M_{BA} = 3656\ \text{N - m}$

$M_{BD} = \text{-}5273.1\ \text{N - m}$ $\qquad M_{DB} = 8413.5\ \text{N - m}$

$M_{BC} = 1617\ \text{N - m}$ $\qquad M_{CB} = 808.5\ \text{N - m}$

Check: $\overset{(+)}{\sum} M_B = 3656 - 5273.1 + 1617 \approx 0$ OK.

Step 5: Compute the member end shears, and the support reactions (see Fig. E5.2.6(c)):

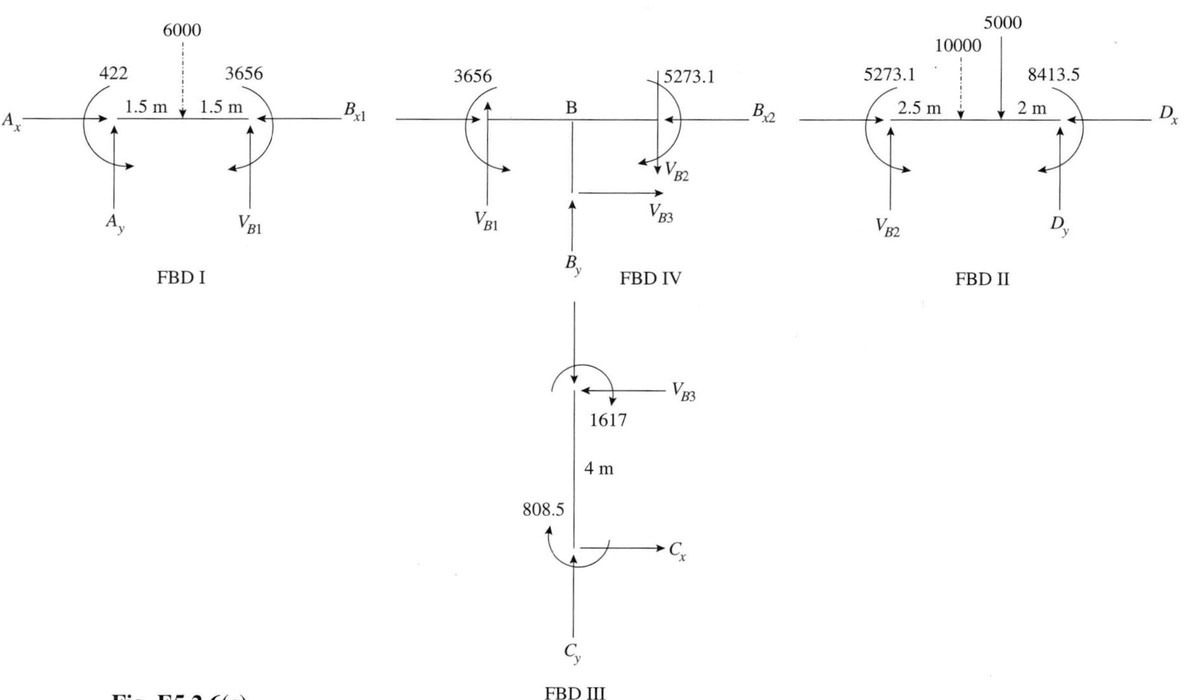

Fig. E5.2.6(c)

FBD I: $\overset{\curvearrowleft +}{\sum} M_B = 0 = 422 + 6000(1.5) - 3656 - A_y(3) \Rightarrow A_y = 1922\,\text{N}(\uparrow)$

$$\overset{\uparrow +}{\sum} F_y = 0 = A_y - 6000 + V_{B1} \Rightarrow V_{B1} = 4078\,\text{N}(\uparrow)$$

FBD II: $\overset{\curvearrowleft +}{\sum} M_B = 0 = 5273.1 - 10000(2.5) - 5000(3) - 8413.5 + D_y(5) \Rightarrow D_y = 8628.1\,\text{N}(\uparrow)$

$$\overset{\uparrow +}{\sum} F_y = 0 = V_{B2} - 15000 + D_y \Rightarrow V_{B2} = 6371.9\,\text{N}(\uparrow)$$

FBD III: $\overset{\curvearrowleft +}{\sum} M_B = 0 = -1617 - 808.5 + C_x(4) \Rightarrow C_x = 606.4\,\text{N}(\rightarrow)$

$$\overset{\uparrow +}{\sum} F_x = 0 \Rightarrow V_{B3} = 606.4\,\text{N}(\leftarrow)$$

Finally, from FBD IV, $\overset{\uparrow +}{\sum} F_y = 0 = B_y - V_{B1} - V_{B2} \Rightarrow B_y = 10450\,\text{N}(\uparrow)$, and from FBD III, $\overset{\uparrow +}{\sum} F_y = 0 = -B_y + C_y \Rightarrow C_y = 10450\,\text{N}(\uparrow)$.

Observation: We cannot evaluate either A_x or D_x. This is because of the inherent assumption in the slope-deflection method that the axial effects are ignored. Solve this problem using the GS-USA computer program and compare the results. What do you observe?

5.2.3 Frames with Sidesway

In this section we look at frames that displace sideways. A sidesway will occur if (a) the frame geometry and loading are unsymmetrical, and (b) sidesway is not prevented due to supports. Figs. 5.2.3.1 and 5.2.3.2 show examples of frames that move sideways.

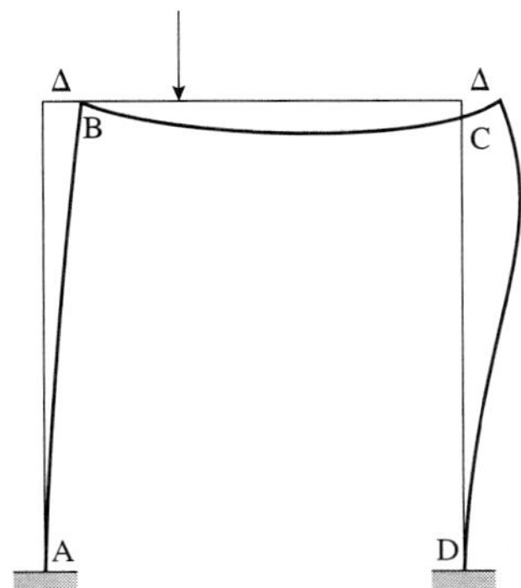

Fig. 5.2.3.1 Sidesway due to unsymmetry.

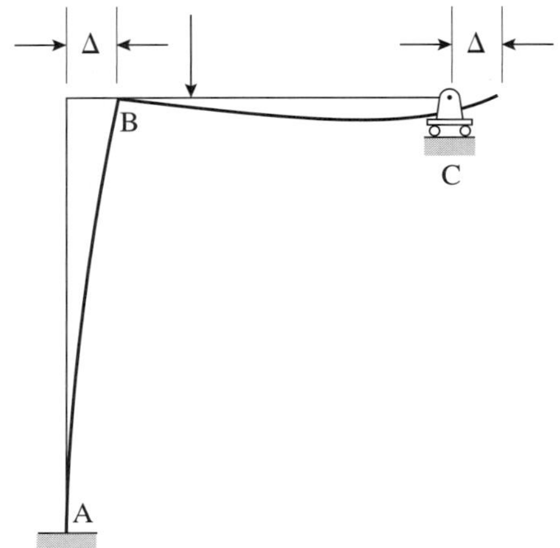

Fig. 5.2.3.2
Frame not restrained to prevent sidesway.

In the frame in Fig. 5.2.3.1, there are three degrees of freedom: θ_B, θ_C, and Δ. Joints B and C deflect the same amount to the right since axial deformations are not accounted for in this method. The frame moves sideways because of the asymmetrical load. Similarly, in the frame in Fig. 5.2.3.2, there are three degrees of freedom: θ_B, θ_C, and Δ. The frame moves to the right since there are no restraints in the horizontal direction to prevent beam BC moving horizontally.

As with beams and frames without sidesway, each segment generates two slope-deflection equations. The rest of the equations deal with (a) the equilibrium of the joints with the kinematic unknowns, and (b) equilibrium of the frame in the direction of the sidesway. We saw examples of condition (a) before (see Example 5.2.4).

Consider the frame in Fig. 5.2.3.1. To solve for the three kinematic unknowns we need three equations of equilibrium. The first two are the usual equilibrium of joints B and C. The FBDs of the two joints are shown in Fig. 5.2.3.3 and the equilibrium equations are

$$\overset{\curvearrowleft +}{\sum} M_B = 0 = M_{BA} + M_{BC} \tag{5.2.3.1}$$

$$\overset{\curvearrowleft +}{\sum} M_C = 0 = M_{CB} + M_{CD} \tag{5.2.3.2}$$

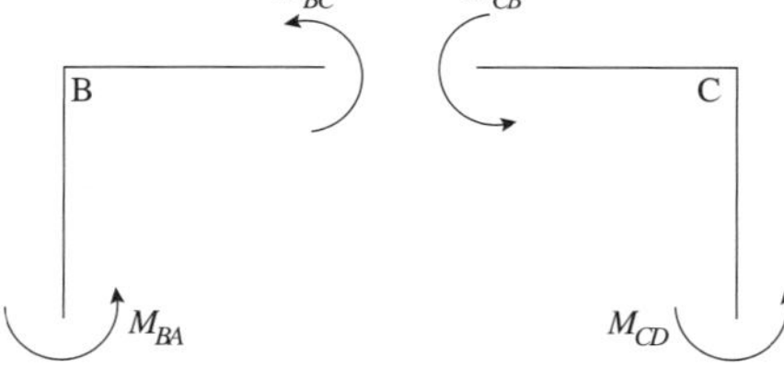

Fig. 5.2.3.3
Joint equilibrium FBDs.

The final equilibrium equation is the horizontal equilibrium of the frame. Taking the FBD of the frame in Fig. 5.2.3.4, we have

$$\overset{\rightarrow +}{\sum} F_x = 0 = A_x + D_x \tag{5.2.3.3}$$

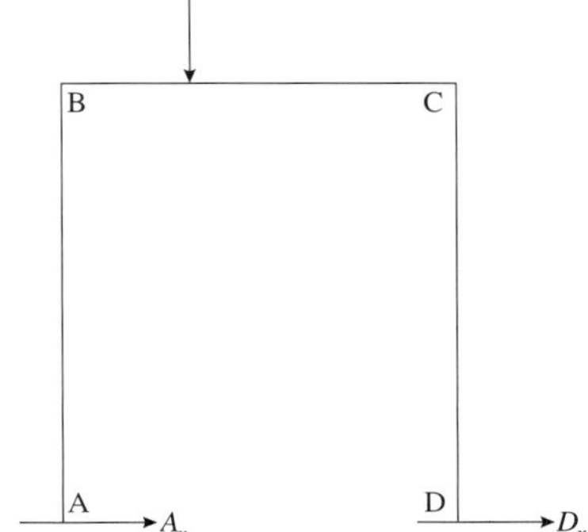

Fig. 5.2.3.4 FBD for computing the horizontal equilibrium of the frame.

Equation (5.2.3.3) by itself is not of use unless we change its components to include the member end moments. We relate the reactions at the base of the two columns to the internal moments using the FBDs of the two columns. From Fig. 5.2.3.5, taking moments about the top of the columns yields

$$\overset{\curvearrowright +}{\sum} M_B = 0 = (A_x)(L_{AB}) - M_{AB} - M_{BA} \Rightarrow A_x = \frac{M_{AB} + M_{BA}}{L_{AB}} \tag{5.2.3.4}$$

$$\overset{\curvearrowright +}{\sum} M_C = 0 = (D_x)(L_{CD}) - M_{CD} - M_{DC} \Rightarrow D_x = \frac{M_{CD} + M_{DC}}{L_{CD}} \tag{5.2.3.5}$$

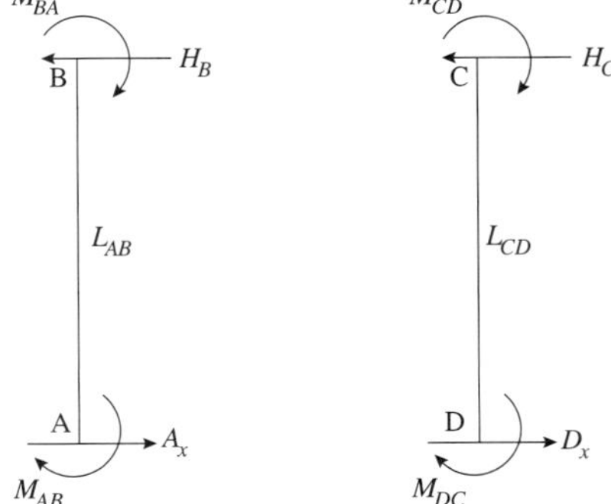

Fig. 5.2.3.5 Column FBDs.

Substituting these two results in Eq. (5.2.3.3), we have

$$\overset{\rightarrow +}{\sum} F_x = 0 = A_x + D_x = \frac{M_{AB} + M_{BA}}{L_{BA}} + \frac{M_{CD} + M_{DC}}{L_{CD}} \tag{5.2.3.6}$$

Into Eqs. (5.2.3.1), (5.2.3.2), and (5.2.3.6) we substitute the appropriate slope-deflection equations so as generate three equations entirely in terms of the kinematic unknowns θ_B, θ_C, and Δ. These equations can then be solved simultaneously to yield their values.

In the following examples, we see how to use the latter equilibrium condition to generate the last of the equations necessary to solve for the unknowns.

EXAMPLE 5.2.7 ***Statically Indeterminate Frame***

Consider the frame in Fig. E5.2.7(a). *EI* is a constant. Compute the support reactions.

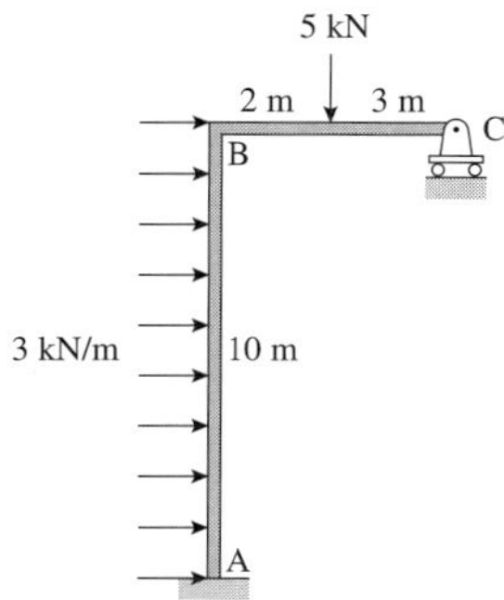

Fig. E5.2.7(a)

SOLUTION

Step 1: The problem units are kN, m. There is sidesway in this problem—the frame will sway to the right. As a result, the chord rotation exists only in member AB. The kinematic unknowns in this problem are θ_B, θ_C, and Δ. The chord rotation in AB is $\psi_{AB} = (\Delta/10)$.

$$FEM_{AB} = -\frac{wL^2}{12} = -\frac{(3)(10^2)}{12} = -25\,\text{kN - m} = -FEM_{BA}$$

$$FEM_{BC} = -\frac{Pab^2}{L^2} = -\frac{5(2)(3^2)}{25} = -3.6\,\text{kN - m}$$

$$FEM_{CB} = \frac{Pa^2b}{L^2} = \frac{5(2^2)(3)}{25} = 2.4\,\text{kN - m}$$

Step 2: Develop the slope-deflection equations:

$$M_{AB} = \frac{2EI}{10}\left(0+\theta_B-3\left(\frac{\Delta}{10}\right)\right)-25 = \frac{2EI\theta_B}{10}-\frac{6EI\Delta}{100}-25 \tag{1}$$

$$M_{BA} = \frac{2EI}{10}\left(2\theta_B+0-3\left(\frac{\Delta}{10}\right)\right)+25 = \frac{4EI\theta_B}{10}-\frac{6EI\Delta}{100}+25 \tag{2}$$

$$M_{BC} = \frac{2EI}{5}(2\theta_B+\theta_C-0)-3.6 = \frac{8EI\theta_B}{10}+\frac{4EI\theta_C}{10}-3.6 \tag{3}$$

$$M_{CB} = \frac{2EI}{5}(2\theta_C+\theta_B-0)+2.4 = \frac{4EI\theta_B}{10}+\frac{8EI\theta_C}{10}+2.4 \tag{4}$$

Step 3: Generate the additional equations and solve for the kinematic unknowns. We need to generate three additional equations corresponding to the three kinematic unknowns.

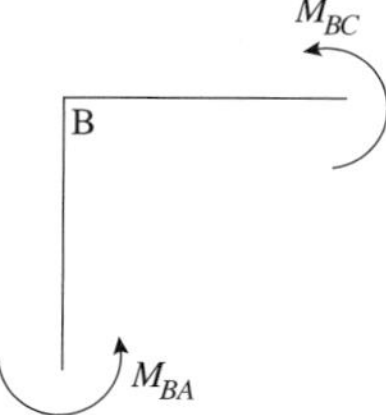

Fig. E5.2.7(b)

From the FBD of joint B in Fig. E5.2.7(b), we have

$$\overset{+}{\curvearrowleft}\sum M_B = 0 = M_{BA} + M_{BC} \tag{5}$$

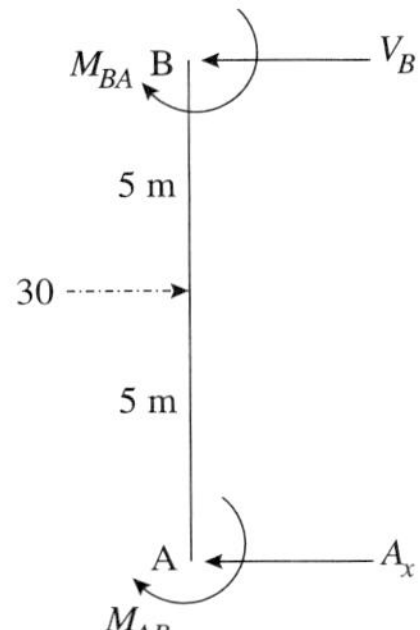

Fig. E5.2.7(c)

From the FBD of column AB (Fig. E5.2.7(c)), we find

$$\overset{\curvearrowleft +}{\sum} M_B = 0 = -M_{AB} - M_{BA} + 30(5) - A_x(10) \quad \text{or} \quad A_x = \frac{150 - M_{AB} - M_{BA}}{10} \tag{6a}$$

From the horizontal equilibrium of the entire frame (see Fig. E5.2.7(d)), we find

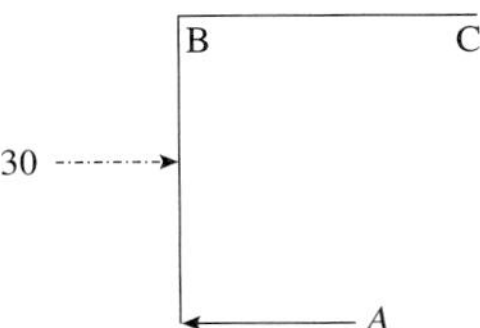

Fig. E5.2.7(d)

$$\overset{\rightarrow +}{\sum} F_x = 0 = 30 - A_x = 150 + M_{AB} + M_{BA} \tag{6b}$$

The third equation arises from the fact that C is a roller support. Hence

$$M_{CB} = 0 \tag{7}$$

Substituting (1)–(4) into (5), (6b), and (7) yields

$$60EI\theta_B + 20EI\theta_C - 3EI\Delta = -1070 \tag{A}$$

$$15EI\theta_B - 3EI\Delta = -3750 \tag{B}$$

$$2EI\theta_B + 4EI\theta_C = -12 \tag{C}$$

Solving, we have

$$EI\theta_B = 78.29 \Rightarrow \theta_B = \frac{78.29}{EI}(\circlearrowright)$$

$$EI\theta_C = -42.14 \Rightarrow \theta_C = \frac{42.14}{EI}(\circlearrowleft)$$

$$EI\Delta = 1641.43 \Rightarrow \Delta = \frac{1641.43}{EI}(\rightarrow)$$

Step 4: Computation of internal moments. Substituting the results into (1)–(4) gives us

$$M_{AB} = -107.8\,\text{kN - m} \qquad M_{BA} = -42.2\,\text{kN - m}$$

$$M_{BC} = 42.2\,\text{kN - m} \qquad M_{CB} = 0$$

As a check, we note that the results satisfy (A)–(C).

Step 5: Computation of the support reactions. Using the FBD of column AB in Fig. E5.2.7(e), we find

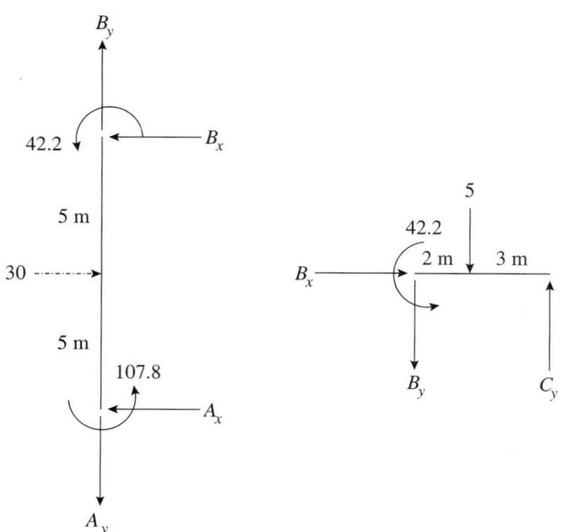

Fig. E5.2.7(e)

$$\overset{\curvearrowleft+}{\sum} M_B = 0 \Rightarrow A_x = 30\,\text{kN}(\leftarrow) \qquad \overset{\rightarrow+}{\sum} F_x = 0 \Rightarrow B_x = 0$$

Using the FBD of beam BC in Fig. E5.2.7(e), we find

$$\overset{\curvearrowleft+}{\sum} M_B = 0 \Rightarrow C_y = 10.44\,\text{kN}(\uparrow) \qquad \overset{\uparrow+}{\sum} F_y = 0 \Rightarrow B_y = 5.44\,\text{kN}(\downarrow)$$

And finally, going back to the FBD of column AB, we have $A_y = 5.44$ kN(↓).

EXAMPLE 5.2.8 *Statically Indeterminate Portal Frame*

Consider the frame in Fig. E5.2.8(a). The member properties are such that $I_{AB} = I_{CD} = I$ and $I_{BC} = 2I$. Compute the support reactions.

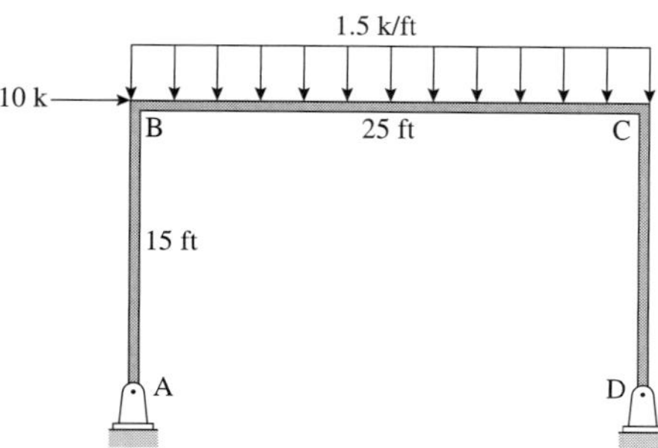

Fig. E5.2.8(a)

SOLUTION

Step 1: An examination of the frame shows that sidesway is possible and that there are five kinematic unknowns: $\theta_A = ?$, $\theta_B = ?$, $\theta_C = ?$, $\theta_D = ?$, and $\Delta = ?$. The chord rotations are a result of the sidesway occurring in columns AB and DC. Let us assume that the sidesway occurs to the right so that the chord rotations are clockwise.

Step 2: Slope-deflection equations:

$$M_{AB} = 2E\left(\frac{I}{15}\right)\left(2\theta_A + \theta_B - 3\left(\frac{\Delta}{15}\right)\right) = \frac{4EI}{15}\theta_A + \frac{2EI}{15}\theta_B - \frac{6EI}{225}\Delta \qquad (1)$$

$$M_{BA} = 2E\left(\frac{I}{15}\right)\left(2\theta_B + \theta_A - 3\left(\frac{\Delta}{15}\right)\right) = \frac{2EI}{15}\theta_A + \frac{4EI}{15}\theta_B - \frac{6EI}{225}\Delta \qquad (2)$$

$$M_{BC} = 2E\left(\frac{2I}{25}\right)(2\theta_B + \theta_C) - 78.125 = \frac{8EI}{25}\theta_B + \frac{4EI}{25}\theta_C - 78.125 \qquad (3)$$

$$M_{CB} = 2E\left(\frac{2I}{25}\right)(2\theta_C + \theta_B) + 78.125 = \frac{4EI}{25}\theta_B + \frac{8EI}{25}\theta_C + 78.125 \qquad (4)$$

$$M_{CD} = 2E\left(\frac{I}{15}\right)\left(2\theta_C + \theta_D - 3\left(\frac{\Delta}{15}\right)\right) = \frac{4EI}{15}\theta_C + \frac{2EI}{15}\theta_D - \frac{6EI}{225}\Delta \qquad (5)$$

$$M_{DC} = 2E\left(\frac{I}{15}\right)\left(2\theta_D + \theta_C - 3\left(\frac{\Delta}{15}\right)\right) = \frac{2EI}{15}\theta_C + \frac{4EI}{15}\theta_D - \frac{6EI}{225}\Delta \qquad (6)$$

Step 3: Additional equations. We need five additional equations, one corresponding to each of the kinematic unknowns. Since A is a pin support, $M_{AB} = 0$. Using (1) gives us

$$60EI\theta_A + 30EI\theta_B - 6EI\Delta = 0 \qquad (A)$$

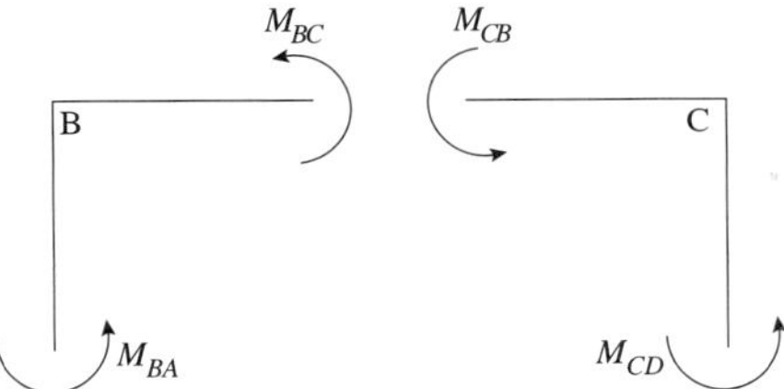

Fig. E5.2.8(b)

The equilibrium of joint B (see Fig. E5.2.8(b)) requires that $M_{BA} + M_{BC} = 0$. Using (2) and (3) yields

$$30EI\theta_A + 132EI\theta_B + 36EI\theta_C - 6EI\Delta = 17578.1 \qquad (B)$$

Similarly, the equilibrium of joint C (Fig. E5.2.8(b)) requires that $M_{CB} + M_{CD} = 0$. Using (4) and (5) yields

$$36EI\theta_B + 132EI\theta_C + 30EI\theta_D - 6EI\Delta = -17578.1 \qquad (C)$$

Since D is a pin support, $M_{DC} = 0$. Using (6), we have

$$30EI\theta_C + 60EI\theta_D - 6EI\Delta = 0 \qquad (D)$$

The final equation deals with horizontal equilibrium of the frame. First, using the FBD of column AB (Fig. E5.2.8(c)) and taking moment about B, we find $A_x = M_{BA}/15$. Similarly, using the FBD of column CD and taking moment about C, we find $D_x = M_{CD}/15$. Finally using the FBD of the entire frame gives us

$$\overset{\rightarrow+}{\sum} F_x = 0 = 10 + A_x + D_x = 10 + \frac{M_{BA}}{15} + \frac{M_{CD}}{15}$$

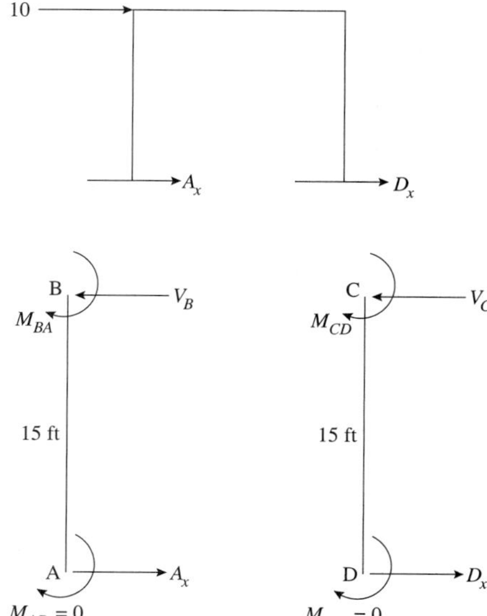

Fig. E5.2.8(c)

Substituting (2) and (4) in the above equation, we have

$$30EI\theta_A + 60EI\theta_B + 60EI\theta_C + 30EI\theta_D - 12EI\Delta = -33750 \tag{E}$$

Solving the five equations (A)–(E) yields

$EI\theta_A = 610.243$ $\quad EI\theta_B = 373.264$ $\quad EI\theta_C = -60.7636$

$EI\theta_D = 827.257$ $\quad EI\Delta = 7968.75$

or

$$\theta_A = \frac{610.243}{EI}(\circlearrowright) \qquad \theta_B = \frac{373.264}{EI}(\circlearrowright) \qquad \theta_C = \frac{60.7636}{EI}(\circlearrowleft)$$

$$\theta_D = \frac{827.257}{EI}(\circlearrowright) \qquad \Delta = \frac{7968.75}{EI}(\rightarrow)$$

Step 4: Computation of the internal moments. Substituting the results in (1)–(6), we have

$M_{AB} = 0$ $\quad M_{BC} = 31.6\,\text{k-ft}$ $\quad M_{CD} = -118.4\,\text{k-ft}$

$M_{BA} = -31.6\,\text{k-ft}$ $\quad M_{CB} = 118.4\,\text{k-ft}$ $\quad M_{DC} = 0$

As a check, these internal moments satisfy (A)–(E).

Step 5: Support reactions and member end shears. Using the FBD of column AB in Fig. E5.2.8(d) gives

$$\overset{\curvearrowleft +}{\sum} M_B = 31.6 - A_x(15) \Rightarrow A_x = 2.11\,\text{k}(\leftarrow)$$

$$\overset{\rightarrow +}{\sum} F_x = 0 = -A_x + B_x \Rightarrow B_x = 2.11\,\text{k}(\rightarrow)$$

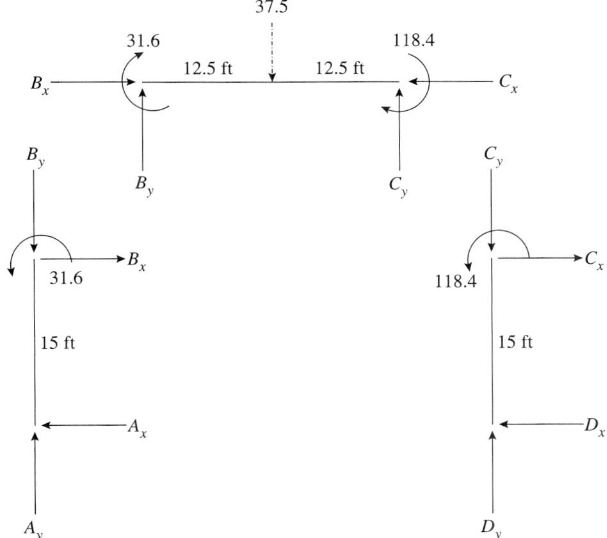

Fig. E5.2.8(d)

Using the FBD of column CD gives

$$\overset{\curvearrowleft +}{\sum} M_C = 118.4 - D_x(15) \Rightarrow D_x = 7.89\,\text{k}(\leftarrow)$$

$$\overset{\rightarrow +}{\sum} F_x = 0 = -D_x + C_x \Rightarrow C_x = 7.89\,\text{k}(\rightarrow)$$

Using the FBD of beam BC gives

$$\overset{\curvearrowleft +}{\sum} M_B = -31.6 - 118.4 - 37.5(12.5) + C_y(25) \Rightarrow C_y = 24.75\,\text{k}(\uparrow)$$

$$\overset{\uparrow +}{\sum} F_y = 0 = B_y - 37.5 + C_y \Rightarrow B_y = 12.75\,\text{k}(\uparrow)$$

Finally, going back to the FBDs of the two columns, we find

$A_y = 12.75\,\text{k}(\uparrow) \qquad D_y = 24.75\,\text{k}(\uparrow)$

The shear force and bending moment diagrams and the elastic curve are shown if Figs. E5.28(e) and (f).

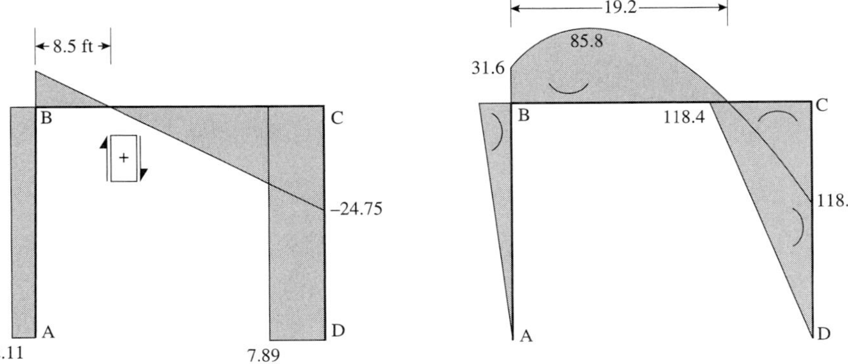

Fig. E5.2.8(e) Shear force (k) and bending moment diagrams (k-ft).

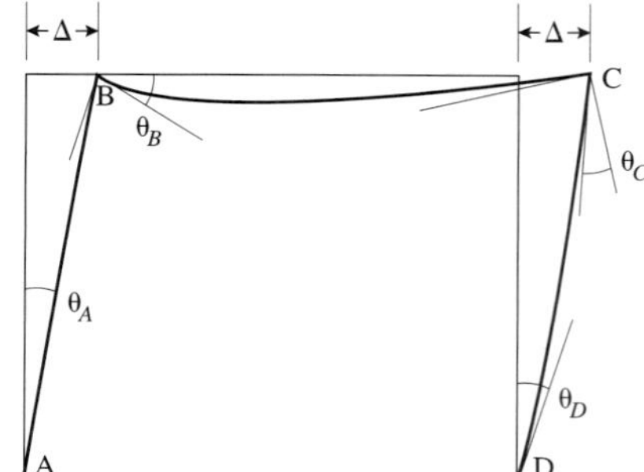

Fig. E5.2.8(f) Deflected shape.

EXAMPLE 5.2.9 ***Statically Indeterminate Frame with Overhangs***

Compute the support reactions for the frame in Fig. E5.2.9(a). The cross-sectional properties are such that the material used is the same for all the members but $I_{AB} = I_{BD} = I_{DF} = I$ and $I_{BC} = I_{DE} = 2I$.

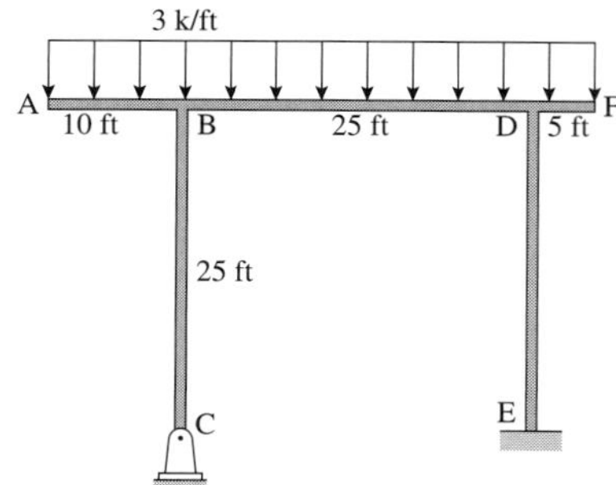

Fig. E5.2.9(a)

SOLUTION

Step 1: An examination of the frame shows that overhanging sections or segments AB and DF are statically determinate. We can remove the two sections and replace them on the main frame CBDE with their equivalent forces. The equivalent problem is shown in Fig. E5.2.9(b).

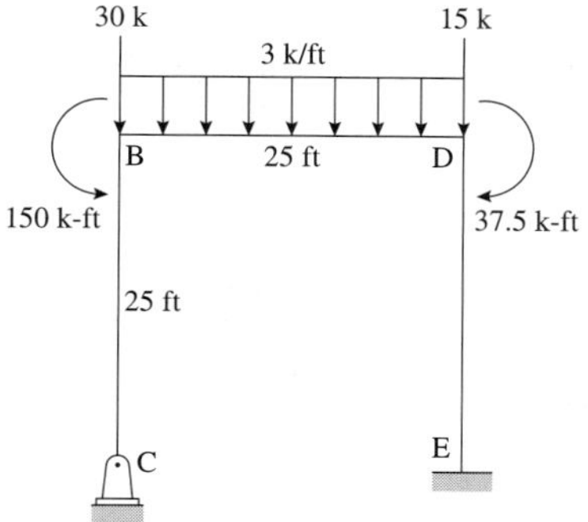

Fig. E5.2.9(b)

The kinematic conditions are: $\theta_C = ?$, $\theta_B = ?$, $\theta_D = ?$, $\Delta = ?$, and $\theta_E = 0$. We assume the sidesway is to the right so that the chord rotations $\psi_{CB} = (\Delta/L_{CB})$ and $\psi_{DE} = (\Delta/L_{DE})$ are clockwise.

$$FEM_{BD} = -\frac{wL^2}{12} = -\frac{3(25)^2}{12} = -156.25 \text{ k - ft} = -FEM_{DB}$$

Step 2: Slope-deflection equations.

$$M_{CB} = 2E\left(\frac{2I}{25}\right)\left(2\theta_C + \theta_B - 3\left(\frac{\Delta}{25}\right)\right) = \frac{4EI}{25}\theta_B + \frac{8EI}{25}\theta_C - \frac{12EI}{625}\Delta \quad (1)$$

$$M_{BC} = 2E\left(\frac{2I}{25}\right)\left(2\theta_B + \theta_C - 3\left(\frac{\Delta}{25}\right)\right) = \frac{8EI}{25}\theta_B + \frac{4EI}{25}\theta_C - \frac{12EI}{625}\Delta \quad (2)$$

$$M_{BD} = 2E\left(\frac{I}{25}\right)(2\theta_B + \theta_D) - 156.25 = \frac{4EI}{25}\theta_B + \frac{2EI}{25}\theta_D - 156.25 \quad (3)$$

$$M_{DB} = 2E\left(\frac{I}{25}\right)(2\theta_D + \theta_B) + 156.25 = \frac{2EI}{25}\theta_B + \frac{4EI}{25}\theta_D + 156.25 \quad (4)$$

$$M_{DE} = 2E\left(\frac{2I}{25}\right)\left(2\theta_D - 3\left(\frac{\Delta}{25}\right)\right) = \frac{8EI}{25}\theta_D - \frac{12EI}{625}\Delta \quad (5)$$

$$M_{ED} = 2E\left(\frac{2I}{25}\right)\left(\theta_D - 3\left(\frac{\Delta}{25}\right)\right) = \frac{4EI}{25}\theta_D - \frac{12EI}{625}\Delta \quad (6)$$

Step 3: Additional equations. We need four additional equations. These are similar to those generated in the previous problems. The joint equilibrium FBDs are shown in Fig. E5.2.9(c).

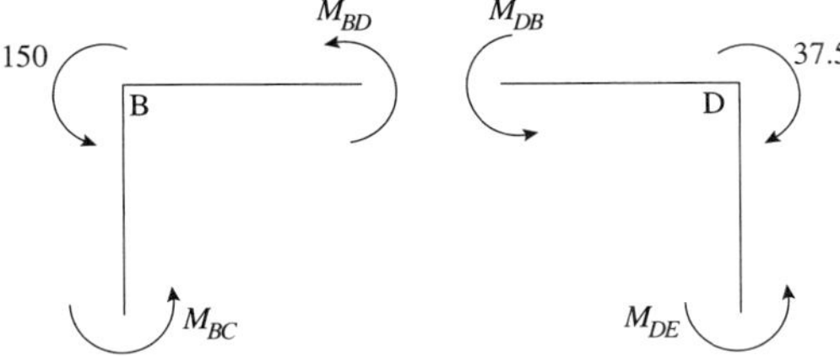

Fig. E5.2.9(c)

Since C is a pin support,

$$M_{CB} = 0$$

or $100EI\theta_B + 200EI\theta_C - 12EI\Delta = 0$ (A)

$$\overset{+}{\sum} M_B = 0 = M_{BD} + M_{BC} + 150$$

or $300EI\theta_B + 100EI\theta_C + 50EI\theta_D - 12EI\Delta = 3906.25$ (B)

$$\overset{+}{\sum} M_D = 0 = M_{DB} + M_{DE} - 37.5$$

or $50EI\theta_B + 300EI\theta_D - 12EI\Delta = 74218.75$ (C)

The final equation deals with the equilibrium of the frame in the direction of sidesway (see Fig. E5.2.9(d)).

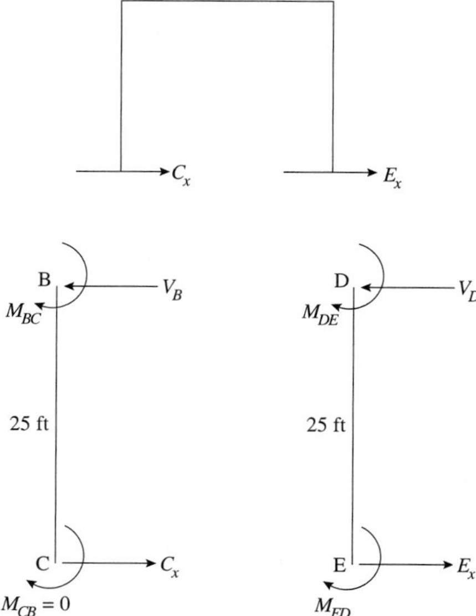

Fig. E5.2.9(d)

From the equilibrium of column CB, we have

$$\overset{\curvearrowleft +}{\sum} M_B = 0 = 25C_x - M_{BC} \Rightarrow C_x = \frac{M_{BC}}{25}$$

From the equilibrium of column ED, we have

$$\overset{\curvearrowleft +}{\sum} M_D = 0 = 25E_x - M_{DE} - M_{ED} \Rightarrow E_x = \frac{M_{DE} + M_{ED}}{25}$$

Since the entire frame is in equilibrium, we have

$$\overset{\rightarrow +}{\sum} F_x = 0 = C_x + E_x = \frac{M_{BC}}{25} + \frac{M_{DE} + M_{ED}}{25}$$

or $200EI\theta_B + 100EI\theta_C + 300\theta_D - 36EI\Delta = 0$ (D)

Step 4: Solution of the equilibrium equations. We can write the above four equations (A)–(D) in a matrix form as follows:

$$\begin{bmatrix} 100 & 200 & 0 & -12 \\ 300 & 100 & 50 & -12 \\ 50 & 0 & 300 & -12 \\ 200 & 100 & 300 & -36 \end{bmatrix} \begin{Bmatrix} EI\theta_B \\ EI\theta_C \\ EI\theta_D \\ EI\Delta \end{Bmatrix} = \begin{Bmatrix} 0 \\ 3906.25 \\ -74218.75 \\ 0 \end{Bmatrix}$$

Solving gives us

$$EI\theta_B = -0.9889 \Rightarrow \theta_B = \frac{0.9889}{EI} (\curvearrowleft)$$

$$EI\theta_C = -247.23 \Rightarrow \theta_C = \frac{247.23}{EI} (\curvearrowleft)$$

$$EI\theta_D = -412.38 \Rightarrow \theta_D = \frac{412.38}{EI} (\curvearrowleft)$$

$$EI\Delta = -4128.76 \Rightarrow \Delta = \frac{4128.76}{EI} (\leftarrow)$$

Step 5: Computation of the member end forces. Substituting these values in (1)–(6), we have the internal moments as follows:

$$M_{CB} = 0 \qquad M_{BC} = 39.4 \text{ k - ft}$$

$$M_{BD} = -189.4 \text{ k - ft} \qquad M_{DB} = 90.19 \text{ k - ft}$$

$$M_{DE} = -52.69 \text{ k - ft} \qquad M_{ED} = 13.29 \text{ k - ft}$$

We should as usual check whether these moments satisfy the additional equations.

Step 6: Computation of support reactions and end shears. The FBDs generated as a result of the computation of the end moments are shown in Fig. E5.2.9(e).

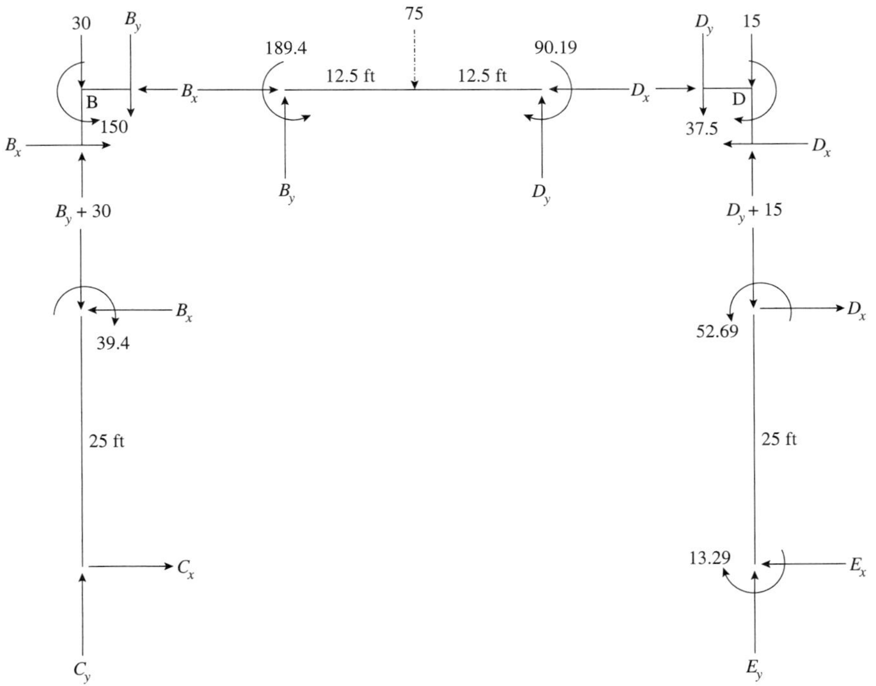

Fig. E5.2.9(e)

From the FBD of column CB, we have

$$\overset{\curvearrowleft +}{\sum} M_B = 0 = 25C_x - 39.4 \Rightarrow C_x = 1.58 \text{ k}(\rightarrow)$$

From the FBD of column DE, we have

$$\overset{\curvearrowleft +}{\sum} M_D = 0 = 52.69 - 13.29 - E_x(25) \Rightarrow E_x = 1.58 \text{ k}(\leftarrow)$$

From the FBD of beam BD, we have

$$\overset{\curvearrowleft +}{\sum} M_D = 0 = 189.4 - 75(12.5) + D_y(25) - 90.19 \Rightarrow D_y = 33.53 \text{ k}(\uparrow)$$

$$\overset{\uparrow +}{\sum} F_y = 0 = B_y - 75 + D_y \Rightarrow B_y = 41.47 \text{ k}(\uparrow)$$

Hence going back to the column FBDs, $C_y = 71.5$ k(↑) and $E_y = 48.5$ k(↑). The final structural FBD is in Fig. E5.2.9(f).

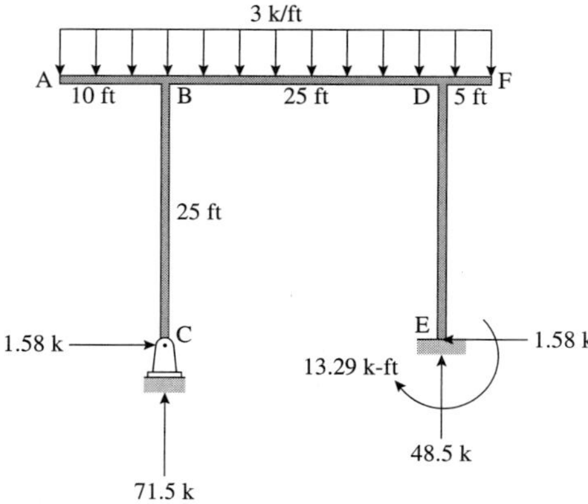

Fig. E5.2.9(f)

EXERCISES

Some of the following problems were solved using the force method. You should compare the two solutions along with the solution obtained from the GS-USA computer program.

Appetizers

5.2.1. Compute all the support reactions in Fig. P5.2.1. EI is a constant.

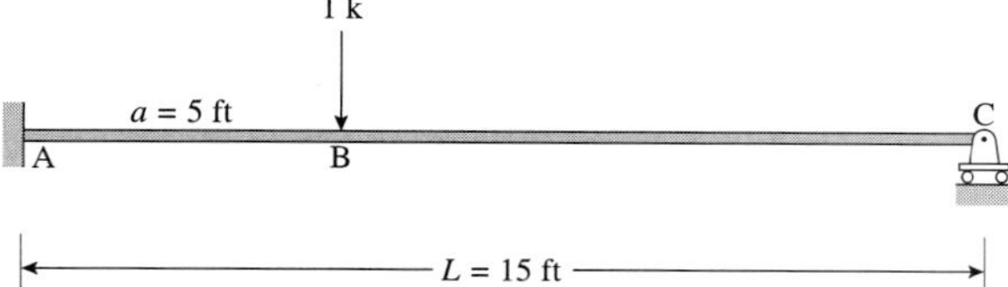

Fig. P5.2.1

5.2.2. First solve for all the support reactions in the system in Fig. P5.2.2 and then draw the shear force and bending moment diagrams. EI is a constant. Take $a = 10$ m and $b = 5$ m.

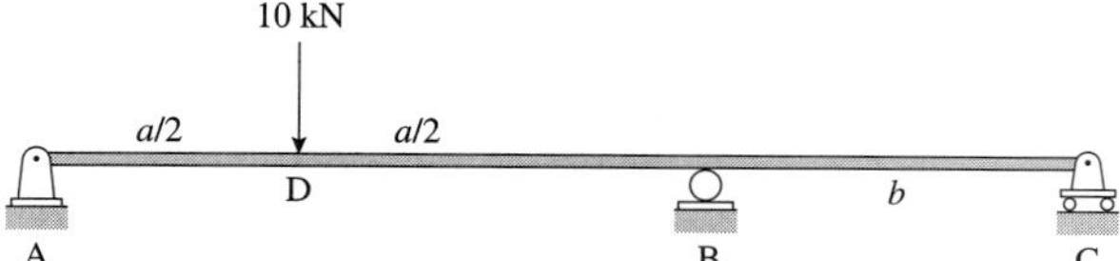

Fig. P5.2.2

5.2.3. First solve for all the support reactions and then draw the shear force and bending moment diagrams for the beam in Fig. P5.2.3. EI is a constant.

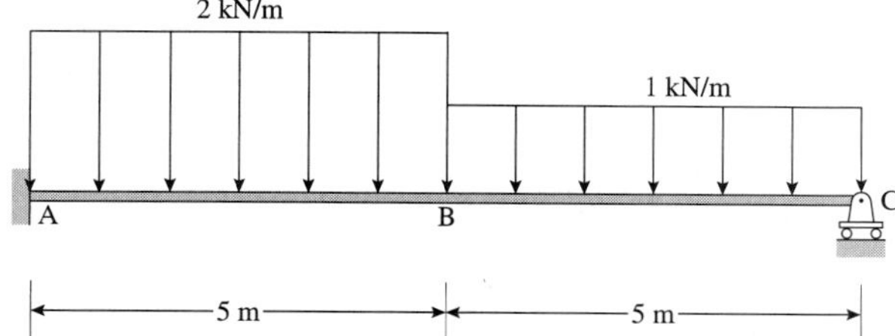

Fig. P5.2.3

5.2.4. Draw the shear force and bending moment diagrams for the frame in Fig. P5.2.4. Assume that *EI* is a constant.

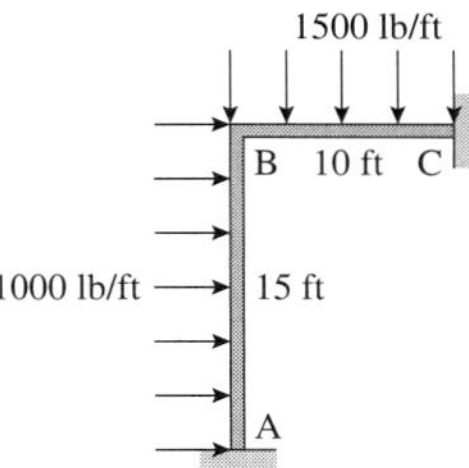

Fig. P5.2.4

5.2.5. Draw the shear force and bending moment diagrams for the frame in Fig. P5.2.5. Assume that *EI* is a constant.

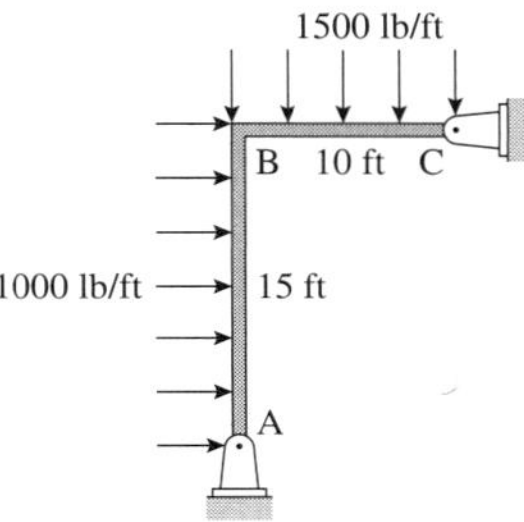

Fig. P5.2.5

Main Course

5.2.6. Draw the shear force and bending moment diagrams for the frame in Fig. P5.2.6. Assume that *EI* is a constant.

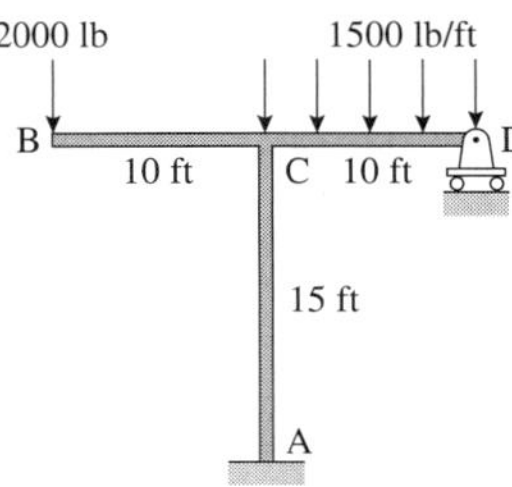

Fig. P5.2.6

5.2.7. For the system in Fig. P5.2.7, take $E = 30000$ ksi, $I = 400$ in^4, $L = 20$ ft, and $w = 0$. Support B settles 1 in. Solve for all the support reactions, and draw the shear force and bending moment diagrams.

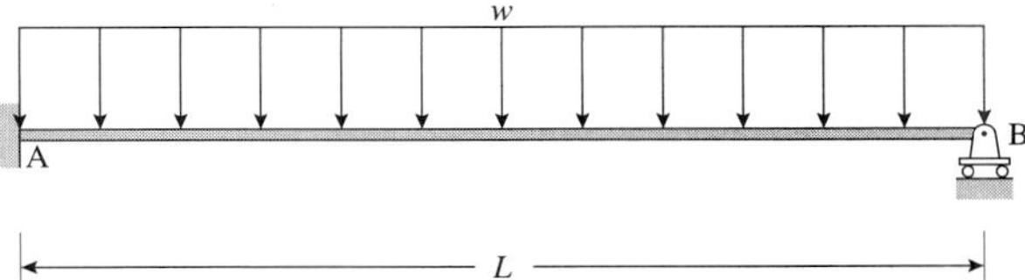

Fig. P5.2.7

Compare the solution with the case where $w = 1000$ lb/ft.

5.2.8. Draw the shear force and bending moment diagrams for the frame in Fig. P5.2.8. Take $I_{AB} = I_{CD} = 2I_{BC} = 2I$.

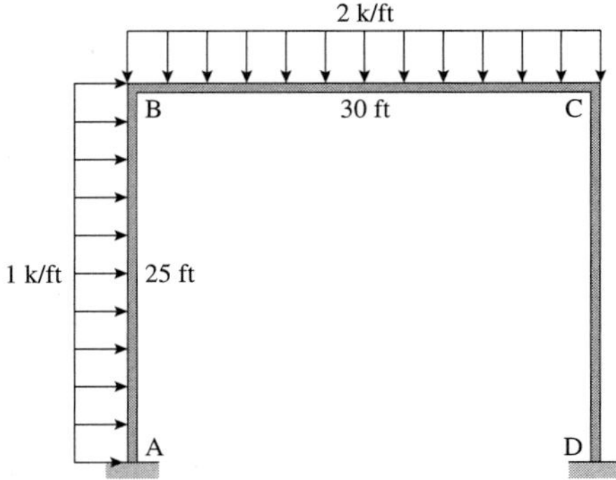

Fig. 5.2.8

5.2.9. Draw the shear force and bending moment diagrams for the frame in Fig. P5.2.9. Assume that EI is a constant.

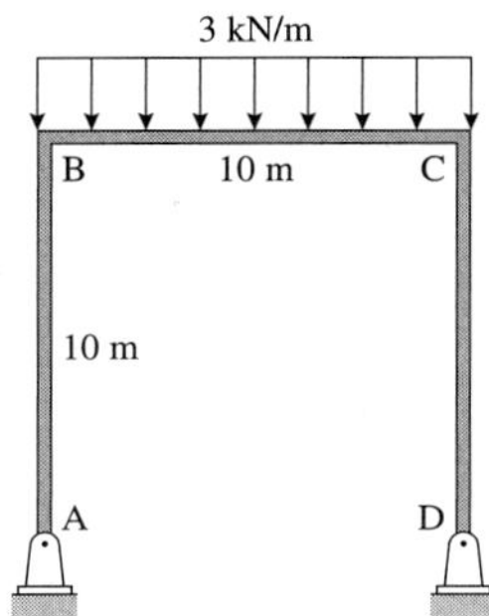

Fig. P5.2.9

5.2.10. Draw the shear force and bending moment diagrams for the frame in Fig. 5.2.10. Assume that EI is a constant.

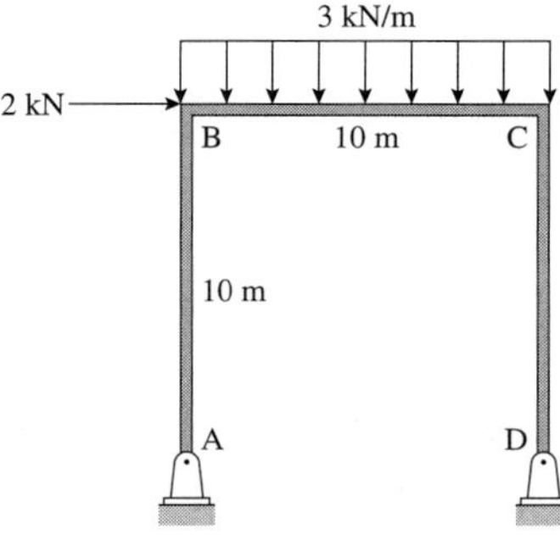

Fig. P5.2.10

5.2.11. Solve for all the support reactions in Fig. P5.2.11, and draw the shear force and bending moment diagrams. EI is a constant.

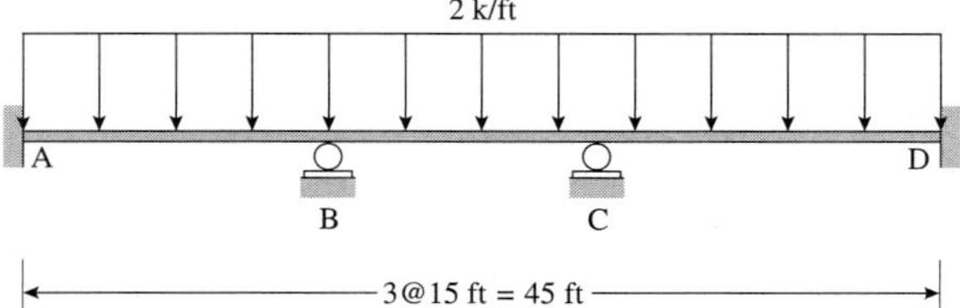

Fig. P5.2.11

5.2.12. For the beam in Fig. P5.2.11 take $E = 30000$ ksi and $I = 400$ in^4. Support B settles 1 in. Solve for all the support reactions, and draw the shear force and bending moment diagrams.

5.2.13. Figure P5.2.13 shows a planar frame. The material is steel, $E = 2(10)^{11}$ Pa and the cross-sectional properties are such that $A = 0.01$ m^2 and $I = 0.0001$ m^4. Solve for the member nodal forces and the support reactions.

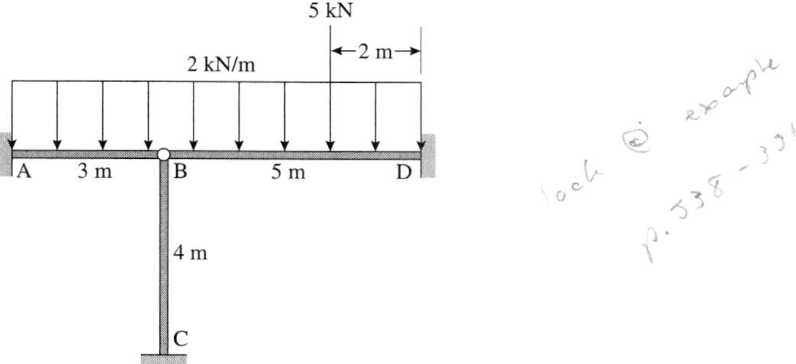

Fig. P5.2.13

Structural Concepts

5.2.14. We will modify the derivation of the slope-deflection equations by considering a beam segment that is an end span AB with B being an end roller, as in Fig. P5.2.14. The implication is that $M_{BA} = 0$. In other words, there is only one effective equation for the end span. Write the two slope-deflection equations for M_{BA} and M_{BA} as usual. Now set $M_{BA} = 0$. Simplify that equation and use the result to simplify the equation for M_{BA}. What are the advantages and disadvantages of this approach?

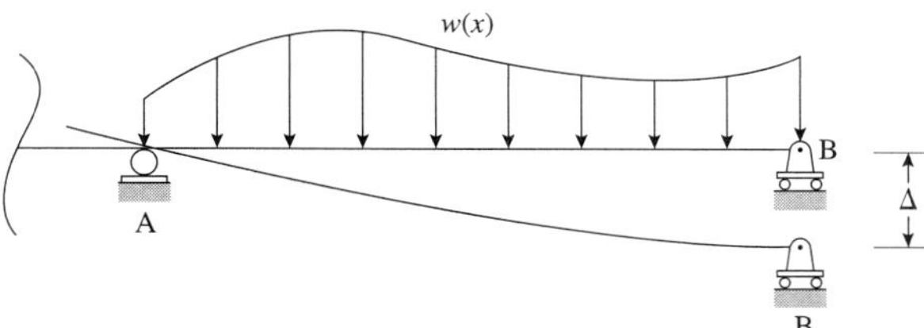

Fig. P5.2.14

SUMMARY

In this chapter we investigated in great detail classical solution techniques to (a) compute the deflections of truss, beam, and frame structures, and (b) solve for the deflections and internal forces in indeterminate structural systems. The background material will prove extremely useful in the later chapters dealing with numerical solution techniques. One of the advantages of looking at the slope-deflection method is that it is a displacement method, a method where the primary unknowns are displacements. In the next chapter we look at numerical methods that also have displacements as the primary unknowns.

You should solve the problems in this chapter using the GS-USA© computer program. The important fact to remember is that for planar beams and frames both the force method and the slope deflection method ignore axial and shear deformations. One should carefully look at the assumptions behind the implementation of the computer program. In the GS-USA© Frame program the shear deformation is ignored but not the axial deformation. Hence to obtain results from the computer program that are close to the results from these classical techniques, it is necessary to use large values for the cross-sectional areas of the members—the axial deformation of a member is inversely proportional to the cross-sectional area.

There is much debate on what constitutes a proper background for a student in structural analysis. While it is clear that basic principles and fundamentals are very important, it is, however, not clear as to what should be the mix between classical techniques and the more modern numerical techniques for structural analysis in a typical curriculum. We believe that classical techniques are necessary in grooming the student—introducing the concepts of modeling, compatibility, and equilibrium. However, we recognize that limitations are associated with these techniques in terms of the assumptions made, and especially in the size of the problems that can be solved. In this chapter we have learned to stand up and walk, and in the next chapter we will learn how to jog.

SUMMARY EXERCISES

Appetizers

Draw the shear force and bending moment diagrams for the beams and frames shown in Fig. P5.1–P5.3.

5.1. EI is a constant. Take $P = 2$ k.

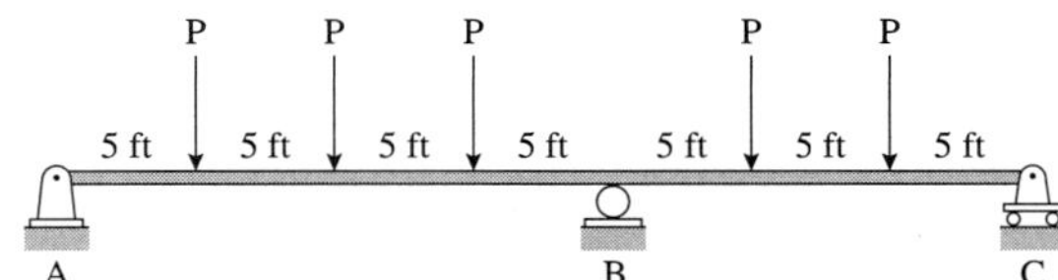

Fig. P5.1

5.2. EI is a constant.

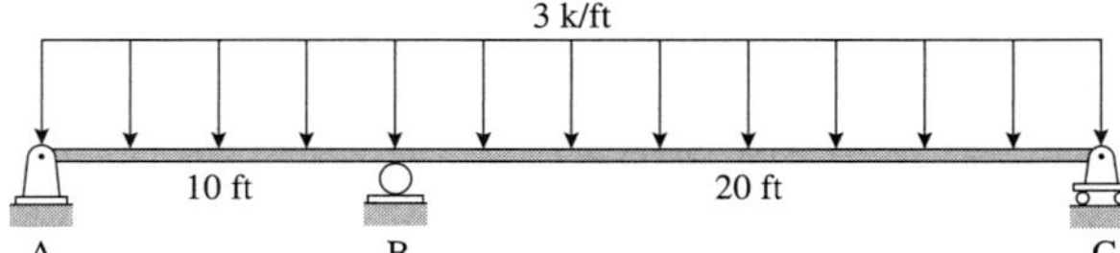

Fig. P5.2

5.3. Take $E = 200$ GPa, $I_{AB} = I$, $I_{BD} = 2I$, $I_{BC} = 1.25I$, and $I = 10^{-4}$ m^4.

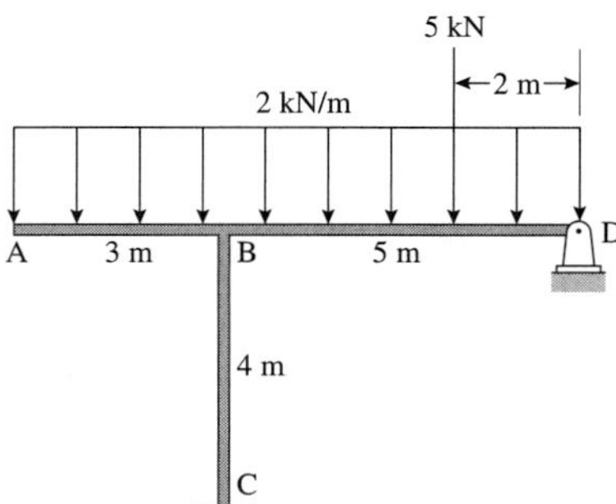

Fig. P5.3

Main Course

5.4. The truss in Fig. P5.4 is made of steel and the member cross-sections are as follows: top chord 1.5 in^2, bottom chord 1.0 in^2, and the web members 0.75 in^2. The vertical displacement at B and C cannot exceed 0.25 in. Does the design satisfy the requirement? If not, make the smallest possible changes to the cross-sectional areas so that the design meets the requirement.

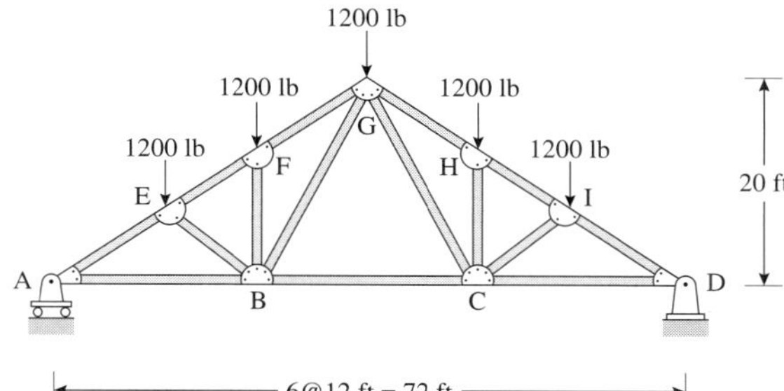

Fig. P5.4

5.5. The steel beam in Fig. P5.5 is subjected to very high loads at the mid-span. The moment of inertia of the beam is $4(10^8)$ mm^4. The span BC is then strengthened. What should the new moment of inertia of BC be so that the largest vertical displacement in the beam is less than 0.11 m?

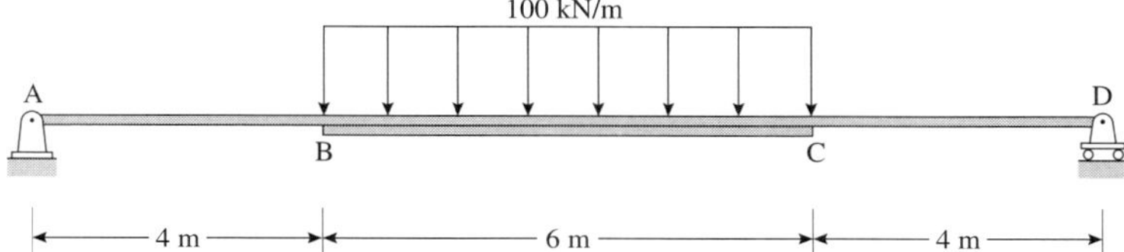

Fig. P5.5

5.6. Compute the support reactions of the beam in Fig. P5.6. *EI* is a constant.

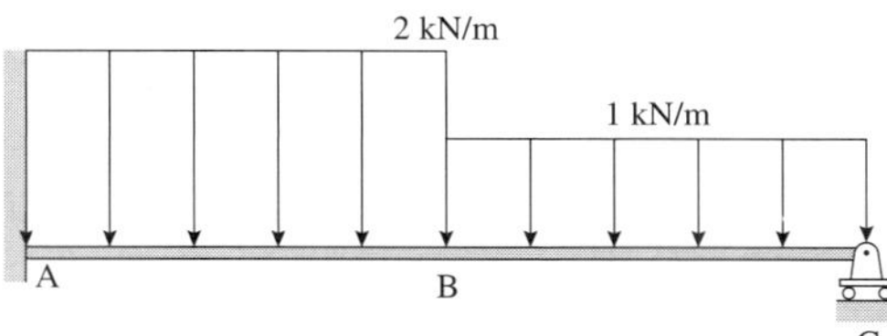

Fig. P5.6

Structural Concepts

The analysis of symmetric structures—beams, frames, and trusses—can be carried out more efficiently by exploiting the symmetric conditions. Consider the beam in Fig. P5(a).

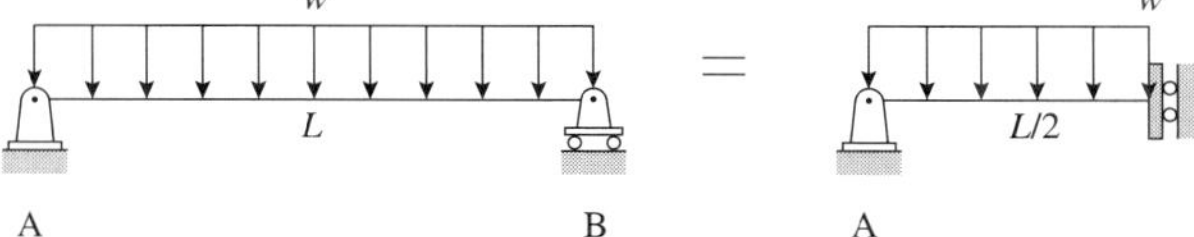

Fig. P5(a)

The determinate structure is symmetric in geometry, member properties, supports, and loading. In other words, there is an axis of symmetry—a mirror image of the structure exists on the other side. Hence the beam deforms so that center of the beam displaces vertically downwards and the slope is zero. The equivalent structure is shown on the right. It consists of the left half of the structure with the support conditions at the axis of symmetry such that the beam is free to displace vertically but is (i) constrained from moving horizontally and (ii) constrained so that the slope is zero.

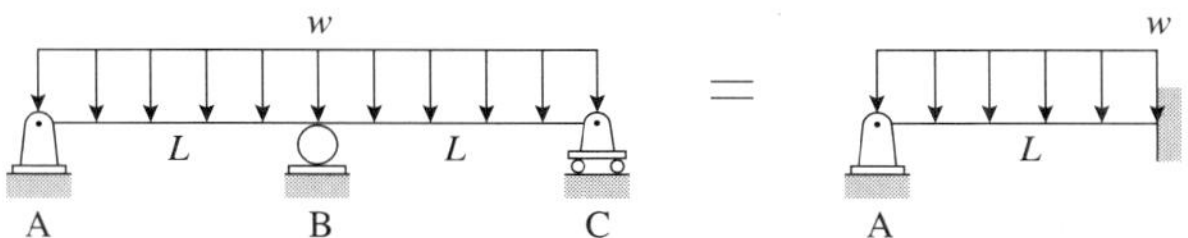

Fig. P5(b)

An indeterminate beam is shown in Fig. P5(b). The beam is symmetric in geometry, member properties, supports, and loading, with the axis of symmetry being the vertical axis through B. In this case, the beam deforms such that center of the beam is a fixed support—B cannot move horizontally and vertically, and the slope is zero. The equivalent structure is shown above. Finally, consider the indeterminate portal frame in Fig. P5(c). The behavior of the beam is similar to the beam in Fig. P5(a). The axis of symmetry is the axis through the center of the frame. The equivalent frame is shown to the right.

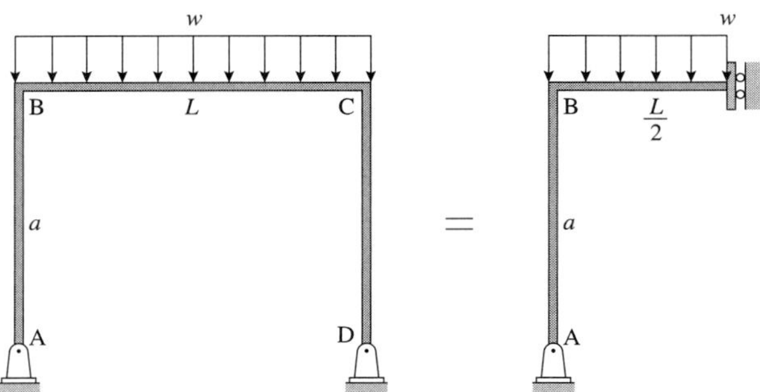

Fig. P5(c)

5.7. Compute the support reactions of the beam in Fig. P5.7. *EI* is a constant.

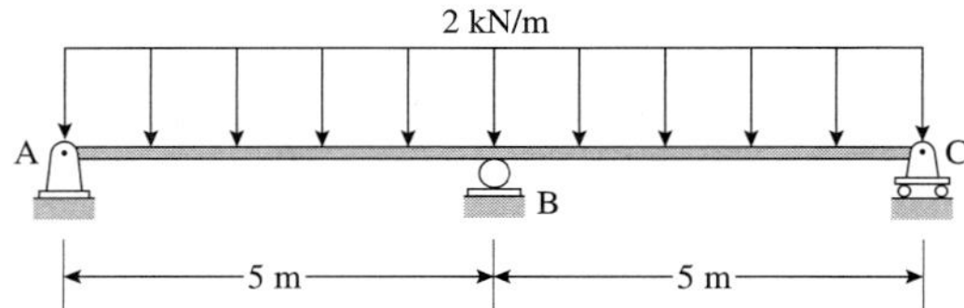

Fig. P5.7

5.8. Compute the support reactions of the frame shown below.

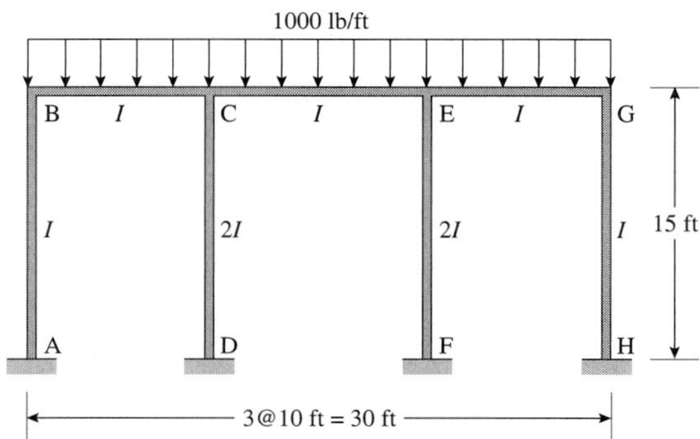

Fig. P5.8

5.9. The beam in Fig. P5.9 is symmetric but the loading is unsymmetrical. *EI* is a constant. How would you use superposition to analyze this beam? Compute the support reactions.

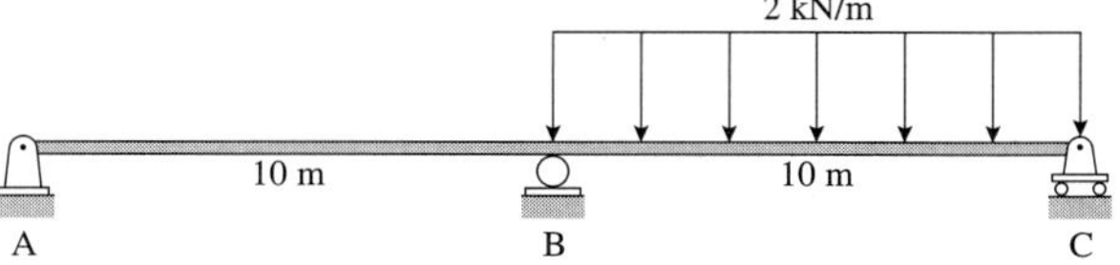

Chapter 6

Matrix-Based Numerical Methods of Structural Analysis

Structural analysis of complex systems such as antenna dish requires the use of computer-based numerical techniques.

"Everything should be made as simple as possible, not simpler." Albert Einstein

"Although the finite element method can make a good engineer better, it can make a poor engineer more dangerous . . . . One can now make mistakes with more confidence than ever before." Robert Cook, author of several texts on finite elements

"Nothing in life is for free." Anon

"I just bought a Mac to help me design the next Cray." Seymour Cray, when informed that Apple Inc. had recently bought a Cray supercomputer to help design the next Mac

In earlier courses in statics and deformable solids (or strength of materials), most of the structural systems were statically determinate. The analyses of these structures were carried out using the concept of static equilibrium. In Chapter 5 we saw some of the classical techniques for finding the structural response of statically determinate and indeterminate structural systems. We also saw the slope-deflection method in which the primary unknowns were linear displacements and rotations of beams and frame structures. These methods can be used, with hand calculations, when the degree of static or kinematic indeterminacy is small.

In this chapter we look at the modern methods of structural analysis. These methods not only overcome some of the drawbacks of the classical techniques but will also provide a mechanism to solve a variety of problem types. Statically determinate and indeterminate problems will be solved in the same way. The basic methodology can be applied to truss, beam, and frame structures uniformly. The methodology can also be extended to handle other classes of structural problems—plane elasticity, plate, and shells, etc. The most important characteristic is the ability to automate the solution process so that implementation in a computer program is possible. These methodologies form the backbone of the commercial programs that are used routinely to analyze a variety of structural systems. Understanding the theory, the limitations, and the solution procedure will also enable us to use these computer programs intelligently!

OBJECTIVES

- To understand the direct stiffness method as applied to planar truss and frame analysis.
- To understand the basics of the finite element method as a more powerful and alternate method for analyzing truss and frame systems with the potential to solve other classes of problems.
- To use the methods developed to solve problems by hand, further reinforcing the ideas introduced in Chapters 2, 4, and 5.

ASSUMPTIONS

- Deformations are assumed to be small. Deformations are small (a) if their magnitudes are small in comparison to the dimensions of the structure and (b) if the relationships among the applied loads and the reactions (supports and internal member forces) are not affected by the deformations of the structure.
- Numerical calculations by nature involve truncation and round-off errors. One must be careful to interpret the results from a numerical procedure. We will see some examples illustrating this aspect.
- Both the GS-USA© Frame program and the UNDO© program can be used to solve problems in this chapter. The former provides a mechanism for printing the intermediate steps during the solution of a truss or a frame. The latter provides a matrix toolbox that can be used to (a) generate the element matrices for different elements, and (b) solve for the nodal displacements, member forces, etc. One can also use a spreadsheet, e.g., Microsoft Excel©, or a computer program for mathematics, e.g., Matlab©, Matcad©, Maple©, etc., as an aid in solving problems.

6.1 FUNDAMENTALS OF MATRIX ALGEBRA

The solution methodology to be discussed in this chapter is numerical in nature. The initial, intermediate, and final steps usually involve matrices. While knowledge of linear algebra is essential in understanding the material in this chapter, we will focus on a narrower topic—matrix algebra. You may wish to read a book on linear algebra to gain a better understanding of the background material.

6.1.1 Definitions

Matrix. A two-dimensional matrix is a rectangular array of numbers. Each number or element of the matrix is identified by its location—a row number and a column number. Consider the following example:

$$\mathbf{A}_{m\times x} = \begin{bmatrix} A_{11} & A_{12} & \cdots & \cdots & A_{1n} \\ A_{21} & A_{22} & \cdots & \cdots & A_{2n} \\ \cdots & \cdots & \cdots & \cdots & \cdots \\ \cdots & \cdots & \cdots & \cdots & \cdots \\ A_{m1} & A_{m2} & \cdots & \cdots & A_{mn} \end{bmatrix} \qquad (6.1.1.1)$$

A typical element of the matrix **A** is designated A_{ij} where i is the row number and j is the column number. We usually, but not always, denote matrices with uppercase letters.

Vector. A vector is a special instance of a matrix. It has either one row or one column. We will usually, but not always, denote vectors with lowercase letters.

Row Vector. A vector with one row is called a row vector. Consider the following example.

$$a_{1\times n} = \{ a_1 \quad a_2 \quad \ldots \quad a_n \} \qquad (6.1.1.2)$$

Column Vector. A vector with one column is called a column vector. Consider the following example.

$$a_{m\times 1} = \begin{Bmatrix} a_1 \\ a_2 \\ \ldots \\ a_m \end{Bmatrix} \qquad (6.1.1.3)$$

Null Vector. A null vector is such that all the elements of the vector are zero, for example

$$a_{1\times n} = \{ 0 \quad 0 \quad \ldots \quad 0 \} \qquad (6.1.1.4)$$

Square Matrix. A square matrix has the same number of rows and columns. For example,

$$\mathbf{B}_{3\times 3} = \begin{bmatrix} 12 & -3 & 1 \\ 5 & 8 & 0 \\ -55 & 1 & 22 \end{bmatrix}$$

is a square matrix with integer elements.

Symmetric Matrix. A square matrix such that $A_{ij} = A_{ji}$ for any i, j is a symmetric matrix. For example,

$$\mathbf{B}_{3\times 3} = \begin{bmatrix} 12 & -3 & 1 \\ -3 & 8 & 0 \\ 1 & 0 & 22 \end{bmatrix} = \begin{bmatrix} 12 & -3 & 1 \\ & 8 & 0 \\ \text{Sym} & & 22 \end{bmatrix}$$

is a symmetric matrix.

Diagonal Matrix. A square matrix such that $A_{ij} = 0$ if $i \neq j$ is a diagonal matrix. For example,

$$\mathbf{B}_{3\times3} = \begin{bmatrix} 12 & 0 & 0 \\ 0 & 8 & 0 \\ 0 & 0 & 22 \end{bmatrix}$$

is a diagonal matrix.

Identity Matrix. A diagonal matrix such that $A_{ii} = 1$ is an identity matrix and is denoted $\mathbf{I}_{n\times n}$. For example, the following is an identity (or unit) matrix of order or size 3:

$$\mathbf{I}_{3\times3} = \begin{bmatrix} 1 & 0 & 0 \\ 0 & 1 & 0 \\ 0 & 0 & 1 \end{bmatrix}$$

Upper Triangular Matrix. A square matrix such that $A_{ij} = 0$ if $i < j$ is an upper triangular matrix. For example,

$$\mathbf{B}_{3\times3} = \begin{bmatrix} 12 & -55 & 0 \\ 0 & 8 & 10 \\ 0 & 0 & 22 \end{bmatrix}$$

is an upper triangular matrix.

Lower Triangular Matrix. A square matrix such that $A_{ij} = 0$ if $i > j$ is a lower triangular matrix. For example,

$$\mathbf{B}_{3\times3} = \begin{bmatrix} 12 & 0 & 0 \\ -55 & 8 & 0 \\ 0 & 10 & 22 \end{bmatrix}$$

is a lower triangular matrix.

6.1.2 Operations

Addition and Subtraction. Two matrices of the same size can be added or subtracted from one another. For example, if

$$\mathbf{A}_{m\times n} = \mathbf{B}_{m\times n} + \mathbf{C}_{m\times n} \quad \text{then} \quad A_{ij} = B_{ij} + C_{ij} \tag{6.1.2.1}$$

and, if

$$\mathbf{A}_{m\times n} = \mathbf{B}_{m\times n} - \mathbf{C}_{m\times n} \quad \text{then} \quad A_{ij} = B_{ij} - C_{ij} \tag{6.1.2.2}$$

Consider the following example. Let

$$\mathbf{B}_{3\times3} = \begin{bmatrix} 12 & -3 & 1 \\ -3 & 8 & 0 \\ 1 & 0 & 22 \end{bmatrix} \qquad \mathbf{C}_{3\times3} = \begin{bmatrix} 0 & 12 & -1 \\ 15 & 8 & 1 \\ 11 & 0 & 7 \end{bmatrix}$$

Then

$$\mathbf{A} = \mathbf{B} + \mathbf{C} = \begin{bmatrix} 12 & 9 & 0 \\ 12 & 16 & 1 \\ 12 & 0 & 29 \end{bmatrix} \qquad \mathbf{A} = \mathbf{B} - \mathbf{C} = \begin{bmatrix} 12 & -15 & 2 \\ -18 & 0 & -1 \\ -10 & 0 & 15 \end{bmatrix}$$

Multiplication. Two matrices can be multiplied as follows:

$$\mathbf{A}_{m\times n} = \mathbf{B}_{m\times o}\mathbf{C}_{o\times n} \tag{6.1.2.3}$$

provided the number of columns in **B** is equal to the number of rows in **C**. This condition makes the two matrices conformable. The resulting matrix **A** has its number of rows equal to the number of rows in **B** and number of columns equal to the number of columns in **C**. To generate the elements of the resulting matrix **A** we need

$$A_{ij} = \sum_{k=1}^{o} B_{ik} C_{kj} \tag{6.1.2.4}$$

In other words, the product of the corresponding elements from row i of **B** with the elements from column j of **C** yields A_{ij}. This operation is similar to computing the dot product.

For example, let

$$\mathbf{B}_{3\times 3} = \begin{bmatrix} 12 & -3 & 1 \\ -3 & 8 & 0 \\ 1 & 0 & 22 \end{bmatrix} \qquad \mathbf{C}_{3\times 2} = \begin{bmatrix} 0 & 12 \\ 15 & 8 \\ 11 & 0 \end{bmatrix}$$

Then $\mathbf{A}_{3\times 2} = \mathbf{B}_{3\times 3}\mathbf{C}_{3\times 2}$ can be computed by writing the three matrices as follows:

$$\begin{array}{cc} & \begin{bmatrix} 0 & 12 \\ 15 & 8 \\ 11 & 0 \end{bmatrix} \\ \begin{bmatrix} 12 & -3 & 1 \\ -3 & 8 & 0 \\ 1 & 0 & 22 \end{bmatrix} = & \begin{bmatrix} A_{11} & A_{12} \\ A_{21} & A_{22} \\ A_{31} & A_{32} \end{bmatrix} \end{array}$$

where A_{11} = the product of the first row of **B** and the first column of **C**

$= (12)(0) + (-3)(15) + (1)(11) = -34$

A_{12} = the product of the first row of **B** and the second column of **C**

$= (12)(12) + (-3)(8) + (1)(0) = 120$

A_{21} = the product of the second row of **B** and the first column of **C**

$= (-3)(0) + (8)(15) + (0)(11) = 120$

A_{22} = the product of the second row of **B** and the second column of **C**

$= (-3)(12) + (8)(8) + (0)(0) = 28$

A_{31} = the product of the third row of **B** and the first column of **C**

$= (1)(0) + (0)(15) + (22)(11) = 242$

A_{32} = the product of the third row of **B** and the second column of **C**

$= (1)(12) + (0)(8) + (22)(0) = 12$

Transpose. The transpose of matrix $\mathbf{A}_{m\times n}$ is denoted $\mathbf{A^T}_{n\times m}$. The transpose matrix is constructed such that

$$A_{ij}^{\mathbf{T}} = A_{ji} \tag{6.1.2.5}$$

As can be seen from Eq. (6.1.2.5), the transpose matrix is obtained by interchanging the rows and columns of the original matrix. Let

$$\mathbf{C}_{3\times 2} = \begin{bmatrix} 0 & 12 \\ 15 & 8 \\ 11 & 0 \end{bmatrix}. \qquad \text{Then } \mathbf{C}_{2\times 3}^{\mathbf{T}} = \begin{bmatrix} 0 & 15 & 11 \\ 12 & 8 & 0 \end{bmatrix}$$

Determinant. The determinant of a square matrix $\mathbf{A}_{n\times n}$ is denoted det(**A**) and is given by

$$\det(\mathbf{A}) = \sum_{j=1}^{n} a_{ij}A_{ij} \qquad \text{for any} \qquad i = 1, 2, ..., n \tag{6.1.2.6a}$$

or

$$\det(\mathbf{A}) = \sum_{i=1}^{n} a_{ij}A_{ij} \qquad \text{for any} \qquad j = 1, 2, ..., n \tag{6.1.2.6b}$$

where minor M_{ij} is the determinant of the $(n-1) \times (n-1)$ submatrix obtained by deleting the ith row and j[th] column, and cofactor a_{ij} associated with M_{ij} is defined to be $a_{ij} = (-1)^{i+j}M_{ij}$. While it will not be necessary for us to compute the determinant of a matrix, we still need to understand the concept. Let

$$\mathbf{A}_{2\times 2} = \begin{bmatrix} 4 & -3 \\ 1 & 6 \end{bmatrix} \qquad \mathbf{B}_{2\times 2} = \begin{bmatrix} 8 & 3 \\ 16 & 6 \end{bmatrix}$$

Then using Eq. (6.1.2.6a) with $i = 1$, we have

$$\det(\mathbf{A}) = \sum_{j=1}^{n} a_{ij}A_{ij} = a_{11}A_{11} + a_{12}A_{12} = 4a_{11} - 3a_{12}$$

$$a_{11} = (-1)^{1+1}\det[6] = (1)(6) = 6 \qquad a_{12} = (-1)^{1+2}\det[1] = (-1)(1) = -1$$

$$\det(\mathbf{A}) = 4a_{11} - 3a_{12} = 4(6) - 3(-1) = 27$$

and

$$\det(\mathbf{B}) = \sum_{j=1}^{n} b_{ij}B_{ij} = b_{11}B_{11} + b_{12}B_{12} = 8b_{11} + 3b_{12}$$

$$b_{11} = (-1)^{1+1}\det[6] = (1)(6) = 6 \qquad b_{12} = (-1)^{1+2}\det[16] = (-1)(16) = -16$$

$$\det(\mathbf{B}) = 8b_{11} + 3b_{12} = 8(6) + 3(-16) = 0$$

Since the determinant of **B** is zero, **B** is known as a *singular* matrix.

Solution of Linear Algebraic Equations. The following set of linear algebraic equations

$$\mathbf{A}_{n\times m}\mathbf{x}_{m\times 1} = \mathbf{b}_{n\times 1} \tag{6.1.2.7}$$

has a unique, nontrivial solution in **x** if and only if

a. the number of equations is equal to the number of unknowns, i.e., $n = m$,
b. the coefficient matrix **A** is a nonsingular matrix, and
c. the right-hand side (RHS) vector **b** is not a null vector.

We will not look at any specific techniques to solve these equations. We assume that the elimination technique discussed in Appendix E can be used effectively for up to three equations and that for larger systems a computer program (e.g., UNDO©) or a programmable calculator is available.

EXERCISES

Appetizers

Solve the problems below given the following matrices:

$$\mathbf{A}_{2\times 2} = \begin{bmatrix} 1 & 6 \\ 6 & -3 \end{bmatrix} \qquad \mathbf{B}_{2\times 2} = \begin{bmatrix} 5 & 7 \\ 33 & 20 \end{bmatrix} \qquad \mathbf{a}_{1\times 2} = \{-4 \quad 10\}$$

6.1.1.

(a) Is **A** symmetric? Is **A** an upper triangular matrix?

(b) Compute **A** + **B**.

(c) Compute **A** – **B**.

(d) Can **Aa** be computed? Can **aA** be computed?

Main Course

Solve the problems below given the following matrices:

$$\mathbf{A}_{3\times2} = \begin{bmatrix} 1 & 6 \\ 0 & -3 \\ 5 & -15 \end{bmatrix} \qquad \mathbf{B}_{3\times2} = \begin{bmatrix} 0 & -5 \\ 11 & 7 \\ -19 & 4 \end{bmatrix} \qquad \mathbf{C}_{2\times3} = \begin{bmatrix} -12 & 8 & 0 \\ 0 & 16 & -9 \end{bmatrix}$$

$$\mathbf{d}_{1\times2} = \{3 \quad -5\} \qquad \mathbf{a}_{3\times1} = \begin{Bmatrix} 6 \\ -11 \\ 7 \end{Bmatrix}$$

6.1.2

(a) Compute **AC**.

(b) Compute **BC**.

(c) Compute $\mathbf{A}^T$ and $\mathbf{C}^T$.

(d) Compute $\mathbf{C}^T\mathbf{A}$.

(e) Compute $\mathbf{dC} - \mathbf{a}^T$.

6.2 DIRECT STIFFNESS METHOD[1]

Tracing the history of the modern numerical methods of structural analysis takes us to the flexibility and stiffness methods as the first methods that were investigated widely. The former is similar to the classical force methods in which the primary unknowns are forces (or redundants). The latter is a displacement method in which the primary unknowns are displacements. The displacement methods have certain advantages over the force methods and hence are more commonly used. It may be appropriate at this stage to review the material from Chapter 5 before proceeding further.

6.2.1 Overview

We begin our investigations with a system consisting of simple, linear springs as shown in Fig. 6.2.1.1. The roller supports ensure that a typical spring moves strictly along the x axis. A single spring is the basic building block. Two *nodes* define a typical spring *element* as shown in Fig. 6.2.1.2.

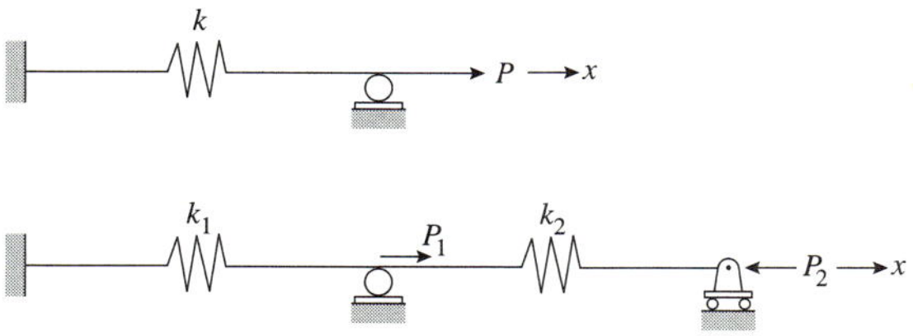

Fig. 6.2.1.1 Linear spring systems with (a) one spring (b) two springs.

[1] The reader may elect to cover Section 6.2 and proceed directly to Section 6.5 without loss of continuity. Time permitting, it is recommended that both the approaches—the direct stiffness method and the finite element method—be covered to gain a proper perspective of the advances made in matrix-based numerical structural analysis.

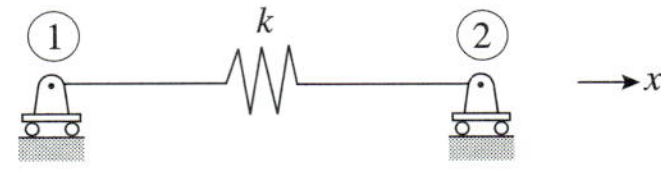

Fig. 6.2.1.2
The basic building block.

The two nodes are arbitrarily labeled 1 and 2. Alternately, we could label them i and j, or L and R. The following parameters describe the behavior of a typical spring: the spring stiffness k, the displacements d_1 and d_2, and the forces f_1 and f_2 at the ends of the spring. The displacements can cause either tension or compression in the spring. At each end of the spring there is a single unknown—the displacement. This is known as the *degree of freedom*, or dof for short. Hence, two degrees of freedom define a typical spring element. Note that the displacements and forces are shown acting in the positive coordinate direction. This is by convention where all quantities are initially assumed to be positive. The solution to the problem will yield the correct direction (a positive value indicating a displacement or force along the positive x direction).

We know that the behavior of a linear spring arises from Hooke's Law. When an elastic spring of stiffness k is subjected to an axial force f, it deforms by an amount d such that

$$d = \frac{f}{k} \tag{6.2.1.1}$$

Using this equation with the typical spring element as shown in Fig. 6.2.1.2, we have

$$f_1 = kd_1 - kd_2 \tag{6.2.1.2}$$

If we assume that the spring is in equilibrium, then $f_2 = -f_1$ and

$$f_2 = -kd_1 + kd_2 \tag{6.2.1.3}$$

We can rewrite the above two equations in matrix form as

$$\begin{bmatrix} k & -k \\ -k & k \end{bmatrix} \begin{Bmatrix} d_1 \\ d_2 \end{Bmatrix} = \begin{Bmatrix} f_1 \\ f_2 \end{Bmatrix} \tag{6.2.1.4}$$

or,

$$\mathbf{k}_{2\times2}\mathbf{d}_{2\times1} = \mathbf{f}_{2\times1} \tag{6.2.1.5}$$

where $\mathbf{k}_{2\times2}$ is the element stiffness matrix,
$\mathbf{d}_{2\times1}$ is the vector of element nodal displacements, and
$\mathbf{f}_{2\times1}$ is the vector of element nodal forces.

This simple example will give us further insight into the direct stiffness method. We have used two very important concepts to derive the equations defining the behavior of a typical element—equilibrium and Hooke's Law. The two equations in Eq. (6.2.1.5) are the equilibrium-compatibility equations at the two nodes of the element. We can interpret the physical significance of the elements of the element stiffness matrix. A typical element k_{ij} (or stiffness coefficient k_{ij}) in the element stiffness matrix is the force required at node i to produce a unit displacement at node j. The stiffness matrix is symmetric.[2]

How do we use the element equations to solve a problem involving one or more springs? Let us look at a system with one spring as shown below with $k = 300$ lb/in and $P = 30$ lb. We have labeled the two nodes as 1 and 2.

[2] The Maxwell–Betti reciprocal theorem states that all stiffness matrices for linear structures referred to orthogonal coordinate systems must be symmetric.

Fig. 6.2.1.3(a) One-spring system.

We can easily construct the element equations for this problem as follows (see Eq. (6.2.1.4)):

$$\begin{bmatrix} 300 & -300 \\ -300 & 300 \end{bmatrix} \begin{Bmatrix} d_1 \\ d_2 \end{Bmatrix} = \begin{Bmatrix} f_1 \\ f_2 \end{Bmatrix} \tag{6.2.1.6}$$

In this example, the element equations are also the *system equations*. The term system equations refers to the equilibrium-compatibility equations for the entire system involving all the spring elements. In general, the number of system equations is equal to the total number of degrees of freedom in the system. The one-spring system in Fig. 6.2.1.1 has a total of two dof and the two-springs system has three dof.

Several problems arise when we try to solve the set of Eq. (6.2.1.6). First, what are f_1 and f_2? The former is the support reaction. The latter is the applied load $P = 30$ lb. At this stage we do not know the support reaction. Second, if we somehow attempt to solve these equations (say, by assuming $f_1 = -30$), the process will fail since the coefficient matrix is singular.

What is necessary in general before we solve the system equations is to impose *boundary conditions* (or nodal fixity conditions) applicable for the problem. From Fig. 6.2.1.3(a) it is clear that $d_1 = 0$. Hence the problem has only one effective dof, d_2. How do we reflect the boundary conditions in Eq. (6.2.1.6)? Since the first equation deals with d_1 whose value is known, the first equation is superfluous. We cannot just delete the first equation since then we will be left with one equation involving d_1 and d_2,

$$\begin{bmatrix} -300 & 300 \end{bmatrix} \begin{Bmatrix} d_1 \\ d_2 \end{Bmatrix} = \{30\} \tag{6.2.1.7}$$

Keeping in mind the nature of matrix multiplication and that fact that $d_1 = 0$, the above equation can be suitably transformed to the following by deleting the first column:

$$[300]\{d_2\} = \{30\} \tag{6.2.1.8}$$

which can be solved to yield $d_2 = 30/300 = 0.1$ in. It is no coincidence that the force term corresponding to the known displacement is an unknown as much as the displacement term corresponding to the known force is an unknown.[3]

In general, imposition of a boundary condition involving $D_i = 0$ involves deleting the ith row and the ith column from the system equations.[4] For example, if $D_2 = 0$ in

$$\begin{bmatrix} K_{11} & K_{12} & K_{13} \\ K_{21} & K_{22} & K_{23} \\ K_{31} & K_{32} & K_{33} \end{bmatrix} \begin{Bmatrix} D_1 \\ D_2 \\ D_3 \end{Bmatrix} = \begin{Bmatrix} F_1 \\ F_2 \\ F_3 \end{Bmatrix}, \quad \text{then} \quad \left[\begin{array}{c|c|c} K_{11} & \not{K}_{12} & K_{13} \\ \hline \not{K}_{21} & \not{K}_{22} & \not{K}_{23} \\ \hline K_{31} & \not{K}_{32} & K_{33} \end{array}\right] \begin{Bmatrix} D_1 \\ \not{D}_2 \\ D_3 \end{Bmatrix} \not= \begin{Bmatrix} F_1 \\ \not{F}_2 \\ F_3 \end{Bmatrix}$$

Simplifying, we get

$$\begin{bmatrix} K_{11} & K_{13} \\ K_{31} & K_{33} \end{bmatrix} \begin{Bmatrix} D_1 \\ D_3 \end{Bmatrix} = \begin{Bmatrix} F_1 \\ F_3 \end{Bmatrix} \tag{6.2.1.9}$$

---

[3] This fact arises from the theory of differential equations. For example, for a problem to be well-posed, known displacements correspond to essential (or Dirichlet) boundary condition and the unknown forces corresponds to natural (or Neumann) boundary condition. In general, at a boundary, either the known condition is essential or natural but not both.

[4] The more curious reader is encouraged to read Section 6.5.3 where further details are provided.

Finally, we can use the computed displacements to solve for the forces at the ends of the spring. This step is accomplished using the left-hand side of Eq. (6.2.1.4). Substituting yields

$$\begin{bmatrix} 300 & -300 \\ -300 & 300 \end{bmatrix} \begin{Bmatrix} 0 \\ 0.1 \end{Bmatrix} = \begin{Bmatrix} -30 \\ 30 \end{Bmatrix}$$

The negative sign implies that the 30 lb force is acting in the negative x-direction. The free-body diagram of the element is given in Fig. 6.2.1.3(b).

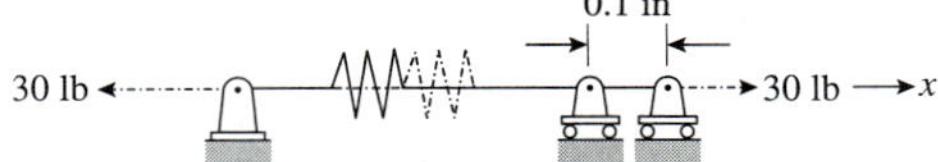

Fig. 6.2.1.3(b) FBD of the one-spring system.

Let us now focus our attention on the two-spring system shown in Fig. 6.2.1.4. The three nodes are arbitrarily labeled 1, 2, and 3. Similarly, the two elements are arbitrarily labeled 1 and 2. A change in notation is necessary. How do we differentiate between the two displacements associated with each element and the three displacements associated with the spring system? We will use (lowercase) d_1 and d_2 to denote the displacements at the first and the second node of a typical element. The displacements at the system level are denoted by (uppercase) D_i where i is the node number. The relationship between the element displacements and the system displacements is shown in Fig. 6.2.1.5.

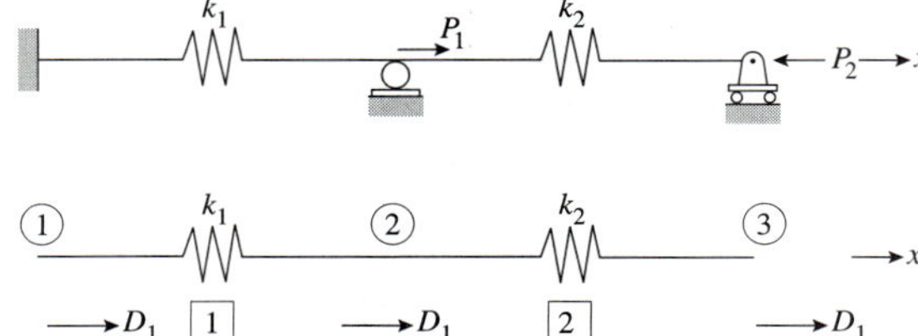

Fig. 6.2.1.4 Two-spring system.

The element nodes 1 and 2 are being labeled left to right for the time being so that the vector from 1 to 2 is along the positive x direction. We will relax this requirement in later sections.

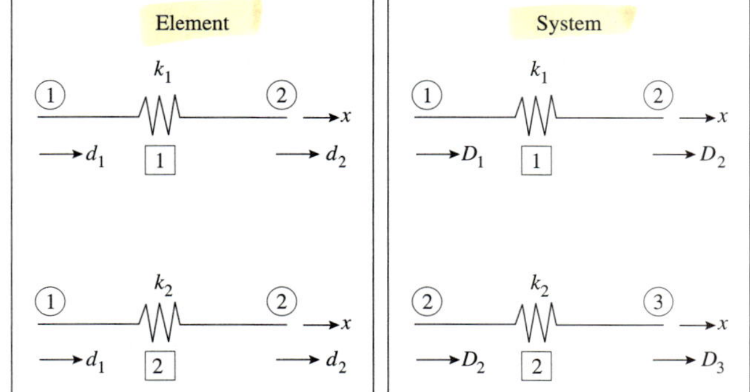

Fig. 6.2.1.5 Element and system degrees of freedom for each element.

The next step in the solution process is to generate the element equations for the two elements. The element equations are

$$\begin{bmatrix} k_1^1 & -k_1^1 \\ -k_1^1 & k_1^1 \end{bmatrix} \begin{Bmatrix} d_1^1 \\ d_2^1 \end{Bmatrix} = \begin{Bmatrix} f_1^1 \\ f_2^1 \end{Bmatrix} \quad \text{or} \quad \begin{bmatrix} k_1^1 & -k_1^1 \\ -k_1^1 & k_1^1 \end{bmatrix} \begin{Bmatrix} D_1 \\ D_2 \end{Bmatrix} = \begin{Bmatrix} f_1^1 \\ f_2^1 \end{Bmatrix} \tag{6.2.1.10}$$

where the superscript 1 indicates association of the quantity with element 1. Similarly, for element 2 we have

$$\begin{bmatrix} k_2^2 & -k_2^2 \\ -k_2^2 & k_2^2 \end{bmatrix} \begin{Bmatrix} d_1^2 \\ d_2^2 \end{Bmatrix} = \begin{Bmatrix} f_1^2 \\ f_2^2 \end{Bmatrix} \quad \text{or} \quad \begin{bmatrix} k_2^2 & -k_2^2 \\ -k_2^2 & k_2^2 \end{bmatrix} \begin{Bmatrix} D_2 \\ D_3 \end{Bmatrix} = \begin{Bmatrix} f_1^2 \\ f_2^2 \end{Bmatrix} \tag{6.2.1.11}$$

Note that the local dof have been replaced with the global dof using Fig. 6.2.1.5. The task of generating the system equations will involve taking the *four* equations in Eqs. (6.2.1.10) and (6.2.1.11) and constructing the *three* system equilibrium-compatibility equations of the form

$$\begin{bmatrix} K_{11} & K_{12} & K_{13} \\ K_{21} & K_{22} & K_{23} \\ K_{31} & K_{32} & K_{33} \end{bmatrix} \begin{Bmatrix} D_1 \\ D_2 \\ D_3 \end{Bmatrix} = \begin{Bmatrix} F_1 \\ F_2 \\ F_3 \end{Bmatrix} \tag{6.2.1.12}$$

The process of taking the element equations and constructing the system equations is known as the *assembly process*. The recommended procedure is to start populating the system equations with element 1 and then updating the equations with the other elements. From Eq. (6.2.1.10) it is clear that the two equations correspond to equilibrium along D_1 and D_2 (the first and the second system equations). Hence the system equations are as follows with the contribution from element 1:

$$\begin{bmatrix} k_1^1 & -k_1^1 & 0 \\ -k_1^1 & k_1^1 & 0 \\ 0 & 0 & 0 \end{bmatrix} \begin{Bmatrix} D_1 \\ D_2 \\ D_3 \end{Bmatrix} = \begin{Bmatrix} f_1^1 \\ f_2^1 \\ 0 \end{Bmatrix} \tag{6.2.1.13}$$

Similarly, for element 2, the two equations correspond to equilibrium along D_2 and D_3 (the second and the third system equations). Hence the updated system equations are as follows with the contribution from element 2:

$$\begin{bmatrix} k_1^1 & -k_1^1 & 0 \\ -k_1^1 & k_1^1 + k_2^2 & -k_2^2 \\ 0 & -k_2^2 & k_2^2 \end{bmatrix} \begin{Bmatrix} D_1 \\ D_2 \\ D_3 \end{Bmatrix} = \begin{Bmatrix} f_1^1 \\ f_2^1 + f_1^2 \\ f_2^2 \end{Bmatrix} \tag{6.2.1.14}$$

or

$$\begin{bmatrix} k_1^1 & -k_1^1 & 0 \\ -k_1^1 & \boxed{k_1^1 + k_2^2} & -k_2^2 \\ 0 & -k_2^2 & k_2^2 \end{bmatrix} \begin{Bmatrix} D_1 \\ D_2 \\ D_3 \end{Bmatrix} = \begin{Bmatrix} F_1 \\ F_2 \\ F_3 \end{Bmatrix} \tag{6.2.1.15}$$

The symbolic way of representing the system equilibrium-compatibility equations is

$$\mathbf{K}_{3\times3}\mathbf{D}_{3\times1} = \mathbf{F}_{3\times1} \tag{6.2.1.16}$$

where $\mathbf{K}_{3\times3}$ is the system (or structural) stiffness matrix,
$\mathbf{D}_{3\times1}$ is the vector of system nodal displacements, and
$\mathbf{F}_{3\times1}$ is the vector of system nodal forces.

Observation: Note that the system stiffness matrix is symmetric. This is a consequence of Betti-Maxwell Theorem (see Problem 4.5.22). The boxed term in Eq. (6.2.1.15) shows the coupling between springs 1 and 2 at the second degree of freedom. There is no coupling at the first and the last degrees of freedom since they are connected to either spring 1 or spring 2 but not both.

What do the three terms in the system force vector represent? The first term is the support reaction at node 1. The second and the third terms are the *net* forces at the second and third nodes. Let $k_1 = 200$ lb/in, $k_2 = 300$ lb/in, $P_1 = 20$ lb, and $P_2 = 40$ lb. Hence the system equations are

$$\begin{bmatrix} 200 & -200 & 0 \\ -200 & 500 & -300 \\ 0 & -300 & 300 \end{bmatrix} \begin{Bmatrix} D_1 \\ D_2 \\ D_3 \end{Bmatrix} = \begin{Bmatrix} F_1 \\ 20 \\ -40 \end{Bmatrix} \qquad (6.2.1.17)$$

As in the previous problem, we cannot solve these equations until we impose the boundary conditions. From the problem data, $D_1 = 0$. Imposing this boundary condition yields

$$\begin{bmatrix} 500 & -300 \\ -300 & 300 \end{bmatrix} \begin{Bmatrix} D_2 \\ D_3 \end{Bmatrix} = \begin{Bmatrix} 20 \\ -40 \end{Bmatrix} \qquad (6.2.1.18)$$

Solving the equations yields $D_2 = -0.1$ in and $D_3 = -0.233333$ in. Finally, we can use the computed displacements to solve for the forces at the ends of the spring. For element 1 we find

$$\begin{bmatrix} 200 & -200 \\ -200 & 200 \end{bmatrix} \begin{Bmatrix} 0 \\ -0.1 \end{Bmatrix} = \begin{Bmatrix} 20 \\ -20 \end{Bmatrix} \qquad (6.2.1.19)$$

and for element 2

$$\begin{bmatrix} 300 & -300 \\ -300 & 300 \end{bmatrix} \begin{Bmatrix} -0.1 \\ -0.233333 \end{Bmatrix} = \begin{Bmatrix} 40 \\ -40 \end{Bmatrix} \qquad (6.2.1.20)$$

The free-body diagrams of the two elements as well as that of node 2 are shown in Fig. 6.2.1.6. From looking at the FBDs it is clear that equilibrium is satisfied and that both elements are in compression.

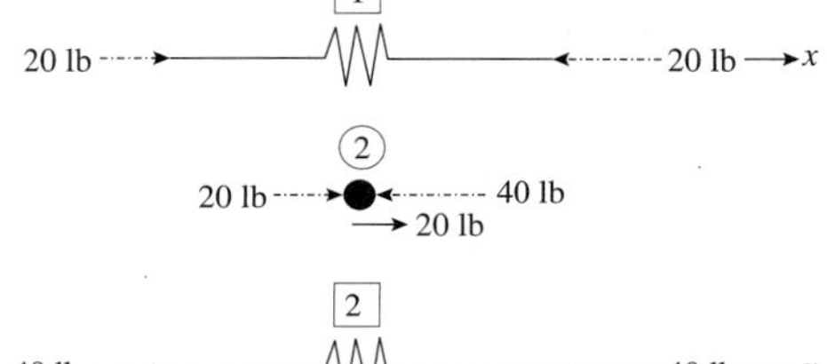

Fig. 6.2.1.6 FBDs of the two elements and node 2.

EXERCISES

Appetizers

6.2.1. For the system of springs in Fig. P6.2.1(a), compute the force in each spring. Use the model in Fig. P6.2.1(b).

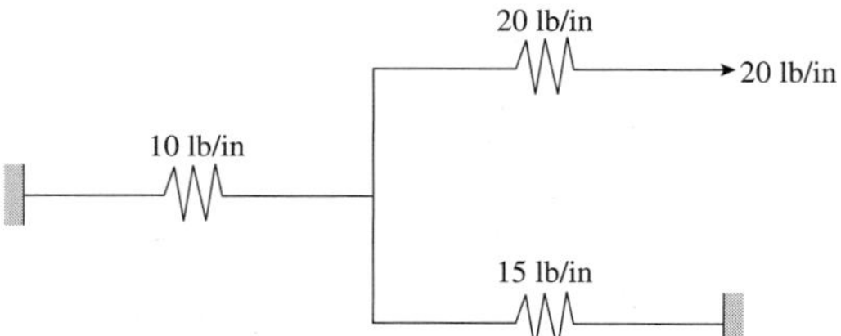

Fig. P6.2.1(a)

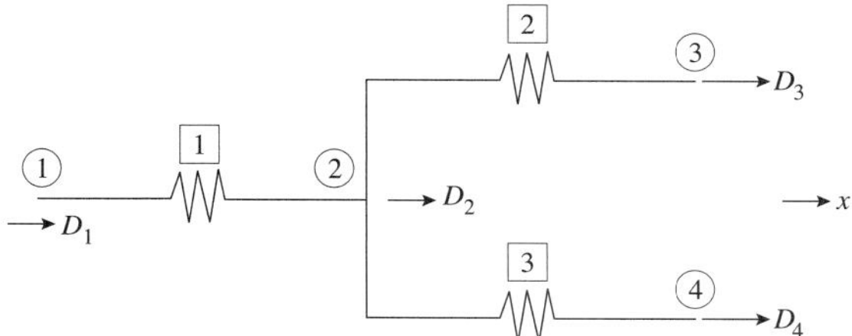

Fig. P6.2.1(b)

6.2.2 Truss Analysis

A truss is a structural system that satisfies the following requirements:

a. The members are straight, slender, and prismatic. The cross-sectional dimensions are small in comparison to the member lengths. The weights of the members are small compared to the applied loads and can be neglected. Also when constructing the truss model for analysis, we treat the members as a one-dimensional entity (having length and negligible cross-sectional dimensions).
b. The joints are assumed to be frictionless pins (or internal hinges).
c. The loads are applied only at the joints in the form of concentrated forces.

As a consequence of these assumptions, the members are two-force members, meaning that they carry only axial forces. In very many ways, a truss member is quite similar to the typical linear spring that we investigated in the previous section. Two nodes define a typical truss element, as shown in Fig. 6.2.2.1. The notation is slightly different from that used for the linear spring. The coordinate system for the typical truss element is now called x'. The prime notation is important so that we can distinguish between the local coordinate system x' and the global coordinate system x-y. Why is this necessary? Unlike the spring examples, a truss system can be composed of several members or elements such that each member has a different local axis or coordinate system (see Fig. 6.2.2.2). However, we can locate the entire truss in one reference frame, the global coordinate system. In other words, while the truss behavior is uniaxial, the truss system is located in a two-dimensional or even three-dimensional spatial system.

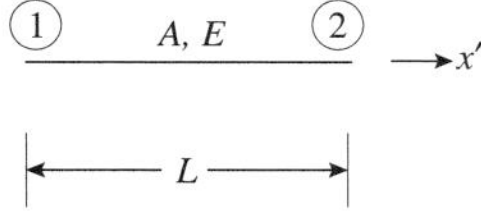

$d'_1, f'_1 \longrightarrow$ ———— $\longrightarrow d'_2, f'_2 \longrightarrow x'$

Fig. 6.2.2.1
A typical truss element (local coordinate system)

The following parameters describe the behavior of a typical truss element: the cross-sectional area A, the modulus of elasticity E, the length L, the local displacements d'_1 and d'_2, and the local forces f'_1 and f'_2 at the ends of the element.

As we saw in Chapter 4, when an elastic bar is subjected to an axial force f, by Hooke's Law it deforms by an amount d such that

$$d = \frac{f}{AE/L} = \frac{fL}{AE} \qquad (6.2.2.1)$$

Using this equation for the truss element as shown in Fig. 6.2.2.1, we have

$$f'_1 = \frac{AE}{L} d'_1 - \frac{AE}{L} d'_2 \qquad (6.2.2.2)$$

Fig. 6.2.2.2
A planar truss system showing the global coordinate system and the three local coordinate systems associated with each element.

If we assume that the element is in equilibrium, then $f_2' = -f_1'$ and

$$f_2' = \frac{-AE}{L} d_1' + \frac{AE}{L} d_2' \tag{6.2.2.3}$$

We can rewrite the above two equations in matrix form as

$$\frac{AE}{L}\begin{bmatrix} 1 & -1 \\ -1 & 1 \end{bmatrix}\begin{Bmatrix} d_1' \\ d_2' \end{Bmatrix} = \begin{Bmatrix} f_1' \\ f_2' \end{Bmatrix} \tag{6.2.2.4}$$

or

$$\mathbf{k}'_{2\times2}\mathbf{d}'_{2\times1} = \mathbf{f}'_{2\times1} \tag{6.2.2.5}$$

where $\mathbf{k}'_{2\times2}$ is the local element stiffness matrix,
$\mathbf{d}'_{2\times1}$ is the vector of local element nodal displacements, and
$\mathbf{f}'_{2\times1}$ is the vector of local element nodal forces.

It should be evident by now that the "spring stiffness" of a truss element is AE/L. While we can generate the element equations, how do we relate the element equations for all the different elements in a truss system? The answer lies in defining these equations in a common reference frame—the global coordinate system that is the same for all the elements. The element shown in Fig., 6.2.2.1 is now placed in the global coordinate system as shown in Fig. 6.2.2.3. The global coordinates of nodes 1 and 2 are (x_1, y_1) and (x_2, y_2), respectively. At node 1, the x and y components of the local displacement d_1' are d_1 and d_2. Similarly, at node 2, the x and y components of the local displacement d_2' are d_3 and d_4. We can relate the local and global displacements at node 1 as

$$(d_1')^2 = (d_1)^2 + (d_2)^2 \tag{6.2.2.6}$$

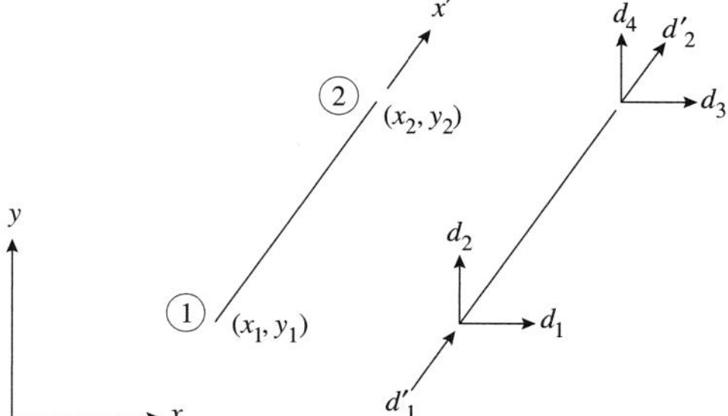

Fig. 6.2.2.3
Displacement local-to-global transformation.

Dividing throughout by d_1', we have

$$d_1' = \frac{d_1}{d_1'} d_1 + \frac{d_2}{d_1'} d_2 = l d_1 + m d_2 \tag{6.2.2.7a}$$

where (l, m) are the direction cosines of the x' axis with respect to the global coordinate system. Similarly, we have

$$d_2' = \frac{d_3}{d_2'} d_3 + \frac{d_4}{d_2'} d_4 = ld_3 + md_4 \tag{6.2.2.7b}$$

The direction cosines are computed using the nodal coordinates as follows:

$$L = \sqrt{(x_2 - x_1)^2 + (y_2 - y_1)^2};\ l = \frac{x_2 - x_1}{L};\ m = \frac{y_2 - y_1}{L} \tag{6.2.2.7c}$$

We can write Eqs. (6.2.2.7a)–(6.2.2.7b) as

$$\begin{Bmatrix} d_1' \\ d_2' \end{Bmatrix} = \begin{bmatrix} l & m & 0 & 0 \\ 0 & 0 & l & m \end{bmatrix} \begin{Bmatrix} d_1 \\ d_2 \\ d_3 \\ d_4 \end{Bmatrix} \tag{6.2.2.8}$$

or

$$\mathbf{d}'_{2\times1} = \mathbf{T}_{2\times4}\mathbf{d}_{4\times1} \tag{6.2.2.9}$$

where $\mathbf{T}_{2\times4}$ is the displacement local-to-global transformation matrix
$\mathbf{d}_{4\times1}$ is the vector of global element nodal displacements.

It should be noted that since we are using the undeformed geometry of the member to compute the direction cosines, the implicit assumption is that the nodal displacements are small; otherwise the equilibrium equations are not valid.

We can relate the local and global forces acting at the two nodes as shown in Fig. 6.2.2.4 in a similar fashion as follows. At node 1, relating the x and y components, we have

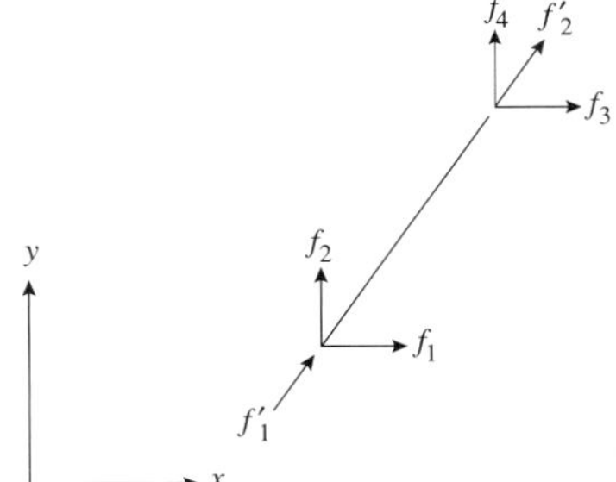

Fig. 6.2.2.4 Force global-to-local transformation.

$$f_1 = lf_1' \qquad f_2 = mf_1' \tag{6.2.2.10a}$$

and at node 2

$$f_3 = lf_2' \qquad f_4 = mf_2' \tag{6.2.2.10b}$$

We can write the two equations as

$$\begin{Bmatrix} f_1 \\ f_2 \\ f_3 \\ f_4 \end{Bmatrix} = \begin{bmatrix} l & 0 \\ m & 0 \\ 0 & l \\ 0 & m \end{bmatrix} \begin{Bmatrix} f_1' \\ f_2' \end{Bmatrix} \tag{6.2.2.11}$$

or

$$\mathbf{f}_{4\times1} = \mathbf{T}^{\mathbf{T}}_{4\times2}\mathbf{f}'_{2\times1} \tag{6.2.2.12}$$

where $\mathbf{T}^{\mathbf{T}}_{4\times2}$ is the force global-to-local transformation matrix
$\mathbf{f}_{4\times1}$ is the vector of global element nodal forces.

We are now ready to generate the element equations in the global coordinate system. We start with Eq. (6.2.2.5). Substituting Eq. (6.2.2.9) for $\mathbf{d}'_{2\times1}$, we have

$$\mathbf{k}'_{2\times2}\mathbf{T}_{2\times4}\mathbf{d}_{4\times1} = \mathbf{f}'_{2\times1} \tag{6.2.2.13}$$

Premultiplying both sides by $\mathbf{T}^{\mathbf{T}}_{4\times2}$ gives us

$$\mathbf{T}^{\mathbf{T}}_{4\times2}\mathbf{k}'_{2\times2}\mathbf{T}_{2\times4}\mathbf{d}_{4\times1} = \mathbf{T}^{\mathbf{T}}_{4\times2}\mathbf{f}'_{2\times1} \tag{6.2.2.14}$$

Using Eq. (6.2.2.12), the above equation can be rewritten as

$$\mathbf{T}^{\mathbf{T}}_{4\times2}\mathbf{k}'_{2\times2}\mathbf{T}_{2\times4}\mathbf{d}_{4\times1} = \mathbf{f}_{4\times1} \tag{6.2.2.15}$$

or

$$\mathbf{k}_{4\times4}\mathbf{d}_{4\times1} = \mathbf{f}_{4\times1} \tag{6.2.2.16}$$

where $\mathbf{k}_{4\times4} = \mathbf{T}^{\mathbf{T}}_{4\times2}\mathbf{k}'_{2\times2}\mathbf{T}_{2\times4}$ is the global element stiffness matrix. These are the element equilibrium-compatibility equations in the global coordinate system. The global element stiffness matrix can be computed by multiplying the three matrices to yield

$$\mathbf{k}_{4\times4} = \frac{AE}{L}\begin{bmatrix} l^2 & lm & -l^2 & -lm \\ lm & m^2 & -lm & -m^2 \\ -l^2 & -lm & l^2 & lm \\ -lm & -m^2 & lm & m^2 \end{bmatrix} \tag{6.2.2.17}$$

Just like the local element stiffness matrix, the global element stiffness matrix is symmetric. It should also be noted that a typical truss element has two dof in the local coordinate system and four dof in the global coordinate system.

We can now summarize the major steps in solving any planar truss problem using the direct stiffness method.

Step 1: Select the problem units. Set up the coordinate system. Identify and label the nodes and the elements. For each element select a start node (node 1) and an end node (node 2). We use an arrow along the member to indicate the direction from the start node to the end node. This establishes the local coordinate system for each element. Label the two global dof at each node starting at node 1 and proceeding sequentially.

Step 2: Construct the equilibrium-compatibility equations for a typical element (Eq. (6.2.2.16)).

Step 3: Using the problem data, construct the element equations from Step 2 for all the elements in the problem.

Step 4: Assemble the element equations into the system equations, $\mathbf{K}_{2j\times2j}\mathbf{D}_{2j\times1} = \mathbf{F}_{2j\times1}$, where j is the number of joints in the truss.

Step 5: Impose the boundary conditions.

Step 6: Solve the system equations $\mathbf{KD} = \mathbf{F}$ for the nodal displacements $\mathbf{D}$.

Step 7: For each element using the nodal displacements, compute the element nodal forces. We first start with Eq. (6.2.2.5) written as

$$\mathbf{f}'_{2\times1} = \mathbf{k}'_{2\times2}\mathbf{d}'_{2\times1}$$

Substituting Eq. (6.2.2.9), we have

$$\mathbf{f}'_{2\times1} = \mathbf{k}'_{2\times2}\mathbf{T}_{2\times4}\mathbf{d}_{4\times1} \tag{6.2.2.18}$$

Since each element is in equilibrium, $f'_1 = -f'_2$ and it is necessary to compute only one of the element local forces. For example,

$$f_1' = \frac{AE}{L}\begin{bmatrix} l & m & -l & -m \end{bmatrix}\begin{Bmatrix} d_1 \\ d_2 \\ d_3 \\ d_4 \end{Bmatrix} \tag{6.2.2.19}$$

From Fig. 6.2.2.4 it should be clear that if f_1' is positive, then the element is in compression.

EXAMPLE 6.2.1

Element Stiffness Matrix

Figure E6.2.1(a) shows a planar truss. The material is steel, $E = 2(10)^{11}$ Pa and the cross-sectional area of both the members is 0.01 m^2. Construct the global element equilibrium equations for each element.

SOLUTION

This example serves to illustrate the first three steps to solving for the response of a planar truss using the direct stiffness method.

Step 1: We select N, m as the problem units. The origin of the global coordinate system will be located at point A. The labeled model is shown in Figs. E6.2.1(b) and (c).

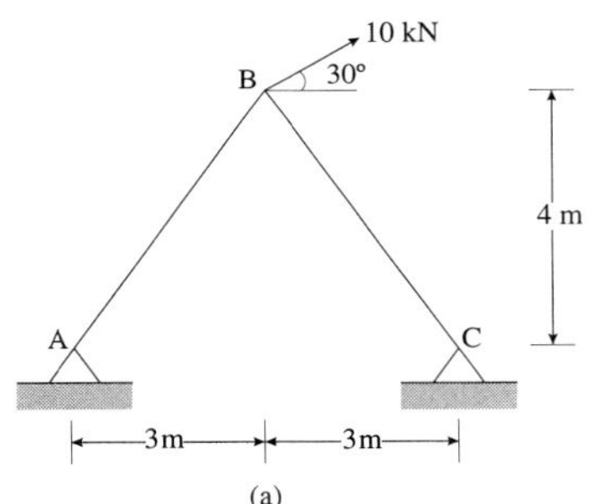

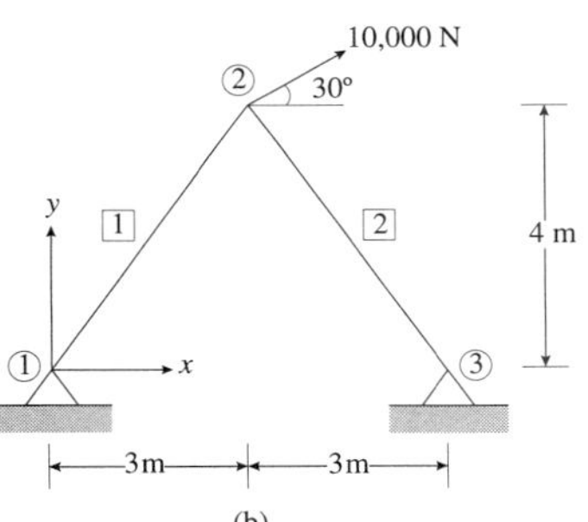

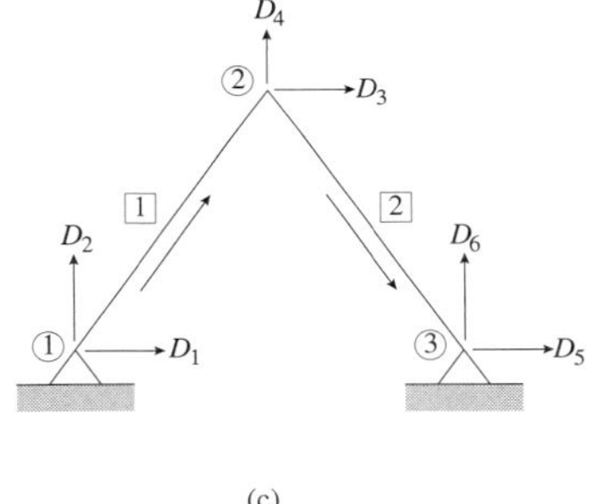

Fig. E6.2.1(a-c)

It helps to create the following table that contains the terms in the global element stiffness matrix.

Member	(x_1, y_1)	(x_2, y_2)	L	l	m	AE/L
1	(0,0)	(3,4)	5	0.6	0.8	$4(10^8)$
2	(3,4)	(6,0)	5	0.6	–0.8	$4(10^8)$

Step 2: We now create the two global element stiffness matrices. For Element 1, we have

$$\mathbf{k}_{4\times4} = \frac{AE}{L}\begin{bmatrix} l^2 & lm & -l^2 & -lm \\ lm & m^2 & -lm & -m^2 \\ -l^2 & -lm & l^2 & lm \\ -lm & -m^2 & lm & m^2 \end{bmatrix} = 4(10^8)\begin{bmatrix} 0.36 & 0.48 & -0.36 & -0.48 \\ 0.48 & 0.64 & -0.48 & -0.64 \\ -0.36 & -0.48 & 0.36 & 0.48 \\ -0.48 & -0.64 & 0.48 & 0.64 \end{bmatrix}$$

Similarly, for Element 2 we have

$$\mathbf{k}_{4\times4} = \frac{AE}{L}\begin{bmatrix} l^2 & lm & -l^2 & -lm \\ lm & m^2 & -lm & -m^2 \\ -l^2 & -lm & l^2 & lm \\ -lm & -m^2 & lm & m^2 \end{bmatrix} = 4(10^8)\begin{bmatrix} 0.36 & -0.48 & -0.36 & 0.48 \\ -0.48 & 0.64 & 0.48 & -0.64 \\ -0.36 & 0.48 & 0.36 & -0.48 \\ 0.48 & -0.64 & -0.48 & 0.64 \end{bmatrix}$$

Finally, the global element equilibrium equations for the two elements can be written as follows.

Element 1:

$$4(10^8)\begin{bmatrix} 0.36 & 0.48 & -0.36 & -0.48 \\ 0.48 & 0.64 & -0.48 & -0.64 \\ -0.36 & -0.48 & 0.36 & 0.48 \\ -0.48 & -0.64 & 0.48 & 0.64 \end{bmatrix}\begin{Bmatrix} D_1 \\ D_2 \\ D_3 \\ D_4 \end{Bmatrix} = \begin{Bmatrix} f_1^1 \\ f_2^1 \\ f_3^1 \\ f_4^1 \end{Bmatrix}$$

Element 2:

$$4(10^8)\begin{bmatrix} 0.36 & -0.48 & -0.36 & 0.48 \\ -0.48 & 0.64 & 0.48 & -0.64 \\ -0.36 & 0.48 & 0.36 & -0.48 \\ 0.48 & -0.64 & -0.48 & 0.64 \end{bmatrix}\begin{Bmatrix} D_3 \\ D_4 \\ D_5 \\ D_6 \end{Bmatrix} = \begin{Bmatrix} f_1^2 \\ f_2^2 \\ f_3^2 \\ f_4^2 \end{Bmatrix}$$

EXAMPLE 6.2.2 ***Structural or System Stiffness Matrix***

For the truss shown in Example 6.2.1, construct the system equilibrium equations.

SOLUTION Using the eight equations from the previous example, we need to construct the six equations that describe the equilibrium of the entire truss. We carry out this task sequentially starting with element 1. The system equations now appear as follows:

$$4(10^8)\begin{bmatrix} 0.36 & 0.48 & -0.36 & -0.48 & 0 & 0 \\ 0.48 & 0.64 & -0.48 & -0.64 & 0 & 0 \\ -0.36 & -0.48 & 0.36 & 0.48 & 0 & 0 \\ -0.48 & -0.64 & 0.48 & 0.64 & 0 & 0 \\ 0 & 0 & 0 & 0 & 0 & 0 \\ 0 & 0 & 0 & 0 & 0 & 0 \end{bmatrix}\begin{Bmatrix} D_1 \\ D_2 \\ D_3 \\ D_4 \\ D_5 \\ D_6 \end{Bmatrix} = \begin{Bmatrix} f_1^1 \\ f_2^1 \\ f_3^1 \\ f_4^1 \\ 0 \\ 0 \end{Bmatrix}$$

Now using the equations from element 2, we get

$$4(10^8)\begin{bmatrix} 0.36 & 0.48 & -0.36 & -0.48 & 0 & 0 \\ 0.48 & 0.64 & -0.48 & -0.64 & 0 & 0 \\ -0.36 & -0.48 & \boxed{0.36+0.36} & \boxed{0.48-0.48} & -0.36 & 0.48 \\ -0.48 & -0.64 & \boxed{0.48-0.48} & \boxed{0.64+0.64} & 0.48 & -0.64 \\ 0 & 0 & -0.36 & 0.48 & 0.36 & -0.48 \\ 0 & 0 & 0.48 & -0.64 & -0.48 & 0.64 \end{bmatrix}\begin{Bmatrix} D_1 \\ D_2 \\ D_3 \\ D_4 \\ D_5 \\ D_6 \end{Bmatrix} = \begin{Bmatrix} f_1^1 \\ f_2^1 \\ f_3^1 + f_1^2 \\ f_4^1 + f_2^2 \\ f_3^2 \\ f_4^2 \end{Bmatrix}$$

The elements within the rectangle show the coupling between the two elements. Simplifying and noting that the components of the applied load at node 2 are 10000 cos (30°) and 10000 sin (30°), we have the system equations as

$$4(10^8)\begin{bmatrix} 0.36 & 0.48 & -0.36 & -0.48 & 0 & 0 \\ 0.48 & 0.64 & -0.48 & -0.64 & 0 & 0 \\ -0.36 & -0.48 & 0.72 & 0 & -0.36 & 0.48 \\ -0.48 & -0.64 & 0 & 1.28 & 0.48 & -0.64 \\ 0 & 0 & -0.36 & 0.48 & 0.36 & -0.48 \\ 0 & 0 & 0.48 & -0.64 & -0.48 & 0.64 \end{bmatrix}\begin{Bmatrix} D_1 \\ D_2 \\ D_3 \\ D_4 \\ D_5 \\ D_6 \end{Bmatrix} = \begin{Bmatrix} F_1 \\ F_2 \\ 8660 \\ 5000 \\ F_5 \\ F_6 \end{Bmatrix}$$

or

$$\mathbf{K}_{6\times6}\mathbf{D}_{6\times1} = \mathbf{F}_{6\times1}$$

We can carry out a few checks to ensure that the results are acceptable. First, the structural stiffness matrix **K** should be symmetric. Second, **K** is *usually* diagonally dominant, meaning that the diagonal element has the largest magnitude in that row and column. As we can see in the equations above, this condition is met by most but not all diagonal elements. However, the largest number is a diagonal element, K_{44}. Note also that all diagonal elements are positive.

EXAMPLE 6.2.3 ***Planar Truss Analysis***

Solve for the nodal displacements and the element forces for the truss in Example 6.2.1.

SOLUTION

We need to implement steps 5, 6, and 7 from the general procedure (see page 362).

Step 5: The boundary conditions for this problem are $D_1 = D_2 = D_5 = D_6 = 0$. Imposing these conditions on the system equations yields the modified system equations:

$$4(10^8)\begin{bmatrix} 0.72 & 0 \\ 0 & 1.28 \end{bmatrix}\begin{Bmatrix} D_3 \\ D_4 \end{Bmatrix} = \begin{Bmatrix} 8660 \\ 5000 \end{Bmatrix}$$

or $\mathbf{K}_{2\times2}\mathbf{D}_{2\times1} = \mathbf{F}_{2\times1}$

We can carry out a few checks, as in the previous problem, to ensure that the equations are acceptable. First, the structural stiffness matrix **K** should be symmetric. Second, **K** is *usually* diagonally dominant, meaning that the diagonal element has the largest magnitude in that row and column. This condition is met by both the equations above. For a structure to be stable, all diagonal elements must be greater than zero.[5]

Step 6: Solving the modified system equations, $D_3 = 3.00694(10^{-5})$ m and $D_4 = 9.76563(10^{-6})$ m. Node 2 moves to the right and up since both the displacements are positive.

Step 7: Using the nodal displacements obtained in Step 6, we can compute the member axial forces (see Eq. (6.2.2.19)). For element 1, we have

$$f_1' = 4(10^8)[0.6 \quad 0.8 \quad -0.6 \quad -0.8][\,0 \quad 0 \quad D_3 \quad D_4\,]^{\mathrm{T}} = -10341.7 \text{ N}$$

The member is in tension. For element 2, we have

$$f_1' = 4(10^8)[0.6 \quad -0.8 \quad -0.6 \quad 0.8][D_3 \quad D_4 \quad 0 \quad 0\,]^{\mathrm{T}} = 4091.65 \text{ N}$$

The member is in compression.

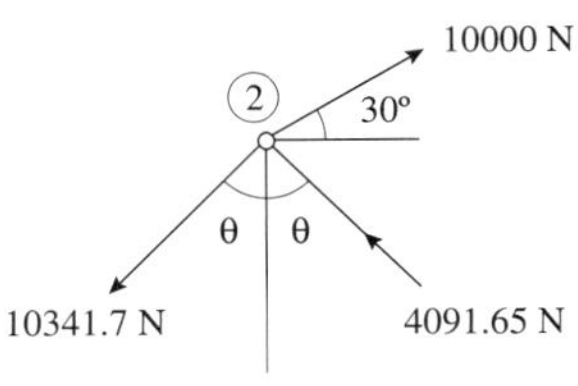

Fig. E6.2.3

We can now check the solution by drawing the FBD of node 2 and verifying that joint equilibrium is satisfied; see Fig. E6.2.3.

$$\overset{\rightarrow+}{\sum} F_x = -10341.7\sin(\theta) - 4091.65\sin(\theta) + 10000\cos(30°) = 0. \qquad \text{OK.}$$

$$\overset{\uparrow+}{\sum} F_x = -10341.7\cos(\theta) + 4091.65\cos(\theta) + 10000\sin(30°) = 0. \qquad \text{OK.}$$

[5] This is a necessary but not sufficient condition. In other words, even if all diagonal elements are positive, the structure could still be unstable. As we see later with the finite element method, the structural stiffness matrix must be symmetric and positive definite.

EXAMPLE 6.2.4 ***A More Efficient Solution Process***

Solve for the nodal displacements and member forces for the truss in Fig. E6.2.4(a). The modulus of elasticity is $30(10^6)$ psi and the cross-sectional area of each member is 1.2 in^2.

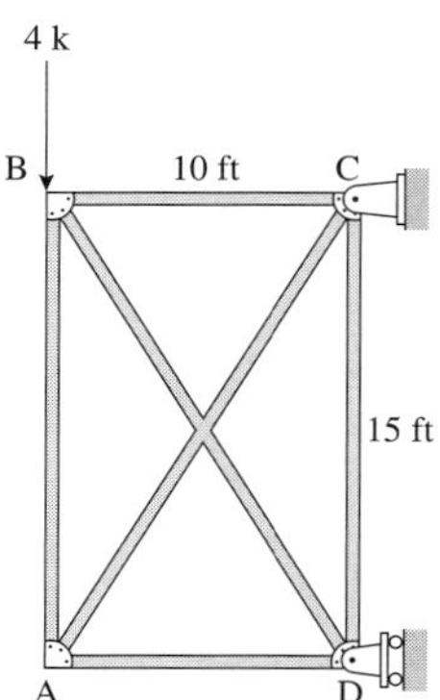

Fig. E6.2.4(a)

SOLUTION

Step 1: The model details are shown in Fig. E6.2.4(b). The problem units are lb, in.

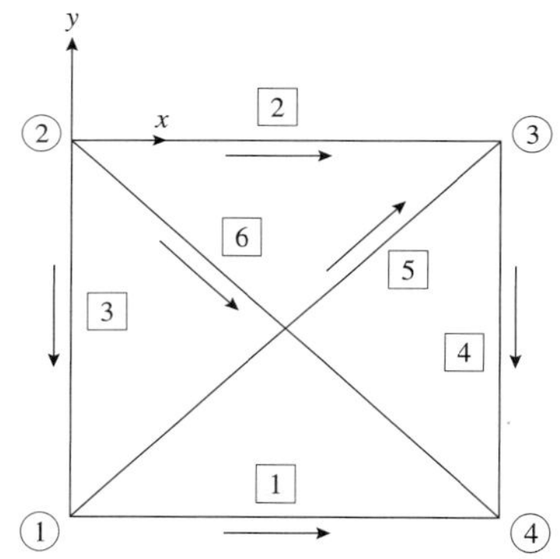

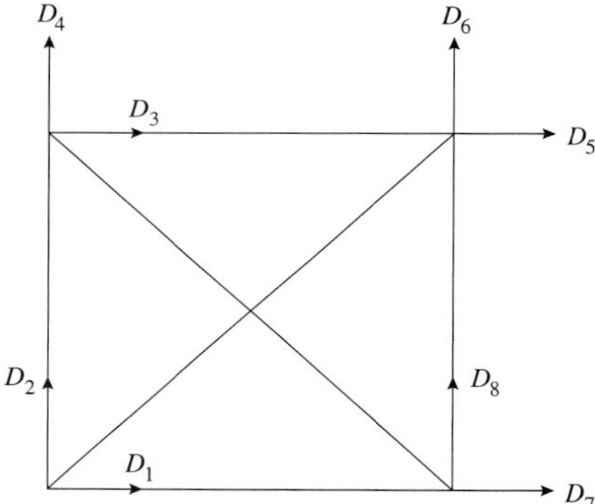

Fig. E6.2.4(b)

One approach to increasing the efficiency of the overall procedure is to construct only the effective equations. In other words, with respect to the current problem, we will not generate the global element and system equations corresponding to D_5, D_6, and D_7 degrees of freedom (since they are zeros). We first construct the table of element-related values.

Member	(x_1, y_1)	(x_2, y_2)	L	l	m	AE/L
1	(0,–180)	(120,–180)	120	1	0	$3(10^5)$
2	(0,0)	(120,0)	120	1	0	$3(10^5)$
3	(0,0)	(0,–180)	180	0	–1	$2(10^5)$
4	(120,0)	(120,–180)	180	0	–1	$2(10^5)$
5	(0,–180)	(120,0)	216.333	0.5547	0.832051	$1.664(10^5)$
6	(0,0)	(120,–180)	216.333	0.5547	–0.832051	$1.664(10^5)$

Step 2: Now we can construct the global element equations using the global element stiffness matrix:

$$\mathbf{k}_{4\times 4} = \frac{AE}{L}\begin{bmatrix} l^2 & lm & -l^2 & -lm \\ lm & m^2 & -lm & -m^2 \\ -l^2 & -lm & l^2 & lm \\ -lm & -m^2 & lm & m^2 \end{bmatrix}$$

We start with element 1 and immediately note that we need not construct the third equation (dealing with D_7) or the terms dealing with it (the third column). At these locations we place an × symbol. Note also that we have placed the appropriate column numbers on top of the four columns.

$$\begin{array}{c} \boxed{1 \quad 2 \quad 7 \quad 8} \\ 10^5\begin{bmatrix} 3 & 0 & \times & 0 \\ 0 & 0 & \times & 0 \\ \times & \times & \times & \times \\ 0 & 0 & \times & 0 \end{bmatrix}\begin{Bmatrix} D_1 \\ D_2 \\ D_7 \\ D_8 \end{Bmatrix} \end{array}$$

Similarly, we can construct the other element equations.

Element 2:

$$\begin{array}{c} \boxed{3 \quad 4 \quad 5 \quad 6} \\ 10^5\begin{bmatrix} 3 & 0 & \times & \times \\ 0 & 0 & \times & \times \\ \times & \times & \times & \times \\ \times & \times & \times & \times \end{bmatrix}\begin{Bmatrix} D_3 \\ D_4 \\ D_5 \\ D_6 \end{Bmatrix} \end{array}$$

Element 3:

$$\begin{array}{c} \boxed{3 \quad 4 \quad 1 \quad 2} \\ 10^5\begin{bmatrix} 0 & 0 & 0 & 0 \\ 0 & 2 & 0 & -2 \\ 0 & 0 & 0 & 0 \\ 0 & \boxed{-2} & 0 & 2 \end{bmatrix}\begin{Bmatrix} D_3 \\ D_4 \\ D_1 \\ D_2 \end{Bmatrix} \end{array}$$

Element 4:

$$\begin{array}{c} \boxed{5 \quad 6 \quad 7 \quad 8} \\ 10^5\begin{bmatrix} \times & \times & \times & \times \\ \times & \times & \times & \times \\ \times & \times & \times & \times \\ \times & \times & \times & 2 \end{bmatrix}\begin{Bmatrix} D_5 \\ D_6 \\ D_7 \\ D_8 \end{Bmatrix} \end{array}$$

Element 5:

$$\begin{array}{c} \boxed{1 \quad 2 \quad 5 \quad 6} \\ 10^5\begin{bmatrix} 0.5120 & 0.7679 & \times & \times \\ 0.7679 & 1.1520 & \times & \times \\ \times & \times & \times & \times \\ \times & \times & \times & \times \end{bmatrix}\begin{Bmatrix} D_1 \\ D_2 \\ D_5 \\ D_6 \end{Bmatrix} \end{array}$$

Element 6:

$$\begin{array}{c} \boxed{3 \quad 4 \quad 7 \quad 8} \\ 10^5\begin{bmatrix} 0.5120 & \boxed{-0.7679} & \times & 0.7679 \\ -0.7679 & 1.1520 & \times & -1.1520 \\ \times & \times & \times & \times \\ 0.7679 & -1.1520 & \times & 1.1520 \end{bmatrix}\begin{Bmatrix} D_3 \\ D_4 \\ D_7 \\ D_8 \end{Bmatrix} \end{array}$$

We explain the significance of the column numbers. The row numbers of the elements of the stiffness matrix are associated with the global displacement associated with that row. For example, for element 6, the first equation deals with D_3 or the third row in the system equations. The column number will help us carry out the assembly process in an automated manner. For example, in element 3, the boxed number, –2, will be placed at row 2 and column 4 of the structural or system stiffness matrix. Similarly, the boxed element in 6 is associated with row 3 and column 4 of the system stiffness matrix.

Step 3: Using the element equations we can construct the system equations. We show the successive snapshots of the system stiffness matrix as we proceed from elements 1 through 6 to assemble the matrix.

$$\textbf{\textit{Element 1:}}\ (10)^5\begin{bmatrix}3&0&0&0&0\\0&0&0&0&0\\0&0&0&0&0\\0&0&0&0&0\\0&0&0&0&0\end{bmatrix}\begin{Bmatrix}D_1\\D_2\\D_3\\D_4\\D_8\end{Bmatrix}\Rightarrow \textbf{\textit{Element 2:}}\ (10)^5\begin{bmatrix}3&0&0&0&0\\0&0&0&0&0\\0&0&3&0&0\\0&0&0&0&0\\0&0&0&0&0\end{bmatrix}\begin{Bmatrix}D_1\\D_2\\D_3\\D_4\\D_8\end{Bmatrix}\Rightarrow$$

$$\textbf{\textit{Element 3:}}\ (10)^5\begin{bmatrix}3&0&0&0&0\\0&2&0&-2&0\\0&0&3&0&0\\0&-2&0&2&0\\0&0&0&0&0\end{bmatrix}\begin{Bmatrix}D_1\\D_2\\D_3\\D_4\\D_8\end{Bmatrix}\Rightarrow$$

$$\textbf{\textit{Element 4:}}\ (10)^5\begin{bmatrix}3&0&0&0&0\\0&2&0&-2&0\\0&0&3&0&0\\0&-2&0&2&0\\0&0&0&0&2\end{bmatrix}\begin{Bmatrix}D_1\\D_2\\D_3\\D_4\\D_8\end{Bmatrix}\Rightarrow$$

$$\textbf{\textit{Element 5:}}\ (10)^5\begin{bmatrix}3.5120&0.7679&0&0&0\\0.7679&3.1520&0&-2&0\\0&0&3&0&0\\0&-2&0&2&0\\0&0&0&0&2\end{bmatrix}\begin{Bmatrix}D_1\\D_2\\D_3\\D_4\\D_8\end{Bmatrix}$$

$$\textbf{\textit{Element 6:}}\ (10)^5\begin{bmatrix}3.5120&0.7679&0&0&0\\0.7679&3.1520&0&-2&0\\0&0&3.5120&-0.7679&0.7679\\0&-2&-0.7679&3.1520&-1.1520\\0&0&0.7679&-1.1520&3.1520\end{bmatrix}\begin{Bmatrix}D_1\\D_2\\D_3\\D_4\\D_8\end{Bmatrix}=\begin{Bmatrix}0\\0\\0\\-4000\\0\end{Bmatrix}$$

Step 4: With this modified procedure, there is no need to impose the boundary conditions at the system level since we have already imposed them at the element level. Solving the above equations, we obtain

$$\{D_1, D_2, D_3, D_4, D_8\} = 10^{-3}\{4.44367, -20.3232, 4.44367, -30.3232, -10\} \text{ in}$$

Step 5: We are finally ready to compute the member forces.

Element 1:

$f_1' = 3(10^5)[1 \quad 0 \quad -1 \quad 0][D_1 \quad D_2 \quad D_7 \quad D_8]^T = 1333$ lb . Member is in compression.

Element 2:

$f_2' = 3(10^5)[1 \quad 0 \quad -1 \quad 0][D_3 \quad D_4 \quad D_5 \quad D_6]^T = -1333$ lb . Member is in tension.

Element 3:

$f_3' = 2(10^5)[0 \quad -1 \quad 0 \quad 1][D_3 \quad D_4 \quad D_1 \quad D_2]^T = 2000$ lb . Member is in compression.

Element 4:

$f_4' = 2\left(10^5\right)\left[0 \quad -1 \quad 0 \quad 1\right]\left[D_5 \quad D_6 \quad D_7 \quad D_8\right]^{\mathbf{T}} = -2000 \text{ lb}$. Member is in tension.

Element 5:

$f_5' = 1.664\left(10^5\right)\left[0.5547 \quad 0.832051 \quad -0.5547 \quad -0.832051\right]\left[D_1 \quad D_2 \quad D_5 \quad D_6\right]^{\mathbf{T}}$

$= -2404$ lb. Member is in tension.

Element 6:

$f_6' = 1.664\left(10^5\right)\left[0.5547 \quad -0.832051 \quad -0.5547 \quad 0.832051\right]\left[D_3 \quad D_4 \quad D_7 \quad D_8\right]^{\mathbf{T}}$

$= 2404$ lb. Member is in compression.

Step 6: We can carry out a few equilibrium checks to ensure that out solution is correct. The FBDs of joint A or 1 and joint B or 2 are shown in Fig. E6.2.4(c). Note that $\theta = \tan^{-1}(15/10) = 56.31°$.

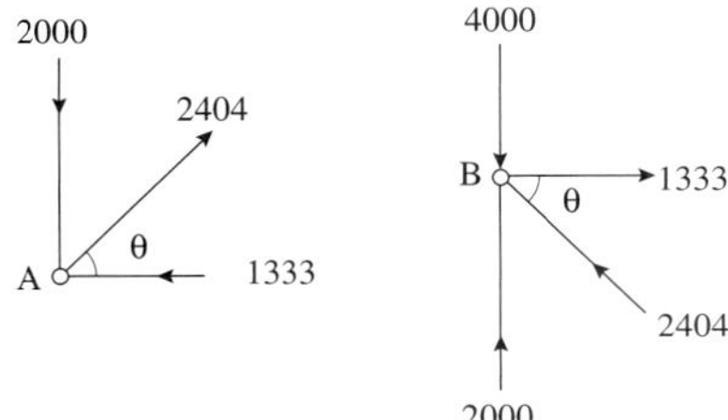

Fig. E6.2.4(c)

$$\text{Joint A:} \quad \overset{\rightarrow +}{\sum} F_x = -1333 + 2404(\cos\theta) = 0 \qquad \text{OK.}$$

$$\overset{\uparrow +}{\sum} F_y = -2000 + 2404(\sin\theta) = 0 \qquad \text{OK.}$$

$$\text{Joint B:} \quad \overset{\rightarrow +}{\sum} F_x = 1333 - 2404(\cos\theta) = 0 \qquad \text{OK.}$$

$$\overset{\uparrow +}{\sum} F_y = -4000 + 2000 + 2404(\sin\theta) = 0 \qquad \text{OK.}$$

Observation: One can start appreciating the power of the direct stiffness method. We have solved for all the nodal displacements and member forces in the truss simultaneously. The truss is statically indeterminate to degree one. Solving for the member forces using the force method would require analysis of two determinate trusses. But solving for all the nodal displacements will require additional analysis of five indeterminate trusses!

Computing the Support Reactions

There are at least two different approaches to computing the support reactions. The force vector **F** in the system equilibrium equations **KD** = **F** represents the net forces along each degree of freedom. Some of these are externally applied forces and others are the support reactions. Hence one could pick the rows from **K** that correspond to the support reactions and multiply the rows with the displacement vector. With respect to the previous example, it would be necessary to construct rows 5, 6, and 7 of $\mathbf{K}_{8\times8}$ and multiply these rows with the displacement vector $\mathbf{D}_{8\times1}$. The three dot products yield the three support reactions.

The other approach is to use the computed member nodal forces to generate the support reactions vector **R** that is initially set to zero. After using Eq. (6.2.2.18) to compute the member nodal forces **f′** for each element, we can update **R** as[6]

[6] The equation is not dimensionally correct but serves to illustrate the basic idea. The correct equation would be $\mathbf{R} = \mathbf{R} + \mathbf{A}\mathbf{T}^{\mathrm{T}}\mathbf{f}'$ where **A** is a Boolean matrix of the appropriate size containing 0s and 1s.

$$\mathbf{R} = \mathbf{R} + \mathbf{T}^{\mathbf{T}}\mathbf{f}' \tag{6.2.2.20}$$

using only the terms corresponding to the support reactions from the product $\mathbf{T}^{\mathbf{T}}\mathbf{f}'$.

Continuing with the previous analysis, the steps to compute the support reactions would yield the following $\left(\mathbf{R} = \{ R_5, R_6, R_7 \}\right)$.

Element 1:

$$\mathbf{T}^{\mathbf{T}}\mathbf{f}' = \{ 1333, 0, -1333, 0 \} \Rightarrow \mathbf{R} = \{ 0, 0, -1333 \}$$

Element 2:

$$\mathbf{T}^{\mathbf{T}}\mathbf{f}' = \{ -1333, 0, 1333, 0 \} \Rightarrow \mathbf{R} = \{ 1333, 0, -1333 \}$$

Element 3:

$$\mathbf{T}^{\mathbf{T}}\mathbf{f}' = \{ 0, -2000, 0, 2000 \} \Rightarrow \mathbf{R} = \{ 1333, 0, -1333 \}$$

Element 4:

$$\mathbf{T}^{\mathbf{T}}\mathbf{f}' = \{ 0, 2000, 0, -2000 \} \Rightarrow \mathbf{R} = \{ 1333, 2000, -1333 \}$$

Element 5:

$$\mathbf{T}^{\mathbf{T}}\mathbf{f}' = \{ -1333.67, -2000, 1333.5, 2000 \} \Rightarrow \mathbf{R} = \{ 2666.67, 4000, -1333 \}$$

Element 6:

$$\mathbf{T}^{\mathbf{T}}\mathbf{f}' = \{ 1333.67, -2000, -1333.5, 2000 \} \Rightarrow \mathbf{R} = \{ 2666.67, 4000, -2666.67 \}$$

The resulting FBD of the truss is shown in Fig. E6.2.4(d). A quick check shows that $\overset{\rightarrow +}{\sum} F_x = 0$, $\overset{\uparrow +}{\sum} F_y = 0$, and $\overset{\curvearrowleft +}{\sum} M = 0$.

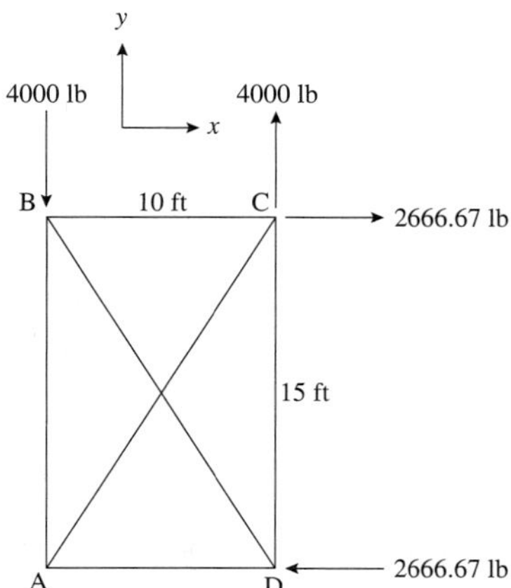

Fig. E6.2.4(d)

Space Truss Analysis

The basic behavior of a truss element was captured early on in this section and specifically by Eq. (6.2.2.5). The element behavior is captured in a local coordinate system that is usually different for different members of the truss. The transformation from local to global system can be carried out in a similar manner to the planar truss when the global coordinate system defines a three-dimensional space. Hence, extending the ideas from planar trusses to the analysis of space trusses involves increasing the dimensionality of the appropriate matrices. There are now three degrees of freedom per node—displacements in the global x, y, z directions, and six degrees of freedom per element, as seen in Fig. 6.2.2.5.

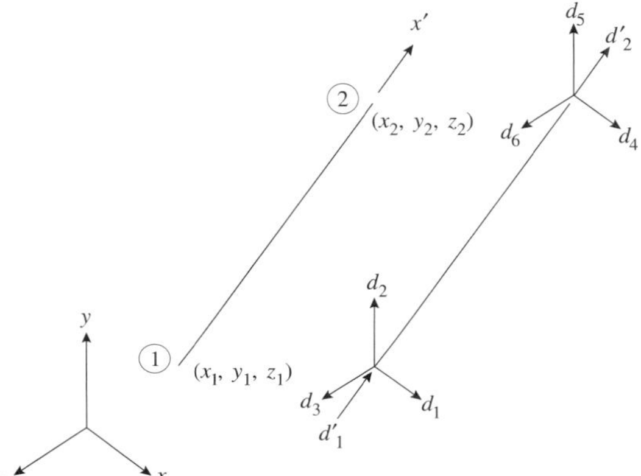

Fig. 6.2.2.5 Displacement global-to-local transformation.

The relevant equations are summarized below. The definitions of the matrices are the same as before.

$$\mathbf{k}'_{2\times2}\mathbf{d}'_{2\times1} = \mathbf{f}'_{2\times1} \tag{6.2.2.21}$$

$$\mathbf{d}'_{2\times1} = \mathbf{T}_{2\times6}\mathbf{d}_{6\times1} \tag{6.2.2.22}$$

$$\mathbf{f}_{6\times1} = \mathbf{T}^{\mathrm{T}}_{6\times2}\mathbf{f}'_{2\times1} \tag{6.2.2.23}$$

$$\mathbf{k}_{6\times6}\mathbf{d}_{6\times1} = \mathbf{f}_{6\times1} \tag{6.2.2.24}$$

where $\mathbf{k}_{6\times6} = \mathbf{T}^{\mathrm{T}}_{6\times2}\mathbf{k}'_{2\times2}\mathbf{T}_{2\times6}$ (6.2.2.25)

$$\mathbf{T}_{2\times6} = \begin{bmatrix} l & m & n & 0 & 0 & 0 \\ 0 & 0 & 0 & l & m & n \end{bmatrix} \tag{6.2.2.26}$$

$$L = \sqrt{(x_2 - x_1)^2 + (y_2 - y_1)^2 + (z_2 - z_1)^2} \tag{6.2.2.27a}$$

$$l = \frac{x_2 - x_1}{L};\ m = \frac{y_2 - y_1}{L};\ n = \frac{z_2 - z_1}{L} \tag{6.2.2.27b}$$

Finally, after the nodal displacements are computed from the solution of the system equations, the member force can be computed using

$$f'_1 = \frac{AE}{L}\begin{bmatrix} l & m & n & -l & -m & -n \end{bmatrix}\begin{Bmatrix} d_1 \\ d_2 \\ d_3 \\ d_4 \\ d_5 \\ d_6 \end{Bmatrix} \tag{6.2.2.28}$$

It should be evident that the major difference between the planar and space truss analysis is that the dimensions of the matrices are larger for space truss. The basic steps are exactly the same.

EXERCISES

Appetizers

For the following problems (a) compute the direction cosines of each element, (b) label the global degrees of freedom at the nodes, and (c) identify the known and unknown global degrees of freedom.

6.2.2. The truss members in Fig. P6.2.2 have the following properties: modulus of elasticity is 29000 ksi and the cross-sectional area is 2.0 in^2.

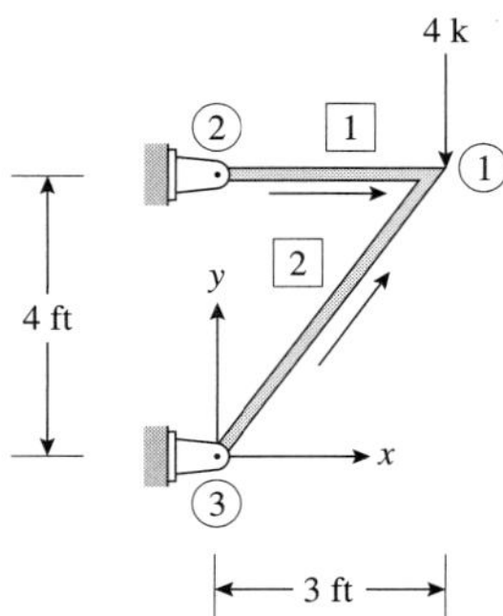

Fig. P6.2.2

6.2.3. All the truss members in Fig. P6.2.3 have the following properties: modulus of elasticity is 200 GPa, $\alpha = 10^{-5}/°C$, and the cross-sectional area is 0.01 m^2.

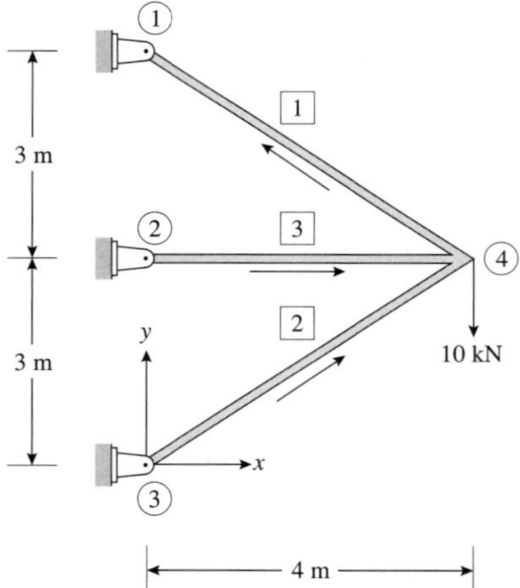

Fig. P6.2.3

6.2.4. The members in the truss in Fig. P6.2.4 have the following cross-sectional areas: $A_1 = A_2 = 0.01$ m^2, $A_6 = A_7 = 0.02$ m^2, and the rest of the members = 0.005 m^2. The material used is steel. Compute all the member forces.

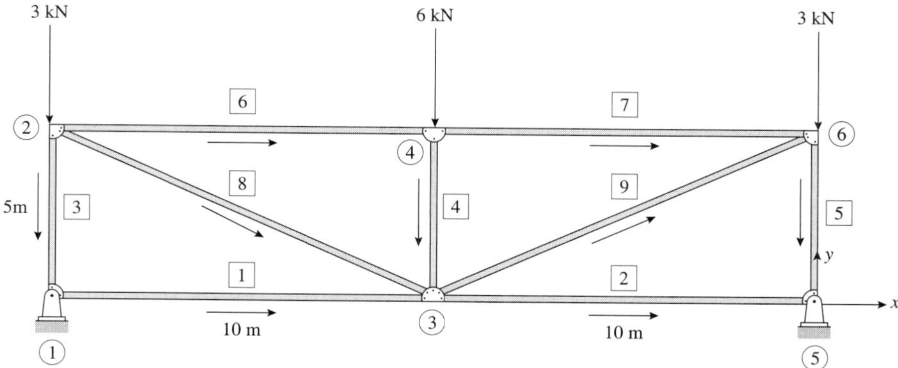

Fig. P6.2.4

6.2.5. The truss members in Fig. P6.2.5 have the following properties: modulus of elasticity is 2 GPa and the cross-sectional area of members 1 and 2 is 0.01 m^2 and of members 3, 4, and 5 is 0.02 m^2.

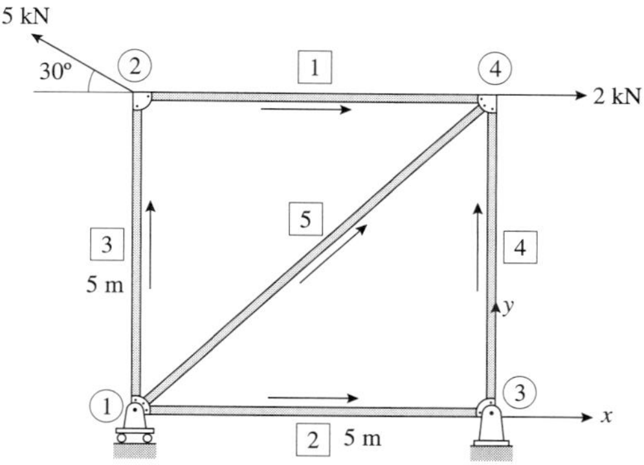

Fig. P6.2.5

Main Course

6.2.6. For the truss in Problem 6.2.2, compute the nodal displacements and the member forces.

6.2.7. For the truss in Problem 6.2.3, compute the nodal displacements and the member forces.

6.2.8. For the truss in Problem 6.2.4, compute the nodal displacements and the member forces.

6.2.9. For the truss in Problem 6.2.5, compute the nodal displacements and the member forces.

6.2.10. Solve Example 7.4.3.

Structural Concepts

6.2.11. Compute the support reactions of the truss structures in Problems 6.2.2 and 6.2.4. Check the equilibrium of the structure.

6.2.12. Compute the support reactions of the truss structure in Problem 6.2.9. Check the equilibrium of the structure.

6.2.3 Frame Analysis

A planar frame is a structural system that satisfies the following requirements:

a. The members are slender and prismatic. They can be straight or curved, vertical, horizontal, or inclined. The cross-sectional dimensions are small in comparison to the member lengths. Also when constructing the frame model, we treat the members as one-dimensional entities (having length and negligible cross-sectional dimensions).

b. The joints can be assumed to be rigid connection, frictionless pins (or internal hinges), or typical connections.

c. The loads can be concentrated forces or moments that act at joints or on the frame members, or distributed forces acting on the members.

In this section, however, we assume that the frame is made of straight members and that the connections are rigid. We develop the element capable of modeling a planar frame in two stages. In the first stage, the flexure effects (due to shear force and bending moments) will be considered. In the second stage, the axial effects will be considered. Using the superposition principle, we can then construct the behavior of a frame element. As before, the superposition is valid only if the displacements are small. In structural analysis terminology, members that are subjected primarily to flexural effects are said to be beams whereas members with combined axial-flexural effects are called beam-columns. By the end of this section we will have developed the element equations for the combined effects that can also be used to model pure beam behavior. *To avoid construction and usage of several different terminology, we refer to this element simply as the* ***beam*** element.

Consider a beam having transverse displacements and rotations as shown in Fig. 6.2.3.1. The dashed line shows the undeformed state of the element; the solid line shows the deformation due to flexural effects.

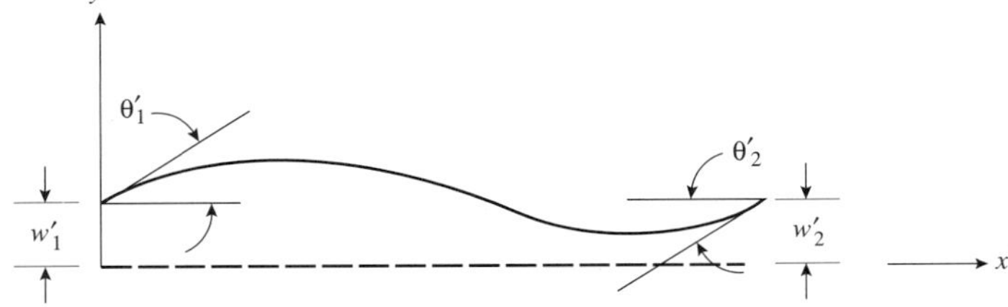

Fig. 6.2.3.1
Planar beam behavior.

There are four degrees of freedom: the transverse displacement w_1' and rotation θ_1' at the start node, and similarly the transverse displacement w_2' and rotation θ_2' at the end node. When the rotations are small, it is possible to use the approximation $\theta \approx \tan\theta$ and say that the rotation and slope are analogous. As we saw with the derivation of the spring and truss element equations, in the direct stiffness method, the stiffness coefficients are derived. To compute the stiffness coefficient k_{ij} while suppressing all degrees of freedom but j, we compute the force required along degree of freedom i to produce a unit displacement along degree of freedom j. Consider the stiffness coefficients associated with the transverse displacement, w_1'. Figure 6.2.3.2 shows the beam with all degrees of freedom restrained except w_1', which is set to unity. We saw in Chapter 5 (section 5.2) how to compute the resulting shear and bending moments at the two ends of the beam. They are shown in the figure.

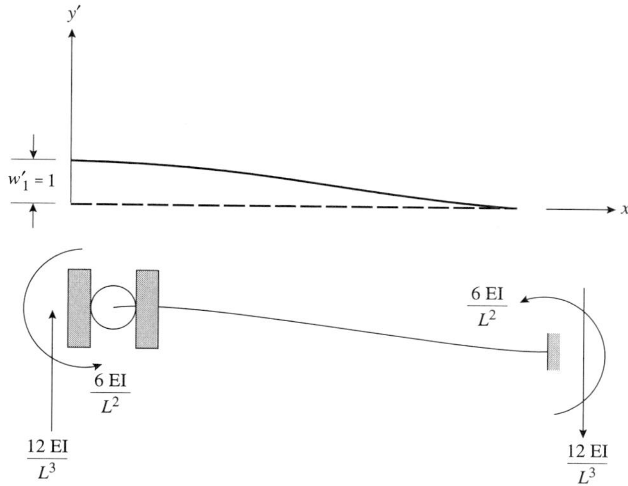

Fig. 6.2.3.2
Stiffness coefficients for transverse displacement w_1.

Hence, with the order of the dof as $\{w'_1, \theta'_1, w'_2, \theta'_2\}$, $k_{11} = 12EI/L^3$, $k_{21} = 6EI/L^2$, $k_{31} = -12EI/L^3$, and $k_{41} = 6EI/L^2$.

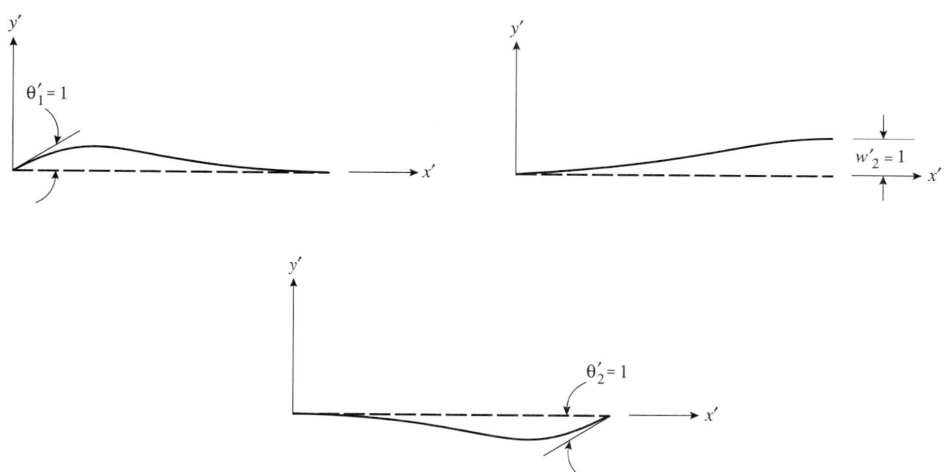

Fig. 6.2.3.3
Deformed shapes to compute the stiffness coefficients for the other three degrees of freedom.

In a similar manner, the stiffness coefficients associated with the other three degrees of freedom can be obtained (see Fig. 6.2.3.3). The element stiffness in the local coordinate system can be written as follows:

$$\mathbf{k}'_{4\times4} = \frac{EI}{L^3}\begin{bmatrix} 12 & 6L & -12 & 6L \\ 6L & 4L^2 & -6L & 2L^2 \\ -12 & -6L & 12 & -6L \\ 6L & 2L^2 & -6L & 4L^2 \end{bmatrix} \tag{6.2.3.1}$$

In a beam or a frame structure, a member may undergo axial deformation in addition to the transverse displacements and rotations. The six degrees of freedom for the complete beam element are shown in Fig. 6.2.3.4. The stiffness coefficients due to the axial effects were computed earlier with the truss element in Section 6.2.2 and given by Eq. (6.2.2.4). Superposing the two element stiffness matrices yields the element equations

$$\begin{bmatrix} \frac{AE}{L} & 0 & 0 & -\frac{AE}{L} & 0 & 0 \\ 0 & \frac{12EI}{L^3} & \frac{6EI}{L^2} & 0 & -\frac{12EI}{L^3} & \frac{6EI}{L^2} \\ 0 & \frac{6EI}{L^2} & \frac{4EI}{L} & 0 & -\frac{6EI}{L^2} & \frac{2EI}{L} \\ -\frac{AE}{L} & 0 & 0 & \frac{AE}{L} & 0 & 0 \\ 0 & -\frac{12EI}{L^3} & -\frac{6EI}{L^2} & 0 & \frac{12EI}{L^3} & -\frac{6EI}{L^2} \\ 0 & \frac{6EI}{L^2} & \frac{2EI}{L} & 0 & -\frac{6EI}{L^2} & \frac{4EI}{L} \end{bmatrix} \begin{Bmatrix} u_1' \\ w_1' \\ \theta_1' \\ u_2' \\ w_2' \\ \theta_2' \end{Bmatrix} = \begin{Bmatrix} f_1' \\ f_2' \\ f_3' \\ f_4' \\ f_5' \\ f_6' \end{Bmatrix} \quad (6.2.3.2a)$$

or

$$\mathbf{k}'_{6\times 6}\mathbf{d}'_{6\times 1} = \mathbf{f}'_{6\times 1} \quad (6.2.3.2b)$$

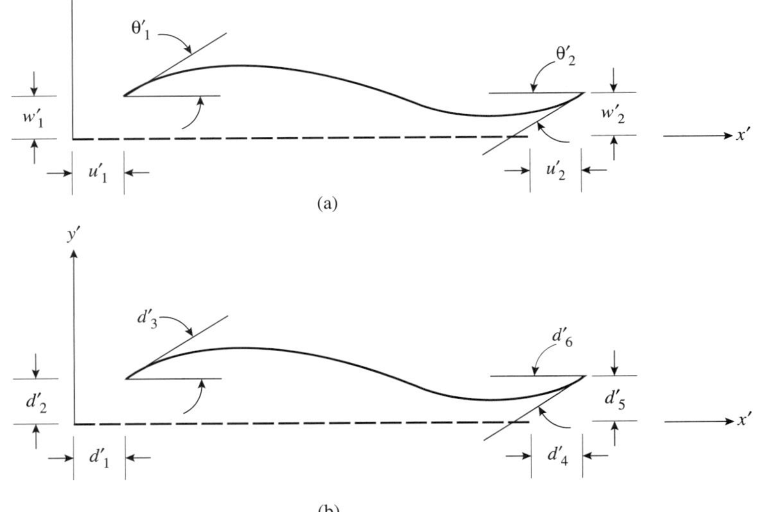

Fig. 6.2.3.4 The complete beam element with axial and flexural effects: (a) displacement components at the two nodes, and (b) the six local degrees of freedom.

While Eq. (6.2.3.2) represents the equilibrium compatibility of the beam element in the local coordinate system, we still need to generate the element equations in the global coordinate system as we did for the truss element. Figure 6.2.3.5 shows the local and global coordinate systems and the degrees of freedom at the two nodes of the element. This planar element is assumed to lie in the x'–y' and x–y plane. In other words, the z' and z axes are the same.

The local coordinate system is defined as follows. The (positive) x' axis is from node 1 to node 2 of the element. The z' axis points away from the page towards the viewer. Noting that the cross product of the x' axis and the y' axis defines the z' axis (for a right-handed coordinate system), we have the y' axis defined by taking the cross product of the z' axis with the x' axis. The global coordinate system is such that the beam element lies in the x–y plane with the z axis pointing out of the page.

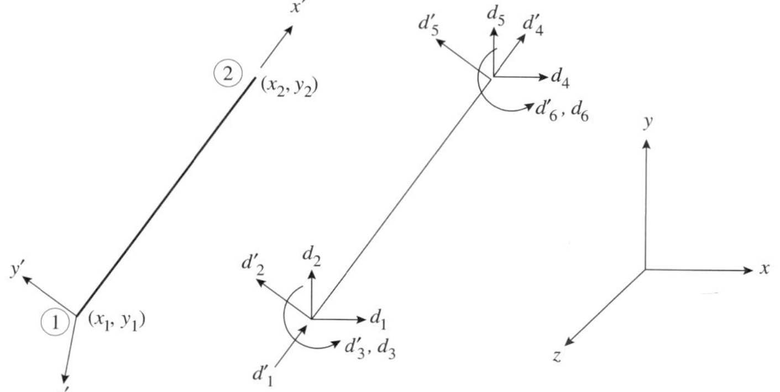

Fig. 6.2.3.5 Beam element's local and global coordinate systems and degrees of freedom. z' and z axes point out of the page.

First we construct the direction cosines of the local axes. Let $(l_{x'}, m_{x'}, n_{x'})$ be the direction cosines of the x'-axis and $(l_{y'}, m_{y'}, n_{y'})$ be the direction cosines of the y'-axis in terms of the global coordinate system. Hence

$$L = \sqrt{(x_2 - x_1)^2 + (y_2 - y_1)^2};\ l_{x'} = \frac{x_2 - x_1}{L};\ m_{x'} = \frac{y_2 - y_1}{L};\ n_{x'} = 0 \tag{6.2.3.3a}$$

The direction cosines of the local z' axis are such that $l_{z'} = 0$, $m_{z'} = 0$, and $n_{z'} = 1$. Since we have the y' axis defined by taking the cross product of the z' axis with the x' axis, we can construct the direction cosines of the y' axis as

$$(0i + 0j + 1k) \times (l_{x'}i + m_{x'}j + 0k) = (-m_{x'}i + l_{x'}j + 0k) \tag{6.2.3.3b}$$

Or, we can simply say that the direction cosines of the x'-axis are (l, m) and those of the y' axis are $(-m, l)$. We can ignore the third component, which is zero, for planar beams.

With respect to the two displacements and rotation at the first node, the relationship between the local and the global displacements can be expressed as

$$d'_1 = ld_1 + md_2 \tag{6.2.3.4a}$$

$$d'_2 = -md_1 + ld_2 \tag{6.2.3.4b}$$

$$d'_3 = d_3 \tag{6.2.3.4c}$$

Similarly, at the second node of the element we have

$$d'_4 = ld_4 + md_5 \tag{6.2.3.4d}$$

$$d'_5 = -md_4 + ld_5 \tag{6.2.3.4e}$$

$$d'_6 = d_6 \tag{6.2.3.4f}$$

These six equations can be written in the matrix form as

$$\begin{Bmatrix} d'_1 \\ d'_2 \\ d'_3 \\ d'_4 \\ d'_5 \\ d'_6 \end{Bmatrix} = \begin{bmatrix} l & m & 0 & 0 & 0 & 0 \\ -m & l & 0 & 0 & 0 & 0 \\ 0 & 0 & 1 & 0 & 0 & 0 \\ 0 & 0 & 0 & l & m & 0 \\ 0 & 0 & 0 & -m & l & 0 \\ 0 & 0 & 0 & 0 & 0 & 1 \end{bmatrix} \begin{Bmatrix} d_1 \\ d_2 \\ d_3 \\ d_4 \\ d_5 \\ d_6 \end{Bmatrix} \tag{6.2.3.5a}$$

or

$$\mathbf{d}'_{6\times1} = \mathbf{T}_{6\times6}\mathbf{d}_{6\times1} \tag{6.2.3.5b}$$

In a similar we can construct the relationship between the local nodal forces and the global nodal forces. Figure 6.2.3.6(a)–(b) shows the local and global components at the two nodes, and Figure 6.2.3.6(c) shows the equivalence between f_1, the global x-force at node 1, and the two local force components, f_1' and f_2'. The other relationships can be formed similarly. Hence,

$$f_1 = lf_1' - mf_2' \tag{6.2.3.6a}$$

$$f_2 = mf_1' + lf_2' \tag{6.2.3.6b}$$

$$f_3 = f_3' \tag{6.2.3.6c}$$

Similarly, at the second node of the element

$$f_4 = lf_4' - mf_5' \tag{6.2.3.6d}$$

$$f_5 = mf_4' + lf_5' \tag{6.2.3.6e}$$

$$f_6 = f_6' \tag{6.2.3.6f}$$

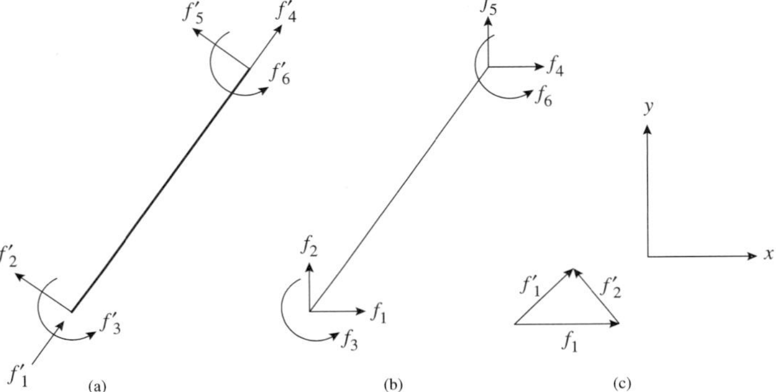

Fig. 6.2.3.6 Beam elements: (a) local, (b) global nodal forces, and (c) force components.

The three forces in the local coordinate system at each node can be identified as follows: f_1', f_4' are the axial forces, f_2', f_5' are the shear forces, and f_3', f_6' are the bending moments. These six equations can be written in matrix form as

$$\begin{Bmatrix} f_1 \\ f_2 \\ f_3 \\ f_4 \\ f_5 \\ f_6 \end{Bmatrix} = \begin{bmatrix} l & -m & 0 & 0 & 0 & 0 \\ m & l & 0 & 0 & 0 & 0 \\ 0 & 0 & 1 & 0 & 0 & 0 \\ 0 & 0 & 0 & l & -m & 0 \\ 0 & 0 & 0 & m & l & 0 \\ 0 & 0 & 0 & 0 & 0 & 1 \end{bmatrix} \begin{Bmatrix} f_1' \\ f_2' \\ f_3' \\ f_4' \\ f_5' \\ f_6' \end{Bmatrix} \tag{6.2.3.7a}$$

or

$$\mathbf{f}_{6\times1} = \mathbf{T}^{\mathbf{T}}_{6\times6}\mathbf{f}'_{6\times1} \tag{6.2.3.7b}$$

We are now ready to generate the element equations in the global coordinate system. We start with Eq. (6.2.3.2b). Substituting Eq. (6.2.3.5b) for $\mathbf{d}'_{6\times1}$, we have

$$\mathbf{k}'_{6\times6}\mathbf{T}_{6\times6}\mathbf{d}_{6\times1} = \mathbf{f}'_{6\times1} \tag{6.2.3.8}$$

Premultiplying both sides by $\mathbf{T}^{\mathbf{T}}_{6\times6}$ yields

$$\mathbf{T}^{\mathbf{T}}_{6\times6}\mathbf{k}'_{6\times6}\mathbf{T}_{6\times6}\mathbf{d}_{6\times1} = \mathbf{T}^{\mathbf{T}}_{6\times6}\mathbf{f}'_{6\times1} \tag{6.2.3.9}$$

Using Eq. (6.2.3.7b), the above equation can be rewritten as

$$\mathbf{T}^{\mathbf{T}}_{6\times6}\mathbf{k}'_{6\times6}\mathbf{T}_{6\times6}\mathbf{d}_{6\times1} = \mathbf{f}_{6\times1} \tag{6.2.3.10}$$

or

$$\mathbf{k}_{6\times6}\mathbf{d}_{6\times1} = \mathbf{f}_{6\times1} \tag{6.2.3.11}$$

where $\mathbf{k}_{6\times6} = \mathbf{T}^{\mathbf{T}}_{6\times6}\mathbf{k}'_{6\times6}\mathbf{T}_{6\times6}$ is the global element stiffness matrix. These are the element equilibrium-compatibility equations in the global coordinate system. The global element stiffness matrix can be computed by multiplying the three matrices to yield the following result: with

$$a = \frac{AE}{L} \qquad b = \frac{12EI}{L^3} \qquad c = \frac{6EI}{L^2} \qquad d = \frac{2EI}{L}$$

we have

$$\mathbf{k}_{6\times6} = \begin{bmatrix} (al^2+bm^2) & (a-b)lm & -cm & -(al^2+bm^2) & -(a-b)lm & -cm \\ & (am^2+bl^2) & cl & -(a-b)lm & -(am^2+bl^2) & cl \\ & & 2d & cm & -cl & d \\ & & & (al^2+bm^2) & (a-b)lm & cm \\ & \text{Symmetric} & & & (am^2+bl^2) & -cl \\ & & & & & 2d \end{bmatrix} \tag{6.2.3.12}$$

Just like the local element stiffness matrix, the global element stiffness matrix is also symmetric. It should also be noted that a typical beam element (unlike the truss element) has six dof in the local and global coordinate systems.

Handling Element Loads. As we have seen several times before in beam and frame structures, the loads acting on the beam or frame can be either nodal loads or loads acting on the member itself. The latter are known as element loads. We will assume that the element loads—concentrated forces or linearly distributed loads acting on the entire length of a member—act along the local y' direction. Later we will see how to handle loads that act in different directions. The question at hand is: how do we handle the effects due to element loads? The first thing to note is that the system equations **KD** = **F** that are solved to compute the nodal displacements are the equilibrium equations at the nodes. The direct stiffness method converts the distributed properties of the structure into equivalent properties along the degrees of freedom at the nodes. With the element loads, we must compute their equivalent forces that act along the degrees of freedom at the element nodes. These forces must be added to the nodal forces to create the complete nodal force vector **F**. Second, when the nodal displacements are used to compute the element nodal forces, these effects due to element loads must again be considered; otherwise the element will not be in equilibrium.

These concepts are illustrated in Fig. 6.2.3.7. The element load (uniformly distributed load w) acting on the beam is replaced with its nodal equivalent force system—a force and a moment at each end of the member. The equivalent force system (also called the equivalent joint forces) is nothing other than the *opposite* of the fixed-end forces that were used with the slope-deflection method! The frame in Fig. 6.2.3.7(b) is analyzed and system equations **KD** = **F** are solved for the nodal displacements.

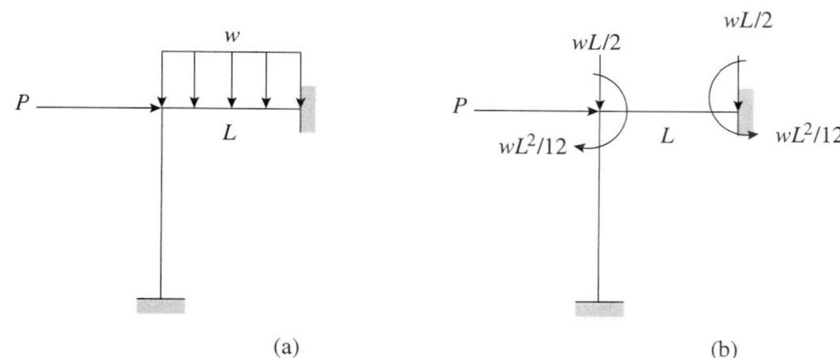

Fig. 6.2.3.7
(a) Actual frame.
(b) Equivalent nodal loads.

Once the element nodal forces are computed using the nodal displacements, we must take into account the element loads acting on the element. It is important to understand why this is necessary. Figure. 6.2.3.8(a) shows the original beam. We replaced the loading with its equivalent nodal loads as shown in Fig. 6.2.3.8(b). However, to get back the original beam, we must reintroduce the element load and the fixed-end moments that are equal and opposite to the equivalent joint loads. Only then does the superposition of (b) and (c) give back the original beam in (a).

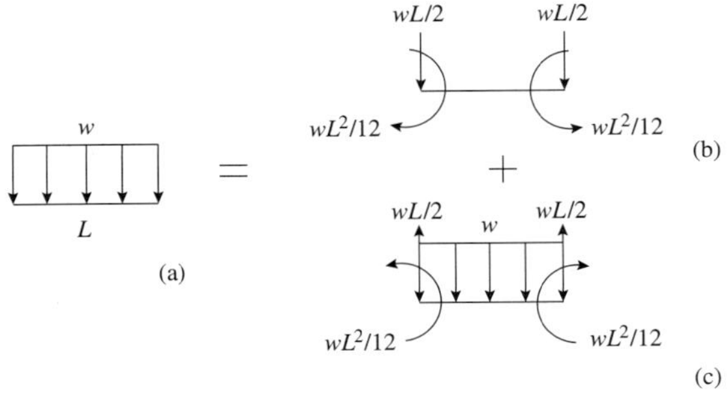

Fig. 6.2.3.8 For a uniformly distributed load: (a) Original beam, (b) equivalent joint loads, and (c) equivalent loading system showing the fixed-end moments.

We present the two most common element loads and their equivalent joint loads in Fig. 6.2.3.9. The results are familiar since we computed them in Chapter 5. The same procedure can be used with other types of loading such as a triangular loading, trapezoidal loading, etc.

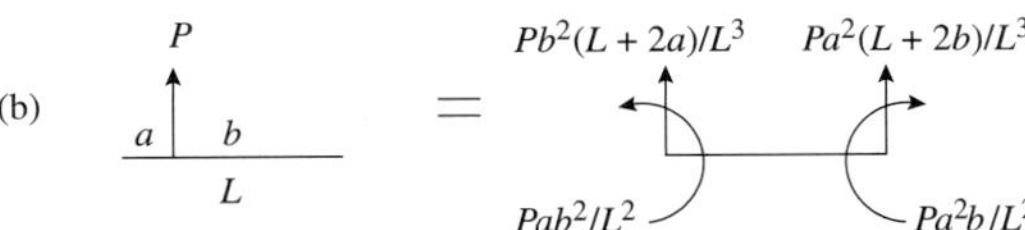

Fig. 6.2.3.9 Equivalent joint loads for (a) uniformly distributed load, (b) concentrated load.

We can now summarize the major steps in solving any planar frame problem using the direct stiffness method.

Step 1: Select the problem units. Set up the coordinate system. Identify and label the nodes and the elements. For each element select a start node (node 1) and an end node (node 2). We use an arrow along the member to indicate the direction from the start node to the end node. This establishes the local coordinate system for each element. Label the three global dof at each node starting at node 1 and proceeding sequentially.

Step 2: Construct the equilibrium-compatibility equations for a typical element (Eqs. (6.2.3.11) and (6.2.3.12)).

Step 3: Using the problem data, construct the element equations from Step 2 for all the elements in the problem. If there are element loads, compute the equivalent joint loads (EJL) $\mathbf{q}'_{6\times1}$ and transform them to the global coordinate system using Eq. 6.2.3.7 as

$$\mathbf{q}_{6\times1} = \mathbf{T}^{\mathrm{T}}_{6\times6}\mathbf{q}'_{6\times1} \tag{6.2.3.13a}$$

or $$\mathbf{q}_{6\times1} = \left[(lq'_1 - mq'_2) \quad (mq'_1 + lq'_2) \quad q'_3 \quad (lq'_4 - mq'_5) \quad (mq'_4 + lq'_5) \quad q'_6\right]^{\mathrm{T}} \tag{6.2.3.13b}$$

Note that if there is more than one element load acting on an element, $\mathbf{q}_{6\times1}$ represents the linear superposition (algebraic sum) of all the element loads acting on that element.

Step 4: Assemble the element equations into the system equations $\mathbf{K}_{3j\times3j}\mathbf{D}_{3j\times1} = \mathbf{F}_{3j\times1}$, where j is the number of nodes in the frame.

Step 5: Impose the boundary conditions.

Step 6: Solve the system equations $\mathbf{KD} = \mathbf{F}$ for the nodal displacements $\mathbf{D}$.

Step 7: For each element, using the nodal displacements, compute the element nodal forces using

$$\mathbf{f}'_{6\times1} = \mathbf{k}'_{6\times6}\mathbf{T}_{6\times6}\mathbf{d}_{6\times1} - \mathbf{q}'_{6\times1} \tag{6.2.3.14a}$$

or

$$\begin{Bmatrix} f_1' \\ f_2' \\ f_3' \\ f_4' \\ f_5' \\ f_6' \end{Bmatrix} = \begin{Bmatrix} a\left[l(d_1-d_4)+m(d_2-d_5)\right] \\ b\left[l(d_2-d_5)-m(d_1-d_4)\right]+c(d_3+d_6) \\ c\left[l(d_2-d_5)-m(d_1-d_4)\right]+d(2d_3+d_6) \\ -a\left[l(d_1-d_4)+m(d_2-d_5)\right] \\ -b\left[l(d_2-d_5)-m(d_1-d_4)\right]-c(d_3+d_6) \\ c\left[l(d_2-d_5)-m(d_1-d_4)\right]+d(d_3+2d_6) \end{Bmatrix} - \begin{Bmatrix} q_1' \\ q_2' \\ q_3' \\ q_4' \\ q_5' \\ q_6' \end{Bmatrix} \tag{6.2.3.14b}$$

where the last term represents the adjustment to the element nodal forces discussed in Fig. 6.2.3.8. The support reactions can be computed using the procedure discussed in Section 6.2.2 and Eq. (6.2.2.21).

EXAMPLE 6.2.5

Continuous Beam

Figure E6.2.5(a) shows a continuous beam. The material is steel, $E = 2(10)^{11}$ Pa and the cross-sectional properties are such that A = 0.01 m^2 and I = 0.0001 m^4. Construct the global element equilibrium equations for each element.

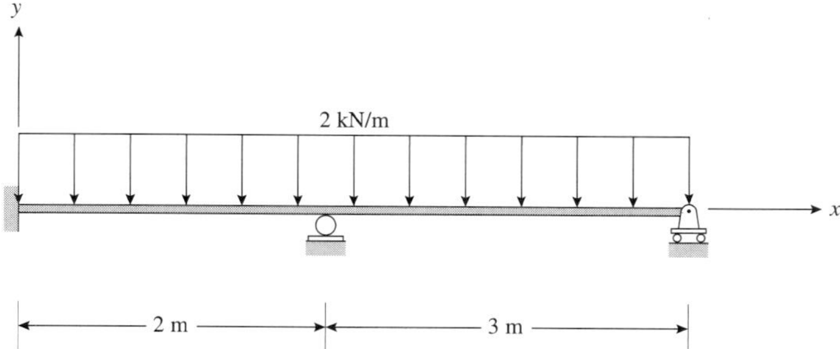

Fig. E6.2.5(a) Continuous beam.

SOLUTION

Step 1: The model details are shown in Fig. E6.2.5(b). The problem units are N, m.

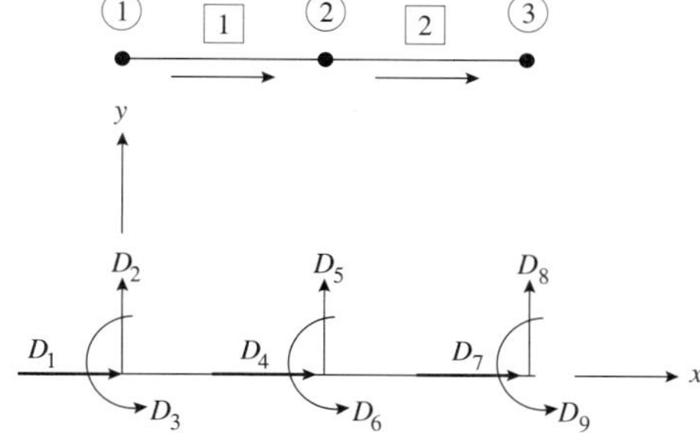

Fig. E6.2.5 (b) Model details showing node and element numbers and the global degrees of freedom.

Steps 2 and 3: It would be too tedious to generate the element equations by hand. We will use a spreadsheet to illustrate how this can be done using a computer tool. The shaded cells have the input values cross-sectional area, moment of inertia, modulus of elasticity, the (x, y) coordinates of the two nodes that are labeled Node 1 (start node) and Node 2 (end node), the uniformly distributed load w, and the concentrated force P and its distance from the start node. The rest of the cells have the formula relating to Eqs. (6.2.3.12) and (6.2.3.13).

The row and column headers represent the degree of freedom associated with that row and column. These are generated as follows. The degrees of freedom at node i are ($3i - 2$, $3i - 1$, $3i$).

Element 1:

	A	0.01	**E**	2.00E+11	**x1**	0
	I	1.00E–04			**x2**	2
					y1	0
	Node 1	1	**Node 2**	2	**y2**	0
w	–2000					
q'	0	–2000	–666.666667	0	–2000	666.666667
P		**a**		b	**0**	
q'	0	0	0	0	**0**	0
Sum	0	–2000	–666.666667	0	–2000	666.666667
	L	2	**l**	1	**m**	0
	a	1.00E+09	**b**	3.00E+07	**c**	3.00E+07
	d	2.00E+07				

k

	1	**2**	**3**	**4**	**5**	**6**
1	1.00E+09	0.00E+00	0.00E+00	–1.00E+09	0.00E+00	0.00E+00
2	0.00E+00	3.00E+07	3.00E+07	0.00E+00	–3.00E+07	–3.00E+07
3	0.00E+00	3.00E+07	4.00E+07	0.00E+00	–3.00E+07	2.00E+07
4	–1.00E+09	0.00E+00	0.00E+00	1.00E+09	0.00E+00	0.00E+00
5	0.00E+00	–3.00E+07	–3.00E+07	0.00E+00	3.00E+07	3.00E+07
6	0.00E+00	–3.00E+07	2.00E+07	0.00E+00	3.00E+07	4.00E+07

q

0	–2000	–666.666667	0	–2000	666.666667

Pay particular attention to the sign associated with the distributed load w. The formula assumes that a positive w acts in the positive y' direction. The same comments apply to the concentrated load P.

Element 2:

	A	0.01	**E**	2.00E+11	**x1**	2
	I	1.00E–04			**x2**	5
					y1	0
	Node 1	**2**	**Node 2**	**3**	**y2**	0
w	–2000					
q'	0	–3000	–1500	0	–3000	1500
P		**a**		b	**0**	
q'	0	0	0	0	0	0
Sum	0	–3000	–1500	0	–3000	1500

L	3	**l**	1	**m**	0
a	6.67E+08	**b**	8.89E+06	**c**	1.33E+07
d	1.33E+07				

k

	4	**5**	**6**	**7**	**8**	**9**
4	6.67E+08	0.00E+00	0.00E+00	–6.67E+08	0.00E+00	0.00E+00
5	0.00E+00	8.89E+06	1.33E+07	0.00E+00	–8.89E+06	–1.33E+07
6	0.00E+00	1.33E+07	2.67E+07	0.00E+00	–1.33E+07	1.33E+07
7	–6.67E+08	0.00E+00	0.00E+00	6.67E+08	0.00E+00	0.00E+00
8	0.00E+00	–8.89E+06	–1.33E+07	0.00E+00	8.89E+06	1.33E+07
9	0.00E+00	–1.33E+07	1.33E+07	0.00E+00	1.33E+07	2.67E+07

q

0	–3000	–1500	0	–3000	1500

EXAMPLE 6.2.6

System Equilibrium Equations

For the continuous beam in Fig. E6.2.5(a), construct the system equilibrium equations.

SOLUTION

In the previous example, we constructed the element global equilibrium equations.

Steps 4 and 5: We now construct the effective rsystem equilibrium equations. Considering the beam it is clear that $D_1 = D_2 = D_3 = D_5 = D_8 = 0$. Hence assembling the equations from element 1, we have the current snapshot of $\mathbf{KD} = \mathbf{F}$ as

$$\begin{bmatrix} 1(10^9) & 0 & 0 & 0 \\ 0 & 4(10^7) & 0 & 0 \\ 0 & 0 & 0 & 0 \\ 0 & 0 & 0 & 0 \end{bmatrix} \begin{Bmatrix} D_4 \\ D_6 \\ D_7 \\ D_9 \end{Bmatrix} = \begin{Bmatrix} 0 \\ 666.67 \\ 0 \\ 0 \end{Bmatrix}$$

After assembling the equations from element 2, we have the final system equations $\mathbf{KD} = \mathbf{F}$ as

$$(10^7)\begin{bmatrix} 166.7 & 0 & -66.7 & 0 \\ 0 & 6.67 & 0 & 1.33 \\ -66.7 & 0 & 66.7 & 0 \\ 0 & 1.33 & 0 & 2.67 \end{bmatrix} \begin{Bmatrix} D_4 \\ D_6 \\ D_7 \\ D_9 \end{Bmatrix} = \begin{Bmatrix} 0 \\ -833.33 \\ 0 \\ 1500 \end{Bmatrix}$$

EXAMPLE 6.2.7 ***Nodal Displacements and Member Forces***

For the continuous beam in Fig. E6.2.5(a), solve for the member nodal forces.

SOLUTION We continue with the solution from the previous examples.

Step 6: Solving the system equations yields

$$\{D_4, D_6, D_7, D_9\} = \{0, -2.63092(10^{-5})\,\text{rad}, 0, 6.92851(10^{-5})\,\text{rad}\}$$

It should come as no surprise that the x-displacements are zero.

Step 7: The final step in the procedure is to compute the members' nodal forces using Eq. (6.2.3.14b). We augment the spreadsheet shown in Example 6.2.5 with additional data on the nodal displacements associated with the element. The results for the two elements are shown in Figs. E6.2.7(a) and E6.2.7(b).

Element 1:

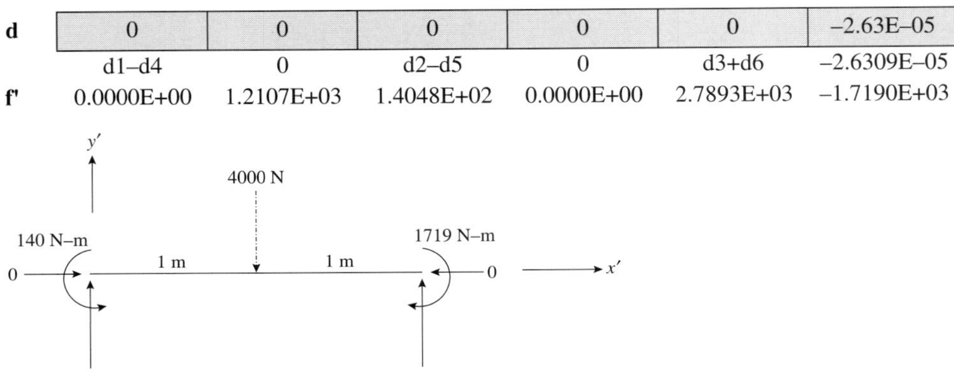

d	0	0	0	0	0	–2.63E–05
	d1–d4	0	d2–d5	0	d3+d6	–2.6309E–05
f'	0.0000E+00	1.2107E+03	1.4048E+02	0.0000E+00	2.7893E+03	–1.7190E+03

Fig. E6.2.7(a)

Check: $\overset{\rightarrow +}{\sum} F_x = 0$ OK.

$$\overset{\uparrow +}{\sum} F_y = 1211 - 4000 + 2789 = 0 \qquad \text{OK.}$$

$$\overset{\curvearrowleft +}{\sum} M_1 = 140 - 4000(1) - 1719 + 2789(2) \approx 0 \qquad \text{OK.}$$

Element 2:

d	0	0	–2.63E–05	0	0	6.93E–05
	d1–d4	0	d2–d5	0	d3+d6	4.2976E–05
f'	0.0000E+00	3.5730E+03	1.7222E+03	0.0000E+00	2.4270E+03	–3.1880E+00

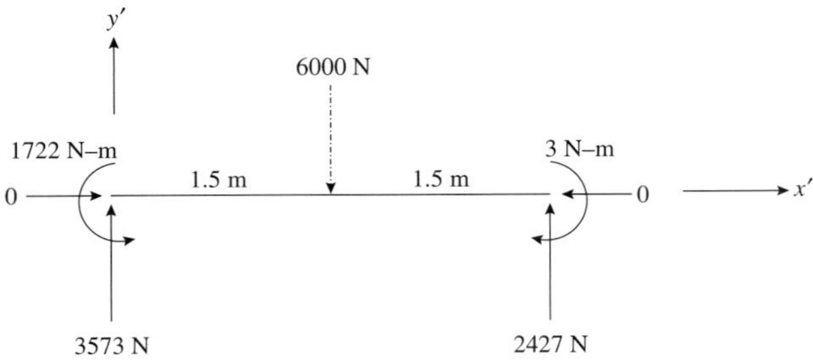

Fig. E6.2.7(b)

$$\text{Check: } \overset{\rightarrow+}{\sum} F_x = 0 \quad \text{OK.}$$

$$\overset{\uparrow+}{\sum} F_y = 3573 - 6000 + 2427 = 0 \qquad \text{OK.}$$

$$\overset{\curvearrowleft+}{\sum} M_2 = 1722 - 6000(1.5) - 3 + 2427(3) = 0 \qquad \text{OK.}$$

The moment at node 2 of element 1 should be equal and opposite to the moment at node 1 of element 2, i.e., 1719 N-m versus 1722 N-m. The moment at node 3 should be zero but is computed as 3 N-m. These are the numerical errors alluded to earlier but note that they are extremely small—compute the error as a fraction of the maximum moment in each element. If care is exercised by using sufficiently large precision, the numerical errors will be small.

EXAMPLE 6.2.8

A Planar Frame (see Example 5.2.6)

Figure E6.2.8(a) shows a planar frame. The material is steel, $E = 2(10)^{11}$ Pa and the cross-sectional properties are such that $A = 0.01$ m^2 and $I = 0.0001$ m^2. Solve for the member nodal forces and the support reactions.

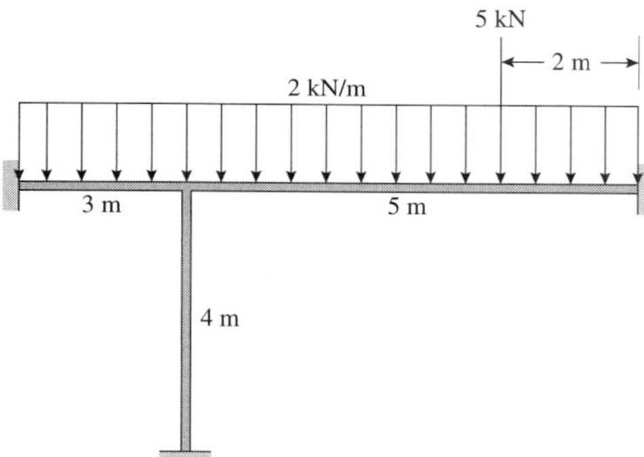

Fig. E6.2.8(a)

SOLUTION

Step 1: The problem units are N, m. The model details are shown in Fig. E6.2.8(b).

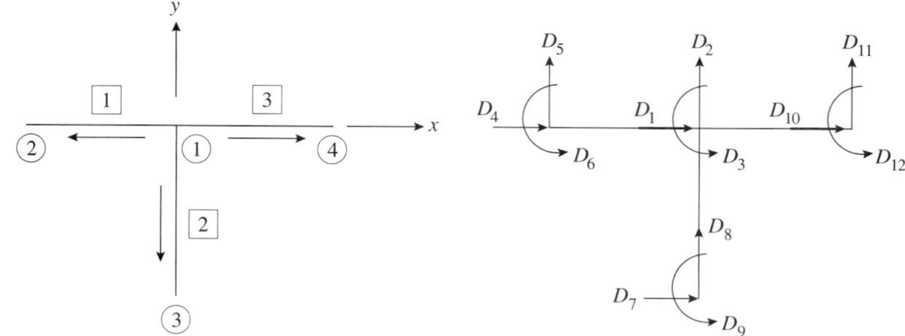

Fig. E6.2.8(b)

Steps 2 and 3: We construct the element equations using the spreadsheet template.

Element 1:

	A	0.01	E	2.00E+11	x1	0
	I	1.00E–04			x2	–3
					y1	0
	Node 1	1	Node 2	2	y2	0
w	2000					
q'	0	3000	1500	0	3000	–1500
P	0	a	0	b	0	
q'	0	0	0	0	0	0
Sum	0	3000	1500	0	3000	–1500

L	3	l	–1	m	0
a	6.67E+08	b	8.89E+06	c	1.33E+07
d	1.33E+07				

k

	1	2	3	4	5	6
1	6.67E+08	0.00E+00	0.00E+00	–6.67E+08	0.00E+00	0.00E+00
2	0.00E+00	8.89E+06	–1.33E+07	0.00E+00	–8.89E+06	1.33E+07
3	0.00E+00	–1.33E+07	2.67E+07	0.00E+00	1.33E+07	1.33E+07
4	–6.67E+08	0.00E+00	0.00E+00	6.67E+08	0.00E+00	0.00E+00
5	0.00E+00	–8.89E+06	1.33E+07	0.00E+00	8.89E+06	–1.33E+07
6	0.00E+00	1.33E+07	1.33E+07	0.00E+00	–1.33E+07	2.67E+07

q

0	–3000	1500	0	–3000	–1500

Note the sign associated with w, the uniformly distributed load acting on the element.

Element 2:

	A	0.01	E	2.00E+11	x1	0
	I	1.00E–04			x2	0
					y1	0
	Node 1	1	Node 2	3	y2	–4
w	0					
q'	0	0	0	0	0	0
P	0	a	0	b	0	
q'	0	0	0	0	0	0
Sum	0	0	0	0	0	0

L	4	l	0	m	–1
a	5.00E+08	b	3.75E+06	c	7.50E+06
d	1.00E+07				

k

	1	2	3	7	8	9
1	3.75E+06	0.00E+00	7.50E+06	–3.75E+06	0.00E+00	7.50E+06
2	0.00E+00	5.00E+08	0.00E+00	0.00E+00	–5.00E+08	0.00E+00
3	7.50E+06	0.00E+00	2.00E+07	–7.50E+06	0.00E+00	1.00E+07
7	–3.75E+06	0.00E+00	–7.50E+06	3.75E+06	0.00E+00	–7.50E+06
8	0.00E+00	–5.00E+08	0.00E+00	0.00E+00	5.00E+08	0.00E+00
9	7.50E+06	0.00E+00	1.00E+07	–7.50E+06	0.00E+00	2.00E+07

q

0	0	0	0	0	0

Element 3:

	A	0.01	**E**	2.00E+11	**x1**	0
	I	1.00E–04			**x2**	5
					y1	0
	Node 1	1	**Node 2**	4	**y2**	0
w	–2000					
q'	0	–5000	–4166.666667	0	–5000	4166.666667
P	–5000	**a**	**3**	**b**	**2**	
q'	0	–1760	–2400	0	**–3240**	3600
Sum	0	–6760	–6566.666667	0	–8240	7766.666667
	L	5	**l**	1	**m**	0
	a	4.00E+08	**b**	1.92E+06	**c**	4.80E+06
	d	8.00E+06				

k

	1	2	3	10	11	12
1	4.00E+08	0.00E+00	0.00E+00	–4.00E+08	0.00E+00	0.00E+00
2	0.00E+00	1.92E+06	4.80E+06	0.00E+00	–1.92E+06	–4.80E+06
3	0.00E+00	4.80E+06	1.60E+07	0.00E+00	–4.80E+06	8.00E+06
10	–4.00E+08	0.00E+00	0.00E+00	4.00E+08	0.00E+00	0.00E+00
11	0.00E+00	–1.92E+06	–4.80E+06	0.00E+00	1.92E+06	4.80E+06
12	0.00E+00	–4.80E+06	8.00E+06	0.00E+00	4.80E+06	1.60E+07

q

0	–6760	–6566.666667	0	–8240	7766.666667

Steps 4 and 5: The boundary conditions are:

$$D_4 = D_5 = D_6 = D_7 = D_8 = D_9 = D_{10} = D_{11} = D_{12} = 0$$

Hence there are three effective degrees of freedom and the assembly process yields the following system equations:

$$10^7\begin{bmatrix} 107.075 & 0 & 0.75 \\ 0 & 51.081 & -0.85 \\ 0.75 & -0.85 & 6.27 \end{bmatrix}\begin{Bmatrix} D_1 \\ D_2 \\ D_3 \end{Bmatrix} = \begin{Bmatrix} 0 \\ -9760 \\ -5066.7 \end{Bmatrix}$$

Step 6: Solving the system equations yields

$$D_1 = 5.85975(10^{-7})\text{ m},\; D_2 = -2.0499(10^{-5})\text{ m},\; D_3 = -8.36577(10^{-5})\text{ rad}$$

Step 7: From the nodal displacements the member forces can be computed, as in Figs. E6.2.8(c)–(e).

Element 1:

d	5.86E–07	–2.05E–05	–8.37E–05	0	0	0.00E+00
	d1–d4	5.85975E–07	d2–d5	–2.0499E–05	d3+d6	–8.3658E–05
f'	–3.9065E+02	–3.9332E+03	–3.4576E+03	3.9065E+02	–2.0668E+03	6.5788E+02

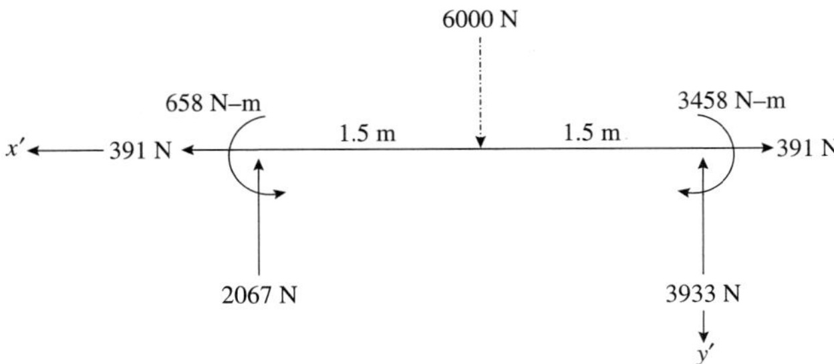

Fig. E6.2.8(c)

Element 2:

d	5.86E–07	–2.05E–05	–8.37E–05	0	0	0.00E+00
	d1–d4	5.85975E–07	d2–d5	–2.0499E–05	d3+d6	–8.3658E–05
f'	1.0250E+04	–6.2524E+02	–1.6688E+03	–1.0250E+04	6.2524E+02	–8.3218E+02

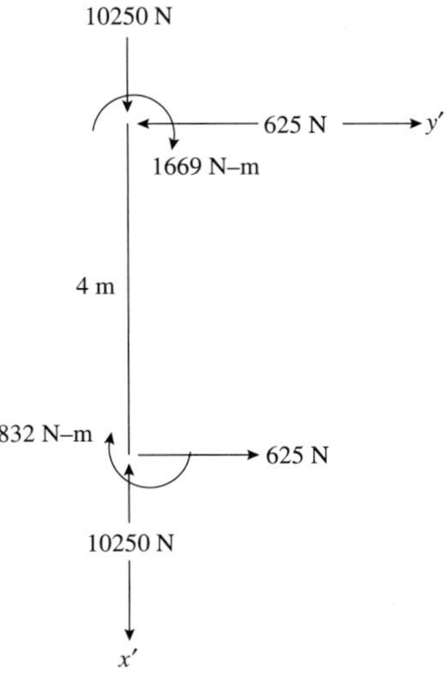

Fig. E6.2.8(d)

Element 3:

d	5.86E–07	–2.05E–05	–8.37E–05	0	0	0.00E+00
	d1–d4	5.85975E–07	d2–d5	–2.0499E–05	d3+d6	–8.3658E–05
f'	2.3439E+02	6.3191E+03	5.1297E+03	–2.3439E+02	8.6809E+03	–8.5343E+03

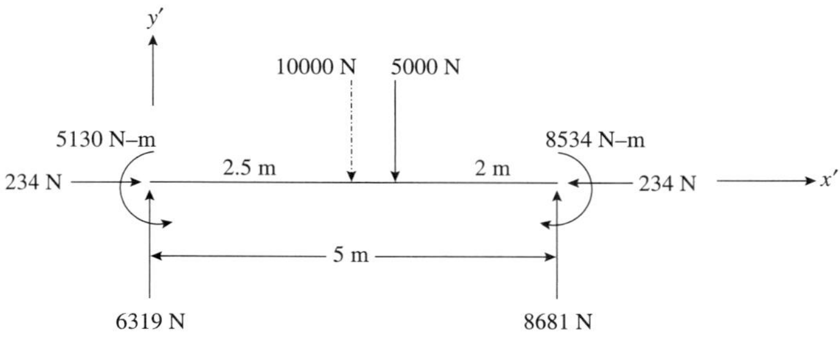

Fig. E6.2.8(e)

One should check the equilibrium of each element and the equilibrium of each joint to ensure that the computations are correct. The FBD of node 1 is shown in Fig. E6.2.8(f).

Check: $\overset{\rightarrow +}{\sum} F_x = -391 + 625 - 234 = 0$ OK.

$$\overset{\uparrow +}{\sum} F_y = -3933 + 10250 - 6319 \approx 0 \qquad \text{OK.}$$

$$\overset{\circlearrowleft +}{\sum} M_1 = 3457 + 1669 - 5130 \approx 0 \qquad \text{OK.}$$

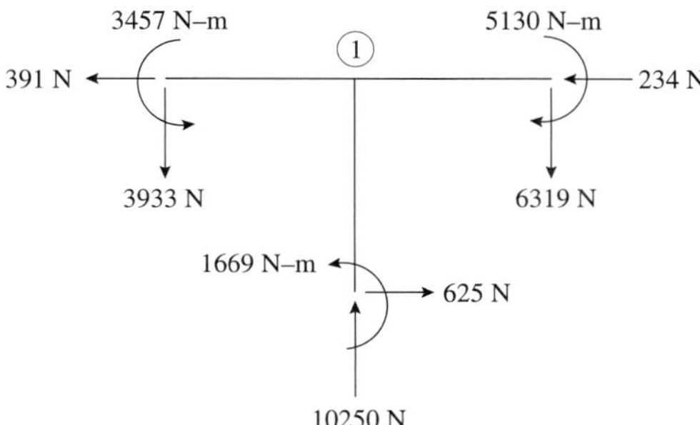

Fig. E6.2.8(f)

EXERCISES

In all the problems in this section, unless otherwise stated, compute the nodal displacements and the element nodal forces.

Appetizers

6.2.14. Solve the beam in Fig. P6.2.14. Take E = 200 GPa and $I = 4(10^6)$ mm^4.

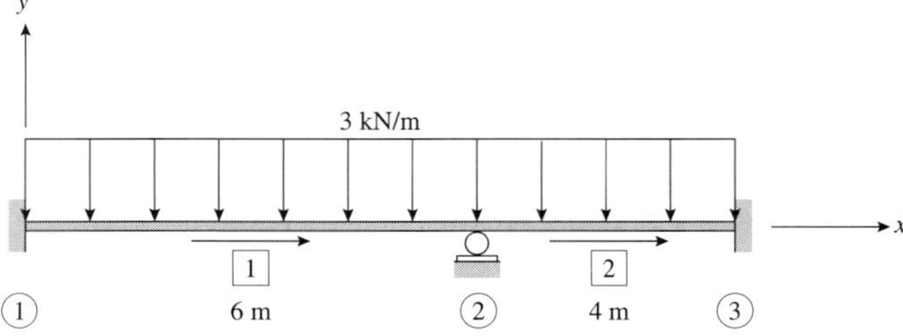

Fig. P6.2.14

6.2.15. Solve Example 7.4.1.

6.2.16. Solve the frame in Example 5.2.4. However, take the loading as in Fig. P6.2.16.

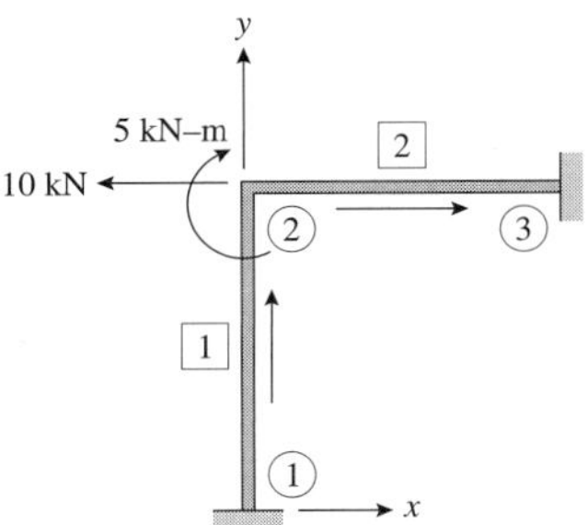

Fig. P6.2.16

Main Course

6.2.17. Solve Problem 5.2 using the model in Fig. P6.2.17. Take E = 29000 ksi, A = 10 in^2 and I = 50 in^4.

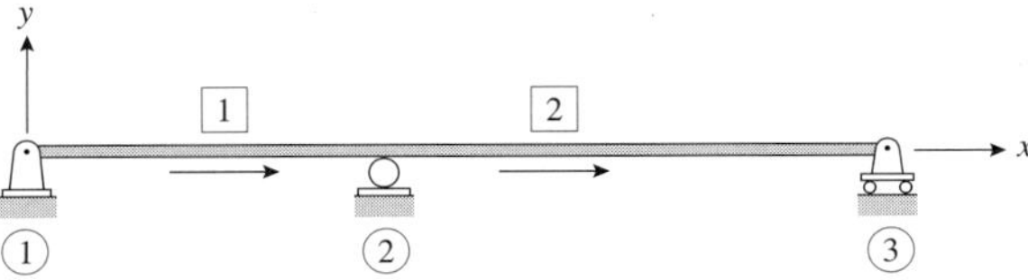

Fig. P6.2.17

6.2.18. Solve Example 5.2.4. The model is defined in Fig. P6.2.18. Take A = 0.01m^2.

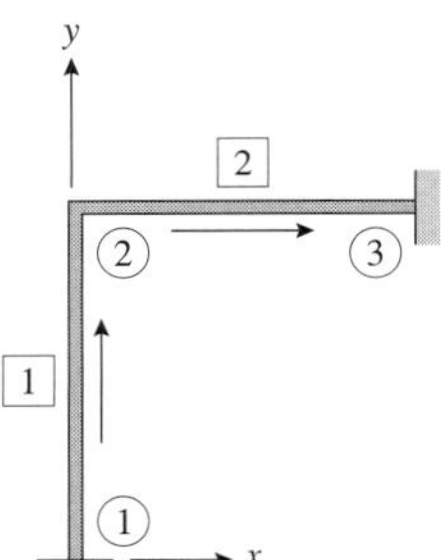

Fig. P6.2.18

6.2.19. Compute the nodal displacements and member nodal forces for the wooden beam (E = 1500 ksi) in Fig. P6.2.19. The cross-section is 12 in × 8 in rectangular section. Ignore self-weight.

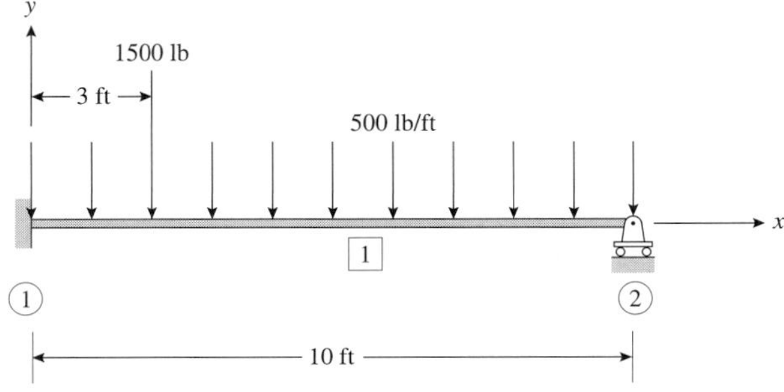

Fig. P6.2.19

6.2.20. Compute the member nodal forces for the frame in Fig. P6.2.20. The members are W21 × 44. Ignore self-weight.

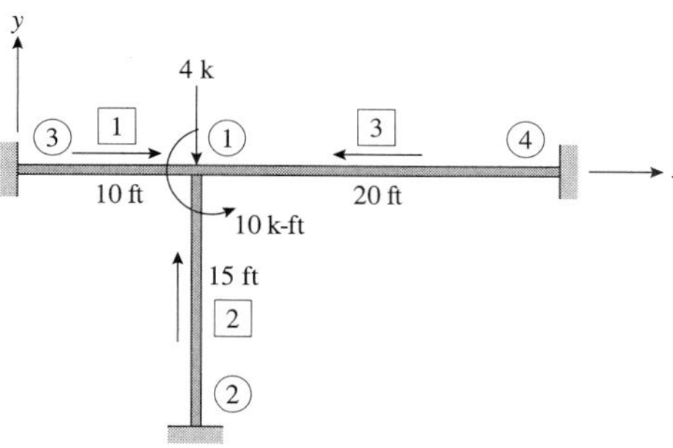

Fig. P6.2.20

Structural Concepts

6.2.21. The frame in Fig. P6.2.21 is made of steel. The two members have the cross-sectional properties $A = 0.01\ \text{m}^2$ and $I = 10^{-4}\ \text{m}^4$. Compute the support reactions. Draw the shear force and bending moment diagrams.

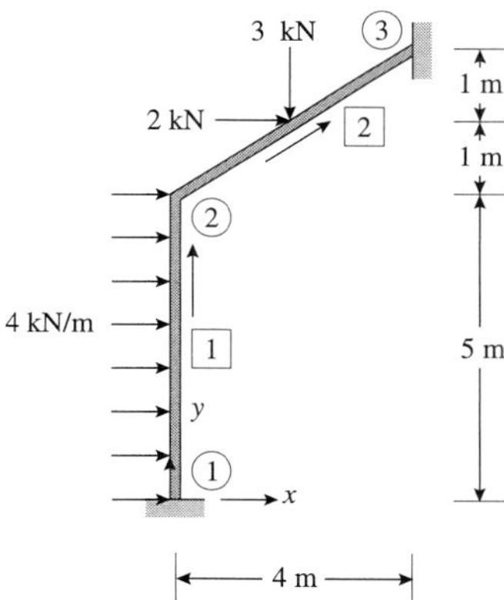

Fig. P6.2.21

6.3 THEOREM OF MINIMUM POTENTIAL ENERGY

While the direct stiffness method was used to derive the appropriate equations for the truss and beam elements, more useful and complex structures cannot be studied effectively using this approach. The limitations of the stiffness method were recognized decades ago and were overcome by the development of the finite element method. In this section we lay the foundation for the finite element approach most commonly used to solve structural analysis problems.

> The theorem of minimum potential energy states that for a *conservative* system, amongst all *admissible configurations* those that satisfy the equations of *equilibrium* make the potential energy stationary with respect to small variations of displacement. If the stationary condition is a *minimum*, the equilibrium state is *stable*.

Pay particular attention to the italicized terms to understand the applicability and limitations of the theorem. Review, if necessary, the material from Section 4.4.

Consider the following situation. Let Π denote the total potential energy of the system. Let the potential energy be a function of a set of displacements $\mathbf{D} = \{D_1, D_2, \ldots, D_n\}$. If the displacements satisfy the boundary conditions such that the system is in stable equilibrium, then the following conditions must be satisfied:

$$\frac{\partial \Pi}{\partial \mathbf{D}} = 0 \quad i = 1, 2, \ldots n \tag{6.3.1}$$

and can be used to compute the displacements. Figure 6.3.1 shows the state of equilibrium of a sphere resting on different surfaces.

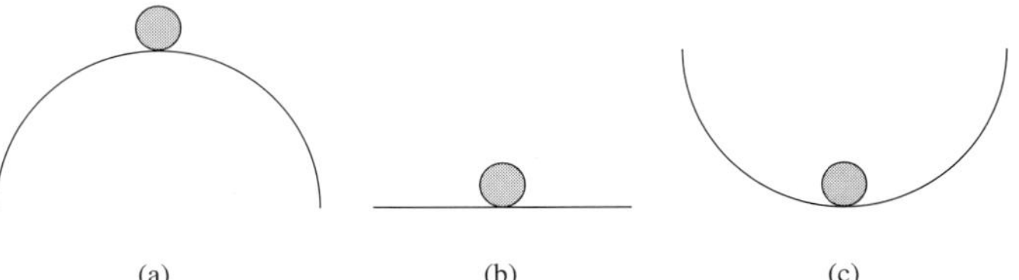

Fig. 6.3.1
(a) Unstable equilibrium.
(b) Neutral equilibrium.
(c) Stable equilibrium.

The unstable equilibrium corresponds to the state of highest potential energy of the sphere, the neutral equilibrium state shows no change in the potential energy when the sphere is perturbed slightly, and the stable equilibrium state corresponds to the state of lowest potential energy of the sphere (in the immediate neighborhood). The same comments apply to structural systems. We are interested in finding the equilibrium state that has the minimum potential energy so that the structural system is stable.

Total Potential Energy

We revisit the topic of the total potential energy introduced in Chapter 4. The total potential energy of a linearly elastic system is given by

$$\begin{aligned} \Pi &= \text{strain energy} + \text{work potential} \\ &= \int_V U_0 \, dV - \int_V \mathbf{f}^T \mathbf{F} \, dV - \int_S \mathbf{f}^T \mathbf{\Phi} \, dS - \mathbf{D}^\mathbf{T} \mathbf{P} \end{aligned} \tag{6.3.2}$$

where

- U_0 strain energy per unit volume
- $\mathbf{f}$ displacement field, e.g. $\{u \; v \; w\}$ in three dimensions
- $\mathbf{F}$ body forces per unit volume
- $\mathbf{\Phi}$ surface traction per unit area
- $\mathbf{D}$ nodal displacements
- $\mathbf{P}$ concentrated forces

An example of body force is self-weight. Surface traction refers to distributed loads acting on the surface of the body. The strain energy density is given by

$$U_0 = \frac{1}{2}\{\varepsilon\}^T \mathbf{E}\{\varepsilon\} - \{\varepsilon\}^T \mathbf{E}\{\varepsilon_0\} + \{\varepsilon\}^T \{\sigma_0\} \tag{6.3.3}$$

where

- $\{\varepsilon\}$ strain components
- $\{\varepsilon_0\}$ initial strain components (e.g., fabrication errors)

$\{\sigma_0\}$ initial stress components (e.g., thermal stresses)

E material matrix relating strains and stresses (see Eq. (4.1.4.1))

$$\{\sigma\} = \mathbf{E}\{\varepsilon\} - \mathbf{E}\{\varepsilon_0\} + \{\sigma_0\} \tag{6.3.4}$$

Let us now look at an example in which we compute the total potential energy and find the equilibrium state of the system.

EXAMPLE 6.3.1 ***Axially Loaded Bar***

Consider a bar of constant cross-section A, length L, and modulus of elasticity E subjected to a constant axial force P at the right tip and fixed at the left end. Compute the tip displacement and the state of stress in the bar.

SOLUTION Let the tip displacement be D. This is the sole unknown or degree of freedom in this problem. Using Eqs. (6.3.2) and (6.3.3), we have

$$\varepsilon_x = D/L \qquad dV = Adx$$

$$U_0 = \int_0^L \left\{\frac{1}{2}\left(\frac{D}{L}\right)E\left(\frac{D}{L}\right)\right\} Adx$$

$$\Pi(D) = \int_V U_0\, dV - PD = \frac{D^2EA}{2L} - PD \tag{6.3.5}$$

Using the theorem of minimum potential energy, we have

$$\frac{d\Pi}{dD} = 0 = \frac{DEA}{L} - P \tag{6.3.6}$$

$$\text{or} \quad D = \frac{PL}{AE} \tag{6.3.7}$$

Hence

$$\varepsilon_x = D/L = \frac{P}{AE} \tag{6.3.8a}$$

and

$$\sigma = E\varepsilon_x = \frac{P}{A} \tag{6.3.8b}$$

Let's examine the process and the solution. We assumed that the entire problem could be described by a single unknown D at the tip of the bar. Is this correct? With the displacement at the left end assumed to be zero (since it is fixed) and the right end displacement as D, the net effect of the assumptions is that the bar has a linear displacement field; i.e., a linear function with a value zero at $x = 0$ and D at $x = L$ describes the deformation of the bar. This assumption is certainly valid for this problem but is not true if the loading on the bar is changed, for example. How can we overcome this *ad hoc* nature and formalize the solution process? We see the approach in the next section.

Rayleigh-Ritz Technique

Engineering systems are usually described by a system of partial differential equations. These problems are referred to as boundary value problems. With the generality associated with practical systems, it is virtually impossible to find the exact solution—a solution that satisfies the differential equations at every interior point of the domain *and* the boundary conditions. The trick is to *assume the form of solution.* For example, for structural problems one would assume the form of displacement field and then use the theorem of minimum potential energy to find the unknown parameters. This is the Rayleigh-Ritz technique.

Step 1: The first step is to assume the solution. The assumed form will have one or more unknown parameters or degrees of freedom **a**. A typical example is a polynomial. The assumed form must be able to satisfy the essential boundary conditions for the problem.

Step 2: Using the assumed form of the solution, construct the total potential energy using Eq. (6.3.2). The potential energy will now be in terms of **a**, i.e., $\Pi = \Pi(\mathbf{a})$.

Step 3: The final step is to use the theorem of minimum potential energy and minimize the total potential energy. In other words,

$$\frac{\partial \Pi}{\partial \mathbf{a}} = 0 \tag{6.3.9}$$

The above condition leads to a set of linear algebraic equations that can be then be solved using any solution technique for the unknown parameters **a**.

EXAMPLE 6.3.2 ***Axially Loaded Bar***

Resolve Example 6.3.1.

SOLUTION

The displacement field **f** is described by a single displacement component u in the bar. Let us assume a linear displacement field of the form

$$u(x) = a_0 + a_1 x \tag{6.3.10}$$

The essential boundary condition for this problem is

$$u(x=0) = 0 \tag{6.3.11}$$

Substituting Eq. (6.3.11) into (6.3.10), we have

$$u(x=0) = 0 = a_0 \tag{6.3.12}$$

Hence, Eq. (6.3.10) can be rewritten as

$$u(x) = a_1 x \tag{6.3.13}$$

The axial strain ε_x is given by

$$\varepsilon_x = \frac{du}{dx} = a_1 \tag{6.3.14}$$

Substituting Eqs. (6.3.13) and (6.3.14) into Eq. (6.3.2) gives us

$$\Pi(a_1) = \int_V U_0 \, dV - PD = \int_0^L \frac{1}{2}(a_1)(E)(a_1)A\,dx - P(a_1 L)$$

$$= \frac{1}{2} a_1^2 EAL - Pa_1 L \tag{6.3.15}$$

Using the theorem of minimum potential energy (Eq. 6.3.9), we have

$$\frac{d\Pi}{da_1} = 0 = a_1 EAL - PL \tag{6.3.16}$$

$$\text{or } a_1 = \frac{P}{AE} \tag{6.3.17}$$

Hence,

$$u(x) = \frac{Px}{AE}$$

$$\varepsilon_x = \frac{du}{dx} = \frac{P}{AE} \tag{6.3.18}$$

and

$$\sigma = E\varepsilon_x = \frac{P}{A} \tag{6.3.19}$$

The results are the same as the previous example. It should, however, be noted that we have solved a simple problem where the solution is smooth. Let us look at a slightly different problem.

EXAMPLE 6.3.3 ***Axially Loaded Bar***

Consider a bar of unit length that is fixed at both ends and is loaded by a unit point at the center of the bar, as in Fig. E6.3.3(a). Assume that $AE = 1$. Find the displacement and the stresses in the bar.

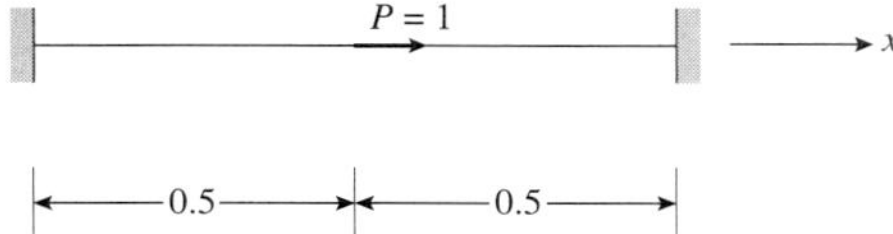

Fig. E6.3.3(a)

SOLUTION Again we assume a polynomial as the solution as[7]

$$u(x) = a_0 + a_1 x + a_2 x^2 \tag{6.3.20}$$

The essential boundary conditions for this problem are

$$u(x = 0) = 0 \tag{6.3.21a}$$

$$u(x = 1) = 0 \tag{6.3.21b}$$

Substituting Eq. (6.3.21) into (6.3.20), we have

$$u(x = 0) = 0 = a_0$$

$$u(x = 1) = 0 = a_1 + a_2 \Rightarrow a_2 = -a_1 \tag{6.3.22}$$

Hence, Eq. (6.3.20) can be rewritten as

$$u(x) = a_1 x - a_1 x^2 \tag{6.3.23}$$

$$\frac{du}{dx} = a_1 - 2a_1 x \tag{6.3.24}$$

---

[7] This is the lowest-order polynomial that can be assumed as the solution. For example, we cannot assume $u(x) = a_0 + a_1 x$. Imposing the conditions in Eq. (6.3.21) would leave no free parameters.

Now constructing the total potential energy, we have

$$\Pi(a_1) = \int_V U_0\, dV - PD = \int_0^1 \frac{1}{2}(a_1 - 2a_1 x)^2\, dx - (1)(0.5a_1 - 0.25a_1)$$
$$= \frac{a_1^2}{6} - 0.25a_1 \tag{6.3.25}$$

Using the theorem of minimum potential energy (Eq. 6.3.9), we have

$$\frac{d\Pi}{da_1} = 0 = \frac{a_1}{3} - 0.25 \tag{6.3.26}$$

or $a_1 = 0.75$ (6.3.27)

Hence,

$$u(x) = 0.75\left(x - x^2\right) \tag{6.3.28}$$

$$\varepsilon_x = \frac{du}{dx} = 0.75 - 1.5x \tag{6.3.29}$$

$$\sigma = E\varepsilon_x = 0.75 - 1.5x \tag{6.3.30}$$

Figure E6.3.3(b) shows the comparison between the Rayleigh-Ritz (RR) solution and the exact (mechanics of materials) solution.

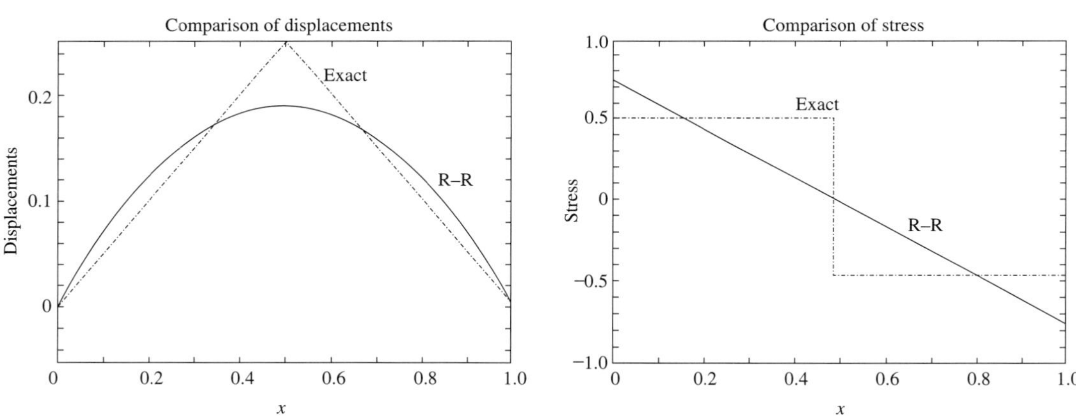

Fig. E6.3.3(b) Comparison of results.

The exact displacement is piecewise linear and the exact stress is discontinuous. Hence it should be clear that increasing the order of the polynomial while yielding better solutions will never yield the exact solution to this problem.

EXERCISES

Appetizers

6.3.1. Consider a bar of constant cross-section A, length L, and modulus of elasticity E subjected to a constant axial force P as shown in Fig. P6.3.1. Compute the tip displacement and the state of stress in the bar.

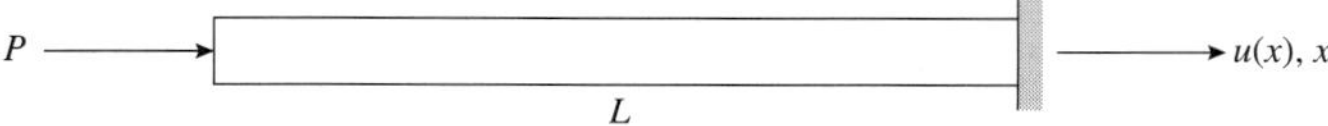

Fig. P6.3.1

Main Course

6.3.2. Consider a bar of constant cross-section A, length L, and modulus of elasticity E subjected to a uniform axial loading, q as shown in Fig. P6.3.2. Compute the displacement field $u(x)$ and the state of stress in the bar.

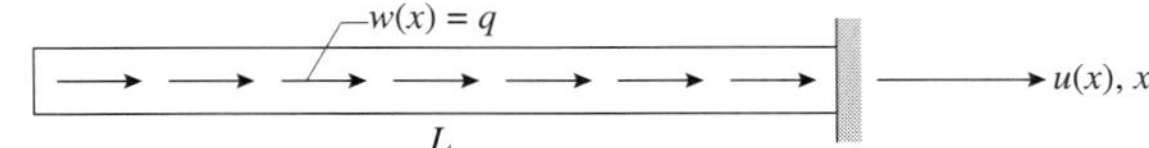

Fig. P6.3.2

6.4 FINITE ELEMENT METHOD

The Rayleigh-Ritz technique, while powerful, has severe limitations as a practical tool. First, the assumed solution is valid for the entire problem domain. As we saw in the last section, very simple solutions such as piecewise linear cannot be handled with a single function. Second, the assumed solution lacks physical meaning. For example, if the assumed solution is a linear polynomial, what do the two coefficients represent?

Two elegant modifications can be made to the above procedure. The Rayleigh-Ritz concept (of assuming an approximate solution) can be used over an element instead of the entire problem domain. Second, the approximate solution can be transformed and related to the unknown nodal values using the concept of interpolation.

The finite element method (FEM) has evolved over a long time. The basic building blocks and ideas originated in the 1940s. With the advent of digital computers in the 1950s, the ideas were converted into matrix form, making for a practical implementation. Today engineers have recognized the power of this very practical tool and finite elements are routinely used to solve very diverse problems in all engineering areas—civil, aerospace, mechanical, biomedical, electrical, chemical, etc. There are tens of books devoted exclusively to the treatment of finite elements. In this text, and especially in this section, we introduce the very basic ideas. However, these ideas are very powerful and can easily be extended to treat other classes of problems.

Finally, a definition of FEM before we look at the details. The finite element method is a computer-aided mathematical technique for obtaining approximate numerical solutions to the abstract equations of calculus that predict the response of physical systems subjected to external influences.[8] Now to the details.

EXAMPLE 6.4.1 ***Finite Element Solution***

Resolve Example 6.3.3.

SOLUTION

The basic approach is now to discretize the domain into finite elements. Let us use two elements, one from $x = 0$ to $x = 0.5$ and the other from $x = 0.5$ to $x = 1.0$, as in Fig. E6.4.1(a).

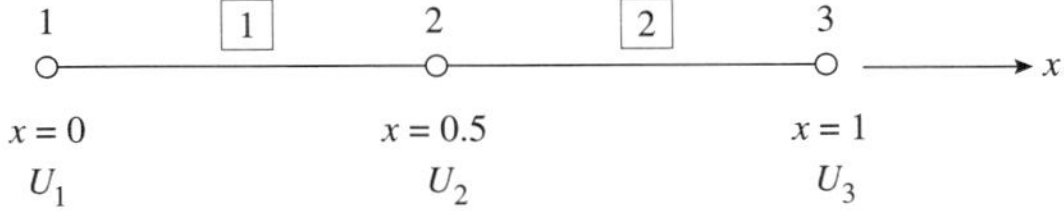

Fig. E6.4.1(a)

We could assume the trial solution for element 1 as

$$u_1(x) = a_1 + a_2 x^2 \tag{6.4.1}$$

[8] David S. Burnett, Finite Element Analysis: From Concepts to Applications, Addison-Wesley, 1988.

and for element 2 as

$$u_2(x) = b_1 + b_2 x^2 \tag{6.4.2}$$

To ensure that the displacement is continuous at the element interface, i.e., at $x = 0.5$, we could enforce the following constraint:

$$u_1(x = 0.5) = u_2(x = 0.5) \tag{6.4.3}$$

This approach is laborious, especially as the size of the problem increases. Instead, we can convert the trial solution to a form involving the nodal values via the concept of interpolation. For a typical element as in Fig. E6.4.1(b), we can assume (s is a local coordinate system with the same sense as x) the displacement as

$$u(s) = a_1 + a_2 s \tag{6.4.4}$$

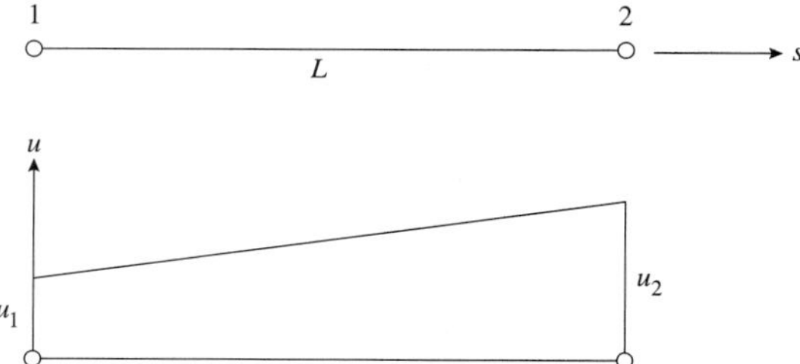

Fig. E6.4.1(b) A typical element.

Using the nodal conditions $u(s = 0) = u_1$ and $u(s = L) = u_2$, we can rewrite the above equation as

$$u(s) = a_1 + a_2 s = \frac{L - s}{L} u_1 + \frac{s}{L} u_2 = \phi_1 u_1 + \phi_2 u_2 \tag{6.4.5}$$

where ϕ_1 and ϕ_2 are the shape functions. Hence, the strain and stress in the element can be expressed as

$$\varepsilon = \frac{du}{ds} = -\frac{1}{L} u_1 + \frac{1}{L} u_2 = \frac{1}{L}(u_2 - u_1) = \begin{bmatrix} -\frac{1}{L} & \frac{1}{L} \end{bmatrix}_{1\times 2} \begin{Bmatrix} u_1 \\ u_2 \end{Bmatrix}_{2\times 1} = \mathbf{B}_{1\times 2}\mathbf{d}_{2\times 1} \tag{6.4.6}$$

$$\sigma = E\varepsilon \tag{6.4.7}$$

$\mathbf{B}_{1\times 2}$ is known as the strain-displacement matrix since it relates strain ε (left-hand side) to the displacements $\mathbf{d}_{2\times 1}$ (right-hand side). Note that the strain and the stress within the element are constants. Hence, the strain energy in a typical element can be written as

$$U = \int_V U_0 \, dV = \frac{1}{2}\int_V \varepsilon\sigma \, dV = \frac{1}{2}\varepsilon E \varepsilon A L \tag{6.4.8}$$

Substituting Eqs. (6.4.6) and (6.4.7) into (6.4.8), we have

$$U = \frac{1}{2}\mathbf{d}^T_{1\times 2}\mathbf{B}^T_{2\times 1}(EAL)_{1\times 1}\mathbf{B}_{1\times 2}\mathbf{d}_{2\times 1} \tag{6.4.9}$$

$$U = \frac{1}{2}\mathbf{d}^T_{1\times 2}\,\mathbf{k}_{2\times 2}\,\mathbf{d}_{2\times 1} \tag{6.4.10}$$

where $\mathbf{k}_{2\times 2}$ is the element stiffness matrix given as

$$k_{2\times 2} = \mathbf{B}^T_{2\times 1}(EAL)_{1\times 1}\mathbf{B}_{1\times 2} = \frac{AE}{L}\begin{bmatrix} 1 & -1 \\ -1 & 1 \end{bmatrix} \tag{6.4.11}$$

Since there are no body forces or surface tractions in this problem, the work potential term need not be computed. In general, however, we have to compute the work potential. The total potential energy in a typical element is given by

$$\Pi_e(\mathbf{d}) = \frac{1}{2}\mathbf{d}_{1\times2}^T\,\mathbf{k}_{2\times2}\,\mathbf{d}_{2\times1} + \text{work potential} \tag{6.4.12}$$

Now using the numerical data for the problem, we have the following.

Element 1: $EA = 1,\ L = 0.5,\ \mathbf{d}_{1\times2}^T = \{U_1, U_2\}$.

Element 2: $EA = 1,\ L = 0.5,\ \mathbf{d}_{1\times2}^T = \{U_2, U_3\}$.

Hence the total potential energy in the system can be written as

$$\Pi(\mathbf{D}) = \frac{1}{2}\mathbf{D}_{1\times3}^T\begin{bmatrix} 2 & -2 & 0 \\ -2 & 2 & 0 \\ 0 & 0 & 0 \end{bmatrix}\mathbf{D}_{3\times1} + \frac{1}{2}\mathbf{D}_{1\times3}^T\begin{bmatrix} 0 & 0 & 0 \\ 0 & 2 & -2 \\ 0 & -2 & 2 \end{bmatrix}\mathbf{D}_{3\times1} - (1)U_2 \tag{6.4.13}$$

where $\mathbf{D}^T = \{U_1\ U_2\ U_3\}$ is the vector of (system) nodal displacements; the first term is due to the strain energy in element 1, the second term is due to the strain energy in element 2, and the last term is the work potential due to the concentrated force $P = 1$ acting at $x = 0.5$.

Using the theorem of minimum potential energy by finding the stationary point of $\Pi(\mathbf{D})$, we have

$$\frac{\partial \Pi}{\partial U_1} = 0 = 2U_1 - 2U_2$$

$$\frac{\partial \Pi}{\partial U_2} = 0 = -2U_1 + 4U_2 - 2U_3 - 1 \tag{6.4.14}$$

$$\frac{\partial \Pi}{\partial U_3} = 0 = -2U_2 + 2U_3$$

The three equations can be written in matrix form as

$$\begin{bmatrix} 2 & -2 & 0 \\ -2 & 4 & -2 \\ 0 & -2 & 2 \end{bmatrix}\begin{Bmatrix} U_1 \\ U_2 \\ U_3 \end{Bmatrix} = \begin{Bmatrix} 0 \\ 1 \\ 0 \end{Bmatrix} \tag{6.4.15}$$

or $\mathbf{K}_{3\times3}\mathbf{D}_{3\times1} = \mathbf{F}_{3\times1}$ (6.4.16)

These are the system equations. The process of obtaining these equations was a bit involved. We could have generated the elements equations and gone through the assembly process as we did with the direct stiffness method.

Now imposing the boundary conditions, $U_1 = U_3 = 0$, we have effectively a single equation to solve:

$$4U_2 = 1 \Rightarrow U_2 = 0.25 \tag{6.4.17}$$

which is the exact solution! Now the strains and stresses can be computed in each element using the equations developed earlier.

Observations: This example illustrates the basic idea behind the theorem of minimum potential energy as used with the finite element approach. It is applicable to any system provided the assumptions are not violated. Following up on Eq. (6.4.12), the total potential energy of a structural system is given as

$$\Pi(\mathbf{D}) = \frac{1}{2}\mathbf{D}^\mathbf{T}\mathbf{K}\mathbf{D} - \mathbf{D}^\mathbf{T}\mathbf{F} \tag{6.4.18}$$

(continued)

Applying the theorem yields

$$\frac{\partial \Pi}{\partial \mathbf{D}} = 0 = \mathbf{KD} - \mathbf{F} \tag{6.4.19}$$

or $\mathbf{KD} = \mathbf{F}$ (6.4.20)

The potential energy is a minimum since $\frac{\partial^2 \Pi}{\partial \mathbf{D}^2} = \mathbf{K} > 0$ since $\mathbf{K}$ is a positive definite matrix (a square matrix in which all the eigenvalues are positive).

6.4.1 Truss Analysis

In Fig. 6.4.1.1, the displacement $u(s)$ in the truss element can be assumed as a linear polynomial, and using the nodal conditions seen in the previous section (see Eq. (6.4.5)), we have

$$u(s) = \phi_1(s)\,d_1' + \phi_2(s)\,d_2' = \frac{L-s}{L}d_1' + \frac{s}{L}d_2' \tag{6.4.1.1}$$

Hence the strain ε in the element is constant as

$$\varepsilon = \frac{du}{ds} = \frac{d}{ds}\begin{bmatrix}\phi_1 & \phi_2\end{bmatrix}\begin{Bmatrix}d_1' \\ d_2'\end{Bmatrix} = \begin{bmatrix}-\frac{1}{L} & \frac{1}{L}\end{bmatrix}\begin{Bmatrix}d_1' \\ d_2'\end{Bmatrix} = \mathbf{B}_{1\times 2}\,\mathbf{d}'_{2\times 1} \tag{6.4.1.2}$$

The term $\mathbf{B}$ is usually called the strain-displacement "matrix." The stress-strain relationship is given as (a scalar relationship)

$$\sigma = E\varepsilon \tag{6.4.1.3}$$

Hence the strain energy in the truss element can be written as

$$U = \int_V U_0\,dV = \int_0^L \frac{1}{2}\varepsilon\,\sigma\,A\,ds = \int_0^L \frac{1}{2}[\mathbf{B}_{1\times 2}\,\mathbf{d}'_{2\times 1}]^T E[\mathbf{B}_{1\times 2}\,\mathbf{d}'_{2\times 1}]A\,ds \tag{6.4.1.4}$$

Simplifying yields

$$U = [\mathbf{d}']^T_{1\times 2}\left[\int_0^L \mathbf{B}^T_{2\times 1}(EA)_{1\times 1}\mathbf{B}_{1\times 2}\,ds\right][\mathbf{d}']_{2\times 1} = [\mathbf{d}']^T_{1\times 2}[\mathbf{k}']_{2\times 2}[\mathbf{d}']_{2\times 1} \tag{6.4.1.5}$$

where

$$[\mathbf{k}']_{2\times 2} = \int_0^L \mathbf{B}^T_{2\times 1}(EA)_{1\times 1}\mathbf{B}_{1\times 2}\,ds = \frac{AE}{L}\begin{bmatrix}1 & -1 \\ -1 & 1\end{bmatrix} \tag{6.4.1.6}$$

is the element stiffness matrix and is identical to the one derived in Eq. (6.2.2.5).

The work potential takes place due to concentrated forces acting at the ends of the element and can be written as

$$W = -[\mathbf{d}']^T_{2\times 1}[\mathbf{f}']_{2\times 1} \tag{6.4.1.10}$$

Using the theorem of minimum potential energy, we have

$$\Pi(\mathbf{d}') = U + W$$

and minimizing Π we can express the element equations as

$$\frac{AE}{L}\begin{bmatrix}1 & -1 \\ -1 & 1\end{bmatrix}\begin{Bmatrix}d_1' \\ d_2'\end{Bmatrix} = \begin{Bmatrix}f_1' \\ f_2'\end{Bmatrix} \tag{6.4.1.11}$$

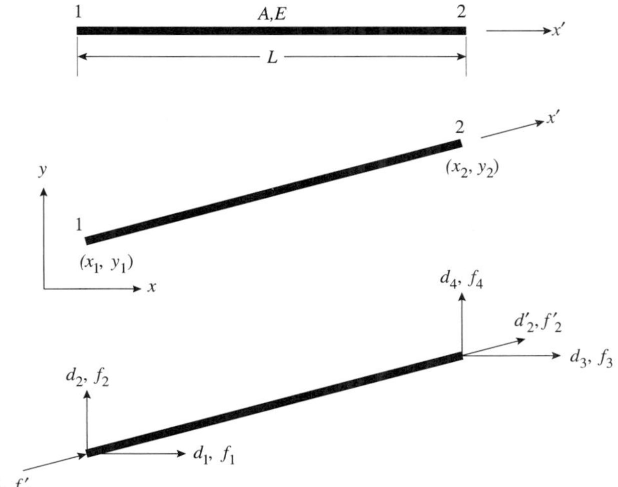

Fig. 6.4.1.1
Description of the planar truss element.

or
$$\mathbf{k}'_{2\times 2}\mathbf{d}'_{2\times 1} = \mathbf{f}'_{2\times 1} \tag{6.4.1.12}$$

where A, E, and L are the element cross-sectional area, modulus of elasticity, and the length respectively, and d'_1, d'_2 and f'_1, f'_2 are the element nodal displacement and element nodal forces, respectively, along the x' (or, axial) direction at nodes 1 and 2.

A few comments are in order.

a. The behavior of the truss element is fundamentally a one-dimensional phenomenon. However, since the different elements in a truss can have different orientations, we need to define the element behavior in a local coordinate system x'. Quantities that are appropriately described in a local coordinate system are denoted as *primed* (′) quantities. To describe the behavior of the truss system, which is now a collection of two or more truss elements, we need to define a global coordinate system x–y that is same for all the elements. Quantities that are appropriately described in a global coordinate system are denoted without any primes.
b. Since the truss element is located in the x–y plane, there are two displacements and two force components at each node of the element. In other words, there are two degrees of freedom at each node leading to a total of four degrees of freedom per element.
c. The displacement in a typical element is linear. Hence the strain and stress in each element are constants.
d. The $\mathbf{k}'$, $\mathbf{d}'$, and $\mathbf{f}'$ are the element stiffness matrix, element nodal displacement vector, and element nodal force vector in the local coordinate system.
e. Note how the displacements (and forces) are numbered: the x-displacement at a node is numbered first followed by the y-displacement.

Equation (6.4.1.12) describes the equilibrium-compatibility of a typical element in the local coordinate system. However, each element in a truss typically can have different local coordinate systems. The underlying question is: how do we relate the element equations for all the different elements in a truss system? The answer lies in defining these equations in a common reference frame—the global coordinate system that is the same for all the elements.

Our next task is to transform Eq. (6.4.1.12) from the local coordinate system to the common reference frame, the global coordinate system. This can be done by first relating the local and global displacements and forces. Note that

$$d_1' = \sqrt{(d_1)^2 + (d_2)^2} \tag{6.4.1.13a}$$

or

$$(d_1')^2 = (d_1)^2 + (d_2)^2 \tag{6.4.1.13b}$$

or

$$d_1' = \frac{d_1}{d_1'} d_1 + \frac{d_2}{d_1'} d_2 \tag{6.4.1.13c}$$

or

$$d_1' = l_{x'} d_1 + m_{x'} d_2 \tag{6.4.1.13d}$$

where $(l_{x'}, m_{x'})$ are the direction cosines of the x' coordinate system with respect to the global coordinate system. Similarly, we can write the equation for the other local displacement as

$$d_2' = l_{x'} d_3 + m_{x'} d_4 \tag{6.4.1.13e}$$

Note that the direction cosines can be computed as

$$l_{x'} = \frac{x_2 - x_1}{L}, \quad m_{x'} = \frac{y_2 - y_1}{L}, \quad L = \sqrt{(x_2 - x_1)^2 + (y_2 - y_1)^2} \tag{6.4.1.13f}$$

Combining Eqs. (6.4.1.13d) and (6.4.1.13e), we have

$$\mathbf{d}'_{2\times1} = \begin{bmatrix} l_{x'} & m_{x'} & 0 & 0 \\ 0 & 0 & l_{x'} & m_{x'} \end{bmatrix} \begin{Bmatrix} d_1 \\ d_2 \\ d_3 \\ d_4 \end{Bmatrix} = \mathbf{T}_{2\times4} \mathbf{d}_{4\times1} \tag{6.4.1.14}$$

Similarly, we can relate the nodal forces at the ends of the member as

$$\begin{Bmatrix} f_1 \\ f_2 \\ f_3 \\ f_4 \end{Bmatrix} = \begin{bmatrix} l_{x'} & 0 \\ m_{x'} & 0 \\ 0 & l_{x'} \\ 0 & m_{x'} \end{bmatrix} \begin{Bmatrix} f_1' \\ f_2' \end{Bmatrix} \quad \Rightarrow \quad \mathbf{f}_{4\times1} = \mathbf{T}^{\mathbf{T}}_{4\times2} \, \mathbf{f}'_{2\times1} \tag{6.4.1.15}$$

Substituting Eqs. (6.4.1.14) and (6.4.1.15) into (6.4.1.12), we have

$$\mathbf{k}_{4\times4} \mathbf{d}_{4\times1} = \mathbf{f}_{4\times1} \tag{6.4.1.16a}$$

where

$$\mathbf{k}_{4\times4} = \mathbf{T}^{\mathbf{T}}_{4\times2} \mathbf{k}'_{2\times2} \mathbf{T}_{2\times4} \tag{6.4.1.16b}$$

is the element stiffness matrix in the global coordinate system.

After the structural equations are solved for the nodal displacements, the strain ε, stress σ, and axial force N in a typical element is computed by first using Eq. (6.4.1.14) to obtain $\mathbf{d}'$ and then

$$\varepsilon = \frac{du}{ds} = \frac{d}{ds}(\phi_1 d_1' + \phi_2 d_2') = \frac{d_2' - d_1'}{L} \tag{6.4.1.17}$$

$$\sigma = E\varepsilon \tag{6.4.1.18}$$

$$N = \sigma A \tag{6.4.1.19}$$

Space Truss Element

The space truss element has only minor differences from the planar truss element. The equations in the local coordinate system do not change. However, there are three degrees of freedom per node and six degrees of freedom per element in the global coordinate system:

$$\mathbf{d}'_{2\times1} = \mathbf{T}_{2\times6}\mathbf{d}_{6\times1} \tag{6.4.1.20}$$

$$\mathbf{f}_{6\times1} = \mathbf{T}^{\mathbf{T}}_{6\times2}\,\mathbf{f}'_{2\times1} \tag{6.4.1.21}$$

$$\mathbf{k}_{6\times6}\mathbf{d}_{6\times1} = \mathbf{f}_{6\times1} \tag{6.4.1.22}$$

$$\mathbf{k}_{6\times6} = \mathbf{T}^{\mathbf{T}}_{6\times2}\mathbf{k}'_{2\times2}\mathbf{T}_{2\times6} \tag{6.4.1.23}$$

$$l_{x'} = \frac{x_2 - x_1}{L},\ m_{x'} = \frac{y_2 - y_1}{L},\ n_{x'} = \frac{z_2 - z_1}{L} \tag{6.4.1.24}$$

$$L = \sqrt{(x_2 - x_1)^2 + (y_2 - y_1)^2 + (z_2 - z_1)^2} \tag{6.4.1.25}$$

$$\mathbf{T}_{2\times6} = \begin{bmatrix} l_{x'} & m_{x'} & n_{x'} & 0 & 0 & 0 \\ 0 & 0 & 0 & l_{x'} & m_{x'} & n_{x'} \end{bmatrix} \tag{6.4.1.26}$$

EXAMPLE 6.4.2 ***Space Truss***

For the space truss shown in Fig. E6.4.2(a), compute the nodal displacements and the force in each member. The modulus of elasticity is 29000 ksi and the cross-sectional area of each member is 2.5 in^2.

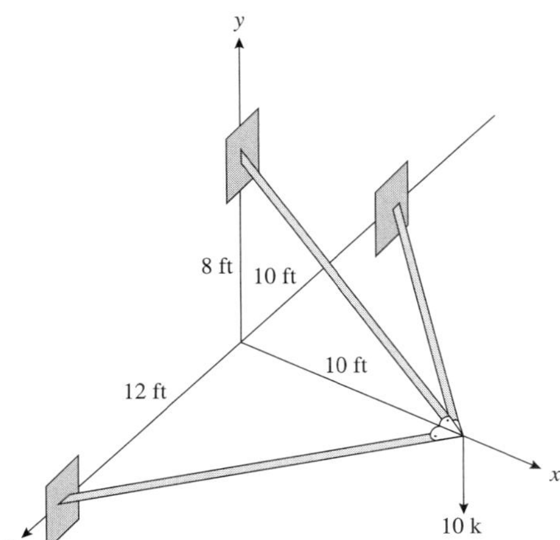

Fig. E6.4.2(a)

SOLUTION

Step 1: The problem units are lb, in. The model details are shown in Fig. E6.4.2(b).

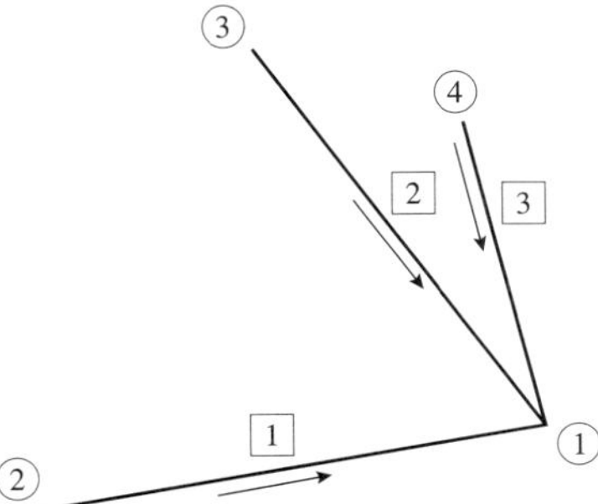

Fig. E6.4.2(b)

The nodal boundary conditions are $D_4 = \ldots = D_{12} = 0$. In other words, there are only three effective degrees of freedom (see Fig. E6.4.2(c)).

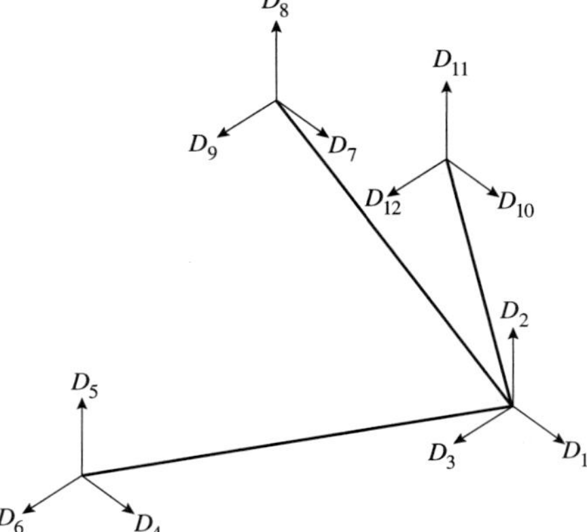

Fig. E6.4.2(c)

We omit some of the details here, but it should be noted that the steps are essentially the same as those used for planar trusses.

Step 2: Element equations:

Element 1:

$$10^5 \begin{bmatrix} 1.5852 & 0 & -1.9022 & -1.5852 & 0 & 1.9022 \\ & 0 & 0 & 0 & 0 & 0 \\ & & 2.2826 & 1.9022 & 0 & -2.2826 \\ & & & 1.5852 & 0 & -1.9022 \\ & & & & 0 & 0 \\ \text{Sym} & & & & & 2.2826 \end{bmatrix} \begin{Bmatrix} D_4 \\ D_5 \\ D_6 \\ D_1 \\ D_2 \\ D_3 \end{Bmatrix} = \begin{Bmatrix} f_1^1 \\ f_2^1 \\ f_3^1 \\ f_4^1 \\ f_5^1 \\ f_6^1 \end{Bmatrix}$$

Element 2:

$$10^5 \begin{bmatrix} 2.8767 & -2.3013 & 0 & -2.8767 & 2.3013 & 0 \\ & 1.8411 & 0 & 2.3013 & -1.8411 & 0 \\ & & 0 & 0 & 0 & 0 \\ & & & 2.8767 & -2.3013 & 0 \\ & & & & 1.8411 & 0 \\ \text{Sym} & & & & & 0 \end{bmatrix} \begin{Bmatrix} D_7 \\ D_8 \\ D_9 \\ D_1 \\ D_2 \\ D_3 \end{Bmatrix} = \begin{Bmatrix} f_1^2 \\ f_2^2 \\ f_3^2 \\ f_4^2 \\ f_5^2 \\ f_6^2 \end{Bmatrix}$$

Element 3:

$$10^5 \begin{bmatrix} 2.1361 & 0 & 2.1361 & -2.1361 & 0 & -2.1361 \\ & 0 & 0 & 0 & 0 & 0 \\ & & 2.1361 & -2.1361 & 0 & -2.1361 \\ & & & 2.1361 & 0 & 2.1361 \\ & & & & 0 & 0 \\ \text{Sym} & & & & & 2.1361 \end{bmatrix} \begin{Bmatrix} D_{10} \\ D_{11} \\ D_{12} \\ D_1 \\ D_2 \\ D_3 \end{Bmatrix} = \begin{Bmatrix} f_1^3 \\ f_2^3 \\ f_3^3 \\ f_4^3 \\ f_5^3 \\ f_6^3 \end{Bmatrix}$$

A few observations: a row (and the corresponding column) with all zero elements indicates that the element has zero stiffness along that degree of freedom. Every element in the truss has a zero row (and column). Element 1 lies in the x–z plane and hence has zero stiffness in the y-direction (rows 3 and 6). Similarly, element 2 has zero stiffness in the z-direction. Element 3 is similar to element 1—zero stiffness in the y-direction. While certain rows and columns can be zero in an element stiffness matrix, we cannot have a zero row or column in the structural stiffness matrix after the boundary conditions are imposed. If it did, it would indicate a zero stiffness along that degree of freedom. In other words, we would have an unstable structure.

Step 3: Assembly of the system equations:

$$10^5 \begin{bmatrix} 6.5979 & -2.3013 & 0.23386 \\ -2.3013 & 1.8411 & 0 \\ 0.23386 & 0 & 4.4187 \end{bmatrix} \begin{Bmatrix} D_1 \\ D_2 \\ D_3 \end{Bmatrix} = \begin{Bmatrix} 0 \\ -10^4 \\ 0 \end{Bmatrix}$$

Step 4: Solution of the system equations. Solving, we have

$$D_1 = -3.3703(10^{-2}) \text{ in} \qquad D_2 = -9.6445(10^{-2}) \text{ in} \qquad D_3 = 1.7838(10^{-3}) \text{ in}$$

Step 5: Element nodal forces. The computation of the element nodal forces can be carried out using Eq. (6.2.2.27).

Element 1:

$$f_1' = 386778[\,0.64 \quad 0 \quad -0.768 \quad -0.64 \quad 0 \quad 0.768\,] \begin{Bmatrix} 0 \\ 0 \\ 0 \\ -0.0337 \\ -0.09644 \\ 0.00178 \end{Bmatrix} = 8875 \text{ lb} \quad \text{(C)}$$

Element 2:

$$f_1' = 471775[\,0.781 \quad -0.625 \quad 0 \quad -0.781 \quad 0.625 \quad 0\,] \begin{Bmatrix} 0 \\ 0 \\ 0 \\ -0.0337 \\ -0.09644 \\ 0.00178 \end{Bmatrix} = -16008 \text{ lb} \quad \text{(T)}$$

Element 3:

$$f_1' = 427210[\,0.707 \quad 0 \quad -0.707 \quad -0.707 \quad 0 \quad 0.707\,] \begin{Bmatrix} 0 \\ 0 \\ 0 \\ -0.0337 \\ -0.09644 \\ 0.00178 \end{Bmatrix} = 9642 \text{ lb} \quad \text{(C)}$$

A check of the FBD of node 1 shows that the node is in equilibrium.

6.4.2 Frame Analysis

The beam behavior illustrated in this section is one that includes axial, shear, and moment effects. This beam is also referred to as the Euler-Bernoulli beam. Figure 6.4.2.1 shows a simply supported beam subjected to transverse loads. From elementary beam theory (compression is negative), we have

$$\sigma_x = -\frac{M_z y}{I_z} \tag{6.4.2.1a}$$

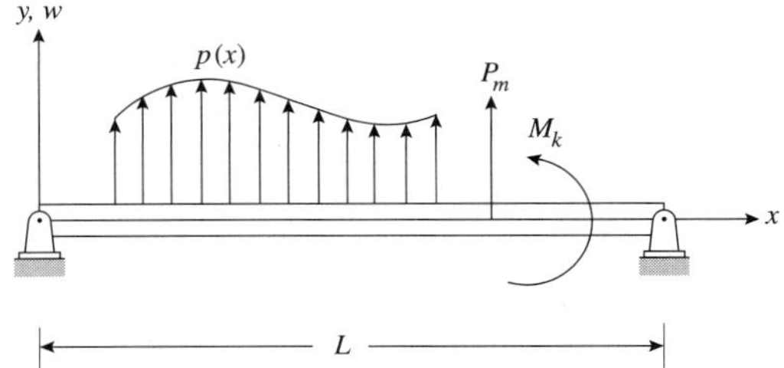

Fig. 6.4.2.1 Planar beam subjected to different loads.

$$\sigma_x = E\varepsilon_x \tag{6.4.2.1b}$$

$$\frac{d^2w(x)}{dx^2} = \frac{M_z}{EI_z} \tag{6.4.2.1c}$$

where M_z is the moment, E is the modulus of elasticity, w is the transverse deflection of the centroidal axis, and I_z is the moment of inertia about the centroidal axis. We drop the subscripts in the next step. The strain energy in the beam is given by

$$U = \int_V U_0 \, dV = \int_0^L \int_A \frac{1}{2} \varepsilon \sigma \, dA \, dx = \frac{1}{2} \int_0^L \left[\frac{M^2}{EI^2} \int_A y^2 \, dA \right] dx \tag{6.4.2.2}$$

Noting that $I = \int_A y^2 \, dA$, we have

$$U = \frac{1}{2} \int_0^L EI \left(\frac{d^2w}{dx^2} \right)^2 dx \tag{6.4.2.3}$$

The total potential energy in the beam is given by

$$\Pi = \frac{1}{2} \int_0^L EI \left(\frac{d^2w}{dx^2} \right)^2 dx - \int_0^L pw \, dx - \sum_m P_m w_m - \sum_k M_k \frac{dw}{dx} \tag{6.4.2.4}$$

where the last three terms are the work potential terms due to distributed element loads, concentrated forces, and concentrated moments, respectively. We are now ready to build a typical finite element and computer its potential energy. Figure 6.4.2.2 shows the two degrees of freedom at any point on the beam (and the beam element).

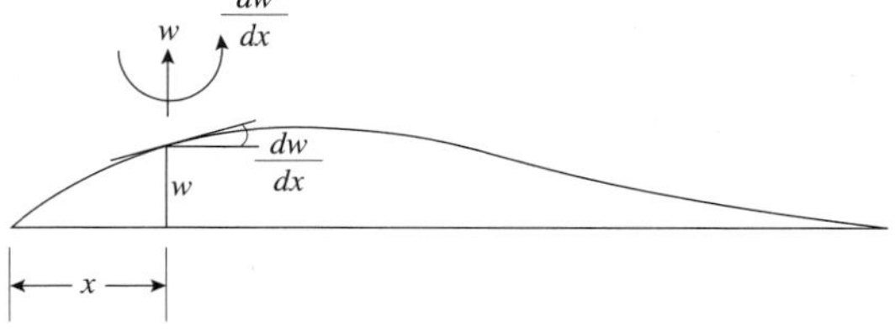

Fig. 6.4.2.2 Deformation of the neutral axis, showing the two dof at any point.

Fig. 6.4.2.3 shows a typical beam element. It is described by two nodes and four degrees of freedom.

Since four nodal conditions are known—a displacement and a slope at each node—the lowest-order polynomial we can use to characterize the transverse displacement is a cubic polynomial:

$$w(x) = a_1 + a_2 x + a_3 x^2 + a_4 x^3 \tag{6.4.2.5}$$

Fig. 6.4.2.3 Typical beam element description.

The nodal conditions are

$$w(x=0)=w_1 \qquad w(x=L)=w_2 \tag{6.4.2.6a}$$

$$\frac{dw}{dx}(x=0)=\theta_1 \qquad \frac{dw}{dx}(x=L)=\theta_2 \tag{6.4.2.6b}$$

After substituting these conditions in Eq. (6.4.2.5) and solving for the four coefficients, we can write the transverse displacement as

$$w(x)=\phi_1 w_1+\phi_2\theta_1+\phi_3 w_2+\phi_4\theta_2 \tag{6.4.2.7}$$

where the shape functions are (see Fig. 6.4.2.4)

$$\phi_1=1-\frac{3x^2}{L^2}+\frac{2x^3}{L^3} \qquad \phi_3=\frac{3x^2}{L^2}-\frac{2x^3}{L^3} \tag{6.4.2.8a}$$

$$\phi_2=x-\frac{2x^2}{L}+\frac{x^3}{L^2} \qquad \phi_4=-\frac{x^2}{L}+\frac{x^3}{L^2} \tag{6.4.2.8b}$$

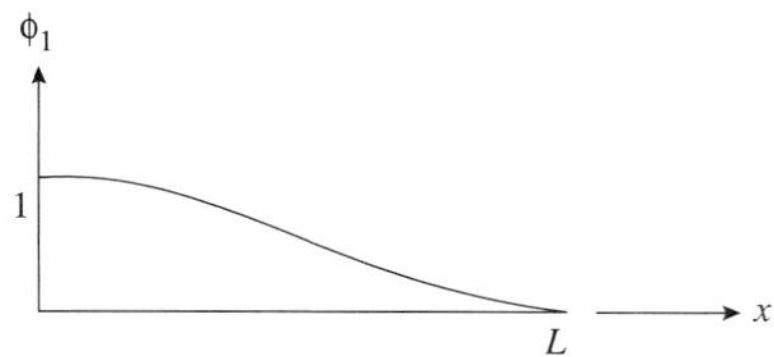

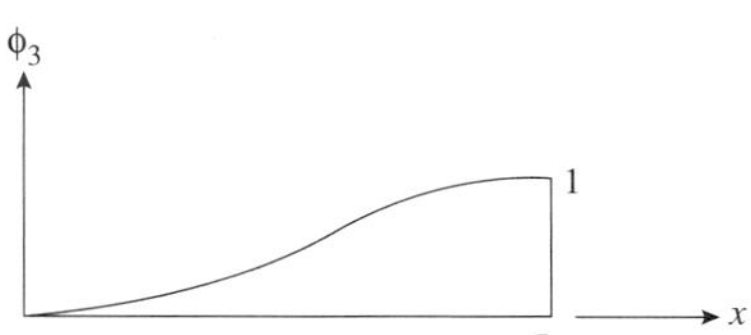

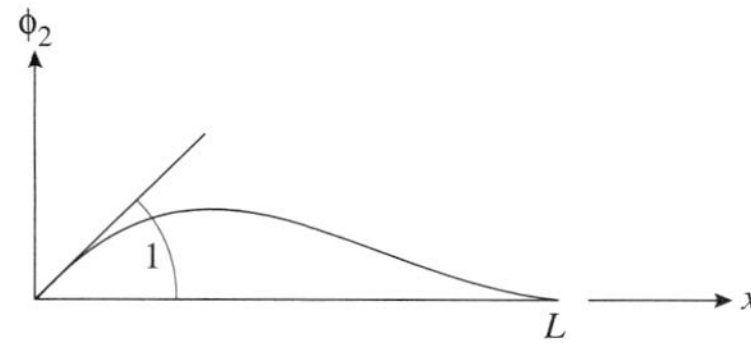

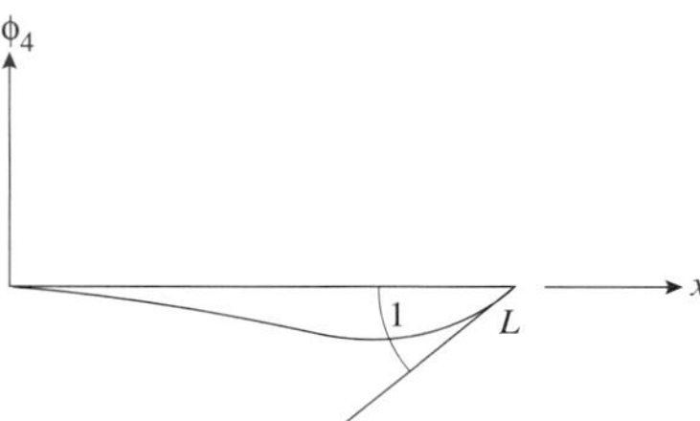

Fig. 6.4.2.4 The four shape functions.

Now using Eq. (6.4.2.4) to compute the strain energy in a typical element, we note that

$$\frac{d^2w}{dx^2}=\left[-\frac{6}{L^2}+\frac{12x}{L^3} \;\middle|\; -\frac{4}{L}+\frac{6x}{L^2} \;\middle|\; \frac{6}{L^2}-\frac{12x}{L^3} \;\middle|\; -\frac{2}{L}+\frac{6x}{L^2}\right]_{1\times 4}\mathbf{d}_{4\times 1} \tag{6.4.2.9a}$$

or

$$\frac{d^2w}{dx^2}=\mathbf{B}_{1\times 4}\mathbf{d}_{4\times 1} \tag{6.4.2.9b}$$

Hence substituting in Eq. (6.4.2.4), we have

$$U = \frac{1}{2}\mathbf{d}^{\mathrm{T}}\left[\int_0^L \mathbf{B}^T\, EI\, \mathbf{B}\, dx\right]\mathbf{d} = \frac{1}{2}\mathbf{d}^{\mathrm{T}}_{1\times 4}\mathbf{k}_{4\times 4}\mathbf{d}_{4\times 1} \tag{6.4.2.10}$$

Hence,

$$\mathbf{k}_{4\times 4} = \frac{EI}{L^3}\begin{bmatrix} 12 & 6L & -12 & 6L \\ & 4L^2 & -6L & 2L^2 \\ Sym & & 12 & -6L \\ & & & 4L^2 \end{bmatrix} \tag{6.4.2.11}$$

While the prime notation has not been used, the above derivation is for the quantities in the local coordinate system. Moreover, it does *not* include axial effects. The inclusion of axial effects is quite simple. Figure 6.4.2.5 shows the general beam element. With the assumptions made at the beginning of this section, the axial effects are independent of the bending effects. The general beam element is the linear superposition of the axial behavior captured by truss element and the bending behavior captured by the beam element. The following should be noted about the element description.

a. The element lies in the x–y plane. The coordinate systems are such that the local z' and the global z coincide. They are obtained by taking the cross product of the local x' and y' axes. To find the direction cosines of the x' axis we can employ the following expressions:

$$l_{x'} = \frac{x_2 - x_1}{L}, \quad m_{x'} = \frac{y_2 - y_1}{L}, \quad L = \sqrt{(x_2 - x_1)^2 + (y_2 - y_1)^2} \tag{6.4.2.12}$$

Similarly, the direction cosines of the y' axis can be written as

$$l_{y'} = -m_{x'}, \quad m_{y'} = l_{x'} \tag{6.4.2.13}$$

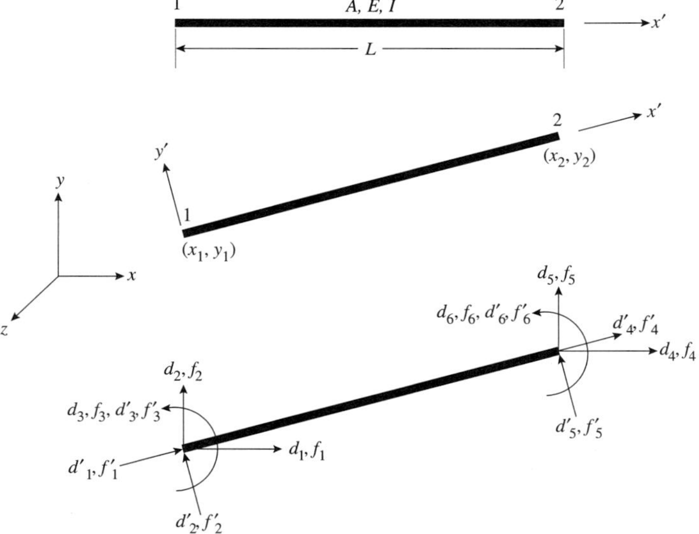

Fig. 6.2.4.5 General beam element description.

b. There are six degrees of freedom in the element in the local (d_1', d_2', d_3', d_4', d_5', d_6') and the global (d_1, d_2, d_3, d_4, d_5, d_6) coordinate systems. The three nodal forces at each node in the local coordinate system refer to the axial force (f_1', f_4'), the shear force (f_2', f_5') and the bending moment (f_3', f_6'). The global forces, in general, cannot be classified. The local and global displacements and forces are related to each other as follows:

$$\mathbf{d}'_{6\times1} = \begin{bmatrix} l_{x'} & m_{x'} & 0 & 0 & 0 & 0 \\ l_{y'} & m_{y'} & 0 & 0 & 0 & 0 \\ 0 & 0 & 1 & 0 & 0 & 0 \\ 0 & 0 & 0 & l_{x'} & m_{x'} & 0 \\ 0 & 0 & 0 & l_{y'} & m_{y'} & 0 \\ 0 & 0 & 0 & 0 & 0 & 1 \end{bmatrix} \begin{Bmatrix} d_1 \\ d_2 \\ d_3 \\ d_4 \\ d_5 \\ d_6 \end{Bmatrix} = \mathbf{T}_{6\times6}\mathbf{d}_{6\times1} \tag{6.4.2.14}$$

$$\mathbf{f}_{6\times1} = \mathbf{T}^{\mathbf{T}}_{6\times6}\mathbf{f}_{6\times1} \tag{6.4.2.15}$$

c. The element local stiffness matrix is obtained by combining Eq. (6.4.2.11) and (6.4.1.6) and is given as

$$\mathbf{k}'_{6\times6} = \begin{bmatrix} \frac{EA}{L} & 0 & 0 & 0 & 0 & 0 \\ & \frac{12EI}{L^3} & \frac{6EI}{L^2} & 0 & -\frac{12EI}{L^3} & \frac{6EI}{L^2} \\ & & \frac{4EI}{L} & 0 & -\frac{6EI}{L^2} & \frac{2EI}{L} \\ & Sym & & \frac{EA}{L} & 0 & 0 \\ & & & & \frac{12EI}{L^3} & -\frac{6EI}{L^2} \\ & & & & & \frac{4EI}{L} \end{bmatrix} \tag{6.4.2.16}$$

The element global stiffness matrix is obtained similarly to the truss element:

$$\mathbf{k}_{6\times6} = \mathbf{T}^{\mathbf{T}}_{6\times6}\mathbf{k}'_{6\times6}\mathbf{T}_{6\times6} \tag{6.4.2.17}$$

d. The equivalent nodal forces due to loads acting on the element can be found from

$$\mathbf{q}_{6\times1} = \mathbf{T}^{\mathbf{T}}_{6\times1}\mathbf{q}'_{6\times1} \tag{6.4.2.18}$$

For example, if a uniformly distributed load of intensity p acts in the positive y' direction, the equivalent nodal forces are computed as

$$q_i = \int_0^L p(x)\phi_i(x)\,dx = p\int_0^L \phi_i(x)\,dx \quad i = 1, 2, 3, 4 \tag{6.4.2.19}$$

where the shape functions ϕ_i are given by Eqs. (6.4.2.8). Evaluating the above, we have

$$\mathbf{q}'_{6\times1} = \left[0, \frac{pL}{2}, \frac{pL^2}{12}, 0, \frac{pL}{2}, -\frac{pL^2}{12}\right]^{\mathbf{T}} \tag{6.4.2.20}$$

These equivalent nodal forces can then be transformed from the local to the global coordinate system:

$$\mathbf{q}_{6\times1} = \mathbf{T}^{\mathbf{T}}_{6\times6}\mathbf{q}'_{6\times1} \tag{6.4.2.21}$$

and added to the system nodal force vector.

e. Once the system or global equilibrium equations are solved for the nodal displacements, the member nodal forces can be computed as

$$\mathbf{d}'_{6\times1} = \mathbf{T}_{6\times6}\,\mathbf{d}_{6\times1} \tag{6.4.2.22}$$

$$\mathbf{f}'_{6\times1} = \mathbf{k}'_{6\times6}\mathbf{d}'_{6\times1} - \sum_i \left(\mathbf{q}'_{6\times1}\right)_i \tag{6.4.2.23}$$

where the summation is over all the element loads acting on the element. The last term is necessary to satisfy element equilibrium since the element is subjected to element loads. Note that the strains and stresses can be computed only if the cross-sectional shape is known.

EXAMPLE 6.4.3 ***Planar Frame***

Consider the frame in Fig. E6.4.3(a). The modulus of elasticity is $2(10^{11})$ Pa, the cross-sectional area is 0.01 m^2 and the moment of inertia is 0.0001 m^4 for both the members. Compute the member nodal forces.

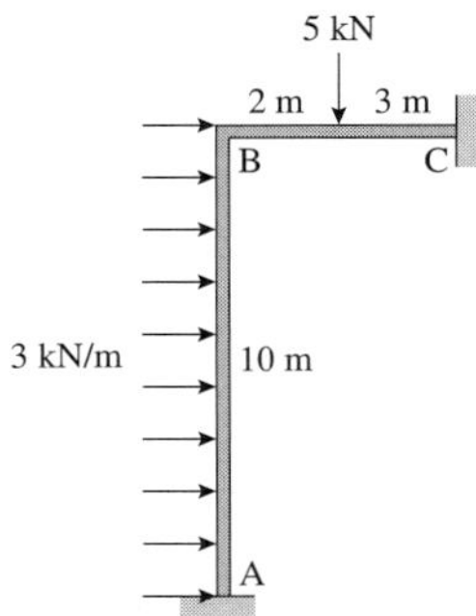

Fig. E6.4.3(a)

SOLUTION

Step 1: The problem units are N, m. We select the origin of the coordinate system at A. The node and element numbers are shown in Fig. E6.4.3(b). We also number the global degrees of freedom at the nodes. As can be seen from the figure, there is a total of nine degrees of freedom in the frame. However, the boundary conditions of the frame are such that

$$D_1 = D_2 = D_3 = D_7 = D_8 = D_9 = 0$$

Instead of using Eqs. (6.4.2.17) and (6.4.2.21) to generate the element equilibrium equations, we use the derived form presented in Eqs. (6.2.3.14b) and (6.2.3.13b).

Element	(l, m)	(a, b, c, d)
1	(0,1)	(2e8, 240000, 1.2e6, 4e6)
2	(1,0)	(4e8, 1.92e6, 4.8e6, 8e6)

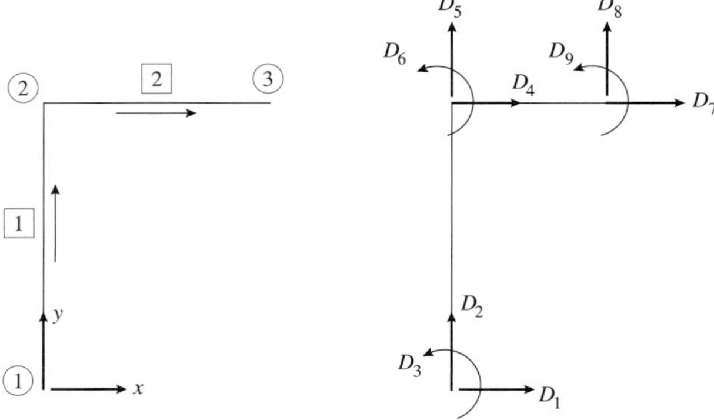

Fig. E6.4.3(b)

Element load on element 1: With $w = -3000$ N/m (see Fig. 6.2.3.9), we have

$$\mathbf{q}'_{6\times1} = \left\{0, \frac{wL}{2}, \frac{wL^2}{12}, 0, \frac{wL}{2}, -\frac{wL^2}{12}\right\} = \{0, 15000, -25000, 0, 15000, 25000\}$$

Element load on element 2: With $P = -5000$ N, $a = 2$ m, $b = 3$ m (see Fig. 6.2.3.9), we have

$$\mathbf{q}'_{6\times1} = \left\{0, \frac{Pb^2(L+2a)}{L^3}, \frac{Pab^2}{L^2}, 0, \frac{Pa^2(L+2b)}{L^3}, -\frac{Pa^2b}{L^2}\right\} = \{0, -3240, -3600, 0, -1760, 2400\}$$

These loads need to be transformed to the global coordinate system using Eq. (6.4.2.21).

Step 2: The element equilibrium equations. We can use the results from step 1 to generate the element equilibrium equations for each element.

Element 1:

$$10^5\begin{bmatrix} 24 & 0 & -12 & -2.4 & 0 & -12 \\ 0 & 2000 & 0 & 0 & -2000 & 0 \\ -12 & 0 & 80 & 12 & 0 & 40 \\ -2.4 & 0 & 12 & 2.4 & 0 & 12 \\ 0 & -2000 & 0 & 0 & 2000 & 0 \\ -12 & 0 & 40 & 12 & 0 & 80 \end{bmatrix}\begin{Bmatrix} D_1 \\ D_2 \\ D_3 \\ D_4 \\ D_5 \\ D_6 \end{Bmatrix} = \begin{Bmatrix} 15000 \\ 0 \\ -25000 \\ 15000 \\ 0 \\ 25000 \end{Bmatrix}$$

Element 2:

$$10^5\begin{bmatrix} 4000 & 0 & 0 & -4000 & 0 & 0 \\ 0 & 19.2 & 48 & 0 & -19.2 & 48 \\ 0 & 48 & 160 & 0 & -48 & 80 \\ -4000 & 0 & 0 & 4000 & 0 & 0 \\ 0 & -19.2 & -48 & 0 & 19.2 & -48 \\ 0 & 48 & 80 & 0 & -48 & 160 \end{bmatrix}\begin{Bmatrix} D_4 \\ D_5 \\ D_6 \\ D_7 \\ D_8 \\ D_9 \end{Bmatrix} = \begin{Bmatrix} 0 \\ -3240 \\ -3600 \\ 0 \\ -1760 \\ 2400 \end{Bmatrix}$$

Step 3: Assembly of the system equations $\mathbf{K}_{3\times3}\mathbf{D}_{3\times1} = \mathbf{F}_{3\times1}$. We assemble only the effective equations:

$$10^5\begin{bmatrix} 4002.4 & 0 & 12 \\ 0 & 2019.2 & 48 \\ 12 & 48 & 240 \end{bmatrix}\begin{Bmatrix} D_4 \\ D_5 \\ D_6 \end{Bmatrix} = \begin{Bmatrix} 15000 \\ -3240 \\ 21400 \end{Bmatrix}$$

Note that both the element stiffness matrix and the system stiffness matrix are symmetric.

Step 4: Solution of the equilibrium equations. Solving the three equations, we have

$$D_4 = 3.48(10^{-5})\text{ m} \qquad D_5 = -3.74(10^{-5})\text{ m} \qquad D_6 = 8.97(10^{-4})\text{ rad}$$

Step 5: Computation of element nodal forces. Using Eqs. (6.4.2.22) and (6.4.2.23), we can compute the element nodal forces. The details (intermediate steps) are not shown here.

Element 1:

$$\mathbf{f}'_{6\times1} = \{7476, 16085, 28631, -7476, 13915, -17779\}\text{ N}$$

Element 2:

$$\mathbf{f}'_{6\times1} = \{13915, 7476, 17779, -13915, -2476, 4600\}\text{ N}$$

The element FBDs are shown in Fig. E6.4.3(c).

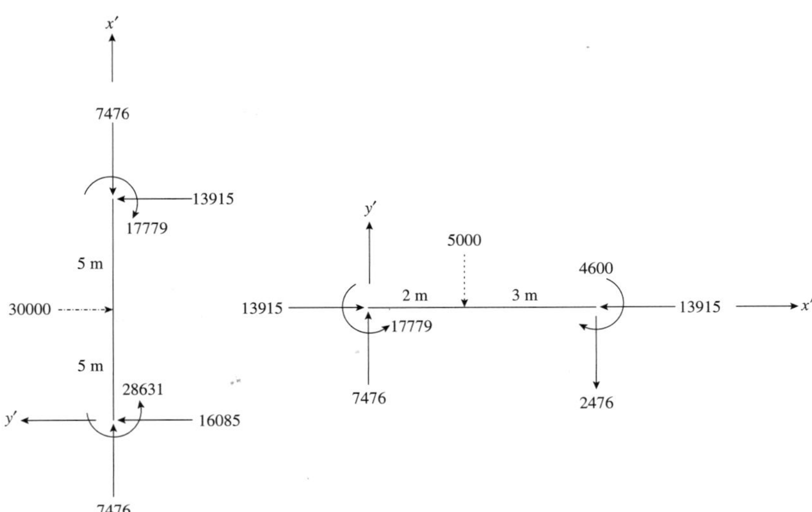

Fig. E6.4.3(c)

Space Beam Element

The space beam element is quite a bit different from the planar beam element. There are 12 degrees of freedom in the element, with 6 degrees of freedom per node. The element is shown in Fig. 6.4.2.6. The composite behavior of the element is a superposition of the following effects:

a. Axial deformation along x'
b. Bending about the y' and z' axes
c. Torsional deformation (rotation) about the x' axis

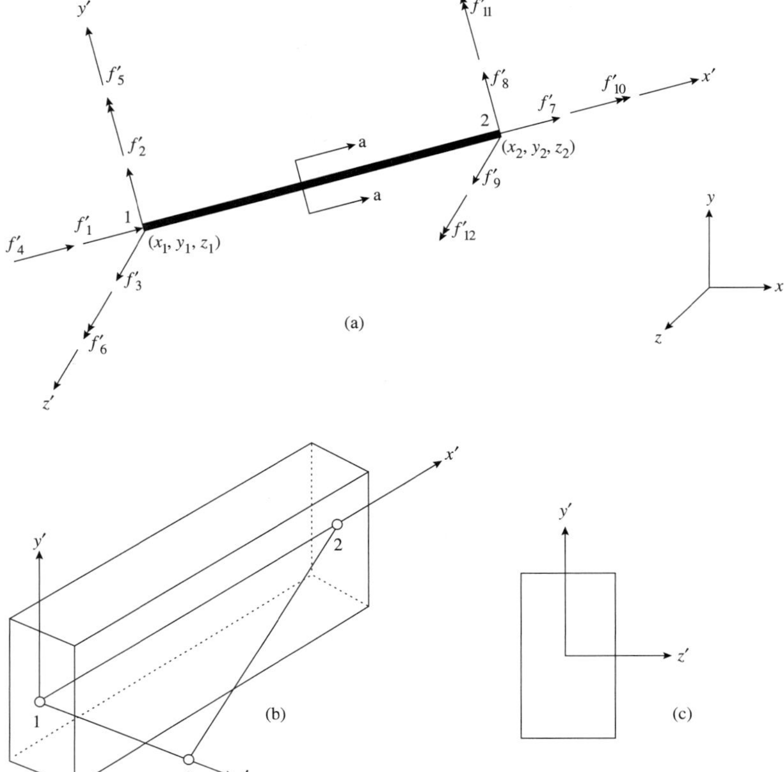

Fig. 6.2.4.6
(a) Orientation. (b) Local coordinate system.
(c) View a-a.

The x' axis is the centroidal axial axis. The y' and z' axes are the principal axes with the $x'-z'$ plane as the major principal plane of bending and $x'-y'$ plane as the minor principal plane of bending. (f_1', f_7') are the axial forces, (f_4', f_8') are the torsional moments, (f_2', f_8') are the shear forces in the y' direction, (f_3', f_9') are the shear forces in the z' direction, (f_5', f_{11}') are the bending moments about the y' axis, and (f_6', f_{12}') are the bending moments about the z' axis. To define the orientation of the element, we need an additional point. Node 3 is known as the reference point. In this formulation, the purpose of specifying this third point is to define the (major) principal plane of bending. While point 3 can be placed anywhere on the principal plane, we place point 3 on the z' axis so that 1–3 points in the positive z' direction.

We now look at an aspect that was ignored for the planar beam element. The total transverse (or lateral) deflection is given as

$$w = w_b + w_s \tag{6.4.2.24}$$

where w_b is the deflection due to bending strains and w_s is the deflection due to shearing strains. The former was considered earlier. The latter is such that

$$\frac{dw_s}{dx} = -\frac{V}{GA_s} \tag{6.4.2.25}$$

where G is the shear modulus and A_s is the beam cross-sectional area effective in shear. This beam is referred to as the Timoshenko beam. The equation can then be used to compute the shearing strain energy. The element stiffness matrix can now be computed and expressed as follows:

$$\mathbf{k}'_{12\times12} = \left[\begin{array}{c|c} \mathbf{k}_{11} & \mathbf{k}_{12} \\ \hline \mathbf{k}_{21} & \mathbf{k}_{22} \end{array}\right]_{12\times12} \tag{6.4.2.26}$$

where

$$\mathbf{k}_{11} = \begin{bmatrix} \dfrac{EA}{L} & 0 & 0 & 0 & 0 & 0 \\ & \dfrac{12EI_z}{L^3\alpha_y} & 0 & 0 & 0 & \dfrac{6EI_z}{L^2\alpha_y} \\ & & \dfrac{12EI_y}{L^3\alpha_z} & 0 & -\dfrac{6EI_y}{L^2\alpha_z} & 0 \\ & & & \dfrac{GJ}{L} & 0 & 0 \\ & \text{Sym} & & & \dfrac{\beta_z EI_y}{L\alpha_z} & 0 \\ & & & & & \dfrac{\beta_z EI_z}{L\alpha_y} \end{bmatrix} \tag{6.4.2.27a}$$

$$\mathbf{k}_{22} = \begin{bmatrix} \dfrac{EA}{L} & 0 & 0 & 0 & 0 & 0 \\ & \dfrac{12EI_z}{L^3\alpha_y} & 0 & 0 & 0 & -\dfrac{6EI_z}{L^2\alpha_y} \\ & & \dfrac{12EI_y}{L^3\alpha_z} & 0 & \dfrac{6EI_y}{L^2\alpha_z} & 0 \\ & & & \dfrac{GJ}{L} & 0 & 0 \\ & \text{Sym} & & & \dfrac{\beta_z EI_y}{L\alpha_z} & 0 \\ & & & & & \dfrac{\beta_z EI_z}{L\alpha_y} \end{bmatrix} \tag{6.4.2.27b}$$

where $\alpha_y = (1+\Phi_y)$, $\alpha_z = (1+\Phi_z)$, $\beta_y = (4+\Phi_y)$, and $\beta_z = (4+\Phi_z)$:

$$\Phi_y = \frac{12EI_z}{GA_{s_y}L^2} \quad \text{and} \quad \Phi_z = \frac{12EI_y}{GA_{s_z}L^2}$$ [9]

$$\mathbf{k}_{12} = \mathbf{k}_{21}^{\mathbf{T}} = \begin{bmatrix} -\frac{EA}{L} & 0 & 0 & 0 & 0 & 0 \\ 0 & -\frac{12EI_z}{L^3\alpha_y} & 0 & 0 & 0 & \frac{6EI_z}{L^2\alpha_y} \\ 0 & 0 & -\frac{12EI_y}{L^3\alpha_z} & 0 & -\frac{6EI_y}{L^2\alpha_z} & 0 \\ 0 & 0 & 0 & -\frac{GJ}{L} & 0 & 0 \\ 0 & 0 & \frac{6EI_y}{L^2\alpha_z} & 0 & \frac{\gamma_z EI_y}{L\alpha_z} & 0 \\ 0 & -\frac{6EI_z}{L^2\alpha_y} & 0 & 0 & 0 & \frac{\gamma_y EI_z}{L\alpha_y} \end{bmatrix}$$

where $\gamma_y = (2-\Phi_y)$ $\quad \gamma_z = (2-\Phi_z)$ (6.4.2.28)

The local-to-global transformation matrix $\mathbf{T}_{12\times12}$ can be constructed as

$$\mathbf{T}_{12\times12} = \begin{bmatrix} \mathbf{\Lambda} & & & \\ & \mathbf{\Lambda} & & \\ & & \mathbf{\Lambda} & \\ & & & \mathbf{\Lambda} \end{bmatrix} \qquad \mathbf{\Lambda}_{3\times3} = \begin{bmatrix} l_{x'} & m_{x'} & n_{x'} \\ l_{y'} & m_{y'} & n_{y'} \\ l_{z'} & m_{z'} & n_{z'} \end{bmatrix} \tag{6.4.2.29}$$

Let $\mathbf{e}_{x'}$, $\mathbf{e}_{y'}$, $\mathbf{e}_{z'}$ be the unit vectors along the local x, y, z axes. Then

$$L = \sqrt{(x_2-x_1)^2 + (y_2-y_1)^2 + (z_2-z_1)^2} \tag{6.4.2.30a}$$

$$\mathbf{e}_{x'} = [l_{x'} \quad m_{x'} \quad n_{x'}] \Rightarrow l_{x'} = \frac{x_2-x_1}{L}, \quad m_{x'} = \frac{y_2-y_1}{L}, \quad n_{x'} = \frac{z_2-z_1}{L} \tag{6.4.2.30b}$$

$$\mathbf{e}_{13} = \frac{x_3-x_1}{L_{13}}\hat{i} + \frac{y_3-y_1}{L_{13}}\hat{j} + \frac{z_3-z_1}{L_{13}}\hat{k} \tag{6.4.2.30c}$$

$$L_{13} = \sqrt{(x_3-x_1)^2 + (y_3-y_1)^2 + (z_3-z_1)^2} \tag{6.4.2.30d}$$

$$\mathbf{e}_{y'} = [l_{y'} \quad m_{y'} \quad n_{y'}] \Rightarrow \mathbf{e}_{y'} = \mathbf{e}_{13} \times \mathbf{e}_{x'} \tag{6.4.2.30e}$$

$$\mathbf{e}_{z'} = [l_{z'} \quad m_{z'} \quad n_{z'}] \Rightarrow \mathbf{e}_{z'} = \mathbf{e}_{13} \tag{6.4.2.30f}$$

The rest of the computations, including computation of the equivalent nodal forces, nodal forces (or element stress resultants), etc., are carried out in a manner similar to those described for the planar beam element.

[9] $A_{s_y} = (5/6)A = A_{s_z}$ for a rectangular cross-section.

EXAMPLE 6.4.4 ***Space Frame***

A cantilever frame (Fig. E6.4.4(a)) is made of a material whose $E = 1600$ ksi and $\nu = 0.2$. The member cross-sections are rectangular with the width 10 in and height 16 in. Compute the nodal displacements and member nodal forces.

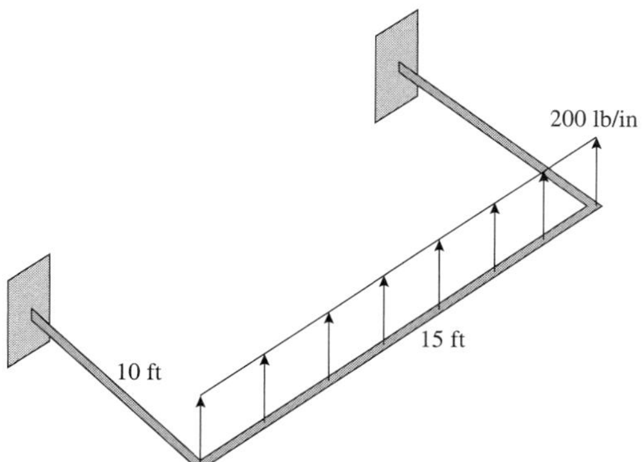

Fig. E6.4.4(a)

SOLUTION ***Step 1:*** The problem units are lb, in. The model details are shown in Fig. E6.4.4(b).

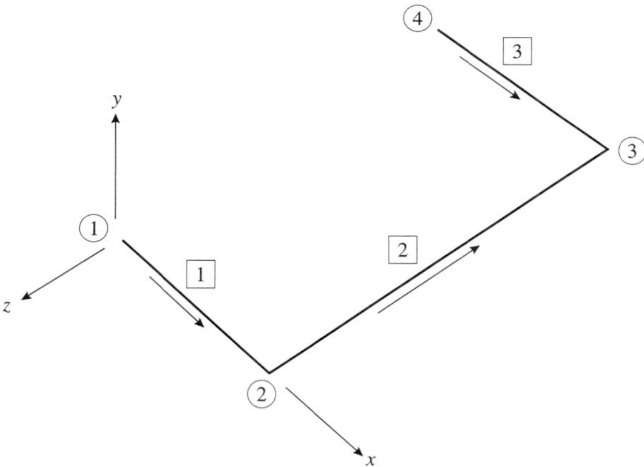

Fig. E6.4.4(b)

There are six degrees of freedom per node for a total of 24 dof. However, based on the manner in which the frame is supported, there are effectively 12 dof (see Fig. E6.4.4(c)). In other words,

$$D_1 = D_2 = D_3 = D_4 = D_5 = D_6 = 0$$

$$D_{19} = D_{20} = D_{21} = D_{22} = D_{23} = D_{24} = 0$$

Step 2: Element equations. For space frames, it is much more convenient to use a computerized tool to carry out the intermediate calculations.

Element 1: Stiffness Matrix

```
ROW :     1
     2.1333E+06    0.0000E+00    0.0000E+00    0.0000E+00    0.0000E+00    0.0000E+00
    -2.1333E+06    0.0000E+00    0.0000E+00    0.0000E+00    0.0000E+00    0.0000E+00
ROW :     2
     0.0000E+00    3.6079E+04    0.0000E+00    0.0000E+00    0.0000E+00    2.1647E+06
     0.0000E+00   -3.6079E+04    0.0000E+00    0.0000E+00    0.0000E+00    2.1647E+06
```

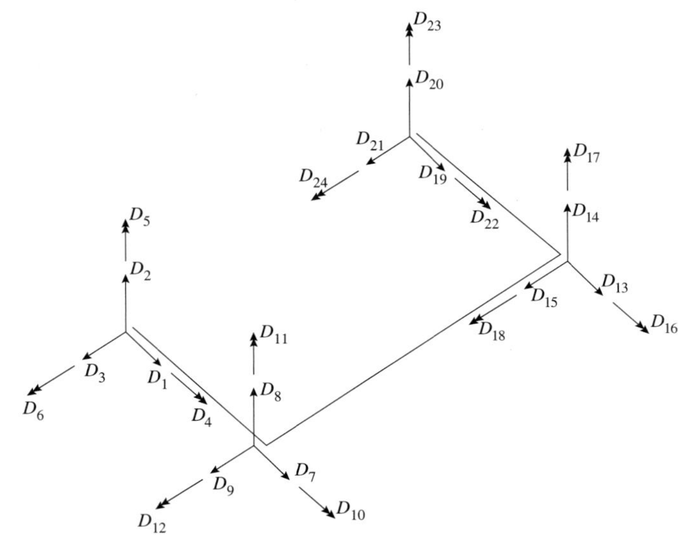

Fig. E6.4.4(c)

```
ROW :     3
   0.0000E+00   0.0000E+00   1.4524E+04   0.0000E+00  -8.7146E+05   0.0000E+00
   0.0000E+00   0.0000E+00  -1.4524E+04   0.0000E+00  -8.7146E+05   0.0000E+00
ROW :     4
   0.0000E+00   0.0000E+00   0.0000E+00   1.8111E+07   0.0000E+00   0.0000E+00
   0.0000E+00   0.0000E+00   0.0000E+00  -1.8111E+07   0.0000E+00   0.0000E+00
ROW :     5
   0.0000E+00   0.0000E+00  -8.7146E+05   0.0000E+00   7.0065E+07   0.0000E+00
   0.0000E+00   0.0000E+00   8.7146E+05   0.0000E+00   3.4510E+07   0.0000E+00
ROW :     6
   0.0000E+00   2.1647E+06   0.0000E+00   0.0000E+00   0.0000E+00   1.7539E+08
   0.0000E+00  -2.1647E+06   0.0000E+00   0.0000E+00   0.0000E+00   8.4372E+07
ROW :     7
  -2.1333E+06   0.0000E+00   0.0000E+00   0.0000E+00   0.0000E+00   0.0000E+00
   2.1333E+06   0.0000E+00   0.0000E+00   0.0000E+00   0.0000E+00   0.0000E+00
ROW :     8
   0.0000E+00  -3.6079E+04   0.0000E+00   0.0000E+00   0.0000E+00  -2.1647E+06
   0.0000E+00   3.6079E+04   0.0000E+00   0.0000E+00   0.0000E+00  -2.1647E+06
ROW :     9
   0.0000E+00   0.0000E+00  -1.4524E+04   0.0000E+00   8.7146E+05   0.0000E+00
   0.0000E+00   0.0000E+00   1.4524E+04   0.0000E+00   8.7146E+05   0.0000E+00
ROW :    10
   0.0000E+00   0.0000E+00   0.0000E+00  -1.8111E+07   0.0000E+00   0.0000E+00
   0.0000E+00   0.0000E+00   0.0000E+00   1.8111E+07   0.0000E+00   0.0000E+00
ROW :    11
   0.0000E+00   0.0000E+00  -8.7146E+05   0.0000E+00   3.4510E+07   0.0000E+00
   0.0000E+00   0.0000E+00   8.7146E+05   0.0000E+00   7.0065E+07   0.0000E+00
ROW :    12
   0.0000E+00   2.1647E+06   0.0000E+00   0.0000E+00   0.0000E+00   8.4372E+07
   0.0000E+00  -2.1647E+06   0.0000E+00   0.0000E+00   0.0000E+00   1.7539E+08
```

Element 2 Stiffness Matrix

```
ROW :     1
   4.3509E+03   0.0000E+00   0.0000E+00   0.0000E+00  -3.9158E+05   0.0000E+00
  -4.3509E+03   0.0000E+00   0.0000E+00   0.0000E+00  -3.9158E+05   0.0000E+00
```

```
ROW :     2
     0.0000E+00   1.0987E+04   0.0000E+00   9.8886E+05   0.0000E+00   0.0000E+00
     0.0000E+00  -1.0987E+04   0.0000E+00   9.8886E+05   0.0000E+00   0.0000E+00
ROW :     3
     0.0000E+00   0.0000E+00   1.4222E+06   0.0000E+00   0.0000E+00   0.0000E+00
     0.0000E+00   0.0000E+00  -1.4222E+06   0.0000E+00   0.0000E+00   0.0000E+00
ROW :     4
     0.0000E+00   9.8886E+05   0.0000E+00   1.1934E+08   0.0000E+00   0.0000E+00
     0.0000E+00  -9.8886E+05   0.0000E+00   5.8656E+07   0.0000E+00   0.0000E+00
ROW :     5
    -3.9158E+05   0.0000E+00   0.0000E+00   0.0000E+00   4.7094E+07   0.0000E+00
     3.9158E+05   0.0000E+00   0.0000E+00   0.0000E+00   2.3390E+07   0.0000E+00
ROW :     6
     0.0000E+00   0.0000E+00   0.0000E+00   0.0000E+00   0.0000E+00   1.2074E+07
     0.0000E+00   0.0000E+00   0.0000E+00   0.0000E+00   0.0000E+00  -1.2074E+07
ROW :     7
    -4.3509E+03   0.0000E+00   0.0000E+00   0.0000E+00   3.9158E+05   0.0000E+00
     4.3509E+03   0.0000E+00   0.0000E+00   0.0000E+00   3.9158E+05   0.0000E+00
ROW :     8
     0.0000E+00  -1.0987E+04   0.0000E+00  -9.8886E+05   0.0000E+00   0.0000E+00
     0.0000E+00   1.0987E+04   0.0000E+00  -9.8886E+05   0.0000E+00   0.0000E+00
ROW :     9
     0.0000E+00   0.0000E+00  -1.4222E+06   0.0000E+00   0.0000E+00   0.0000E+00
     0.0000E+00   0.0000E+00   1.4222E+06   0.0000E+00   0.0000E+00   0.0000E+00
ROW :    10
     0.0000E+00   9.8886E+05   0.0000E+00   5.8656E+07   0.0000E+00   0.0000E+00
     0.0000E+00  -9.8886E+05   0.0000E+00   1.1934E+08   0.0000E+00   0.0000E+00
ROW :    11
    -3.9158E+05   0.0000E+00   0.0000E+00   0.0000E+00   2.3390E+07   0.0000E+00
     3.9158E+05   0.0000E+00   0.0000E+00   0.0000E+00   4.7094E+07   0.0000E+00
ROW :    12
     0.0000E+00   0.0000E+00   0.0000E+00   0.0000E+00   0.0000E+00  -1.2074E+07
     0.0000E+00   0.0000E+00   0.0000E+00   0.0000E+00   0.0000E+00   1.2074E+07
```

Element 2: Load Vector

```
     0.0000E+00   0.0000E+00   1.8000E+04   0.0000E+00   5.4000E+05   0.0000E+00
     0.0000E+00   0.0000E+00   0.0000E+00   1.8000E+04   0.0000E+00  -5.4000E+05
```

Element 3: Stiffness Matrix

```
ROW :     1
     2.1333E+06   0.0000E+00   0.0000E+00   0.0000E+00   0.0000E+00   0.0000E+00
    -2.1333E+06   0.0000E+00   0.0000E+00   0.0000E+00   0.0000E+00   0.0000E+00
ROW :     2
     0.0000E+00   3.6079E+04   0.0000E+00   0.0000E+00   0.0000E+00   2.1647E+06
     0.0000E+00  -3.6079E+04   0.0000E+00   0.0000E+00   0.0000E+00   2.1647E+06
ROW :     3
     0.0000E+00   0.0000E+00   1.4524E+04   0.0000E+00  -8.7146E+05   0.0000E+00
     0.0000E+00   0.0000E+00  -1.4524E+04   0.0000E+00  -8.7146E+05   0.0000E+00
ROW :     4
     0.0000E+00   0.0000E+00   0.0000E+00   1.8111E+07   0.0000E+00   0.0000E+00
     0.0000E+00   0.0000E+00   0.0000E+00  -1.8111E+07   0.0000E+00   0.0000E+00
ROW :     5
     0.0000E+00   0.0000E+00  -8.7146E+05   0.0000E+00   7.0065E+07   0.0000E+00
     0.0000E+00   0.0000E+00   8.7146E+05   0.0000E+00   3.4510E+07   0.0000E+00
ROW :     6
     0.0000E+00   2.1647E+06   0.0000E+00   0.0000E+00   0.0000E+00   1.7539E+08
     0.0000E+00  -2.1647E+06   0.0000E+00   0.0000E+00   0.0000E+00   8.4372E+07
```

```
ROW :     7
   -2.1333E+06   0.0000E+00   0.0000E+00   0.0000E+00   0.0000E+00   0.0000E+00
    2.1333E+06   0.0000E+00   0.0000E+00   0.0000E+00   0.0000E+00   0.0000E+00
ROW :     8
    0.0000E+00  -3.6079E+04   0.0000E+00   0.0000E+00   0.0000E+00  -2.1647E+06
    0.0000E+00   3.6079E+04   0.0000E+00   0.0000E+00   0.0000E+00  -2.1647E+06
ROW :     9
    0.0000E+00   0.0000E+00  -1.4524E+04   0.0000E+00   8.7146E+05   0.0000E+00
    0.0000E+00   0.0000E+00   1.4524E+04   0.0000E+00   8.7146E+05   0.0000E+00
ROW :    10
    0.0000E+00   0.0000E+00   0.0000E+00  -1.8111E+07   0.0000E+00   0.0000E+00
    0.0000E+00   0.0000E+00   0.0000E+00   1.8111E+07   0.0000E+00   0.0000E+00
ROW :    11
    0.0000E+00   0.0000E+00  -8.7146E+05   0.0000E+00   3.4510E+07   0.0000E+00
    0.0000E+00   0.0000E+00   8.7146E+05   0.0000E+00   7.0065E+07   0.0000E+00
ROW :    12
    0.0000E+00   2.1647E+06   0.0000E+00   0.0000E+00   0.0000E+00   8.4372E+07
    0.0000E+00  -2.1647E+06   0.0000E+00   0.0000E+00   0.0000E+00   1.7539E+08
```

Step 3: Assembly of structural equilibrium equations. In an attempt to increase storage efficiency as well as reduce the computational effort, the structural stiffness matrix is not stored as a full matrix. The first improvement that can be made is to store the matrix as a banded upper triangular matrix. Some of the details are presented at the end of this chapter (see Problem 6.7). Below we present the values for the stiffness matrix in this storage scheme.

Structural Stiffness Matrix, $K_{12\times12}$

```
ROW :     1
    2.1377E+06   0.0000E+00   0.0000E+00   0.0000E+00  -3.9158E+05   0.0000E+00
   -4.3509E+03   0.0000E+00   0.0000E+00   0.0000E+00  -3.9158E+05   0.0000E+00
ROW :     2
    4.7066E+04   0.0000E+00   9.8886E+05   0.0000E+00  -2.1647E+06   0.0000E+00
   -1.0987E+04   0.0000E+00   9.8886E+05   0.0000E+00   0.0000E+00   0.0000E+00
ROW :     3
    1.4367E+06   0.0000E+00   8.7146E+05   0.0000E+00   0.0000E+00   0.0000E+00
   -1.4222E+06   0.0000E+00   0.0000E+00   0.0000E+00   0.0000E+00   0.0000E+00
ROW :     4
    1.3745E+08   0.0000E+00   0.0000E+00   0.0000E+00  -9.8886E+05   0.0000E+00
    5.8656E+07   0.0000E+00   0.0000E+00   0.0000E+00   0.0000E+00   0.0000E+00
ROW :     5
    1.1716E+08   0.0000E+00   3.9158E+05   0.0000E+00   0.0000E+00   0.0000E+00
    2.3390E+07   0.0000E+00   0.0000E+00   0.0000E+00   0.0000E+00   0.0000E+00
ROW :     6
    1.8747E+08   0.0000E+00   0.0000E+00   0.0000E+00   0.0000E+00   0.0000E+00
   -1.2074E+07   0.0000E+00   0.0000E+00   0.0000E+00   0.0000E+00   0.0000E+00
ROW :     7
    2.1377E+06   0.0000E+00   0.0000E+00   0.0000E+00   3.9158E+05   0.0000E+00
    0.0000E+00   0.0000E+00   0.0000E+00   0.0000E+00   0.0000E+00   0.0000E+00
ROW :     8
    4.7066E+04   0.0000E+00  -9.8886E+05   0.0000E+00  -2.1647E+06   0.0000E+00
    0.0000E+00   0.0000E+00   0.0000E+00   0.0000E+00   0.0000E+00   0.0000E+00
ROW :     9
    1.4367E+06   0.0000E+00   8.7146E+05   0.0000E+00   0.0000E+00   0.0000E+00
    0.0000E+00   0.0000E+00   0.0000E+00   0.0000E+00   0.0000E+00   0.0000E+00
```

```
ROW :    10
     1.3745E+08   0.0000E+00   0.0000E+00   0.0000E+00   0.0000E+00   0.0000E+00
     0.0000E+00   0.0000E+00   0.0000E+00   0.0000E+00   0.0000E+00   0.0000E+00
ROW :    11
     1.1716E+08   0.0000E+00   0.0000E+00   0.0000E+00   0.0000E+00   0.0000E+00
     0.0000E+00   0.0000E+00   0.0000E+00   0.0000E+00   0.0000E+00   0.0000E+00
ROW :    12
     1.8747E+08   0.0000E+00   0.0000E+00   0.0000E+00   0.0000E+00   0.0000E+00
     0.0000E+00   0.0000E+00   0.0000E+00   0.0000E+00   0.0000E+00   0.0000E+00
```

Load Vector, $F_{12\times 1}$

```
     0.0000E+00   1.8000E+04   0.0000E+00   5.4000E+05   0.0000E+00   0.0000E+00
     0.0000E+00   1.8000E+04   0.0000E+00  -5.4000E+05   0.0000E+00   0.0000E+00
```

Step 4: Solution of the structural equations. Solving the structural equilibrium equations $\mathbf{K}_{12\times12}\mathrm{D}_{12\times1} = \mathbf{F}_{12\times1}$, we obtain the nodal displacements as

$$\mathbf{D}_{12\times1} = \{0,\ 1.923",0,0.006853 \text{ rad},0,0.02373 \text{ rad}\}$$
$$\{0,\ 1.923",0,-0.006853 \text{ rad},0,0.02373 \text{ rad}\}$$

The table below relates this displacements to the values at the four nodes.

NODAL DISPLACEMENTS

NODE	X DISP (in)	Y DISP (in)	Z DISP (in)	X ROT (RAD)	Y ROT (RAD)	Z ROT (RAD)
1	0.000E+00	0.000E+00	0.000E+00	0.000E+00	0.000E+00	0.000E+00
2	0.000E+00	1.923E+00	0.000E+00	6.853E-03	0.000E+00	2.373E-02
3	0.000E+00	1.923E+00	0.000E+00	-6.853E-03	0.000E+00	2.373E-02
4	0.000E+00	0.000E+00	0.000E+00	0.000E+00	0.000E+00	0.000E+00

Step 5: Element member nodal forces. We use the computed displacements, Eqs. (6.4.2.22) and (6.4.2.23), suitably modified as

$$\mathbf{d}'_{12\times1} = \mathbf{T}_{12\times12}\,\mathbf{d}_{12\times1} \qquad \mathbf{f}'_{12\times1} = \mathbf{k}'_{12\times12}\mathbf{d}'_{12\times1} - \sum_i \left(\mathbf{q}'_{12\times1}\right)_i$$

The results are shown below in tabular form with the components identified as axial force, shear force and bending/torsional moment at the two ends of each member.

MEM	NODE	AXIAL (lb)	SHEAR Y (lb)	SHEAR Z (lb)	TORSION (lb -in)	BENDING Y (lb -in)	BENDING Z (lb -in)
1	1	0.000E+00	1.800E+04	0.000E+00	-1.241E+05	0.000E+00	2.160E+06
	2	0.000E+00	-1.800E+04	0.000E+00	1.241E+05	0.000E+00	8.429E-02
2	2	0.000E+00	1.800E+04	0.000E+00	-1.151E-11	0.000E+00	1.241E+05
	3	0.000E+00	1.800E+04	0.000E+00	1.151E-11	0.000E+00	-1.241E+05
3	4	0.000E+00	-1.800E+04	0.000E+00	1.241E+05	0.000E+00	-2.160E+06
	3	0.000E+00	1.800E+04	0.000E+00	-1.241E+05	0.000E+00	-8.429E-02

EXERCISES

The problems in this section are similar to those in Section 6.2. Unless otherwise stated, for each problem compute the nodal displacements, the member nodal forces and the support reactions.

Appetizers

6.4.1. The nodal displacements for the truss shown below are given in Fig. P6.4.1. $E = 30000$ ksi and member cross-section is circular with a radius of 1 in. Compute the member forces in members 2, 6, 9, and 13.

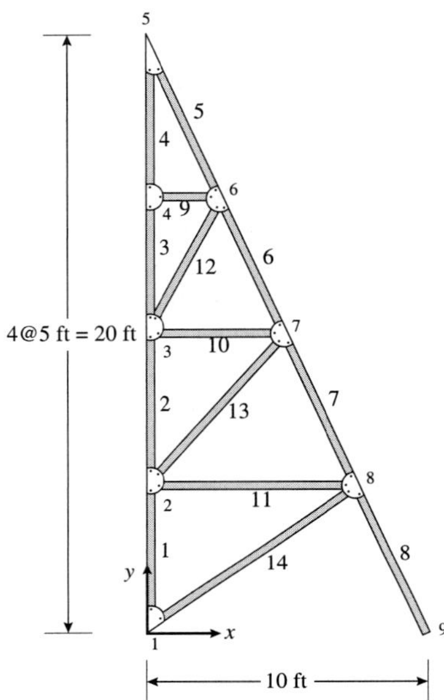

Fig. P6.4.1

Node	X Disp (in)	Y Disp (in)
1	0	0
2	0.022217	0.00763944
3	0.0451335	0.0127324
4	0.0721605	0.0152789
5	0.106827	0.0178254
6	0.0708873	0.00341432
7	0.0413138	-0.0042548
8	0.0145776	-0.00694647
9	0	0

6.4.2. For the beam in Fig. P6.4.2, take $E = 29000$ ksi, $I = 200$ in^4, and $A = 10$ in^2.

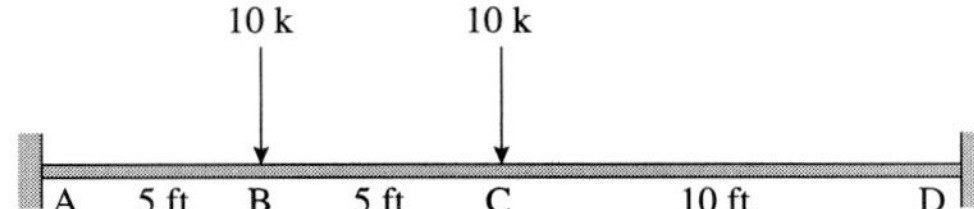

Fig. P6.4.2

6.4.3. For the beam in Fig. P6.4.3, take $E = 29000$ ksi, $I_{AB} = 200$ in^4, and $I_{BC} = 400$ in^4.

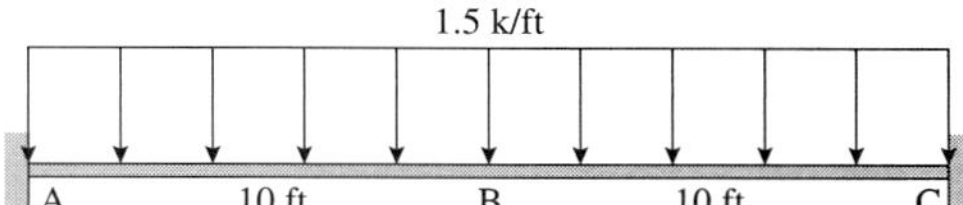

Fig. P6.4.3

6.4.4. The steel frame in Fig. P6.4.4 is constructed such that the properties of the member AB are A = 25 in^2 and I = 1500 in^4, and member BC are A = 30 in^2 and I = 1600 in^4.

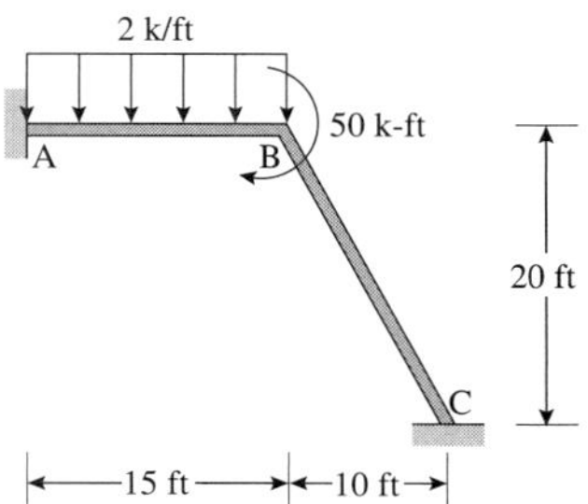

Fig. P6.4.4

Main Course

6.4.5. Using Eq. (6.4.2.19), compute the equivalent nodal forces for the element loads in Fig. P6.4.5.

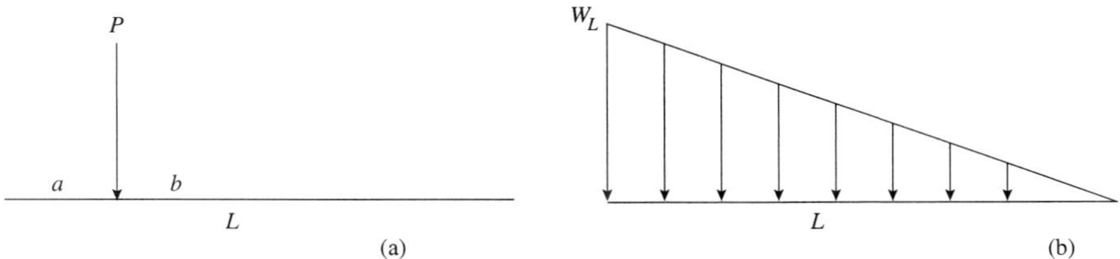

Fig. P6.4.5

6.4.6. For the structure in Fig. P6.4.6, take $E = 40000$ ksi, $A = 50$ in^2, and $I = 800$ in^4.

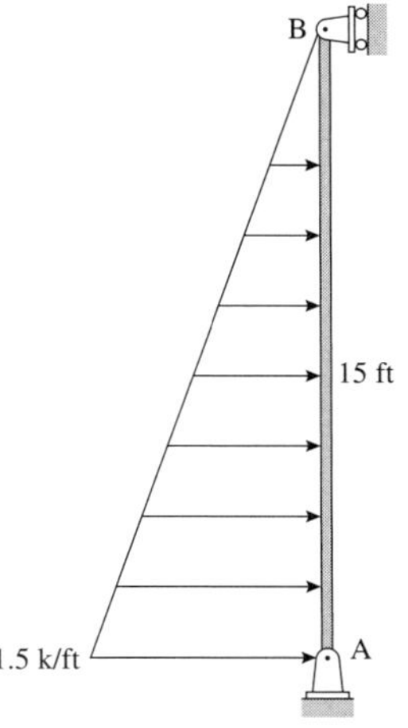

Fig. P6.4.6

6.4.7. The frame in Fig. P6.4.7 has the following properties: A = 20 in^2, I = 1500 in^4, and E = 30000 ksi. Compute the support reactions.

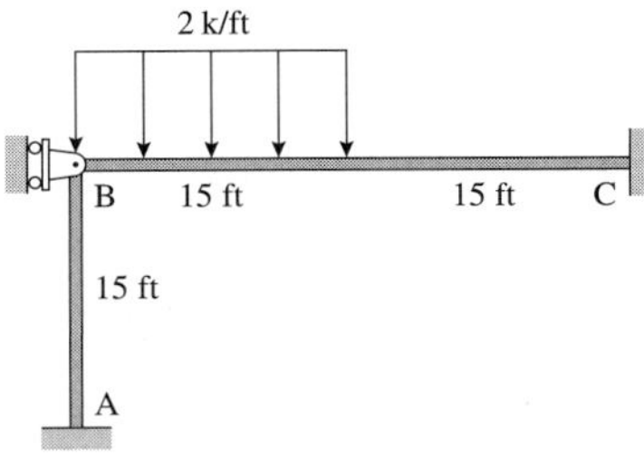

Fig. P6.4.7

6.4.8. The frame in Fig. P6.4.8 has the following properties: for the columns A = 20 in^2 and I = 1500 in^4, for the beams A = 30 in^2 and I = 2000 in^4. For all the members, E = 30000 ksi. The 10 k loads are placed at the center of beams BC and CD.

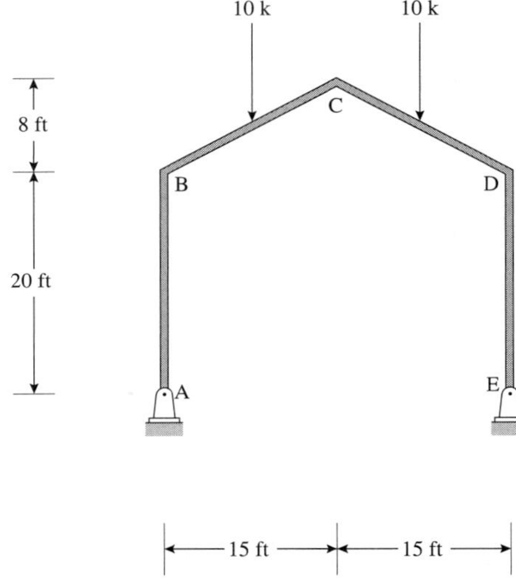

Fig. P6.4.8

6.4.9. The members in the truss in Fig. P6.4.9 are made of aluminum and have a cross-sectional area of 2 in^2. Member 1 connects nodes 2 and 1, member 2 connects 4 and 1, and member 3 connects 3 and 1. Compute all the member forces.

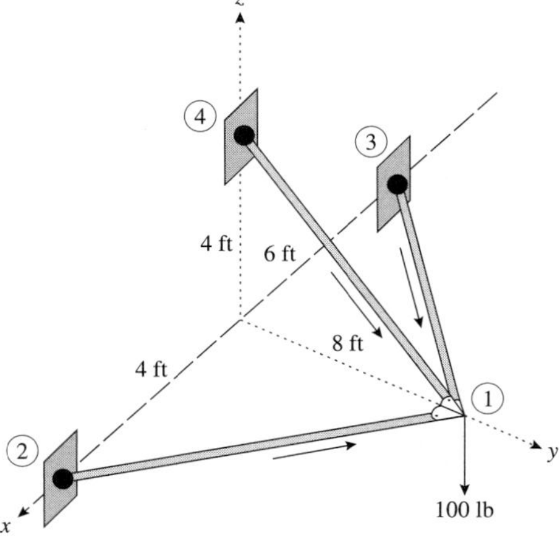

Fig. P6.4.9

6.4.10. The space frame in Fig. P6.4.10 has the following properties: beams are W18 × 65 and the column is W14 × 311. The 6 k force acts at the center of the member AB.

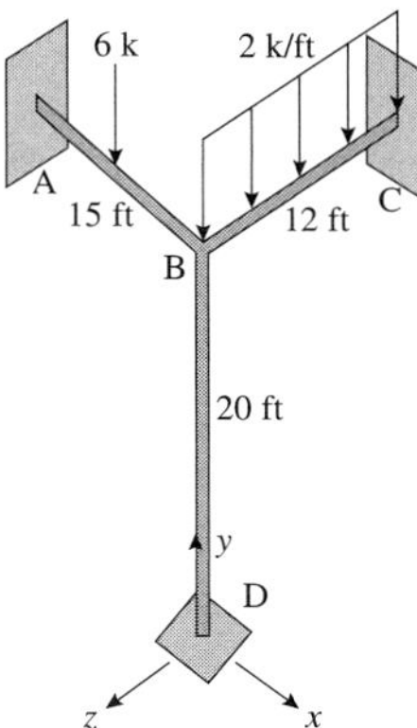

Fig. P6.4.10

6.5 ADVANCED TOPICS

In the earlier sections we looked at the different steps in the solution process. The objective was to compute the nodal displacements and use them to compute the member forces. In this section will be look at special cases that affect one or more steps of the overall process.

6.5.1 Internal Hinge

Internal hinges or moment-release hinges are locations where the bending moment is equal to zero. Figure 6.5.1.1 shows a planar frame. Node 2 of the frame is an internal hinge. The implication is that moments for all the three elements meeting at node 2 at the end corresponding to that node are zero. Also, there is no unique rotation at the node. Each element at that node can potentially have a different rotation value.

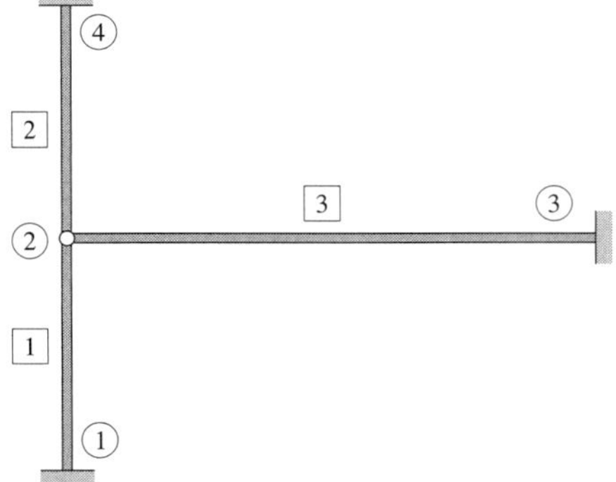

Fig. 6.5.1.1 Frame with node 2 as an internal hinge or moment-release connection.

Figure 6.5.1.2 shows a slightly different situation. Node 2 is not an internal hinge. However, the end of element 3 corresponding to node 2 is moment-free. The other two elements can potentially have nonzero moments (however, they must satisfy $\Sigma M = 0$). These two elements have a unique rotation value at the node but element 3 can potentially have a different rotation value. Both these cases must be treated differently.[10]

[10] An extreme case is to have an internal hinge at both ends of the member. If the structure is supported by pin and/or roller supports and there are no element loads, we have a truss structure! In fact, in the GS-USA program, the default structure is a frame. A truss is created by converting all the joints to an internal hinge while other truss assumptions are satisfied.

We now derive the element equations. Consider the beam element shown in Fig. 6.5.1.3. The moment-release hinge is located at the start node of the element. As we saw in Section 6.2.3, the element equations for a planar beam element subjected to element loads are given by

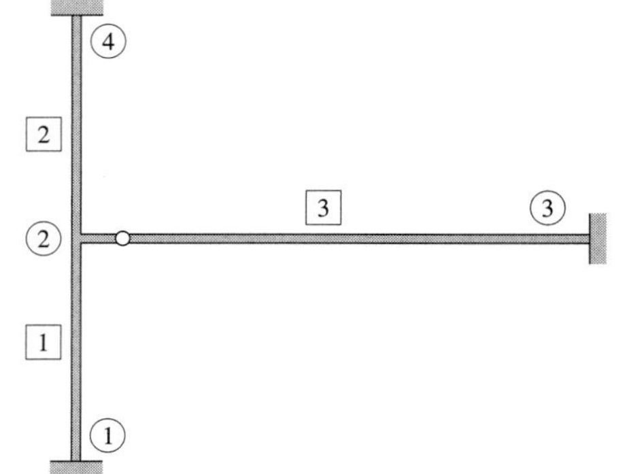

Fig. 6.5.1.2 Element 3 with end at Node 2 as a moment-release connection.

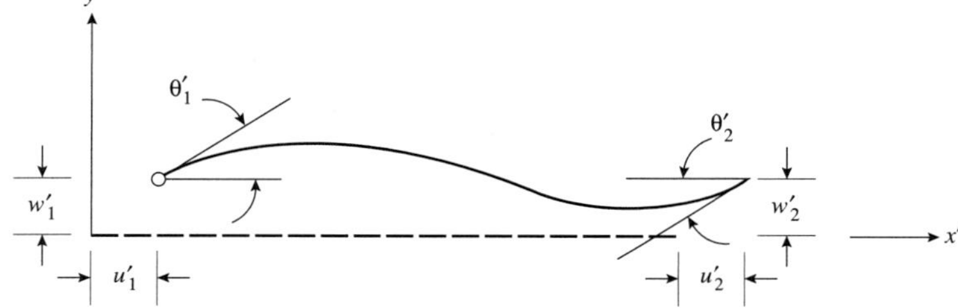

Fig. 6.5.1.3 Planar beam element with an internal hinge at the start node of the element.

$$\mathbf{f}'_{6\times1} = \mathbf{k}'_{6\times6}\mathbf{d}'_{6\times1} - \mathbf{q}'_{6\times1} \tag{6.5.1.1}$$

Expanding the equations, we have

$$\frac{AE}{L}(u'_1 - u'_2) - q'_1 = f'_1 \tag{6.5.1.2a}$$

$$\frac{EI}{L^3}\left(12w'_1 + 6L\theta'_1 - 12w'_2 + 6L\theta'_2\right) - q'_2 = f'_2 \tag{6.5.1.2b}$$

$$\frac{EI}{L^3}\left(6Lw'_1 + 4L^2\theta'_1 - 6Lw'_2 + 2L^2\theta'_2\right) - q'_3 = f'_3 \tag{6.5.1.2c}$$

$$\frac{AE}{L}(-u'_1 + u'_2) - q'_4 = f'_4 \tag{6.5.1.2d}$$

$$\frac{EI}{L^3}\left(-12w'_1 - 6L\theta'_1 + 12w'_2 - 6L\theta'_2\right) - q'_5 = f'_5 \tag{6.5.1.2e}$$

$$\frac{EI}{L^3}\left(6Lw'_1 + 2L^2\theta'_1 - 6Lw'_2 + 4L^2\theta'_2\right) - q'_6 = f'_6 \tag{6.5.1.2f}$$

Since the hinge is at the start node of the element, $f'_3 = 0$. Using this condition with Eq. (6.5.1.2c) and solving for the rotation corresponding to the moment, we find

$$\theta'_1 = \frac{3}{2L}\left(-w'_1 + w'_2\right) - \frac{1}{2}\theta'_2 + \frac{L}{4EI}q'_3 \tag{6.5.1.3}$$

This condition can be used in the other equations to represent θ_1' in terms of the other degrees of freedom shown above. The implication is that θ_1' is not a degree of freedom but can be found if the degrees of freedom on the right-hand side of Eq. (6.5.1.3) are known. After all the algebraic manipulations, the following modified equations are obtained:

$$\begin{bmatrix} \frac{AE}{L} & 0 & 0 & -\frac{AE}{L} & 0 & 0 \\ 0 & \frac{3EI}{L^3} & 0 & 0 & -\frac{3EI}{L^3} & \frac{3EI}{L^2} \\ 0 & 0 & 0 & 0 & 0 & 0 \\ -\frac{AE}{L} & 0 & 0 & \frac{AE}{L} & 0 & 0 \\ 0 & -\frac{3EI}{L^3} & 0 & 0 & \frac{3EI}{L^3} & -\frac{3EI}{L^2} \\ 0 & \frac{3EI}{L^2} & 0 & 0 & -\frac{3EI}{L^2} & \frac{3EI}{L} \end{bmatrix} \begin{bmatrix} u_1' \\ w_1' \\ \theta_1' \\ u_2' \\ w_2' \\ \theta_2' \end{bmatrix} - \begin{Bmatrix} q_1' \\ q_2' - \frac{3q_3'}{2L} \\ 0 \\ q_4' \\ q_5' + \frac{3q_3'}{2L} \\ q_6' - \frac{q_3'}{2} \end{Bmatrix} = \begin{Bmatrix} f_1' \\ f_2' \\ f_3' \\ f_4' \\ f_5' \\ f_6' \end{Bmatrix} \tag{6.5.1.4}$$

$$\mathbf{k}'_{6\times6}\mathbf{d}'_{6\times1} - \mathbf{q}'_{6\times1} = \mathbf{f}'_{6\times1} \tag{6.5.1.5}$$

When the hinge is located at the end node of the element, the same procedure can be used, but now, $f_6' = 0$. Using this condition with Eq. (6.5.1.2f) and solving for the rotation corresponding to the moment yields

$$\theta_2' = \frac{3}{2L}(-w_1' + w_2') - \frac{1}{2}\theta_1' + \frac{L}{4EI}q_6' \tag{6.5.1.6}$$

$$\begin{bmatrix} \frac{AE}{L} & 0 & 0 & -\frac{AE}{L} & 0 & 0 \\ 0 & \frac{3EI}{L^3} & \frac{3EI}{L^2} & 0 & -\frac{3EI}{L^3} & 0 \\ 0 & \frac{3EI}{L^2} & \frac{3EI}{L} & 0 & -\frac{3EI}{L^2} & 0 \\ -\frac{AE}{L} & 0 & 0 & \frac{AE}{L} & 0 & 0 \\ 0 & -\frac{3EI}{L^3} & -\frac{3EI}{L^2} & 0 & \frac{3EI}{L^3} & 0 \\ 0 & 0 & 0 & 0 & 0 & 0 \end{bmatrix} \begin{bmatrix} u_1' \\ w_1' \\ \theta_1' \\ u_2' \\ w_2' \\ \theta_2' \end{bmatrix} - \begin{Bmatrix} q_1' \\ q_2' - \frac{3q_6'}{2L} \\ q_3' - \frac{q_6'}{2} \\ q_4' \\ q_5' + \frac{3q_6'}{2L} \\ 0 \end{Bmatrix} = \begin{Bmatrix} f_1' \\ f_2' \\ f_3' \\ f_4' \\ f_5' \\ f_6' \end{Bmatrix} \tag{6.5.1.7}$$

$$\mathbf{k}'_{6\times6}\mathbf{d}'_{6\times1} - \mathbf{q}'_{6\times1} = \mathbf{f}'_{6\times1} \tag{6.5.1.8}$$

When the element has internal hinges at both ends, $f_3' = 0$ and $f_6' = 0$. Hence,

$$\theta_1' = \frac{1}{L}(-w_1' + w_2') + \frac{L}{6EI}(2q_3' - q_6') \tag{6.5.1.9a}$$

$$\theta_2' = \frac{1}{L}(-w_1' + w_2') + \frac{L}{6EI}(2q_6' - q_3') \tag{6.5.1.9b}$$

Using these with Eq. (6.5.1.2f), we have

$$\begin{bmatrix} \frac{AE}{L} & 0 & 0 & -\frac{AE}{L} & 0 & 0 \\ 0 & 0 & 0 & 0 & 0 & 0 \\ 0 & 0 & 0 & 0 & 0 & 0 \\ -\frac{AE}{L} & 0 & 0 & \frac{AE}{L} & 0 & 0 \\ 0 & 0 & 0 & 0 & 0 & 0 \\ 0 & 0 & 0 & 0 & 0 & 0 \end{bmatrix} \begin{Bmatrix} u_1' \\ w_1' \\ \theta_1' \\ u_2' \\ w_2' \\ \theta_2' \end{Bmatrix} - \begin{Bmatrix} q_1' \\ q_2' - \frac{1}{L}(q_3' + q_6') \\ 0 \\ q_4' \\ q_2' + \frac{1}{L}(q_3' + q_6') \\ 0 \end{Bmatrix} = \begin{Bmatrix} f_1' \\ f_2' \\ f_3' \\ f_4' \\ f_5' \\ f_6' \end{Bmatrix} \quad (6.5.1.10)$$

$$\mathbf{k}'_{6\times 6}\mathbf{d}'_{6\times 1} - \mathbf{q}'_{6\times 1} = \mathbf{f}'_{6\times 1} \quad (6.5.1.11)$$

It should come as no surprise that the stiffness matrix in the above equation corresponds to the stiffness matrix for a truss element. The equivalent joint loads for two loads are shown in Fig. 6.5.1.4.

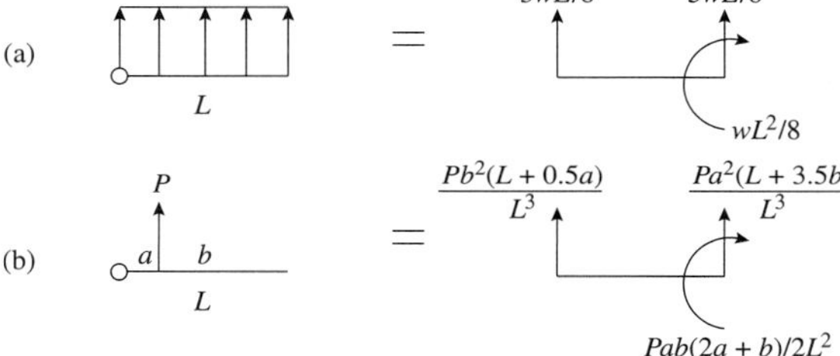

Fig. 6.5.1.4 Equivalent joint loads for (a) uniformly distributed load and (b) concentrated load, with a moment-release connection at the start node.

We can now summarize the major steps in planar frame analysis with or without internal hinges.

Step 1: Select the problem units. Set up the coordinate system. Identify and label the nodes and the elements. For each element, select a start node (node 1) and an end node (node 2). We use an arrow along the member to indicate the direction from the start node to the end node. This establishes the local coordinate system for each element. Label the three global dof at each node starting at node 1 and proceeding sequentially.

Step 2: Construct the equilibrium-compatibility equations for a typical element (Eqs. (6.2.3.11) and (6.2.3.12)). If the element has a moment-release hinge use Eqs. (6.5.1.5) or (6.5.1.8) or (6.5.1.11), but first transform the equations to the global coordinate system.

Step 3: Using the problem data, construct the element equations from Step 2 for all the elements in the problem. If there are element loads, compute the equivalent joint loads (EJL) $\mathbf{q}'_{6\times 1}$ and transform them to the global coordinate system using Eq. (6.2.3.7) as

$$\mathbf{q}_{6\times 1} = \mathbf{T}^{\mathbf{T}}_{6\times 6}\mathbf{q}'_{6\times 1} \quad (6.5.1.12a)$$

or $$\mathbf{q}_{6\times 1} = \left[(lq_1' - mq_2') \quad (mq_1' + lq_2') \quad q_3' \quad (lq_4' - mq_5') \quad (mq_4' + lq_5') \quad q_6'\right]^{\mathbf{T}} \quad (6.5.1.12b)$$

Note that if there is more than one element load, $\mathbf{q}_{6\times 1}$ represents the linear superposition (algebraic sum) of all the element loads acting on that element. Note that $\mathbf{q}'_{6\times 1}$ should reflect whether the element has rigid connections or moment-release connection(s) at the ends.

Step 4: Assemble the element equations into the system equations $\mathbf{K}_{3j\times 3j}\mathbf{D}_{3j\times 1} = \mathbf{F}_{3j\times 1}$, where j is the number of joints in the frame.

Step 5: Impose the boundary conditions. For nodes that correspond to a moment-release connection (see Fig. 6.5.1.2), delete the equation corresponding to the rotational degree of freedom since it is not a free degree of freedom. This is necessary to ensure that we can solve the system equations; otherwise, the equations are linearly dependent. An examination of the transformed equations (Eqs. (6.5.1.4) or (6.5.1.7) or (6.5.1.10)), shows that one or more rows and columns have only zeros in them! This can be achieved by imposing the nodal condition $\theta = 0$ as we do with other fixed degrees of freedom.

Step 6: Solve the system equations $\mathbf{KD} = \mathbf{F}$ for the nodal displacements $\mathbf{D}$.

Step 7: For each element compute the nodal displacements as

$$\mathbf{d}'_{6\times1} = \mathbf{T}_{6\times6}\mathbf{d}_{6\times1} \qquad (6.5.1.13a)$$

After this step, for elements with one or more moment-release hinges, use the appropriate equation to compute the rotation at the hinge—Eq. (6.5.1.3) or (6.5.1.6) or (6.5.1.9a,b). Finally, compute the member end forces:

$$\mathbf{f}'_{6\times1} = \mathbf{k}'_{6\times6}\mathbf{T}_{6\times6}\mathbf{d}_{6\times1} - \mathbf{q}'_{6\times1} \qquad (6.5.1.14)$$

EXAMPLE 6.5.1

Beam with an Internal Hinge

For the beam in Fig. E6.5.1(a), compute the support reactions. Take $E = 2(10^{11})$ N/m² and $I = 10^{-4}$ m².

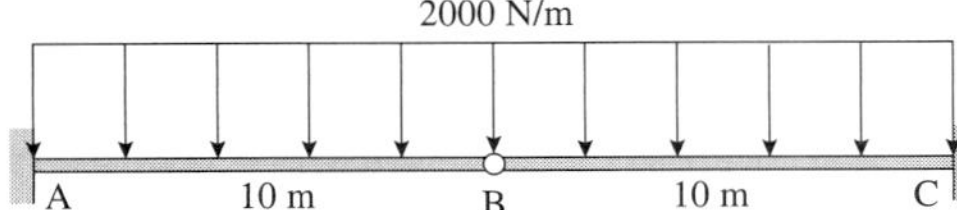

Fig. E6.5.1(a)

SOLUTION

Step 1: Let us select N, m as the problem units. The model is shown in Fig. E6.5.1(b). From the problem data, $D_1 = D_2 = D_3 = D_7 = D_8 = D_9 = 0$. However, the rotation labeled at node 2 as D_6 is incorrect. In fact, there are two independent rotations, D_{61} and D_{62}, corresponding to the rotation of the two elements that meet at node 2. As we will see later, the degrees of freedom at node 2 will be eliminated.

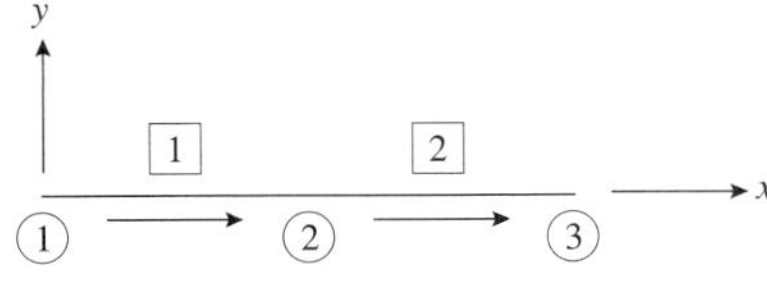

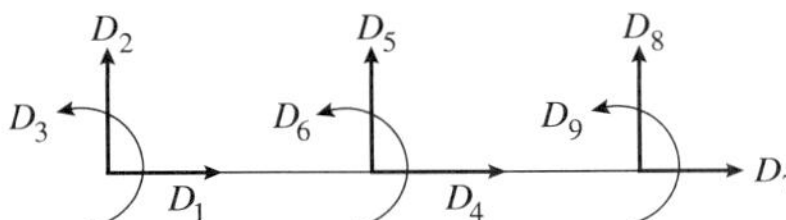

Fig. E6.5.1(b)

Let us assume that the cross-sectional area is 1.0 m².

Step 2: Element equilibrium equations. The element equations for the two elements are presented below. Recognizing that the internal hinge at the end node of the element and using Eq. (6.5.1.7) for element 1, we have

$$10^4\begin{bmatrix} 2(10^6) & 0 & 0 & -2(10^6) & 0 & 0 \\ 0 & 6 & 60 & 0 & -6 & 0 \\ 0 & 60 & 600 & 0 & -60 & 0 \\ -2(10^6) & 0 & 0 & 2(10^6) & 0 & 0 \\ 0 & -6 & -60 & 0 & 6 & 0 \\ 0 & 0 & 0 & 0 & 0 & 0 \end{bmatrix}\begin{Bmatrix} D_1 \\ D_2 \\ D_3 \\ D_4 \\ D_5 \\ D_{61} \end{Bmatrix} = \begin{Bmatrix} 0 \\ -12500 \\ -25000 \\ 0 \\ -7500 \\ 0 \end{Bmatrix}$$

Now recognizing that the internal hinge is at the start node of element 2 and using Eq. (6.5.1.4), we have

$$10^4\begin{bmatrix} 2(10^6) & 0 & 0 & -2(10^6) & 0 & 0 \\ 0 & 6 & 0 & 0 & -6 & 60 \\ 0 & 0 & 0 & 0 & 0 & 0 \\ -2(10^6) & 0 & 0 & 2(10^6) & 0 & 0 \\ 0 & -6 & 0 & 0 & 6 & -60 \\ 0 & 60 & 0 & 0 & -60 & 600 \end{bmatrix}\begin{Bmatrix} D_4 \\ D_5 \\ D_{62} \\ D_7 \\ D_8 \\ D_9 \end{Bmatrix} = \begin{Bmatrix} 0 \\ -7500 \\ 0 \\ 0 \\ -12500 \\ 25000 \end{Bmatrix}$$

Note that we are using the appropriately modified equations to compute the equivalent nodal forces for the element loads.

Step 2: Assembly of the system equations. The assembled equations will not involve any of the rotational degrees of freedom at node 2 since they were eliminated in step 1:

$$10^4\begin{bmatrix} 4(10^6) & 0 \\ 0 & 12 \end{bmatrix}\begin{Bmatrix} D_4 \\ D_5 \end{Bmatrix} = \begin{Bmatrix} 0 \\ -15000 \end{Bmatrix}$$

Step 3: Solution of the system equations. Solving, we have $D_4 = 0$ and $D_5 = -0.125$ m.

Step 4: Computation of the element nodal forces. Since both the elements are tied to an internal hinge, we must recover the rotation at the hinge before the element nodal forces can be computed. Using Eqs. (6.5.1.3) and (6.5.1.6), we have for element 1

$$\theta'_2 = -0.01667 \text{ rad}$$

and for element 2

$$\theta'_1 = 0.01667 \text{ rad}$$

Now substituting the nodal displacements in Eq. (6.5.1.14), we have for element 1

$$\mathbf{f}'_{6\times1} = \{0, 20000\text{ N}, 100000\text{ N - m}, 0, 0, 0, 0\}$$

and for element 2

$$\mathbf{f}'_{6\times1} = \{0, 0, 0, 0, 20000\text{ N}, -100000\text{ N - m}\}$$

The element FBDs are shown in Figs. E6.5.1(c) and (d).

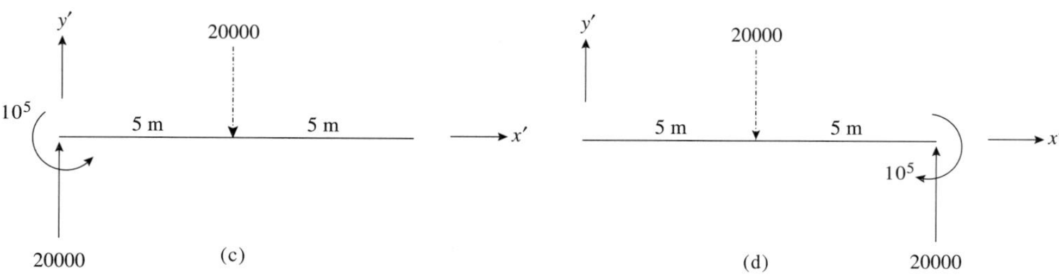

Fig. E6.5.1(c) and (d)
(c) Element 1 FBD.
(d) Element 2 FBD.

EXAMPLE 6.5.2 ***Frame with an Internal Hinge (Example 5.2.5)***

Consider the frame in Fig. E6.5.2(a). The modulus of elasticity is $2(10^{11})$ Pa, the cross-sectional area is $0.01\ m^2$ and the moment of inertia is $0.0001\ m^4$ for both the members. Compute the support reactions.

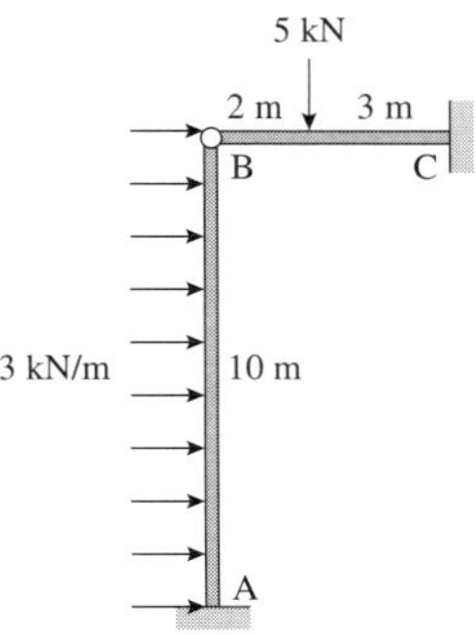

Fig. E6.5.2(a)

SOLUTION

Step 1: The problem is similar to Example 6.4.3 except that here B is an internal hinge. As before, we select N, m as the problem units. The model is shown in Fig. E6.5.2(b). The boundary conditions of the frame are such that

$$D_1 = D_2 = D_3 = D_7 = D_8 = D_9 = 0$$

However, there are two rotations at node 2 corresponding to the rotations of the two elements that meet at node 2, D_{61} and D_{62}. We eliminate these degrees of freedom at the element level by using the modified element equations.

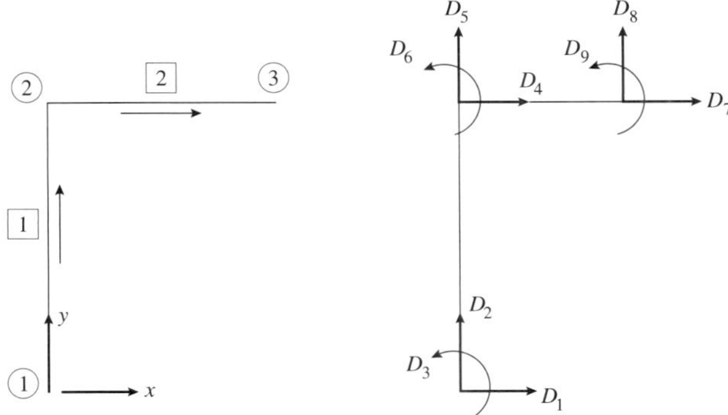

Fig. E6.5.2(b)

Step 2: The element equilibrium equations. Using the modified element equilibrium equations and noting that the end node of Element 1 and the start node of Element 2 are internal hinges, we have the following element equations:

Element 1:

$$10^5\begin{bmatrix} 0.6 & 0 & -6 & -0.6 & 0 & 0 \\ 0 & 2000 & 0 & 0 & -2000 & 0 \\ -6 & 0 & 60 & 6 & 0 & 0 \\ -0.6 & 0 & 6 & 0.6 & 0 & 0 \\ 0 & -2000 & 0 & 0 & 2000 & 0 \\ 0 & 0 & 0 & 0 & 0 & 0 \end{bmatrix}\begin{Bmatrix} D_1 \\ D_2 \\ D_3 \\ D_4 \\ D_5 \\ D_{61} \end{Bmatrix} = \begin{Bmatrix} 18750 \\ 0 \\ -37500 \\ 11250 \\ 0 \\ 0 \end{Bmatrix}$$

Element 2:

$$10^5\begin{bmatrix} 4000 & 0 & 0 & -4000 & 0 & 0 \\ 0 & 4.8 & 0 & 0 & -4.8 & 24 \\ 0 & 0 & 0 & 0 & 0 & 0 \\ -4000 & 0 & 0 & 4000 & 0 & 0 \\ 0 & -4.8 & 0 & 0 & 4.8 & -24 \\ 0 & 24 & 0 & 0 & -24 & 120 \end{bmatrix}\begin{Bmatrix} D_4 \\ D_5 \\ D_{62} \\ D_7 \\ D_8 \\ D_9 \end{Bmatrix} = \begin{Bmatrix} 0 \\ -2160 \\ 0 \\ 0 \\ -2840 \\ 4200 \end{Bmatrix}$$

Step 3: Assembly of the system equations:

$$10^5\begin{bmatrix} 4000.6 & 0 \\ 0 & 2004.8 \end{bmatrix}\begin{Bmatrix} D_4 \\ D_5 \end{Bmatrix} = \begin{Bmatrix} 11250 \\ -2160 \end{Bmatrix}$$

Step 4: Solution of the equilibrium equations. Solving the two equations, we have

$$D_4 = 2.812(10^{-5})\text{ m} \qquad D_5 = -1.0774(10^{-5})\text{ m}$$

Step 5: Computation of element nodal forces. Before the element nodal forces can be computed, we need to recover the element rotations at the internal hinge.

Element 1:

$$\theta_2' = 3.12(10^{-3})\text{ rad}$$

$$\mathbf{f}'_{6\times1} = \{2155\text{ N}, 18752\text{ N}, 37517\text{ N-m}, -2155\text{ N}, 11248\text{ N}, 0\}$$

Element 2:

$$\theta_1' = -2.218(10^{-4})\text{ rad}$$

$$\mathbf{f}'_{6\times1} = \{11248\text{ N}, 2155\text{ N}, 0, -11248\text{ N}, 2845\text{ N}, -4226\text{ N-m}\}$$

The element FBDs are shown in Fig. E6.5.2(c).

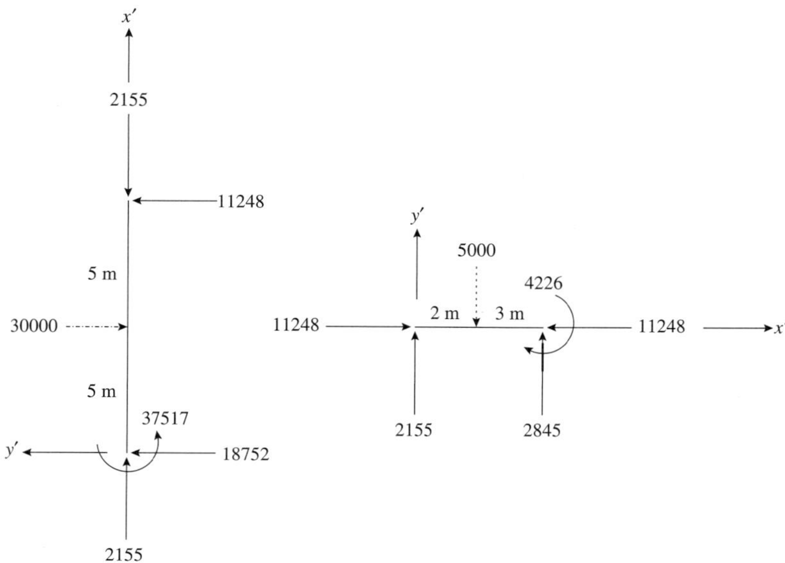

Fig. E6.5.2(c)

6.5.2 Skew Supports

Skew supports, or inclined rollers (see Fig. 6.5.2.1) can be handled as a special case of what are generally referred to as multi-point constraints (MPCs). Consider a case where there exists a known relationship between two different degrees of freedom D_i and D_j. Let the equation that represents the relationship be

$$c_i D_i + c_j D_j = c \tag{6.5.2.1}$$

where c_i, c_j, and c are known constants. The total potential energy for a structure with the MPC along the lines of Eq. (6.4.18) is

$$\Pi(\mathbf{D}) = \frac{1}{2}\mathbf{D}^{\mathrm{T}}\mathbf{K}\mathbf{D} - \mathbf{D}^{\mathrm{T}}\mathbf{F} + \frac{1}{2}C\left(c_i D_i - c_j D_j - c\right)^2 \tag{6.5.2.2}$$

where C is a large number.

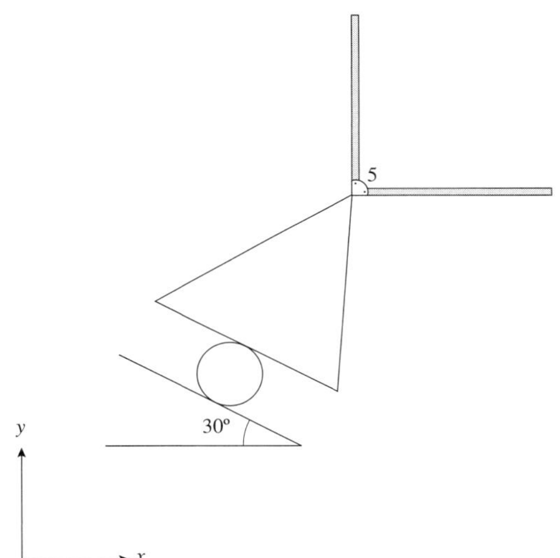

Fig. 6.5.2.1 Skew roller support.

Note that Π takes on a minimum value when $c_i D_i - c_j D_j - c$ is zero (or, numerically very small). Using the theorem of minimum potential energy, $\partial\Pi/\partial D = 0$, yields the usual equilibrium equations except for the rows and columns dealing with D_i and D_j. The usual and the modified terms are:

$$\begin{bmatrix} K_{ii} & K_{ij} \\ K_{ji} & K_{jj} \end{bmatrix} \rightarrow \begin{bmatrix} K_{ii} + Cc_i^2 & K_{ij} + Cc_i c_j \\ K_{ji} + Cc_i c_j & K_{jj} + Cc_j^2 \end{bmatrix} \tag{6.5.2.3}$$

$$\begin{Bmatrix} F_i \\ F_j \end{Bmatrix} \rightarrow \begin{Bmatrix} F_i + Ccc_i \\ F_j + Ccc_j \end{Bmatrix} \tag{6.5.2.4}$$

Note that the modified equations $\mathbf{KD} = \mathbf{F}$ containing terms from Eqs. (6.5.2.3) and (6.5.2.4) are still symmetric and positive definite. The only question left to answer is: what is the suitable value for the large number C? A popular choice that seems to work effectively is to make the constant a function of the largest element in the structural stiffness matrix:

$$C = 10^4 \max\left|K_{pq}\right|, \quad 1 \le p,\ q \le n \tag{6.5.2.5}$$

EXAMPLE 6.5.3 ***Simply Supported Beam with Skew Support (Example 2.5.2)***

A simply supported beam (E = 273600 k/ft^2) with an inclined roller support at one end is shown in Fig. E6.5.3(a). The cross-section is rectangular of width 8 in by height 20 in. Compute all the support reactions.

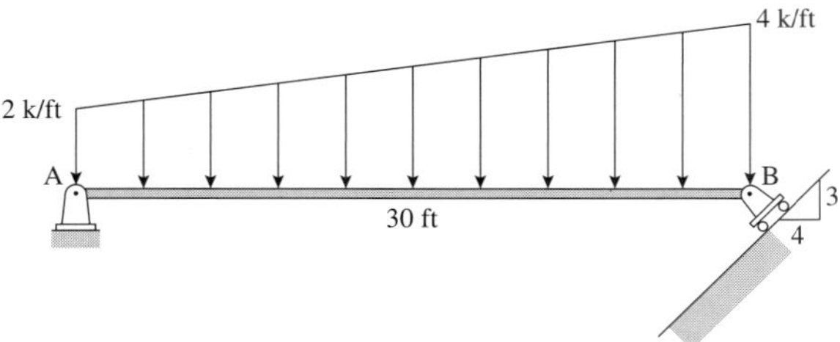

Fig. E6.5.3(a)

SOLUTION

Step 1: The problem units are k, ft. The model is shown in Fig. E6.5.3(b).

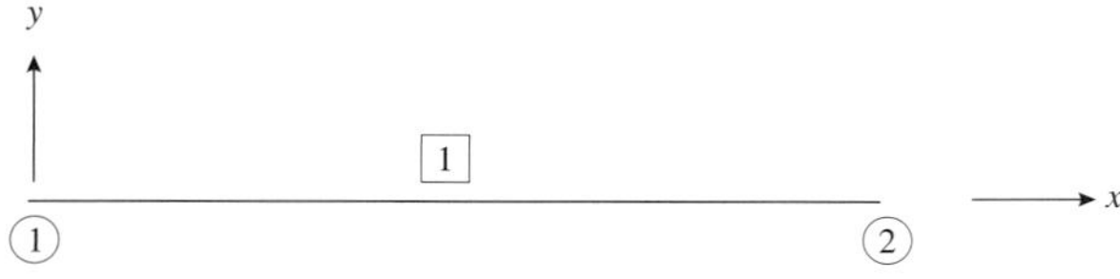

Fig. E6.5.3(b)

The boundary conditions are $D_1 = D_2 = 0$. The inclined or skew support is such that

$$\frac{D_5}{D_4} = \frac{3}{4} \Rightarrow 3D_4 - 4D_5 = 0 \tag{1}$$

Step 2: Element equilibrium equations. The element equilibrium equations are

$$10^2 \begin{bmatrix} 101.33 & 0 & 0 & -101.33 & 0 & 0 \\ 0 & 0.3128 & 4.6914 & 0 & -0.3128 & 4.6914 \\ 0 & 4.6914 & 93.827 & 0 & -4.6914 & 46.914 \\ -101.33 & 0 & 0 & 101.33 & 0 & 0 \\ 0 & -0.3128 & -4.6914 & 0 & 0.3128 & -4.6914 \\ 0 & 4.6914 & 46.914 & 0 & -4.6914 & 93.827 \end{bmatrix} \begin{Bmatrix} D_1 \\ D_2 \\ D_3 \\ D_4 \\ D_5 \\ D_6 \end{Bmatrix} = \begin{Bmatrix} 0 \\ -39 \\ -210 \\ 0 \\ -51 \\ 240 \end{Bmatrix}$$

Step 3: Assembly of the system equations:

$$10^2 \begin{bmatrix} 93.827 & 0 & -4.6914 & 46.914 \\ & 101.33 & 0 & 0 \\ & & 0.3128 & -4.6914 \\ \text{Sym} & & & 93.827 \end{bmatrix} \begin{Bmatrix} D_3 \\ D_4 \\ D_5 \\ D_6 \end{Bmatrix} = \begin{Bmatrix} -210 \\ 0 \\ -51 \\ 240 \end{Bmatrix}$$

These equations cannot be solved until we impose the constraint equation, Eq. (1). Note that $\max|K_{pq}| = 1.0133(10^4)$. Hence

$$C = (10^4)(1.0133 \times 10^4) = 1.0133(10^8)$$

Using Eqs. (6.5.2.3) and (6.5.2.4), we have $c_i = c_4 = 3$, $c_j = c_5 = -4$ and $c = 0$. Hence

$$10^2 \begin{bmatrix} 93.827 & 0 & -4.6914 & 46.914 \\ & 9.1201(10^6) & -1.216(10^7) & 0 \\ & & 1.6213(10^7) & -4.6914 \\ \text{Sym} & & & 93.827 \end{bmatrix} \begin{Bmatrix} D_3 \\ D_4 \\ D_5 \\ D_6 \end{Bmatrix} = \begin{Bmatrix} -210 \\ 0 \\ -51 \\ 240 \end{Bmatrix}$$

Step 4: Solution of the system equations. Solving, we obtain

$D_3 = -0.04699$ rad $\qquad D_4 = -0.0037$ ft

$D_5 = -0.00278$ ft $\qquad D_6 = 0.0489$ rad

Step 5: Element nodal forces. The member nodal forces are obtained in the usual manner using Eqs. (6.4.2.22) and (6.4.2.23). Using the equations, we have

$\mathbf{f}'_{6\times 1} = \{37.5 \text{ k}, 40 \text{ k}, 0, -37.5 \text{ k}, 50 \text{ k}, 0\}$

The element FBD is shown in Fig. E6.5.3(c).

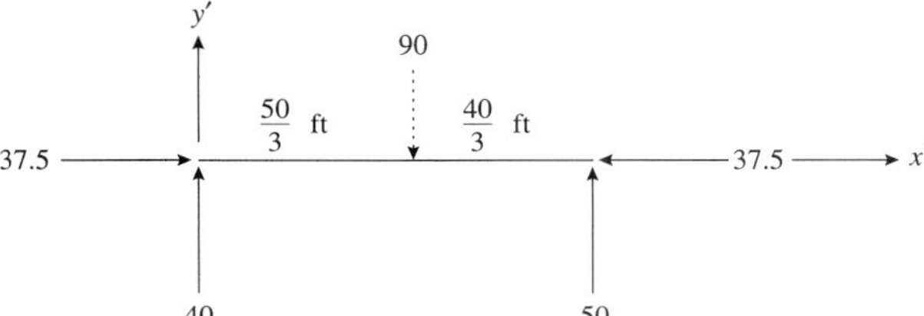

Fig. E6.5.3(c)

6.5.3 Support Settlements

As we saw in Chapter 5, support settlement effects are quite often considered during the analysis of structural systems and the forces induced in the structure due to these settlements can be large. Consideration of support settlement in the overall solution process is quite simple and affects the imposition of the essential boundary conditions before the system equations **KD** = **F** are solved.

Following the discussions in Section 6.2, consider the set of three equations representing the system equations as

$$\begin{bmatrix} K_{11} & K_{12} & K_{13} \\ K_{21} & K_{22} & K_{23} \\ K_{31} & K_{32} & K_{33} \end{bmatrix} \begin{Bmatrix} D_1 \\ D_2 \\ D_3 \end{Bmatrix} = \begin{Bmatrix} F_1 \\ F_2 \\ F_3 \end{Bmatrix} \qquad (6.5.3.1)$$

The system equations are singular unless the proper essential boundary conditions are imposed. Consider the boundary condition $D_2 = c$, where c is a known constant. In other words, the displacement D_2 has a known value that could be either zero or nonzero. A zero value would represent a restrained displacement whereas a nonzero value would represent a specified displacement such as a support settlement.

We rewrite Eq. (6.5.3.1) as three separate equations:

$$K_{11}D_1 + K_{12}D_2 + K_{13}D_3 = F_1 \qquad (6.5.3.2)$$

$$K_{21}D_1 + K_{22}D_2 + K_{23}D_3 = F_2 \qquad (6.5.3.3)$$

$$K_{31}D_1 + K_{32}D_2 + K_{33}D_3 = F_3 \qquad (6.5.3.4)$$

Using the condition $D_2 = c$ in the above equations, we have

$$K_{11}D_1 + (0)D_2 + K_{13}D_3 = F_1 - K_{12}c \qquad (6.5.3.5)$$

$$(0)D_1 + (1)D_2 + (0)D_3 = c \tag{6.5.3.6}$$

$$K_{31}D_1 + (0)D_2 + K_{33}D_3 = F_3 - K_{32}c \tag{6.5.3.7}$$

Note carefully that Eqs. (6.5.3.2) and (6.5.3.5) are the same if $D_2 = c$. We have merely taken the D_2 term to the right-hand side where it belongs, since D_2 is strictly no longer an unknown. Similar comments are valid for Eqs. (6.5.3.4) and (6.5.3.7). In order to (a) not change the number of equations, and (b) recognize that $D_2 = c$, suitable changes have been made to Eq. (6.5.3.3), and Eq. (6.5.3.6) is merely $D_2 = c$. We can rewrite Eqs. (6.5.3.5)–(6.5.3.7) as

$$\begin{bmatrix} K_{11} & 0 & K_{13} \\ 0 & 1 & 0 \\ K_{31} & 0 & K_{33} \end{bmatrix} \begin{Bmatrix} D_1 \\ D_2 \\ D_3 \end{Bmatrix} = \begin{Bmatrix} F_1 - K_{12}c \\ c \\ F_3 - K_{32}c \end{Bmatrix} \tag{6.5.3.8}$$

This approach is known as the elimination approach.[11] The advantage of the equations in the above form is that the number of equations remains the same and that the stiffness matrix is still symmetric. From a viewpoint of implementing the solution procedure in the form of a computer program, these are desirable properties.

General Procedure. If displacement $D_j = c$ is to be imposed, implement the following three steps.

Step 1: Modify the right-hand side vector (or load vector) as $F_i = F_i - K_{ij}c$, $i = 1, \ldots, n$.

Step 2: Modify the coefficient matrix (or stiffness matrix) as

$$K_{ij} = 0, \; i = 1,\ldots,n \qquad K_{ji} = 0, \; i = 1,\ldots,n$$

Step 3: Set $K_{jj} = 1$.

This three-step process must be applied for each displacement D_j that has a known value.

EXAMPLE 6.5.4 ***Support Settlement (Example 5.2.3)***

For the beam in Fig. E6.5.4(a), compute the support reactions. Support B settles 0.5 in. Take $E = 30000$ ksi and $I = 1500$ in^4.

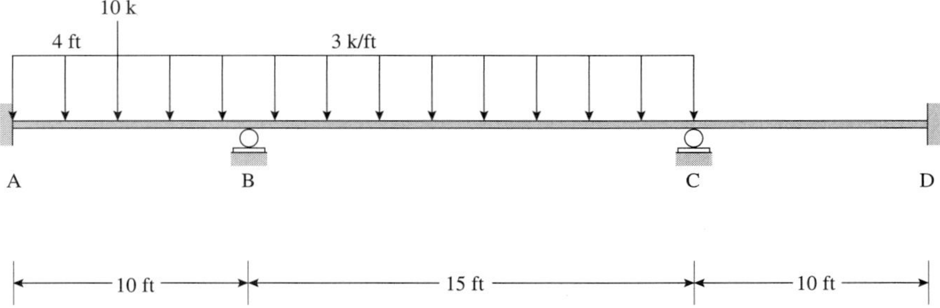

Fig. E6.5.4(a)

SOLUTION

Step 1: We use k, ft as the problem units. The model is shown in Fig. E6.5.4(b).

The cross-sectional area is not given and should not affect the results. For the sake of convenience, we assume $A = 1$ ft^2. Note that the boundary conditions are $D_1 = D_2 = D_3 = D_8 = D_{10} = D_{11} = D_{12} = 0$, $D_5 = -(0.5/12) = -0.04167$, and there are effectively five degrees of freedom (including D_5).

[11] A different way is to use the penalty approach, which was used to handle skew supports in Section 6.5.2.

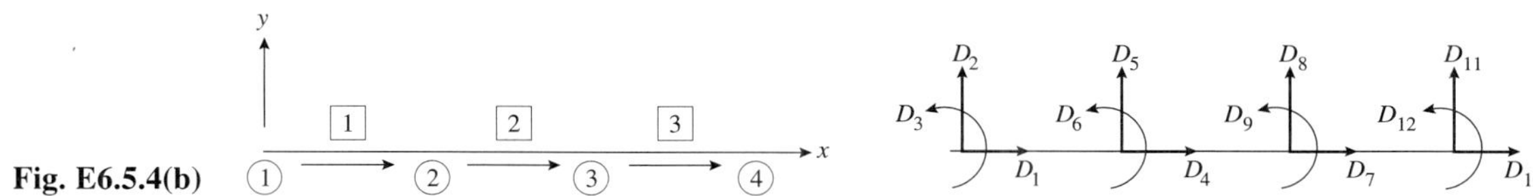

Fig. E6.5.4(b)

Step 2: Element equilibrium equations

Element 1:

$$10^3\begin{bmatrix} 432 & 0 & 0 & -432 & 0 & 0 \\ 0 & 3.75 & 18.75 & 0 & -3.75 & 18.75 \\ 0 & 18.75 & 125 & 0 & -18.75 & 62.5 \\ -432 & 0 & 0 & 432 & 0 & 0 \\ 0 & -3.75 & -18.75 & 0 & 3.75 & -18.75 \\ 0 & 18.75 & 62.5 & 0 & -18.75 & 125 \end{bmatrix}\begin{bmatrix} D_1 \\ D_2 \\ D_3 \\ D_4 \\ D_5 \\ D_6 \end{bmatrix} = \begin{Bmatrix} 0 \\ -21.48 \\ -39.4 \\ 0 \\ -18.52 \\ 34.6 \end{Bmatrix}$$

Element 2:

$$10^3\begin{bmatrix} 288 & 0 & 0 & -288 & 0 & 0 \\ 0 & 1.111 & 8.333 & 0 & -1.111 & 8.333 \\ 0 & 8.333 & 83.33 & 0 & -8.333 & 41.667 \\ -288 & 0 & 0 & 288 & 0 & 0 \\ 0 & -1.111 & -8.333 & 0 & 1.111 & -8.333 \\ 0 & 8.333 & 41.667 & 0 & -8.333 & 83.33 \end{bmatrix}\begin{bmatrix} D_4 \\ D_5 \\ D_6 \\ D_7 \\ D_8 \\ D_9 \end{bmatrix} = \begin{Bmatrix} 0 \\ -22.5 \\ -56.25 \\ 0 \\ -22.5 \\ 56.25 \end{Bmatrix}$$

Element 3:

$$10^3\begin{bmatrix} 432 & 0 & 0 & -432 & 0 & 0 \\ 0 & 3.75 & 18.75 & 0 & -3.75 & 18.75 \\ 0 & 18.75 & 125 & 0 & -18.75 & 62.5 \\ -432 & 0 & 0 & 432 & 0 & 0 \\ 0 & -3.75 & -18.75 & 0 & 3.75 & -18.75 \\ 0 & 18.75 & 62.5 & 0 & -18.75 & 125 \end{bmatrix}\begin{bmatrix} D_7 \\ D_8 \\ D_9 \\ D_{10} \\ D_{11} \\ D_{12} \end{bmatrix} = \begin{Bmatrix} 0 \\ 0 \\ 0 \\ 0 \\ 0 \\ 0 \end{Bmatrix}$$

Step 3: Assembly (system equations):

$$10^3\begin{bmatrix} 720 & 0 & 0 & -288 & 0 \\ & 4.8611 & -10.417 & 0 & 8.333 \\ & & 208.33 & 0 & 41.667 \\ \text{Sym} & & & 720 & 0 \\ & & & & 208.33 \end{bmatrix}\begin{bmatrix} D_4 \\ D_5 \\ D_6 \\ D_7 \\ D_9 \end{bmatrix} = \begin{Bmatrix} 0 \\ -41.02 \\ -21.65 \\ 0 \\ 56.25 \end{Bmatrix}$$

Step 4: Imposition of the boundary condition. Next we need to impose the boundary condition $D_5 = -0.04167$. Using the steps discussed in the elimination technique, we can reduce the equations to the final form:

$$10^3\begin{bmatrix} 720 & 0 & 0 & -288 & 0 \\ & 10^{-3} & 0 & 0 & 0 \\ & & 208.33 & 0 & 41.667 \\ \text{Sym} & & & 720 & 0 \\ & & & & 208.33 \end{bmatrix}\begin{bmatrix} D_4 \\ D_5 \\ D_6 \\ D_7 \\ D_9 \end{bmatrix} = \begin{Bmatrix} 0 \\ -4.1667(10^{-2}) \\ -455.68 \\ 0 \\ 403.47 \end{Bmatrix}$$

Step 5: Solution of the system equilibrium equations. Solving the system equations, we obtain

$$D_4 = 0,\ D_5 = -0.041667 \text{ ft}, \quad D_6 = -2.6819(10^{-3}) \text{ rad} \quad D_7 = 0,\ D_9 = 2.473(10^{-3}) \text{ rad}$$

Step 6: Member nodal forces. The member nodal forces are obtained in the usual manner using Eqs. (6.4.2.22) and (6.4.2.23). The details are not presented here but the final results are summarized.

The FBDs are shown in Figs. E6.5.4(c)–(e).

Element 1: $\mathbf{f}'_{6\times1} = \{0, 127 \text{ k}, 653 \text{ k-ft}, 0, -87 \text{ k}, 411.4 \text{ k-ft}\}$

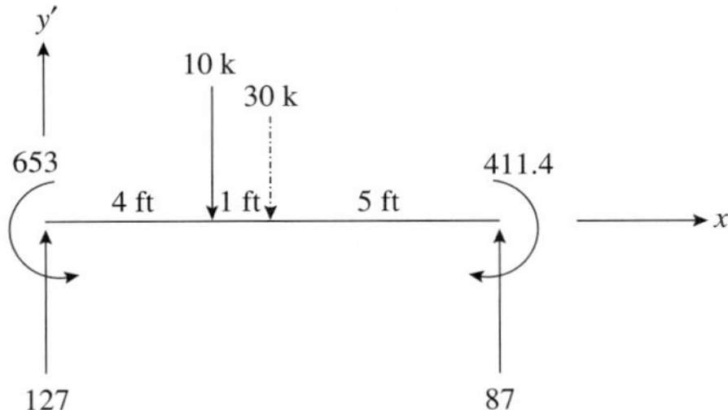

Fig. E6.5.4(c)

Element 2: $\mathbf{f}'_{6\times1} = \{0, -25 \text{ k}, -411.4 \text{ k-ft}, 0, 71 \text{ k}, -309 \text{ k-ft}\}$

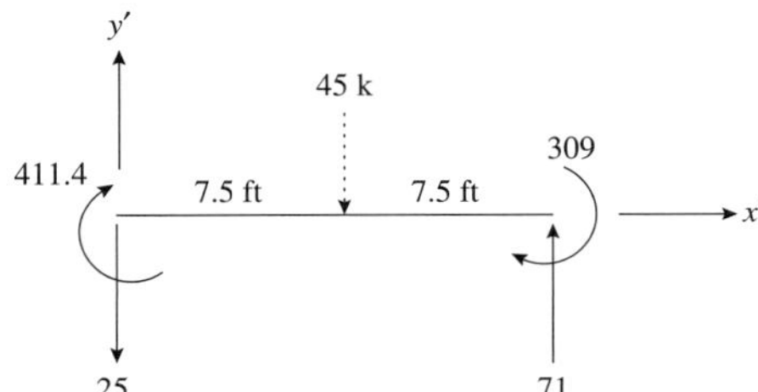

Fig. E6.5.4(d)

Element 3: $\mathbf{f}'_{6\times1} = \{0, 46 \text{ k}, 309 \text{ k-ft}, 0, -46 \text{ k}, 154.6 \text{ k-ft}\}$

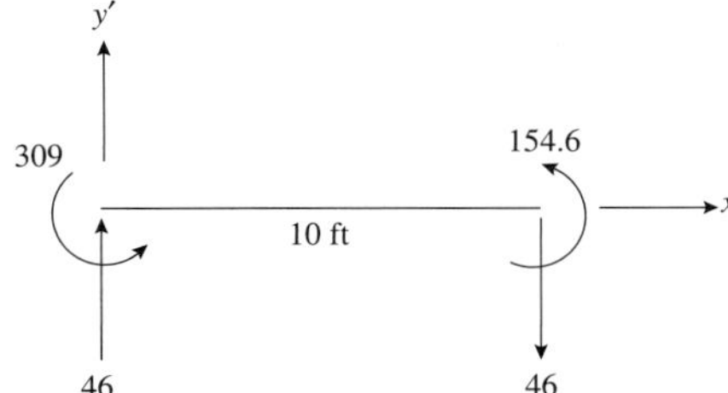

Fig. E6.5.4(e)

6.5.4 Typical Connection

In Chapter 2 we briefly introduced a typical connection—connections that are neither completely rigid nor an internal hinge. The rigidity is a function of several parameters such as the stiffness and the geometry of the individual members, the bolt or screw layout, etc. A torsional spring is used to depict the rigidity of a typical connection. Experiments can be conducted to estimate the torsional stiffness constant. Throughout the text we have looked at either rigid connections or an internal hinge. If the torsional spring constant is infinity, then the connection behaves as a rigid connection. On the other hand, if the spring constant is zero, then the connection behaves as an internal hinge.

We now describe the torsional spring element that can be used to model a typical connection. The element is shown in Fig. 6.5.4.1. The element has two degrees of freedom—rotations θ_1 and θ_2 at the ends of the spring. The element property is defined by the spring constant k_θ. The element equations are

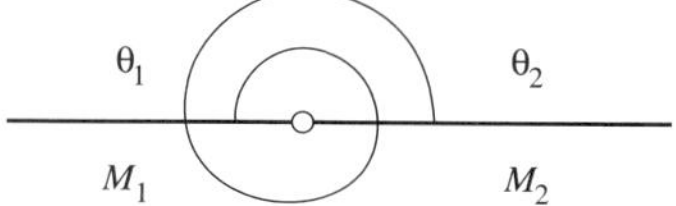

Fig. 6.5.4.1 Torsional spring element.

$$k_\theta \begin{bmatrix} 1 & -1 \\ -1 & 1 \end{bmatrix} \begin{Bmatrix} \theta_1 \\ \theta_2 \end{Bmatrix} = \begin{Bmatrix} M_1 \\ M_2 \end{Bmatrix} \tag{6.5.4.1}$$

A word about modeling with this element. If two members are connected to each other via a known joint, one end of this spring is connected to one member and the other end of the spring is connected to the second member. From Fig. 6.5.4.2, at the joint connecting the two members there are four degrees of freedom, not three. The four dof are the x-displacement Δ_x, the y-displacement Δ_y, θ_1 the rotation associated with element 1 at the joint, and θ_2 the rotation associated with element 2 at the joint. This situation is no different if an internal hinge connected members 1 and 2.

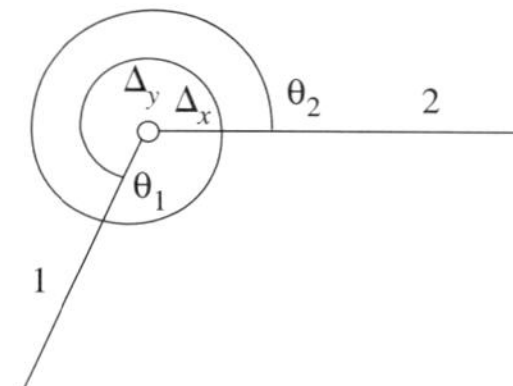

Fig. 6.5.4.2 A torsional spring connecting two members.

EXAMPLE 6.5.5 ***Frame with a Typical Connection (Example 5.2.5)***

Consider the frame in Fig. E6.5.5(a). The modulus of elasticity is $2(10^{11})$ Pa, the cross-sectional is 0.001 m^2 and the moment of inertia is 0.0001 m^4 for both the members. Draw the shear force and bending moment diagrams.

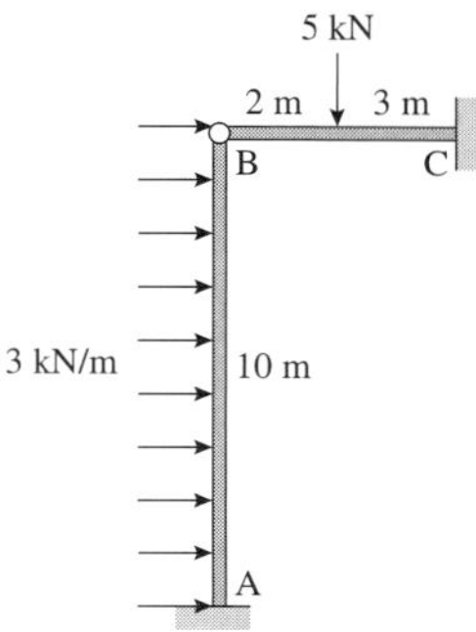

Fig. E6.5.5(a)

SOLUTION

Step 1: The problem characteristics lie between Example 6.4.3, where B is a rigid connection, and Example 6.5.2, where B is an internal hinge. As before, the problem units are N, m. Similarly to Example 6.5.2, there are a total of ten degrees of freedom in the frame (see Fig. E6.5.5(b)). The boundary conditions of the frame are such that

$$D_1 = D_2 = D_3 = D_8 = D_9 = D_{10} = 0$$

However, we have to retain all the other degrees of freedom. Recall that we condensed or eliminated the rotational dof at the hinge in Example 6.5.2. Most of the calculations are similar to Example 6.4.3 and we present them again for easy reference.

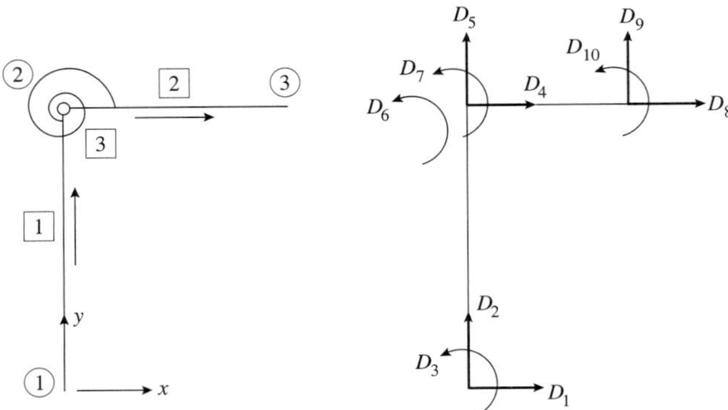

Fig. E6.5.5(b)

Element load on element 1: With $w = -3000$ N/m (see Fig. 6.2.3.9), we have

$$\mathbf{q}'_{6\times 1} = \left\{0, \frac{wL}{2}, \frac{wL^2}{12}, 0, \frac{wL}{2}, -\frac{wL^2}{12}\right\} = \{0, 15000, -25000, 0, 15000, 25000\}$$

Element load on element 2: With $P = -5000$ N, $a = 2$ m, $b = 3$ m (see Fig. 6.2.3.9), we have

$$\mathbf{q}'_{6\times 1} = \left\{0, \frac{Pb^2(L+2a)}{L^3}, \frac{Pab^2}{L^2}, 0, \frac{Pa^2(L+2b)}{L^3}, -\frac{Pa^2b}{L^2}\right\}$$
$$= \{0, -3240, -3600, 0, -1760, 2400\}$$

These loads need to be transformed to the global coordinate system using Eq. (6.4.2.21).

Step 2: The element equilibrium equations. We can use the results from step 1 to generate the element equilibrium equations for each element:

Element 1:

$$10^5 \begin{bmatrix} 2.4 & 0 & -12 & -2.4 & 0 & -12 \\ 0 & 2000 & 0 & 0 & -2000 & 0 \\ -12 & 0 & 80 & 12 & 0 & 40 \\ -2.4 & 0 & 12 & 2.4 & 0 & 12 \\ 0 & -2000 & 0 & 0 & 2000 & 0 \\ -12 & 0 & 40 & 12 & 0 & 80 \end{bmatrix} \begin{Bmatrix} D_1 \\ D_2 \\ D_3 \\ D_4 \\ D_5 \\ D_6 \end{Bmatrix} = \begin{Bmatrix} 15000 \\ 0 \\ -25000 \\ 15000 \\ 0 \\ 25000 \end{Bmatrix}$$

Element 2:

$$10^5 \begin{bmatrix} 4000 & 0 & 0 & -4000 & 0 & 0 \\ 0 & 19.2 & 48 & 0 & -19.2 & 48 \\ 0 & 48 & 160 & 0 & -48 & 80 \\ -4000 & 0 & 0 & 4000 & 0 & 0 \\ 0 & -19.2 & -48 & 0 & 19.2 & -48 \\ 0 & 48 & 80 & 0 & -48 & 160 \end{bmatrix} \begin{Bmatrix} D_4 \\ D_5 \\ D_7 \\ D_8 \\ D_9 \\ D_{10} \end{Bmatrix} = \begin{Bmatrix} 0 \\ -3240 \\ -3600 \\ 0 \\ -1760 \\ 2400 \end{Bmatrix}$$

Element 3:

$$\begin{bmatrix} k_\theta & -k_\theta \\ -k_\theta & k_\theta \end{bmatrix} \begin{Bmatrix} D_6 \\ D_7 \end{Bmatrix} = \begin{Bmatrix} 0 \\ 0 \end{Bmatrix}$$

We will use the numerical value for the spring later in the example.

Step 3: Assembly of the system equations. We assemble only the effective equations.

$$10^5 \begin{bmatrix} 4002.4 & 0 & 12 & 0 \\ 0 & 2019.2 & 0 & 48 \\ 12 & 0 & 80 + k_\theta & -k_\theta \\ 0 & 48 & -k_\theta & 160 + k_\theta \end{bmatrix} \begin{Bmatrix} D_4 \\ D_5 \\ D_6 \\ D_7 \end{Bmatrix} = \begin{Bmatrix} 15000 \\ -3240 \\ 25000 \\ -3600 \end{Bmatrix}$$

Step 4: Solution of the equilibrium equations. Solving the four equations, we have the following results:

a. For $k_\theta = 0$ (internal hinge):

$$D_4 = 2.812(10^{-5})\text{ m} \qquad D_5 = -1.077(10^{-5})\text{ m}$$

$$D_6 = 3.121(10^{-3})\text{ rad} \qquad D_7 = -2.218(10^{-4})\text{ rad}$$

These values are the same as those obtained in Examples 6.5.2 and 5.2.5.

b. For $k_\theta = \infty$ (rigid connection). Using a numerical value of 10^{10} for k_θ, we have

$$D_4 = 3.48(10^{-5})\text{ m} \qquad D_5 = -3.74(10^{-5})\text{ m}$$

$$D_6 = 8.97(10^{-4})\text{ rad} \qquad D_7 = 8.97(10^{-4})\text{ rad}$$

These values are the same as those obtained in Example 6.4.3.

c. For $k_\theta = 10^4$:

$$D_4 = 2.813(10^{-5})\text{ m} \qquad D_5 = -1.082(10^{-5})\text{ m}$$

$$D_6 = 3.117(10^{-3})\text{ rad} \qquad D_7 = -2.2(10^{-4})\text{ rad}$$

The displacements are closer to the internal hinge case.

Step 5: Computation of element nodal forces. Using Eqs. (6.4.2.22) and (6.4.2.23), we can compute the element nodal forces. The details (intermediate steps) are not shown here.

Element 1:

$$\mathbf{f}'_{6\times1} = \{2164\text{ N}, 18734\text{ N}, 37502\text{ N-m}, -2164\text{ N}, 11253\text{ N}, -30\text{ N-m}\}$$

Element 2:

$$\mathbf{f}'_{6\times1} = \{11253\text{ N}, 2164\text{ N}, 30\text{ N-m}, -11253\text{ N}, 2836\text{ N}, -4212\text{ N-m}\}$$

The element FBDs are shown in Fig. E6.5.5(c).

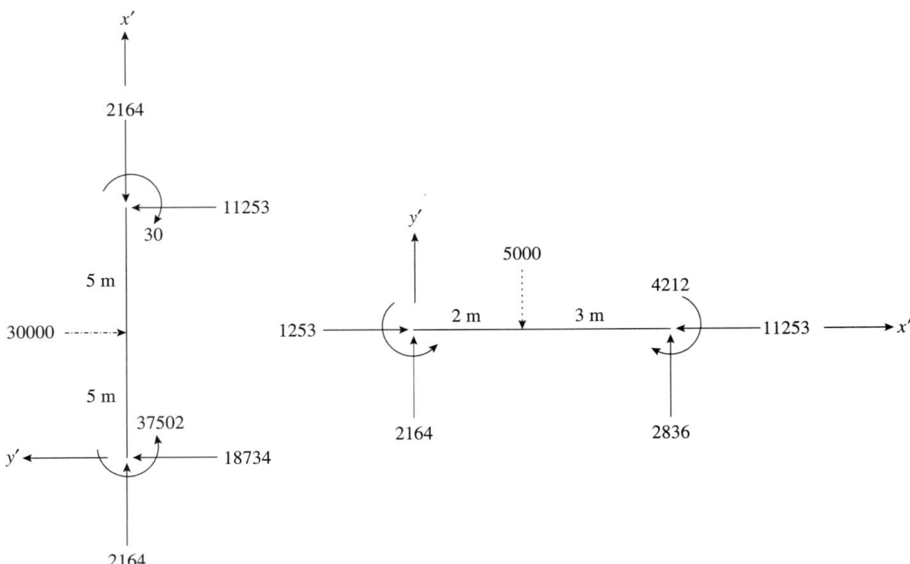

Fig. E6.5.5(c)

6.5.5 Thermal Loads

When temperature effects are considered, additional calculations need to be done. The thermal load vector needs to be generated and added to the overall load vector **F**. Once the nodal displacements **D** are computed, the thermal strains ε_0 must be deducted from the computed strain to calculate the element stress and force.

Truss. Consider a truss element subjected to a uniform temperature change ΔT. First, the load vector must be modified by the addition of the thermal loads:

$$\varepsilon_0 = \alpha \Delta T \tag{6.5.5.1}$$

$$(\mathbf{q}_t')_{2\times 1} = EA\varepsilon_0 \begin{Bmatrix} -1 \\ 1 \end{Bmatrix} \qquad \mathbf{q}_t = \mathbf{T}^{\mathrm{T}} \mathbf{q}_t' \tag{6.5.5.2}$$

where ε_0 represents the initial (or thermal) strains, α is the coefficient of thermal expansion, $\mathbf{q}_t'$ is the thermal load vector in the local coordinate system, and $\mathbf{q}_t$ is in the global coordinate system. Second, after the displacements are computed, the stress in the element is now

$$\sigma = E(\varepsilon - \varepsilon_0) = E\left[\frac{d_2' - d_1'}{L} - \alpha(\Delta T)\right] \tag{6.5.5.3}$$

Beam. The beam behavior under thermal loading is essentially the same as the truss behavior unless there are eccentricities in the connections. The following equations are general equations and it should be clear why the previous statement is true:

Axial Forces, q_1' and q_7': $(q_1')_t = -EA\alpha(\Delta T) = -(q_7')_t$ (6.5.5.4a)

Shear Forces, q_2' and q_8': $(q_2')_t = 0 = (q_8')_t$ (6.5.5.4b)

Shear Forces, q_3' and q_9': $(q_3')_t = 0 = (q_9')_t$ (6.5.5.4c)

Bending Moments, q_5' and q_{11}': $(q_5')_t = \int_A \alpha E(\Delta T)\, z\, dA = -(q_{11}')_t$ (6.5.5.4d)

Bending Moments, q_6' and q_{12}': $(q_6')_t = -\int_A \alpha E(\Delta T)\, y\, dA = (q_{12}')_t$ (6.5.5.4e)

Torsional Moments, q_4' and q_{10}': $(q_4')_t = 0 = (q_{10}')_t$ (6.5.5.4f)

In a manner similar to Eq. (6.5.5.3), the final member nodal forces are given by

$$\mathbf{f}' = \mathbf{k}'\mathbf{d}' - \sum_i (\mathbf{q}')_i - \mathbf{q}_t' \tag{6.5.5.5}$$

where the second term on the right is the summation over all element loads and the final term represents the thermal load component.

As mentioned earlier, thermal loads can be classified either as initial strains or initial stresses. The solution process in this section can be used to model other types of initial strains. For example, fabrication errors can be modeled as an initial strain using a fictitious temperature change in the element. The details are left as an exercise.

EXAMPLE 6.5.6 ***Thermal Loading on a Truss***

Solve for the nodal displacements and member forces for the truss in Fig. E6.5.6(a). The modulus of elasticity is $30(10^6)$ psi, the cross-sectional area of each member is 1.2 in^2, the coefficient of thermal expansion is 1/150000 per °F, and the temperature change in members AB and BC is 100° F.

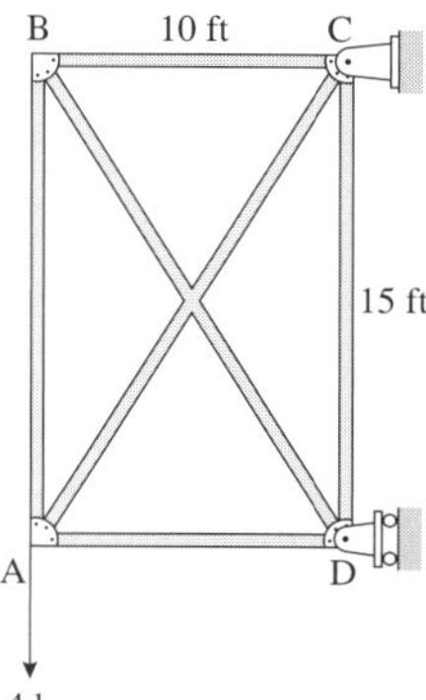

Fig. E6.5.6(a)

SOLUTION ***Step 1:*** The problem units are lb, in. The model is shown in Fig. E6.5.6(b).

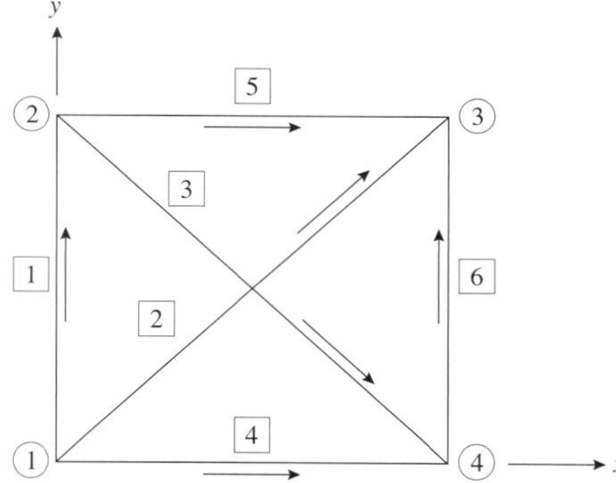

Fig. E6.5.6(b)

The global degrees of freedom are shown in Fig. E6.5.6(c).

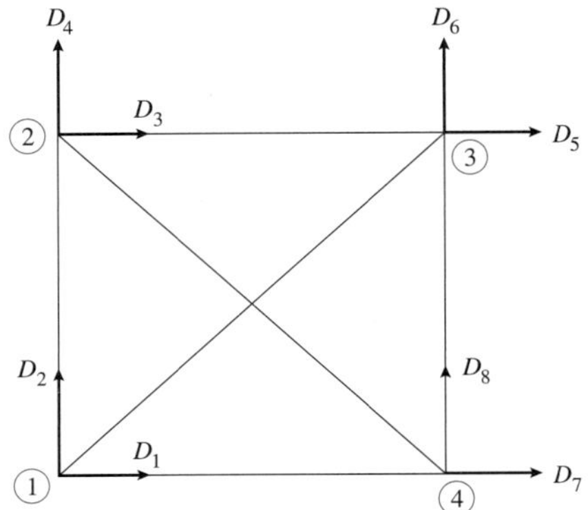

Fig. E6.5.6(c)

The details of the stiffness matrix generation are not presented here. However, we focus on the generation of the load vector due to the temperature changes.

Element 1: Using the given data, we have

$(l, m) = (0.0, 1.0) \qquad EA = 36(10^6) \text{ psi} \qquad \varepsilon_0 = \alpha(\Delta T) = 6.66667(10^{-4})$

Hence, $(\mathbf{q}'_t)_{2\times1} = EA\varepsilon_0 \begin{Bmatrix} -1 \\ 1 \end{Bmatrix} = 24000 \begin{Bmatrix} -1 \\ 1 \end{Bmatrix} = \begin{Bmatrix} -24000 \\ 24000 \end{Bmatrix}$

Since $\mathbf{T}_{2\times4} = \begin{bmatrix} 0 & 1 & 0 & 0 \\ 0 & 0 & 0 & 1 \end{bmatrix}$, using Eq. (6.5.5.2) we have

$$(\mathbf{q}_t)_{4\times1} = \begin{Bmatrix} 0 \\ -24000 \\ 0 \\ 24000 \end{Bmatrix} \text{lb} = \begin{Bmatrix} F_1 \\ F_2 \\ F_3 \\ F_4 \end{Bmatrix}$$

Element 5: In a similar manner, we have

$$(\mathbf{q}_t)_{4\times1} = \begin{Bmatrix} -24000 \\ 0 \\ 24000 \\ 0 \end{Bmatrix} \text{lb} = \begin{Bmatrix} F_3 \\ F_4 \\ F_5 \\ F_6 \end{Bmatrix}$$

Taking these two load vectors, we can construct the structural nodal load vector as

$$\mathbf{F}_{8\times1} = \begin{Bmatrix} 0 \\ -24000 \\ -24000 \\ 24000 \\ 24000 \\ 0 \\ 0 \\ 0 \end{Bmatrix} \text{ and using the external load we have } \mathbf{F}_{8\times1} = \begin{Bmatrix} 0 \\ -28000 \\ -24000 \\ 24000 \\ 24000 \\ 0 \\ 0 \\ 0 \end{Bmatrix}$$

Step 2: Assembly of the system equations:

$$10^5\begin{bmatrix} 3.512 & 0.76805 & 0 & 0 & 0 \\ & 3.1521 & 0 & -2 & 0 \\ & & 3.512 & -0.76805 & 0.76805 \\ Sym & & & 3.5121 & -1.1521 \\ & & & & 3.1521 \end{bmatrix}\begin{Bmatrix} D_1 \\ D_2 \\ D_3 \\ D_4 \\ D_8 \end{Bmatrix} = \begin{Bmatrix} 0 \\ -28000 \\ -24000 \\ 24000 \\ 0 \end{Bmatrix}$$

Step 3: Solution of the system equations. Solving the system equilibrium equations, we have

$$D_1 = 0.0186'' \qquad D_2 = -0.0851'' \qquad D_3 = -0.0703''$$

$$D_4 = 0.0130'' \qquad D_8 = 0.0219''$$

Step 4: Element nodal forces. Using Eq. (6.5.5.3) we can compute the net force in each member by multiplying the stress with the element area. We show the calculations for elements 1 and 5 that have a nonzero second term.

Element 1:

Since $\mathbf{d}_{4\times1} = \{0.0186, -0.0851, -0.0703, 0.0130\}^T$

$$\mathbf{d}'_{2\times1} = \mathbf{T}_{2\times4}\mathbf{d}_{4\times1} = \begin{Bmatrix} -0.0851 \\ 0.0130 \end{Bmatrix} \text{in}$$

$$\sigma = 30(10^6)\left[\frac{0.0130-(-0.0851)}{180} - 6.6667(10^{-4})\right] = -3650 \text{ psi}$$

The negative sign indicates that the element is in compression:

$$f = \sigma A = 3650(1.2) = 4380 \text{ lb} \quad \text{(C)}$$

Element 5:

Since $\mathbf{d}_{4\times1} = \{-0.0703, 0.0130, 0, 0\}^T$

$$\mathbf{d}'_{2\times1} = \mathbf{T}_{2\times4}\mathbf{d}_{4\times1} = \begin{Bmatrix} -0.0703 \\ 0 \end{Bmatrix} \text{in}$$

$$\sigma = 30(10^6)\left[\frac{0-(-0.0703)}{120} - 6.6667(10^{-4})\right] = -2425 \text{ psi}$$

$$f = \sigma A = 2425(1.2) = 2910 \text{ lb} \quad \text{(C)}$$

EXERCISES

Appetizers

6.5.1. Consider the truss from Problem 6.2.3 (see Fig. P6.5.1). Compute the member forces if, in addition to the external load, there is a temperature decrease of 50° C in all the members.

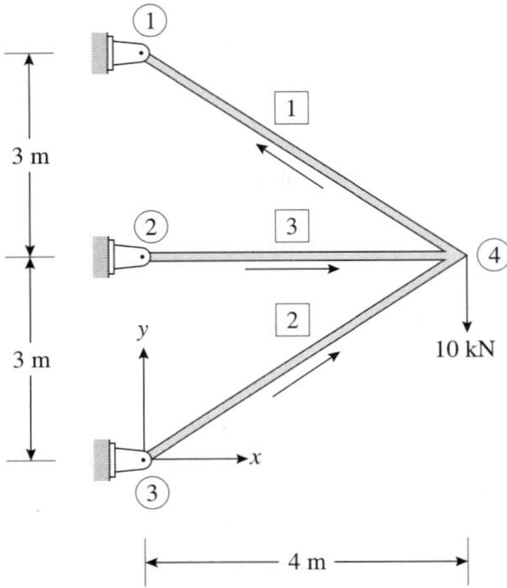

Fig. P6.5.1

6.5.2. Consider the truss in Problem 6.2.5 (see Fig. P6.5.2). In addition to the external loads, there is a support settlement of 0.01 m at node 1. Compute all the member forces.

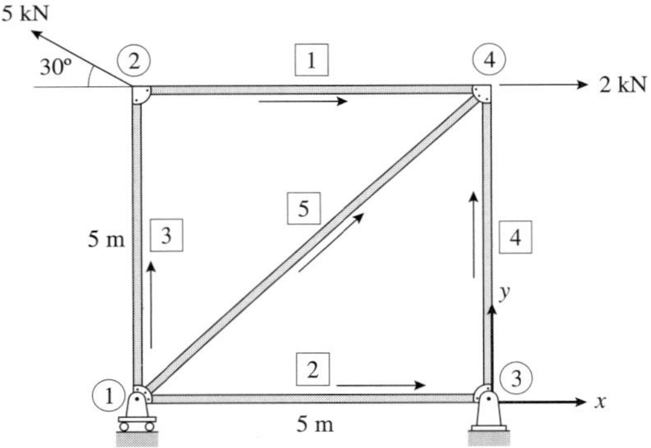

Fig. P6.5.2

6.5.3. Consider the steel beam in Problem 6.2.14 (see Fig. P6.5.3). Take $I = 10^{-4}\,\text{m}^4$ and $A = 0.01\ \text{m}^2$. In addition to the external loads, there is a support settlement of 0.05m at node 2. Draw the shear force and bending moment diagrams.

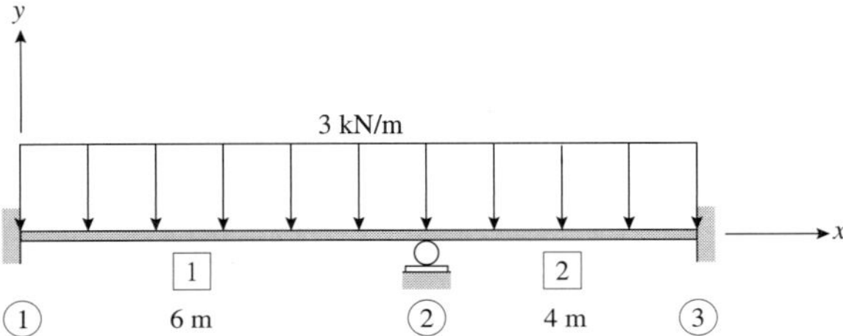

Fig. P6.5.3

Main Course

6.5.4. Compute the nodal displacements and the support reactions of the steel beam in Fig. P6.5.4. The members have the cross-sectional properties $A = 5\ \text{in}^2$ and $I = 400\ \text{in}^4$.

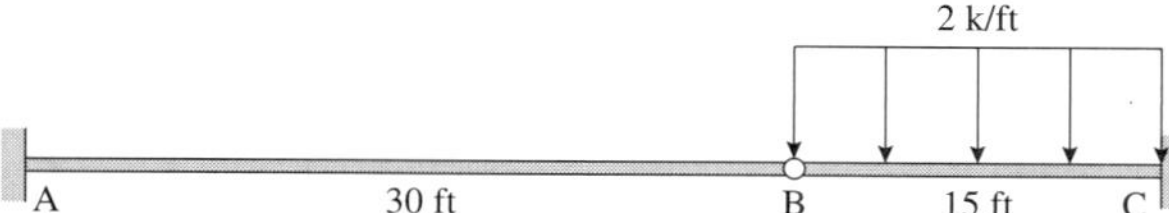

Fig. P6.5.4

6.5.5. Compute the nodal displacements and the support reactions of the beam in Fig. P6.5.5. The member is built using AISC W36 × 210. Ignore self-weight.

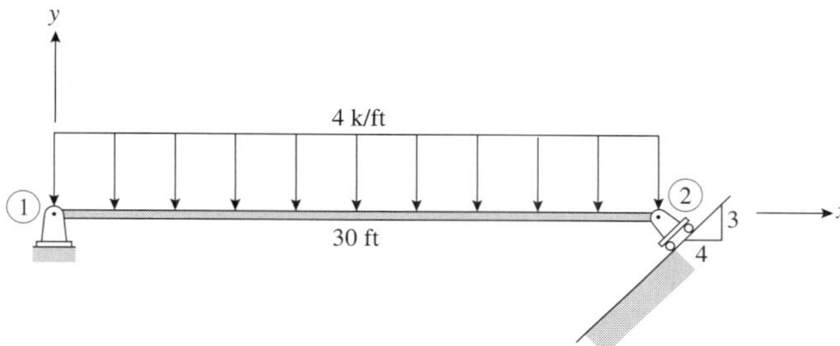

Fig. P6.5.5

6.5.6. Compute the nodal displacements and the member end forces in the beam in Fig. P6.5.6. Take $E = 200$ GPa and $I = 4(10^6)\ \text{mm}^4$ and the torsional spring constant as 100 kN-m/rad.

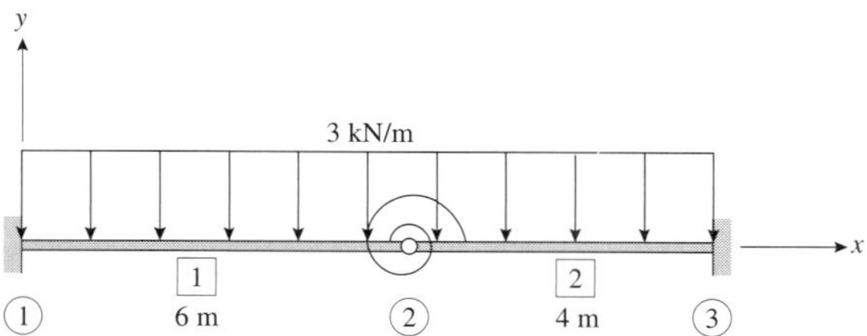

Fig. P6.5.6

6.5.7. Compute the nodal displacements and the support reactions of the frame in Problem 7.1 (see Fig. P6.5.7). The members are AISC W24 × 84. Ignore self-weight.

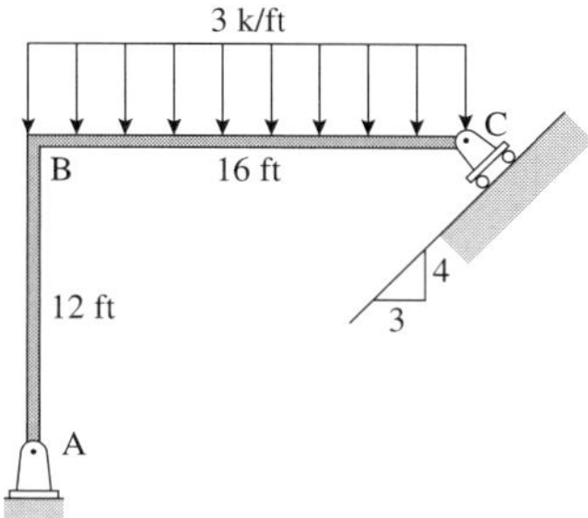

Fig. P6.5.7

Structural Concepts

6.5.8. Fabrication errors can be modeled as an initial strain using a fictitious temperature change in the element as an equivalent load. Solve Problem 6.5.1 with the only source of loading in the truss being a fabrication error in member 3 that is 0.01m too short.

SUMMARY

In this chapter we looked at two different numerical approaches to solving for the response of discrete structural systems—the direct stiffness method and the finite element method. Both these approaches can be used to generate the equilibrium-compatibility equations for a typical element in a system. We derived these equations for the truss and beam elements. These elements can then be used in the modeling of a structure that fits the definition of a truss and beam-column.

There are tremendous advantages to these numerical techniques. They give an engineer a powerful tool to analyze determinate and indeterminate, small or large, simple or complex systems. The computed responses include the displacement, force, stress and strain distribution over the structural elements. We can pose the question that is most often asked today by students—"If there is this powerful technique that is implemented in the form of a user-friendly computer program, why do I need to learn and use the tedious, classical techniques?" The contents of this chapter should make the answer obvious—(a) we simply could not have learnt these concepts if we did not know the fundamentals offered by the classical techniques,[12] and (b) there is tremendous potential to misuse the computer tools if we are not clear about their assumptions, limitations, strengths, and weaknesses.

The finite element method is a very powerful technique. While we have barely scratched the surface of its capabilities, the fundamental ideas introduced in this chapter are the very ideas used to solve a wide variety of problems. These ideas include the concept of using an assumed solution, interpolation, minimizing the total potential energy, generating the element equations, the assembly process, imposition of boundary conditions, the solution of the primary unknowns (the nodal displacements) and computation of the secondary response quantities (element nodal forces and support reactions).

When a powerful analysis tool is combined with a powerful design tool, we have the makings of an efficient practical tool. We explore this tool in the next two chapters.

SUMMARY EXERCISES

Appetizers

6.1. Solve for the nodal displacements, member end force and the support reactions for the beam in Fig. P6.1 below for the following two cases. Take $E = 2(10^{11})$ N/m^2, $A = 0.03$ m^2, and $I = 10^{-4}$ m^4.

(a) Assume that only the external loads are acting on the beam.

(b) Assume that in addition to the external loads, support B settles 0.05 m.

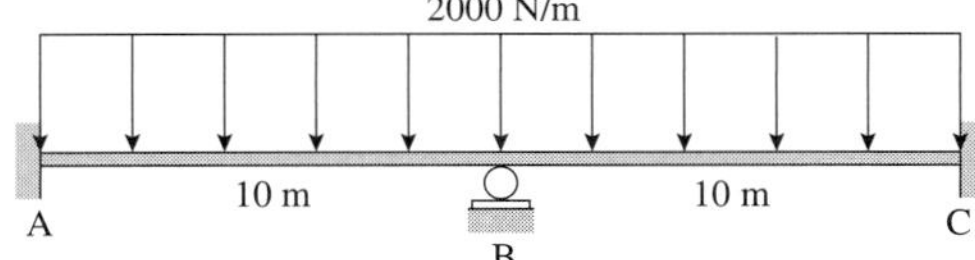

Fig. P6.1

6.2. Solve for the nodal displacements, member end force, and the support reactions for the frame in Fig. P6.2. The member cross-sections made of steel, and are circular hollow with an internal radius of 1 in and a wall thickness of 0.1 in.

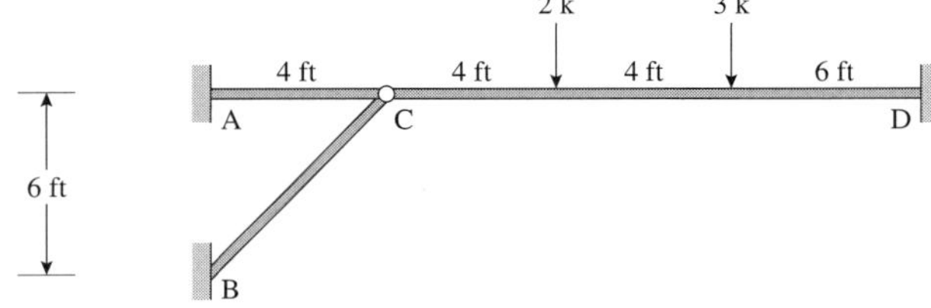

Fig. P6.2

---

[12] The author has found bugs in the GS-USA© computer programs using classical techniques (not hand calculations of the finite element method).

Main Course

6.3. The planar truss in Fig. P6.3 is made of steel. Members AB and BD have a cross-sectional area of 2 in^2 while the rest of the members have a cross-sectional area of 1.25 in^2. Support D settles 0.5 in. In addition, member BC is 0.4 in too long. Compute the member forces and the support reactions identifying the contributions from each of the three effects—external load, support settlement, and fabrication error.

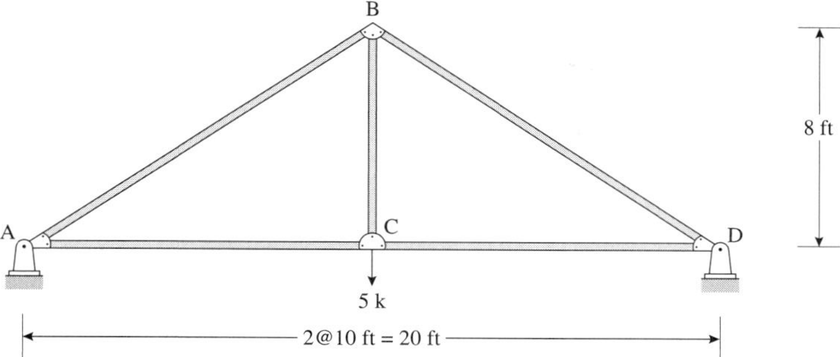

Fig. P6.3

6.4. Redo Problem 6.3 but assume that support at D is a roller support that allows horizontal displacement. Compare the results from the two problems.

6.5. The space truss in Fig. P6.5 is made of aluminum. The members are hollow square tubes with outside dimension of 2 in and wall thickness of 0.2 in. Compute the member forces and the support reactions.

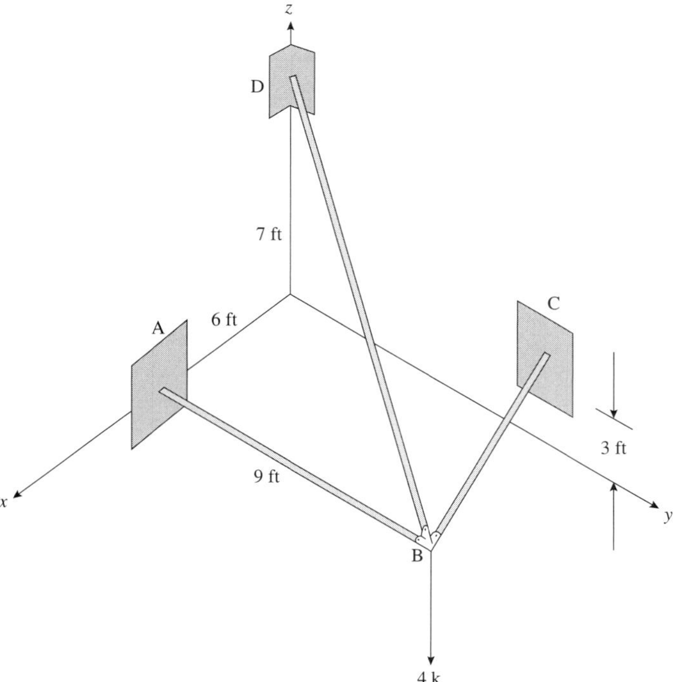

Fig. P6.5

6.6. Redo Problem 6.5 assuming that the structure is a space frame. The connection at B is a rigid connection and all the supports are fixed supports.

Structural Concepts

6.7. The step that takes up most of the computational time in a finite element computer program, especially for larger problems, is the generation and solution of the structural equilibrium equations **KD** = **F**. There are at least two issues in alleviating this problem. First is to minimize the amount of storage space required to store the information in **K**. The structural stiffness matrix is usually sparse. Typically the nonzero entries make up a few percent (1–10%) of the entire matrix. Researchers have devised several storage schemes to minimize the amount of storage space needed by recognizing that **K** is sparse and that the locations of the nonzero entries are known once the structural model is defined. The second issue is to devise a solution algorithm to solve the equilibrium equations with minimal access and mathematical operations. In this problem we address the first issue.

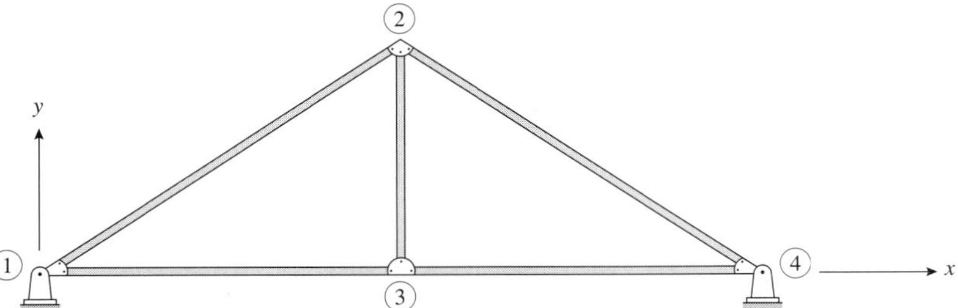

Fig. P6.7

The **K** matrix is symmetric and banded. A banded matrix can be though of as a special case of a sparse matrix. Consider the planar truss in Fig. P6.7. There are a total of eight degrees of freedom. If we go through the process of constructing the element equations and forming the system equations (symbolically, not numerically), we can generate the following map for the **K** matrix where an × marks the locations of all the nonzero entries:

1	2	3	4	5	6	7	8	
×	×	×	×	×	×			1
×	×	×	×	×	×			2
×	×	×	×	×	×	×	×	3
×	×	×	×	×	×	×	×	4
×	×	×	×	×	×	×	×	5
×	×	×	×	×	×	×	×	6
		×	×	×	×	×	×	7
		×	×	×	×	×	×	8

We can now compute an important property of the matrix known as the half-bandwidth (HBW). If we scan every row in the upper triangular portion of the matrix, we can compute the *distance* between the diagonal entry and the last nonzero entry in that row. For example, in the first row, the last nonzero entry is in column 6. Hence the distance is (6 – 1 + 1) = 6, or (last column – column corresponding to the diagonal entry + 1). This value for every row is displayed beside the matrix. The largest of these values is the half-bandwidth of the matrix. Hence the HBW of the given matrix is 6. A solid line now can be constructed that passes through the diagonal entries of the matrix and encompasses all the nonzero entries in the upper triangular matrix. Storing only these numbers is sufficient to extract all the information we need about **K**:

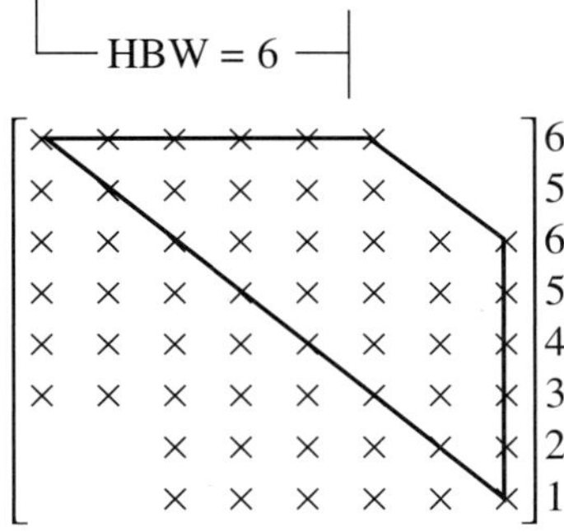

Construct the symbolic map of **K** by interchanging the locations of nodes 3 and 4, and compute the HBW of the matrix. Try this exercise with other problems in this section.

Chapter 7

Computer-Based Structural Analysis

Complex systems such as a water treatment plant are modeled on a computer. The database contains a variety of information including details of the structural model. The advantages of such a computerized model are numerous.

"On two occasions I have been asked [by members of Parliament], 'Pray, Mr. Babbage, if you put into the machine wrong figures, will the right answers come out?' I am not able rightly to apprehend the kind of confusion of ideas that could provoke such a question." Charles Babbage

"There is no reason for any individual to have a computer in their house." Kenneth Olsen, founder of Digital Equipment Corp.

"640K ought to be enough for anyone." Bill Gates

There is a philosophical reason why we are studying the use of computers and computer programs to analyze structural systems (possibly this early on in the course). Computers are wonderful tools for solving problems. They are a tool much like the log tables, slide rules, and basic calculators that engineers have used in the past. But they are much more versatile and powerful. It is imperative that we use this powerful tool in an intelligent fashion. We must recognize its strengths and weaknesses (yes, weaknesses!). When used with caution, computer programs can provide an engineer with powerful tools to carry out mundane calculations very quickly and accurately, to understand structural behavior through visual examinations, to investigate several modeling alternatives, to improve and design better structural components and systems, etc. No rational engineer has claimed that computer programs should supplant the systematic study of structural analysis or that they provide totally automated solutions. What no one can deny is the fact that today computer usage is very common and has led to tremendous gains in productivity. Consider the following quote: "The introduction of powerful digital computers and powerful computer programs has revolutionized the design of lateral load resisting systems. Now the widespread availability of powerful microcomputers and frame analysis programs provides a means for the small consulting firms to use advanced design concepts in moderate size projects" [Allison, 1991].

One obvious question is whether the computer program should be introduced before or after studying its theoretical basis. We feel that computer usage and literacy has progressed to the point that a student can use computer-based tools with minimal (but necessary) supervision and guidance. Whenever possible, we will use classical techniques that you have learned in prior courses, or in the previous chapters and will be introduced to in later chapters, to verify computer-generated results.

In this chapter, the GS-USA© (Graphics-based System for Understanding Structural Analysis) computer program is introduced.[1] This program is one of the several programs on the CD-ROM included with the text. GS-USA is based on the same theory as hundreds of commercial programs. Your interaction with very powerful commercial programs (costing thousands of dollars) will be eased tremendously once you learn how to use the GS-US program effectively.[2]

We will use GS-USA to analyze a mathematical model. What is a mathematical model? The usual starting point for most structural analysis is a physical description of the structure—the material(s) used, the geometry and layout of the members, their dimensions, and the manner in which the structure is expected to be supported and loaded. In the previous chapter, we constructed the mathematical model consistent with the assumptions made. The material behavior, the member, joint and support characteristics, the relationship among displacements, and the application of static equilibrium equations are based on certain assumptions. Being cognizant of the fact that an exact analysis is virtually impossible, how closely are these assumptions satisfied by the physical structure? This closeness determines the extent and magnitude of the errors. The idealized model that is used during structural analysis is the mathematical model.

---

[1] The GS-USA©, SlideTray©, UNDO©, and UCSD© programs are designed to run on Windows 95©/ Windows 98©/Windows NT© or later systems.

[2] I have mentioned the following in the Preface and will do so again: This book is *not* built around the capabilities of the aforementioned software. If you prefer another program with similar or even better capabilities, feel free to use them. The theories and problems discussed in this text can be solved using any competent program.

OBJECTIVES

- To understand the terminology used in computer-based structural analysis.
- To recognize the different components of a mathematical model.
- To build and analyze simple beams, trusses, and frames, and compare the program results with the results from hand calculations made using the theory from the previous chapter.
- To understand the limitations of computer-based structural analysis and learn how to effectively use computer programs as an analysis tool.

ASSUMPTIONS (AND TIPS)

- A computer program for structural analysis solves a mathematical model that the program supports and a user of the program creates a model for. Errors in modeling do occur and errors in computer programs (bugs) do exist. Every computer output must be examined for correctness.
- Numerical solutions to engineering problems suffer from common numerical errors—truncation and round-off. While computer systems can be made to compute in precision that is unavailable with other tools, one must be careful to interpret the numbers.
- **Learning a new tool can be an expensive and often intimidating process. But it should not be. Do not hesitate to use all available resources—online help, program documentation, sample problems, teachers, peers, practicing engineers, e-mail, and Web site—to have your questions answered and to improve your understanding of computer-based structural analysis.**

7.1 OVERVIEW

The basic idea behind computer-based structural analysis is to

a. take the problem statement (typically the physical description of the structural model) and construct a mathematical model of the structural system
b. communicate the mathematical model to the computer program in a language that the computer program understands
c. create and edit the mathematical model, analyze and examine the results

The results are the same response measures that we have seen in the text—member internal forces, deflections, stresses, and support reactions.

We will use the GS-USA program to achieve these objectives. Here is a partial list of its capabilities.

- It can create, edit, analyze, display, print, and plot a planar truss or frame in a windowed environment.
- The structure can be made up of one or several different materials. Similarly, there can be one or several different member cross-sections comprising the structure.
- The loads can be concentrated forces and moments, linearly distributed forces, self-weight, or thermal in nature. There can be more than one load case, if necessary.

- The connections can be rigid or an internal hinge. The supports can be a roller, pin, or fixed support. Effects due to support settlements can be studied. Inclined supports (or skew supports) can be handled.
- The computed responses include displacements, member internal forces, and support reactions. Deformed shape, shear force, and bending moment diagrams can be displayed.

How are these capabilities different from those of a commercial program? The differences are similar to the differences between a simple and a more sophisticated word processor. If you are writing an English report, an appropriate expertise in the English language is assumed. The manner in which you create and edit the document may be different with different word processors, but the language is the same. The more sophisticated program will have more bells and whistles, e.g., multiple-column formatting, creating a table within the document, ability to embed a picture, etc. In a similar fashion, the manner in which the mathematical model is created and edited is going to be different but the components needed to define the model and the final results are going to be essentially the same. While the commercial program may have more capabilities, its mission is to make the program as user-friendly as possible. The mission of the GS-USA program is to provide an environment conducive to learning and understanding structural analysis and design systematically, efficiently, and proficiently.

In the next few sections, we will see how to use the computer program to analyze a structural system. First (since the theory on which GS-USA is based is discussed in Chapter 6), we deal with the mathematical modeling language in the next section. After that, we look at the steps necessary to create the mathematical model. Finally, we look at modeling and solving several problems. To gain confidence in the computer modeling techniques, we first solve some of the same problems we solved in Chapter 2. More advanced examples are solved later in the chapter.

7.2 TERMINOLOGY

We begin with the definitions of some of the most commonly used terms.

Problem Units. Since the problem units are not hard-coded in the GS-USA program, the user must ensure that all problem data are expressed in consistent units. Four quantities are used: mass, force, length, and temperature. Examples of consistent units include (i) kg, N, m and °C, (ii) lbm, lb, in, °F, etc. The unit for time is taken as seconds (s).

Global Coordinate System. In order to carry out the analysis, the program must know the locations of the joints, members, etc., in a structure. The user expresses these locations in terms of (X, Y) coordinates. These coordinates are known as the global coordinates. The user must first define and locate the global coordinate system. While the frame lies in the global X–Y plane, the (global) Z axis is normal to this plane.

Node. A joint is that part of the structure where two or more members meet, or where the structure is supported. Every joint is a node. Sometimes additional locations on members need to be specified. These locations are also called nodes. Every node is identified by an unique node number. Node numbers start at 1 and are numbered consecutively.

Nodal Fixity Conditions. The nodal fixity conditions describe how the nodes move (or deform) when the structure is subjected to external loads.

Element. A member spans two joints. All members are elements. An element spans two nodes. As we see later, there are situations where all elements are not members, e.g., they are part of a member. Every element is identified by an unique element number. Element numbers start at 1 and are numbered consecutively. Note that every node must be associated with at least one element.

Local Coordinate System. The internal forces in frame members need to be defined in a local coordinate system so as to identify the axial force, the shear force, and the bending moment acting at the ends of the member. Usually (but not always) the local coordinate system of each member in a frame is different from the global coordinate system.

Load Case. A load case denotes a set of loads that act on the structure simultaneously. A structure may be subjected to several different sets of loads.

Nodal Loads. The external loads that act exactly at a node are called nodal loads. Note that on a frame these external loads can be concentrated loads acting along the global X–Y directions or bending moments acting along the global Z direction.

Element Loads. The external loads (concentrated forces or linearly distributed forces) that act on an element are known as element loads.

Thermal Loads. The external loads on structures can be due to temperature changes in some or all elements. These are the thermal loads.

Specified Displacements. The internal forces on structures can be due to effects such as support settlements. These are called specified displacements since the displacements (at these nodes) are specified or known.

Multi-Point Constraints. There are situations in which the structure deforms in an unusual but known manner. Multi-point constraints (MPCs) can be used to model inclined roller supports, rigid links, etc.

7.3 BUILDING A MATHEMATICAL MODEL

In this preliminary treatment of this topic, we assume that the starting point is a typical problem statement as seen in Chapter 2.

Figure 7.3.1(a) shows a beam (a propped cantilever) subjected to a tip load. The material and cross-section are constant over the length of the beam. The problem is to construct a model so that the support reactions can be computed. To create the model, two preliminary steps are necessary. First, a set of consistent units must be selected. We express lengths in m and the forces in kN. Secondly, the geometry and topology of the structure needs to be defined in a global coordinate system. In this example, we select A as the origin of the coordinate system with the global x aligned with the beam and y pointing upwards. The topology contains the definition of the nodes (or joints/supports) and elements (or members). Generally speaking, elements define the structure and nodes are used to define the elements.[3] The beam could physically be made of a single piece that is 15 m long and is supported at A and B. However, in the mathematical model, we define two elements—one from A to B, and the other from B to C. This is necessary because the beam is supported at B and that information must be captured in the model. The basic model is presented in Fig. 7.3.1(b).

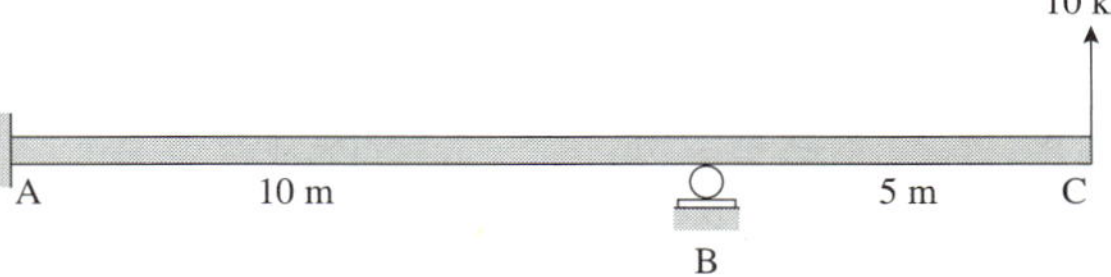

Fig. 7.3.1(a) Propped cantilever.

[3] The exceptions should not concern us in the context of the problems addressed in this text.

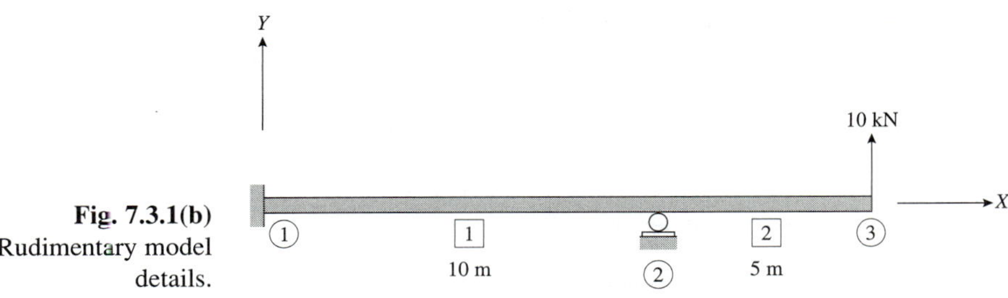

Fig. 7.3.1(b) Rudimentary model details.

Hence, the structure (or beam) is described as being made up of three nodes and two elements. Nodes are identified through *node numbers*. In other words, every node is arbitrarily assigned a number tag starting at 1. Pictorially, a circle enclosing a number is used to identify and tag a node. Similarly, every element is identified through *element numbers*, which are also arbitrarily assigned starting at 1. Pictorially, a square enclosing a number is used to identify and tag an element.

We select the origin of the global coordinate system (x–y) to be at node 1. This selection is made arbitrarily: we could have selected any other location as the origin. Now, node 1 is at (0,0), node 2 at (10,0), and node 3 at (15,0). Node 1 is a fixed support, node 2 is a roller support such that the node (or, joint) cannot move in the y direction, and node 3 is a free end. Element 1 is defined as spanning node 1 to node 2, whereas element 2 spans nodes 2 and 3. The nodal load acts in the y-direction at node 3.

Figure 7.3.2 shows a larger structure—a planar frame. The structure is made up of seven nodes and seven elements. Element 1 is subjected to a linearly distributed load. Element 5 is subjected to two element loads—two concentrated forces. Nodes 2, 3, 4, and 6 are *rigid* connections. They are *free* to move in the X and Y directions and *free* to rotate about the Z axis. Nodes 1 and 5 are fixed supports. Node 7 is a pin support.

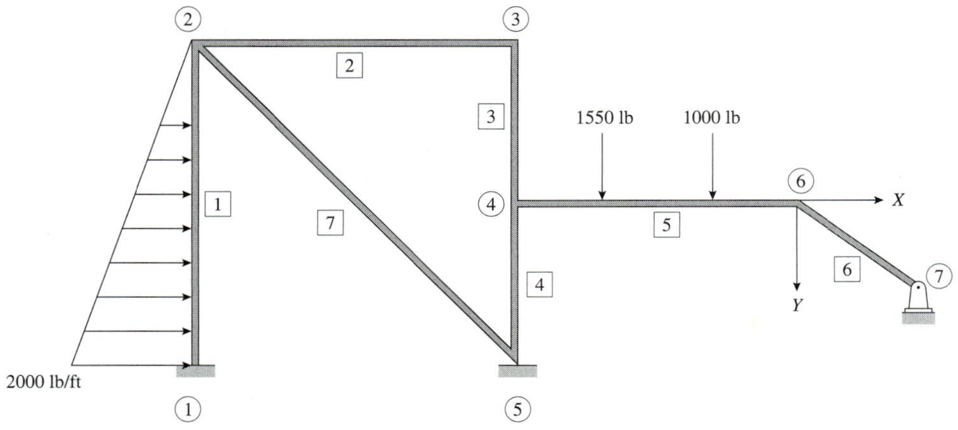

Fig. 7.3.2 Model of a planar frame.

To summarize, every mathematical model (that is used by a computer program such as GS-USA) of a structural system has at least the following components:

1. *Nodal data with their fixity conditions known.* The locations of nodes are denoted using their (X, Y) coordinates. In addition, fixity codes indicate to the analysis programs how the structure is constructed and supported. However, unlike the thought process in statics where supports and connections are associated with support reactions and joint forces, the computer-based structural analysis programs (usually called finite element analysis programs) require support and joint specifications in

terms of *joint displacements*, or whether a joint can or cannot move or rotate. If the joint or support can displace along a direction (x, y, or z) or rotate about an axis then it is said to be *free*. It is *fixed* otherwise.

Let us assume that the structure (truss or frame) is in the X–Y plane with the supports aligned along the X–Y directions for all cases except skew roller support. Connection or support points in a truss are classified as a regular joint (two or more members connected by a pin), a roller support, a pin support, and a skew support. Note that a roller support can be free to move either in the X or the Y direction. Connections or supports in a frame are classified as a rigid joint (the angle between the members that meet at a joint before and after the application of the loads remains the same), a roller support, a pin support, a fixed support, an *internal hinge* (where the moment is released or is zero), and skew support. Do not confuse a rigid joint with a fixed support. Table 7.3.1 shows the appropriate fixity codes that you must specify for different types of joints and supports. To keep the discussions clear, it will be necessary to distinguish joints from supports. The table shows all possible nodal fixity conditions or codes.

We briefly mentioned in Chapter 6 that one could conceivably treat a truss as a special case of frame. Each joint then would be an internal hinge. The supports will be pin or roller supports. There would be no element loads—only concentrated forces applied at the joints. This is the strategy used in the GS-USA program. The default structure is a frame. The conversion to a truss takes place as stated above.

Table 7.3.1 Fixity Codes for Planar Trusses and Frames

Frame (the default structure)			
Support or joint	x displacement	y displacement	z rotation
rigid joint	Free	Free	Free
internal hinge	Free	Free	Hinge
roller support	Free	Fixed	Free
roller support	Fixed	Free	Free
pin support	Fixed	Fixed	Free
fixed support	Fixed	Fixed	Fixed
skew support	Free	Free	Free

Truss			
Support or joint	x displacement	y displacement	z rotation
joint	Free	Free	Hinge
roller support	Free	Fixed	Hinge
roller support	Fixed	Free	Hinge
pin support	Fixed	Fixed	Hinge
skew support	Free	Free	Hinge

2. *Element data including material and cross-sectional properties.* An element is described in terms of the nodes that physically locate the element, the material properties of the material making up the element, and the cross-sectional properties of the element. Since an element spans two nodes, the element definition is in terms of a *start node* and an *end node*. These nodes are arbitrarily designated *start* and *end*. For the beam in Fig. 7.3.1(b), element 1 can either be defined with node 1 as start node and node 2 as end node or vice versa. Similarly, element 7 in Fig. 7.3.2 can either be defined with node 2 as start node and node 5 as end node or vice versa.

One of the more important properties of an element definition is its local coordinate system. Local coordinates are necessary not only for defining the loads that act on a beam/frame element but also for interpreting the forces that act at the ends of those members (see Fig. 7.3.3). The local coordinate system is described as follows. The local x axis is from the start node to the end node (hence the local axis can change if the start and end nodes are interchanged) of the member. *The local z axis points towards you,* assuming that you are looking at a piece of paper that contains the drawing of the structure. Using the right-hand rule, the cross product of the (local) z axis and the (local) x axis gives the orientation of the (local) y axis. The concentrated load or distributed load that acts on a member *may* act along the local y axis. Hence the load value is positive if it acts in the positive y axis.

Once the analysis is complete, GS-USA prints the internal forces and moment acting on the two ends of the member. These internal forces are the axial force (acting along the local x), the shear force (acting along the local y), and the bending moment (acting along the local z axis). A positive value indicates that the force or moment acts in the positive local axis direction.

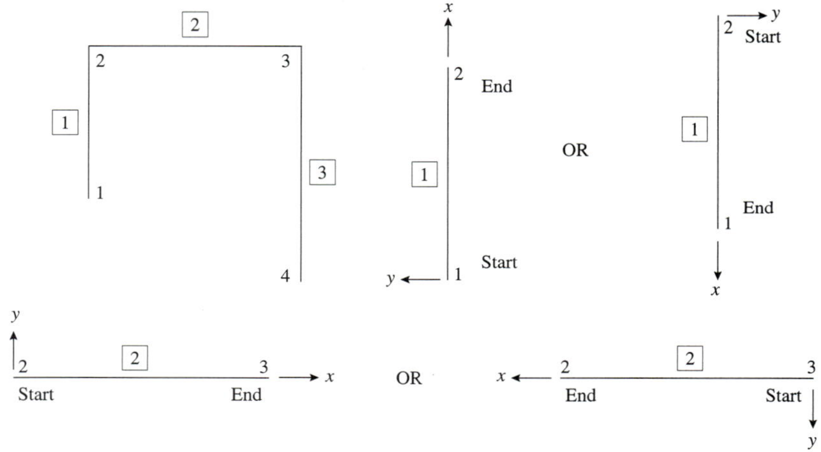

Fig. 7.3.3 Examples of local coordinate system for different elements of a frame.

3. *Loads acting on the structure.* Loads acting on a structure can be classified in two ways. The first classification is *nodal load*, which is a concentrated load that acts at a joint. Both trusses and frames can be subjected to concentrated nodal forces but only frames can also have concentrated nodal moment. The second is *element load*, which is a concentrated load and/or a distributed load that act(s) on the member. Only frames are subjected to element loads. In the program, distributed element loads are assumed to act on the entire element, i.e., from the start node to the end node.

7.4 STEPS IN MODELING THE STRUCTURE

Step 1: Choose the units for the problem. These units must be consistent.

Step 2: Setup the coordinate system. Tip: Select one of the nodes as the origin. If self-weight is to be included, then the Y direction must be the gravitational direction.

Step 3: Number the nodes (or joints) and elements (or members). Both these numbers start at 1.

Step 4: Make a table showing all the different materials used in the structure. The material group number is the identification tag associated with the set of material properties.

Group	Modulus of elasticity	Coeff. of thermal expansion
1	...	
2	...	

Step 5: Make a table showing all the different cross-sections used in the structure. The property group number is the identification tag associated with the set of cross-sectional properties.

Group	Area	Moment of inertia
1	...	
2	...	

Step 6: Make a table showing the nodal properties.

Node	X coordinate	Y coordinate	X fixity	Y fixity	Z fixity
1	...				
2	...				

Step 7: Make a table showing the element properties.

Element	Start node	End node	Material group	Property group
1	...			
2	...			

Step 8: Show the nodal loads, if any, in a tabular form.

Node	X force	Y force	Z moment
1	...		
2	...		

Step 9: Show the concentrated force element loads, if any, in a tabular form. The element load ID is the identification tag associated with the element loads. The load type is local y, or global X, or global Y.

Element load ID	Element	Load type	Distance from start node	Load value
1	...			
2	...			

Step 10: Show the linearly distributed element loads, if any, in a tabular form. The element load ID is the identification tag associated with the element loads. The load type is local y, or global x, or global y, or projected global x, or projected global y.

Element load ID	Element	Load type	Intensity at start node	Intensity at end node
...	...			
...	...			

Once you have gone through these ten steps, you are ready to use the computer program. With practice, you will be able to bypass these steps and create and edit the problem data interactively.

EXAMPLE 7.4.1 ***Simply Supported Beam (Example 2.5.1)***

Figure E7.4.1(a) shows a simply supported beam. Determine the support reactions.

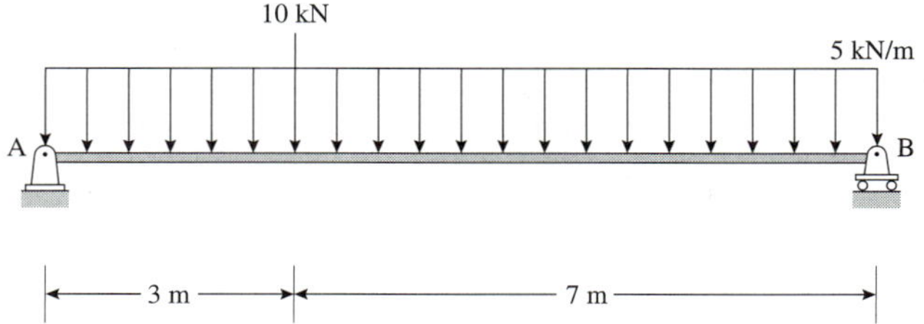

Fig. E7.4.1(a)
Simply supported beam.

SOLUTION

Step 1: Construct the model. The mathematical model showing the nodes and elements is given in Fig. E7.4.1(b).

As units, select m for length and kN for force. The tables from the previous section will be filled using the problem data and our model definition.

Fig. E7.4.1(b)
Mathematical model.

Material data: The support reactions for this problem are not affected by the choice of material properties. (Is this true of all structural systems?) Hence they are not a part of the problem statement. However, we assume the material for two reasons. First, the program will not execute without the material data. Second, it is in our best interest to start a process of associating the structural systems with commonly used structural materials. Let us assume that the material is 0.2% carbon hot-rolled steel.

Group	Modulus of elasticity	Coeff. of thermal expansion
1	$200(10^6)$ kN/m^2	$12(10^{-6})$m/m – °C

Member cross-sectional properties: Once again, the support reactions for this problem are not affected by the choice of member cross-section. We assume an I-section with the following properties (the rough dimensions are: depth of section 75 mm, flange width 50 mm, web thickness 4 mm, flange thickness 6 mm):

Group	Area	Moment of inertia
1	$8(10^{-4})$ m^2	$70(10^{-8})$ m^4

Nodal data: Note that node 1 is a pin support and node 2 is a (x) roller support.

Node	*X* coordinate	*Y* coordinate	*X* fixity	*Y* fixity	*Z* fixity
1	0	0	Fixed	Fixed	Free
2	10	0	Free	Fixed	Free

Element data: There is only one material and one element property group.

Element	Start node	End node	Material group	Property group
1	1	2	1	1
2	2	3	1	1

Nodal loads: There are no nodal loads.
Element concentrated loads: There is one concentrated load acting on an element.

Element load ID	Element	Load type	Distance from start node	Load value
1	1	Local *y*	3 m	–10 kN

Element distributed loads: There is one distributed load acting on an element.

Element load ID	Element	Load type	Intensity at start node	Intensity at end node
2	1	Local *y* distributed	–5 kN/m	–5 kN/m

These steps complete the tabular description of the problem.

Step 2: Execute the GS-USA program. Armed with the GS-USA tutorial and the tables from step 1, execute the program.

Step 3: Examine the results file.

Using annotations, we explain the different sections of the results file. The first section contains general information about the model.

```
        ----------------------------------------
              Graphics-based System for
           Understanding Structural Analysis
         PLANE FRAME ANALYSIS & DESIGN PROGRAM
                COPYRIGHT, 1991-2000
                    S. D. RAJAN
           DEPARTMENT OF CIVIL ENGINEERING
              ARIZONA STATE UNIVERSITY
                  TEMPE, AZ 85287
              VERSION 8.10 : AUG 1999
        ----------------------------------------

 Date: 08/20/99    Time: 10:11 AM

 Project Name : p2-5-1
 Creation Date and Time : 08/20/1999 19:22

 PROBLEM SIZE
 ------------
                   Structure Type: FRAME
                  Number of Nodes:     2
               Number of Elements:     1
        Number of Material Groups:     1
Number of Element Property Groups:     1
                  Number of MPC's:     0
             Number of Load Cases:     1
           Total number of unknowns:   3
```

```
   Total mass of structure:          62.8 kg
  Total weight of structure:            0 kN
  Total volume of structure:        0.008 m ^3
     Length of all members:            10 m
   Total cost of structure: $ 0
```

```
     PROBLEM Notes
     -------------
Project Title:
     Analyst:
  Start Date:
    End Date:

Project Information
-------------------
```

Summary of the Input Data

Section 2 contains the material and cross-sectional data.

MATERIAL PROPERTIES

Group	Young's Modulus (kN/ m ^2)	Mass Density (kg / m ^3)	CTE (m / m - C)	Cost ($ / m ^3)
1	2e+008	7850	1.2e-005	0

CROSS-SECTIONAL DIMENSIONS

Group	Type	Cost ($ / m)	Dimensions (m)
1	General	0	0.0008 7e-007 1 1

CROSS-SECTIONAL PROPERTIES

Group	Area (m ^2)	Moment of Inertia (m ^4)	Shear Factor (m ^2)	Section Modulus (m ^3)
1	~~0.0008~~ 0.8	~~7e-007~~ 0.7	1	1

Section 3 contains the nodal data.

NODAL INFORMATION

Node	Cost ($)	X Coor (m)	Y Coor (m)	X-Fixity	Y-Fixity	Z-Fixity
1	0	0	0	FIXED	FIXED	FREE
2	0	10	0	FREE	FIXED	FREE

Section 4 contains the element data.

```
----------------------------------------------------------
                   ELEMENT  INFORMATION
Elem   Matl Grp    Prop Grp    Start Node   End Node   Delta T
                                                        ( C )
----------------------------------------------------------
1      1           1           1            2           0
```

Section 5 contains the loading information. Note that both the element loads could have been defined as GLOBAL Y LOADs.

```
  -------------------------------------------------
              NODAL LOADS for LOAD CASE  1
  Node       X Force          Y Force          Z Moment
              (kN)             (kN)             (kN- m )
  -------------------------------------------------
                There are no nodal loads

--------------------------------------------------------
           ELEMENT CONCENTRATED LOADS for LOAD CASE  1
Elem  Load Type   Dist from Start Node    Load Intensity
                         ( m )                 (kN)
--------------------------------------------------------
1     LOCAL Y     3                        -10

----------------------------------------------------------
           ELEMENT DISTRIBUTED LOADS for LOAD CASE  1
Elem  Load Type     Int at Start Node     Int at End Node
                       (kN/ m )              (kN/ m )
----------------------------------------------------------
1     LOCAL Y       -5                    -5
```

Section 6 contains additional input data such as MPC data.

```
---------------------------------------------------------------------------
                       MULTI-POINT CONSTRAINTS
MPC Node 1  Local Dof  Coef        Node 2  Local Dof  Coef        Value
---------------------------------------------------------------------------
               There are no MPC's
```

Summary of the Structural Response

Section 7 contains the displacements at the nodes. There are no displacements at node1 (A) and node 2 (B). The rotation at node 1 is in the negative Z direction (clockwise) and at node 2 is positive (counterclockwise).

```
-------------------------------------------------------
          NODAL DISPLACEMENTS for LOAD CASE  1
Node      X Disp          Y Disp          Z Rotations
           ( m )           ( m )             (rad)
-------------------------------------------------------
1         0               0               -1.9131e-006
2         0               0               1.8131e-006

Min:      0               0               -1.9131e-006
at  :     1               1               1
Max:      0               0               1.8131e-006
at  :     1               1               2
```

Section 8 contains the element nodal forces, the min-max internal forces in every member, etc. The element nodal forces are in the local coordinate system and hence can be identified as axial, shear, and bending moment. The first line for each element shows the nodal forces at the start node of the element; the second line is for the end node of the element. A positive value indicates a force or moment along the positive coordinate direction. For this problem the element nodal shear forces are acting along the positive y direction, which is also the same as the global Y direction. While the axial forces are zero, the bending moments appear not to be. However, check their magnitudes!

```
--------------------------------------------------------------------
              ELEMENT NODAL FORCES for LOAD CASE  1
                local coordinate system
Elem    Axial           Shear           Moment          Rotation
        (kN)            (kN)            (kN- m )
--------------------------------------------------------------------
1       0               32              -2.92925e-006
        0               28              -2.02951e-006

  Min:  0               28              -2.92925e-006
  at :  1               1               1
  Max:  0               32              -2.02951e-006
  at :  1               1               1
```

To ensure that we understand these numbers, we draw the FBD of element 1 using the problem data and the results in Fig. E7.4.1(c). The coordinate system is local. You should check to see if the element is in equilibrium.

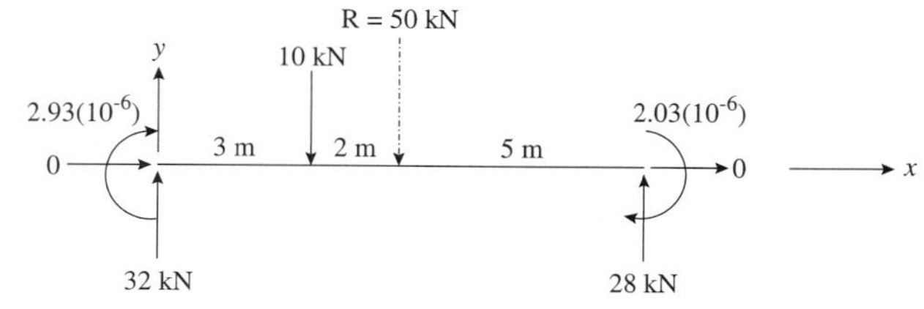

Fig. E7.4.1(c)

		MIN-MAX INTERNAL MEMBER FORCES for LOAD CASE 1		
	Element	Axial Force (kN)	Shear Force (kN)	Bending Moment (kN- m)
Min	1	0	-32	-2.02951e-006
At		10	0	10
Max		0	28	78.3798
At		0	10	4.4898

Section 9 contains the support reactions. The answers are the same as those obtained in Example 2.5.1.

	SUPPORT REACTIONS for LOAD CASE 1		
Node	X Force (kN)	Y Force (kN)	Z Moment (kN- m)
1	0	32	0
2	0	28	0
Min:	0	28	0
at :	1	2	1
Max:	0	32	0
at :	1	1	1
Sum:	0	60	0

Finally, we draw the structural FBD using the problem data and the computed support reactions in Fig. E7.4.1(d). The coordinate system is global. Check the equilibrium of the structure.

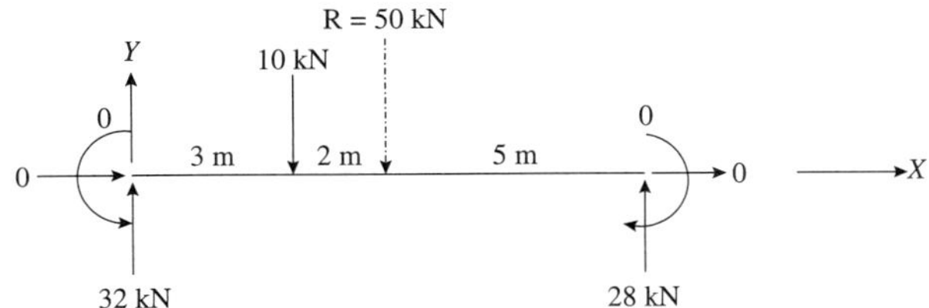

Fig. E7.4.1(d)

A logical question is whether the mathematical model is unique for this problem. The answer is a qualified yes. While it is possible to number the nodes and elements differently, the number of elements and nodes in the model is the *smallest* that we can define. In an alternate model, we can have three nodes (at A, at the 10 kN load, and at C) and hence two elements. The 10 kN force is a nodal load in this model. Try this model and you will discover that the answers are the same as before.

Tip: Having defined the characteristics of the model, in this text, we will always attempt to use the smallest number of nodes and elements necessary to solve the problem.

EXAMPLE 7.4.2 ***Frame with Moment-Release Hinge (Example 2.5.3)***

A planar frame is shown in Fig. E7.4.2(a) with a moment-release hinge at B. Compute all the support reactions and pin forces.

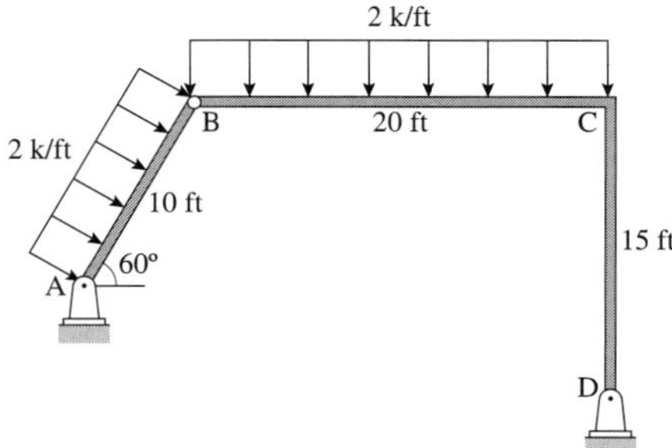

Fig. E7.4.2(a) Planar frame.

SOLUTION

Step 1: Construct the model (Fig. E7.4.2(b)), using arrows to show the direction from start node to end node of each element. As units, select ft for length and k for force. The tables from the previous section will be filled using the problem data and our model definition.

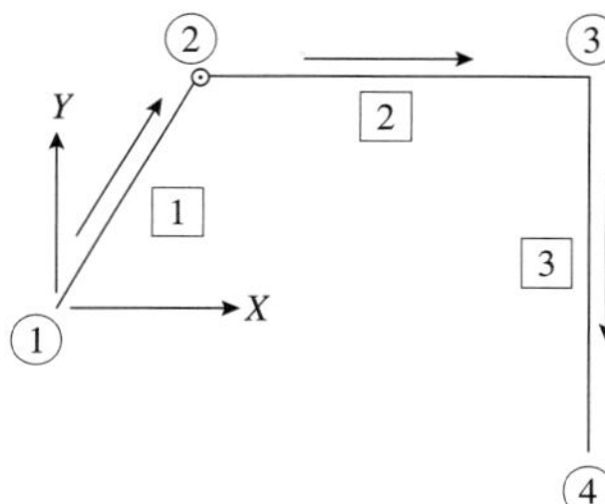

Fig. E7.4.2(b) Mathematical model.

Material data: As in the previous problem, the support reactions for this problem are not affected by the choice of material properties. Let us assume that the material is 0.2% carbon hot-rolled steel.

Group	Modulus of elasticity	Coeff. of thermal expansion
1	$4.176(10^6)$ k/ft^2	$6.67(10^{-6})$ ft/ft–°F

Member cross-sectional properties: Once again, the support reactions for this problem are not affected by the choice of member cross-section. We assume a symmetric I-section with a web height of 18 in, flange width of 11 in, web thickness of 0.4 in and flange thickness of 0.7 in.

Group	Type	Dimension 1	Dimension 2	Dimension 3	Dimension 4
1	Symmetric I-section	1.5 ft	0.033 ft	0.917 ft	0.0583 ft

Nodal data: The values of coordinates are entered as expressions just as one would when using a calculator. Most dialog boxes in the GS-USA program have input fields (which have by default a yellow background color) into which the user can type an expression.

Node	*X* coordinate	*Y* coordinate	*X* fixity	*Y* fixity	*Z* fixity
1	0	0	Fixed	Fixed	Free
2	10 cos(60)	10 sin(60)	Free	Free	Hinge
3	10 cos(60) + 20	10 sin(60)	Free	Free	Free
4	10 cos(60) + 20	10 sin(60) – 15	Fixed	Fixed	Free

Element data: There is only one material and one element property group.

Element	Start node	End node	Material group	Property group
1	1	2	1	1
2	2	3	1	1
3	3	4	1	1

Nodal loads: There are no nodal loads.
Element concentrated loads: There are no element concentrated loads.
Element distributed loads: There are two element distributed loads.

Element load ID	Element	Load type	Intensity at start node	Intensity at end node
1	1	Local *y* distributed	–2 k/ft	–2 k/ft
2	2	Global *y* distributed	–2 k/ft	–2 k/ft

These steps complete the tabular description of the problem.

Step 2: Execute the GS-USA program.

Step 3: Examine the results file.

Summary of the Input Data

```
----------------------------------------
       Graphics-based System for
   Understanding Structural Analysis
 PLANE FRAME ANALYSIS & DESIGN PROGRAM
         COPYRIGHT, 1991-2000
            S. D. RAJAN
    DEPARTMENT OF CIVIL ENGINEERING
       ARIZONA STATE UNIVERSITY
           TEMPE, AZ 85287
        VERSION 8.10 : AUG 1999
----------------------------------------

Date: 08/21/99    Time: 03:28 PM

Project Name : p2-5-3
Creation Date and Time : 08/21/1999 19:32

PROBLEM SIZE
------------
              Structure Type: FRAME
             Number of Nodes:     4
          Number of Elements:     3
   Number of Material Groups:     1
Number of Element Property Groups:     1
              Number of MPC's:     0
```

```
          Number of Load Cases:     1
      Total number of unknowns:     7
       Total mass of structure:          7.0625 km
     Total weight of structure:               0  k
     Total volume of structure:          7.0625 ft^3
         Length of all members:              45 ft
       Total cost of structure: $ 0

    PROBLEM Notes
    -------------
Project Title:
      Analyst:
   Start Date:
     End Date:

Project Information
-------------------
```

MATERIAL PROPERTIES

Group	Young's Modulus (k/ft^2)	Mass Density (km/ft^3)	CTE (ft/ft- F)	Cost ($ /ft^3)
1	4.176e+006	1	1	0

The dimensions of the cross-section are listed below.

CROSS-SECTIONAL DIMENSIONS

Group	Type	Cost ($ /ft)	Dimensions (ft)
1	Sym I	0	1.5 0.0333333 0.916667 0.0583333

The cross-sectional properties of the cross-section are listed below.

CROSS-SECTIONAL PROPERTIES

Group	Area (ft^2)	Moment of Inertia (ft^4)	Shear Factor (ft^2)	Section Modulus (ft^3)
1	0.156944	0.0743314	0.0315047	0.0919563

Note the HINGE Z-Fixity condition for node 2.

NODAL INFORMATION

Node	Cost ($)	X Coor (ft)	Y Coor (ft)	X-Fixity	Y-Fixity	Z-Fixity
1	0	0	0	FIXED	FIXED	FREE
2	0	5	8.66025	FREE	FREE	HINGE
3	0	25	8.66025	FREE	FREE	FREE
4	0	25	-6.33975	FIXED	FIXED	FREE

ELEMENT INFORMATION

Elem	Matl Grp	Prop Grp	Start Node	End Node	Delta T (F)
1	1	1	1	2	0
2	1	1	2	3	0
3	1	1	3	4	0

NODAL LOADS for LOAD CASE 1

Node	X Force (k)	Y Force (k)	Z Moment (k-ft)

There are no nodal loads

ELEMENT CONCENTRATED LOADS for LOAD CASE 1

Elem	Load Type	Dist from Start Node (ft)	Load Intensity (k)

There are no concentrated element loads

The element distributed load for element 1 must be in the local y direction. However, for element 2 it can be either local y or global Y.

ELEMENT DISTRIBUTED LOADS for LOAD CASE 1

Elem	Load Type	Int at Start Node (k/ft)	Int at End Node (k/ft)
1	LOCAL Y	-2	-2
2	GLOBAL Y	-2	-2

MULTI-POINT CONSTRAINTS

MPC Node 1	Local Dof	Coef	Node 2	Local Dof	Coef	Value

There are no MPC's

Summary of the Structural Response

The Z rotation at node 2 is not zero. Since each element that meet at node 2 will have its own rotation at that end, those values are shown in the element table.

NODAL DISPLACEMENTS for LOAD CASE 1

Node	X Disp (ft)	Y Disp (ft)	Z Rotations (rad)
1	0	0	-0.00873543
2	0.0732123	-0.0425319	0
3	0.0727205	-0.000734363	-0.000954197
4	0	0	-0.00679496

```
Min:         0            -0.0425319     -0.00873543
at  :        1             2              1
Max:         0.0732123     0              0
at  :        2             1              2
```

Note the element Z rotations for elements 1 and 2. Once again you can disregard the small bending moment values and effectively take them as zero.

```
-----------------------------------------------------------------
              ELEMENT NODAL FORCES for LOAD CASE  1
                local coordinate system
Elem   Axial          Shear          Moment          Rotation
       ( k)           ( k)           ( k-ft)
-----------------------------------------------------------------
1      14.9109        10             -3.31525e-005
       -14.9109       10             2.27374e-013    -0.0081958
2      16.1157        7.91322        -1.06581e-014   0.00258036
       -16.1157       32.0868        -241.736
3      32.0868        16.1157        241.736
       -32.0868       -16.1157       -9.77359e-006

  Min: -32.0868       -16.1157       -241.736
  at : 3              3              2
  Max: 32.0868        32.0868        241.736
  at : 3              2              3
```

The FBD of element 1 is shown in Fig. E7.4.2(c), and we should check whether the element is in equilibrium (as we should all the elements). The moments are not shown as they are zero at both ends of the element.

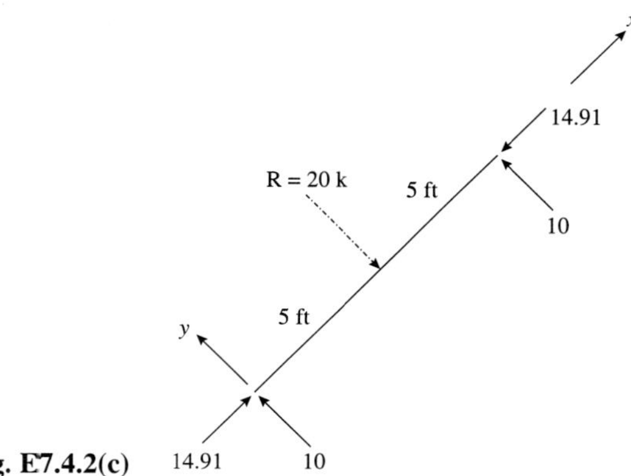

Fig. E7.4.2(c)

```
-------------------------------------------------------------
      MIN-MAX INTERNAL MEMBER FORCES for LOAD CASE  1
Element      Axial Force         Shear Force       Bending Moment
               ( k)                 ( k)              ( k-ft)
-------------------------------------------------------------
Min   1        -14.9109            -10               2.27374e-013
At             10                  0                 10
Max            -14.9109            10                24.9896
```

```
At            0             10              4.89796
Min    2      -16.1157      -7.91322        -241.736
At            20            0               20
Max           -16.1157      32.0868         15.6391
At            0             20              4.08163
Min    3      -32.0868      -16.1157        -241.736
At            15            15              0
Max           -32.0868      -16.1157        -9.77359e-006
At            0             0               15
```

```
        -------------------------------------------------
                SUPPORT REACTIONS for LOAD CASE  1
        Node     X Force         Y Force          Z Moment
                  ( k)            ( k)            ( k-ft)
        -------------------------------------------------
        1        -1.20478       17.9132          0
        4        -16.1157       32.0868          0

    Min:         -16.1157       17.9132          0
    at :         4              1                1
    Max:         -1.20478       32.0868          0
    at :         1              4                1
    Sum:         -17.3205       50               0
```

As a final check, the structural FBD is shown in Fig. E7.4.2(d), from which the equilibrium checks can be carried out.

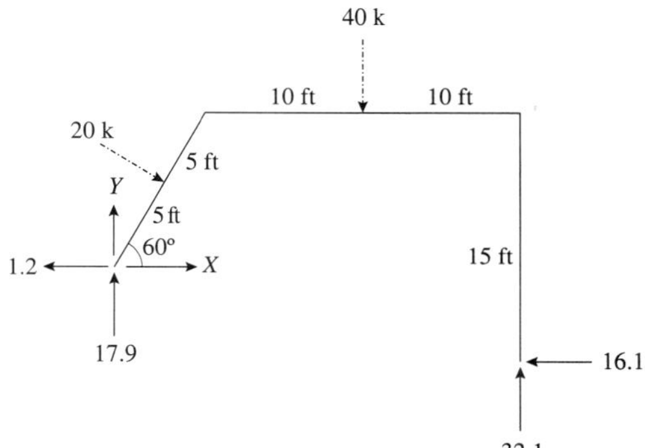

Fig. E7.4.2(d)

Tip: Make it a habit to check the results from any computer program using the concepts of equilibrium and compatibility.

EXAMPLE 7.4.3 ***Cantilevered Truss (Example 2.7.1)***

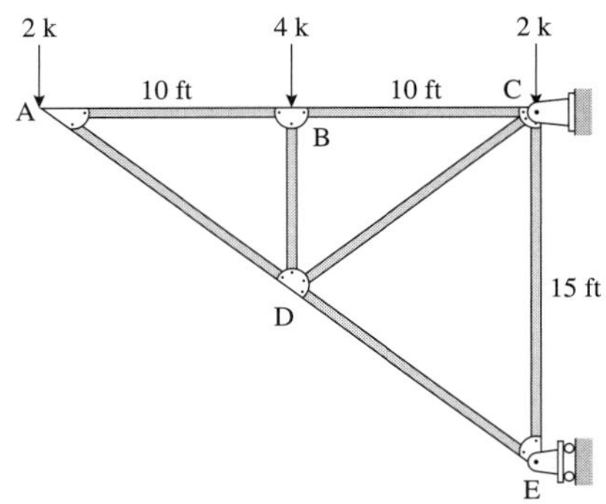

Fig. E7.4.3(a)
Cantilevered truss.

SOLUTION

Step 1: Construct the model for the truss in Fig E. 7.4.3(a), as in Fig. E7.4.3(b).

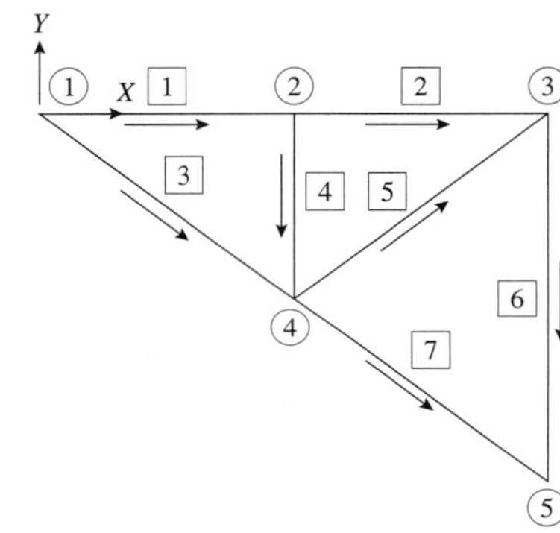

Fig. E7.4.3(b)
Mathematical model.

As units, select ft for length and k for force.

Material data: Let us assume that the material is 0.2% carbon hot-rolled steel.

Group	Modulus of elasticity	Coeff. of thermal expansion
1	$4.176(10^6)$ k/ft^2	$6.67(10^{-6})$ ft/ft–°F

Member cross-sectional properties: We assume a circular hollow cross-section with the properties inner radius 3 in, wall thickness 0.5 in.

Group	Type	Dimension 1	Dimension 2
1	Circular hollow	0.25 ft	0.0417 ft

Nodal data: Since this is a truss, we specify only X and the Y fixity conditions. Converting to truss in the program automatically converts all the joints to pin connections (internal hinge).

Node	X coordinate	Y coordinate	X fixity	Y fixity
1	0	0	Free	Free
2	10	0	Free	Free
3	20	0	Fixed	Fixed
4	10	–7.5	Free	Free
5	20	–15	Fixed	Free

Element data: There is only one material and one element property group.

Element	Start node	End node	Material group	Property group
1	1	2	1	1
2	2	3	1	1
3	1	4	1	1
4	2	4	1	1
5	4	3	1	1
6	3	5	1	1
7	4	5	1	1

Nodal loads: There are three nodal loads.

Node	*X* force	*Y* force
1	0	–2 k
2	0	–4 k
3	0	–2 k

Element concentrated loads: There are no element loads in a truss.
Element distributed loads: There are no element loads in a truss.
These steps complete the tabular description of the problem.

Step 2: Execute the GS-USA program. Remember to convert the structure to a truss.

Step 3: Examine the results file.

As a general note, all the Z components are not applicable for a planar truss. The nodal Z fixity conditions are all HINGE, moments in the Z direction are not allowed, and every node has X and Y displacements but no Z rotation. The member forces are axial (there is no shear force and bending moment), and the reactive forces are X and Y forces only.

```
----------------------------------------
       Graphics-based System for
   Understanding Structural Analysis
 PLANE FRAME ANALYSIS & DESIGN PROGRAM
          COPYRIGHT, 1991-2000
             S. D. RAJAN
     DEPARTMENT OF CIVIL ENGINEERING
        ARIZONA STATE UNIVERSITY
           TEMPE, AZ 85287
        VERSION 8.10 : AUG 1999
----------------------------------------

Date: 08/21/99    Time: 04:21 PM

Project Name : p2-7-1
Creation Date and Time : 08/21/1999 16:44

PROBLEM SIZE
------------
              Structure Type: TRUSS
             Number of Nodes:      5
          Number of Elements:      7
```

```
        Number of Material Groups:     1
Number of Element Property Groups:     1
                Number of MPC's:     0
             Number of Load Cases:     1
          Total number of unknowns:     7
           Total mass of structure:          5.67232 km
         Total weight of structure:                0  k
         Total volume of structure:          5.67232 ft^3
             Length of all members:               80 ft
           Total cost of structure: $ 0

     PROBLEM Notes
     -------------
Project Title:
      Analyst:
   Start Date:
     End Date:

Project Information
-------------------
```

MATERIAL PROPERTIES

Group	Young's Modulus (k/ft^2)	Mass Density (km/ft^3)	CTE (ft/ft- F)	Cost ($ /ft^3)
1	4.176e+006	1	1	0

CROSS-SECTIONAL DIMENSIONS

Group	Type	Cost ($ /ft)	Dimensions (ft)
1	Cir Hollow	0	0.25 0.0416667

CROSS-SECTIONAL PROPERTIES

Group	Area (ft^2)	Moment of Inertia (ft^4)	Shear Factor (ft^2)	Section Modulus (ft^3)
1	0.070904	0.00261582	0.0355916	0.00896851

NODAL INFORMATION

Node	Cost ($)	X Coor (ft)	Y Coor (ft)	X-Fixity	Y-Fixity	Z-Fixity
1	0	0	0	FREE	FREE	
2	0	10	0	FREE	FREE	
3	0	20	0	FIXED	FIXED	
4	0	10	-7.5	FREE	FREE	
5	0	20	-15	FIXED	FREE	

ELEMENT INFORMATION

Elem	Matl Grp	Prop Grp	Start Node	End Node	Delta T (F)
1	1	1	1	2	0
2	1	1	2	3	0
3	1	1	1	4	0
4	1	1	2	4	0
5	1	1	4	3	0
6	1	1	3	5	0
7	1	1	4	5	0

NODAL LOADS for LOAD CASE 1

Node	X Force (k)	Y Force (k)	Z Moment (k-ft)
1	0	-2	
2	0	-4	
3	0	-2	

ELEMENT CONCENTRATED LOADS for LOAD CASE 1

Elem	Load Type	Dist from Start Node (ft)	Load Intensity (k)

There are no concentrated element loads

ELEMENT DISTRIBUTED LOADS for LOAD CASE 1

Elem	Load Type	Int at Start Node (k/ft)	Int at End Node (k/ft)

There are no distributed element loads

MULTI-POINT CONSTRAINTS

MPC	Node 1	Local Dof	Coef	Node 2	Local Dof	Coef	Value

There are no MPC's

NODAL DISPLACEMENTS for LOAD CASE 1

Node	X Disp (ft)	Y Disp (ft)	Z Rotations (rad)
1	-0.000180122	-0.0011464	
2	-9.00612e-005	-0.000554439	
3	0	0	
4	0.000163939	-0.00045312	
5	0	-0.000202638	
Min:	-0.000180122	-0.0011464	
at :	1	1	
Max:	0.000163939	0	
at :	4	3	

To interpret the element nodal force table, draw the FBD of a typical element (see Fig. E7.4.3(c)). As a rule of thumb, if the axial force at the start node is POSITIVE then the truss member is in COMPRESSION.

```
-----------------------------------------------------------------
              ELEMENT NODAL FORCES for LOAD CASE  1
                local coordinate system
Elem   Axial           Shear           Moment          Rotation
       ( k)            ( k)            ( k-ft)
-----------------------------------------------------------------
1      -2.66667
       2.66667
2      -2.66667
       2.66667
3      3.33333
       -3.33333
4      4
       -4
5      -3.33333
       3.33333
6      -4
       4
7      6.66667
       -6.66667

  Min: -6.66667
  at : 7
  Max: 6.66667
  at : 7

      -------------------------------------------------
                SUPPORT REACTIONS for LOAD CASE  1
      Node      X Force          Y Force          Z Moment
                 ( k)             ( k)            ( k-ft)
      -------------------------------------------------
      3         5.33333          8
      5         -5.33333         0

  Min:          -5.33333         0
  at :          5                5
  Max:          5.33333          8
  at :          3                3
  Sum:          0                8
```

It is also essential that we check the equilibrium of the different joints as it is necessary to check the overall structural equilibrium. The structural equilibrium check is left as an exercise. The check for node 3 is:

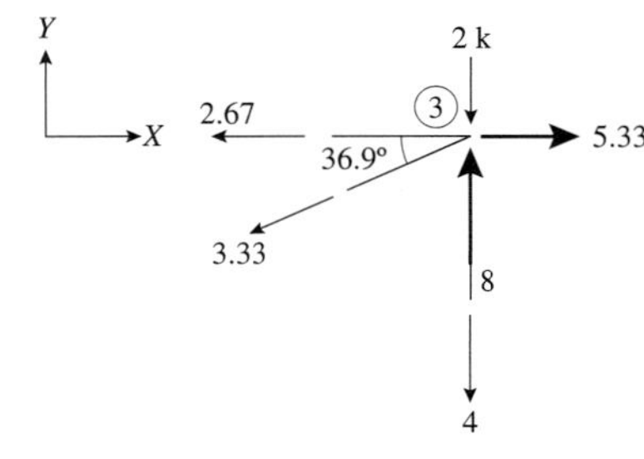

Fig. E7.4.3(c)

$$\overset{\rightarrow+}{\sum} F_x = 0 = -2.67 - 3.33\cos(36.9°) + 5.33 \approx 0 \qquad \text{OK.}$$

$$\overset{\uparrow+}{\sum} F_y = 0 = -3.33\sin(36.9°) - 2 + 8 - 4 \approx 0 \qquad \text{OK.}$$

EXAMPLE 7.4.4 ***Portal Frame with Triangular Loading (Example 2.8.9)***

For the frame in Fig. E7.4.4(a), draw the shear force and bending moment diagrams.

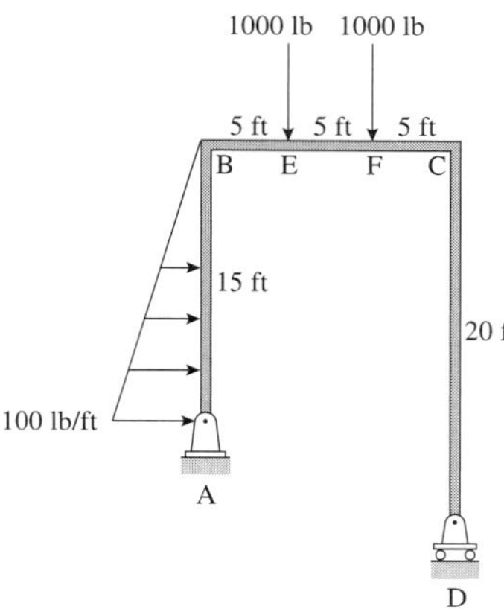

Fig. E7.4.4(a)
Planar frame.

SOLUTION

Step 1: Construct the model (see Fig. E7.4.4(b)), using arrows to show the direction from start node to end node of each element.

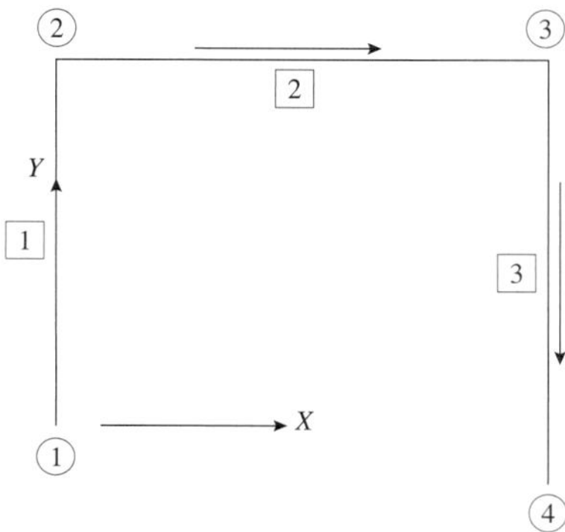

Fig. E7.4.4(b)
Mathematical model.

As units, select ft for length and lb for force. The tables from the previous section will be filled using the problem data and our model definition.

Material data: Let us assume that the material is medium-strength concrete.

Group	Modulus of elasticity	Coeff. of thermal expansion
1	$6.7(10^6)$ lb/ft^2	$6.7(10^{-6})$ ft/ft–°F

Member cross-sectional properties: We assume a rectangular section with an effective height of 8 in and width 4 in (steel reinforcements are ignored here).

Group	Type	Dimension 1	Dimension 2
1	Solid rectangular	0.67 ft	0.33 ft

Nodal data:

Node	*X* coordinate	*Y* coordinate	*X* fixity	*Y* fixity	*Z* fixity
1	0	0	Fixed	Fixed	Free
2	0	15	Free	Free	Free
3	15	15	Free	Free	Free
4	15	–5	Free	Fixed	Free

Element data: There is only one material and one element property group.

Element	Start node	End node	Material group	Property group
1	1	2	1	1
2	2	3	1	1
3	3	4	1	1

Nodal loads: There are no nodal loads.

Element concentrated loads: There are two element concentrated loads acting on element 2.

Element load ID	Element	Load type	Distance from start node	Load value
1	2	Local *y* concentrated	5 ft	–1000 lb
2	2	Local *y* concentrated	10 ft	–1000 lb

Element distributed loads: There is one element distributed load.

Element load ID	Element	Load type	Intensity at start node	Intensity at end node
3	1	Local *y* distributed	–100 lb/ft	0

These steps complete the tabular description of the problem.

Step 2: Execute the GS-USA program.

Step 3: Examine the results file.

```
----------------------------------------
        Graphics-based System for
    Understanding Structural Analysis
  PLANE FRAME ANALYSIS & DESIGN PROGRAM
          COPYRIGHT, 1991-2000
              S. D. RAJAN
      DEPARTMENT OF CIVIL ENGINEERING
         ARIZONA STATE UNIVERSITY
              TEMPE, AZ 85287
         VERSION 8.10 : AUG 1999
----------------------------------------

Date: 08/26/99    Time: 08:02 PM
Project Name : p2-8-9
Creation Date and Time : 08/26/1998 19:55
```

```
      PROBLEM SIZE
      ------------
                         Structure Type: FRAME
                        Number of Nodes:     4
                     Number of Elements:     3
              Number of Material Groups:     1
  Number of Element Property Groups:     1
                         Number of MPC's:     0
                   Number of Load Cases:     1
                Total number of unknowns:     9
                 Total mass of structure:          11.055 slg
               Total weight of structure:               0 lb
               Total volume of structure:          11.055 ft^3
                   Length of all members:              50 ft
                 Total cost of structure: $ 0

      PROBLEM Notes
      -------------
Project Title:
      Analyst:
   Start Date:
     End Date:

Project Information
-------------------
```

MATERIAL PROPERTIES

Group	Young's Modulus (lb/ft^2)	Mass Density (slg/ft^3)	CTE (ft/ft- F)	Cost ($ /ft^3)
1	6.7e+007	1	1	0

CROSS-SECTIONAL DIMENSIONS

Group	Type	Cost ($ /ft)	Dimensions (ft)
1	Rect Solid	0	0.67 0.33

CROSS-SECTIONAL PROPERTIES

Group	Area (ft^2)	Moment of Inertia (ft^4)	Shear Factor (ft^2)	Section Modulus (ft^3)
1	0.2211	0.00827098	0.1474	0.0246895

NODAL INFORMATION

Node	Cost ($)	X Coor (ft)	Y Coor (ft)	X-Fixity	Y-Fixity	Z-Fixity
1	0	0	0	FIXED	FIXED	FREE
2	0	0	15	FREE	FREE	FREE
3	0	15	15	FREE	FREE	FREE
4	0	15	-5	FREE	FIXED	FREE

ELEMENT INFORMATION

Elem	Matl Grp	Prop Grp	Start Node	End Node	Delta T (F)
1	1	1	1	2	0
2	1	1	2	3	0
3	1	1	3	4	0

NODAL LOADS for LOAD CASE 1

Node	X Force (lb)	Y Force (lb)	Z Moment (lb-ft)

There are no nodal loads

ELEMENT CONCENTRATED LOADS for LOAD CASE 1

Elem	Load Type	Dist from Start Node (ft)	Load Intensity (lb)
2	LOCAL Y	5	-1000
2	LOCAL Y	10	-1000

ELEMENT DISTRIBUTED LOADS for LOAD CASE 1

Elem	Load Type	Int at Start Node (lb/ft)	Int at End Node (lb/ft)
1	LOCAL Y	-100	0

MULTI-POINT CONSTRAINTS

MPC	Node 1	Local Dof	Coef	Node 2	Local Dof	Coef	Value

There are no MPC's

NODAL DISPLACEMENTS for LOAD CASE 1

Node	X Disp (ft)	Y Disp (ft)	Z Rotations (rad)
1	0	0	-0.15514
2	1.87033	-0.000759432	-0.0790108
3	1.87033	-0.00168763	0.0619694
4	3.10971	0	0.0619694
Min:	0	-0.00168763	-0.15514
at :	1	3	1
Max:	3.10971	0	0.0619694
at :	4	1	3

```
--------------------------------------------------------------------
              ELEMENT NODAL FORCES for LOAD CASE  1
                local coordinate system
Elem   Axial          Shear          Moment          Rotation
       (lb)           (lb)          (lb-ft)
--------------------------------------------------------------------
1      750            750            -0.000716313
       -750           7.69393e-005   3750
2      0              750            -3750
       0              1250           -0.000385191
3      1250           9.90908e-005   0.000990908
       -1250          -9.90908e-005  0.000990908

  Min: -1250          -9.90908e-005  -3750
  at : 3              3              2
  Max: 1250           1250           3750
  at : 3              2              1
```

```
-----------------------------------------------------------------
     MIN-MAX INTERNAL MEMBER FORCES for LOAD CASE  1
Element       Axial Force         Shear Force       Bending Moment
                 (lb)                (lb)              (lb-ft)
-----------------------------------------------------------------
Min   1       -750                -750              0.000716313
At            15                  0                 0
Max           -750                7.69393e-005      3750
At            0                   15                15
Min   2       0                   -750              -0.000385191
At            15                  0                 15
Max           0                   1250              7500
At            0                   15                5
Min   3       -1250               -9.90908e-005     -0.000990908
At            20                  20                0
Max           -1250               -9.90908e-005     0.000990908
At            0                   0                 20
```

```
     ---------------------------------------------------
          SUPPORT REACTIONS for LOAD CASE  1
     Node      X Force        Y Force        Z Moment
                (lb)           (lb)          (lb-ft)
     ---------------------------------------------------
     1         -750           750            0
     4         0              1250           0

Min:           -750           750            0
at :           1              1              1
Max:           0              1250           0
at :           4              4              1
Sum:           -750           2000           0
```

Once again, we should not only check the equilibrium of every element and the structural equilibrium but also the equilibrium of every node. We will check the equilibrium of element 2 (Fig. E7.4.4(c)) and node 2 (Fig. E7.4.4(d)).

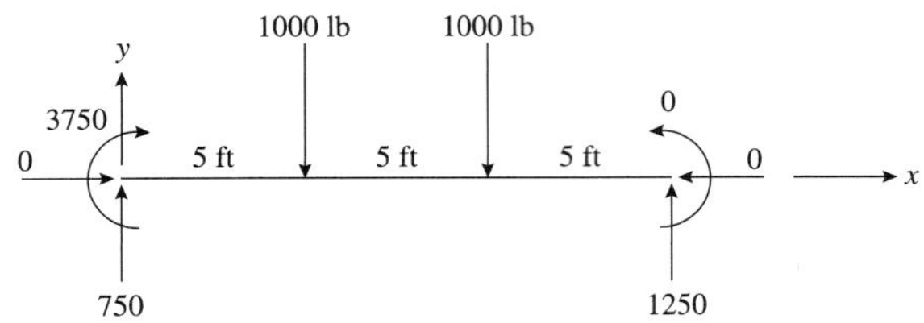

Fig. E7.4.4(c)
FBD of element 2.

$$\overset{\rightarrow+}{\sum} F_x = 0 - 0 = 0 \qquad \text{OK.}$$

$$\overset{\uparrow+}{\sum} F_y = 750 - 1000 - 1000 + 1250 = 0 \qquad \text{OK.}$$

$$\overset{\curvearrowleft+}{\sum} M_{\text{origin}} = -3750 - 1000(5) - 1000(10) + 1250(15) = 0 \qquad \text{OK.}$$

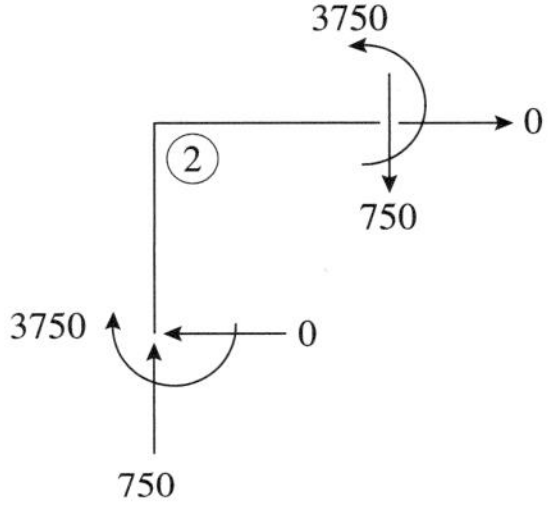

Fig. E7.4.4(d)
FBD of node 2.

$$\overset{\rightarrow+}{\sum} F_x = -0 + 0 = 0 \qquad \text{OK.}$$

$$\overset{\uparrow+}{\sum} F_y = 750 - 750 = 0 \qquad \text{OK.}$$

$$\overset{\curvearrowleft+}{\sum} M_2 = -3750 + 3750 = 0 \qquad \text{OK.}$$

Observation: The only difference between the diagrams in Fig. E7.4.4(e) and (f) and the results from the previous chapter is with respect to the shear force diagram. A RON segment is used in constructing the SF and BM diagrams. In other words, a cut is made so that the exposed face has an outward normal along the local x-axis (the left part of the member from the start node to the cut is used). A positive shear force indicates that the shear force at the cut acts along the positive local y direction. Similarly, a positive bending moment indicates that the bending moment is along the positive local z axis.

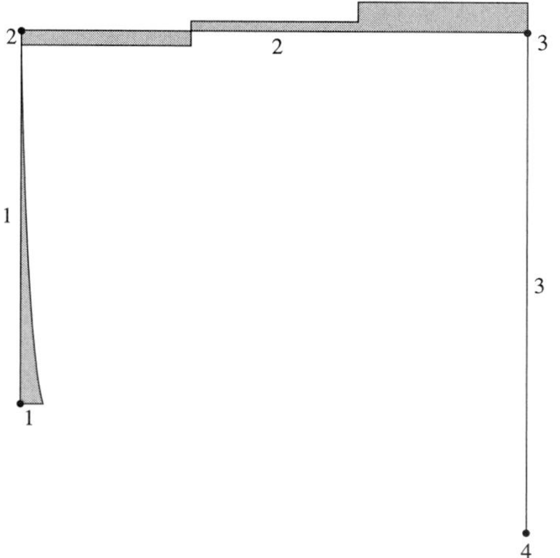

Fig. E7.4.4(e)
Shear force diagram.

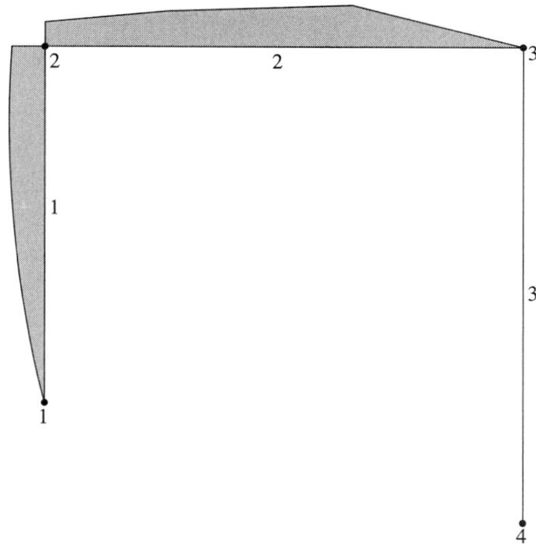

Fig. E7.4.4(f)
Bending moment diagram.

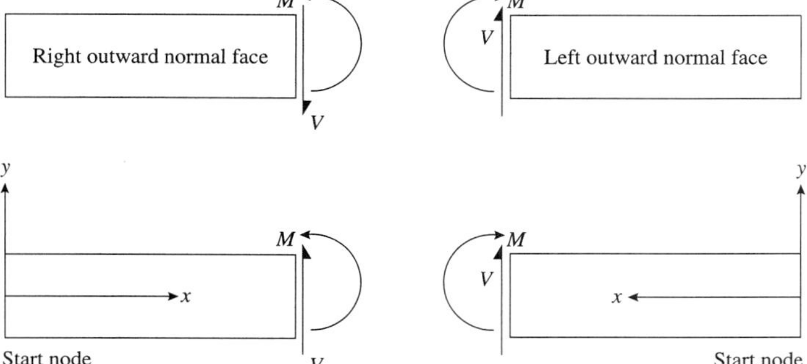

Fig. E7.4.4(g)
Sign convention.

Fig. E7.4.4(g) shows the sign convention (positive quantities are shown in the figure) used in the previous chapter (top) and the GS-USA program (bottom).

7.5 SOME BASIC CHECKS

Visually examine the structure before, during, and after entering the data for the following:

Consistent Units. Make sure that the units of the data are all consistent in system as well as magnitude. For example, if you choose to enter data in terms of inches, pounds, pound-mass, and Fahrenheit, you must still convert feet to inches and kips to pounds.

Stability. Before entering data into the program, check for stability of the structure. The structure must be properly supported and satisfy the equations of equilibrium. For trusses, check for internal as well as external stability.

Nodal Fixity Conditions. Check that the nodal fixity conditions correctly reflect the boundary conditions of the structure. Be sure to check for support conditions and internal hinges.

Materials and Properties. Identify members of the structure that have different materials and properties. Different materials and properties are assigned group numbers, so check that the members are assigned correct group numbers.

EXAMPLE 7.5.1 ***Unstable Truss***

Find the member forces in the truss in Fig. E7.5.1.

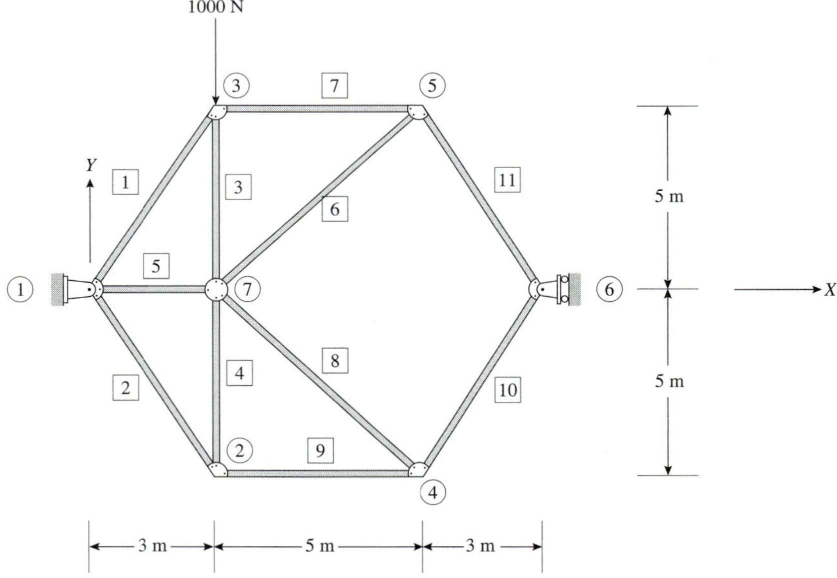

Fig. E7.5.1

SOLUTION

Step 1: Most of the preliminary details are not presented here. Assume that the truss is made of steel with square hollow sections (inner height and width 0.1 m, wall thickness 0.01 m).

Step 2: If we try to analyze the truss, the following message is displayed:

Unstable structure. Instability detected at Node: 6. Direction: Y Disp.

In this case the instability is detected quite well by the program. There may be situations where the program may not display the error message but the error may still manifest itself in a different form. While solving the equilibrium equations, the default stiffness diagonal tolerance (a term discussed in detail in Section 6.2) of 10^{-6} may be too small to detect the instability. However, it is quite likely that the resulting displacements are very large compared to the dimensions of the structure. Clearly the results are invalid in this case.

To summarize, instability (in the context of materially linear, small-displacement analysis) is a function of the structural topology—the manner in which the joints are placed and connected to each other via the members, and the manner in which the structure is supported.

EXERCISES

Appetizers

7.5.1. For each of the problems in Figs. E7.5.1(a)–(c) construct the nodal data table. Select the problem units and the origin of the coordinate system first.

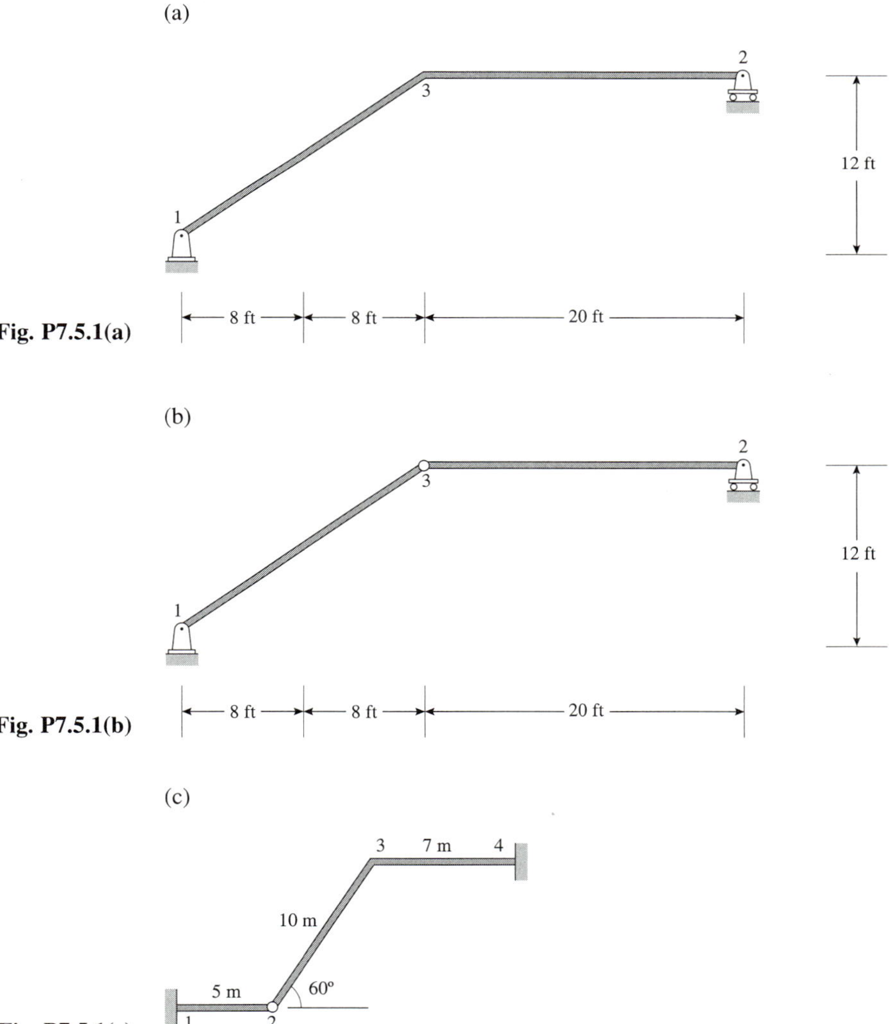

Fig. P7.5.1(a)

Fig. P7.5.1(b)

Fig. P7.5.1(c)

7.5.2. Solve Example 2.5.4 using the GS-USA program and compare the results.

7.5.3. Solve Example 2.7.2 using the GS-USA program and compare the results.

Main Course

In each of the problems below there are one or more modeling errors, either in the structure and/or in the mathematical model. Identify and correct the errors. Solve the (corrected) problems both by hand and using a computer program. Compare the two sets of answers.

7.5.4. The frame in Fig. P7.5.4 is made of steel. The cross-sections are hollow circular tubes (inner radius 1 in and wall thickness 0. in). As units, select ft for length and lb for force.

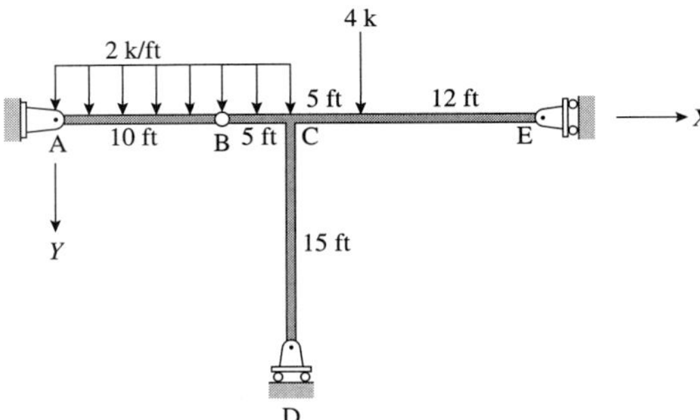

Fig. P7.5.4

The tables that will be used to generate the model are given below.

Material data:

Group	Modulus of elasticity	Coeff. of thermal expansion
1	$4.176(10^6)$ k/ft^2	$6.67(10^{-6})$ ft/ft–°F

Member cross-sectional properties:

Group	Type	Dimension 1	Dimension 2
1	Hollow circular	1"	0.2"

Nodal data:

Node	*X* coordinate	*Y* coordinate	*X* fixity	*Y* fixity	*Z* fixity
1	0	0	Fixed	Fixed	Free
2	10	0	Free	Free	Free
3	15	0	Free	Free	Free
4	15	15	Free	Fixed	Free
5	20	0	Free	Free	Free
6	32	0	Free	Fixed	Free

Element data:

Element	Start node	End node	Material group	Property group
1	1	2	1	1
2	2	3	1	1
3	3	4	1	1
4	3	6	1	1

Nodal loads:

Node	*X* force	*Y* force	*Z* moment
5	0	–4000	0

Element distributed loads:

Element load ID	Element	Load type	Intensity at start node	Intensity at end node
1	1	Global *Y* distributed	–2000	–2000
2	2	Global *Y* distributed	–2000	–2000

7.5.5. The truss in Fig. P7.5.5 is made of steel. The cross-sections are hollow circular tubes (inner radius 10 cm and wall thickness 1 cm). As units, select m for length and kN for force.

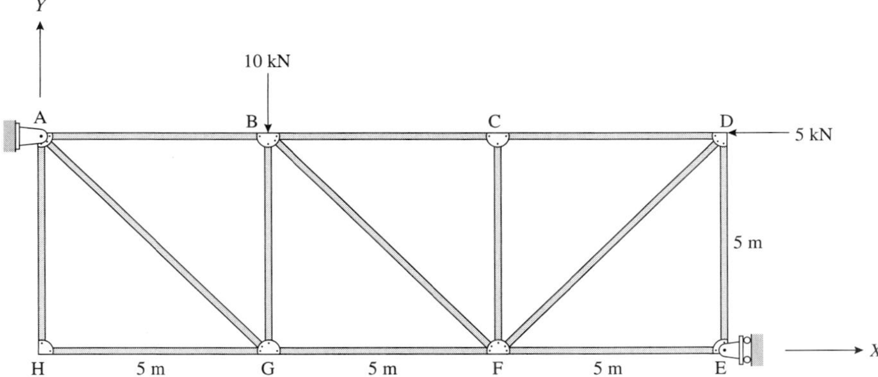

Fig. P7.5.5

Material data:

Group	Modulus of elasticity
1	200 kN/m^2

Member cross-sectional properties:

Group	Type	Dimension 1	Dimension 2
1	Hollow circular	0.1 m	0.01 m

Nodal data:

Node	*X* coordinate	*Y* coordinate	*X* fixity	*Y* fixity	*Z* fixity
1	0	5	Fixed	Fixed	Free
2	5	5	Free	Free	Free
3	10	5	Free	Free	Free
4	15	5	Free	Free	Free
5	15	0	Free	Fixed	Free
6	10	0	Free	Free	Free
7	5	0	Free	Free	Free
8	0	0	Free	Free	Free

Element data:

Element	Start node	End node	Material group	Property group
1	1	2	1	1
2	2	3	1	1
3	3	4	1	1
4	8	7	1	1
5	7	6	1	1
6	6	5	1	1
7	7	1	1	1
8	2	6	1	1
9	4	6	1	1
10	8	1	1	1
11	7	2	1	1
12	6	3	1	1
13	5	4	1	1

Nodal loads:

Node	*X* force	*Y* force	*Z* moment
2	0	10	0
4	5	0	0

Structural Concepts

7.5.6. Reanalyze the truss shown in Example 7.5.1 but change the support at node 6 from a *Y*-roller to a *X*-roller. Is the truss stable? Do you agree that the only stable configuration for a truss is a collection of triangles?

One of the strengths of computer-based analysis is to be able to carry out efficient parametric studies. The mundane calculations are performed in the computer program. In the following problems, compute the required response and plot the stated graph.

7.5.7. Figure P7.5.7 shows a planar truss as the supporting structure. Plot the variations of member forces in DF, DC, CF, and CE with respect to a. Vary a between 4 ft and 10 ft.

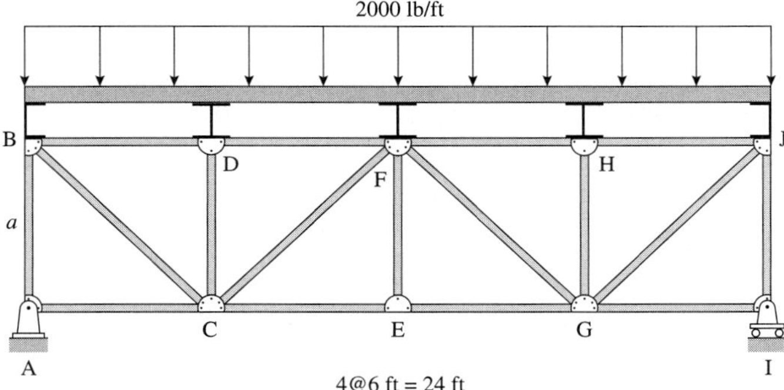

Fig. P7.5.7

7.5.8. Figure P7.5.8 shows a planar truss as the supporting structure. Plot the variations of member forces in DF, DC, DE, and CE with respect to a. Vary a between 4 and 10 ft.

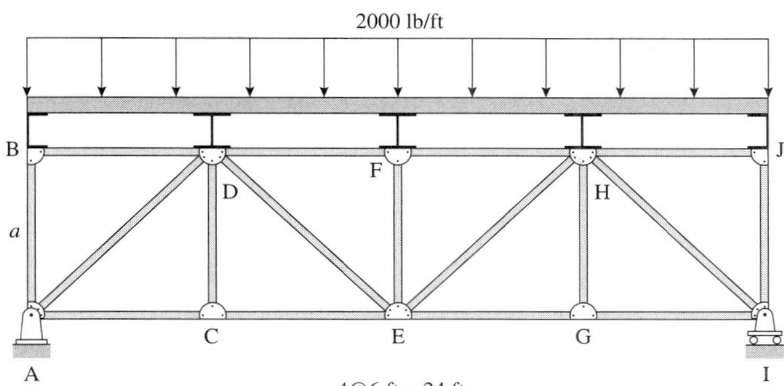

Fig. P7.5.8

7.5.9. For the system in Fig. P7.5.9, plot the variation of $M_{\max}$, the maximum moment in the frame, with respect to a. Vary a between 0 and 12 m.

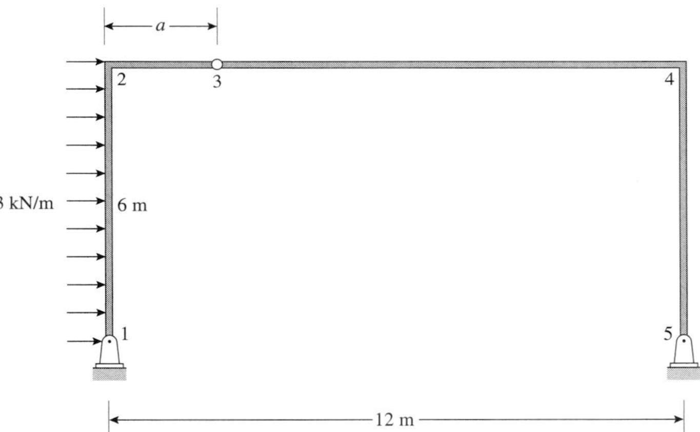

Fig. P7.5.9

7.6 MORE EXAMPLES

A brief review is in order now. The mathematical modeling concepts we have seen so far make it possible to construct models with the following characteristics:

- Modeling of rigid connections and internal hinges.
- Modeling of fixed, pin, and roller supports.
- Modeling of trusses subjected to nodal loads.
- Modeling of beams and frames subjected to nodal loads and element loads.

We now look at some more topics that will make it possible to solve a wider class of problems. First, we will learn more about multi-point constraints (MPCs). This will be useful for modeling skew or inclined supports. We will also see how to use the program to analyze structural systems that are subjected to multiple load cases.

Inclined Roller Support

Imagine a (skew) roller support at joint 5 in the structure shown in Fig. 7.6.1. The positive X axis is from left to right, the positive Y axis from bottom to top. Let the skew support be such that the inclined surface makes an angle of 30° with respect to the horizontal as shown in the figure (or the normal to the support surface makes an angle of 60° with the global X axis when measured counterclockwise from the X axis). If the roller moves down the inclined support surface, the displacement translates into a positive displacement in the X direction and a negative displacement in the Y direction, or vice versa if it moves up. However, the magnitudes of these X and Y displacements are not independent of each other. The relationship between the two can be expressed in the form of a constraint equation as

$$\tan(30) = \frac{Y_{\text{disp}}}{-X_{\text{disp}}} \qquad \tan(30) = \frac{-Y_{\text{disp}}}{X_{\text{disp}}}$$

$$\text{or} \quad (0.577)X_{\text{disp}} + (1.0)Y_{\text{disp}} = 0 \qquad \text{or} \quad (0.577)X_{\text{disp}} + (1.0)Y_{\text{disp}} = 0$$

Note that the coefficient values are not as important as the ratio of the two coefficients.

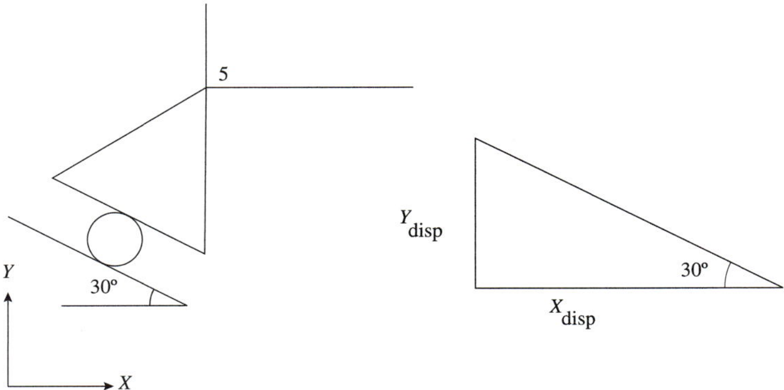

Fig. 7.6.1
Inclined roller support.

Multiple Load Cases

It is rare to design a structure just for one set of loads. Structural systems typically are subjected to quite different sets of loads that may or may not act simultaneously. The implication from a structural analysis viewpoint is that the structure must be analyzed for different sets of load cases. The design implications will be studied in subsequent chapters.

Consider the industrial frame in Figs. 7.6.2 and 7.6.3. In the first figure, the loading on the frame is due to dead and live loads (recall that we looked at the different types of loads and how to estimate their values in Section 3.6). In the second figure, the loading on the frame is due to dead load and wind load (acting left to right). While the structure is the same in both the figures, the loads are not. To facilitate the computer-based analysis, it is necessary to define the model (material and element cross-sectional properties, nodal and element data) just once except for the loads. The different sets of loads are identified in the GS-USA program by load case numbers. This tag is similar to other identification tags such as node number or element number. Load case numbering starts at 1 and is incremented sequentially.

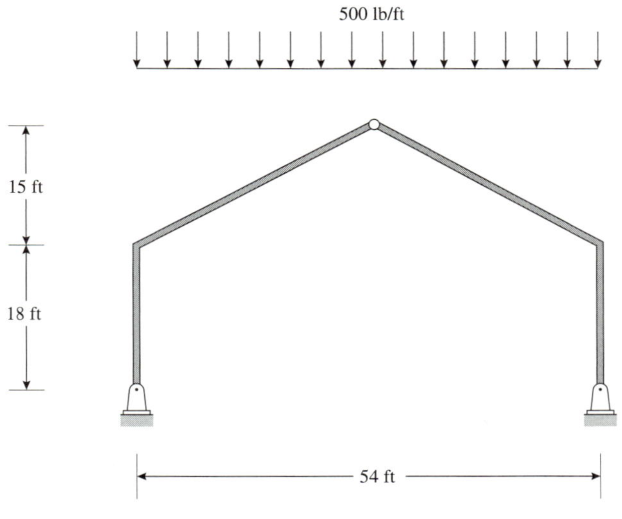

Fig. 7.6.2
Dead plus live load on the frame.

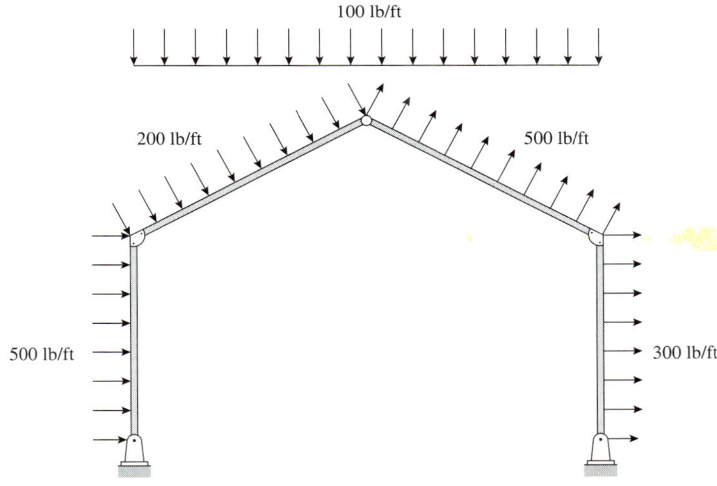

Fig. 7.6.3
Dead plus wind load—pressure (left) and suction (right) loads.

EXAMPLE 7.6.1 ***Handling Skew Supports (Example 2.5.2)***

A simply supported beam with an inclined roller support at one end is shown in Fig. E7.6.1(a). Compute all the support reactions.

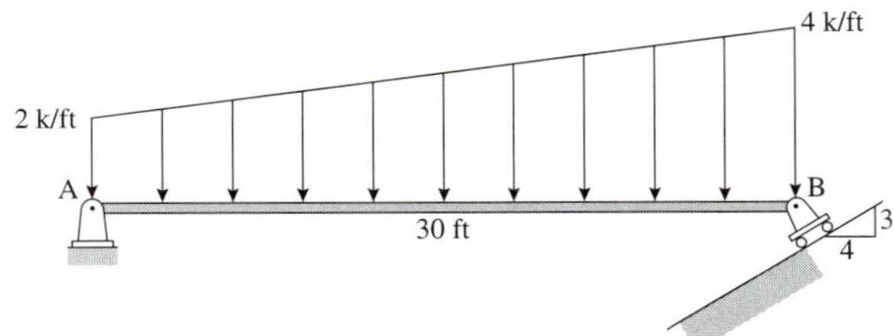

Fig. E7.6.1(a)

SOLUTION

Step 1: As units, select ft for length and k for force. The model is shown in Fig. E7.6.1(b).

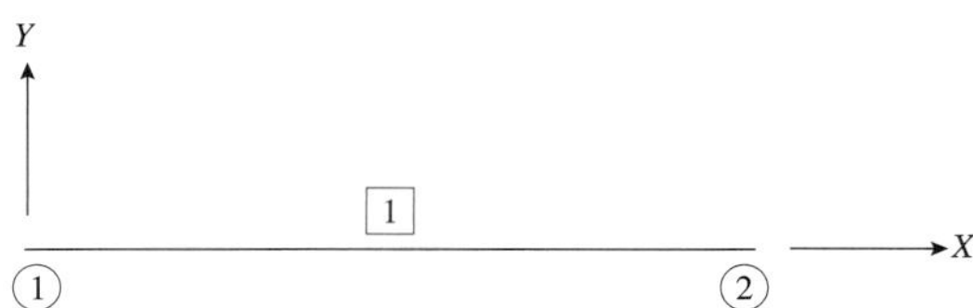

Fig. E7.6.1(b)

Step 2: The tables that will be used to generate the model are given below.

Material data: Let us assume that the beam is made of wood.

Group	Modulus of elasticity
1	273600 k/ft^2

Member cross-sectional properties: Let the cross-section be rectangular (20 in deep × 8 in wide).

Group	Type	Dimension 1	Dimension 2
1	Rectangular	1.67 ft	0.67 ft

Note that the nodal fixity conditions for the node with the inclined support is free in the X, Y, and Z directions.

Nodal data:

Node	X coordinate	Y coordinate	X fixity	Y fixity	Z fixity
1	0	0	Fixed	Fixed	Free
2	30	0	Free	Free	Free

Element data:

Start node	End node	Material group	Property group
1	2	1	1

Element distributed loads:

Element load ID	Element	Load type	Intensity at start node	Intensity at end node
1	1	Global y distributed or local y distributed	−2	−4

Multi-point constraint: With the defined coordinate system, the inclined surface is such that if the roller support moves up the surface, it has moved in the positive X and positive Y directions such that

$$\frac{3}{4} = \frac{Y_{\text{disp}}}{X_{\text{disp}}}$$

$$\text{or } (3.0)X_{\text{disp}} + (-4.0)Y_{\text{disp}} = 0$$

Step 2: Execute the GS-USA program.

Step 3: Examine the results file.

```
              ----------------------------------------
                  Graphics-based System for
                Understanding Structural Analysis
              PLANE FRAME ANALYSIS & DESIGN PROGRAM
                      COPYRIGHT, 1991-2000
                          S. D. RAJAN
                 DEPARTMENT OF CIVIL ENGINEERING
                    ARIZONA STATE UNIVERSITY
                         TEMPE, AZ 85287
                     VERSION 8.10 : AUG 1999
              ----------------------------------------

     Date: 08/16/99    Time: 07:31 PM

     Project Name : p2-5-2
     Creation Date and Time : 08/16/1999 19:27


     PROBLEM SIZE
     ------------
                   Structure Type: FRAME
                  Number of Nodes:     2
               Number of Elements:     1
         Number of Material Groups:    1
Number of Element Property Groups:     1
                   Number of MPC's:    1
              Number of Load Cases:    1
          Total number of unknowns:    4
           Total mass of structure:          33.3333 km
         Total weight of structure:                0  k
         Total volume of structure:          33.3333 ft^3
             Length of all members:               30 ft
           Total cost of structure: $ 0

     PROBLEM Notes
     -------------
Project Title:
      Analyst:
   Start Date:
     End Date:

Project Information
-------------------


   ---------------------------------------------------------------------
                          MATERIAL PROPERTIES
   Group   Young's Modulus    Mass Density    CTE              Cost
             ( k/ft^2)         (km/ft^3)     (ft/ft- F)     ( $ /ft^3)
   ---------------------------------------------------------------------
   1       273600             1               1                0
```

CROSS-SECTIONAL DIMENSIONS

Group	Type	Cost ($ /ft)	Dimensions (ft)
1	Rect Solid	0	1.66667 0.666667

CROSS-SECTIONAL PROPERTIES

Group	Area (ft^2)	Moment of Inertia (ft^4)	Shear Factor (ft^2)	Section Modulus (ft^3)
1	1.11111	0.257202	0.740741	0.308642

NODAL INFORMATION

Node	Cost ($)	X Coor (ft)	Y Coor (ft)	X-Fixity	Y-Fixity	Z-Fixity
1	0	0	0	FIXED	FIXED	FREE
2	0	30	0	FREE	FREE	FREE

ELEMENT INFORMATION

Elem	Matl Grp	Prop Grp	Start Node	End Node	Delta T (F)
1	1	1	1	2	0

NODAL LOADS for LOAD CASE 1

Node	X Force (k)	Y Force (k)	Z Moment (k-ft)

There are no nodal loads

ELEMENT CONCENTRATED LOADS for LOAD CASE 1

Elem	Load Type	Dist from Start Node (ft)	Load Intensity (k)

There are no concentrated element loads

ELEMENT DISTRIBUTED LOADS for LOAD CASE 1

Elem	Load Type	Int at Start Node (k/ft)	Int at End Node (k/ft)
1	LOCAL Y	-2	-4

MULTI-POINT CONSTRAINTS

MPC	Node 1	Local Dof	Coef	Node 2	Local Dof	Coef	Value
1	2	X Disp	3	2	Y Disp	-4	0

```
-----------------------------------------------------
          NODAL DISPLACEMENTS for LOAD CASE  1
Node      X Disp          Y Disp          Z Rotations
           (ft)            (ft)             (rad)
-----------------------------------------------------
1         0               0               -0.0469873
2         -0.00370066     -0.00277552     0.0489338

Min:      -0.00370066     -0.00277552     -0.0469873
at :      2               2               1
Max:      0               0               0.0489338
at :      1               1               2
```

```
------------------------------------------------------------------
             ELEMENT NODAL FORCES for LOAD CASE  1
               local coordinate system
Elem   Axial          Shear          Moment          Rotation
       ( k)           ( k)          ( k-ft)
------------------------------------------------------------------
1      37.5           40             1.41257e-005
       -37.5          50             1.52987e-005

  Min: -37.5          40             1.41257e-005
  at : 1              1              1
  Max: 37.5           50             1.52987e-005
  at : 1              1              1
```

```
---------------------------------------------------------------
     MIN-MAX INTERNAL MEMBER FORCES for LOAD CASE  1
Element      Axial Force         Shear Force       Bending Moment
                ( k)                ( k)              ( k-ft)
---------------------------------------------------------------
Min   1         -37.5               -40               -1.41257e-005
At              30                  0                 0
Max             -37.5               50                338.522
At              0                   30                15.9184
```

```
-------------------------------------------------
          SUPPORT REACTIONS for LOAD CASE  1
Node      X Force         Y Force         Z Moment
           ( k)            ( k)           ( k-ft)
-------------------------------------------------
1         37.5            40              0
2         -37.5           50              0

Min:      -37.5           40              0
at :      2               1               1
Max:      37.5            50              0
at :      1               2               1
Sum:      37.5            40              0
```

EXAMPLE 7.6.2 ***Multiple Load Cases***

Draw the shear force and bending moment diagrams of the industrial frame in Section 7.6 for both the load cases.

SOLUTION

Step 1: The model is shown in Fig. E7.6.2(a).

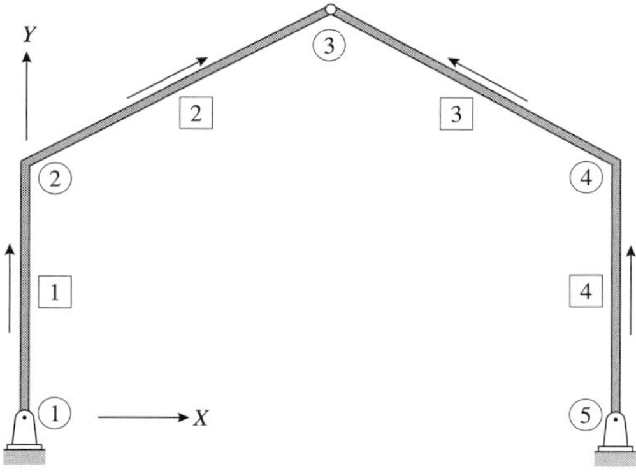

Fig. E7.6.2(a) Model details.

As units, select in for length and lb for force.

Material data: Let us assume that the material is steel.

Group	Modulus of elasticity
1	$2.9(10^7)$ psi

Member cross-sectional properties: We assume AISC Wide Flange Section.

Group	Type	ID
1	Predefined AISC W-section	W21 × 50

Nodal data:

Node	*X* coordinate	*Y* coordinate	*X* fixity	*Y* fixity	*Z* fixity
1	0	0	Fixed	Fixed	Free
2	0	18*12	Free	Free	Free
3	(54/2)*12	(18+15)*12	Free	Free	Hinge
4	54*12	18*12	Free	Free	Free
5	54*12	0	Fixed	Fixed	Free

Element data: There is only one material and one element property group.

Element	Start node	End node	Material group	Property group
1	1	2	1	1
2	2	3	1	1
3	4	3	1	1
4	5	4	1	1

Nodal loads: There are no nodal loads.

Element concentrated loads: There are no element concentrated loads.

Element distributed loads: There are several element distributed loads.

Element load ID	Load case	Element	Load type	Intensity at start node	Intensity at end node
1	1	2	Projected *X* distributed	–500/12	–500/12
2	1	3	Projected *X* distributed	–500/12	–500/12

Element load ID	Load case	Element	Load type	Intensity at start node	Intensity at end node
3	2	2	Projected *X* distributed	–100/12	–100/12
4	2	3	Projected *X* distributed	–100/12	–100/12
5	2	1	Projected *Y* distributed	500/12	500/12
6	2	2	Local *y* distributed	–200/12	–200/12
7	2	3	Local *y* distributed	–500/12	–500/12
8	2	4	Projected *Y* distributed	300/12	300/12

Step 2: Execute the GS-USA program.

Step 3: Examine the results.

Note that each load case is identified in each of the tables below as LOAD CASE 1 and LOAD CASE 2.

```
              ----------------------------------------
                    Graphics-based System for
                Understanding Structural Analysis
              PLANE FRAME ANALYSIS & DESIGN PROGRAM
                      COPYRIGHT, 1991-2000
                          S. D. RAJAN
                 DEPARTMENT OF CIVIL ENGINEERING
                    ARIZONA STATE UNIVERSITY
                        TEMPE, AZ 85287
                    VERSION 8.10 : AUG 1999
              ----------------------------------------

Date: 08/26/99    Time: 07:25 AM

Project Name : e7-6-2
Creation Date and Time : 08/26/1999 06:25

PROBLEM SIZE
------------
              Structure Type: FRAME
             Number of Nodes:     5
          Number of Elements:     4
   Number of Material Groups:     1
Number of Element Property Groups:     1
             Number of MPC's:     0
        Number of Load Cases:     2
     Total number of unknowns:    10
      Total mass of structure:          17007.8 lbm
    Total weight of structure:                0 lb
    Total volume of structure:          17007.8 in^3
        Length of all members:          1173.29 in
      Total cost of structure: $ 0
```

PROBLEM Notes

Project Title:
Analyst:
Start Date:
End Date:

Project Information

MATERIAL PROPERTIES

Group	Young's Modulus (lb/in^2)	Mass Density (lbm/in^3)	CTE (in/in- F)	Cost ($ /in^3)
1	2.9e+007	1	1	0

CROSS-SECTIONAL DIMENSIONS

Group	Type	Cost ($ /in)	Dimensions (in)
1	AISC	0	W21X50

CROSS-SECTIONAL PROPERTIES

Group	Area (in^2)	Moment of Inertia (in^4)	Shear Factor (in^2)	Section Modulus (in^3)
1	14.4959	963.963	6.8686	92.5553

NODAL INFORMATION

Node	Cost ($)	X Coor (in)	Y Cocr (in)	X-Fixity	Y-Fixity	Z-Fixity
1	0	0	0	FIXED	FIXED	FREE
2	0	0	216	FREE	FREE	FREE
3	0	324	396	FREE	FREE	HINGE
4	0	648	216	FREE	FREE	FREE
5	0	648	0	FIXED	FIXED	FREE

ELEMENT INFORMATION

Elem	Matl Grp	Prop Grp	Start Node	End Node	Delta T (F)
1	1	1	1	2	0
2	1	1	2	3	0
3	1	1	4	3	0
4	1	1	5	4	0

```
-------------------------------------------------
            NODAL LOADS for LOAD CASE  1
Node      X Force          Y Force          Z Moment
           (lb)            (lb)             (lb-in)
-------------------------------------------------
           There are no nodal loads

-------------------------------------------------
            NODAL LOADS for LOAD CASE  2
Node      X Force          Y Force          Z Moment
           (lb)            (lb)             (lb-in)
-------------------------------------------------
           There are no nodal loads

--------------------------------------------------------
        ELEMENT CONCENTRATED LOADS for LOAD CASE  1
Elem  Load Type   Dist from Start Node    Load Intensity
                          (in)                 (lb)
--------------------------------------------------------
         There are no concentrated element loads

-----------------------------------------------------------
        ELEMENT DISTRIBUTED LOADS for LOAD CASE  1
Elem  Load Type     Int at Start Node    Int at End Node
                       (lb/in)              (lb/in)
-----------------------------------------------------------
2     PROJECTED X   -41.6667             -41.6667
3     PROJECTED X   -41.6667             -41.6667

--------------------------------------------------------
        ELEMENT CONCENTRATED LOADS for LOAD CASE  2
Elem  Load Type   Dist from Start Node    Load Intensity
                          (in)                 (lb)
--------------------------------------------------------
         There are no concentrated element loads

-----------------------------------------------------------
        ELEMENT DISTRIBUTED LOADS for LOAD CASE  2
Elem  Load Type     Int at Start Node    Int at End Node
                       (lb/in)              (lb/in)
-----------------------------------------------------------
2     PROJECTED X   -8.33333             -8.33333
3     PROJECTED X   -8.33333             -8.33333
1     PROJECTED Y   41.6667              41.6667
2     LOCAL Y       -16.6667             -16.6667
3     LOCAL Y       -41.6667             -41.6667
4     PROJECTED Y   25                   25

---------------------------------------------------------------------------
                     MULTI-POINT CONSTRAINTS
MPC Node 1  Local Dof  Coef       Node 2  Local Dof  Coef        Value
---------------------------------------------------------------------------
               There are no MPC's
```

NODAL DISPLACEMENTS for LOAD CASE 1

Node	X Disp (in)	Y Disp (in)	Z Rotations (rad)
1	0	0	0.00421505
2	-0.578628	-0.00693656	-0.000393592
3	8.62255e-015	-1.06318	0
4	0.578628	-0.00693656	0.000393592
5	0	0	-0.00421505
Min:	-0.578628	-1.06318	-0.00421505
at :	2	3	5
Max:	0.578628	0	0.00421505
at :	4	1	1

NODAL DISPLACEMENTS for LOAD CASE 2

Node	X Disp (in)	Y Disp (in)	Z Rotations (rad)
1	0	0	-0.0194634
2	3.51013	0.00204671	-0.0104509
3	3.48383	0.0617034	0
4	3.4442	-0.000659401	-0.0108967
5	0	0	-0.0186574
Min:	0	-0.000659401	-0.0194634
at :	1	4	1
Max:	3.51013	0.0617034	0
at :	2	3	3

ELEMENT NODAL FORCES for LOAD CASE 1
local coordinate system

Elem	Axial (lb)	Shear (lb)	Moment (lb-in)	Rotation
1	13500	-5522.73	0.118022	
	-13500	5522.73	-1.19291e+006	
2	11383.9	9119.05	1.19291e+006	
	-4827.71	2682.07	0	-0.00467717
3	11383.9	-9119.05	-1.19291e+006	
	-4827.71	-2682.07	0	0.00467717
4	13500	5522.73	-0.118022	
	-13500	-5522.73	1.19291e+006	
Min:	-13500	-9119.05	-1.19291e+006	
at :	1	3	1	
Max:	13500	9119.05	1.19291e+006	
at :	1	2	2	

```
-----------------------------------------------------------------
              ELEMENT NODAL FORCES for LOAD CASE  2
                local coordinate system
Elem   Axial          Shear          Moment         Rotation
       (lb)           (lb)          (lb-in)
-----------------------------------------------------------------
1      -3983.33       13800          -0.820843
       3983.33        -4799.99       2.0088e+006
2      -6130.52       -1150.97       -2.0088e+006
       7441.75        9688.58        0              0.0054882
3      5606           1646.33        -1.8144e+006
       -4294.76       11436.9        0              0.00514982
4      1283.33        11100          -0.163338
       -1283.33       -5700          1.8144e+006

  Min: -6130.52       -5700          -2.0088e+006
  at : 2              4              2
  Max: 7441.75        13800          2.0088e+006
  at : 2              1              1

  -------------------------------------------------------------
       MIN-MAX INTERNAL MEMBER FORCES for LOAD CASE  1
  Element      Axial Force         Shear Force      Bending Moment
                  (lb)                (lb)             (lb-in)
  -------------------------------------------------------------
Min    1         -13500             5522.73          -1.19291e+006
At               216                216              216
Max              -13500             5522.73          -0.118022
At               0                  0                0
Min    2         -11383.9           -9119.05         -1.19291e+006
At               0                  0                0
Max              -4827.71           2682.07          112948
At               370.643            370.643          287.437
Min    3         -11383.9           -2682.07         -112948
At               0                  370.643          287.437
Max              -4827.71           9119.05          1.19291e+006
At               370.643            0                0
Min    4         -13500             -5522.73         0.118022
At               216                216              0
Max              -13500             -5522.73         1.19291e+006
At               0                  0                216

  -------------------------------------------------------------
       MIN-MAX INTERNAL MEMBER FORCES for LOAD CASE  2
  Element      Axial Force         Shear Force      Bending Moment
                  (lb)                (lb)             (lb-in)
  -------------------------------------------------------------
Min    1         3983.33            -13800           0.820843
At               216                0                0
Max              3983.33            -4799.99         2.0088e+006
At               0                  216              216
Min    2         6130.52            1150.97          0
At               0                  0                370.643
Max              7441.75            9688.58          2.0088e+006
At               370.643            370.643          0
Min    3         -5606              -1646.33         0
```

```
At              0               0               370.643
Max             -4294.76        11436.9         1.85276e+006
At              370.643         370.643         45.3848
Min     4       -1283.33        -11100          0.163338
At              216             0               0
Max             -1283.33        -5700           1.8144e+006
At              0               216             216
```

```
        ------------------------------------------------
                 SUPPORT REACTIONS for LOAD CASE  1
        Node       X Force        Y Force        Z Moment
                    (lb)           (lb)          (lb-in)
        ------------------------------------------------
        1          5522.73        13500          0
        5          -5522.73       13500          0

    Min:           -5522.73       13500          0
    at :           5              1              1
    Max:           5522.73        13500          0
    at :           1              1              1
    Sum:           0              27000          0
```

```
        ------------------------------------------------
                 SUPPORT REACTIONS for LOAD CASE  2
        Node       X Force        Y Force        Z Moment
                    (lb)           (lb)          (lb-in)
        ------------------------------------------------
        1          -13800         -3983.33       0
        5          -11100         1283.33        0

    Min:           -13800         -3983.33       0
    at :           1              1              1
    Max:           -11100         1283.33        0
    at :           5              5              1
    Sum:           -24900         -2700          0
```

The shear force and bending moment diagrams are shown next.

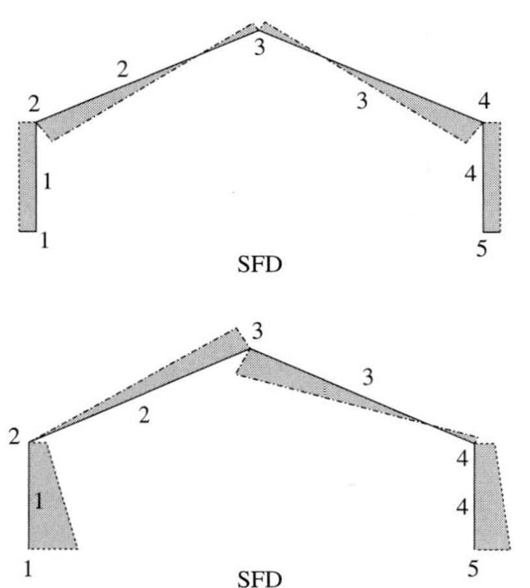

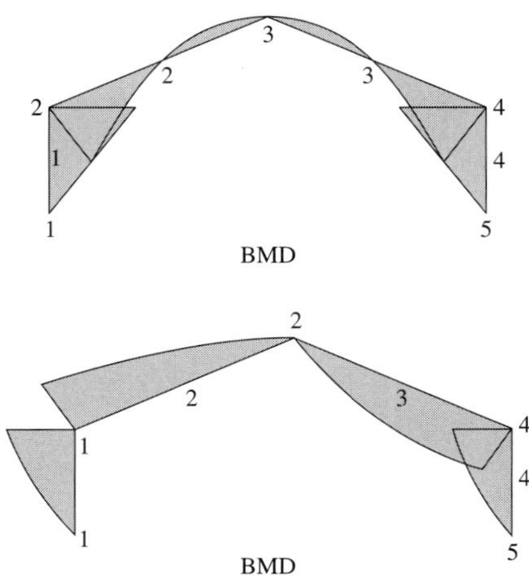

Checks:

a. Applied loads equal to sum of reactions? (Structural FBD in equilibrium?)

Load Case	Forces X direction	Forces Y direction
1	Applied: 0 Sum of the X reactions: –3314 + 3314 = 0	Applied: (300)(54) = –16200 Sum of the Y reactions: 8100 + 8100 = 16200
2	Applied: (200 + 50)(18 + 15) = 8250 Sum of the X reactions: –4534 – 3716 = –8250	Applied: (75)(54) = –4050 Sum of the X reactions: –496 + 4546 = 4050

b. Zero moments at the pin supports and internal hinge?

c. Joint equilibrium of nodes 2, 3, and 4?

SUMMARY

To facilitate computer-based structural analysis, understanding the terminology is the first step. In this chapter we looked at the concepts associated with nodes, elements, nodal fixity conditions, element and nodal loads. To illustrate these concepts, problems from Chapter 2 were solved first. These examples served to illustrate the steps required to create the mathematical model as well as the manner in which to read and interpret the program results.

Some advanced concepts associated with inclined roller supports (multi-point constraints) and multiple loading conditions were also discussed in some detail. The concepts will be useful in other chapters.

The important lesson to be learned is that computer programs are a very versatile and powerful tool. This tool must be used with caution—we can and should use the structural analysis theory to check the results. In the other chapters we will see how to leverage the strengths of computer-based tools efficiently to analyze, understand, and design structural components and systems.

SUMMARY EXERCISES

For those problems requiring the use of a computer program, it is recommended that you use the GS-USA program or a program with similar capabilities.

Appetizers

7.1. Compute the support reactions for the frame in Fig. P7.1 using a computer program. Draw the shear force and bending moment diagrams.

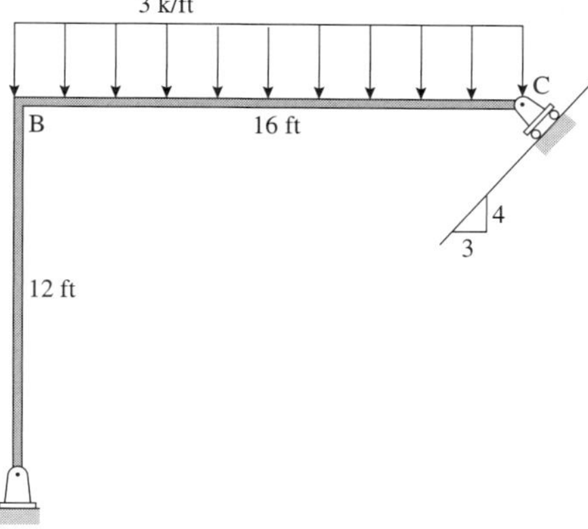

Fig. P7.1

Main Course

7.2. Solve Problem 2.1 using a computer program.

7.3. Solve Problem 2.2 using a computer program.

Structural Concepts

7.4. The truss in Fig. P7.4 (8 joints and 16 members) is made of steel. The cross-sections are hollow circular tubes (inner radius 10 cm and wall thickness 1 cm). The output from a computer program is given below. Systematically check the answers.

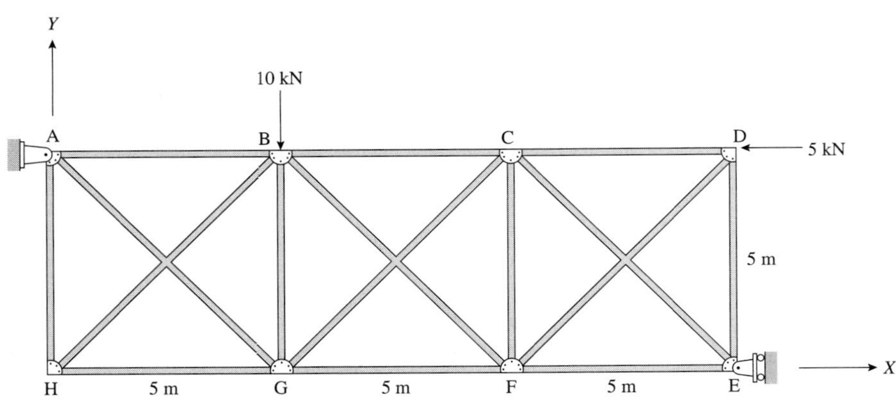

Fig. P7.4

```
---------------------------------------
    ACME PLANE TRUSS ANALYSIS PROGRAM
         VERSION 0.9 : JAN 2000
---------------------------------------
```

MATERIAL PROPERTIES

Group	Young's Modulus (N / m ^2)	Mass Density (kg / m ^3)	CTE (m / m - C)	Cost ($ / m ^3)
1	2e+011	7850	1.2e-005	0

CROSS-SECTIONAL PROPERTIES

Group	Area (m ^2)	Moment of Inertia (m ^4)	Shear Factor (m ^2)	Section Modulus (m ^3)
1	0.00659735	3.64503e-005	0.00330365	0.000331367

NODAL INFORMATION

Node	Cost ($)	X Coor (m)	Y Coor (m)	X-Fixity	Y-Fixity
1	0	0	5	FIXED	FIXED
2	0	5	5	FREE	FREE
3	0	10	5	FREE	FREE
4	0	15	5	FREE	FREE
5	0	15	0	FIXED	FREE
6	0	10	0	FREE	FREE
7	0	5	0	FREE	FREE
8	0	0	0	FREE	FREE

```
--------------------------------------------------------
                   ELEMENT INFORMATION
Elem  Matl Grp   Prop Grp   Start Node  End Node  Delta T
                                                    ( C )
--------------------------------------------------------
1     1          1          1           2         0
2     1          1          2           3         0
3     1          1          3           4         0
4     1          1          8           7         0
5     1          1          7           6         0
6     1          1          6           5         0
7     1          1          8           1         0
8     1          1          7           2         0
9     1          1          6           3         0
10    1          1          5           4         0
11    1          1          1           7         0
12    1          1          8           2         0
13    1          1          2           6         0
14    1          1          7           3         0
15    1          1          3           5         0
16    1          1          6           4         0

    ------------------------------------------------
             NODAL LOADS for LOAD CASE  1
    Node      X Force        Y Force        Z Moment
               ( N )          ( N )          ( N - m )
    ------------------------------------------------
    2        0              -10000
    4        -5000          0

    ---------------------------------------------------
             NODAL DISPLACEMENTS for LOAD CASE  1
    Node     X Disp          Y Disp
              ( m )           ( m )
    ---------------------------------------------------
    1        0              0
    2        -2.00503e-006  -0.000153996
    3        -2.05368e-005  -0.000188296
    4        -4.14888e-005  -0.000175435
    5        0              -0.00017343
    6        -3.5889e-005   -0.000186706
    7        -7.41983e-005  -0.00013346
    8        -9.11403e-005  -1.6942e-005

Min:         -9.11403e-005  -0.000188296
at :         8              3
Max:         0              0
at :         1              1
```

```
------------------------------------------------------------------
              ELEMENT NODAL FORCES for LOAD CASE  1
                local coordinate system
Elem   Axial          Shear          Moment          Rotation
       ( N )          ( N )          ( N - m )
------------------------------------------------------------------
1      529.115
       -529.115
2      4890.42
       -4890.42
3      5529.12
       -5529.12
4      -4470.89
       4470.89
5      -10109.6
       10109.6
6      -9470.89
       9470.89
7      -4470.89
       4470.89
8      5419.53
       -5419.53
9      419.534
       -419.534
10     529.113
       -529.113
11     -7819.35
       7819.35
12     6322.79
       -6322.79
13     154.975
       -154.975
14     154.975
       -154.975
15     -748.281
       748.281
16     -748.281
       748.281

  Min: -10109.6
  at : 5
  Max: 10109.6
  at : 5

--------------------------------------------------------------
    MAX MEMBER STRESSES for LOAD CASE  1
Element  Compressive Stress   Tensile Stress  Pcr(Euler Buck)
             ( N / m ^2)        ( N / m ^2)        ( N )
--------------------------------------------------------------
1         80201.1             0                2.878e+006
2         741270              0                2.878e+006
3         838082              0                2.878e+006
4         0                   677679           2.878e+006
5         0                   1.53237e+006     2.878e+006
6         0                   1.43556e+006     2.878e+006
7         0                   677679           2.878e+006
```

```
8         821472              0                  2.878e+006
9         63591.3             0                  2.878e+006
10        80200.8             0                  2.878e+006
11        0                   1.18523e+006       1.439e+006
12        958384              0                  1.439e+006
13        23490.6             0                  1.439e+006
14        23490.5             0                  1.439e+006
15        0                   113422             1.439e+006
16        0                   113422             1.439e+006

          ------------------------------------------------
                    SUPPORT REACTIONS for LOAD CASE  1
          Node      X Force           Y Force           Z Moment
                     ( N )             ( N )            ( N - m )
          ------------------------------------------------
          1          -5000             10000
          5          10000             0
```

7.5. Influence lines are graphs of some structural response versus a unit load whose location is varied along the structure. The idea is to find the location(s) of the load that would cause the worst effect on the supporting structure.

Consider a simply supported bridge as shown in Fig. P7.5. The two concentrated loads represent the axle loads from a moving vehicle.

(a) Plot the vertical reaction at A versus a.

(b) Plot the maximum bending moment, M_{max} versus a.

In both cases vary a between 0 and 30 ft.

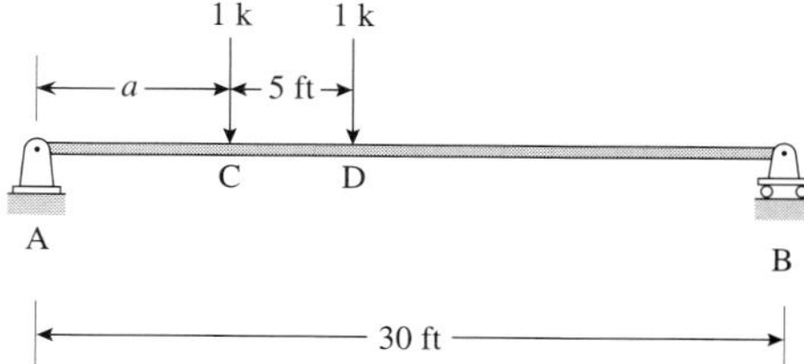

Fig. P7.5

Chapter 8

Optimum Structural Design

Savings of even a few percent in large structural systems such as an auditorium or mass-produced systems such as a roof truss, can translate into millions of dollars. Computerized tools such as numerical optimization can help immensely in achieving lower overall project cost.

"A great pleasure in life is doing what people say you cannot do." Anon

"The difference between theory and implementation is the difference between a soup recipe and a bowl of hot soup."

"Experience is the name everyone gives to their mistakes." Oscar Wilde

In Chapter 3 we looked at some of the major components of structural design problems—material behavior, computation of stresses and strains, failure theories, computation and modeling of loads, and computation of critical structural responses. The present chapter can be divided into two parts. The first covers the general area of optimization techniques. The second part links the optimization techniques with the solution of structural design problems.

In the first few sections, we explore the basic concepts associated with a formal, numerical approach to finding the optimal or best design. We first look at the common terminology. For example, the term *engineering design optimization* (or *design optimization* for short), encompasses the study of mathematical programming with the objective of finding the optimal design of engineering problems. Then we explore the basic ideas starting with one-dimensional problems before moving onto multidimensional problems. After gaining an understanding of the issues, we look at a spectrum of engineering design possibilities. Clearly we are neither prepared nor have the focus to examine all these different problems. To solve these problems, researchers have devised a variety of solution techniques. Some of these have entered mainstream usage in programs such as Microsoft Excel. In this chapter we look at one technique that has recently gained prominence and is powerful and versatile—the genetic algorithm (GA).

As we saw in Chapter 3, there is an intimate link between structural analysis and structural design. Throughout the text we saw several design examples in which structural analysis was used to compute the required structural response. These examples dealt with the design of components or simple systems involving few design variables and constraints. In this chapter we build on those ideas. There are two major objectives. The first is to understand whether and how structural design requirements can be formalized. This process is known as problem formulation. The second is to link structural analysis with optimization technique so that not only do we find the best possible design but we also automate the mundane calculations required to find the best design. In this chapter we will see several open-ended problems. This is the nature of most design problems. While there are correct and incorrect solutions, there is no (one) best design.

Finally, a few words about the placement and coverage of material in this chapter. The formal study of engineering design is becoming more and more important. While some of the background material for this chapter and some of the material in it may appear ambitious, the basic intent in presenting the material is twofold. First, it is imperative to think of engineering design as a systematic, organized activity. An engineer must learn to think of "design problem formulation"—how to pose the design problem in some concrete manner. It should come as no surprise that several tools, including popular computer programs, help the user pose and answer "what-if" scenarios. This is an important component in engineering design. Second, having posed the design problem in a concrete or mathematical form, the design engineer must be able to select an appropriate solution tool. This chapter attempts to answer both these questions in the best possible manner given the time and space constraints. Experience has shown that an understanding and usage of this design-related material is a most satisfying exercise. Industry has realized the potential of computer-based tools including design optimization and is increasingly reliant on these tools to solve design problems efficiently, accurately, and in a timely manner.

OBJECTIVES

- To become familiar with the language of design optimization.
- To understand how to formulate and solve analytically simple unconstrained and constrained minimization problems.
- To learn the basics of GA.
- To learn how to use the UNDO (Understanding Design Optimization) program for solving constrained optimization problems using the genetic algorithm.
- To learn how to formulate and solve structural design problems using the GS-USA program.

ASSUMPTIONS

- Structural analysis will be carried out using the usual linear, small-displacement, small-strain behavior assumptions.
- We encourage the use of numerical analysis techniques discussed in Chapter 6. Both bending and axial effects will be considered.
- The functions that define the design optimization problems will be assumed to have the form and properties amenable to using the genetic algorithm discussed in this chapter.

8.1 BACKGROUND

The simplest form of a mathematical programming problem is

$$\text{Find} \quad \mathbf{x} \in R^n \tag{8.1.1a}$$

$$\text{to minimize} \quad f(\mathbf{x}) \tag{8.1.1b}$$

Mathematics provides us a very powerful language to express our ideas. In the above equations, $\mathbf{x}$ represents the vector of *design variables*. These are variables we seek in order to complete the design process. The notation $\mathbf{x} \in R^n$ indicates that the design variables are real-valued and that there are n variables, $x_1, x_2,\ldots,x_n$. The function $f(\mathbf{x})$ is the *objective function*. This is the function that drives the design process and is either directly or indirectly a function of the n design variables. Such a problem is called an *unconstrained minimization* problem.

Consider the following problem.

Find x

to minimize $f(x) = (x - 10)^2 + 1$

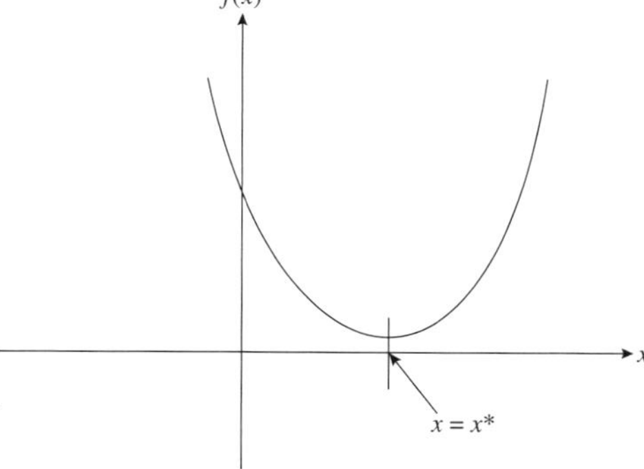

Fig. 8.1.1 Minimum point for a unimodal function.

Since there is only one design variable, the problem is referred to as a *one-dimensional unconstrained minimization problem.*

From calculus, we know for continuous differentiable functions, the necessary condition to find the minimum of a function $f(x)$ is $(df/dx) = 0$. This condition yields the stationary point(s). Therefore, $(df/dx) = 0 = 2(x - 10)$. Solving, we have $x = 10$. The sufficient condition that this point $x = x^*$ corresponds to a minimum is $\dfrac{d^2 f(x = x^*)}{dx^2} > 0$. Checking, $\dfrac{d^2 f(x = x^*)}{dx^2} = 2 > 0$. Hence, the solution to the above problem is: the *optimal solution* is at $x^* = 10$ and the lowest value of the objective function is $f(x^*) = 1$. Fig. 8.1.1 shows the problem and the solution graphically.

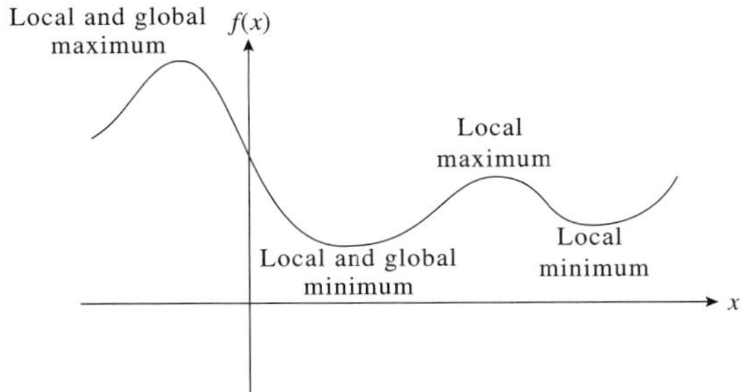

Fig. 8.1.2
Local and global minima.

There are other characteristics of even simple problems that we must be aware of. For example, in Fig. 8.1.3, the function $f(x) = x^4 - 3x^2 + x$ has multiple *local minima*. Such a function is called a *multimodal function* (the function in Fig. 8.1.1 is a *unimodal function*). Each trough or valley captures a minimum that is local. Among all these local minima, there are one or more points that have the absolute minimum value. These points are called the *global minima* (Fig. 8.1.2).

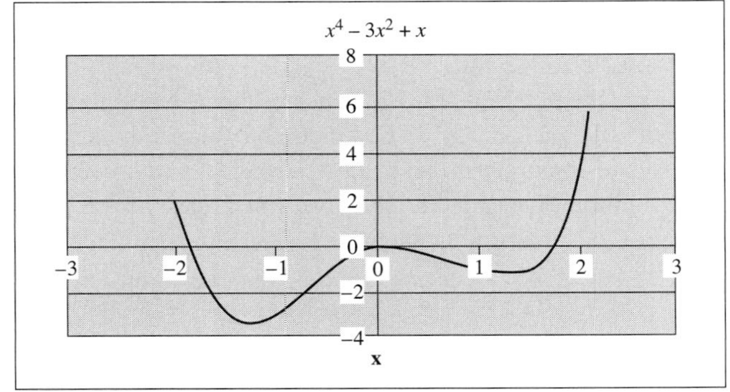

Fig. 8.1.3
A multimodal function $f(x) = x^4 - 3x^2 + x$.

We can use the calculus theorems seen earlier to solve this problem.

$$\frac{df}{dx} = 0 = 4x^3 - 6x + 1 \Rightarrow x = -1.300839567,\ 0.1699384435,\ 1.130901123$$

Using the sufficient condition $(d^2f/dx^2) = 12x^2 - 6$, we find that

$$(d^2f(x = -1.300839567))/dx^2 > 0$$

$$(d^2f(x = 0.1699384435))/dx^2 < 0$$

$$(d^2f(x = 1.130901123))/dx^2 > 0$$

Hence the points $x^* = -1.3008$ and $x^* = 1.1309$ are local minima points. Moreover, the point $x^* = -1.3008$ is also a global minimum point. The function has the lowest value at this point.

As the functions become "more nonlinear," the task of analytically finding the minimum value becomes more difficult. One must then resort to a numerical technique.

Let us now turn our attention to two-variable problems. Since there is now more than one design variable, the problem is referred to as a *multidimensional unconstrained minimization problem*. As an example consider the following problem.

Find $\{x_1, x_2\}$

to minimize[1] $f(x_1, x_2) = 100(x_2 - x_1^2)^2 + (1 - x_1)^2$

The function is shown in Fig. 8.1.4. To find the minimum, we can use the first-order condition[2] as

$$\frac{\partial f}{\partial x_1} = 0 = -400x_1(x_2 - x_1^2) - 2(1 - x_1)$$

$$\frac{\partial f}{\partial x_2} = 0 = 200(x_2 - x_1^2)$$

Solving, we have $\{x_1, x_2\} = \{1, 1\}$. The second-order condition involves computing **H**, the Hessian matrix that contains the second-order (partial) derivatives of $f(\mathbf{x})$:

$$\mathbf{H}_{2\times 2} = \begin{bmatrix} \dfrac{\partial^2 f}{\partial x_1^2} & \dfrac{\partial^2 f}{\partial x_1 \partial x_2} \\ \dfrac{\partial^2 f}{\partial x_2 \partial x_1} & \dfrac{\partial^2 f}{\partial x_2^2} \end{bmatrix} = \begin{bmatrix} 1200x_1^2 - 400x_2 + 2 & -400x_1 \\ -400x_1 & 200 \end{bmatrix}$$

At the point (1, 1), we have

$$\mathbf{H} = \begin{bmatrix} 802 & -400 \\ -400 & 200 \end{bmatrix}$$

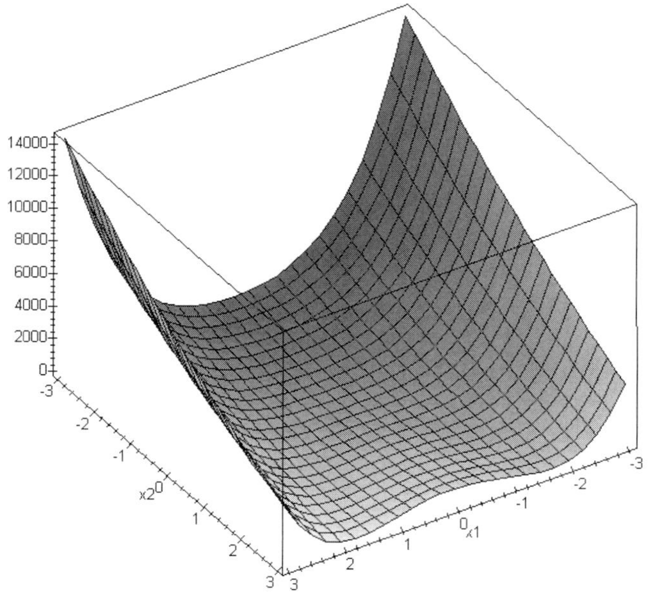

Fig. 8.1.4 Two-variable unconstrained minimization problem.

[1] This function is popularly known as Rosenbrock's function.

[2] The syntactically correct condition is $\nabla f(x) = 0$ where $\nabla f(\mathbf{x}) = \left[\dfrac{\partial f}{\partial x_1} \;\; \dfrac{\partial f}{\partial x_2} \;\; \cdots \;\; \dfrac{\partial f}{\partial x_n}\right]$ is called the gradient vector.

This matrix is positive definite (all the eigenvalues are positive), implying that the point (1, 1) is a minimum point.

The graphical solution, while providing a visual look of the design space, is difficult to use in locating the precise minimum. The purpose of this example is to show that the complexity of a problem increases exponentially with increasing dimension of the design space and that obtaining an analytical solution is cumbersome if not impractical.

It is difficult to formulate most engineering problems as unconstrained minimization problems. Typical engineering design problems posed in the mathematical programming format are usually of the following form:

$$\text{Find} \quad \mathbf{x} \in R^n \tag{8.1.2a}$$

$$\text{to minimize} \quad f(\mathbf{x}) \tag{8.1.2b}$$

$$\text{subject to} \quad g_i(\mathbf{x}) \le 0, \quad i = 1, 2, \ldots, l \tag{8.1.2c}$$

$$h_j(\mathbf{x}) = 0, \quad j = 1, 2, \ldots, m \tag{8.1.2d}$$

$$x_k^L \le x_k \le x_k^U, \quad k = 1, 2, \ldots, n \tag{8.1.2e}$$

Performance requirements, manufacturing constraints, or even the permissible range of values for the design variables can be specified only through *constraints*. The constraints $g_i(\mathbf{x})$ are *inequality constraints* while $h_j(\mathbf{x})$ are *equality constraints*. The constraint functions, similar to the objective function, can be linear or nonlinear, continuous, discontinuous, or piecewise continuous, differentiable, or nondifferentiable. Eq. (8.1.2e) establish the lower and upper bounds on the permissible values of the n design variables. These constraints are usually referred to as *bound constraints* or *side constraints*. A problem posed in the above form is called a *constrained minimization problem*.

An example of the design space for a two-variable constrained problem is shown in Fig. 8.1.5. Here the design variables are (x_1, x_2). There are two inequality constraints, g_1 and g_2. The side constraints are $x_1 \ge 0$ and $x_2 \ge 0$. The hatch marks on the lines and curves representing these four constraints indicate the constraint boundary marking the barrier between the *feasible* domain and the *infeasible* domain. A feasible domain contains all the design points that satisfy all the constraints. In Fig. 8.1.5, the feasible domain is bounded by these

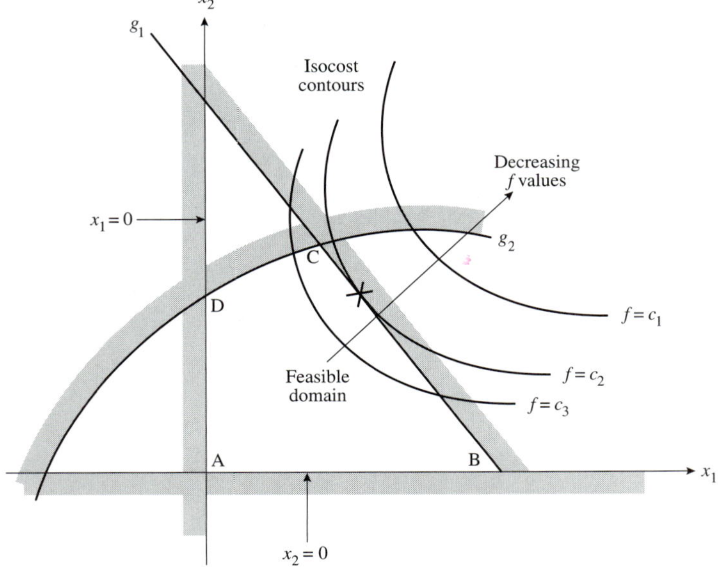

Fig. 8.1.5 Constrained design problem.

four lines. The vertices of the feasible domain are labeled A-B-C-D. The rest of the design space is infeasible. The objective function f is represented in the figure as isocost contours or curves that have a constant f value. As this example indicates, by sliding the isocost contours in the direction of decreasing objective function value, we can locate the optimal solution. The constraint g_1 controls the design and the optimal solution is indicated with an $\times$. The objective function has a value c_2 at this point.

In Chapter 3 we saw four different techniques to solve simple constrained minimization problems: exhaustive search, graphical, trial and error, and "constraint-controlled" optimal design. Exhaustive search is too expensive for solving practical problems. The graphical technique can work at most for two-variable problems. The trial-and-error approach is tedious and unlikely to find a local optimum. The "constraint controlled" approach was ad hoc. We will formalize the approach in Section 8.3, but before doing so, we look at different types of constrained minimization problems.

8.2 TYPES OF MATHEMATICAL PROGRAMMING PROBLEMS

It would be easy to state that there is one standard type of mathematical programming (MP) problem. It would then perhaps be easy to create one or more solution techniques to solve that problem. The reality is that there are numerous types of MP problems. We categorize the problems now.

Types of Design Variables. The design variables commonly encountered in engineering applications can be of different types.

Continuous design variables are those that vary continuously. For example, the height and width of a concrete beam can be taken as continuous design variables since they can, in theory, be cast in any size. Similarly, plate girders can be manufactured to any practical dimensions.

Discrete design variables are those that are available in discrete or predefined values. For example, steel I-beams are usually manufactured in predefined size and dimensions. The AISC sections are examples of such beams.

Integer design variables assume only integer values. An example of an integer design variable is the number of panel points in a roof truss. This number is a positive integer (greater than zero).

Zero-one design variables, as the name suggests, have either a zero or one value. In structural design, one can interpret the zero-one values as being equivalent to present-absent state. For example, we can designate the presence or absence of a member in a truss as being a zero-one design variable. If the design variable value is one, then the member is assumed to be a part of the truss.

Note that the properties of the design space or domain are dictated by the design variable type. With continuous design variables, the design space can be continuous. With the other design variable types, the design space is discontinuous.

Types of Functions. The functions that are used to designate the objective and the constraint functions can be of several types. Note that the independent parameters are the design variables.

The simplest function is a *linear* function.

Nonlinear functions are those where the relationship between the function and the independent variable (e.g., design variable) is nonlinear. Note that a nonlinear function is not necessarily a polynomial. For example, $f(x) = x^2 - 3x$ is nonlinear and so is $f(\mathbf{x}) = e^{x_1}/\sin(x_2)$.

Posynomial functions have a special form given by

$$f(\mathbf{x}) = C_1 x_1^{a_{11}} x_2^{a_{12}} \ldots x_n^{a_{1n}} + C_2 x_1^{a_{21}} x_2^{a_{22}} \ldots x_n^{a_{2n}} + \ldots + C_s x_1^{a_{s1}} x_2^{a_{s2}} \ldots x_n^{a_{sn}} \quad (8.2.1)$$

where $C_i > 0$, a_{ij}, is a known coefficient, and $x_i > 0$.

Types of Constraints. There are two types of constraints.

Equality constraints are used to express the relationship between two or more design variables using the equality operator. The function used to describe an equality constraint can be of any one of the forms defined earlier.

Inequality constraints are used to express the relationship between two or more design variables using a greater than, less than, greater than or equal to, and less than or equal to operators. The function used to describe an inequality constraint can be of any one of the forms defined earlier.

Types of Objective Functions. There are two major types of objective functions, single objective and multi-objective functions.

Single objective. Design problems discussed in this text are driven primarily by a single, primary objective.

Multi-objective. However, there are engineering problems in which more than one objective is important and needs to be minimized simultaneously. For example, an unconstrained minimization problem with multiple objectives can be stated as

$$\text{Find} \qquad \mathbf{x} \in R^n \tag{8.2.2}$$

$$\text{to minimize} \qquad [f_1(\mathbf{x}), f_2(\mathbf{x}), \ ..., f_r(\mathbf{x})] \tag{8.2.3}$$

An example of a multi-objective structural design problem is

Find The design variables

to minimize [project cost, construction time]

There are two objective functions—the project cost and the construction time. In this example, it should be recognized that minimizing construction time can lead to an increase in the overall project cost.

Using the different types of objective and constraint functions, and design variables, different types of mathematical programming problems can be described.

Linear Programming (LP) Problem. Consider the following problem

$$\text{Find} \qquad \{\mathbf{x}\} \tag{8.2.4a}$$

$$\text{to maximize} \qquad \sum_{k=1}^{n} c_k x_k \tag{8.2.4b}$$

$$\text{subject to} \qquad a_{i1}x_1 + a_{i2}x_2 + ... + a_{in} \le b_i \qquad i = 1, \ ..., \ l \tag{8.2.4c}$$

$$a_{i1}x_1 + a_{i2}x_2 + ... + a_{in} \ge b_i \qquad i = l+1, \ ..., \ m \tag{8.2.4d}$$

$$a_{i1}x_1 + a_{i2}x_2 + ... + a_{in} = b_i \qquad i = m+1, \ ..., \ r \tag{8.2.4e}$$

$$x_1 \ge 0, x_2 \ge 0, ..., x_n \ge 0 \tag{8.2.4f}$$

where the coefficients c_k and a_{ij} are constant coefficients and b_i are fixed real constants that are required to be nonnegative. As you can see from the above problem formulation, the objective and the constraint functions are linear functions of the design variables; hence the problem is called a linear programming problem. To solve the above problem, the problem definition is transformed so that all the constraints are equality constraints and $\mathbf{b} \ge 0$. The standard LP problem is then

Find	$\{\mathbf{x}\}$	(8.2.5a)
to maximize	$\mathbf{c}^{\mathbf{T}}\mathbf{x}$	(8.2.5b)
subject to	$\mathbf{A}_{r\times n}\mathbf{x}_{n\times 1} = \mathbf{b}_{r\times 1}$	(8.2.5c)
	$\mathbf{x} \geq \mathbf{0}$	(8.2.5d)

Plastic designs (encountered in design of steel structural systems) can, under restrictive conditions, be posed as an LP problem.

Dynamic Programming (DP) Problem. DP methodology is applicable to engineering problems that can be broken into stages and exhibit the Markovian[3] property. While not widely used in the area of structural design, one can set up a problem that could be solved effectively as a DP problem. Consider a building system consisting of a roof system, a set of floors consisting of beams and columns, and a foundation system. The load path typically starts at the roof and is finally transmitted to the foundation via the floor system, starting at the top floor and progressing to the bottommost floor. Hence one could design the building in stages, starting at the roof and progressing to the foundation.

Nonlinear Programming (NLP) Problem. As mentioned earlier in this chapter, most engineering problems require that constraints be satisfied. These constraints, and frequently the objective function, are nonlinear, leading to an NLP problem:

Find	$\mathbf{x} \in R^n$		(8.2.6a)
to minimize	$f(\mathbf{x})$		(8.2.6b)
subject to	$g_i(\mathbf{x}) \leq 0$	$i = 1, 2, \ldots, l$	(8.2.6c)
	$h_j(\mathbf{x}) = 0$	$j = 1, 2, \ldots, m$	(8.2.6d)
	$x_k^L \leq x_k \leq x_k^U$	$k = 1, 2, \ldots, n$	(8.2.6e)

Special cases of this formulation include those where the design variables are integers (*integer programming problem*), or Boolean (*zero-one programming problem*), or discrete (*discrete programming problem*). Perhaps the biggest difference between the NLP and LP problems is the likelihood of multiple solutions with NLP problems and the difficulty of finding the solution effectively.

We focus in this chapter on NLP problems. Simple NLP problems can be solved "by hand," However, most problems including structural design problems must be solved numerically.

8.3 NONLINEAR PROGRAMMING (NLP) PROBLEM

As we saw in the previous section, the nonlinear programming problem is by far the most commonly encountered structural design problem. We restate the nonlinear programming problem as:

Find	$\mathbf{x} \in R^n$	(8.3.1a)
to minimize	$f(\mathbf{x})$	(8.3.1b)

[3] A Markovian property is encompassed in a process if the decisions for optimal return at a stage in the process depend only on the current state of the system and subsequent decisions.

$$\text{subject to} \quad g_i(\mathbf{x}) \le 0, \quad i = 1, 2, ..., l \tag{8.3.1c}$$

$$h_j(\mathbf{x}) = 0, \quad j = 1, 2, ..., m \tag{8.3.1d}$$

$$x_k^L \le x_k \le x_k^U, \quad k = 1, 2, ..., n \tag{8.3.1e}$$

In addition, we assume that (i) $\mathbf{x}$ is continuous and real-valued and (ii) $f(\mathbf{x})$, $g_i(\mathbf{x})$, and $h_j(\mathbf{x})$ are continuous and differentiable. The objective is to find $\mathbf{x} = \mathbf{x}^*$ so that the objective function has the lowest possible value without violating the constraints.

Mathematical Background. In Chapter 4, we examined the concept of derivatives or gradients looking at these quantities as scalar values. We must now extend this idea into multi-dimensions. The building block is a vector. For example, in two-dimensional space a vector can be expressed as $\mathbf{g}_{2\times 1} = [g_x \ g_y]^T$. An example of vector in two dimensions is a vector of direction cosines (or unit vector) $\mathbf{g}_{2\times 1} = [0.6 \ -0.8]^T$. If $f = f(\mathbf{x})$, then the gradient vector of $f(\mathbf{x})$ is written as $\nabla \mathbf{f}(\mathbf{x})$. For example, let $f(\mathbf{x}) = x_1^2 - 2x_1x_2 + x_2^3$. Then $\partial f/\partial x_1 = 2x_1 - 2x_2$ and $\partial f/\partial x_2 = -2x_1 + 3x_2^2$, and we can write the gradient of $f(\mathbf{x})$ as

$$\nabla \mathbf{f}(\mathbf{x})_{2\times 1} = \left[2x_1 - 2x_2 \quad -2x_1 + 3x_2^2\right]^T.$$

Consider the problem of minimizing $f(\mathbf{x})$ subject to $h_j(\mathbf{x}) = 0$. A point $\mathbf{x}^*$ is a *regular point* provided $\mathbf{h}(\mathbf{x}^*) = 0$ and the gradients of all the constraints at $\mathbf{x}^*$ are linearly independent. The linear independence arises from the fact that no two gradients are parallel to each other nor is it possible to write a gradient as a linear combination of two or more of the other gradients. For example, let $h_1(x_1, x_2) = 2x_1^2 - 3x_2$ and $h_2(x_1, x_2) = -12x_1^2 + 18x_2$. Then $\nabla \mathbf{h}_1 = [4x_1 \ -3]^T$ and $\nabla \mathbf{h}_2 = [-24x_1 \ 18]^T$. Let $x^* = \{2, 1\}$. Then $\nabla \mathbf{h}_1(\mathbf{x}^*) = [4 \ -3]^T$ and $\nabla \mathbf{h}_2(\mathbf{x}^*) = [-24 \ 18]^T$. The two vectors $\nabla \mathbf{h}_1$ and $\nabla \mathbf{h}_2$ are parallel to (or linearly dependent on) each other since $\nabla \mathbf{h}_2 = -6\nabla \mathbf{h}_1$.

8.3.1 Kuhn–Tucker Conditions

Let us consider the following problem:

$$\text{Find} \quad \mathbf{x} \in R^n \tag{8.3.1.1a}$$

$$\text{to minimize} \quad f(\mathbf{x}) \tag{8.3.1.1b}$$

$$\text{subject to} \quad h_j(\mathbf{x}) = 0 \quad j = 1, 2, ..., m \tag{8.3.1.1c}$$

Lagrange is credited with developing a simple but effective way of solving the problem. He suggested that a function L (called the Lagrangian) be developed as

$$L = f(\mathbf{x}) + \sum_{j=1}^{m} \lambda_j h_j(\mathbf{x}) \tag{8.3.1.2}$$

where λ_j is called the Lagrange multiplier. The regular point $\mathbf{x}^*$ is a stationary point if

$$\frac{\partial L}{\partial x_i} = 0 = \frac{\partial f}{\partial x_i} + \sum_{j=1}^{m} \lambda_j \frac{\partial h_j(\mathbf{x})}{\partial x_i} \quad i = 1, 2, ..., n \tag{8.3.1.3}$$

$$h_j(\mathbf{x}) = 0 \quad j = 1, 2, ..., m \tag{8.3.1.4}$$

In other words, solving Eqs. (8.3.1.3) and (8.3.1.4) is equivalent to solving the original problem given by Eqs. (8.3.1.1a)–(8.3.1.1c)). Note that there are $(m + n)$ unknowns in these $(m + n)$ equations. However, the form of these equations is dependent on the form of the objective function and the equality constraints and the equations are usually not linear equations. These conditions are known as first-order Kuhn–Tucker necessary conditions. The second-order Kuhn–Tucker sufficient conditions establish whether $\mathbf{x}^*$ is a local minimum or not. Treatment of these second-order conditions is outside the scope of this text (see a book on optimal design; a comprehensive list is provided in the Bibliography).

EXAMPLE 8.3.1 ***Constrained Minimization with Equality Constraint***

Find $\{x_1, x_2\}$

to minimize $4x_1 - x_2$

subject to $2x_1^2 + x_2^2 = 1$

SOLUTION

Step 1: Form the Lagrangian, we find

$$L = 4x_1 - x_2 + \lambda_1\left(2x_1^2 + x_2^2 - 1\right)$$

Step 2: Using Eq. (8.3.1.3), we have

$$\frac{\partial L}{\partial x_1} = 0 = 4 - 4\lambda_1 x_1 \Rightarrow x_1 = \frac{1}{\lambda_1}$$

$$\frac{\partial L}{\partial x_1} = 0 = -1 + 2\lambda_1 x_2 \Rightarrow x_2 = \frac{1}{2\lambda_1}$$

Substituting in Eq. (8.3.1.4) or the constraint equation, we have

$$2\left(\frac{1}{\lambda_1}\right)^2 + \left(\frac{1}{2\lambda_1}\right)^2 = 1$$

Solving, yields $\lambda_1 = \pm(3/2)$. For each root, we have

$$\lambda_1 = \frac{3}{2} : x_1^* = \frac{2}{3} \quad x_2^* = \frac{1}{3}; \; f(\mathbf{x}^*) = \frac{7}{3}$$

$$\lambda_1 = -\frac{3}{2} : x_1^* = -\frac{2}{3}; \; x_2^* = -\frac{1}{3}; \; f(\mathbf{x}^*) = -\frac{7}{3}$$

Obviously the minimum is at $x_1^* = -2/3$, $x_2^* = -1/3$, and $f(\mathbf{x}^*) = -\frac{7}{3}$.

Now let us consider the following problem with equality and inequality constraints:

Find $\mathbf{x} \in R^n$ (8.3.1.5a)

to minimize $f(\mathbf{x})$ (8.3.1.5b)

subject to $h_j(\mathbf{x}) = 0, \quad j = 1, 2, \ldots, m$ (8.3.1.5c)

$g_i(\mathbf{x}) \le 0, \quad i = 1, 2, \ldots, l$ (8.3.1.5d)

The Lagrangian function for this problem can be defined as

$$L = f(\mathbf{x}) + \sum_{j=1}^{m} \lambda_j h_j(\mathbf{x}) + \sum_{i=1}^{l} \mu_i g_i(\mathbf{x}) \tag{8.3.1.6}$$

where λ_j and μ_i are the Lagrange multipliers. The regular point $\mathbf{x}^*$ is a stationary point if

$$\frac{\partial L}{\partial x_k} = 0 = \frac{\partial f}{\partial x_k} + \sum_{j=1}^{m} \lambda_j \frac{\partial h_j(\mathbf{x})}{\partial x_k} + \sum_{i=1}^{l} \mu_i \frac{\partial g_i(\mathbf{x})}{\partial x_k} \qquad k = 1, 2, \ldots, n \tag{8.3.1.7}$$

$$h_j(\mathbf{x}) = 0 \qquad j = 1,\ 2,\ \ldots,\ m \tag{8.3.1.8}$$

$$g_i(\mathbf{x}) \le 0 \qquad i = 1,\ 2,\ \ldots,\ l \tag{8.3.1.9}$$

$$\mu_i g_i(\mathbf{x}) = 0 \qquad i = 1,\ 2,\ \ldots,\ l \tag{8.3.1.10}$$

$$\mu_i \ge 0 \qquad i = 1,\ 2,\ \ldots,\ l \tag{8.3.1.11}$$

In other words, solving Eqs. (8.3.1.7)–(8.3.1.11) is equivalent to solving the original problem given by Eqs. (8.3.1.5a)–(8.3.1.5d). We illustrate the usage of these conditions with an example.

EXAMPLE 8.3.2 ***Constrained Minimization with Inequality Constraint***

Find $\{x_1, x_2\}$

to minimize $(x_1 - 3)^2 + (x_2 - 4)^2$

subject to $x_1 + x_2 - 5 \le 0$

$x_1 \ge 0, x_2 \ge 0$

SOLUTION

Step 1: We do not use the last two constraints requiring that the design variable be positive but will use them to select the optimal solution. Form the Lagrangian as

$$L = (x_1 - 3)^2 + (x_2 - 4)^2 + \mu_1(x_1 + x_2 - 5)$$

Step 2: Hence, the necessary conditions to be satisfied are

$$\frac{\partial L}{\partial x_1} = 0 = 2x_1 - 6 + \mu_1 \tag{1}$$

$$\frac{\partial L}{\partial x_2} = 0 = 2x_2 - 8 + \mu_1 \tag{2}$$

$$\mu_1(x_1 + x_2 - 5) = 0 \tag{3}$$

$$x_1 + x_2 - 5 \le 0 \tag{4}$$

$$\mu_1 \ge 0 \tag{5}$$

Step 3: The key to solving these conditions is to recognize the importance of Eq. (3). Equation (8.3.1.10) represents the switching conditions since from those equations either $\mu_i = 0$ or $g_i = 0$. Using the switching conditions, the possibilities for this problem are discussed below.

Case 1: $\mu_1 = 0$ (meaning that the inequality constraint g_1 is not active). From (1) and (2), $2x_1 - 6 = 0$ and $2x_2 - 8 = 0$, yielding $x_1 = 3$ and $x_2 = 4$. However, this solution does not satisfy Eq. (4). Hence this is an unacceptable case.

Case 2: $g_1 = 0$ (meaning that the inequality constraint g_1 is active). Hence,

$$2x_1 - 6 + \mu_1 = 0$$

$$2x_2 - 8 + \mu_1 = 0$$

$$x_1 + x_2 - 5 = 0$$

Solving these linear simultaneous equations gives us $x_1 = 2$, $x_2 = 3$, and $\mu_1 = 2$. These values satisfy Eqs. (4), (5), and the requirements that $x_1 \ge 0$, $x_2 \ge 0$. Hence $\mathbf{x}^* = \{2, 3\}$. For these values, the objective function is $f(\mathbf{x}^*) = 2$.

EXAMPLE 8.3.3 ***Constrained Minimization with Inequality Constraints***

Find $\{x_1, x_2\}$

to maximize $-4x_1^2 - 2x_2$

subject to $-2x_1 + x_2 \le 4$

$x_1 + 2x_2 \ge 2$

$x_1 \ge 0, x_2 \ge 0$

SOLUTION

Step 1: The first thing to notice about the problem is that it is not in the standard form. We must first convert the problem to a minimization problem. Maximizing $f(\mathbf{x})$ is equivalent to minimizing $-f(\mathbf{x})$. Hence $f(\mathbf{x}) = 4x_1^2 + 2x_2$. Second, the second constraint must be transformed to $a \le 0$ constraint. Note that $g(\mathbf{x}) \ge a$ is equivalent to $-g(\mathbf{x}) \le -a$. Hence $g_2 = -x_1 - 2x_2 + 2 \le 0$. As in the previous problem, we do not use the last two constraints requiring that the design variable be positive, but will use them to select the optimal solution. Form the Lagrangian as

$$L = 4x_1^2 + 2x_2 + \mu_1(-2x_1 + x_2 - 4) + \mu_2(-x_1 - 2x_2 + 2)$$

Step 2: Hence, the necessary conditions to be satisfied are

$$\frac{\partial L}{\partial x_1} = 0 = 8x_1 - 2\mu_1 x_1 - \mu_2 \tag{1}$$

$$\frac{\partial L}{\partial x_2} = 0 = 2 + \mu_1 - 2\mu_2 \tag{2}$$

$$\mu_1(-2x_1 + x_2 - 4) = 0 \tag{3}$$

$$\mu_2(-x_1 - 2x_2 + 2) = 0 \tag{4}$$

$$-2x_1 + x_2 - 4 \le 0 \tag{5}$$

$$-x_1 - 2x_2 + 2 \le 0 \tag{6}$$

$$\mu_1 \ge 0,\ \mu_2 \ge 0 \tag{7}$$

Step 3: Using the switching conditions, there are four possibilities.

Case 1: $\mu_1 > 0$, $\mu_2 > 0$ (meaning that the inequality constraints are active, $g_1 = 0$ and $g_2 = 0$). Hence

$$-2x_1 + x_2 - 4 = 0$$

$$-x_1 - 2x_2 + 2 = 0$$

$$8x_1 - 2\mu_1 x_1 - \mu_2 = 0$$

$$2 + \mu_1 - 2\mu_2 = 0$$

From the first two equations, $x_1 = -6/5$. This is an unacceptable solution.

Case 2: $\mu_1 > 0$, $\mu_2 = 0$ (meaning that the inequality constraint $g_1 = 0$). Hence

$$-2x_1 + x_2 - 4 = 0$$

$$8x_1 - 2\mu_1 x_1 = 0$$

$$2+\mu_1=0$$

From the last equation, $\mu_1 = -2 < 0$, and is an unacceptable solution.

Case 3: $\mu_1 = 0$, $\mu_2 = 0$ (meaning that the inequality constraints are inactive). Hence

$$8x_1=0$$

$$2=0$$

The second equation expresses an invalid condition, and the solution is unacceptable.

Case 4: $\mu_1 = 0$, $\mu_2 > 0$ (meaning that the inequality constraint $g_2 = 0$). Hence

$$-x_1-2x_2+2=0$$

$$8x_1-\mu_2=0$$

$$2-2\mu_2=0$$

Solving, we have $x_1 = 1/8$, $x_2 = 15/16$, and $\mu_2 = 1$. The constraint g_1 is satisfied. Hence $\mathbf{x}^* = \{(1/8), (15/16)\}$ and $f(\mathbf{x}^*) = 1.9375$.

The two examples above illustrate the strengths and weaknesses of the Kuhn–Tucker (K–T) conditions. Now, a few notes about the first-order K–T conditions.

a. A regular point is a candidate K–T point. In other words, a nonregular point cannot be used to satisfy the K–T conditions.
b. Points that satisfy K–T conditions are local minimum points. However, all local minimum points are not K–T points. In other words, K–T conditions may not be able to locate all local minimum points.
c. It may not be possible to obtain an analytical solution to K–T conditions. This is because the resulting equations may be nonlinear equations requiring a numerical method.

EXERCISES

Appetizers

8.3.1. An aluminum can manufacturer needs to design a (closed) cylindrical soft drink can. The volume of the can must be 400 ml. The height h of the can must be at least twice the diameter d and cannot be more than three times the diameter. Packaging considerations restrict the height to 25 cm and require the use of the smallest amount of sheet metal. Formulate the optimal design problem by clearly identifying the design variables, objective function, and constraints.

8.3.2. A box frame is made of steel angle and hollow tube sections as shown in Fig. P8.3.2. It must be designed for least cost. The angle sections cost \$200/m and the tube sections cost \$500/m. The box frame must enclose a space of 1000 m^3. A side of the box cannot be less than 2 m. Formulate the optimal design problem by clearly identifying the design variables, objective function, and constraints.

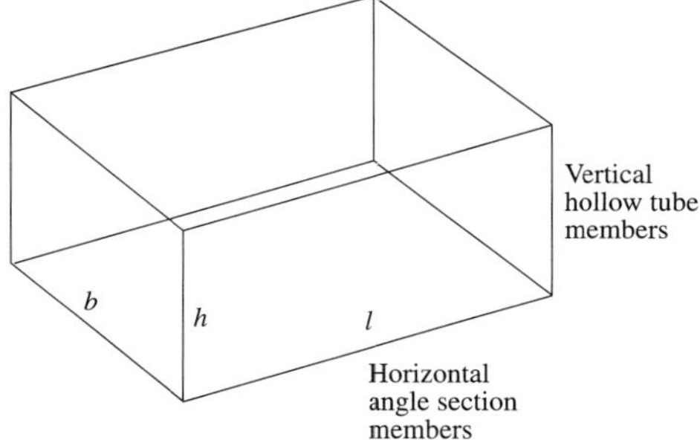

Fig. P8.3.2

8.3.3. A factory makes two products, A and B. The manufacturing of these products takes place in two stages, Stage One and Stage Two. Departments 1 and 2 handle Stage One and Stage Two respectively. Product A requires 2 hours for Stage One and 2.25 hours for Stage Two. Product B requires 2 hours for Stage One and 1.75 hours for Stage Two. Each of the two departments can be operational for a total of 20 hours per day, seven days a week. The profit from sale of product A is \$1.35 per unit and for product B is \$1.20 per unit. How many units of A and B should be produced per week so as to maximize the profit?

Main Course

8.3.4. Solve Problem 8.3.1 using the graphical technique. Verify your answer using K–T conditions.

8.3.5. Solve Problem 8.3.2 using the K–T conditions.

8.3.6. Solve Problem 8.3.3 using the graphical technique. Verify your answer using K–T conditions.

Structural Concepts

8.3.7. Answer True or False. If False, state the reason(s) why.

(a) A feasible design is the one with the lowest objective function value.

(b) An optimal design problem must have constraints.

(c) The number of inequality constraints in a problem cannot be greater than the number of design variables for a problem to be well posed.

(d) Every optimal design problem has one or more optimum solutions.

(e) An inequality constraint $h(\mathbf{x}) = 0$ can be expressed as two inequality constraints, $h(\mathbf{x}) \leq 0$ and $h(\mathbf{x}) \geq 0$.

8.3.2 Numerical Solution Techniques

As you may have seen in a course in numerical analysis, problems such as root finding or the solution of linear algebraic equations can be carried out by more than one solution technique. Different techniques have different assumptions, restrictions, strengths, and weaknesses. Similarly, several numerical techniques can potentially be used to solve NLP problems. They are broadly classified as either gradient-based techniques or direct-search techniques.

Gradient-Based Techniques. The nonlinear nature of the problem makes it difficult to solve directly. Instead, the solution is obtained iteratively. The original nonlinear problem is transformed into a more manageable subproblem whose solution can be obtained numerically. Solution of these subproblems requires not only that the function values be known but also that their gradients (or derivatives) with respect to the design variables be known. Some of the popular solution techniques are the sequential linear programming (SLP) technique, the sequential quadratic programming (SQP) technique, the feasible directions method, the generalized reduced gradient (GRG) method, the augmented Lagrangian technique, the sequential unconstrained minimization technique (SUMT), etc. Treatment of one or more of these techniques can be found in any book on optimization techniques.

Direct Search Techniques. The direct search techniques do not require derivatives. Hence they have the advantage of being able not only to solve problems where the derivatives are discontinuous but also to find the global minimum. This advantage is offset by an increase in computational requirement—usually the function values are required at a very large number of locations in the design space. Some of the popular solution techniques are the Hooke and Jeeves Method, Powell's method of conjugate directions, simulated annealing (SA), genetic algorithm (GA), etc.

In the next section we look at the genetic algorithm in some detail to understand how to use the method for solving constrained minimization problems.

8.4 GENETIC ALGORITHM

The genetic algorithm (GA) is a search strategy based on the rules of natural genetic evolution. Even before the traits of genetic systems were used in solving optimization problems, biologists had used digital computers to perform simulations of genetic system by the early 1950s. The application of genetic algorithms for adaptive systems was first proposed by John Holland (University of Michigan) in 1962, and the term "genetic algorithms" was first used in his student's dissertation.

GAs have been used to solve a variety of structural design problems, including the optimal design of all the structural systems we have seen in this text and more. Because of their discrete nature, GAs lend themselves well to the process of automating the design of skeletal structures. GAs do not require gradient or derivative information. For this reason alone, they have been applied by researchers to solve discrete, nondifferentiable, combinatory, and global optimization engineering problems, such as transient optimization of gas pipeline, topology design of general elastic mechanical system, time scheduling, circuit layout design, composite panel design, pipe network optimization, and several hundred others. GAs are recognized as different from traditional gradient-based optimization techniques in the following four major ways [Goldberg, 1989]:

1. GAs work with a coding of the design variables and parameters in the problem, rather than with the actual parameters themselves.
2. GAs make use of population-type search. Many different design points are evaluated during each iteration instead of sequentially moving from one point to the next.
3. GAs need only a fitness or objective function value. No derivatives or gradients are necessary.
4. GAs use probabilistic transition rules to find new design points for exploration rather than using deterministic rules based on gradient information to find these new points.

The idea behind GA is to simulate the behavior of natural evolutionary selection. Although there exist many different variations of GAs, the basic structure is that shown in Fig. 8.4.1. We explain the basic flow in Fig. 8.4.1 next.

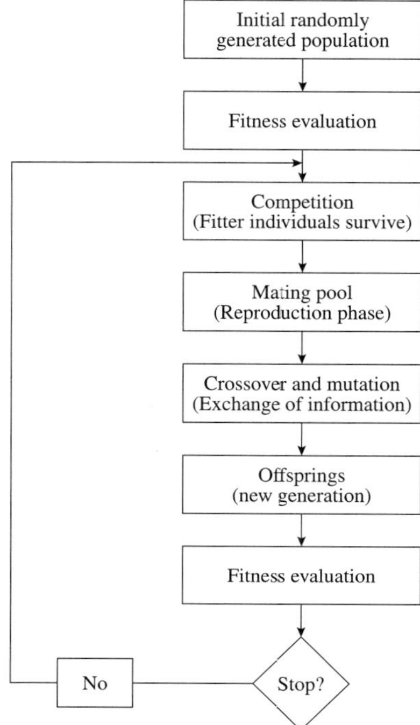

Fig. 8.4.1 Flow in a simple genetic algorithm (SGA).

8.4.1 The Basic Algorithm

The genetic algorithm is used to solve the following problem.

$$\text{to minimize} \quad \hat{f}(\mathbf{x}) \tag{8.4.1.1a}$$

$$\text{subject to} \quad x_k^L \le x_k \le x_k^U, \qquad k = 1, 2, \ldots, n \tag{8.4.1.1b}$$

Note that the problem is primarily an unconstrained minimization problem with lower and upper bounds on the design variables. We first present the necessary background before detailing the algorithm. In the language of GA, we are computing $\hat{f}(\mathbf{x})$, the fitness function, not $f(\mathbf{x})$, the objective function. The two functions are related and the distinction between the two will be made later.

Binary Encoding and Decoding of Design Variables

Binary encoding is the most popular way of encoding the design variables. A binary number is represented as $(b_m \ldots b_1 b_0)_2$ where b_i is either 0 or 1. For example, $(101)_2$ is a three-digit or three-bit binary number. To understand this concept we first explain the relationship between binary and decimal numbers.

$$(b_m \ldots b_1 b_0)_2 = (2^0 b_0 + 2^1 b_1 + \ \ldots \ + 2^m b_m)_{10} \tag{8.4.1.2}$$

Hence, as an example

$$(101)_2 = \left(2^0 \cdot 1 + 2^1 \cdot 0 + 2^2 \cdot 1\right)_{10} = (5)_{10}$$

In other words, the binary number $(101)_2$ is equal to decimal 5.

The process of taking a decimal number and constructing its binary representation (not value) is called encoding. Decoding is the inverse process of taking the binary encoded value and constructing its decimal equivalent.

Continuous Design Variables. A design variable x_i is between x_i^L and x_i^U. Note that x_i is a decimal number. If m bits are available to represent x_i, then the precision p_i with which the number is represented is given by

$$p_i = \frac{x_i^U - x_i^L}{2^m - 1} \tag{8.4.1.3}$$

To understand the term in the denominator, let us look at the example with 3 bits. The possible binary representations with 3 bits are 000, 001, 010, 011, 100, 101, 110, and 111, or 8 possible combinations. Similarly, with 4 bits we have 0000, 0001, 0010, 0011, 0100, 0101, 0110, 0111, 1000, 1001, 1010, 1011, 1100, 1101, 1110, and 1111, or 16 possible combinations. In other words, if there are m bits then there are 2^m combinations or $2^m - 1$ intervals.

The range of values between x_i^L and x_i^U is divided into $2^m - 1$ intervals. For example, if $x^L = 1$, $x^U = 8$, and $m = 3$, then $p = (8 - 1)/7 = 1$. The following table shows the relationship between the binary representation and their decimal equivalents with this example.

Binary representation $(b_2 b_1 b_0)_2$	Decimal equivalent $x = x^L + p(b_2 b_1 b_0)_2$
000	1.0
001	2.0
010	3.0
011	4.0
100	5.0
101	6.0
110	7.0
111	8.0

The decoding is achieved using

$$x = x^L + p(b_m \dots b_1 b_0)_2 \tag{8.4.1.4}$$

where decimal values are used. The previous example is quite straightforward since the bounds and the precision are integers. Now consider the following example. Let $x^L = 1$, $x^U = 10$ and $m = 3$. Then $p = (10 - 1)/7 = 1.28571$ if we assume that the precision is finite. The new relationship is as follows.

Binary representation $(b_2b_1b_0)_2$	Decimal equivalent $x = x^L + p(b_2b_1b_0)_2$
000	1.0
001	2.25571
010	3.57143
011	4.85714
100	6.14286
101	7.42857
110	8.71429
111	10.0

Integer Design Variable. The above example illustrates the encoding and decoding problems when the range is not a multiple of $2^m - 1$. With integer design variables, one approach is to apply Eq. (8.4.1.3) with the precision p being 1 and compute the smallest number of bits required to achieve the precision. The number of bits obviously is an integer. For example, let $x^L = 1$, $x^U = 10$, and $p = 1$. Using

$$2^m - 1 = \frac{x_i^U - x_i^L}{p} \Rightarrow m = \log\left(x_i^U - x_i^L + 1\right)\Big/\log(2) \tag{8.4.1.5}$$

we have $m = \log(10 - 1 + 1)/\log(2) = 3.32193$. We round 3.32193 to the next highest integer, 4. Hence, the new precision with 4 bits is $p = (10 - 1)/(2^4 - 1) = 0.6$. Once we compute the decimal equivalent, we can either truncate or round the value to an integer.

Binary representation $(b_3b_2b_1b_0)_2$	Decimal equivalent $x = x^L + p(b_3b_2b_1b_0)_2$	Integer equivalent (rounded value)
0000	1.0	1
0001	1.6	2
0010	2.2	2
0011	2.8	3
0100	3.4	3
0101	4.0	4
0110	4.6	5
0111	5.2	5
1000	5.8	6
1001	6.4	6
1010	7.0	7
1011	7.6	8
1100	8.2	8
1101	8.8	9
1110	9.4	9
1111	10.0	10

While problems exist with the procedure (note that the integers 2, 3, 5, 6, 8, 9 appear more than once), experience has shown that the procedure works quite well for most problems.

Discrete Design Variable. The representation is similar to integer design variables with $x^L = 1$ and $x^U = q$ where there are q possible discrete values. The discrete values are usually stored in a table in some sorted manner and the integer value between 1 and q is used as an index to obtain the corresponding value(s) from the table.

Zero-One (Binary) Design Variable. Nothing special needs to be done since we need exactly one bit to represent a zero-one design variable.

Chromosome. To represent all the design variables in a problem, we need to create the chromosome for the problem. A chromosome is a concatenated binary string of all the binary representations of the design variables. If there are n design variables with $m = 3$ to represent each design variable, then the chromosome looks as shown in Fig. 8.4.1.1, where x is 0 or 1. The number of bits need not be equal for all the design variables, nor need the design variables be ordered from 1 to n in the chromosome.

xxx xxx xxx xxx
x_1 x_2 x_3 x_n

Fig. 8.4.1.1 Possible chromosome (or gene).

The basic steps in the GA algorithm are discussed next.

Initial Population. The first step is to create the initial population. Unlike gradient-based methods, where the search for the optimal solution takes place by moving from one point to the next, in a GA the traits of a population (of members) are used to move from one generation to the next. Figure 8.4.1.2 shows an initial population consisting of z members. The initial population is usually created randomly.

Member 1 xxx xxx xxx xxx
x_1 x_2 x_3 x_n

Member 2 xxx xxx xxx xxx
x_1 x_2 x_3 x_n

......

Member z xxx xxx xxx xxx
x_1 x_2 x_3 x_n

Fig. 8.4.1.2 Initial population.

With the example in Fig. 8.4.1.2, the size of the chromosome is $3n$ bits. A random number generator can be used to generate a random number between 0.0 and 1.0. Invoking the random number generator $3n$ times, we can generate each member of the population as follows—if the random number is ≤ 0.5 then a 0 is assigned to that bit, otherwise if the number is > 0.5 a 1 is assigned to that bit.

We study the effect of the size of the population z at the end of this section.

Fitness Evaluation

Once the initial population is generated, the actual search process starts. The chromosome is decoded to obtain the values of the design variables x and the fitness function value is computed for each member of the population. In other words, z fitness values $\hat{f}(\mathbf{x})$ are calculated.

Reproduction

To generate the members of the next generation, the reproduction phase has at least three distinct steps. First the mating pool is created. Typically, the weaker members (higher fitness values) are replaced by stronger members (lower fitness values). To produce offspring, two members from the mating pool are selected and a crossover operation is carried out to create the chromosome of the offspring. Finally, to bring diversity into the population, the mutation operation is carried out.

Mating Pool. The mating pool is constructed by selecting members from the population. We describe two commonly used methods. In *roulette-wheel selection*, the chance of being selected is based on the fitness value. The individual members of the population are mapped to segments of a line such that the length of the segment is related to its fitness value. Consider the example shown below, where the population size is 13.

Individual	1	2	3	4	5	6	7	8	9	10	11	12	13	Sum
Fitness	1.20	1.50	1.70	2.20	2.50	2.70	3.90	4.50	4.70	5.20	5.50	6.10	7.80	49.50
Scaled Fitness	41.25	33.00	29.12	22.50	19.80	18.33	12.69	11.00	10.53	9.52	9.00	8.11	6.35	231.21
Selection Probability	0.18	0.14	0.13	0.10	0.09	0.08	0.05	0.05	0.05	0.04	0.04	0.04	0.03	1.00
Cumulative Value	0.18	0.32	0.45	0.54	0.63	0.71	0.76	0.81	0.86	0.90	0.94	0.97	1.00	

Once the fitness values are computed, the other required values can be computed. The sum of the fitness values is $S = \sum_{i=1}^{13} \hat{f}_i = 49.5$. A scaled fitness value[4] is created as $\hat{f}_{is} = \dfrac{S}{\hat{f}_i}$. Let $S_S = \sum_{i=1}^{13} \hat{f}_{is} = 231.21$. The selection probability $p_i = \dfrac{\hat{f}_{is}}{S_S}$. As can be seen from the table above, the length of the segment is greater for lower fitness values than for larger fitness values. For selecting the individual into the mating pool, a random number between 0 and 1.0 is generated. For example, if 7 random numbers are generated as {0.79, 0.10, 0.33, 0.01, 0.99, 0.51, 0.83}, then the individuals selected are (8, 1, 3, 1, 13, 4, 9).

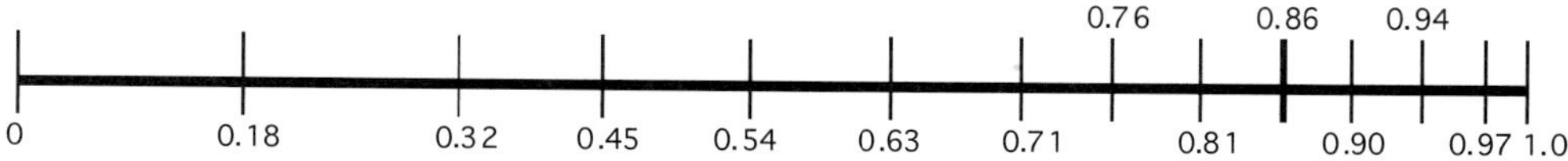

In *tournament selection*, using a random number generator, two members of the population are selected. Their fitness values are compared head-to-head and the one with the lower fitness value is put into the mating pool. This is done z times to create the mating pool of size z. In a "double elimination" tournament selection method, all the individuals in the population are placed in a bag. Two individuals are chosen at random. Their fitness values are compared head-to-head and the one with the lower fitness value is put into the mating pool. These two individuals are then eliminated from the bag and the process is repeated until the bag is empty. This will occur when the mating pool is half full. To complete the mating pool, the process is repeated once again.

In a simple GA, once the mating pool is constructed, two parents are selected and the reproduction process is carried out using the crossover and mutation operators.

Crossover. There are several types of crossover operators. We illustrate the three most commonly discussed.

[4] To generate a general procedure the fitness values are transformed, if necessary, so that all the values are greater than zero.

One-point crossover. Consider two chromosomes selected randomly from the mating pool. They are labeled Parent 1 and Parent 2 in Fig. 8.4.1.3. Based on a predetermined probability, a single crossover point in chosen. If the length of the chromosome is n_c bits, then a random number is generated between 1 and n_c and used as the crossover point. Two offspring are formed and they become the part of the next generation. The first offspring is formed by taking the front or left section of Parent 1 and the rear or right section of Parent 2. The second offspring is formed by taking the front or left section of Parent 2 and the rear or right section of Parent 1. The results are shown in Fig. 8.4.1.4.

Parent 1 1 0 0 0 1 0 0 1
Parent 2 0 0 1 1 0 1 1 1

Fig. 8.4.1.3 Parents selected for the crossover operation.

Parent 1 1 0 0 | 0 1 0 0 1
Parent 2 0 0 1 | 1 0 1 1 1

Offspring 1 1 0 0 | 1 0 1 1 1
Offspring 2 0 0 1 | 0 1 0 0 1

Fig. 8.4.1.4 Offspring resulting from one-point crossover operation occurring at location 3.

Two-point crossover. The idea of the single-point crossover can be extended to include multi-point crossover locations. The section between the first variable and the first crossover point is not exchanged. However, the bits between every other successive crossover point are exchanged between the two parents. This process is illustrated with the two-point crossover example in Fig. 8.4.1.5.

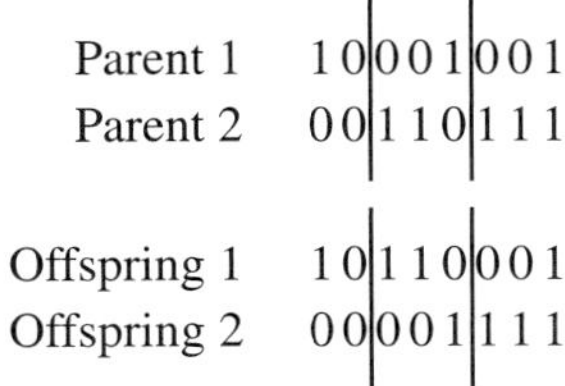

Fig. 8.4.1.5 Offspring resulting from two-point crossover operation occurring at locations 2 and 5.

Uniform crossover. In uniform crossover, every location is a potential crossover point. First, a crossover mask is created randomly. This mask has the same length as the chromosome and the bit value (parity) is used to select which parent will supply the offspring with the bit. If the mask value is 0 then the bit is taken from the first parent; otherwise the bit is taken from parent 2. If two offspring are needed, the mask is used with the parents to create the first offspring and the inverse of the mask is used to create the second offspring (see Fig. 8.4.1.6).

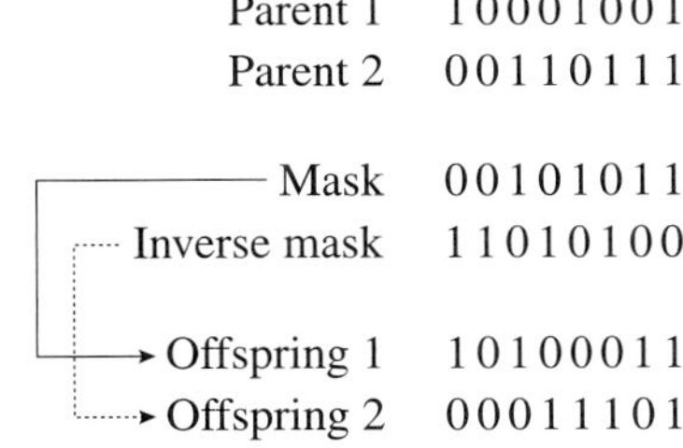

Fig. 8.4.1.6 Example showing uniform crossover

Mutation. This operator occurs much less frequently both in nature and in GA. Offspring variables are mutated by small random changes with a low probability. The basic idea is to introduce some diversity into the population—in other words, to delay the situation in which all the population becomes so homogenous that no further improvement is possible. If the length of the chromosome is n_c bits, then a random number is generated between 1 and n_c. The bit at that location is switched. An example is shown in Fig. 8.4.1.7.

Fig. 8.4.1.7
Example showing mutation taking place at location 4.

Before	1 0 0 1 0 1 1 1
After	1 0 0 0 0 1 1 1

Next Generation

The new generation is formed when sufficient offspring are generated in the reproduction phase. The whole process of fitness evaluation and reproduction starts all over again with this new population. Obviously, somewhere in the evolutionary procedure the iterative process is stopped. Typically this is done if a predetermined number of iterations have been completed or if the fitness function does not change appreciably. Unlike most gradient-based techniques, there is no convergence criterion for the iterative process associated with the GA.

8.4.2 Problem Formulation

GAs were developed to tackle unconstrained optimization problems. However, as mentioned before, most engineering and structural design problems are constrained optimization problems. The standard approach is to transform the original constrained problem to an unconstrained problem as follows.

$$\begin{aligned} &\text{Find} && x \\ &\text{to minimize} && \hat{f}(\mathbf{x}) = f(\mathbf{x}) + \sum_{i=1}^{l} c_i \cdot \max(0,\ g_i) + \sum_{j=1}^{m} c_j \cdot \left|h_j\right| \qquad (8.4.2.1) \\ &\text{subject to} && x_k^L \le x_k \le x_k^U, \qquad k = 1,\ 2,\ \ldots,\ n \end{aligned}$$

where c_i and c_j are penalty parameters. The selection of appropriate penalty weights c_i and c_j is always problematic even in traditional NLP schemes. Typically, a large value is used initially for these parameters. These values are then reduced as the design iterations continue. This feature is implemented in the UNDO program.

8.5 DESIGN EXAMPLES

In this section we solve several design problems using the genetic algorithm as implemented in the UNDO program.[5] Default values are chosen for the parameters associated with the GA with the option of changing these values if required. The GA parameters under the control of the user are shown in Fig. 8.5.1.

---

[5] The UNDO© Tutorial and User's Manual are in the *manual* subdirectory of the directory in which the program is installed.

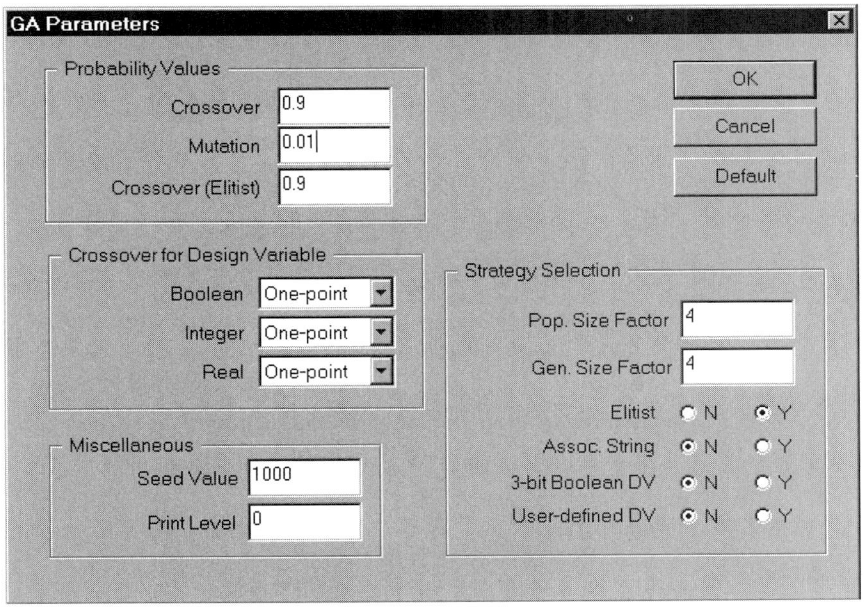

Fig. 8.5.1
User-controllable GA parameters.

Probability Values

Crossover: The probability value used to determine if crossover should or should not take place.

Mutation: The probability value used to determine if mutation should or should not take place.

Crossover (elitist): The probability value used to determine if crossover should or should not take place if the elitist strategy is used.

Crossover for Design Variables

There are three types of crossovers—one-point, two-point, and uniform.

Boolean: Type of crossover for Boolean (zero-one) design variables.

Integer (including discrete): Type of crossover for integer design variables.

Real: Type of crossover for floating (or real/continuous) design variables.

Strategy Selection

Elitist: The best member of current population is carried over to the next generation.

Association string: The program internally tries to find the association between different design variables by using an association string.

3-Bit design variables: Three bits are used to store the representation of Boolean design variables.

User-defined DV: Initial guess provided by the user is used to create a member of the initial population.

Miscellaneous Values

Seed value: Seed value used in random number generation.

Print code: Use a nonzero value so that the program creates output text files (GA_optm.out and GA_optm.hst) that contain additional GA-related output.

We present a few guidelines for formulating and solving problems using GAs.

1. It is a good idea to start solving a problem with as few design variables as possible. It is easier to debug the problem formulation with a manageable number of design variables.
2. The selection of the lower and upper bounds must be done with care. It is necessary to have some prior knowledge of the possible range of values the design variables can assume. One approach is to start with a wide range and obtain the solution. Once a solution is obtained, one can reduce the range by increasing the lower bound or decreasing the upper bound or both.
3. The penalty approach to handling constrained optimization problems works best if the constraints are normalized. For example, consider a problem where $0 \le x_1 \le 10$ and $-10 \le x_2 \le 5$. Instead of writing the following two constraints as

$$g_1(\mathbf{x}) = x_1^2 - 4x_2^3 + 12000 \le 0$$

$$g_2(\mathbf{x}) = 40x_1x_2^2 - \frac{1000}{x_2 + 20} \le 0$$

one can rewrite them as

$$g_1(\mathbf{x}) = \frac{x_1^2 - 4x_2^3}{12000} + 1 \le 0$$

$$g_2(\mathbf{x}) = \frac{\left(x_1x_2^2\right)\left(x_2 + 20\right)}{10^4} - \frac{1}{400} \le 0$$

The basic idea is to avoid very large positive and negative values.

4. Avoid using equality constraints. More often than not, equality constraints can be rewritten expressing one design variable in terms of the others. In other words, a design variable can be eliminated from the problem. Consider a problem where an equality constraint is

$$24x_3 - 4x_4 + 36 = 0$$

The constraint can be rewritten as

$$x_4 = 6x_3 + 9$$

and x_4 can be eliminated as a problem parameter.

5. Changing the default GA parameters: When the GA leads to an infeasible or unsatisfactory solution, it may be worthwhile changing the default values of the GA parameters.

EXAMPLE 8.5.1 ***Constrained Minimization (Box Design Problem)***

Find the optimal solution to the following problem.

Find $\{x_1, x_2, x_3\}$

to minimize $f(\mathbf{x}) = -x_1x_2x_3$

subject to $g_1(\mathbf{x}) \equiv 42 - x_1 \ge 0$

$$g_2(\mathbf{x}) \equiv 42 - x_2 \ge 0$$

$$g_3(\mathbf{x}) \equiv 42 - x_3 \ge 0$$

$$g_4(\mathbf{x}) \equiv x_1 + 2x_2 + 2x_3 \geq 0$$

$$g_5(\mathbf{x}) \equiv 72 - (x_1 + 2x_2 + 2x_3) \geq 0$$

$$0 \leq x_i \leq 100 \qquad i = 1, 2, 3$$

SOLUTION

Step 1: We rewrite the problem as follows, normalizing the constraints.

Find $\{x_1,\ x_2,\ x_3\}$

to minimize $f(\mathbf{x}) = -x_1 x_2 x_3$

subject to

$$g_1(\mathbf{x}) \equiv 1 - x_1/42 \geq 0$$

$$g_2(\mathbf{x}) \equiv 1 - x_2/42 \geq 0$$

$$g_3(\mathbf{x}) \equiv 1 - x_3/42 \geq 0$$

$$g_4(\mathbf{x}) \equiv \frac{x_1 + 2x_2 + 2x_3}{500} \geq 0$$

$$g_5(\mathbf{x}) \equiv 1 - \frac{(x_1 + 2x_2 + 2x_3)}{72} \geq 0$$

$$0 \leq x_i \leq 100 \qquad i = 1, 2, 3$$

Step 2: Based on the problem formulation, the following variables and functions are necessary.

Variable name	Remarks
x1	Design variable x1
x2	Design variable x2
x3	Design variable x3

Function expression	Remarks
–x1*x2*x3	Objective function
1–x1/42	Constraint g1
1–x2/42	Constraint g2
1–x3/42	Constraint g3
(x1+2*x2+2*x3)/500	Constraint g4
1–(x1+2*x2+2*x3)/72	Constraint g5

Design variable	Lower bound	Upper bound	Precision
x1	0	100	Auto
x2	0	100	Auto
x3	0	100	Auto

Step 3: The result of executing the GA option in the UNDO program is shown below after 70 generations.

The solution obtained is $\mathbf{x} = \{9.29, 28.5, 2.25\}$. To see whether we can obtain a better solution, we reduce the upper bound of all the design variables to 30.

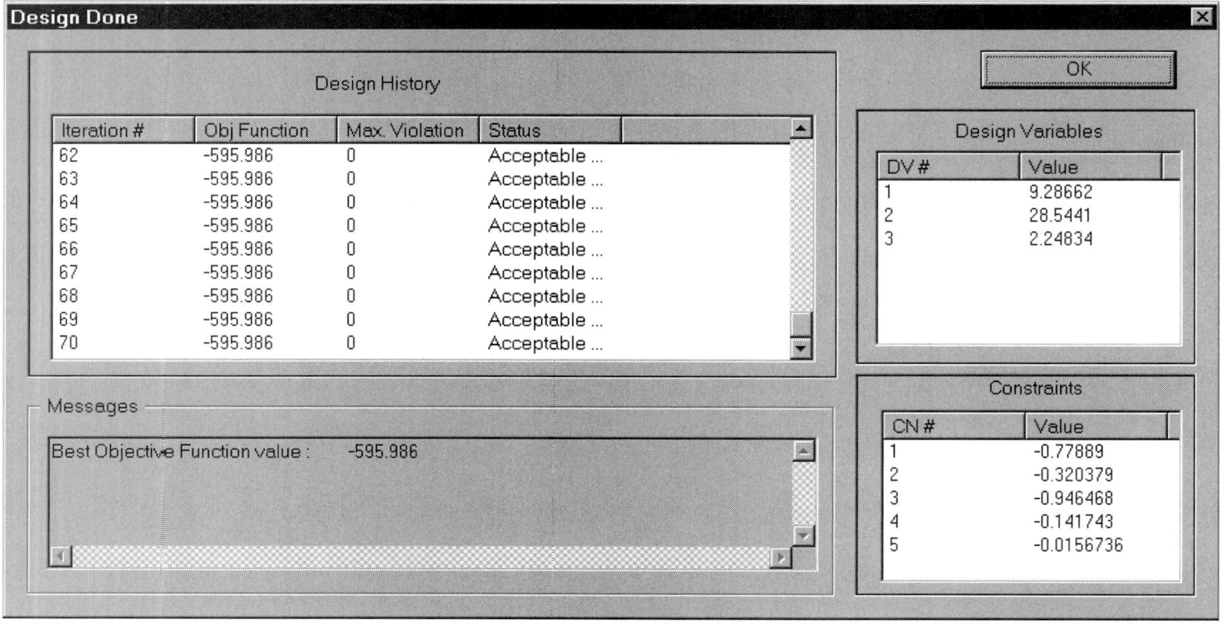

Step 4: With the new upper bound, the result of executing the GA option in the UNDO program is shown below after another 70 generations.

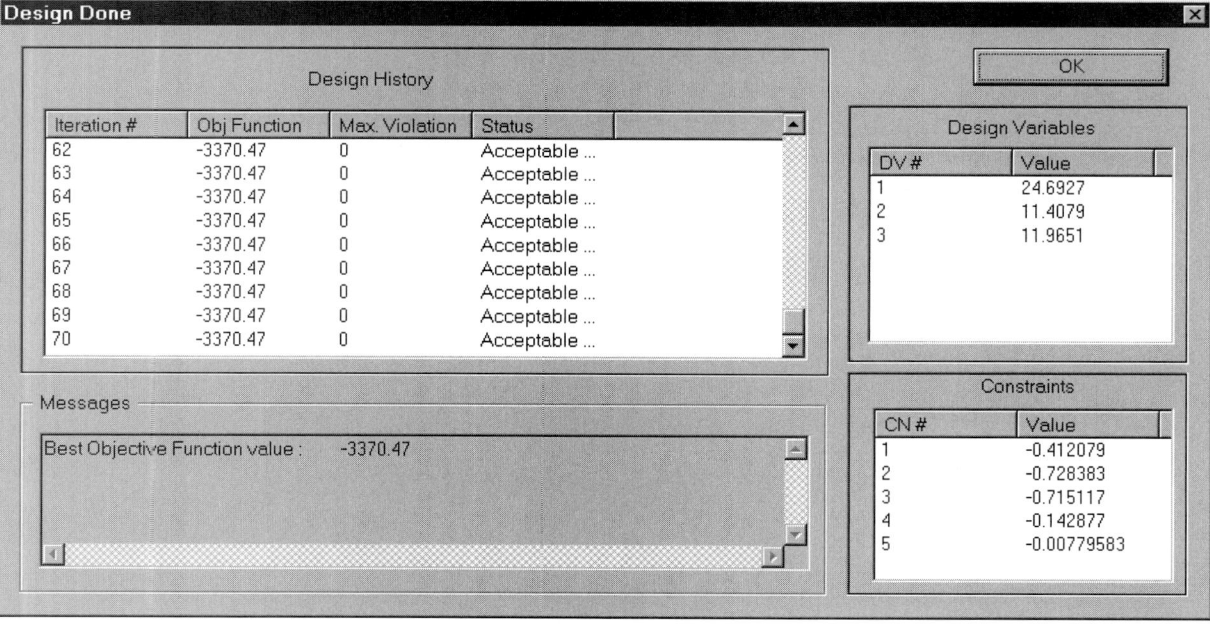

The solution obtained is $\mathbf{x} = \{24.7, 11.4, 12\}$ with $f(\mathbf{x}) = -3370$. The optimal solution is $\mathbf{x} = \{24, 12, 12\}$ with $f(\mathbf{x}) = -3456$. The fifth constraint (g_5) controls the design, i.e., is active at the optimum.

EXAMPLE 8.5.2

Column Design

Figure E8.5.2 shows a column with a rectangular cross-section that must support an axial force of 100 kN. The column must not fail due to axial stress or Euler buckling (in the x–y plane). The allowable axial stress is 20 kN/cm^2. The modulus of elasticity of the material is 1 GPa. The primary objective is to minimize the material volume.

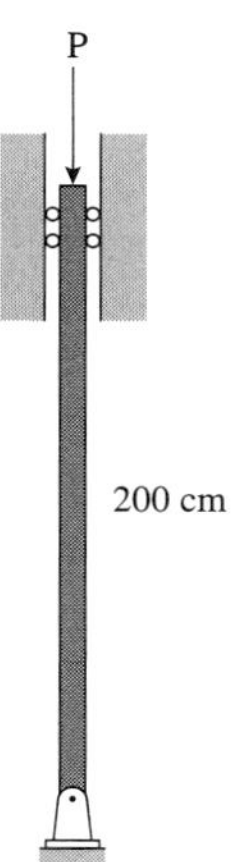

Fig. E8.5.2
Column design.

SOLUTION

Step 1: The design problem can be formulated as follows. Converting all quantities to cm and kN,

a. the volume of material can be expressed as $200bh$ cm^3,
b. the axial stress in the member is $100/bh \leq 20$ kN/cm^2, and
c. the Euler buckling requirement can be stated as $P_{cr} = (\pi^2 EI/L^2) > 100$ kN, or $1 - (\pi^2 bh^3/480000) \leq 0$

Hence we have

Find $\quad \mathbf{x} = \{b, h\}$

to minimize $\quad f(\mathbf{x}) = 200bh$

subject to

$$g_1(\mathbf{x}) = \frac{5}{bh} - 1 \leq 0$$

$$g_2(\mathbf{x}) = 1 - 2.056(10^{-5})bh^3 \leq 0$$

$$g_3(\mathbf{x}) = \frac{b}{h} - 1 \leq 0$$

$$b, h \geq 0$$

Note that all constraints are normalized.

Step 2: Based on the problem formulation, the following variables and functions are necessary:

Variable name	Remarks
b	Cross-section width
h	Cross-section height

Function expression	Remarks
200*b*h	Objective function: column volume
5/(b*h)-1	Axial stress constraint
1-2.056e-5*b*h**3	Euler buckling constraint
b/h-1	Cross-section shape

Design variable	Lower bound	Upper bound	Precision
b	1	20	Auto
h	1	20	Auto

The choice of lower and upper bounds should be based on some knowledge of the problem. In this example, the precision is the smallest (or finest) precision that the program will allow. One should *also* ask the question—When the column is fabricated or constructed, what is the precision (or tolerance) with which it will be made? The basic strategy is to solve the problem in stages. If at the end of first stage, a refined solution is needed, then you can increase the lower bound, decrease the upper bound, or both. The net effect is that you can then reduce the precision value and (you hope) obtain a better solution.

Step 3: The result of executing the GA option in the UNDO program is shown below after 70 generations. The solution obtained is $b = 1$ cm and $h = 7.9$ cm and $f(\mathbf{x}) = 1574$ cm^3. This solution is very close to the optimal solution.

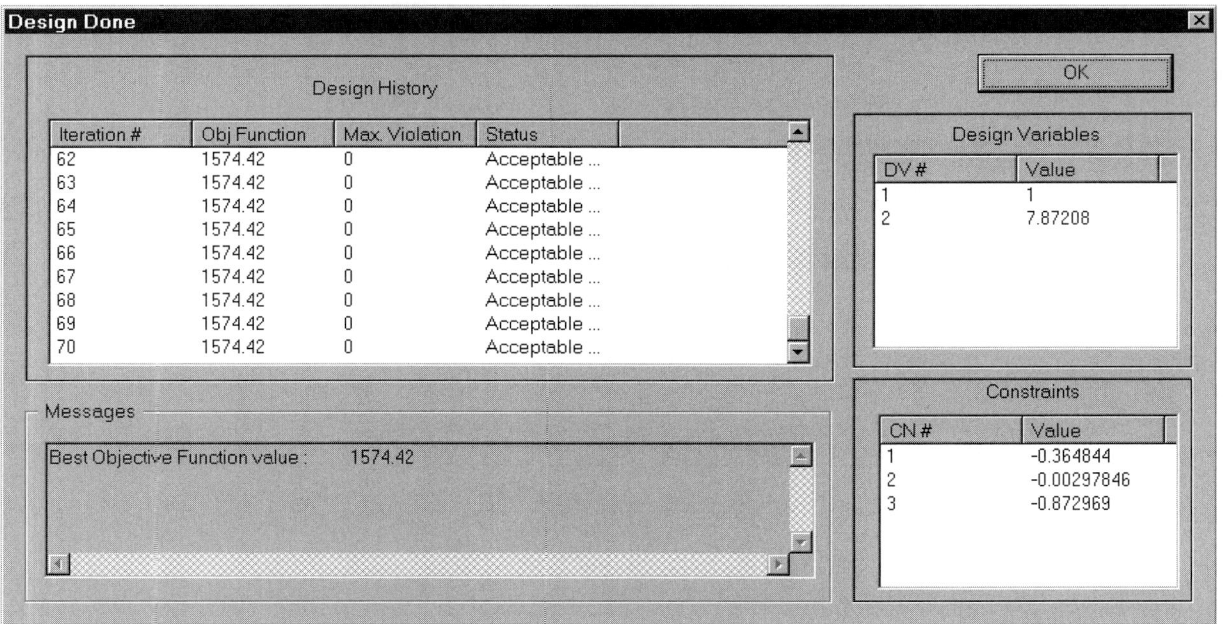

The second constraint (g_2) controls the design, i.e., it is active at the optimum.

EXAMPLE 8.5.3 ***Beam Design (Example 3.7.3)***

Figure E8.5.3 shows a simply supported beam. The beam is made of Douglas fir (modulus of elasticity = 1800 ksi, mass density = 1.0 slugs/ft^3). Find the width b and height h so that the maximum normal stress is less than 2 ksi and the shear stress is less than 0.1 ksi, so that the resulting beam is the lightest beam. The height of the beam should not exceed twice the width. Neglect self-weight of the beam.

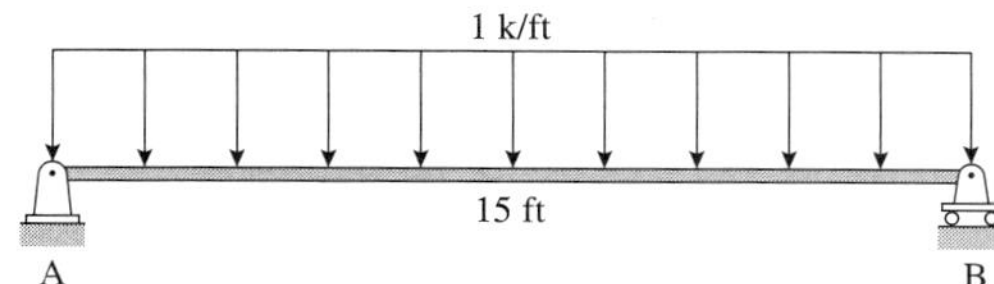

Fig. E8.5.3

SOLUTION

Step 1: Using lb, in as the units, the design problem can be formulated as follows:

Find $\{b, h\}$

To minimize $f(b,h) = 180bh$

$$g_1(\mathbf{x}) = \frac{1.0085(10^3)}{bh^2} - 1 \le 0$$

$$g_2(\mathbf{x}) = \frac{112.05}{bh} - 1 \le 0$$

$$g_3(\mathbf{x}) = 1 - \frac{2b}{h} \le 0$$

$$7'' < h, b < 15''$$

Step 2: Based on the problem formulation, the following variables and functions are necessary:

Variable name	Remarks
b	Cross-section width
h	Cross-section height

Function expression	Remarks
180*b*h	Objective function : column volume
1008.5/(b*h**2)-1	Normal stress constraint
112.05/(b*h)-1	Shear stress constraint
1-(2*b/h)	Cross-section shape

Design variable	Lower bound	Upper bound	Precision
b	7	15	Auto
h	7	15	Auto

Step 3: The result of executing the GA option in the UNDO program is shown below after 50 generations.

The solution obtained is b = 13.2 in, h = 8.8 in, and $f(\mathbf{x})$ = 20928 in^3. This solution is very close to the optimal solution of b = 12.45 in, h = 9.0 in, and $f(\mathbf{x})$ = 20169 in^3. The first constraint (g_1) and the second constraint (g_2) control the design.

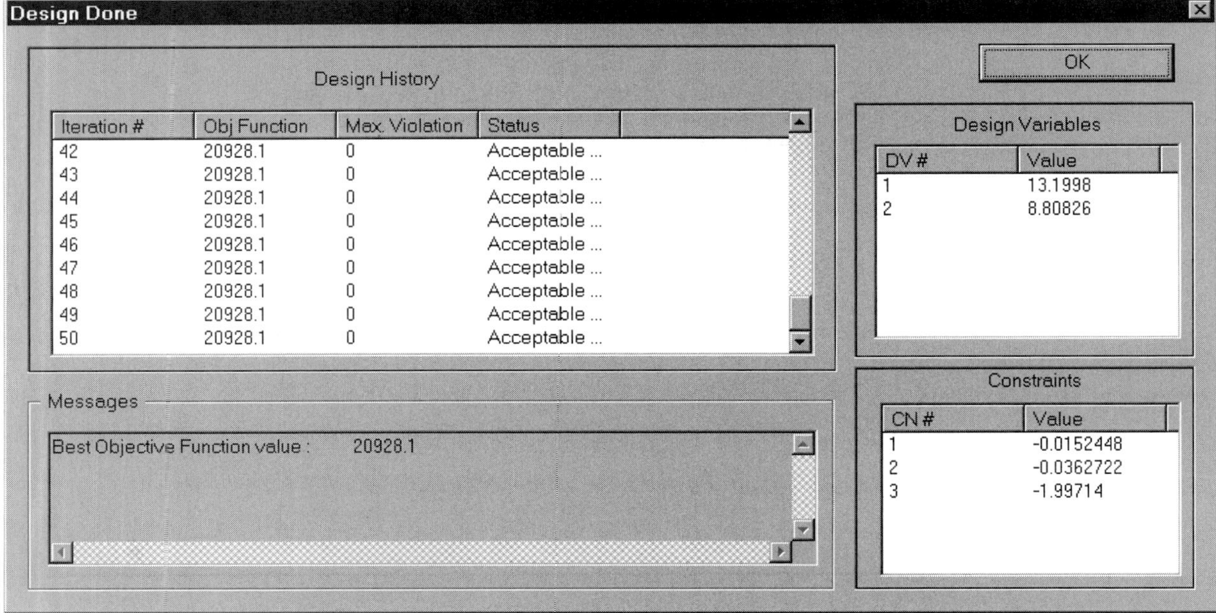

EXAMPLE 8.5.4

Truss Member Design

Figure E8.5.4 shows a two-bar truss. Each member is made of a solid square cross-section. The modulus of elasticity E is $30(10^6)$ psi. The allowable normal stress is 10000 psi. The axial stress in the members should be less than the allowable normal stress and the members should satisfy the Euler buckling requirement. The primary objective is to minimize the material volume.

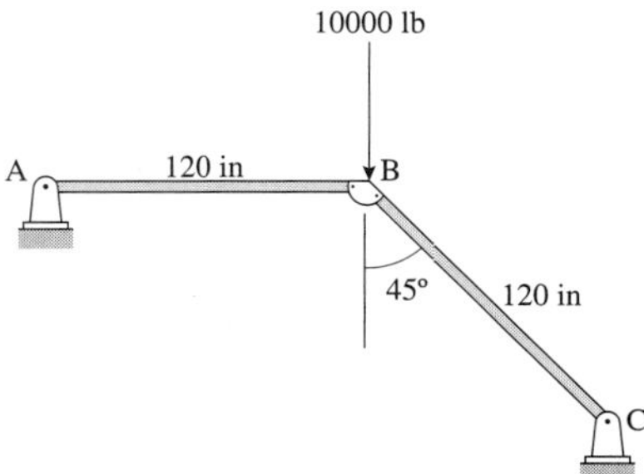

Fig. E8.5.4
Two-bar truss design.

SOLUTION

Step 1: Using lb and in as the problem units, the optimal design problem can be stated as follows:

Find $\mathbf{x} = \{a_{AB}, a_{BC}\}$

to minimize $f(\mathbf{x}) = V = 120(a_{AB}^2 + a_{BC}^2)$

subject to $g_1(\mathbf{x}) = \dfrac{P_{AB}}{10000\, a_{AB}^2} - 1 \le 0$

$$g_2(\mathbf{x}) = \frac{P_{BC}}{10000\, a_{BC}^2} - 1 \le 0$$

$$g_3(\mathbf{x}) = 1 - \frac{(P_{cr})_{AB}}{P_{AB}} \le 0$$

$$g_4(\mathbf{x}) = 1 - \frac{(P_{cr})_{BC}}{P_{BC}} \le 0$$

$$a_{AB}, a_{BC} \ge 0$$

where a_{AB}, a_{BC} are cross-section sides for members AB and BC, P_{AB}, P_{BC} are the magnitude of the axial forces in members AB and BC, and $P_{cr} = (\pi^2 EI/L^2)$ represents the Euler buckling capacity of the member in compression. Using the method of joints, $P_{AB} = 10000$ lb and $P_{BC} = 14150$ lb with both the members in compression.

Step 2: Based on the problem formulation, the following variables and functions are necessary:

Variable name	Type	Expression	Remarks
a1	Simple		Cross-section side for member AB
a2	Simple		Cross-section side for member BC
E	Derived	30e6	Modulus of elasticity
pi	Derived	3.1415926	Pi
I1	Derived	a1**4/12	Moment of inertia (member AB)
I2	Derived	a2**4/12	Moment of inertia (member BC)
stress1	Derived	10000/a1**2	Stress in member AB
stress2	Derived	14150/a2**2	Stress in member BC

Function expression	Remarks
120*(a1**2+a2**2)	Objective function: Truss volume
stress1/10000-1	Axial stress constraint: member AB
Stress2/10000-1	Axial stress constraint: member BC
1-(pi**2*E*I1)/(120**2*10000)	Euler buckling: member AB
1-(pi**2*E*I2)/(120**2*14150)	Euler buckling: member BC

Design variable	Lower bound	Upper bound	Precision
a1	1	2	Auto
a2	1	2	Auto

This example illustrates how to use simple and derived variables to reduce the amount of hand-calculation necessary to formulate the design problem. We could have used more derived variables than shown above.

Step 3: The result of executing the GA option in the UNDO program is shown below after 70 generations.

The solution obtained is $a_1 = 1.56$ in, $a_2 = 1.70$ in, and $f(\mathbf{x}) = 641.5$ in^3. This solution is very close to the optimal solution. The third constraint (g_3) and the fourth constraint (g_4) control the design.

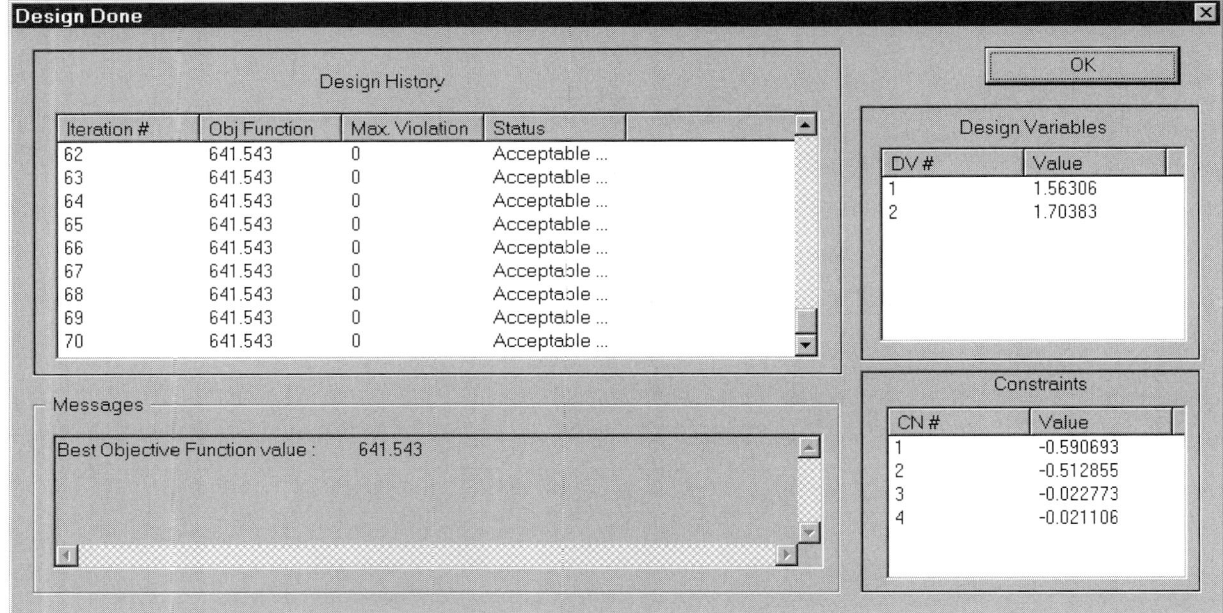

EXERCISES

Appetizers

8.5.1. Figure P8.5.1 shows a determinate planar beam. The cross-section is rectangular (height h and width w). The allowable normal stress is 12 MPa and the shear stress is 5 MPa. Design the minimum-volume beam so that the height is not more than twice the width.

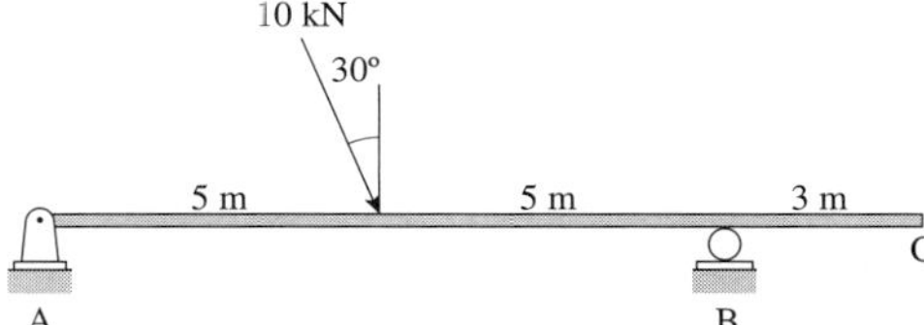

Fig. P8.5.1

8.5.2. How would you formulate and solve the previous problem if the weight density of the beam material was given as 6000 N/m^3?

Main Course

8.5.3. A steel pipe is moved to place by a crane using the system shown in Fig. P8.5.3. The inner pipe diameter is 40 in and the wall thickness is 0.375 in. The length of the pipe is 24 ft. The maximum axial load capacity of the cable is 2000 lb and its length is restricted to 24 ft. Find the distances d and h between the lifting points to minimize the maximum bending stress in the pipe.

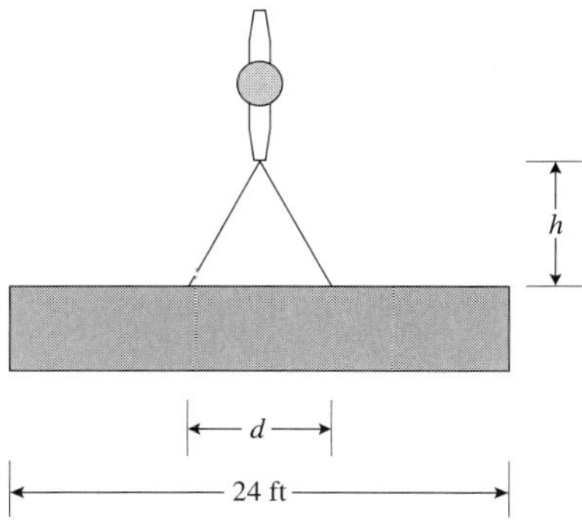

Fig. P8.5.3

8.5.4. Figure P8.5.4 shows a planar two-bar truss. Support C is located directly above A. The allowable stress in member AB is 100 MPa and in member CB is 200 MPa. Find the cross-sectional areas of members AB and BC and the distance x to design the minimum-volume truss.

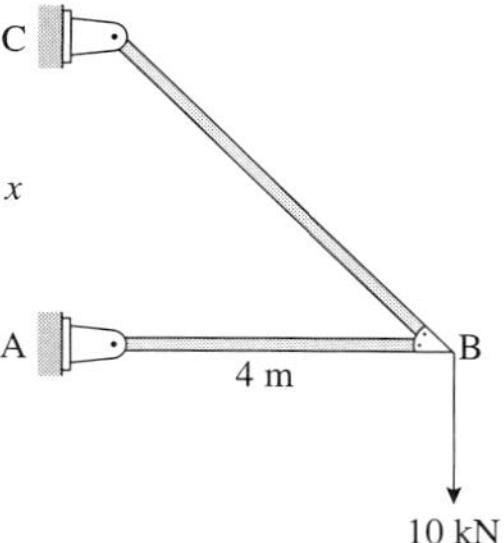

Fig. P8.5.4

Structural Concepts

8.5.5. It is required to design a support bracket as shown in Fig. P8.5.5. Member ADC is W16 × 31. Member BD has a circular hollow cross-section. Supports A and B are pin supports and connection at D is a pin connection. Design the lightest steel member BD while simultaneously reducing the bending stress in member ADC so that the normal stress in the member is less than 10 ksi and Euler buckling is prevented with a safety factor of 2. The wall thickness of the pipe cannot be less than 15% of the inner radius. Also find the optimal values for x and d.

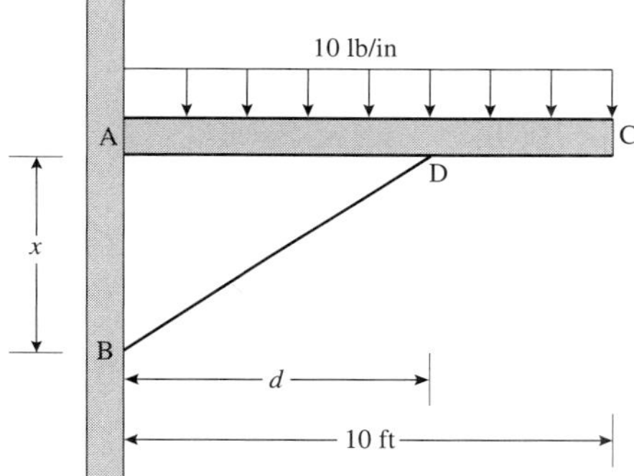

Fig. P8.5.5

8.6 LINKING ANALYSIS AND DESIGN TECHNIQUES

The overall design process was conceptually discussed earlier (see Fig. 8.6.1). Several design tools are developed, linked, and used to help achieve the overall design goals. However, in this section we look at the link between structural analysis and design in one specific stage of the design process, e.g., preliminary or intermediate, and final design.

Once the design problem is formulated as discussed in the previous section, the iterative process can be automated. Figure 8.6.2 shows the localized flow of calculations. There are four major blocks.

Block A. The design process usually starts with an initial design in which depending on the type of design problem (we will see this in the next section), several design decisions have already been made.

Block B. The design problem formulation will determine the type of structural analyses required. For example, it may be necessary to compute (a) the member forces and displacements under the action of static loads, (b) the natural frequency of the structure, (c) the member forces and displacements under the action of dynamic loads, etc.

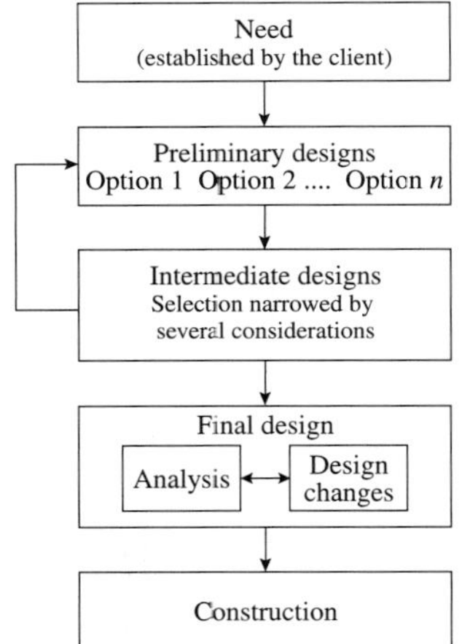

Fig. 8.6.1 Conceptual view of the design process.

Block C. Having computed the structural response in Block B, we can ascertain if the current design meets the design requirements. At this stage, a decision needs to be made whether the current design is acceptable. If acceptable, are design improvements necessary? If not acceptable, how can design changes be made to meet the design requirements?

Block D. If design changes are necessary, what design variables can be changed and by how much? The answer is neither trivial nor easy except in the simplest of design problems.

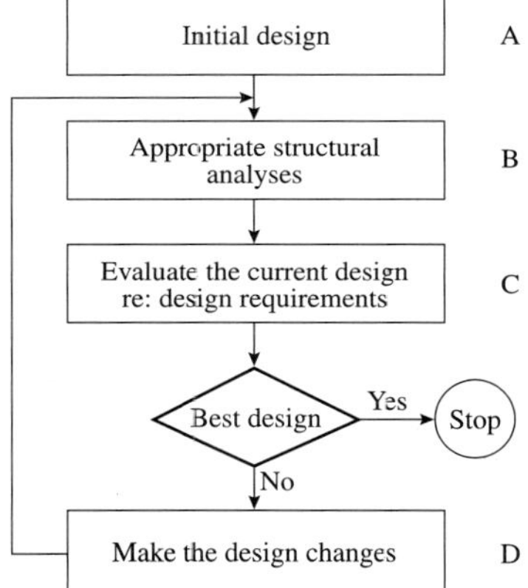

Fig. 8.6.2 Localized optimal design flow.

Block B incorporates structural analysis techniques. On the other hand, Block D relies on optimization technique. We have seen these techniques in the previous chapters. In the following sections we see some of the details of Block A and Block C. In the last section, we link the blocks to find the optimal design.

8.7 STRUCTURAL OPTIMIZATION

To recap, the design problem can be expressed in a formal mathematical form the design problem can be expressed as

Find $\mathbf{x}$

to minimize $f(\mathbf{x})$

subject to $\mathbf{g}(\mathbf{x}) \leq 0$

$$\mathbf{h}(\mathbf{x}) = 0$$

$$\mathbf{x}_L \leq \mathbf{x} \leq \mathbf{x}_U$$

The vector $\mathbf{x}$ is the vector of design variables. The function $f(\mathbf{x})$ is the objective function. The functions $\mathbf{g}(\mathbf{x})$ and $\mathbf{h}(\mathbf{x})$ are the inequality and equality constraint functions, respectively. Since the design variables typically cannot assume any value, each design variable is limited by a lower bound x_L and an upper bound x_U. The final (acceptable) value of the design variable must lie between these limits.

Structural (optimal) design problems can be posed as stated above, and can be classified broadly into three categories. To understand some of the terminology associated with optimal design problems, let us ask the following question: "What components of a structural system (truss and frame) can be used as design variables?" Some of the components are:

a. The material used. We briefly discussed in Section 3.4 the four most commonly used materials—steel, concrete, masonry, and wood.
b. The cross-sectional shape and dimensions of the members.
c. The assemblage of the joints and members.
d. The locations of the joints.
e. The types of the joints and the supports.

Consider the four trusses in Fig. 8.7.1. The two trusses on the left have the same topology but different shapes. The two trusses on the right have the same shape but different topologies.

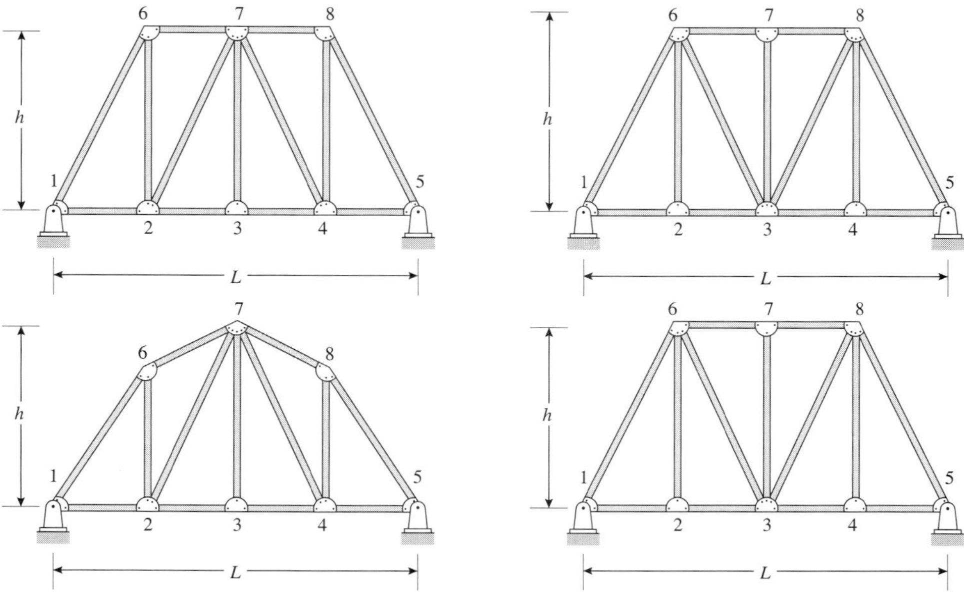

Fig. 8.7.1
(a) Shape changes.
(b) Topology changes.

In the following sections we look at *example design problem formulations*.

8.7.1 Sizing Optimal Design

In this design problem formulation, the structural topology and shape are *not* allowed to vary. The design variables are the cross-sectional properties of some or all the members. We looked at a few examples in Chapter 3, and we now revisit one of those examples—Example 3.7.4 (see Fig. 8.7.1.1).

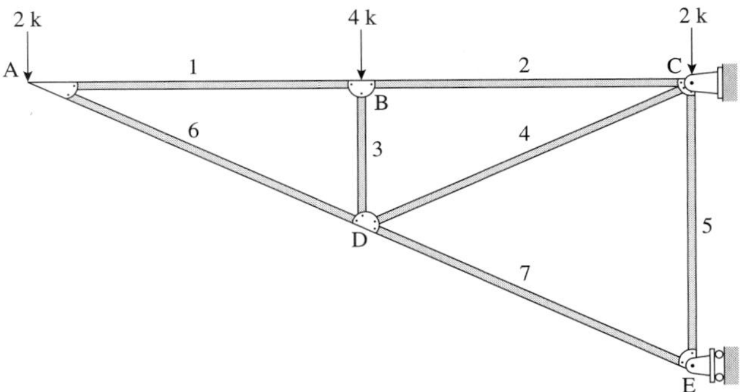

Fig. 8.7.1.1
Cantilever truss.

Problem Statement. The truss members are made of steel and have a circular hollow cross-section. Select three cross-sections for the top chord, the bottom chord, and the rest of the members, respectively. Design the lightest truss so that the normal stress in the members is less than 24 ksi, the Euler buckling criterion is satisfied and the wall thickness t is at least 20% of the inner radius r. The radius cannot be less than 0.5 in and greater than 2 in. The wall thickness cannot be less than 0.1 in.

Problem Formulation. The design problem can be formally stated as follows, using in, lb as the problem units:

Find $$\mathbf{x} = \{t_1, r_1, t_2, r_2, t_3, r_3\} \tag{8.7.1.1}$$

to minimize $$f(\mathbf{x}) = \sum_{i=1}^{7} L_i A_i \rho_i \tag{8.7.1.2}$$

such that $$g_i(\mathbf{x}) \equiv \sigma_i - 24000 \le 0 \qquad i = 1, 2, ..., 7 \tag{8.7.1.3}$$

$$h_i(\mathbf{x}) \equiv P_i - \frac{\pi^2 EI_i}{L_i^2} \le 0 \qquad i = 1, 2, ..., 7 \tag{8.7.1.4}$$

$$i_i(\mathbf{x}) \equiv 0.2r_i - t_i \le 0 \qquad i = 1, 2, 3 \tag{8.7.1.5}$$

and $$0.1 \le t_i \le 1 \qquad i = 1, 2, 3 \tag{8.7.1.6}$$

$$0.5 \le r_i \le 2.0 \qquad i = 1, 2, 3 \tag{8.7.1.7}$$

with $$r_1, t_1 \Rightarrow 1, 2 \tag{8.7.1.8a}$$

$$r_2, t_2 \Rightarrow 3, 4, 5 \tag{8.7.1.8b}$$

$$r_3, t_3 \Rightarrow 6, 7 \tag{8.7.1.8c}$$

Equation (8.7.1.1) shows the vector of design variables. There are six design variables in this problem—three inner radii and three wall thickness. Equations. (8.7.1.6) and (8.7.1.7) establish the range of values that the wall thickness and the radius can assume. We have assumed that these values vary continuously. There are seven members in the truss. How do we relate them to the six design variables? Equations (8.7.1.8a–c) shows the relationship between the design variables and the seven members. For example, both members 1 and 2 have the same inner radius r_1 and the same wall thickness t_1 etc. The objective function, $f(\mathbf{x})$

is the weight of the truss. Equation (8.7.1.2) represents the sum of the weights of the seven members—the weight of a single member is the product of the length L_i, cross-sectional area A_i, and the weight density, ρ_i. Of the three parameters, only the area is related to the design variables as

$$A = \pi\left[(r+t)^2 - r^2\right] = \pi\left[2rt + t^2\right] \tag{8.7.1.9}$$

The axial stress constraints are represented by Eq. (8.7.1.3), one for each member of the truss. This stress is indirectly a function (an implicit function) of the design variables since $\sigma_i = (P_i/A_i)$. Equation (8.7.1.4) is the Euler buckling constraint where P_i is the axial force in the member, E is the modulus of elasticity of steel, I_i is the moment of inertia of member, and L_i is the length of the member. Again, of the four parameters, only the moment of inertia is a function of the design variables:[6]

$$I = \frac{\pi}{4}\left[(r+t)^4 - r^4\right] \tag{8.7.1.10}$$

Finally, Eq. (8.7.1.5) captures the required relationship between the wall thickness and the inner radius. We have included this constraint to avoid the problem of local buckling that could arise with a cross-section with a very large radius and a very small wall thickness.

Observation: For the optimization problem to be well posed, the objective function and all the constraints must be functions of the design variables. Otherwise, changing the design variable values does not affect the values of these functions.

We now draw the relationship between the blocks in Fig. 8.7.1.2 and the problem formulation. We need to set the values of all the structural parameters including the design variables to initiate the design process. The initial values (or guesses) for the design variables should satisfy Eqs. (8.7.1.6) and (8.7.1.7). This is Block A. Block B represents a linear, static analysis so as to compute the member forces. In Block C, we would compute the current values of the objective function and all the constraints—Eqs. (8.7.1.1)–(8.7.1.5). Based on the information from Block C and other pieces of information, the values of the design variables **x** are changed in Block D. The changes that lead to the new values for the design variables must satisfy Eqs. (8.7.1.6) and (8.7.1.7). The iterative process is repeated until some termination criterion is satisfied.

8.7.2 Shape Optimal Design

In this design problem formulation, the structural topology and the member cross-sections are *not* allowed to vary. The design variables are the x–y coordinates of the some or all the nodes. We go back to the previous example (in Section 8.7.1) to formulate the design problem.

Problem Statement. The truss members are made of steel and have a circular hollow cross-section. Select the locations of joints D and E. Design the lightest truss so that the normal stress in the members is less than 24 ksi and the Euler buckling criterion is satisfied. The entire truss must fit in the box or rectangle ACGF. The position and location of the nodes on the top chord cannot be changed. The cross-sectional dimensions of the members are known (see Fig. 8.7.1.2).

[6] This is, strictly speaking, not always correct. It is correct for a determinate truss. For an indeterminate truss, the axial force is also indirectly a function (implicit function) of the cross-sectional dimensions, other parameters being held constant.

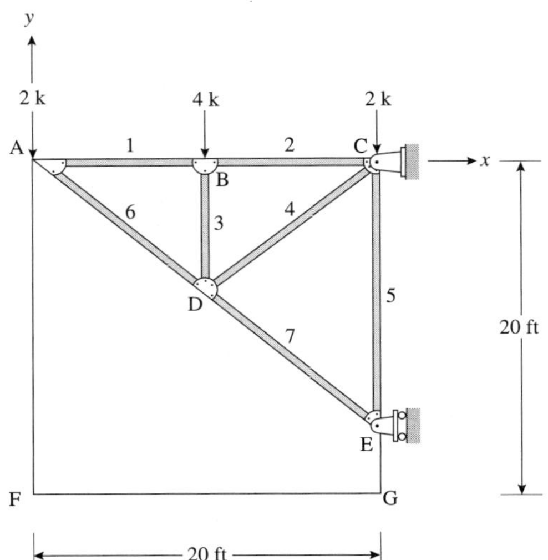

Fig. 8.7.1.2
Shape optimal design.

Problem Formulation. The design problem can be formally stated as follows, using in, lb as the problem units.

Find $\mathbf{x} = \{x_D, y_D, y_E\}$ (8.7.2.1)

to minimize $f(\mathbf{x}) = \sum_{i=1}^{7} L_i A_i \rho_i$ (8.7.2.2)

such that $g_i(\mathbf{x}) \equiv \sigma_i - 24000 \le 0 \qquad i = 1, 2, ..., 7$ (8.7.2.3)

$h_i(\mathbf{x}) \equiv P_i - \dfrac{\pi^2 EI_i}{L_i^2} \le 0 \qquad i = 1, 2, ..., 7$ (8.7.2.4)

and $60 \le x_D \le 180$ (8.7.2.5a)

$-240 \le y_D \le -60$ (8.7.2.5b)

$-240 \le y_E \le -60$ (8.7.2.5c)

We focus on the differences between the sizing and the shape design problem formulations. The objective function in Eq. (8.7.2.1) is the same as before. However, parameter L_i (for members 3–7) is the only parameter that is a function of the design variables. The functions $g_i(\mathbf{x})$ and $h_i(\mathbf{x})$ are both functions of P_i. As stated before, P_i is a function of the locations of the nodes of the element. In addition, $h_i(\mathbf{x})$ is also a function of L_i. Hence varying the design variables in this problem formulation will affect the values of the objective and the constraint functions. To make the design changes valid, the lower and upper bounds on the design variables are selected as shown in Eqs. (8.7.2.5a–c).

Finally, what are the values of the inner radius and the wall thickness? These values need to be selected and then held constant during the design process just as in the sizing optimal design, where the locations of the nodes were initially set and then held constant.

8.7.3 Topology Optimal Design

In this design problem formulation, the member cross-sections and the nodal coordinates are *not* allowed to vary. The design variables are the member connectivity, i.e., the presence and absence of some or all of the members.

Consider the previous example. The initial layout of the truss was selected as shown in Fig. 8.7.1.1 for both the sizing and shape optimal designs. In other words, there are seven members and five nodes, and the members are connected to each other as shown in the figure. Alternately, we could have started with either of the trusses in Fig. 8.7.3.1.

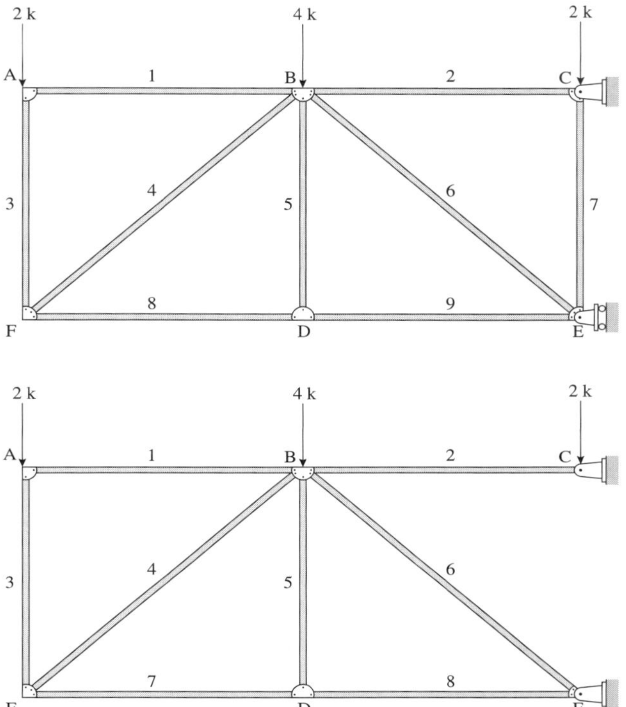

Fig. 8.7.3.1
Possible truss layouts.

However, to obtain the optimal truss configuration or topology, we must start with an initial design and allow the optimization process to remove or add members. The former is much easier to implement. The methodology for adding members (and joints) to an existing truss is much more difficult. We illustrate the procedure for topology optimal design using the same truss example. The process starts with a truss structure that contains all possible nodes that describe the truss and all the possible members that the designer wishes to consider for design. Such a truss, called the *ground structure*, is shown in Fig. 8.7.3.2. Using the ground structure, it is possible to formulate the design problem.

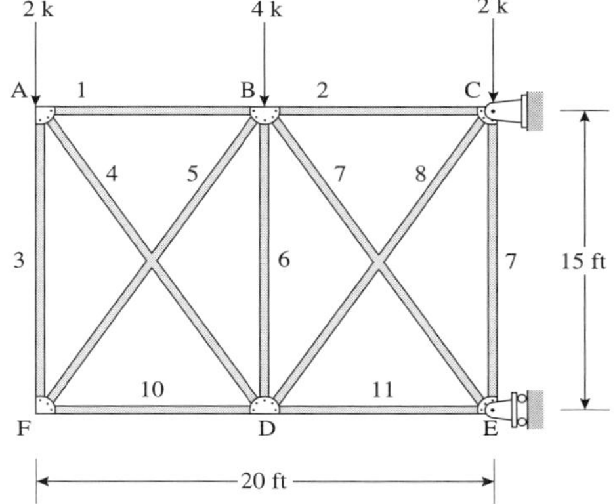

Fig. 8.7.3.2
Ground structure.

Problem Statement. The truss members are made of steel and have a circular hollow cross-section. Select which of the members 4 through 8 should be a part of the lightest truss so that the normal stress in all the members is less than 24 ksi and the Euler buckling criterion is also satisfied. The cross-sectional dimensions of the members are known.

Problem Formulation. The design problem can be formally stated as follows, using in, lb as the problem units.

$$\text{Find} \qquad \mathbf{x} = \{b_4, b_5, b_6, b_7, b_8\} \tag{8.7.3.1}$$

$$\text{to minimize} \qquad f(\mathbf{x}) = \sum_{i=1}^{11} L_i A_i \rho_i \tag{8.7.3.2}$$

$$\text{such that} \qquad g_i(\mathbf{x}) \equiv \sigma_i - 24000 \le 0 \qquad i = 1,\ 2,\ \ldots,\ 11 \tag{8.7.3.3}$$

$$h_i(\mathbf{x}) \equiv P_i - \frac{\pi^2 E I_i}{L_i^2} \le 0 \qquad i = 1,\ 2,\ \ldots,\ 11 \tag{8.7.3.4}$$

$$\text{and} \qquad b_j \in [0,1] \qquad j = 4,\ 5,\ 6,\ 7,\ 8 \tag{8.7.3.5}$$

The design variables in this problem as shown by Eqs. (8.7.3.1) and (8.7.3.5) are Boolean in nature. The subscripts indicate that we wish to consider members 4 through 8 as candidate design variables. The optimization process will determine which of these members should be part of the structure that has the lowest weight and also satisfies the constraints expressed by Eqs. (8.7.3.3) and (8.7.3.4). The hidden danger in this problem formulation is the possibility of encountering unstable structures. The optimization methodology must be able to handle the "discontinuities in the design space." The genetic algorithm that we saw earlier is one such technique that can be used to solve the design problem.

8.7.4 Combination Optimal Design

When two or more of the above design scenarios are combined, we have a combination optimal design problem. This situation is more common as a design problem than just sizing, or shape, or topology design optimization problems. We now formulate an example design problem.

Problem Statement. The truss members are made of steel and have a circular hollow cross-section. Select three cross-sections for the top chord, the bottom chord, and the rest of the members, respectively. Design the lightest truss so that the normal stress in the members is less than 24 ksi, the Euler buckling criterion is satisfied, and the wall thickness t is at least 20% of the inner radius r. The radius cannot be less than 0.5 in or greater than 2 in. The wall thickness cannot be less than 0.1 in. In addition, select the locations of joints D, E, and F. The entire truss must fit in the box or rectangle ACGH. The position and location of the nodes on the top chord cannot be changed. Also, select which of the members 4 through 8 should be a part of the lightest truss, as shown in Fig. 8.7.4.1.

Problem Formulation. The design problem can be formally stated as follows using in, lb as the problem units.

$$\text{Find} \qquad \mathbf{x} = \{t_1,\ r_1,\ t_2,\ r_2,\ t_3,\ r_3,\ x_D,\ y_D,\ y_E,\ x_F,\ y_F,\ b_4,\ b_5,\ b_6,\ b_7,\ b_8\} \tag{8.7.4.1}$$

$$\text{to minimize} \qquad f(\mathbf{x}) = \sum_{i=1}^{11} L_i A_i \rho_i \tag{8.7.4.2}$$

such that $$g_i(\mathbf{x}) \equiv \sigma_i - 24000 \le 0 \qquad i = 1, 2, \ldots, 11 \tag{8.7.4.3}$$

$$h_i(\mathbf{x}) \equiv P_i - \frac{\pi^2 EI_i}{L_i^2} \le 0 \qquad i = 1, 2, \ldots, 11 \tag{8.7.4.4}$$

$$i_i(\mathbf{x}) \equiv 0.2r_i - t_i \le 0 \qquad i = 1, 2, 3 \tag{8.7.4.5}$$

and $$0.1 \le t_i \le 1 \qquad i = 1, 2, 3 \tag{8.7.4.6}$$

$$0.5 \le r_i \le 2.0 \qquad i = 1, 2, 3 \tag{8.7.4.7}$$

$$60 \le x_D \le 180 \tag{8.7.4.8}$$

$$-240 \le y_D \le -60 \tag{8.7.4.9}$$

$$-240 \le y_E \le -60 \tag{8.7.4.10}$$

$$0 \le x_F \le 60 \tag{8.7.4.11}$$

$$-180 \le y_F \le -30 \tag{8.7.4.12}$$

$$b_j \in [0,1] \qquad j = 4, 5, 6, 7, 8 \tag{8.7.4.13}$$

with $$r_1, t_1 \Rightarrow 1, 2 \tag{8.7.4.14a}$$

$$r_2, t_2 \Rightarrow 3, \ldots, 9 \tag{8.7.4.14b}$$

$$r_3, t_3 \Rightarrow 10, 11 \tag{8.7.4.14c}$$

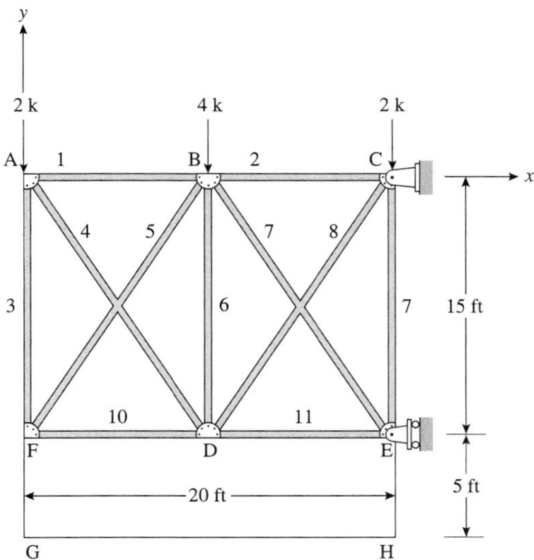

Fig. 8.7.4.1
Sizing shape and topology optimization.

In this section we have looked at different ways of formulating the design problem. The appropriateness of the type of design problem is a function of several factors. We would expect that as the number of design variable increases, the chances of obtaining a better solution also increases. Here are some of the questions that one could ask while formulating the previous design problem.

a. How do we group the different members so that they have the same cross-sectional dimensions? In the examples, we divided the members into three groups—top chord, bottom chord, and the web members. Is this the optimal way of grouping the members? Could we have assumed more than three groups? Should this decision be influenced by construction practices?
b. How do we select which joints locations are design variables? Should this decision be influenced by construction practices?
c. The total load on the truss is apparently 8 k and is applied at three joints. Could we have chosen a different configuration, say with four loaded joints on the top chord?
d. What factors influence the decision to fit the truss in the box ACGH?
e. Could we also have selected the type of support at E (pin vs. roller support) as a design variable? How can we tie in the "cost" of a support with the problem formulation? In the example, the truss was supported at two locations. Could we have selected three support points?
f. Steel was used in this example. Could we have tried other materials such as wood?
g. We considered hollow circular tubes. Could we have used other available shapes?
h. Is there a rational way of bringing cost and reliability into the problem formulation?
i. Is the aesthetics of the structure a valid consideration? How can aesthetics be incorporated in problem formulation?

As we have seen with a relatively simple example, design problems are usually open-ended in terms of both problem formulation and solution. However, the amount of freedom a designer has in terms of formulating the design problem is problem-dependent. We examine some of these issues in the next section.

EXERCISES

For all the problems in this section, formulate (but do *not* solve) the optimal design problem.

Appetizers

8.7.1. A vertical load of 10000 N (representing the weight of a sign) is to be supported by a planar truss (see Fig. P8.7.1). The load is located at a distance of 2 m from the wall on which the truss is to be anchored. The truss is made of steel. The design requirements are as follows.

(1) The truss must lie in the box ABCD.

(2) The truss members must not fail due to axial stress that is limited to 130 MPa. The tip displacement is limited to 0.01 m. Consider the Euler buckling requirement with a factor of safety of 2.

(3) Design the minimum-weight truss.

(4) Restrict the number of different cross-sections used in the truss to three.

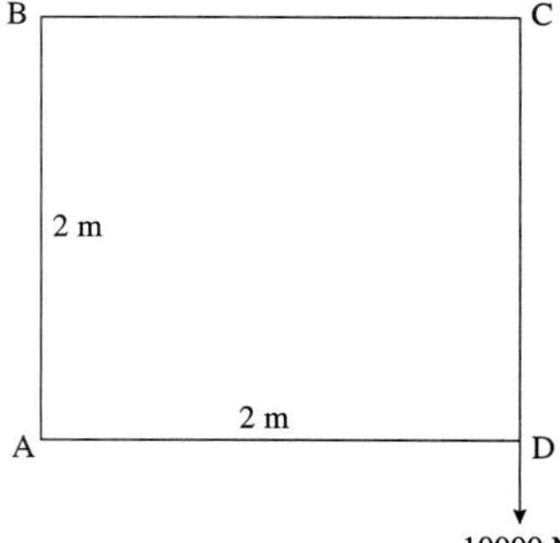

Fig. P8.7.1

8.7.2. Figure P8.7.2 shows a building that is made up of a series of trusses that are 2 feet apart. There is a total of 11 trusses. The trusses support a roof that is subjected to a uniformly distributed vertical load (dead plus live) of 200 psf. The trusses are connected to the two wooden beams (AB and CD) with a pin support on beam AB and a roller support on beam CD. The beams in turn are supported on the columns through a pin support at each end. It is required to design the beams to satisfy the following performance requirements.

(1) The beams must not fail in tension (allowable tensile stress is 1800 psi), compression (allowable compressive stress is 1800 psi), or shear (allowable shear stress is 180 psi).

(2) The maximum vertical deflection must not exceed $L/360$ where L is the span of the beam.

(3) Design the minimum-weight beam assuming that the rectangular cross-section beam is made of Douglas fir. The height of the beam cannot be greater than three times the width of the beam.

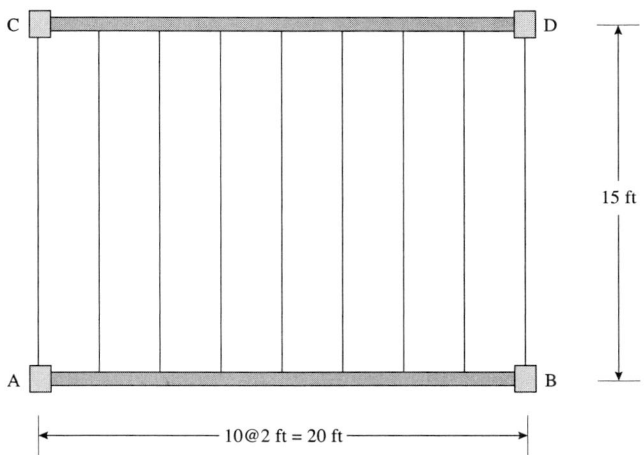

Fig. P8.7.2

Main Course

8.7.3. Figure P8.7.3 shows a plan view of a building covering a 80 ft by 30 ft area. The roof system is such that the total vertical load is 30 psf. The wind load acting normal to the ridge can be assumed to be 10 psf. Five gable frames spaced at 20 ft apart are used as the structural system. Design a minimum-weight interior frame of the building. The normal stress in tension and compression cannot exceed 20000 psi. The shear stress cannot exceed 10000 psi. The maximum horizontal and vertical displacement cannot exceed 1 in. Use AISC W-sections.

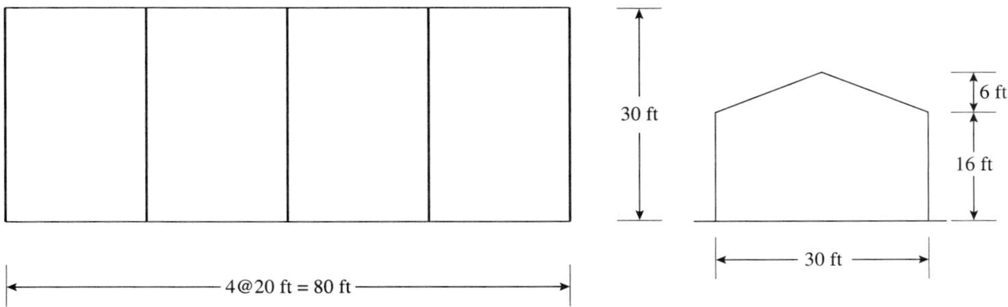

Fig. P8.7.3

8.7.4. Figure P8.7.4 shows a maintenance bay around an industrial building. The bay is 6 ft wide and is to be supported on the exterior columns that are placed 10 ft apart. The design load (dead load plus live load) on the bay floor is 4000 psf. Modeling assumptions dictate that the maintenance bay supporting structure be modeled as a planar truss. It is required to design the maintenance bay supporting structure so as to meet and satisfy the following requirements.

Design the minimum-cost truss. The truss members are made of hollow circular tubes. The material used is 0.2% CR steel (E = 29000 ksi, yield stress = 51000 psi).

The compressive and the tensile stress in each member is to be limited to the allowable value. In addition, members in compression must satisfy the Euler buckling requirement. The factor of safety for each of these three checks is 2.

The truss width is equal to the width of the bay floor.

The maximum vertical deflection must not exceed 0.1 in.

Do not use more than three different cross-sections.

The ratio of the inner radius of the section to the wall thickness cannot exceed 5. The smallest inner radius is 0.5 in.

The cost of steel is $0.4/in^3. The cost of each joint is $50. The cost of each support is $200.

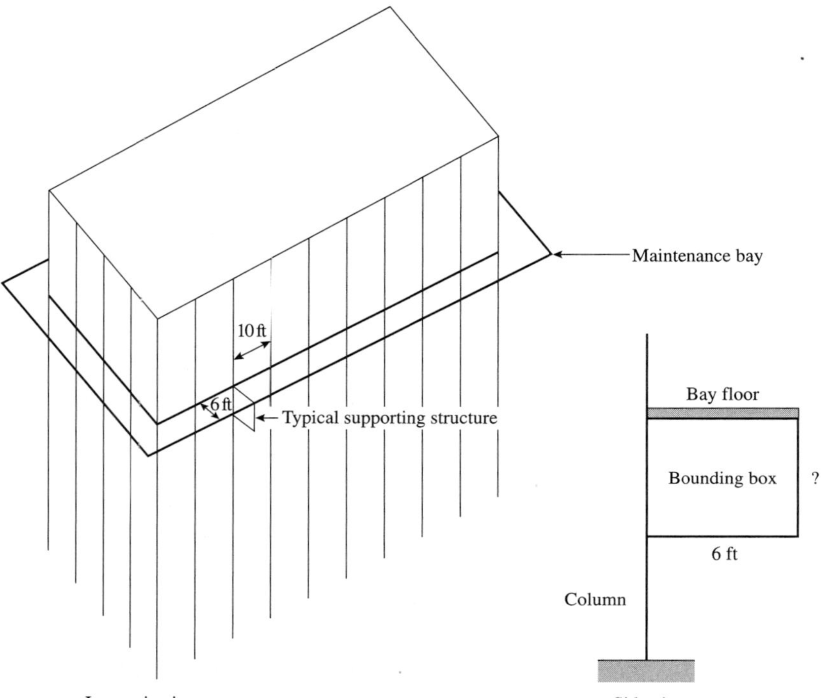

Fig. P8.7.4

Structural Concepts

8.7.5. Consider a design optimization problem that is being solved in stages. In the first stage, only sizing design variables are considered. With the optimized solution so obtained, in the next stage only shape design variables are considered. An optimized solution is obtained. Is this optimal solution better or worse than a solution obtained when both sizing and shape optimal design variables are considered simultaneously?

8.8 DESIGN EXAMPLES

The material in this section is a continuation of the ideas explored in previous section. First, we look at some of the types of commonly encountered design problems. We attempt to classify these problems. Second, we look at how to formulate these design problems and use the optimal design capabilities of the GS-USA program. Finally, we solve a few design problems to illustrate the ideas.

8.8.1 Broad Classification of Structural Forms

Throughout the text we have primarily looked at beams, frames, and trusses, and occasionally at arches and cables. In this section we look at structural systems that can be reasonably approximated as frames and trusses. One way of classifying these systems is to look at how they are supported. It should be noted that typically, a planar structure (or a planar approximation of a three-dimensional structure) needs to fit inside a rectangular bounding box, the dimensions of which are determined by the client and the designer.

Cantilever Structures. These planar structural systems can be supported on one edge of the rectangular box. The load path is such that the "bending moment" increases towards the support. Efficient structural forms can be devised to handle the load path, as we will see in the following examples. Figure 8.8.1.1 shows a signboard and the supporting structure for the signboard. The structure can be approximated as a planar frame with the two supports placed on the left edge of the bounding box. Figures 8.8.1.2 and 8.8.1.3 show two other commonly seen structural systems—a supporting structure for a billboard and a traffic signal cum street light structure. The former is modeled as a cantilevered planar truss and the latter as a planar frame.

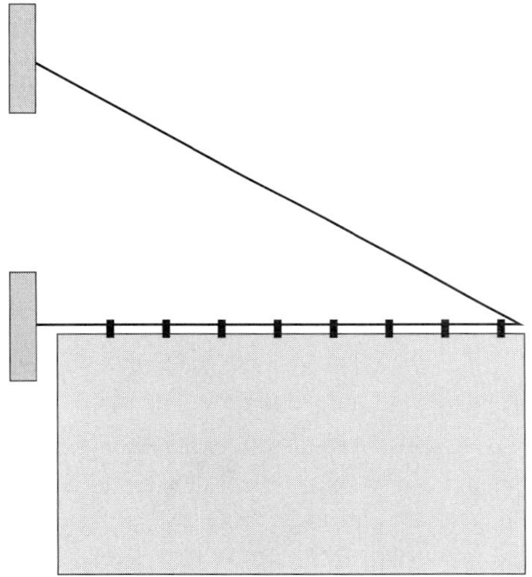

Fig. 8.8.1.1
Signboard structure.

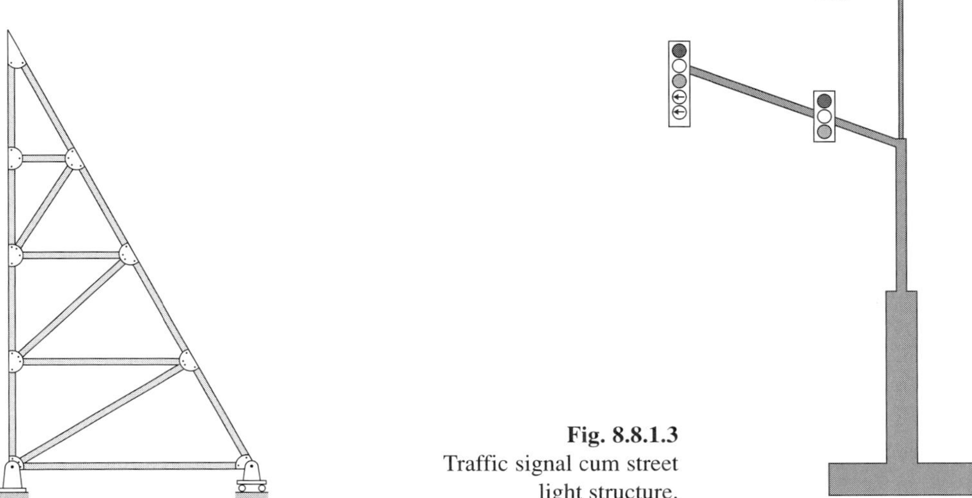

Fig. 8.8.1.2
Billboard supporting structure.

Fig. 8.8.1.3
Traffic signal cum street light structure.

As we can see in each of the above examples, the structural shape and topology is such that the structure resists the loading as efficiently as possible. The primary loading is due to the wind loads. In the case of the billboard supporting structure, the width of the truss increases towards the base. Similarly, with the traffic signal frame, the dimensions of the frame increase towards the base, thereby providing the structure with an increased moment of inertia (or section modulus).

Two-Support Points. These planar structural systems can be supported on two edges of the rectangular box thereby providing continuous open space. Figures 8.8.1.4 through 8.8.1.6 are similar in that the support points are located at the two ends of the structure. These structures can be approximated as planar trusses. However, depending on the modeling of the connections and the loads, they can perhaps be modeled more accurately as planar frames.

The load path with the roof truss and the simply supported bridge structure is such that the one can visualize them as behaving as simply supported beams. The top chord members in the roof truss are in compression whereas the bottom chord members are in tension. The same comments apply to the members in the bridge structure. On the other hand, the second bridge structure behaves more like a fixed-fixed beam.

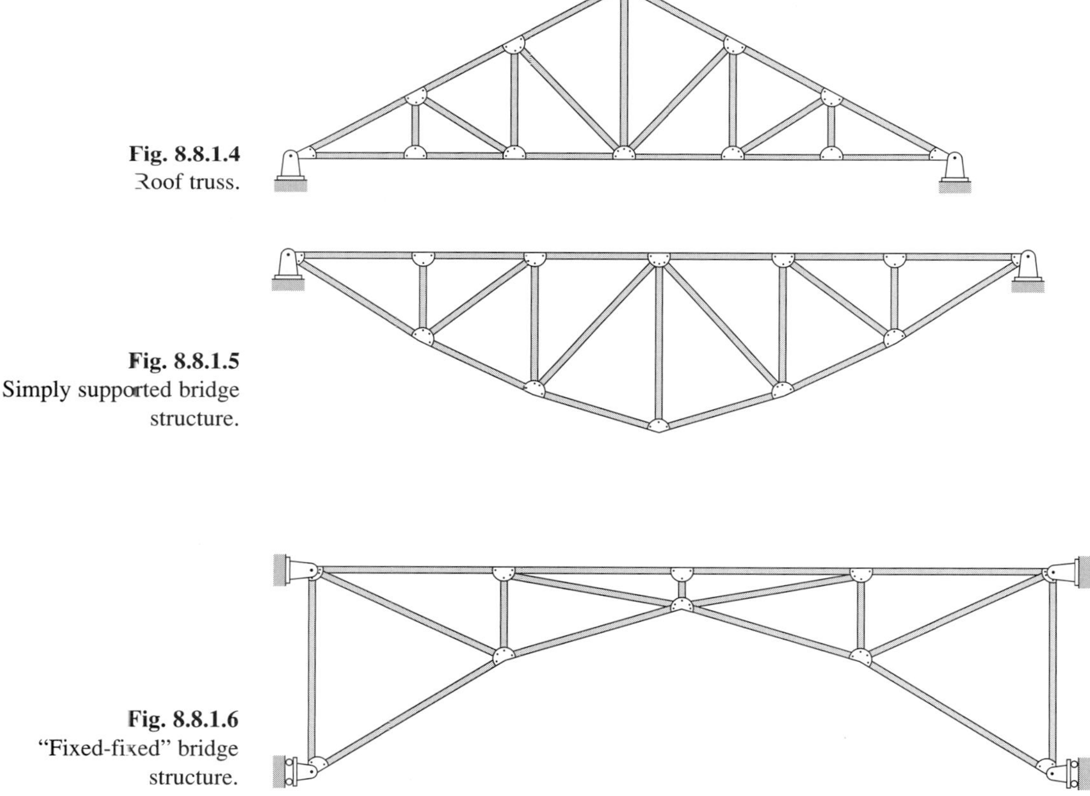

Fig. 8.8.1.4 Roof truss.

Fig. 8.8.1.5 Simply supported bridge structure.

Fig. 8.8.1.6 "Fixed-fixed" bridge structure.

Let's consider two more examples in which the structural systems are utilized to enclose a required area. Figure 8.8.1.7 shows an industrial frame. Once again the support points are at the ends of the structure on the bottom edge of the bounding box. A supporting structure for a greenhouse or a shed is shown in Fig. 8.8.1.8. In this example, the supporting points are on opposite sides of the bounding box.

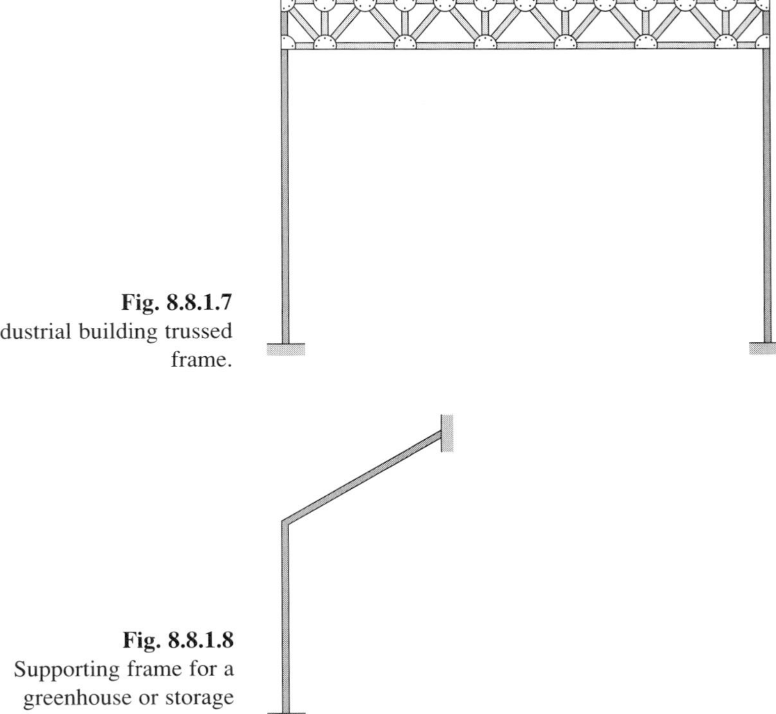

Fig. 8.8.1.7
Industrial building trussed frame.

Fig. 8.8.1.8
Supporting frame for a greenhouse or storage shed.

Figure 8.8.1.9 shows another form of industrial building—a transverse bent with a "knee brace." The brace provides additional rigidity to prevent or minimize excessive sideways deformation of the entire frame, or "racking" of the structure.

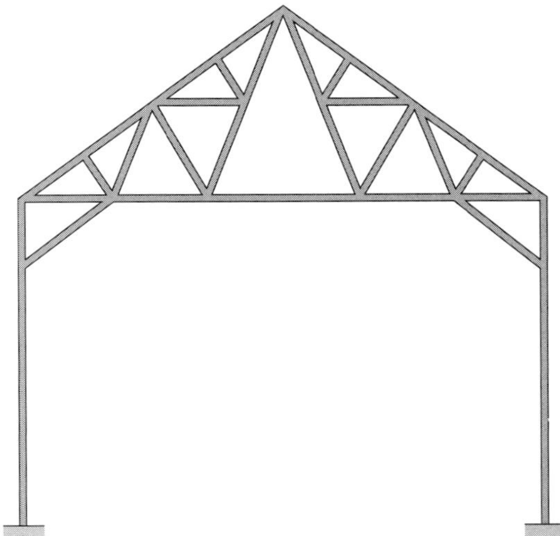

Fig. 8.8.1.9
Transverse bent.

Multiple Support Points. The class of planar structural systems in which the support points can be placed at several locations is quite common. These structural systems are typically used to enclose a large area and are not constrained to provide a large continuous (or open) space. Figure 8.8.1.10 shows a planar frame used in an office building.

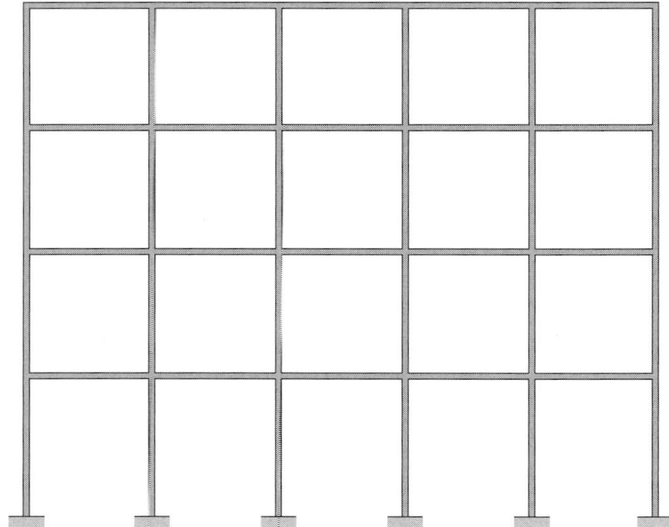

Fig. 8.8.1.10 Multi-storied office building.

Figure 8.8.1.11 shows a multi-bay industrial building where once again a large area is enclosed but intermediate support(s) do not hinder the use of the open space. Both these systems are subjected to gravity and lateral loads, with the load path finally terminating at the base of the structure.

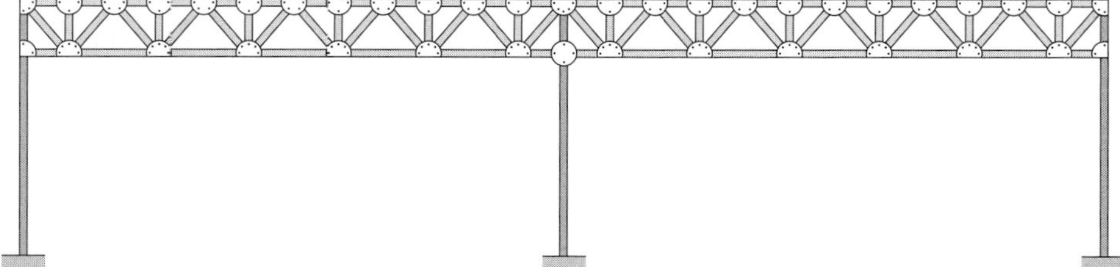

Fig. 8.8.1.11 Multi-bay industrial building.

One could look at a multitude of different structural systems to understand the relationship between the architectural form that is based on the client's need and aesthetics, and the structural form that efficiently resists the loads acting on the structure. Some of these ideas are explored as exercises in this section.

8.8.2 Optimal Design Using the GS-USA© Program

In this section we look at the design capabilities of the GS-USA frame program. In addition to the analysis capabilities discussed in Chapter 7, the program can also be used for sizing, shape, and topology optimal design. Details of the design features can be found in the GS-USA Tutorial contained in the CD-ROM.

Structural Parameters. Structural parameters are those attributes of the structural model that can be used as design variables or in defining constraints. The following structural parameters are supported in the current version of the program.

Entity	Attribute
(A) Cross-sectional properties	(1) Dimension (2) Actual cross-section if AISC cross-sections are used
(B) Nodes	(1) *X* Coordinate (2) *Y* Coordinate (3) *X* Fixity code (4) *Y* Fixity code (5) *Z* Fixity code
(C) Nodes	(1) *X* Displacement (2) *Y* Displacement (3) *Z* Rotation (4) *X* Reaction (5) *Y* Reaction (6) *Z* (moment) Reaction
(D) Elements	(1) Maximum tensile stress (2) Maximum compressive stress (3) Maximum shear stress (4) Euler buckling (5) AISC-ASD89 code (6) Connectivity
(E) Modes	(1) Natural frequency (Hz)

Design Variables. The design variables can be A1, A2, B1, B2, B3, B4, B5, and D6. The constraints can be defined using A1, B1, B2, B3, B4, B5, C1, C2, C3, C4, C5, C6, D1, D2, D3, D4, D5, and E1. The current version of the program restricts the number of design variables to a maximum of 10.

Inequality Constraints. The inequality constraints can include limits on stresses, displacements, natural frequencies, and relationships between cross-sectional parameters.

Equality Constraints. The equality constraints can be used to tie one design parameter to another that is also defined as a design variable. For example, if the x coordinate of node 7 should equal the x coordinate of node 4, then define two design parameters one for each node. Then define either node 4 or node 7 as a design variable. This will ensure that the x coordinate of the selected node is changed during the design process. Now define the equality constraint, for example

$$(1.0)(\text{node } 4\ \text{x}) - (1.0)(\text{node } 7\ \text{x}) = 0$$

The equality constraints are always satisfied (or enforced) in the program during the design process.

Objective Function. The objective function can be one of the following.

1. Mass of the structure
2. Weight of the structure
3. Volume of the structure
4. Cost of the structure
5. Support reaction
6. Nodal displacement

These structural attributes can be either maximized or minimized.

8.8.3 Case Studies

We look at several examples in this section illustrating some of the ideas we have discussed in this chapter. These examples will be solved using the design features in the GS-USA program.

EXAMPLE 8.8.1 ***Design of a Continuous Beam***

Figure E8.8.1(a) shows a continuous beam that is fixed at A with rollers at B and C. The beam is made of steel with the following properties: modulus of elasticity = 29(10^6) psi, allowable normal stress = 10000 psi, allowable shear stress = 8000 psi. The entire beam is defined by a symmetric I section. The design problem is to find the lightest cross-section so that the beam normal stress (due to bending moment) and shear stress (due to shear force) do not exceed the allowable values.

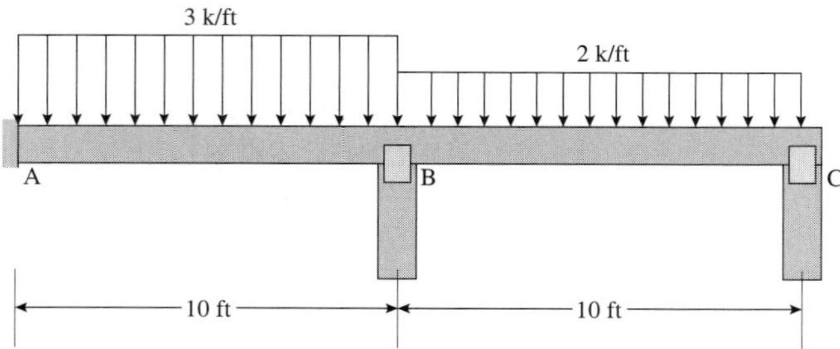

Fig. E8.8.1(a)

SOLUTION

We formulate the problem first and then solve the design problem using the GS-USA program. Note that the structure is statically indeterminate.

Problem Formulation. The design problem now can be stated as

Find $\mathbf{x} = \{w_h, w_t, f_w, f_t\}$ (1)

to minimize $f(\mathbf{x}) \Rightarrow$ volume (2)

such that $\sigma^c_{\max} \le \sigma_a$ (3)

$$\sigma^t_{\max} \le \sigma_a \tag{4}$$

$$\tau_{\max} \le \tau_a \tag{5}$$

$$w_h \le 2f_w \tag{6}$$

$$6'' \le w_h \le 18'' \qquad 4'' \le f_w \le 9'' \tag{7}$$

$$0.1'' \le w_t, f_t \ge 0.7'' \tag{8}$$

Equation (1) shows the four design variables whose values are being sought as the design answers (see Fig. E8.8.1(b)). Equation (2) represents the objective function, the volume of the continuous beam. Equation (3) represents the constraint on the maximum compressive stress in the beam. Similarly, Eq. (4) represents the constraint on the maximum tensile stress, and Eq. (5) the constraint on the maximum shear stress. Since all the cross-sectional dimensions are being varied in this exercise, constraint Eq. (6) ensures that the cross-section that does not become oddly shaped (thin and long). Equations (7) and (8) place lower and upper limits on the four design variables.

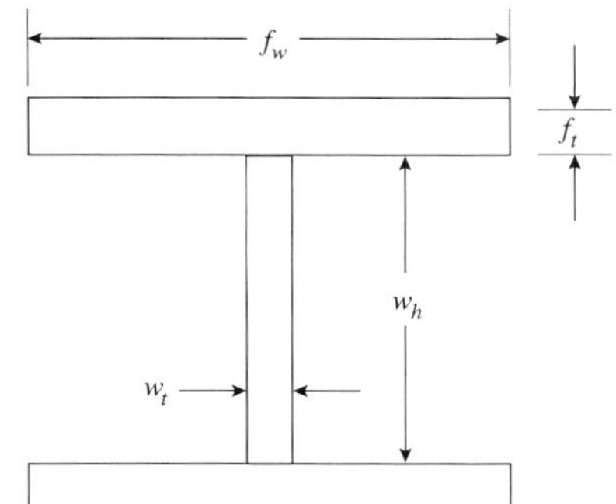

Fig. E8.8.1(b) Beam cross-section design variables

Step 1: Create the analysis model. Use lbm, lb, in, and F as the units. Place the origin of the coordinate system at A. The model will contain one material group, one property group with the cross-section type defined as *symmetric I-section*, three nodes, two elements, and two element loads. Assume the following values as the initial guess for the four design variables : $\mathbf{x} = \{w_h, w_t, f_w, f_t\} = \{6", 0.2", 4", 0.1"\}$.

Step 2: Analyze the structure. Ensure that there are no errors.

Step 3: Define the design model. Click on **Design** > **Define Parameters**. Define the seven parameters required to solve this problem as per the following table.

Parameter	Name	Type, attributes
1	Web height	(X-Sectional Properties, 1, Web height)
2	Web thk	(X-Sectional Properties, 1, Web thickness)
3	Flange width	(X-Sectional Properties, 1, Flange width)
4	Flange thk	(X-Sectional Properties, 1, Flange thickness)
5	Comp Stress	(Element Data, 0, 1, Compressive Stress)
6	Tensile Stress	(Element Data, 0, 1, Tensile Stress)
7	Shear Stress	(Element Data, 0, 1, Shear Stress)

The parameter names are restricted to a maximum of 15 characters.

X-Sectional Properties require two pieces of data—the Property Group number and the dimension associated with that property group. Element Data require three pieces of data—the element number (a 0 value signifies all elements), the load case number (a 0 value signifies all load cases), and the element attribute (compressive stress, tensile stress, etc.).

Step 4: Define the design variables. Click on **Design** > **Define Design Variables**. Select the four design variables one at a time from the Parameter list and then specify the lower bound, upper bound and the precision. Use 0.5 in as the precision for the web height and the flange width. Use 0.1 in as the precision for the web thickness and the flange thickness.

Step 5: Define the design goal. Click on **Design** > **Define Goals**. Select Volume from the Parameter list.

Step 6: Define the constraints. Click on **Design** > **Define Checks**. There are four constraints to be defined. Select Comp Stress from the first parameter box (the first Coef is 1.0). There is no second parameter to select. Type –10000 in the last box. The constraint type is ≤ (the default type). This is because the compressive stress constraint Eq. (3) is expressed as

$$\sigma^c_{\max} - \sigma_a \leq 0 \tag{9}$$

Repeat the same for the tensile stress and the shear stress constraints using the Coef value as 1.0, and –10000 and –8000 respectively, in the last box. Finally, select Web height as the first Parameter (with the first Coef as 1.0), select Flange Width as the second parameter (with the second Coef as -2.0) and 0.0 in the last box. This represents constraint Eq. (6) rewritten as

$$w_h - 2f_w \le 0 \tag{10}$$

Step 7: Now you are ready to design the beam. Click on **Design** > **Design Now**. The program first checks for formulation errors. If there are no errors or warnings, click on the Proceed button.

Step 8: Click on the OK button after the program has run through the 30 design iterations. Now use the **Design** > **View ...** menu options to view the design problem and results. Print the results file and analyze the output.

Summary of the Results. Selected portions of the Results file are shown below.

```
                    DESIGN RESULTS
                    ==============

   PROBLEM SIZE
   ------------
    Number of Design Parameters:      7
     Number of Design Variables:      4
        Number of Design Checks:      4
Minimize Objective Function: Mass. Current Value: 15.4431 slg

------------------------------------------------------------------
                          DESIGN PARAMETERS
Number    Name               Attributes
------------------------------------------------------------------
1       web height       Property Group: 1. Sym I. Web height.
2       web thk          Property Group: 1. Sym I. Web thickness.
3       flange width     Property Group: 1. Sym I. Flange width.
4       flange thk       Property Group: 1. Sym I. Flange thickness.
5       comp stress      Element: 0. LC: 0. Compressive Stress.
6       tensile stress   Element: 0. LC: 0. Tensile Stress.
7       shear stress     Element: 0. LC: 0. Shear Stress.

------------------------------------------------------------------------
                          DESIGN VARIABLES : DEFINITION
Number  Design Variable     Lower Bound      Upper Bound      Precision
------------------------------------------------------------------------
 1          web height       6                18               0.5
 2             web thk       0.1              0.7              0.1
 3        flange width       4                9                0.5
 4          flange thk       0.1              0.7              0.1

               ----------------------------------
                DESIGN VARIABLES : FINAL VALUES
               Number  Design Variable     Value
               ----------------------------------
                1          web height       10.5
                2             web thk       0.2
                3        flange width       6.5
                4          flange thk       0.4
```

```
------------------------------------------------------------------
                      CONSTRAINT VALUES
   Number           Coef * Constraint                    Value
------------------------------------------------------------------
    1            {1 * (Comp. Stress: (Element,LC)(1,1))} -0.0237168
    2            {1 * (Comp. Stress: (Element,LC)(2,1))} -0.0237168
    3            {1 * (Tens. Stress: (Element,LC)(1,1))} -0.0237168
    4            {1 * (Tens. Stress: (Element,LC)(2,1))} -0.0237168
    5            {1 * (Shear Stress: (Element,LC)(1,1))} -0.104064
    6            {1 * (Shear Stress: (Element,LC)(2,1))} -0.252235
    7      {X-Section: 1 * (1,0)}{X-Section: -2 * (1,2)} -0.192308

Max:1            {1 * (Comp. Stress: (Element,LC)(1,1))} -0.0237168

Notes:
A negative constraint value indicates a satisfied constraint.
A positive constraint value indicates an unacceptable design.
```

To summarize, the best solution obtained by the program is as follows.
From Problem Size section:

$$\text{Mass} = 15.4 \text{ slugs}$$

From Design Variables: Final Values section:

$$\text{Web height} = 10.5 \text{ in} \qquad \text{Web thickness} = 0.2 \text{ in}$$
$$\text{Flange width} = 6.5 \text{ in} \qquad \text{Flange thickness} = 0.4 \text{ in}$$

From Constraint Values section:

By looking for the Max value we can conclude that the compressive and tensile stresses in elements 1 and 2 control the design.

EXAMPLE 8.8.2 ***Design of a Roof Truss***

Figure E8.8.2(a) shows the outline of a roof truss. This outline cannot be changed. The roof trusses are placed 2 ft o.c. The design problem is to design the lightest truss. The material is wood with the following properties: modulus of elasticity E = 1500000 psi, weight density ρ = 0.02 lb/in^3, allowable compressive stress $\sigma_a^c = 1500\,\text{psi}$, allowable tensile stress $\sigma_a^t = 2000\,\text{psi}$ and allowable shear stress τ_a = 200 psi. The only loading to be considered is gravity loading on the roof, which has been computed to be 20 psf on the top chord and 5 psf on the bottom chord. The maximum vertical displacement is to be restricted to L/360 or 1 in.

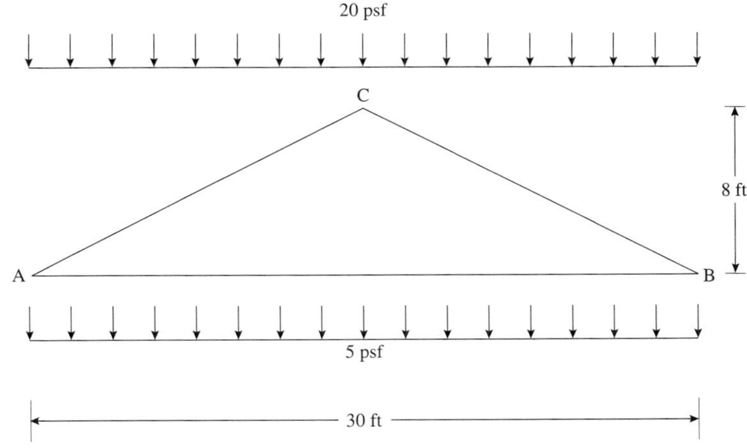

Fig. E8.8.2(a)

Solution. We solve this design problem in stages. In the first stage we will treat the problem as a sizing optimal design problem.

Sizing Optimal Design. Several structural data are not defined and the designer must make decisions to define the missing pieces of information. First, we assume that all the members have a rectangular cross-section (see Fig. E8.8.2(b)). Second, the internal layout of the truss is not defined. We assume a Fink truss (see Fig. E8.8.2(c)). Third, we assume that the top chord members have the same cross-sectional dimensions, the bottom chord members have the same cross-sectional dimensions, and the web members have the same cross-sectional dimensions. Lastly, we assume that the left and right heels (where the truss is supported) are at pin supports. Furthermore, assuming that (i) the loads act directly on the top chord and bottom chord members and (ii) connections are rigid connections, it is necessary for us to model the roof truss as a planar frame. We take lb, in as the problem units. Since the trusses are spaced 2 ft apart, the loading is as follows:

Top chord, $w_{TC} = 20 \text{ lb / ft}^2 \times 2 \text{ ft} = 40 \text{ lb / ft} = 3.33 \text{ lb / in}$

Bottom chord, $w_{BC} = 5 \text{ lb / ft}^2 \times 2 \text{ ft} = 10 \text{ lb / ft} = 0.833 \text{ lb / in}$

Problem Formulation. The design problem now can be stated as

Find $$\mathbf{x} = \{b_{BC}, h_{BC}, b_{TC}, h_{TC}, b_{WEB}, h_{WEB}\} \quad (1)$$

to minimize $$f(\mathbf{x}) \Rightarrow \text{volume} = \sum_{i=1}^{15} A_i L_i \quad (2)$$

such that $$\sigma^c_{\max} \le 1500 \quad (3)$$

$$\sigma^t_{\max} \le 2000 \quad (4)$$

$$\tau_{\max} \le 200 \quad (5)$$

$$|\Delta_j| \le 1'' \qquad j \in \text{y} - \text{displacement of all nodes} \quad (6)$$

$$h_i \ge b_i \qquad i = 1,\ 2,\ 3 \quad (7)$$

$$1'' \le x_i \le 6'' \qquad i = 1,\ \ldots,\ 6 \quad (8)$$

Equation (1) shows the six design variables whose values are being sought as the design answers. Equation (2) represents the objective function, the volume of all the members of the roof truss. The requirements on the maximum compressive, tensile, and shear stress in a member are captured in Eqs. (3)–(5). Equation (6) takes care of the restriction on the vertical displacement. Equation (7) is needed to ensure that the height of the cross-section is in fact greater than the width. Finally, Eq. (8) is used to restrict the dimensions of both the height and the width of the cross-sections.

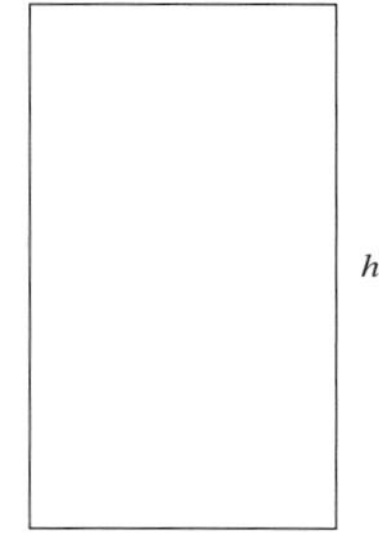

Fig. E8.8.2(b) Beam cross-section design variables.

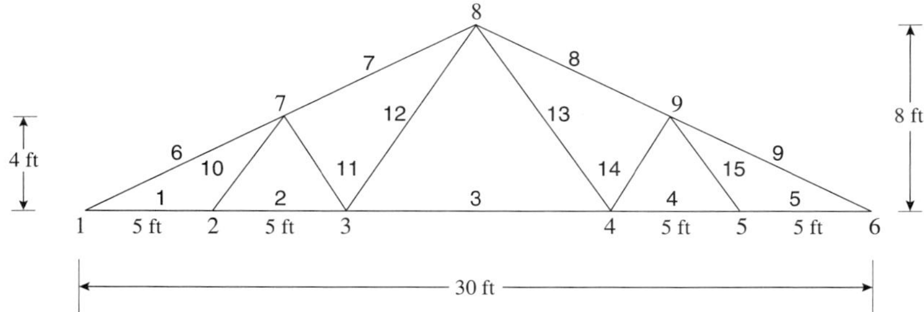

Fig. E8.8.2(c) Initial layout of truss.

We have not imposed the Euler buckling constraint because it is not clear what are the end conditions of the members are. Recall that $P_{cr} = (\pi^2 EI/(kL)^2)$, where k is a function of the end conditions of the member. We address this issue in greater detail in the next chapter. Instead, we have reduced the allowable compressive stress (from 2000 psi to 1500 psi) and restricted all members to a maximum length of 10 ft.

The solution steps in using the GS-USA program are similar to the previous example and are not shown here (the problem file can be found on the CD-ROM).

Summary of the Results. Selected portions of the Results file are shown below.

```
                    DESIGN RESULTS
                    ==============


  PROBLEM SIZE
  ------------
   Number of Design Parameters:    10
    Number of Design Variables:     6
       Number of Design Checks:     7
Minimize Objective Function: Volume. Current Value: 3301.68 in^3

-----------------------------------------------------------------
                    DESIGN PARAMETERS
Number   Name                Attributes
-----------------------------------------------------------------
1      BCHeight         Property Group: 1. Rect Solid. Height.
2      BCWidth          Property Group: 1. Rect Solid. Width.
3      TCHeight         Property Group: 2. Rect Solid. Height.
4      TCWidth          Property Group: 2. Rect Solid. Width.
5      WEBHeight        Property Group: 3. Rect Solid. Height.
6      WEBWidth         Property Group: 3. Rect Solid. Width.
7      CompStress       Element: 0. LC: 0. Compressive Stress.
8      TensStress       Element: 0. LC: 0. Tensile Stress.
9      ShearStress      Element: 0. LC: 0. Shear Stress.
10     VertDisp         Node: 0. LC: 0. Y Displacement.

-----------------------------------------------------------------------
                    DESIGN VARIABLES : DEFINITION
Number  Design Variable    Lower Bound      Upper Bound       Precision
-----------------------------------------------------------------------
 1          BCHeight       1                6                 0.0195508
 2           BCWidth       1                6                 0.0195508
 3          TCHeight       1                6                 0.0195508
 4           TCWidth       1                6                 0.0195508
 5         WEBHeight       1                6                 0.0195508
 6          WEBWidth       1                6                 0.0195508
```

```
------------------------------------
 DESIGN VARIABLES : FINAL VALUES
Number  Design Variable      Value
------------------------------------
 1            BCHeight       2.25125
 2             BCWidth       1
 3            TCHeight       3.03328
 4             TCWidth       1.62563
 5           WEBHeight       1.05865
 6            WEBWidth       1
```

```
-------------------------------------------------------------------------
                        CONSTRAINT VALUES
Number                Coef * Constraint                          Value
-------------------------------------------------------------------------
 1              {1 * (Comp. Stress: (Element,LC)(1,1))} -0.576853
 2              {1 * (Comp. Stress: (Element,LC)(2,1))} -0.459765
 3              {1 * (Comp. Stress: (Element,LC)(3,1))} -0.256944
 4              {1 * (Comp. Stress: (Element,LC)(4,1))} -0.459765
 5              {1 * (Comp. Stress: (Element,LC)(5,1))} -0.576853
 6              {1 * (Comp. Stress: (Element,LC)(6,1))} -0.0244658
 7              {1 * (Comp. Stress: (Element,LC)(7,1))} -0.0701519
 8              {1 * (Comp. Stress: (Element,LC)(8,1))} -0.0701519
 9              {1 * (Comp. Stress: (Element,LC)(9,1))} -0.0244658
 10            {1 * (Comp. Stress: (Element,LC)(10,1))} -0.844018
 11            {1 * (Comp. Stress: (Element,LC)(11,1))} -0.577667
 12            {1 * (Comp. Stress: (Element,LC)(12,1))} -1500
 13            {1 * (Comp. Stress: (Element,LC)(13,1))} -1500
 14            {1 * (Comp. Stress: (Element,LC)(14,1))} -0.577667
 15            {1 * (Comp. Stress: (Element,LC)(15,1))} -0.844018
 16             {1 * (Tens. Stress: (Element,LC)(1,1))} -0.607426
 17             {1 * (Tens. Stress: (Element,LC)(2,1))} -0.536133
 18             {1 * (Tens. Stress: (Element,LC)(3,1))} -0.576613
 19             {1 * (Tens. Stress: (Element,LC)(4,1))} -0.536133
 20             {1 * (Tens. Stress: (Element,LC)(5,1))} -0.607426
 21             {1 * (Tens. Stress: (Element,LC)(6,1))} -0.540883
 22             {1 * (Tens. Stress: (Element,LC)(7,1))} -0.525893
 23             {1 * (Tens. Stress: (Element,LC)(8,1))} -0.525893
 24             {1 * (Tens. Stress: (Element,LC)(9,1))} -0.540883
 25            {1 * (Tens. Stress: (Element,LC)(10,1))} -0.819599
 26            {1 * (Tens. Stress: (Element,LC)(11,1))} -2000
 27            {1 * (Tens. Stress: (Element,LC)(12,1))} -0.676766
 28            {1 * (Tens. Stress: (Element,LC)(13,1))} -0.676766
 29            {1 * (Tens. Stress: (Element,LC)(14,1))} -2000
 30            {1 * (Tens. Stress: (Element,LC)(15,1))} -0.819599
 31             {1 * (Shear Stress: (Element,LC)(1,1))} -0.870324
 32             {1 * (Shear Stress: (Element,LC)(2,1))} -0.890048
 33             {1 * (Shear Stress: (Element,LC)(3,1))} -0.833426
 34             {1 * (Shear Stress: (Element,LC)(4,1))} -0.890048
 35             {1 * (Shear Stress: (Element,LC)(5,1))} -0.870324
 36             {1 * (Shear Stress: (Element,LC)(6,1))} -0.735432
 37             {1 * (Shear Stress: (Element,LC)(7,1))} -0.764794
 38             {1 * (Shear Stress: (Element,LC)(8,1))} -0.764794
 39             {1 * (Shear Stress: (Element,LC)(9,1))} -0.735432
 40            {1 * (Shear Stress: (Element,LC)(10,1))} -0.986513
 41            {1 * (Shear Stress: (Element,LC)(11,1))} -0.992646
```

```
42              {1 * (Shear Stress: (Element,LC)(12,1))} -0.996215
43              {1 * (Shear Stress: (Element,LC)(13,1))} -0.996215
44              {1 * (Shear Stress: (Element,LC)(14,1))} -0.992646
45              {1 * (Shear Stress: (Element,LC)(15,1))} -0.986513
46            {1 * (Displacement: (Node,Dof,LC)(1,2,1))} -1
47            {1 * (Displacement: (Node,Dof,LC)(2,2,1))} -0.906018
48            {1 * (Displacement: (Node,Dof,LC)(3,2,1))} -0.881697
49            {1 * (Displacement: (Node,Dof,LC)(4,2,1))} -0.881697
50            {1 * (Displacement: (Node,Dof,LC)(5,2,1))} -0.906018
51            {1 * (Displacement: (Node,Dof,LC)(6,2,1))} -1
52            {1 * (Displacement: (Node,Dof,LC)(7,2,1))} -0.886944
53            {1 * (Displacement: (Node,Dof,LC)(8,2,1))} -0.923666
54            {1 * (Displacement: (Node,Dof,LC)(9,2,1))} -0.886944
55         {X-Section: 1 * (1,0)}{X-Section: -1 * (1,1)} -0.555803
56         {X-Section: 1 * (2,0)}{X-Section: -1 * (2,1)} -0.464071
57         {X-Section: 1 * (3,0)}{X-Section: -1 * (3,1)} -0.0554029

Max:6            {1 * (Comp. Stress: (Element,LC)(6,1))} -0.0244658

Notes:
A negative constraint value indicates a satisfied constraint.
A positive constraint value indicates an unacceptable design.
```

To summarize, the best solution obtained by the program is as follows.
From Problem Size section:

$$\text{Volume} = 3302 \text{ in}^3$$

From Design Variables: Final Values section:

Section	Height (in)	Width (in)
Top Chord members	3"	1.6"
Bottom Chord members	2.3"	1"
Web members	1.1"	1"

From Constraint Values section:

By looking for the Max value we can conclude that the compressive stress in elements 6 and 9 controls the design.

Sizing and Shape Optimal Design

We now solve the same problem but include shape design variables. In addition to the sizing design variables, we will attempt to find the optimal locations of nodes 2, 3, and 7. We maintain structural symmetry by adjusting the locations of nodes 4, 5, and 9 appropriately. Hence, we have three new design variables, the x coordinates of nodes 2 and 3, and y coordinate of node 7. Additional equality constraints need to be introduced so that the final structural geometry is symmetric. These constraints are

$$X_2 + X_5 - 360 = 0 \tag{9}$$

$$X_3 + X_4 - 360 = 0 \tag{10}$$

$$Y_7 - Y_9 = 0 \tag{11}$$

$$\frac{Y_7}{X_7} = \frac{8}{15} \Rightarrow 15Y_7 - 8X_7 = 0 \tag{12}$$

$$X_7 + X_9 - 360 = 0 \tag{13}$$

Equations (9), (10), and (13) maintain symmetry of nodes 4, 5, and 9 with respect to 3, 2, and 7 respectively about the plane of symmetry ($X = 180$ in). Equation (11) ensures that the y coordinates of nodes 7 and 9 are equal. Equation (12) ensures that node 7 remains on the top chord; i.e., if the y coordinate of node 7 changes, then the x coordinate is suitably adjusted.

We could have recognized that the structure geometry, topology, loads, and boundary conditions are symmetric and used only one half of the structure. However, the intent of this example is to show how structural symmetry (geometry and topology) can be maintained even if the rest of the model, e.g., loading, is not symmetric. The new problem formulation is as follows.

Problem Formulation. The design problem can now be stated as

Find $$\mathbf{x} = \{b_{BC}, h_{BC}, b_{TC}, h_{TC}, b_{WEB}, h_{WEB}, X_2, X_3, Y_7\} \tag{14}$$

to minimize $$f(\mathbf{x}) \Rightarrow \text{volume} = \sum_{i=1}^{15} A_i L_i \tag{15}$$

such that $$\sigma^c_{\max} \leq 1500 \tag{16}$$

$$\sigma^t_{\max} \leq 2000 \tag{17}$$

$$\tau_{\max} \leq 200 \tag{18}$$

$$|\Delta_j| \leq 1'' \qquad j \in \text{y} - \text{displacement of all nodes} \tag{19}$$

$$h_i \geq b_i \qquad i = 1,\ 2,\ 3 \tag{20}$$

$$1'' \leq x_i \leq 6'' \qquad i = 1,\ \ldots,\ 6 \tag{21}$$

$$24'' \leq X_2 \leq 84'' \tag{22}$$

$$85'' \leq X_3 \leq 144'' \tag{23}$$

$$36'' \leq Y_7 \leq 72'' \tag{24}$$

Summary of the Results. Selected portions of the Results file are shown below.

```
                DESIGN RESULTS
                ==============

       PROBLEM SIZE
       ------------
        Number of Design Parameters:    18
         Number of Design Variables:     9
            Number of Design Checks:    13
Minimize Objective Function: Volume. Current Value: 2820.94 in^3
```

DESIGN PARAMETERS

Number	Name	Attributes
1	BCHeight	Property Group: 1. Rect Solid. Height.
2	BCWidth	Property Group: 1. Rect Solid. Width.
3	TCHeight	Property Group: 2. Rect Solid. Height.
4	TCWidth	Property Group: 2. Rect Solid. Width.
5	WEBHeight	Property Group: 3. Rect Solid. Height.
6	WEBWidth	Property Group: 3. Rect Solid. Width.
7	CompStress	Element: 0. LC: 0. Compressive Stress.
8	TensStress	Element: 0. LC: 0. Tensile Stress.
9	ShearStress	Element: 0. LC: 0. Shear Stress.
10	VertDisp	Node: 0. LC: 0. Y Displacement.
11	XCoor2	Node: 2. X Coordinate.
12	XCoor3	Node: 3. X Coordinate.
13	XCoor4	Node: 4. X Coordinate.
14	XCoor5	Node: 5. X Coordinate.
15	YCoor7	Node: 7. Y Coordinate.
16	YCoor9	Node: 9. Y Coordinate.
17	XCoor7	Node: 7. X Coordinate.
18	XCoor9	Node: 9. X Coordinate.

DESIGN VARIABLES : DEFINITION

Number	Design Variable	Lower Bound	Upper Bound	Precision
1	BCHeight	1	6	0.0195508
2	BCWidth	1	6	0.0195508
3	TCHeight	1	6	0.0195508
4	TCWidth	1	6	0.0195508
5	WEBHeight	1	6	0.0195508
6	WEBWidth	1	6	0.0195508
7	XCoor2	24	84	0.234609
8	XCoor3	85	144	0.230699
9	YCoor7	36	72	0.140766

DESIGN VARIABLES : FINAL VALUES

Number	Design Variable	Value
1	BCHeight	1.31281
2	BCWidth	1.15641
3	TCHeight	3.4243
4	TCWidth	1.31281
5	WEBHeight	1
6	WEBWidth	1
7	XCoor2	80.7754
8	XCoor3	138.522
9	YCoor7	45.009

CONSTRAINT INFORMATION

Num	Coef	Parameter	Coef	Parameter	Constant	Type
1	1	CompStress			-1500	<=
2	1	TensStress			-2000	<=

```
 3   1          ShearStress                                  -200        <=
 4   1          VertDisp                                     1           >=
 5   1          BCHeight        -1          BCWidth          0           >=
 6   1          TCHeight        -1          TCWidth          0           >=
 7   1          WEBHeight       -1          WEBWidth         0           >=
 8   1          XCoor2          1           XCoor5           -360        =
 9   1          XCoor3          1           XCoor4           -360        =
10   1          YCoor7          -1          YCoor9           0           =
11   1          XCoor4          -1          XCoor3           -120        <=
12   15         YCoor7          -8          XCoor7           0           =
13   1          XCoor9          1           XCoor7           -360        =

-----------------------------------------------------------------
                         CONSTRAINT VALUES
Number              Coef * Constraint                       Value
-----------------------------------------------------------------
 1                {1 * (Comp. Stress: (Element,LC)(1,1))}  -0.243396
 2                {1 * (Comp. Stress: (Element,LC)(2,1))}  -0.386794
 3                {1 * (Comp. Stress: (Element,LC)(3,1))}  -0.00924331
 4                {1 * (Comp. Stress: (Element,LC)(4,1))}  -0.386794
 5                {1 * (Comp. Stress: (Element,LC)(5,1))}  -0.243396
 6                {1 * (Comp. Stress: (Element,LC)(6,1))}  -0.00345403
 7                {1 * (Comp. Stress: (Element,LC)(7,1))}  -0.0688253
 8                {1 * (Comp. Stress: (Element,LC)(8,1))}  -0.0688252
 9                {1 * (Comp. Stress: (Element,LC)(9,1))}  -0.00345403
10               {1 * (Comp. Stress: (Element,LC)(10,1))}  -0.654977
11               {1 * (Comp. Stress: (Element,LC)(11,1))}  -0.527565
12               {1 * (Comp. Stress: (Element,LC)(12,1))}  -1500
13               {1 * (Comp. Stress: (Element,LC)(13,1))}  -1500
14               {1 * (Comp. Stress: (Element,LC)(14,1))}  -0.527565
15               {1 * (Comp. Stress: (Element,LC)(15,1))}  -0.654977
16                {1 * (Tens. Stress: (Element,LC)(1,1))}  -0.360052
17                {1 * (Tens. Stress: (Element,LC)(2,1))}  -0.472992
18                {1 * (Tens. Stress: (Element,LC)(3,1))}  -0.491535
19                {1 * (Tens. Stress: (Element,LC)(4,1))}  -0.472993
20                {1 * (Tens. Stress: (Element,LC)(5,1))}  -0.360052
21                {1 * (Tens. Stress: (Element,LC)(6,1))}  -0.552176
22                {1 * (Tens. Stress: (Element,LC)(7,1))}  -0.521086
23                {1 * (Tens. Stress: (Element,LC)(8,1))}  -0.521086
24                {1 * (Tens. Stress: (Element,LC)(9,1))}  -0.552176
25               {1 * (Tens. Stress: (Element,LC)(10,1))}  -0.682842
26               {1 * (Tens. Stress: (Element,LC)(11,1))}  -2000
27               {1 * (Tens. Stress: (Element,LC)(12,1))}  -0.706439
28               {1 * (Tens. Stress: (Element,LC)(13,1))}  -0.706439
29               {1 * (Tens. Stress: (Element,LC)(14,1))}  -2000
30               {1 * (Tens. Stress: (Element,LC)(15,1))}  -0.682841
31                {1 * (Shear Stress: (Element,LC)(1,1))}  -0.830777
32                {1 * (Shear Stress: (Element,LC)(2,1))}  -0.879186
33                {1 * (Shear Stress: (Element,LC)(3,1))}  -0.829241
34                {1 * (Shear Stress: (Element,LC)(4,1))}  -0.879186
35                {1 * (Shear Stress: (Element,LC)(5,1))}  -0.830777
36                {1 * (Shear Stress: (Element,LC)(6,1))}  -0.705697
37                {1 * (Shear Stress: (Element,LC)(7,1))}  -0.731889
38                {1 * (Shear Stress: (Element,LC)(8,1))}  -0.731889
39                {1 * (Shear Stress: (Element,LC)(9,1))}  -0.705697
40               {1 * (Shear Stress: (Element,LC)(10,1))}  -0.973597
41               {1 * (Shear Stress: (Element,LC)(11,1))}  -0.992975
```

```
42          {1 * (Shear Stress: (Element,LC)(12,1))}  -0.995771
43          {1 * (Shear Stress: (Element,LC)(13,1))}  -0.995771
44          {1 * (Shear Stress: (Element,LC)(14,1))}  -0.992975
45          {1 * (Shear Stress: (Element,LC)(15,1))}  -0.973597
46        {1 * (Displacement: (Node,Dof,LC)(1,2,1))}  -1
47        {1 * (Displacement: (Node,Dof,LC)(2,2,1))}  -0.894134
48        {1 * (Displacement: (Node,Dof,LC)(3,2,1))}  -0.890892
49        {1 * (Displacement: (Node,Dof,LC)(4,2,1))}  -0.890892
50        {1 * (Displacement: (Node,Dof,LC)(5,2,1))}  -0.894134
51        {1 * (Displacement: (Node,Dof,LC)(6,2,1))}  -1
52        {1 * (Displacement: (Node,Dof,LC)(7,2,1))}  -0.893483
53        {1 * (Displacement: (Node,Dof,LC)(8,2,1))}  -0.920555
54        {1 * (Displacement: (Node,Dof,LC)(9,2,1))}  -0.893483
55     {X-Section: 1 * (1,0)}{X-Section: -1 * (1,1)}  -0.119138
56     {X-Section: 1 * (2,0)}{X-Section: -1 * (2,1)}  -0.616619
57     {X-Section: 1 * (3,0)}{X-Section: -1 * (3,1)}   0
58     {1*(Coordinate:(Node,Dof)(4,0))}
       {-1*(Coordinate:(Node,Dof)(3,0))}              -0.16726

Max:57 {X-Section: 1 * (3,0)}{X-Section: -1 * (3,1)}  0

Notes:
A negative constraint value indicates a satisfied constraint.
A positive constraint value indicates an unacceptable design.
```

To summarize, the best solution obtained by the program is as follows.
From Problem Size section:

$$\text{Volume} = 2821 \text{ in}^3$$

From Design Variables: Final Values section:

Section	Height (in)	Width (in)
Top Chord members	3.4"	1.3"
Bottom Chord members	1.3"	1.6"
Web members	1"	1"
	Coordinate (in)	
X Coordinate Node 2	81"	
X Coordinate Node 3	139"	
Y Coordinate Node 7	45"	

From Constraint Values section:

By looking at the values in this section, we can conclude that the compressive stress in elements 6 and 9, as well as the relationship between the height and the width of the web members, control the design. The final truss shape and layout are shown in Fig. E8.8.2(d).

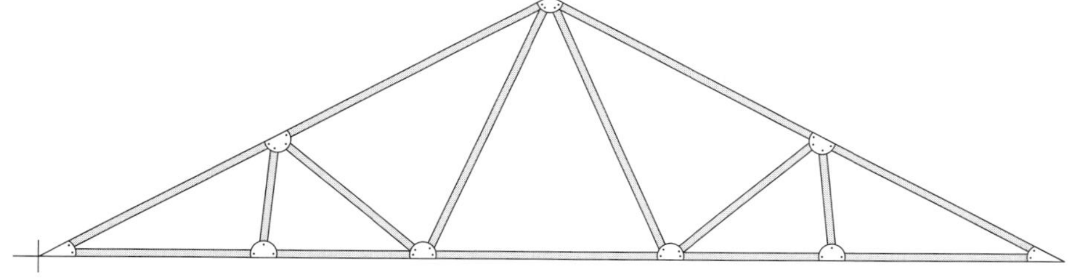

Fig. E8.8.2(d)

Issues Connected with the Design Problem Formulation

We now examine the problem formulation and the solution obtained.

1. *Structural modeling*: Is the structure a truss or a frame? An examination of the structural response (as seen in the results file) should show that the structure is closer to a frame than a truss. The bending moments in the chord members cannot be ignored. The normal stress due to the bending moment is of the same magnitude as the axial force. Are the joints correctly modeled? We had assumed that the joints are rigid connections. In reality, the joints are between a rigid connection and an internal hinge. Experience has shows that the connection details typically used in roof trusses makes the connections behave closer to a rigid connection.
2. *Self-weight*: We have not considered the self-weight. One should check the self-weight of the members. In fact, including them in the computer model does not increase the complexity of the solution algorithm. We will consider the self-weight in the next example.
3. *Truss topology*: Is a Fink truss the best? For a general design problem of the type considered in this example, there is no conclusive answer. We could have tried other types of truss layout. This is left as an exercise (see Problem 8.8.8).
4. *Buckling*: One of the design requirements that controls the design is buckling, both Euler and local. We will address these issues in the next chapter. Euler buckling is considered in the next example.
5. *Multiple load cases*: We did not consider different load cases, as an actual design would require. This is left as an exercise (see Problem 8.8.8).
6. *Cost values*: We designed the truss for minimum volume. A better objective function would have been to minimize the cost of the truss (Rajan et al., 1999).

Tip: Finally, a few words about the design process. Start with the simplest case. Solve a problem first as a sizing optimal design problem. Having found the optimal solution, the next step is to treat the problem as a combined sizing and shape optimal design problem. And, finally, combine the sizing, shape, and topology design variables, if necessary. Too often good solutions and problem characteristics can be missed by not solving a problem in stages.

EXAMPLE 8.8.3 ***Design of a Pedestrian Bridge Truss***

A walkway 30 ft long and 15 ft wide connects two buildings (Fig. E8.8.3(a)). Several components constitute the structure—concrete flooring, transverse beams transferring the load from the concrete floor to three identical trusses, and the trusses that are supported at the ends as pin or roller supports. The truss spacing is 7.5 ft. The concrete floor is designed to transmit a total uniform live load of 100 psf. Figure E8.8.3(b) shows a sample supporting truss.

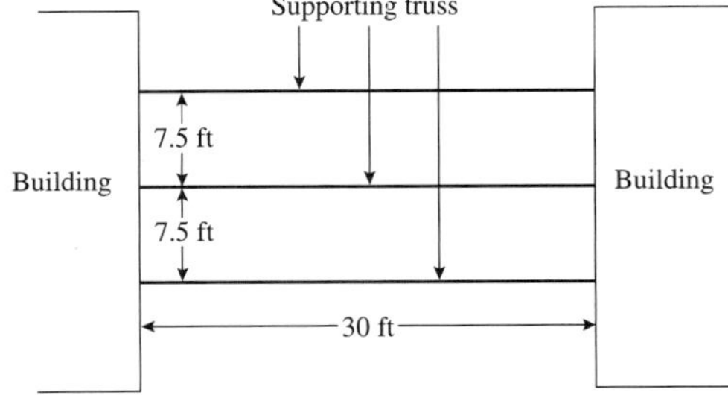

Fig. E8.8.3(a) Plan view showing the walkway and the three supporting trusses.

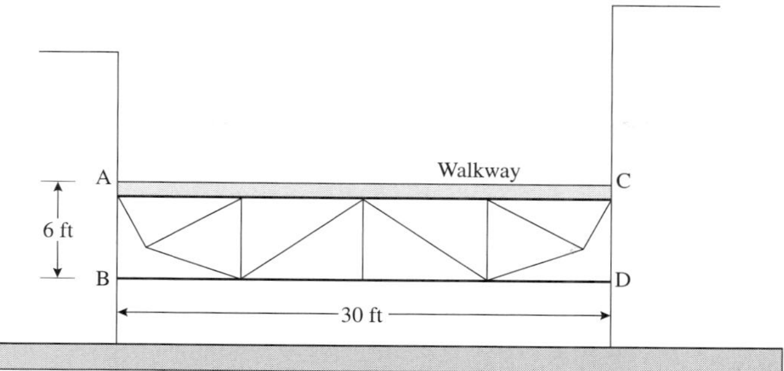

Fig. E8.8.3(b) Side view showing the bounding box ABCD and a sample supporting truss.

The thickness of the concrete flooring (10 psf/") is given as follows. The slab thickness has been adjusted to include the effects of other dead loads acting on the walkway. The transverse beam spacing is limited to 8 ft.

Slab thickness (in)	Transverse beam spacing
8"	< 3'
10"	3' to < 4'
12"	4' to < 5'
14"	5' to < 6'
20"	6' to < 8'

It is required to design a typical supporting truss so as to meet and satisfy the following requirements.

1. Design the minimum-cost truss. The cross-section of the truss members is hollow circular. The material used is 0.2%C CR Steel (E = 29000 ksi, yield stress = 51000 psi).
2. The compressive stress and the tensile stress in each member are to be limited to the allowable value. In addition, members in compression must satisfy the Euler buckling requirement. The factor of safety for each of these three checks is 2.
3. The truss must fit in the bounding box ABCD. The supports may be placed only on the edges defined by AB and CD.
4. The maximum vertical deflection must not exceed 1 in.
5. Do not use more than three different cross-sections.
6. The ratio of the inner radius of the section to the wall thickness cannot exceed 5. The smallest inner radius is 0.5 in.
7. The cost of steel is $0.4/in^3. The cost of each joint is $100. The cost of each support is $500.

SOLUTION

As in the previous example, several structural data are not defined and the designer must make decisions to define the missing pieces.

1. The truss topology is not defined.
2. The truss support conditions are not defined.
3. Once the truss layout is defined, the members must be grouped together so that members with the same cross-sectional properties belong to the same group.

Note that the thickness of the concrete floor is a function of the number of transverse beams, or the number of panel points in the truss. We assume that there are three panel points and the initial truss layout is as in Fig. E8.8.3(c). We design the center truss assuming that the supports at the two ends of the truss are pin supports.

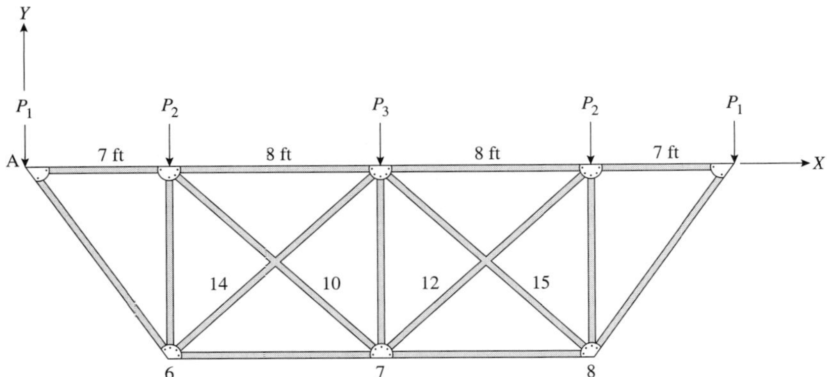

Fig. E8.8.3(c) Initial shape and layout.

From the initial layout, the required slab thickness is 20 in. With the truss spacing of 7.5 ft and the concept of tributary area loading, the loads acting on the top chord of the truss can be computed as follows:

$$P_1 = \left[(20 \text{ in})(10 \text{ lb / ft}^2 \text{ - in}) + 100 \text{ lb / ft}^2\right](7.5 \text{ ft})(3.5 \text{ ft}) = 7875 \text{ lb}$$

$$P_2 = \left[(20 \text{ in})(10 \text{ lb / ft}^2 \text{ - in}) + 100 \text{ lb / ft}^2\right](7.5 \text{ ft})(7.5 \text{ ft}) = 16875 \text{ lb}$$

$$P_3 = \left[(20 \text{ in})(10 \text{ lb / ft}^2 \text{ - in}) + 100 \text{ lb / ft}^2\right](7.5 \text{ ft})(8.0 \text{ ft}) = 18000 \text{ lb}$$

For the sake of brevity, the initial sizing optimal design details are not presented. Based on this preliminary design, the following optimal design problem is formulated. We assume lb, in as the problem units. We also include the self-weight of the members.

Problem Formulation. The design problem now can be stated as

Find $\quad \mathbf{x} = \{r_{BC}, r_{TC}, r_{WEB}, X_6, Y_6, Y_7, e_{10}, e_{12}, e_{14}, e_{15}\}$ (1)

to minimize $\quad f(\mathbf{x}) \Rightarrow \text{cost} = \sum_{i=1}^{13} A_i L_i c_i$ (2)

such that $\quad \sigma^c_{\max} \le 25500$ (3)

$$\sigma^t_{\max} \le 25500 \quad (4)$$

$$P_{member} \le \frac{P_{cr}}{2} \quad (5)$$

$$|\Delta_j| \le 1'' \qquad j \in \text{y} - \text{displacement of all nodes} \quad (6)$$

$$0.5'' \le r \le 2'' \quad (7)$$

$$40'' \le X_6 \le 120'' \quad (8)$$

$$-72'' \le Y_6 \le -36'' \quad (9)$$

$$-72'' \le Y_7 \le -36'' \quad (10)$$

$$e_i \in (0,\ 1) \quad (11)$$

There are 10 design variables. The members are grouped so that all the top chord members have the same inner radii r_{TC} and wall thickness. The same applies for the bottom chord members and the web members. These are the three sizing design variables. The wall thickness is assumed to be exactly 20% of the inner radius, and hence is not considered a design variable. The x and y coordinates of node 6, X_6 and Y_6, and the y coordinate of node 7, Y_7, are taken as the shape design variables. Equality constraints are introduced in the problem to maintain symmetry of the truss. In other words, the coordinates of node 8 are adjusted to obtain structural symmetry. Finally, topology design variables e_{10}, e_{12}, e_{14}, e_{15} are introduced so that we can ascertain which of these four members are necessary to effectively meet the design requirements. The objective function expressed in Eq. (2) is the cost of the truss where c_i is the unit cost of the material. Equations (3)–(4) take care of the compressive and tensile stress requirements. The Euler buckling requirement is expressed in Eq. (5). The 2 factor in the constraint is the safety factor. The displacement limit is expressed in Eq. (6). The rest of the equations establish the lower and upper bounds on the design variables.

Summary of the Results. The final truss shape and layout are shown in Fig. E8.8.3(d). Selected portions of the Results file are shown below.

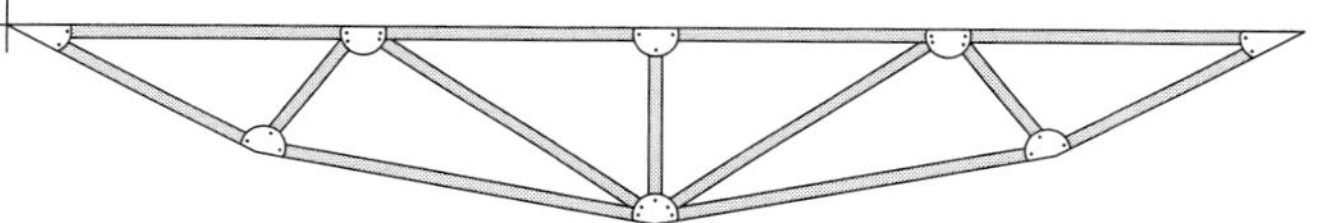

Fig. E8.8.3(d) Final truss shape and topology.

```
                    DESIGN RESULTS
                    ==============

   PROBLEM SIZE
   ------------
    Number of Design Parameters:    19
     Number of Design Variables:    10
        Number of Design Checks:     9
Minimize Objective Function: Cost. Current Value: 2196.84  $

-----------------------------------------------------------------
                       DESIGN PARAMETERS
Number   Name                Attributes
-----------------------------------------------------------------
1       Radius1          Property Group: 1. Cir Hollow. Inner radius.
2       Radius2          Property Group: 2. Cir Hollow. Inner radius.
3       Radius3          Property Group: 3. Cir Hollow. Inner radius.
4       T_Stress         Element: 0. LC: 0. Tensile Stress.
5       C_Stress         Element: 0. LC: 0. Compressive Stress.
6       E_Buckling       Element: 0. LC: 0. Euler Buckling.
7       Y_displacement   Node: 0. LC: 0. Y Displacement.
8       Depth1           Node: 6. Y Coordinate.
9       Depth2           Node: 7. Y Coordinate.
10      Depth3           Node: 8. Y Coordinate.
11      xcoor6           Node: 6. X Coordinate.
12      xcoor8           Node: 8. X Coordinate.
13      Wallthk1         Property Group: 1. Cir Hollow. Wall thickness.
14      Wallthk2         Property Group: 2. Cir Hollow. Wall thickness.
15      Wallthk3         Property Group: 3. Cir Hollow. Wall thickness.
16      E10              Element: 10. LC: 0. Connectivity.
```

```
17      E14           Element: 14. LC: 0. Connectivity.
18      E12           Element: 12. LC: 0. Connectivity.
19      E15           Element: 15. LC: 0. Connectivity.
```

DESIGN VARIABLES : DEFINITION

Number	Design Variable	Lower Bound	Upper Bound	Precision
1	Radius1	0.5	1	0.01
2	Radius2	0.5	2	0.01
3	Radius3	0.5	2	0.01
4	Depth1	-72	-36	1
5	Depth2	-72	-36	1
6	xcoor6	40	80	1
7	E10	0	0	1
8	E14	0	0	1
9	E12	0	0	1
10	E15	0	0	1

DESIGN VARIABLES : FINAL VALUES

Number	Design Variable	Value
1	Radius1	0.77
2	Radius2	1.22
3	Radius3	0.88
4	Depth1	-38
5	Depth2	-60
6	xcoor6	60
7	E10	1
8	E14	0
9	E12	1
10	E15	0

CONSTRAINT INFORMATION

Num	Coef	Parameter	Coef	Parameter	Constant	Type
1	1	T_Stress			-25500	<=
2	1	C_Stress			-25500	<=
3	2	E_Buckling			0	<=
4	1	Y_displacement			1	>=
5	1	Depth1	-1	Depth3	0	=
6	1	xcoor6	1	xcoor8	-360	=
7	0.2	Radius1	-1	Wallthk1	0	=
8	0.2	Radius2	-1	Wallthk2	0	=
9	0.2	Radius3	-1	Wallthk3	0	=

CONSTRAINT VALUES

Number	Coef * Constraint	Value
1	{1 * (Tens. Stress: (Element,LC)(1,1))}	-0.749172
2	{1 * (Tens. Stress: (Element,LC)(2,1))}	-25500
3	{1 * (Tens. Stress: (Element,LC)(3,1))}	-25500
4	{1 * (Tens. Stress: (Element,LC)(4,1))}	-0.749172

```
 5    {1 * (Tens. Stress: (Element,LC)(5,1))} -0.0715825
 6    {1 * (Tens. Stress: (Element,LC)(6,1))} -0.000263274
 7    {1 * (Tens. Stress: (Element,LC)(7,1))} -0.000263274
 8    {1 * (Tens. Stress: (Element,LC)(8,1))} -0.0715825
 9    {1 * (Tens. Stress: (Element,LC)(9,1))} -25500
10    {1 * (Tens. Stress: (Element,LC)(10,1))} -25500
11    {1 * (Tens. Stress: (Element,LC)(11,1))} -25500
12    {1 * (Tens. Stress: (Element,LC)(12,1))} -25500
13    {1 * (Tens. Stress: (Element,LC)(13,1))} -25500
14    {1 * (Tens. Stress: (Element,LC)(14,1))} -1.00004
15    {1 * (Tens. Stress: (Element,LC)(15,1))} -1.00004
16    {1 * (Comp. Stress: (Element,LC)(1,1))} -25500
17    {1 * (Comp. Stress: (Element,LC)(2,1))} -0.780526
18    {1 * (Comp. Stress: (Element,LC)(3,1))} -0.780526
19    {1 * (Comp. Stress: (Element,LC)(4,1))} -25500
20    {1 * (Comp. Stress: (Element,LC)(5,1))} -25500
21    {1 * (Comp. Stress: (Element,LC)(6,1))} -25500
22    {1 * (Comp. Stress: (Element,LC)(7,1))} -25500
23    {1 * (Comp. Stress: (Element,LC)(8,1))} -25500
24    {1 * (Comp. Stress: (Element,LC)(9,1))} -0.28373
25    {1 * (Comp. Stress: (Element,LC)(10,1))} -0.973776
26    {1 * (Comp. Stress: (Element,LC)(11,1))} -0.339761
27    {1 * (Comp. Stress: (Element,LC)(12,1))} -0.973776
28    {1 * (Comp. Stress: (Element,LC)(13,1))} -0.28373
29    {1 * (Comp. Stress: (Element,LC)(14,1))} -1.00004
30    {1 * (Comp. Stress: (Element,LC)(15,1))} -1.00004
31    {2 * (Euler Buck: (Element,LC)(1,1))} -1
32    {2 * (Euler Buck: (Element,LC)(2,1))} -0.00347658
33    {2 * (Euler Buck: (Element,LC)(3,1))} -0.00347658
34    {2 * (Euler Buck: (Element,LC)(4,1))} -1
35    {2 * (Euler Buck: (Element,LC)(5,1))} -1
36    {2 * (Euler Buck: (Element,LC)(6,1))} -1
37    {2 * (Euler Buck: (Element,LC)(7,1))} -1
38    {2 * (Euler Buck: (Element,LC)(8,1))} -1
39    {2 * (Euler Buck: (Element,LC)(9,1))} -0.454235
40    {2 * (Euler Buck: (Element,LC)(10,1))} -0.873225
41    {2 * (Euler Buck: (Element,LC)(11,1))} -0.103437
42    {2 * (Euler Buck: (Element,LC)(12,1))} -0.873225
43    {2 * (Euler Buck: (Element,LC)(13,1))} -0.453914
44    {2 * (Euler Buck: (Element,LC)(14,1))} -1
45    {2 * (Euler Buck: (Element,LC)(15,1))} -1
46    {1 * (Displacement: (Node,Dof,LC)(1,2,1))} -1
47    {1 * (Displacement: (Node,Dof,LC)(2,2,1))} -0.655854
48    {1 * (Displacement: (Node,Dof,LC)(3,2,1))} -0.596348
49    {1 * (Displacement: (Node,Dof,LC)(4,2,1))} -0.655854
50    {1 * (Displacement: (Node,Dof,LC)(5,2,1))} -1
51    {1 * (Displacement: (Node,Dof,LC)(6,2,1))} -0.755507
52    {1 * (Displacement: (Node,Dof,LC)(7,2,1))} -0.631199
53    {1 * (Displacement: (Node,Dof,LC)(8,2,1))} -0.755507

Max:6     {1 * (Tens. Stress: (Element,LC)(6,1))} -0.000263274

Notes:
A negative constraint value indicates a satisfied constraint.
A positive constraint value indicates an unacceptable design.
```

To summarize, the best solution obtained by the program is as follows.
From Problem Size section:

Cost = $2196

From Design Variables: Final Values section:

Section	Inner radius (in)
Top Chord members	0.77
Bottom Chord members	1.22
Web members	0.88
	Coordinate (in)
X Coordinate Node 6	60"
Y Coordinate Node 6	–38"
Y Coordinate Node 7	–60"
	Status
Element 10	Should exist
Element 12	Should exist
Element 14	Not needed
Element 15	Not needed

From Constraint Values section

By looking at the values in this section, we can conclude that the tensile stress in members 6 and 7 controls the design.

Issues Connected with the Design Problem Formulation

We now examine the problem formulation and the solution obtained.

1. *Structural modeling*. We have modeled the structure as a planar truss. The load is transferred from the floor slab to the nodes on the top chord (panel points) via the transverse beams.
2. *Self-weight*: We have considered the self-weight.
3. *Truss topology*: Is the considered topology (base structure) the best truss? Since the loading is a function of the transverse beam spacing, we cannot move the locations of the panel points automatically in the design program and recompute the loads. Current methodology in the GS-USA program requires that for this problem, the user find the optimal design for a fixed number of panel points. In other words, the number of panel points cannot be a design variable when the structure is a truss.
4. *Buckling*: We considered Euler buckling. The local buckling issue was handled indirectly using a minimum ration of wall thickness to inner radius.
5. *Cost values*: The cost values were assumed. Note that the final optimal design is greatly affected by the choice of the cost values.

When the above problem is solved using AISC pipe sections (with minor changes to the constraints), the lowest obtained cost is $2403 and the final shape and topology of the truss is as in Fig. E8.8.3(d). The details and results are available on the CD-ROM.

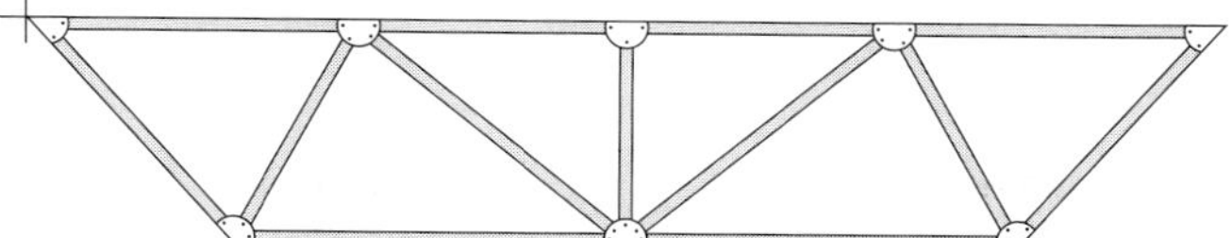

Fig. E8.8.3(d) Final truss shape and topology when AISC pipe sections are used.

With the specified problem formulation, are these designs the best design? The answer is "No." No theory available today provides a mechanism for evaluating or finding the best design when sizing, shape, and topology design variables are considered simultaneously.

EXERCISES

Appetizers

8.8.1. Solve Problem 8.7.1.

8.8.2. Solve Problem 8.7.2.

8.8.3. Solve Problem 8.7.3.

Main Course

8.8.4. Solve Problem 8.7.4.

8.8.5. A retaining wall 15 ft high is constructed of horizontal wood planks. A wooden frame assembly is then used to transfer the loads from the planks to the supporting structure. Modeling assumptions dictate that the supporting structure be modeled as independent planar trusses. The truss spacing is 5 ft. It is required to design a typical truss so as to meet and satisfy the following requirements (see Figs. P8.8.5(a) and (b)).

Design the minimum cost truss. The cross-section of the truss members can be either solid circular or hollow circular. The material used is 0.2%C CR steel (E = 29000 ksi, yield stress = 51000 psi).

The compressive stress and the tensile stress in each member are to be limited to the allowable value. In addition, members in compression must satisfy the Euler buckling requirement. The factor of safety for each of these three checks is 2.

The truss must fit in the bounding box ABCD. The truss supports can be placed only on AB.

The maximum horizontal deflection must not exceed 0.1 in.

Do not use more than three different cross-sections. All the cross-sections must have the same shape.

If hollow circular cross-sections are used, the ratio of the inner radius of the section to the wall thickness cannot exceed 5. The smallest inner radius is 0.5 in. If solid circular cross-sections are used, the smallest radius is 0.5 in.

The cost of steel is \$0.4/in^3. The cost of each joint is \$50. The cost of each support is \$200.

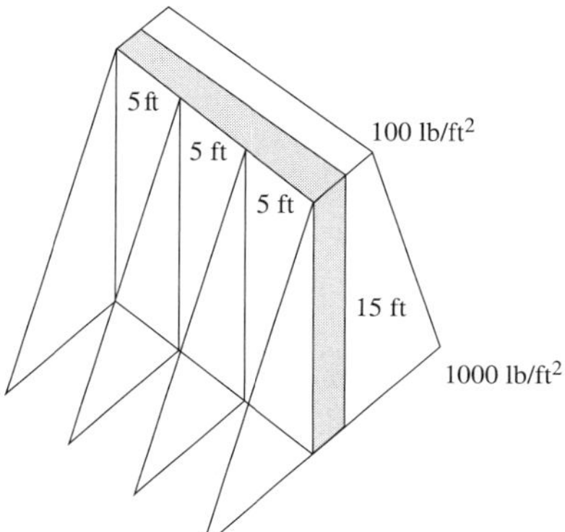

Fig. P8.8.5(a)

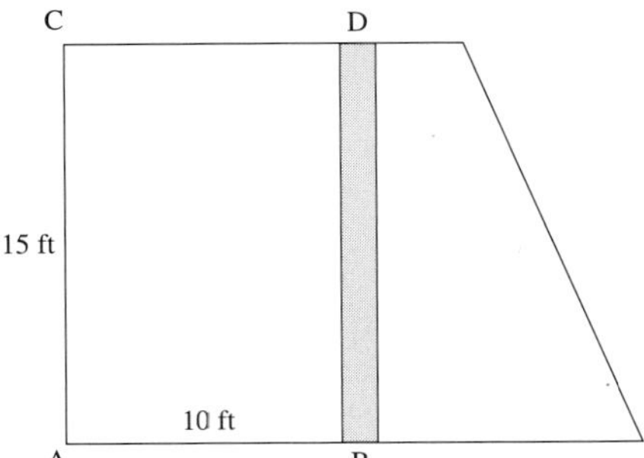

Fig. P8.8.5(b)

8.8.6. A boat ramp is to be constructed of concrete slabs. A supporting structure is a steel frame that is to be used to transfer loads from the concrete deck to the foundations. There are three similar planar frames that form the supporting structure and each of them is assumed to act independently. The frame spacing is 12 ft and the height of the frame at its highest point is 20 ft. The depth of the concrete deck slab is 12 in. The weight of the vehicles acting on the slab can be approximated as a uniform live load of intensity 250 psf.

It is required to design a typical frame (see Fig. P8.8.6) so as to meet and satisfy the following design requirements:

Design the minimum-cost frame. The cross-section of the frame members can be either solid circular or hollow circular.

The material to be used is 0.2%C CR Steel ($E = 29000$ psi, yield stress = 51, 000 psi).

The compressive stress and the tensile stress in each member are to be limited to the allowable value that is to be calculated as yield stress over the factor of safety. The required factor of safety is 2.

The critical Euler buckling load in the member is also to be satisfied with a safety factor of 2.

The frame supports can be placed only at A and D and the frame must fit in the box ABCD.

Do not use more than three cross-sections. All top chord members should have the same cross-section, all bottom chord members should have the same cross-section, and all web members should have the same cross-section.

If hollow circular cross-sections are used, the smallest inner radius is 0.5 in. The minimum wall thickness is to be restricted to 0.2 times the inner radius. If solid circular sections are to be used, the smallest radius is 0.5 in.

The cost of steel is \$0.4/in^3. The cost of each joint is \$50. The cost of each support is \$200. The total cost of a typical frame that is to be computed is the material cost plus the cost of the joints plus the cost of the supports.

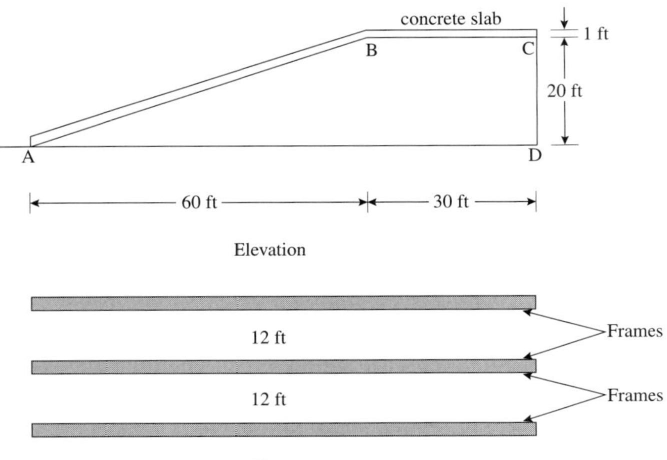

Fig. P8.8.6

Structural Concepts

8.8.7. Chapter 15 of Ref. 10 (Structural Analysis) (see the Bibliography section) presents an informative overview of different structural systems—wood, concrete and steel. Read the chapter and write a summarized report.

8.8.8.

(a) Resolve Example 8.8.2 using a different layout of the web members. Also, study the effect of the cost of the supports and joints on the final answer.

(b) Resolve Example 8.8.2 by considering the effects of dead loads, live loads, wind loads, and snow loads.

8.8.9. Design a typical interior frame from Example 3.6.2.

SUMMARY

In this chapter we discussed introductory issues connected with the general area of optimization and how they can be used for structural design (design optimization). The design problems can be either at the component level (designing one beam, column, etc.) or at the system level (a truss system, a frame system, etc.). The coverage should convince the reader of at least two observations. First, the optimization tools are very powerful. They can help the designer in examining design alternatives as well as automating the mundane calculations that go with iterative procedures. Second, design problems are mostly open-ended. One should use optimization tools to examine different design alternatives. As we have seen in this chapter, the most difficult step in the design process is formulating the design problem, e.g., cost figures are sometimes difficult to obtain.

The design problems in this chapter were "strength- or mechanics-based." In the next chapter, we look at design of steel and concrete structures where the design requirements are specified by design codes. These problems too can be solved very effectively using design optimization techniques.

Chapter 9

Design of Steel and Concrete Structures

Large structural systems are mostly built of steel and concrete, the two most popular building materials.

"If we really understand the problem, the answer will come out of it, because the answer is not separate from the problem." Jiddu Krishnamurthi

"I strive for structural simplicity. . . . The technical man mustn't be lost in his own technology." Fazlur Khan

"To the optimist, the glass is half full. To the pessimist, the glass is half empty. To the engineer, the glass is twice as big as it needs to be." Anon

This chapter provides an introduction to the design of steel and reinforced concrete structures. As we saw in Chapter 3, steel and concrete are the two most commonly used materials for civil engineering structural systems. The treatment of this material in this chapter is narrow and largely introductory. It is expected that those who are interested in a career in structural design will take a second course in steel design and concrete design.

Finally, a few words about the differences between the previous chapters and this one. So far we have assumed that the material behavior is linearly elastic. From a practical viewpoint, we cannot require that steel and concrete structures strictly satisfy this requirement. The AISC and ACI Design Codes allow the structures components to yield. More economical and safer structures result from the use of this philosophy. However, does this mean that we need to change our structural analysis procedures to accommodate "plastic analysis"? No. Both codes allow a linear elastic analysis, while the design requirements and checks assume that the material may have yielded.

OBJECTIVES

- To understand the basics of steel design.
- To understand the basics of concrete design.
- Whenever possible, to formulate design problems as optimization problems using the concepts studied in Chapter 8.
- To understand and use the UCSD© (Understanding Concrete and Steel Design) computer program.

ASSUMPTIONS

- Structural analysis will be carried out using the usual linear, small displacement, small strain behavior assumptions.
- Steel structures will behave in such a way that plane sections remain plane. The residual stresses in the material will not be ignored. The stress-strain curve is the same in tension and compression.
- Reinforced concrete structures will behave in such a way that (a) the strain in concrete and steel is the same before failure, (b) plane sections remain plane, and (c) tensile strength of concrete can be neglected.

ADDITIONAL REFERENCES

The following manuals, which govern the design of steel and concrete structures, are necessary for a proper understanding of the subject material.

Manual of Steel Construction—Load and Resistance Factor Design, Vol. 1, Second Edition, American Institute of Steel Construction, 1994.

Building Code Requirements for Structural Concrete, ACI 318-99, American Concrete Institute, 1999.

9.1 INTRODUCTION TO DESIGN OF STEEL STRUCTURES

Steel is a remarkable man-made material. Structural steel can be either cold-rolled or hot-rolled steel. Cold-rolled steel members are manufactured from steel sheet or strip material. They are available in different compositions and strength. The yield strength can vary from about 30 ksi to as high as 65 ksi. The AISI Code can be used for the design of structural systems made with cold-formed steel. Hot-rolled steel is perhaps more commonly used for structural systems with more demanding load levels and performance requirements. The AISC Code can be used for the design of structural systems made of hot-rolled steel. In this section we look at some of the provisions of the 1994 *AISC LRFD Manual of Steel Construction.*

Steel Properties

The mechanical properties of steel can be found through a series of tests. Figure 9.1.1(a) shows the stress-strain curve obtained through a tension test for carbon steel. The diagram looks similar if a compressive load is applied (with buckling prevented). For the purposes of design, however, the steel behavior is taken as elastic-perfectly plastic (Fig. 9.1.1(b))

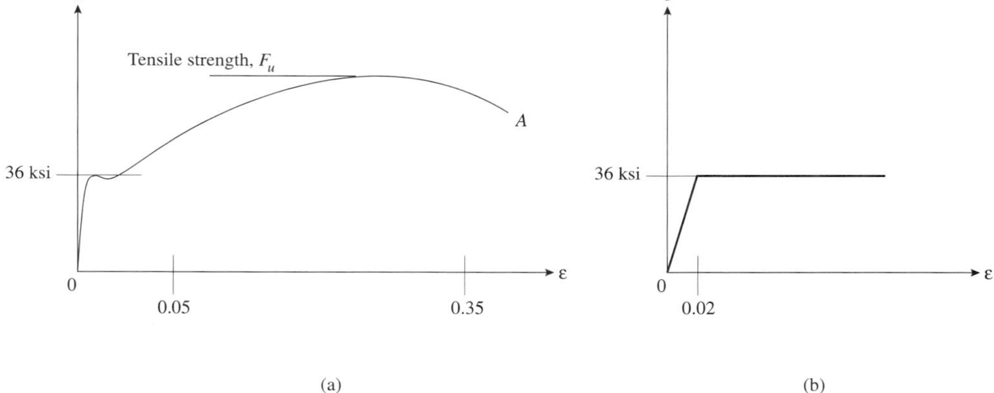

Fig. 9.1.1 Typical (a) stress-strain curve for carbon steel (b) assumed elasto-plastic behavior.

In Chapter 3, we saw the stress-strain curve for a typical ductile material (see Fig. 3.1.2.1). The corresponding material properties—modulus of elasticity, Poisson's ratio, shear modulus, yield stress, and ultimate stress—were discussed. For structural steel, the modulus of elasticity is roughly 29000 ksi. Poisson's ratio is 0.3 when exhibiting elastic behavior. This value increases as steel begins to yield and deform. Using these two values one can compute the value for the shear modulus, which is about 11000 ksi. The ductility of steel permits steel to undergo localized yielding without catastrophic failure.

Hot-rolled structural steel is available in different forms—carbon steels, high-strength steels, and alloy steels. The properties of carbon steel can be varied by varying the amount of carbon used. For example, in low-carbon steel, the percentage of carbon used is less than 0.15%. The other forms of steel are similarly composed—mild carbon (0.15–0.29%), medium carbon (0.30–0.59%) and high carbon steel(0.60–1.70%). High-strength low-alloy steels achieve their properties by the addition of other metals such as copper, manganese, nickel, vanadium, chromium, etc. A higher strength is obtained when low-alloy steels are quenched and tempered. The yield strength of hot-rolled steel can vary from as low as 24 ksi to as high as 100 ksi (see Table 9.1.1).

The A36 steel is one of the most commonly used steels, mainly for buildings. A572 is used in structural shapes, though the use of A709 is becoming more common. A852 is used in harsh environments where corrosion resistance is required.

Table 9.1.1 Properties of Some Commonly Used Structural Steel

ASTM designation	Minimum yield stress, ksi	Tensile strength, ksi	Max. thickness for plates, in
A36	32	58–80	Over 8
	36	58–80	< 8
A570 Grade 40	40	55	
Grade 45	45	60	
Grade 50	50	65	
A572 Grade 42	42	60	< 6
Grade 50	50	65	< 4
Grade 60	60	75	< 1.5
Grade 65	65	80	< 1.25
A709 Grade 36	36	58–80	< 4
Grade 50	50	65	< 4
Grade 70W	70	90–110	< 4
Grade 100/100W	100	110–130	< 2.5
A852	70	90–110	< 4

Standard Shapes

Typical rolled shapes, as designated by AISC and ASTM, are available in different forms. The wide-flange shape (see Fig. 9.1.2) is designated W$d \times w$, where d is the nominal depth of the section in inches and w is the weight per unit length in lb/ft.

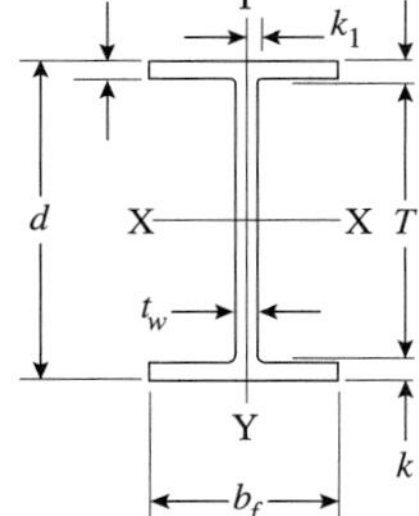

Fig. 9.1.2 AISC wide-flange section.

The American Standard beam (S-beam) is also shaped as the wide-flange beam but is narrower and has thicker web and sloping flanges. The other commonly used shapes include the channel, the angle, structural tee (made by cutting W or S shapes in half), pipe, and tube sections (see Fig. 9.1.3).

Design Philosophy

The design of steel structures is governed by two different design philosophies—working stress design (or allowable stress design, ASD) and limit states design (or load and resistance factor design, LRFD). Irrespective of the design philosophy employed, the major objectives of structural design are: to provide a safe structure under design loads, to provide enough stiffness that the structure does not deform excessively, and to provide economical initial and operating costs for the intended life of the structure. In the ASD methodology, the structural elements are proportioned so that the computed stresses do not exceed the allowable stresses. These allowable stresses are a fraction of the stress levels at failure. In other words, the factor of safety (FS) is built into the allowable values:

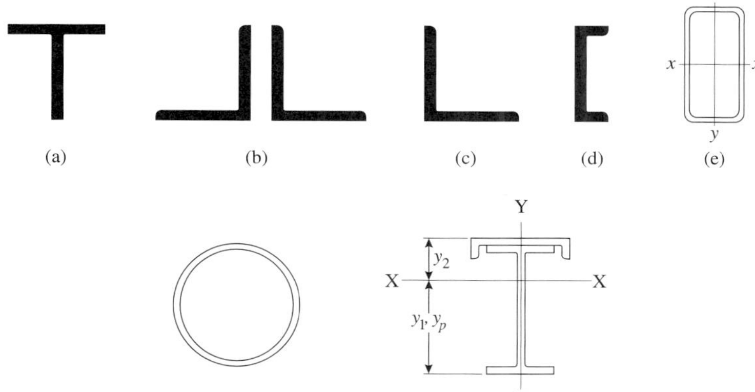

Fig. 9.1.3 Commonly used structural shapes. (a) Structural tee. (b) Double angle. (c) Single angle. (d) Channel. (e) Rectangular tubing. (f) Pipe. (g) Combination W and channel section.

$$\text{Allowable stress} = \frac{\text{Stress limit}}{\text{FS}} \tag{9.1.1}$$

The stress limit is a measure of appropriate stress value associated with the mode of failure. The factor of safety is based on experiments and experience from design and use of a class of structures over a period of time. Several factors affect the safe design of a structure, among them (a) material properties, (b) member properties, (c) quality of construction, (d) load values, and (e) accuracy of structural analysis. Manufacturing practices affect the variability in material properties, the shape and size of member cross-sections, the amount of residual stresses in these members, etc. The quality of construction affects the material properties (e.g., cast-in-place concrete), member locations and dimensions, the efficiency of connections and joints, etc. As we have seen before, loads are even more difficult to estimate and locate on the structure. Such variability in the factors that affect the design make the actual factor of safety an unknown parameter when the ASD design philosophy is used.

The philosophy used in the limit states design is quite different. Limit states are those conditions under which the structure ceases to fulfill its intended function. The limit states can be categorized as strength-based or serviceability-based. The former refers to behavior phenomena such as buckling, fracture, plastic strength, fatigue, etc.; the latter deals with the occupancy of structures, e.g., cracking, deflections, vibrations, etc. The limit states approach is felt to be more rational because it provides an estimate of the actual factor of safety. A probability-based reliability analysis is used to establish the margin of safety as

$$\phi R_n \geq \sum_i \gamma_i Q_i \tag{9.1.2}$$

where ϕ is the strength reduction factor, R_n is the nominal resistance, γ_i are the overload factors, and Q_i represent the different loads acting on the structure. In other words, the left side of the equation represents the resistance offered by the component or system and the right side represents the load expected to be carried by the structure.

The resistance factor ϕ for different members is given by the following:

Tension members:

$$\phi_t = 0.90 \text{ for yielding limit state} \tag{9.1.3a}$$

$$\phi_t = 0.75 \text{ for fracture limit state} \tag{9.1.3b}$$

Compression members:

$$\phi_c = 0.85 \tag{9.1.3c}$$

Beams:

$$\phi_b = 0.90 \text{ for flexure} \tag{9.1.3d}$$

$$\phi_v = 0.90 \text{ for shear} \tag{9.1.3e}$$

Welds:

$$\phi \text{ depends on nature of force} \tag{9.1.3f}$$

Fasteners:

$$\phi_c = 0.75 \tag{9.1.3g}$$

The overload factors were discussed in Chapter 3 in the context of load factors and are presented here for completeness. The ASCE 7-95 standard prescribes that: Structures, components, and foundations shall be designed so that their design strength equals or exceeds the effects of the factored loads in the following combinations:

1. $1.4D$
2. $1.2(D+F+T)+1.6(L+H)+0.5(L_r \text{ or } S \text{ or } R)$
3. $1.2D+1.6(L_r \text{ or } S \text{ or } R)+(0.5L \text{ or } 0.8W)$ (3.6.7.1)
4. $1.2D+1.3W+0.5L+0.5(L_r \text{ or } S \text{ or } R)$
5. $1.2D \pm 1.0E+0.5L+0.2S$
6. $0.9D \pm (1.3W \text{ or } 1.0E)$

In the next two sections we look at issues in the design of compressive members and the design of beams that are laterally supported.

9.1.1 Design of Tension Members

We start our study of steel structures by looking at members subjected primarily to tensile forces. When steel members are subjected to increasing tensile forces, one of two situations occurs—(a) the member yields away from a connection, and (b) fracture occurs through the bolt holes at a connection. In case (a), the nominal strength T_n is given by

$$T_n = F_y A_g \tag{9.1.1.1}$$

where F_y is the yield stress and A_g is the gross cross-sectional area.

The presence of bolt holes leads to stress concentration such that the stresses in the immediate vicinity of the holes are several times larger than the stresses away from the holes. When the localized yielding leads to fracture through the holes, the nominal strength is given by

$$T_n = F_u A_e \tag{9.1.1.2}$$

where F_u is the minimum tensile strength (see Fig. 9.1.1) and A_e is the effective net area.

The effective area is a function of the net area A_n (gross area minus the areas of the bolt holes)

$$A_e = UA_n \tag{9.1.1.3}$$

where U is the efficiency factor or reduction coefficient.

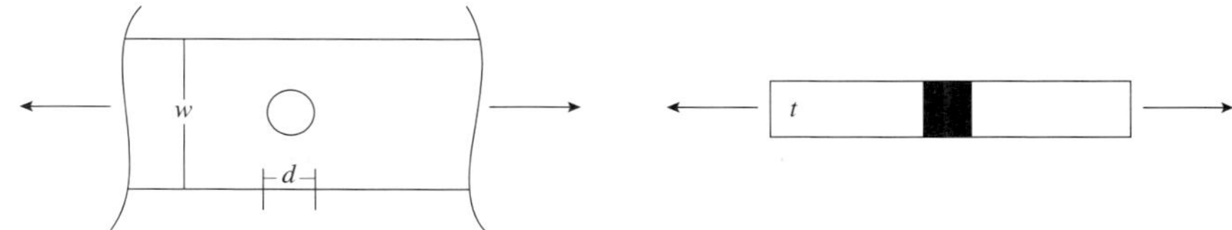

Fig. 9.1.1.1(a) Single bolt hole (plan and elevation).

In bolt assembly in Fig. 9.1.1.1(a), where the width of the plate is w and its thickness t, the gross cross-sectional area $A_g = wt$. It should be noted that the diameters of the bolt holes are nominally larger than the diameter of the bolts. Typically, a hole 1/16 in larger than the diameter of the bolt is punched. Due to damage on the circumference of the hole, a larger value is used to compute the amount of material that has been removed. In other words, the total amount of material removed is taken as the diameter of the bolt (or fastener) plus 1/8 in. The net cross-sectional area is then $A_n = A_g - (d + 0.125)t$, with the units being inches. Another method of creating the holes is to drill holes that are 1/32 in larger than the diameter of the bolt.

In Fig. 9.1.1.1(b) two different bolt assemblies are shown. When two or more bolt holes that are aligned along a line, the failure path is 1–2. When a staggered assembly of bolts is used, the failure line is the one that leads to the largest stress on an effective net area. Consider the staggered bolt assembly. One failure path is 1–2, where one bolt area is to be deducted to compute the net area. The other failure path, 1–3, while longer, is also a distinct possibility since two bolt areas need to be deducted to compute the net area.

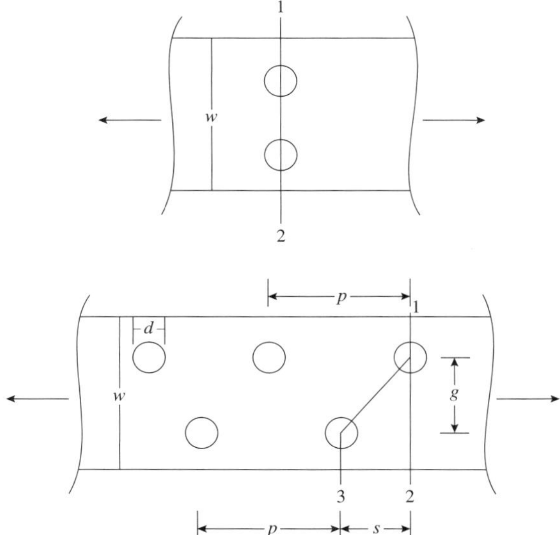

Fig. 9.1.1.1(b) Different failure paths.

An empirical analysis of staggered bolt assemblies proposed by Cochrane (see Bibliography) is used in the AISC LRFD and ASD code. For the staggered bolt assembly,

$$\text{Length of failure path 1–2} = w - (d + 0.0625) \tag{9.1.1.4}$$

$$\text{Length of failure path 1–3} = w - 2(d + 0.0625) + \frac{s^2}{4g} \tag{9.1.1.5}$$

The last term in Eq. (9.1.1.5) is the length correction factor proposed by Cochrane. The smaller of the two values is then used in computing the net area.

Equation (9.1.1.3) is applicable to both bolted and welded connections.[1] In the former case, U is a function of the eccentricity of the loading as

$$U = 1 - \frac{\bar{x}}{l_c} \leq 0.9 \tag{9.1.1.6}$$

where $\bar{x}$ is the distance from the centroid of the element being connected to the plane of load transfer and l_c is the length of the connection in the direction of the loading. An example involving an I-section and gusset plates is shown in Fig. 9.1.1.2.

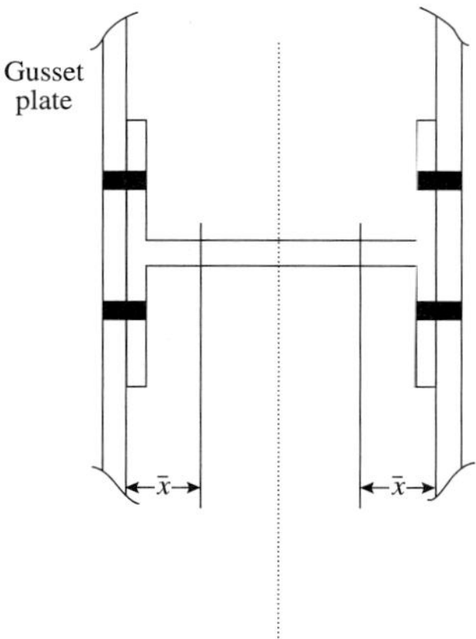

Fig. 9.1.1.2 Eccentricity used in the computation of the reduction coefficient for an I-section.

Design of Tension Members

The design strength $\phi_t T_n$ is taken as the smaller of

a. Yielding failure of the gross cross-section (Eq. (9.1.1.1)):

$$\phi_t T_n = \phi_t F_y A_g = 0.90 F_y A_g \tag{9.1.1.7}$$

b. Fracture of the net section (Eq. (9.1.1.2)):

$$\phi_t T_n = \phi_t F_u A_e = 0.75 F_u A_e \tag{9.1.1.8}$$

where the resistance factor ϕ_t is 0.90 for yielding and 0.75 for fracture. In addition, AISC-LRFD also requires other considerations as follows.

c. Shear rupture:

$$\phi V_n = \phi(0.6 F_u) A_{nv} \tag{9.1.1.9}$$

d. Tension rupture:

$$\phi T_n = \phi F_u A_{nt} \tag{9.1.1.10}$$

e. Shear-tension combination:

i. When $F_u A_{nt} \geq 0.6 F_u A_{nv}$

$$\phi R_{bs} = \phi(0.6 F_y A_{gv} + F_u A_{nt}) \tag{9.1.1.11}$$

ii. When $0.6 F_u A_{nv} \geq F_u A_{nt}$

$$\phi R_{bs} = \phi(0.6 F_u A_{nv} + F_y A_{gt}) \tag{9.1.1.12}$$

[1] Discussion of connections, including connection detailing, is outside the scope of the text.

where A_{gv} is the gross area acted upon by shear, A_{gt} is the gross area acted upon by tension, A_{nv} is the net area acted upon by shear, A_{nt} is the net area acted upon by tension, and ϕ is 0.75 for fracture.

EXAMPLE 9.1.1 ***Design of a Tension Member***

A member in a truss is subjected to tensile forces as follows: (a) 50 k dead load and (b) 10 k live load. The truss is built with W shape A572 Grade 50 steel members. The member is 10 ft long. Typical joints in the truss use 3/4 in diameter bolts with two bolts per gage line. Design the member.

SOLUTION

We use k, in as the problem units.

Step 1: Factored load:

$$T_u = 1.2(50) + 1.6(10) = 76 \text{ k} \qquad \text{or} \qquad T_u = 1.4(50) = 75 \text{ k}$$

Hence, the factored load to be considered is 76 k.

Step 2: Check failure mode:

$$\text{Yielding:} \quad \phi_t F_y A_g = 0.90(50)A_g = 45A_g \Rightarrow A_g = \frac{76}{45} = 1.69 \text{ in}^2$$

$$\text{Fracture:} \quad \phi_t F_u A_e = 0.75(65)A_e = 48.75A_e \Rightarrow A_e = \frac{76}{48.75} = 1.56 \text{ in}^2$$

Since the exact connection details are not known, we estimate the reduction factor as 0.8. Hence,

$$A_n = \frac{A_e}{U} = \frac{1.56}{0.8} = 1.95 \text{ in}^2$$

Since the net area is greater than the gross area, the fracture limit state governs the design.

Step 3: Member selection. Let's try W6 × 9: A = 2.68 in^2; flange thickness = 0.215 in, $r_{\min}$ = 0.905 in:

Net area = 2.68 – 2(0.215)(0.75 + 0.125) = 2.31 in^2 > 1.95 in^2

Step 4: Verify reduction factor. Figure E9.1.1 shows a possible connection involving the truss member. Since the bolts are placed in the two flanges, they must be treated separately. Hence using, half of the W shape, we can compute the location of the centroid, i.e., the value of $\bar{x}$. Computing yields $\bar{x} = 0.68$ in.

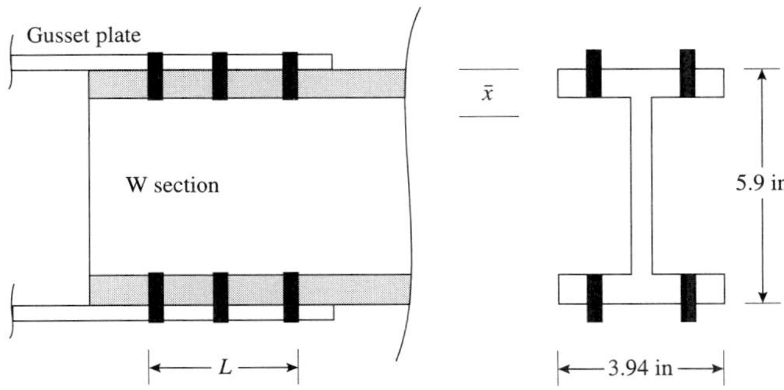

Fig. E9.1.1

From Eq. (9.1.1.6), we have

$$U = 1 - \frac{\bar{x}}{l_c} = 1 - \frac{0.68}{l_c} = 0.8 \Rightarrow l_c = 3.4 \text{ in}$$

In other words, as long as the length of the connection is greater than or equal to 3.4 in, the design is acceptable.

Tip: While not a requirement, it is desirable that the slenderness ratio L/r of the member be less than 240. One of the reasons is to avoid problems if a force reversal occurs in the member. Hence,

$$\min\ r = \frac{L}{240} = \frac{120}{240} = 0.5$$

The selected section W6 × 9 satisfies this requirement.

9.1.2 Design of Compression Members

One of the modes of failure that governs the design of steel structures is buckling—Euler buckling and various forms of local buckling. In Chapter 3 we saw the *elastic* Euler buckling phenomenon. The particular equation of interest is given as

$$P_{cr} = \frac{\pi^2 EI}{(KL)^2} \qquad \text{or} \qquad F_{cr} = \frac{P_{cr}}{A_g} = \frac{\pi^2 E}{(L/r)^2} \tag{9.1.2.1}$$

where L is the effective length, K is the effective length factor, F_{cr} is the average compressive stress, and A_g is the gross cross-sectional area. Recall that

$$I = A_g r^2 \tag{9.1.2.2}$$

where r is the *radius of gyration.*

Test results obtained in the past have consistently failed to reach this critical value for column lengths commonly used in structural design. Two researchers, Engesser and Considere (see Bibliography), recognized that the columns become inelastic before buckling takes place and one cannot use the (elastic) E to compute the buckling load. The modified formula to be used is

$$P_{cr} = \frac{\pi^2 E_t}{(KL/r)^2} A_g \tag{9.1.2.3}$$

where E_t is the tangent modulus of elasticity computed at stress $\frac{P_{cr}}{A_g}$.

For practical design situations, it is understood that a column loaded axially can (a) yield if the column is short and stubby, (b) buckle after a portion of the cross-section has yielded (inelastic buckling), and (c) buckle for longer-length columns (elastic buckling). In addition, one must consider the effects of residual stresses. *Residual stresses* are stresses that are present in an unloaded specimen. The stresses can be attributed to a variety of sources such as uneven cooling for hot-rolled steel, bending during fabrication for cold-rolled steel, and can also be brought about during fabrication (welding, sawing, punching of holes, etc.). The residual stresses affect the Euler buckling formula since the tangent modulus generally varies across the cross-section.

Extensive experiments have been conducted to generate design curves that incorporate the failure modes and effects discussed earlier. The AISC LRFD specifications define a different parameter from the slenderness ratio term we saw in Chapter 3. The slenderness function (or parameter) λ_c is defined as

$$\lambda_c = \frac{KL}{r\pi}\sqrt{\frac{F_y}{E}} \tag{9.1.2.4}$$

where F_y is the yield stress. The Structural Stability Research Council (SSRC)[2] is credited with the equation that is used today for design. The nominal strength P_n of a rolled compression member is given as (LRFD-E2)

$$P_n = A_g F_{cr} \tag{9.1.2.5}$$

where

$$F_{cr} = \left(0.658^{\lambda_c^2}\right) F_y \qquad \text{for } Q\lambda_c \le 1.5 \tag{9.1.2.6a}$$

$$F_{cr} = \left[\frac{0.877}{\lambda_c^2}\right] F_y \qquad \text{for } Q\lambda_c > 1.5 \tag{9.1.2.6b}$$

where Q is a strength reduction factor discussed in more detail in the next section.

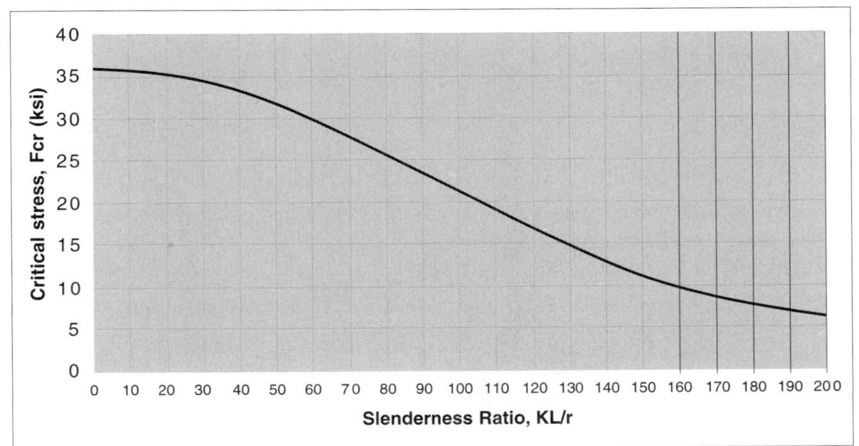

Fig. 9.1.2.1 Critical stress F_{cr} versus slenderness ratio KL/r (for $Q = 1$, $F_y = 36$ ksi).

A sample plot that shows the relationships in Eq. (9.1.2.6) is given in Fig. 9.1.2.1. From the LRFD design perspective, the requirement for design is as follows:

$$\phi_c P_n \ge P_u \tag{9.1.2.7}$$

where ϕ_c can be obtained from Eq. (9.1.3c), P_n from Eq. (9.1.2.5), and P_u is the factored service load. The above equations are valid for cross-sections where the width-to-thickness ratio is not small. The limiting values are given in Table 9.1.2.1 (AISC Table B5.1); here F_y is the yield stress and F_{yf} is the yield stress for the flange.

With thin-walled sections (width-to-thickness ratio is small), local buckling is possible before the section fully develops M_p; hence the load-carrying capacity of the section must be reduced. Per the LRFD Code Section B5,

> Steel sections are classified as compact, noncompact, or slender-element sections. For a section to qualify as *compact*, its flanges must be continuously connected to the web or webs and the width-thickness ratios of its compression elements must not exceed the limiting width-thickness ratios λ_p from Table B5.1 If the width-thickness ratio of one or more compression elements exceeds λ_p, but does not exceed λ_r, the section is *noncompact*. If the width-thickness ratio exceeds λ_r from Table B5.1, the section is referred to as a *slender-element* compression section.

[2] Theodore V. Galambos, ed., *Guide to Stability Design Criteria for Metal Structures*, 4th ed., New York, Wiley, 1988.

Table 9.1.2.1 AISC Table B5.1: Limiting Width-Thickness Ratios for Compression Elements (Abridged Version)

	Description of element	Width-thickness ratio	Limiting width-thickness ratios λ_p (compact)	λ_r (noncompact)
Unstiffened elements	Flanges of I-shaped rolled beams and channels in flexure	b/t	$65/\sqrt{F_y}$ [a]	$141/\sqrt{F_y - 10}$
	Flanges of I-shaped hybrid or welded beams in flexure	b/t	$65/\sqrt{F_{yf}}$	$\dfrac{162}{\sqrt{(F_{yf} - 16.5)/k_c}}$ [b]
	Outstanding legs of pairs of angles in continuous contact; flanges of channels in axial compression; angles and plates projecting from beams or compression members	b/t	NA	$95/\sqrt{F_y}$
Stiffened elements	Webs in flexural compression	h/t_w	$640/\sqrt{F_y}$ [a]	$970/\sqrt{F_y}$ [d]
	Flanges of square and rectangular box and hollow structural sections of uniform thickness subject to bending or compression; flange cover plates and diaphragm plates between lines of fasteners or welds	b/t	$190/\sqrt{F_y}$	$238/\sqrt{F_y}$
	All other uniformly compressed stiffened elements, i.e., supported along two edges	b/t h/t_w	NA	$253/\sqrt{F_y}$
	Circular hollow sections	D/t	[c]	
	In axial compression		NA	$3300/F_y$
	In flexure		$2070/F_y$	$8970/F_y$

[a] Assumes an inelastic rotation capacity of 3. For structures in zones of high seismicity, a greater rotation capacity may be required.

[b] $k_c = \dfrac{4}{\sqrt{h/t_w}}$ but not less than $0.35 \le k_c \le 0.763$.

[c] For plastic design use $1300/F_y$.

[d] For members with unequal flanges, see AISC Appendix B5.1. F_y is the minimum yield stress of the type of steel being used.

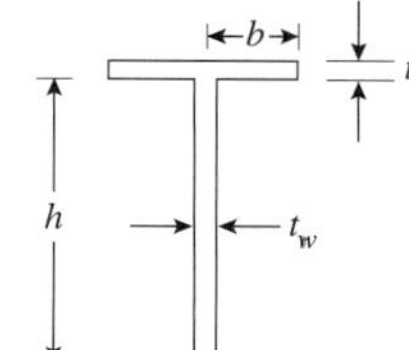

Fig. 9.1.2.2
Details for an I-section.

When the width-thickness ratios are not satisfied as per Table 9.1.2.1 (LRFD Table B5.1), the member capacity must be reduced as follows:

a. For $\lambda_c\sqrt{Q} \leq 1.5$,

$$F_{cr} = Q\left(0.658^{Q\lambda_c^2}\right)F_y \tag{9.1.2.8}$$

b. For $\lambda_c\sqrt{Q} > 1.5$,

$$F_{cr} = \left[\frac{0.877}{\lambda_c^2}\right]F_y \tag{9.1.2.9}$$

where $Q = Q_s Q_a$. (9.1.2.10)

For cross-sections comprised of only unstiffened elements, $Q = Q_s$ ($Q_a = 1.0$).
For cross-sections comprised of only stiffened elements, $Q = Q_a$ ($Q_s = 1.0$).
For cross-sections comprised of both stiffened and unstiffened elements, $Q = Q_s Q_a$.

The reduction factor Q_s for unstiffened compression members whose width-thickness ratio exceeds the limit λ_r is given in LRFD Appendix B. The reduction factor Q_a for stiffened compression members is discussed in Appendix B of the LRFD Manual.

In the next section, we look at design of columns subjected to axial compressive forces. In the section on beams, we investigate further, noncompact and slender sections.

EXAMPLE 9.1.2 ***Column Type and Properties***

Is W12 × 65 locally stable?

SOLUTION

We need to carry out two checks in order to ensure that the section is locally stable. These checks for the flange and the web are discussed in AISC Table B5.1 (Table 9.1.2.1).

From the AISC Properties Table (Table 9.1.1), $b_f = 12.000$ in and $t_f = 0.605$ in. Hence

$$\frac{b_f}{2t_f} = \frac{12.000}{2(0.605)} = 9.9$$

This value is also available in the table. With $F_y = 36$ ksi, we have (from AISC Table B5.1):

$$\frac{95}{\sqrt{F_y}} = \frac{95}{\sqrt{36}} = 15.8 > 9.9. \text{ Hence OK.}$$

Similarly, for the web, from the AISC Properties Table (Table 9.1.1):

$$\frac{h}{t_w} = 24.9$$

Taking the appropriate condition from AISC Table B5.1, we have

$$\frac{253}{\sqrt{F_y}} = \frac{253}{\sqrt{36}} = 42.2 > 24.9. \text{ OK.}$$

Hence the section is locally stable.

9.1.3 Column Design

The end conditions of a column determine its resistance to axial forces—factor K in Eq. (9.1.2.3). In reality, the end restraint condition is difficult to evaluate. For column design, an estimate of the end condition is necessary so that the appropriate section can be selected. The

AISC-LRFD code offers some guidance based on whether the column member is used in a braced or unbraced frame.

An unbraced frame is defined as one in which "lateral stability depends on the bending stiffness of rigidly connected beams and columns." Fig. 9.1.3.1 shows an unbraced frame. The deflected shape of the frame shows that the two columns are such that $KL > 2L$. In other words, $K > 1$.

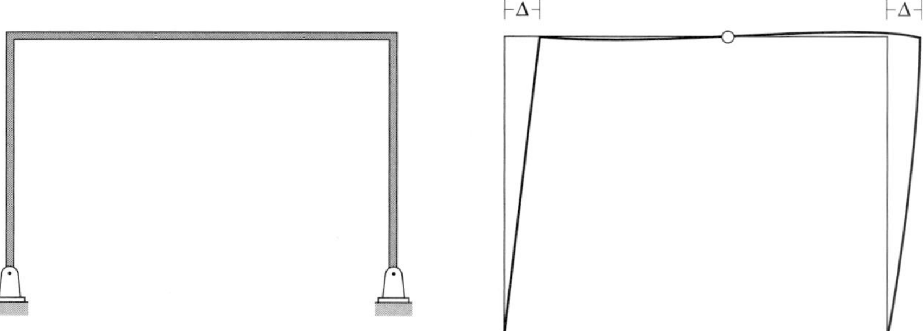

Fig. 9.1.3.1 Unbraced frame.

A braced frame, on the other hand, is defined as one in which "lateral stability is provided by diagonal bracing, shear walls, or equivalent means." Figure 9.1.3.2 shows a braced frame. The bracing provides the required lateral restraint so that the joints do not move laterally. Under this scenario, the effective length factor $K < 1$.

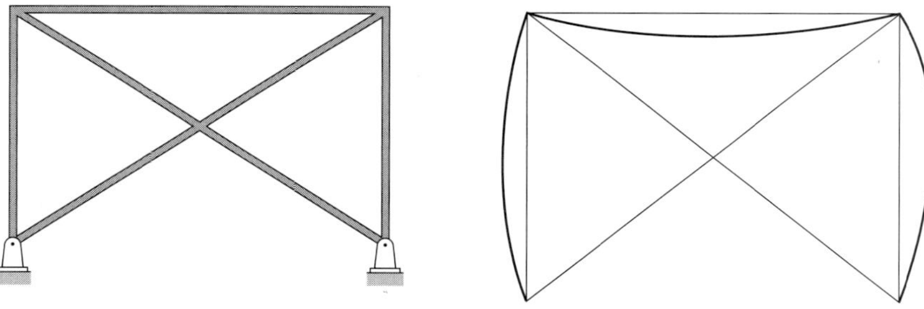

Fig. 9.1.3.2 Braced frame.

In lieu of analyzing the entire frame, the effective length factor can be found by using a chart developed by several researchers. One of the most commonly used charts was developed by O. J. Julian and L. S. Lawrence, and is now adopted by AISC. The chart is shown in Fig. 9.1.3.3. To find the value of K, carry out the following steps.

a. Select the appropriate diagram—sidesway inhibited or sidesway uninhibited.
b. Compute the values of G at the top (say, G_A) and bottom (say, G_B) of the column.
c. Locate these two values on the diagram and connect them by a straight line.
d. The intersection of the straight line with K scale gives the required value.

Design Procedure

1. Compute the factored load P_u.
2. Assume KL/r for the member. AISC-LRFD recommends that this value be limited to 200 for compression members. Obtain the critical stress F_{cr}.
3. Compute the gross area A_g using Eqs. (9.1.2.5) and (9.1.2.7) as $A_g = \dfrac{P_u}{\phi_c F_{cr}}$.

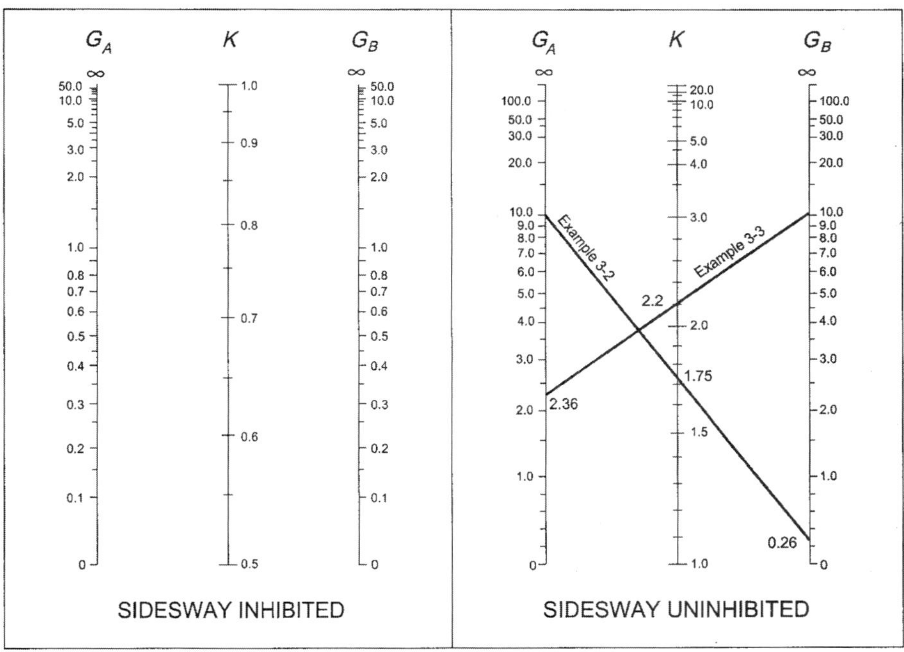

Fig. 9.1.3.3 Alignment charts for effective length of columns in continuous frames (Part 3, Fig. 3.1, AISC Manual) (the examples in the charts are discussed in the AISC manual).

Fig. 3-1. Alignment charts for effective length of columns in continuous frames. The subscripts A and B refer to the joints at the two ends of the column section being considered. G is defined as

$$G = \frac{\Sigma(I_c / L_c)}{\Sigma(I_g / L_g)}$$

in which Σ *indicates a summation of all members rigidly connected to that joint and lying on the plane in which buckling of the column is being considered.* I_c *is the moment of inertia and* L_c *the unsupported length of a column section, and* I_g *is the moment of inertia and* L_g *is the unsupported length of a girder or other restraining member.* I_c *and* I_g *are taken about axes perpendicular to the plane of buckling being considered.*

For column ends supported but not rigidly connected to a footing or foundation, G is theoretically infinity, but, unless actually designed a true friction free pin, may be taken as 10 for practical designs. If the column end is rigidly attached to a properly designed footing, G may be taken as 1.0. Smaller values may be used if justified by analysis.

4. Select a section so that width to thickness ratio is not small (see Table 9.1.2.1).
5. Compute KL/r_x and KL/r_y. With the larger of the two values, compute F_{cr}.
6. Check Eq. (9.1.2.7). If not satisfied, go back to Step 2.

Note: Once again, it cannot be overemphasized that the analysis and design procedures discussed above are applicable to those situations in which the predominant force is axial and in which local buckling is *not* an issue.

Optimal Design Problem Formulation

The optimal design problem for the design of a compression member can be posed as follows.

Find $\mathbf{x}$ = AISC Section

to minimize $f(\mathbf{x})$ = cost or weight (9.1.3.1)

subject to $g_1(\mathbf{x}) \equiv \phi_c P_n \geq P_u$ (9.1.3.2)

$$g_2(\mathbf{x}) \equiv \frac{KL}{r} \leq 200 \tag{9.1.3.3}$$

$$\mathbf{x} \in \mathbf{S} \tag{9.1.3.4}$$

where the design variable $\mathbf{x}$ is an available AISC section, e.g., W section, the set $\{\mathbf{S}\}$ contains the AISC sections to choose from, and the objective function $f(\mathbf{x})$ captures the prescribed cost or weight of the available sections.

There are at least two approaches to solving this problem. As we saw in Chapter 3, an exhaustive search can be carried out using the set $\{\mathbf{S}\}$ to find the section that yields the lowest cost or weight. Alternatively, the genetic algorithm (GA) can be used to find the section that yields the lowest cost or weight. For the design of a single compression member, there is little to choose between these two approaches. However, if the compression member is a part of structural system, then clearly using GA is more efficient and practical than exhaustive search.

EXAMPLE 9.1.3 ***Design of a Column***

Find an economical W section (A36 steel) for a column that is assumed to be pinned at both ends and is 20 ft long. The column carries a factored load of 200 k. Assume that the structure is braced.

SOLUTION We use k, in as the problem units.

Step 1: Factored loads. The factored load is given as 200 k.

Step 2: Estimate of the critical stress and required area. Since the member is pinned at both ends, $K = 1.0$. Also, for rolled W shapes with $F_y = 50$ ksi, $Q = 1$. Since a column 20 ft long can be thought of as a long column, we can estimate KL/r to be 100. From Fig. 9.1.2.1, $F_{cr} = 21$ ksi. Hence the required gross area is

$$A_g = \frac{P_u}{\phi_c F_{cr}} = \frac{200}{0.85(21)} = 11.2 \text{ in}^2$$

Step 3: Member selection. Since W shapes are not equally strong in both the x and y directions, buckling will occur about the weak axis, the y axis. In other words, with KL constant, KL/r_y is larger than KL/r_x. The AISC Column Load Tables (Table 9.1.2.1) can be used to select a suitable section whose area is at least equal to 11.2 in^2. Scanning the table, we find that W8 × 48 has a design axial strength of 214 k for an effective length of $L = 20$ ft.

$$\text{W8} \times 48\text{: } A = 14.1 \text{ in}^2, \quad \frac{KL}{r_y} = \frac{240}{2.08} = 115 \Rightarrow F_{cr} = 18 \text{ ksi}$$

Hence, $\phi_c F_{cr} A_g = 0.85(18)(14.1) = 215.7$ k > 200 k and the section is acceptable.

> **Observation:** The most efficient column under the circumstances discussed in this example is one for which the ratio r_x/r_y has the lowest value. The material distribution is such that the member is "equally" strong in both directions.

EXAMPLE 9.1.4 ***Analysis and Design of a Column in a Planar Frame***

Figure E9.1.4 shows a planar frame that is braced out of plane. A36 steel is used.

i. Compute the effective length factor for Column DC.
ii. Column DC is subjected to a factored axial load of 500 k. Is the design acceptable?

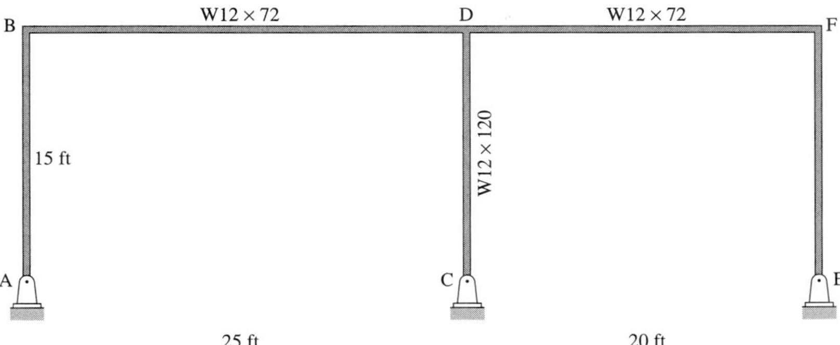

Fig. E9.1.4

SOLUTION

Step 1: Frame definition. Based on the structural geometry provided, the frame is an unbraced frame. For joint D,

$$G = \frac{\sum I_c/L_c}{\sum I_g/L_g} = \frac{1070/15}{597/25 + 597/20} = \frac{71.33}{53.73} = 1.33$$

Note that we are assuming the cross-sections are oriented in such a manner that the largest moment of inertia is available to resist the moment (bending is about the strong axis). In other words, the flanges are normal to the plane of the frame (or the web is in the plane of the frame).

Joint C is at a pin support. Since the exact nature of the support is not known, AISC recommends that for this situation G be taken as 10.0 (value of 1.0 is recommended if the support is fixed).

Step 2: Use of the alignment chart. Using $G_A = 1.33$ and $G_B = 10.0$, we obtain $K_x = 1.95$ for column DC.

Step 3: Estimate of the critical stress and required area. From the given data, for rolled W shapes with $F_y = 36$ ksi, $Q = 1$.

$$\frac{K_x L}{r_x} = \frac{(1.95)(180)}{5.51} = 63.7$$

From Fig. 9.1.2.1, $F_{cr} = 29$ ksi. Hence the required gross area is

$$A_g = \frac{P_u}{\phi_c F_{cr}} = \frac{500}{0.85(29)} = 20.3 \text{ in}^2$$

Note that since the frame is assumed to be adequately braced out of plane, the strength considerations about the y axis do not apply.

Step 4: Member check. From the AISC Column Load Tables, for W12 × 120, the design axial strength value is about 540 k > 500 k for an effective length of $KL = 29.25$ ft. Hence the section is acceptable.

9.1.4 Beam Design

Figure 9.1.4.1 shows the stress distribution in a beam subjected to increasing load. When the beam is laterally supported, one possible mode of failure is through local buckling of the compression flange or the web. By definition, the yield moment M_y is reached when the extreme fiber of the cross-section yields, as shown in Fig. 9.1.4.1(c). In other words,

$$M_y = S_x F_y \tag{9.1.4.1}$$

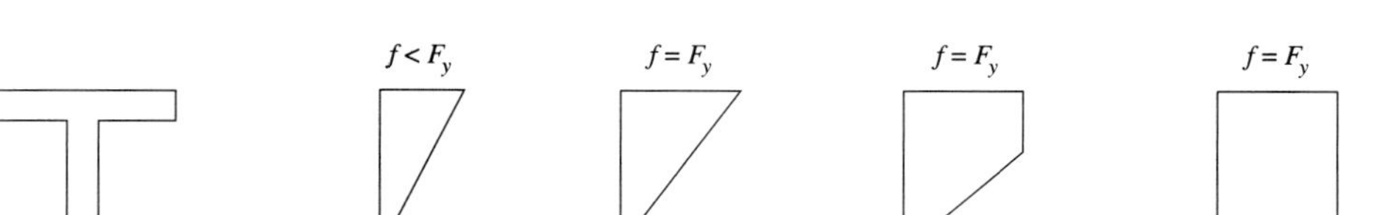
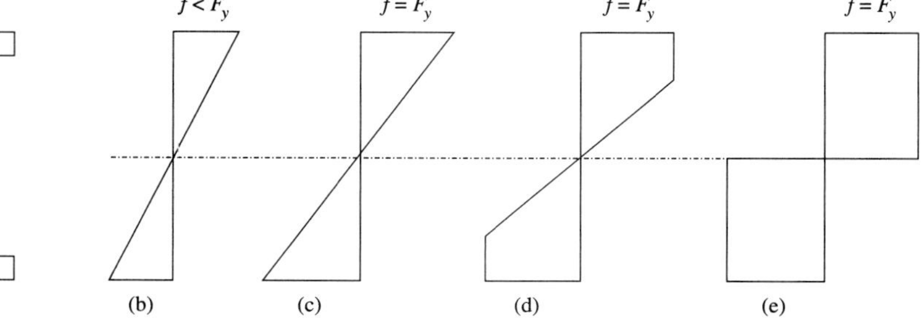

Fig. 9.1.4.1 Stress distribution for a symmetric I-section with increasing load.

The moment corresponding to Fig. 9.1.4.1(e), i.e,. when the entire cross-section has yielded, is called the plastic moment M_p:

$$M_p = F_y \int_A y\,dA = F_y Z \tag{9.1.4.2}$$

where $Z = \int_A y\,dA$ is referred to as the plastic modulus. The ratio of the two moments in Eqs. (9.1.4.1) and (9.1.4.2), designated ξ, is

$$\xi = \frac{M_p}{M_y} = \frac{Z}{S} \tag{9.1.4.3}$$

and is called the shape factor.

The beam member can realize the fully plastic state provided buckling (overall or local), does not take place. Buckling in the AISC sections is taken to be a function of the cross-sectional shape, dimensions, and the grade of steel used. In Section 9.1.2, sections were classified as compact or noncompact. To recap, the width-thickness ratio λ of the section was used in this classification along with λ_p, the upper limit for the compact category, and λ_r, the upper limit for the noncompact category.

Compact Section. A section is compact if $\lambda \le \lambda_p$ and the flange is continuously connected to the web.

Noncompact Section. A section is noncompact if $\lambda_p < \lambda < \lambda_r$.

Slender Section. A section is slender if $\lambda > \lambda_r$.

The AISC Code states that the nominal flexure strength M_n is the lowest value obtained according to the limit states of yielding arising from lateral-torsional buckling (LTB), flange local buckling (FLB), and web local buckling (WLB). Consider the beam in Fig. 9.1.4.2. Assume that the beam cross-section is an I-section. It is possible for the beam to undergo overall buckling (Euler buckling) as shown in Fig. 9.1.4.2(b). Under the action of the applied loads, the top flange is in compression. If the beam is not adequately supported laterally, it is possible for the beam to twist and deflect outwards, as shown in Fig. 9.1.4.2(c).

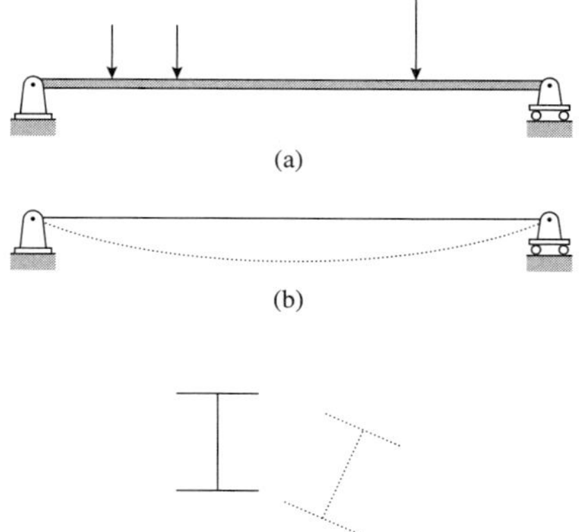

Fig. 9.1.4.2 (a) Simply supported I-beam. (b) Overall buckling mode. (c) Lateral-torsional buckling (LTB) mode. (d) Beam with three lateral support points.

The beam is laterally supported or braced if the compression flange is not allowed to twist and deflect outwards. The lateral bracing can be continuous or discrete (as shown in Fig. 9.1.4.2(d)). For example, continuous lateral support is provided when the compression flange is encased in a concrete floor slab. Discrete support can be provided through cross beams, struts, etc. It should be noted that the unbraced length controls the moment capacity of the beam. It is also possible for the beam to fail if the compression flange buckles or if the web buckles, as shown in Fig. 9.1.4.3.

(a)

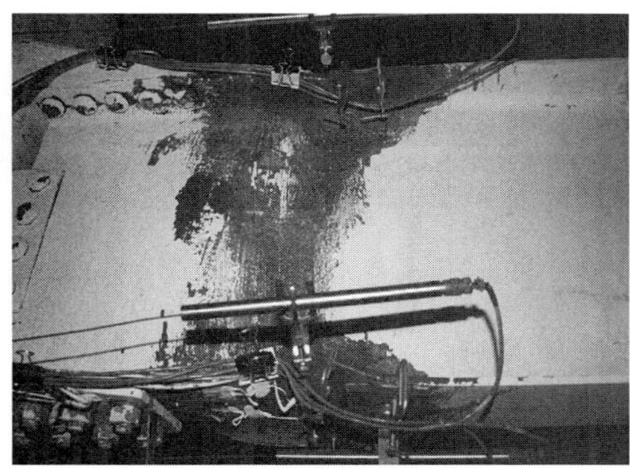
(b)

Fig. 9.1.4.3 (a) Flange local buckling. (b) Web local buckling.

The design of laterally supported beams as per AISC-LRFD code (Appendix F) requires that

$$\phi_b M_n \geq M_u \tag{9.1.4.4}$$

where ϕ_b is the strength reduction factor for flexure (taken as 0.9), M_n is the nominal strength (moment), and M_u is the factored moment.

Compact Sections. For laterally stable compact sections ($\lambda \leq \lambda_p$),

$$M_n = M_p = F_y Z \leq 1.5 M_y \tag{9.1.4.5}$$

The upper limit is placed on M_p so as to avoid excessive deformation.

Lateral-Torsional Buckling. The solution to the differential equation for *elastic* lateral-torsional buckling (Timoshenko and Gere, 1961) is given by

$$M = \frac{\pi}{L_b}\sqrt{\left(\frac{\pi E}{L_b}\right)^2 C_w I_y + EI_y GJ} < M_p \tag{9.1.4.6}$$

where M is the moment
L_b is the unbraced length
C_w is the warping constant
G is the shear modulus
J is the torsional constant
I_y is the moment of inertia about the weak axis

From studies by several researchers, it is generally recognized that the moment strength of a typical I-section subjected to bending is as shown in Fig. 9.1.4.4.

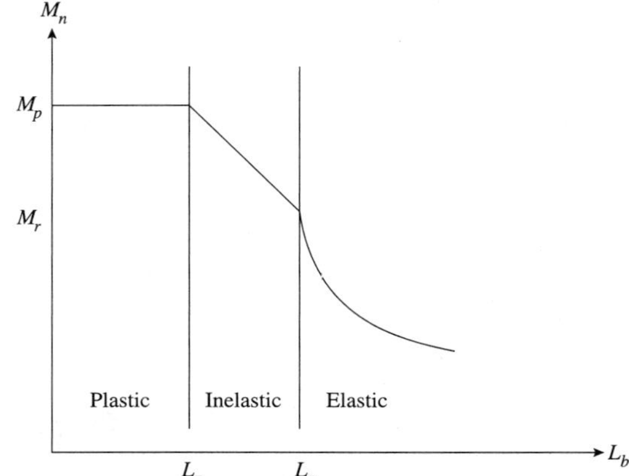

Fig. 9.1.4.4 Moment-length relationship for typical W-sections.

When LTB is not a factor, the beam attains its full plastic strength M_p (Eq. (9.1.4.5)). *Inelastic* LTB occurs if the moment at the instance of LTB is greater than moment at first yield, given by

$$M_r = F_L S_x \tag{9.1.4.7}$$

where F_L is the smaller of $(F_{yf} - F_r)$ and F_{yw}, and F_r is the residual stress. Due to the presence of residual stresses, the flange yield stress is reduced. The prescribed value for F_r is 10 ksi for rolled sections and 16.5 ksi for built-up sections. In the inelastic range, for a uniform bending moment, the nominal strength is given by

$$M_n = M_p - \left(M_p - M_r\right)\left(\frac{L_b - L_p}{L_r - L_p}\right) < M_p \tag{9.1.4.8}$$

where

$$L_p = \frac{300}{\sqrt{F_{yf}}} r_y \tag{9.1.4.9a}$$

To obtain the unbraced length L_r, M_r can be used in Eq. (9.4.1.6) instead of M to yield

$$L_r = \frac{r_y X_1}{\left(F_y - F_r\right)}\sqrt{1 + \sqrt{1 + X_2\left(F_y - F_r\right)^2}} \tag{9.1.4.9b}$$

where

$$X_1 = \frac{\pi}{S_x}\sqrt{\frac{EGJA}{2}} \tag{9.1.4.10}$$

$$X_2 = \frac{4C_w}{I_y}\left(\frac{S_x}{GJ}\right)^2 \tag{9.1.4.11}$$

When the applied moment is not uniform, the modified nominal strength is given by

$$M_n = C_b\left[M_p - \left(M_p - M_r\right)\left(\frac{L_b - L_p}{L_r - L_p}\right)\right] \tag{9.1.4.12}$$

The factor C_b is given as

$$C_b = \frac{12.5M_{\max}}{2.5M_{\max} + 3M_A + 4M_B + 3M_C} \tag{9.1.4.13}$$

where $M_{\max}$ is the absolute maximum of the bending moment within the unbraced length
M_A is the absolute value of the bending moment at the quarter point of the unbraced length
M_B is the absolute value of the bending moment at the midpoint of the unbraced length
M_C is the absolute value of the bending moment at the three-quarter point of the unbraced length

Noncompact Sections. For partially compact (or noncompact) laterally stable sections ($\lambda_p < \lambda \leq \lambda_r$), for the limit states of flange and web local buckling[3] we have

$$M_n = M_p - \left(M_p - M_r\right)\left(\frac{\lambda - \lambda_p}{\lambda_r - \lambda_p}\right) < M_p \tag{9.1.4.14}$$

where $M_r = \left(F_y - F_r\right)S_x$ and F_r is the residual stress.

Slender Sections. For laterally stable slender sections ($\lambda > \lambda_r$), we have

$$M_n = M_{cr} = SF_{cr} \leq M_p \tag{9.1.4.15}$$

where M_{cr} is the buckling moment
M_r is the limiting buckling moment (equal to M_{cr} when $\lambda = \lambda_r$)
F_{cr} is the critical stress
λ is the controlling slenderness parameter and can be taken as $\frac{b_f}{2t_f}$ for flanges of I-shaped members and $\frac{h}{t_w}$ for beam webs

Analysis of I- and H-Sections

We now discuss the steps in the analysis of a typical I-section beam. Given beam shape, dimensions, L_b, and the factored moment, M_u, find the adequacy of the beam section.

Step 1: Classify the section as compact or noncompact. See Table 9.1.2.1 (AISC Table B5.1).

[3] Web local buckling is not possible for the rolled AISC shapes since the webs are compact.

Step 2: For a compact section:

a. Compute L_p using Eq. (9.1.4.9a). If $L_b \le L_p$, then LTB does not take place. Set $M_n = M_p$.
b. Compute L_r using Eq. (9.1.4.9b). If $L_p < L_b < L_r$, there is inelastic LTB. Compute M_n using Eqs. (9.1.4.7) and (9.1.4.8).
c. If $L_b > L_r$, then there is elastic LTB. Compute M_n using Eq. (9.1.4.6).

Step 3: For a noncompact section:

a. Check FLB as follows. If $\lambda \le \lambda_p$, then there is no FLB. Set $M_n = M_p$. If $\lambda_p < \lambda \le \lambda_r$, the flange is noncompact. Compute M_n using Eq. (9.1.4.14).
b. Check WLB as follows. If $\lambda \le \lambda_p$, then there is no WLB. Set $M_n = M_p$. If $\lambda_p < \lambda \le \lambda_r$, the web is noncompact. Compute M_n using Eq. (9.1.4.14).
c. Check LTB as follows. Compute L_p using Eq. (9.1.4.9a). If $L_b \le L_p$, then LTB does not take place. Set $M_n = M_p$. Compute L_r using Eq. (9.1.4.9b). If $L_p < L_b \le L_r$, there is inelastic LTB. Compute M_n using Eqs. (9.1.4.7) and (9.1.4.8). If $L_b > L_r$, then there is elastic LTB. Compute M_n using Eq. (9.1.4.6).

EXAMPLE 9.1.5

Design Checks for a Simply Supported Beam

A simply supported beam of span 35 ft supports a dead load of 400 lb/ft (including the weight of the beam). The service live load is 500 lb/ft. The beam is made of W14 × 53 A36 steel and has continuous lateral support. Does the beam have adequate moment strength?

SOLUTION

Step 1: Compute the factored load and maximum moment. Using k, ft as the units, we get

$$w_u = 1.2w_{DL} + 1.6w_{LL} = 1.28 \text{ k/ft}$$

The maximum moment M_{max} for a simply supported beam occurs at the midspan:

$$M_u = M_{max} = \frac{w_u L^2}{8} = \frac{(1.28)(35)^2}{8} = 196 \text{ k-ft}$$

Step 2: Classify the section. Using the values from the AISC Properties Table for W-Sections gives us

$$\frac{b_f}{2t_f} = 6.1 \quad \text{and} \quad \frac{65}{\sqrt{F_y}} = \frac{65}{\sqrt{36}} = 10.8$$

$$\frac{h}{t_w} = 30.8 \quad \text{and} \quad \frac{640}{\sqrt{F_y}} = \frac{640}{\sqrt{36}} = 106.7$$

Since 10.8 > 6.1 and 106.7 > 30.8, the section is compact.

Step 3: Obtain the design moment strength. For a compact beam that is laterally supported (note the conversion to ft), we have

$$M_n = M_p = F_y Z_x = \frac{36(87.1)}{12} = 261.3 \text{ k-ft}$$

Since $\phi_b M_n = (0.9)261.3 = 235$ k-ft $> M_u$, the section is adequate.

EXAMPLE 9.1.6

Design Checks for a Simply Supported (Compact) Beam

A simply supported beam of span 20 ft (Fig. E9.1.6(a)) supports a dead load of 500 lb/ft (including the weight of the beam). The service live load is 1000 lb/ft. The beam is made of W12 × 35 A36 steel and is laterally supported at the ends. Does the beam have adequate moment strength?

w

20 ft

Fig. E9.1.6(a)

SOLUTION

Step 1: Compute the factored load and maximum moment. Using k, ft as the units, we have

$$w_u = 1.2w_{DL} + 1.6w_{LL} = 2.2 \text{ k/ft}$$

The maximum moment $M_{\max}$ for a simply supported beam occurs at the midspan:

$$M_u = M_{\max} = \frac{w_u L^2}{8} = \frac{(2.2)(20)^2}{8} = 110 \text{ k-ft}$$

Step 2: Classify the section. Using the values from the AISC Properties Table for W-Sections, we have

$$\lambda = \frac{b_f}{2t_f} = 6.3 \qquad \text{and} \qquad \lambda_p = \frac{65}{\sqrt{F_y}} = \frac{65}{\sqrt{36}} = 10.8$$

$$\lambda = \frac{h}{t_w} = 36.2 \qquad \text{and} \qquad \lambda_p = \frac{640}{\sqrt{F_y}} = \frac{640}{\sqrt{36}} = 106.7$$

Since 10.8 > 6.3 and 106.7 > 36.2, the section is compact. Hence,

$$L_p = \frac{300}{\sqrt{F_{yf}}} r_y = \frac{300}{\sqrt{36}}(1.54) = 77 \text{ in} = 6.42 \text{ ft}$$

Furthermore, from the AISC Tables, X_1 = 2420 ksi, X_2 = 0.004340/in^2, and F_r = 10 ksi. Hence,

$$L_r = \frac{r_y X_1}{(F_y - F_r)} \sqrt{1 + \sqrt{1 + X_2 (F_y - F_r)^2}}$$

$$L_r = \frac{(1.54)(2420)}{(36-10)} \sqrt{1 + \sqrt{1 + 0.004340\ (36-10)^2}} = 143.3(1.73) = 247.5 \text{ in} = 20.6 \text{ ft}$$

Since $L_p < L_b = 20$ ft $< L_r$, inelastic LTB controls the design.

Step 3: Compute the moment gradient factor C_b. The critical points for a uniformly loaded simply supported beam are shown in Fig. E9.1.6(b). The bending moments are such that

$$M_{\max} = M_B = \frac{wL^2}{8}; \qquad M_A = M_C = \frac{3wL^2}{32}$$

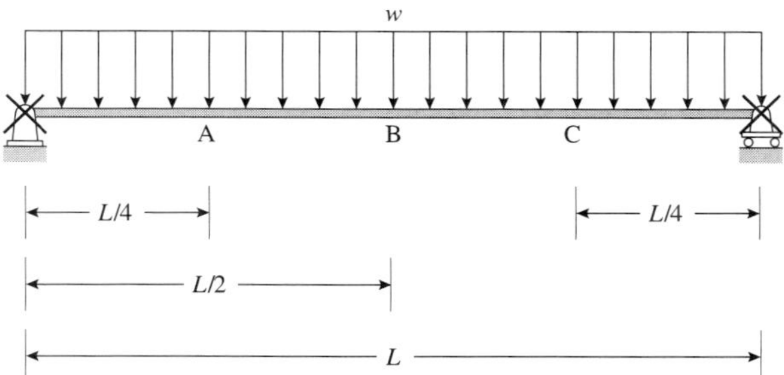

Fig. E9.1.6(b)

Hence

$$C_b = \frac{12.5M_{\max}}{2.5M_{\max} + 3M_A + 4M_B + 3M_C} = 1.14$$

Step 4: Computation of the design moment:

$$M_p = F_y Z_x = 36(51.2) = 1843.2 \text{ k-in} = 153.6 \text{ k-ft}$$

$$M_r = (F_y - F_r)S_x = (36 - 10)45.6 = 1186 \text{ k-in} = 98.8 \text{ k-ft}$$

$$M_n = C_b\left[M_p - (M_p - M_r)\left(\frac{L_b - L_p}{L_r - L_p}\right)\right] = 1.14[153.6 - 54.8(0.96)] = 115.1 \text{ k-ft}$$

$$\phi M_n = 0.9(115.1) = 103.6 \text{ k-ft}$$

Since $\phi M_n < M_u$, the section is inadequate.

EXAMPLE 9.1.7

Design Checks for a Simply Supported (Noncompact) Beam

A simply supported beam of span 20 ft (Fig. E9.1.7) supports a dead load of 1000 lb/ft (including the weight of the beam). The service live load is 2000 lb/ft. The beam is made of W12 × 65 A572 Grade 50 steel and is laterally supported at the ends. Does the beam have adequate moment strength?

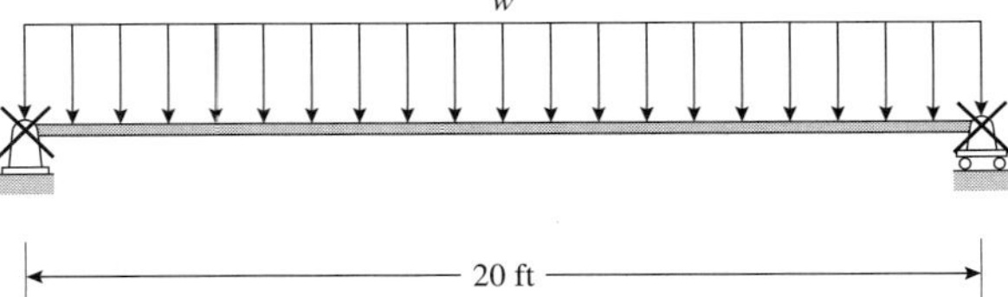

Fig. E9.1.7

SOLUTION

Step 1: Compute the factored load and maximum moment. Using k, ft as the units, we have

$$w_u = 1.2w_{DL} + 1.6w_{LL} = 4.4 \text{ k/ft}$$

The maximum moment $M_{\max}$ for a simply supported beam occurs at the midspan:

$$M_u = M_{\max} = \frac{w_u L^2}{8} = \frac{(4.4)(20)^2}{8} = 220 \text{ k-ft}$$

Step 2: Classify the section. Using the values from the AISC Properties Table for W-Sections gives us

$$\lambda = \frac{b_f}{2t_f} = 9.9 \qquad \lambda_p = \frac{65}{\sqrt{F_y}} = \frac{65}{\sqrt{50}} = 9.19$$

$$\lambda_r = \frac{141}{\sqrt{F_y - F_r}} = \frac{141}{\sqrt{50 - 10}} = 22.3$$

Since $\lambda_p < \lambda < \lambda_r$, the section is noncompact. We need to compute the limit state of FLB and LTB.

Step 3: FLB limit state:

$$M_p = F_y Z_x = 50(96.8) = 4840 \text{ k-in} = 403.3 \text{ k-ft}$$

$$M_r = (F_y - F_r)S_x = (50 - 10)87.9 = 3516 \text{ k-in} = 293 \text{ k-ft}$$

$$M_n = M_p - (M_p - M_r)\left(\frac{\lambda - \lambda_p}{\lambda_r - \lambda_p}\right) = 403.3 - (110.3)(0.054) = 397.3 \text{ k-ft}$$

$$\phi M_n = 0.9(397.3) = 357.6 \text{ k-ft}$$

Step 4: LTB limit state:

$$L_p = \frac{300}{\sqrt{F_{yf}}} r_y = \frac{300}{\sqrt{50}}(3.02) = 128.1 \text{ in} = 10.68 \text{ ft}$$

Furthermore, from the AISC Tables, X_1 = 2940 ksi, X_2 = 0.001720/in^2, and F_r = 10 ksi Hence,

$$L_r = \frac{r_y X_1}{(F_y - F_r)} \sqrt{1 + \sqrt{1 + X_2 (F_y - F_r)^2}}$$

$$L_r = \frac{(3.02)(2940)}{(50-10)} \sqrt{1 + \sqrt{1 + 0.001720\,(50-10)^2}} = 380.4 \text{ in} = 31.7 \text{ ft}$$

Since $L_p < L_b = 20$ ft $< L_r$, inelastic LTB takes place. Hence

$$M_n = C_b\left[M_p - (M_p - M_r)\left(\frac{L_b - L_p}{L_r - L_p}\right)\right] = 1.14\left[403.3 - 110.3\left(\frac{20 - 10.68}{31.7 - 10.68}\right)\right] = 404 \text{ k-ft}$$

$$\phi M_n = 0.9(404) = 363.6 \text{ k-ft}$$

Step 5: Design moment. Since 363.6 k-ft > 357.6 k-ft, FLB controls the design. Since 357.6 k-ft > 220 k-ft, the section is acceptable.

Other Considerations in the Design of Beams

There are other considerations in the design of beams. We address some of the major ones here.

Serviceability. Proper design procedures must incorporate the checks that limit the maximum deflection of the structure. If deflection is not controlled, damage may be caused to nonstructural components. This may prevent the structure from not fulfilling its desired functionality. For example, a beam that deflects excessively may cause the ceiling below to crack.

With regards to deflections, the AISC Code states that "Limiting values of structural behavior to ensure serviceability (e.g., maximum deflections, accelerations, etc.) shall be chosen with due regard to the intended function of the structure. Where necessary, serviceability shall be checked using realistic loads for the appropriate serviceability limit state." Furthermore, Chapter L of the AISC-LRFD Code states that "Deformations in structural members and structural systems due to service loads shall not impair the serviceability of the structure."[4] Finally, it is the responsibility of the design engineer to specify the limiting value for the structure being designed.

In this text we assume the following values as limiting deflection values:

Unplastered floor construction	$L/240$
Unplastered roof construction	$L/180$
Plastered construction	$L/360$

[4] The AISC-ASD Code provides some guidelines on computing the limiting values.

Shear. In Chapter 3 we derived the shear stress formula as

$$\tau = \frac{VQ}{It} \tag{9.1.4.16}$$

The AISC-LRFD design requirement can be expressed as

$$\phi_v V_n \geq V_u \tag{9.1.4.17}$$

where ϕ_v is the strength reduction factor for shear (0.9), V_n is the nominal shear strength, and V_u is the factored shear force.

The nominal shear strength value is usually taken as

$$V_n = 0.6 F_{yw} A_w \tag{9.1.4.18}$$

for beams without transverse stiffeners and not exceeding the h/t_w given by

$$\frac{h}{t_w} = \frac{418}{\sqrt{F_{yw}}} \tag{9.1.4.19}$$

where F_{yw} is the yield stress of the web and A_w is the area of the web. For $418/\sqrt{F_{yw}} < h/t_w \leq 523/\sqrt{F_{yw}}$, inelastic web buckling can occur and

$$V_n = 0.6 F_{yw} A_w \frac{418/\sqrt{F_y}}{h/t_w} \tag{9.1.4.20}$$

For $523/\sqrt{F_{yw}} < h/t_w \leq 260$, inelastic web buckling occurs and

$$V_n = \frac{132000 A_w}{\left(h/t_w\right)^2} \tag{9.1.4.21}$$

One must take into account not only the largest value of the shear force being carried by the beam, but also the reduction in the capacity of the section due to bolt holes or holes through which service pipes pass.

Local Web Yielding. Local web yielding is a possibility when large concentrated loads are applied to beams. The AISC-LRFD requirement can be expressed as

$$\phi R_n \geq R_u \tag{9.1.4.22}$$

where ϕ is the strength reduction factor (taken as 1.0), R_n is the nominal reaction strength, and R_u is the factored reaction.

Recent research has led to better understanding of the behavior of the member when large concentrated loads are applied. A simply supported beam subjected to a concentrated load is shown in Fig. 9.1.4.5. At the support and at the point of application of the load, the load spreads over a distance greater than N, the bearing length. For such a case, R_n is given as follows (computed at the toe of the fillet of a rolled I-section):

For an interior load (point of application of the load is greater than the depth of the member),

$$R_n = (5k + N) F_{yw} t_w \tag{9.1.4.23}$$

For end reactions,

$$R_n = (2.5k + N) F_{yw} t_w \tag{9.1.4.24}$$

where k is the distance from the outer face of the flange to web toe of fillet, F_{yw} is the specified yield stress of the web, N is the bearing length > k, and t_w is the web thickness.

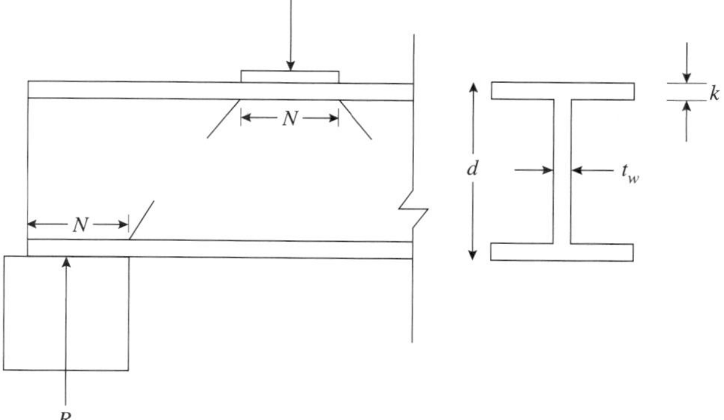

Fig. 9.1.4.5

Web Crippling. Under the action of concentrated loads, the web-crippling criterion controls the stability of the web. The AISC-LRFD Code specifies the nominal reaction strength as follows.

For an interior load (point of application of the load is greater than $d/2$),

$$R_n = 135t_w^2\left[1+3\left(\frac{N}{d}\right)\right]\left(\frac{t_w}{t_f}\right)^{1.5}\sqrt{\frac{F_{yw}t_f}{t_w}} \tag{9.1.4.25}$$

For end reactions when $N/d \le 0.2$,

$$R_n = 68t_w^2\left[1+3\left(\frac{N}{d}\right)\right]\left(\frac{t_w}{t_f}\right)^{1.5}\sqrt{\frac{F_{yw}t_f}{t_w}} \tag{9.1.4.26}$$

For end reactions when $N/d > 0.2$,

$$R_n = 68t_w^2\left(1+\frac{4N}{d}-0.2\right)\left(\frac{t_w}{t_f}\right)^{1.5}\sqrt{\frac{F_{yw}t_f}{t_w}} \tag{9.1.4.27}$$

where t_f is the flange thickness, and d is the overall depth of the beam. The value of ϕ, the strength reduction factor, is taken as 0.75.

EXAMPLE 9.1.8 ***Design Checks for a Simply Supported Beam***

Check the beam in Example 9.1.7 for the following requirements: (a) shear, (b) deflection limit of $L/360$.

SOLUTION

Step 1: Shear check.

a. Compute the maximum shear in the beam. The maximum shear in a simply supported beam subjected to a uniformly distributed load occurs at the supports. Hence

$$V_u = \frac{w_u L}{2} = \frac{4.4(20)}{2} = 44 \text{ k}$$

b. Classify the beam:

$$\frac{h}{t_w} = 24.9 \qquad \frac{418}{\sqrt{F_y}} = \frac{418}{\sqrt{50}} = 59.1$$

Since $h/t_w < 418/\sqrt{F_y}$, there is no web instability. The shear strength is then

$$V_n = 0.6\,F_y A_w = 0.6(50)(12.12 \times 0.39) = 141.8 \text{ k}$$

$$\phi V_n = (0.90)(141.8) = 127.6 \text{ k} > 44 \text{ k}.$$

Hence, the section is acceptable.

Step 2: Deflection check. The largest deflection in a simply supported beam subjected to a uniformly distributed load occurs at the center of the beam and is given as

$$\Delta_{max} = \frac{5w_u L^4}{384EI} = \frac{5(4.4)(20)^4}{384(29 \times 10^3 \times 533)/144} = 0.085 \text{ ft}$$

$$\frac{L}{360} = \frac{20}{360} = 0.056 \text{ ft}$$

Hence, the beam does not satisfy the deflection requirement.

Optimal Design Problem Formulation

The optimal design problem for the design of a beam member can be posed as follows.

Find $\mathbf{x}$ = AISC section

to minimize $f(\mathbf{x})$ = cost or weight (9.1.4.28)

subject to

$$g_1(\mathbf{x}) \equiv \phi_b M_n \geq M_u \quad (9.1.4.29)$$

$$g_2(\mathbf{x}) \equiv \phi_v V_n \geq V_u \quad (9.1.4.30)$$

$$g_3(\mathbf{x}) \equiv \Delta_{max} \leq \Delta_{allowed} \quad (9.1.4.31)$$

$$g_4(\mathbf{x}) \equiv \phi R_n \geq R_u \quad (9.1.4.32)$$

$$\mathbf{x} \in \mathbf{S} \quad (9.1.4.33)$$

where the design variable $\mathbf{x}$ is an available AISC section, e.g., W section, the set $\mathbf{S}$ contains the AISC sections to choose from, and the objective function $f(\mathbf{x})$ captures the prescribed cost or weight of the available sections. Consistent with the comments made about the optimal design of a compression member, there are at least two approaches to solving this problem—exhaustive search and GA. For the design of a single beam member, there is little to choose from between these two approaches. However, if the beam member is a part of structural system, then clearly GA is more efficient and practical than exhaustive search.

Design of Indeterminate Steel Structures

In Chapter 3 we examined in some detail the design process for simple determinate beams. In Chapter 8 we formulated the design problem as a mathematical programming problem. The process was applicable to both determinate and indeterminate structures. In this section we will look in some detail at the trial-and-error procedure for the (sizing) design of indeterminate steel structures.

Typically, the detailed design process starts with the structural geometry (material, members, joints, support conditions) and loading defined. The designer must guess the initial size of the members in the structure. For steel members, one can choose an available section (e.g., from the AISC tables) or configure a built-up section. Designers rely on their experience to make this choice. The experience may take into account factors such as cost and constructability. Having established the member sizes, one must carry out a structural analysis to compute the following response quantities: (a) support reactions, (b) the internal force distribution in the form of the shear force and bending moment diagrams, and (c) key deflections.

The results from the structural analysis are then used to carry out the design checks as we have seen above in some of the examples. It is possible that (a) the checks are satisfied with a large margin of safety, (b) the checks are just about satisfied, or (c) one or more checks are not satisfied. In case (c), a redesign is required. In cases (a) and (b), the designer must decide whether the savings offsets the redesign cost. Note that in the case of indeterminate structures, changing the cross-sectional properties of one member can lead to a redistribution of the internal forces.

EXAMPLE 9.1.9 ***Design of an Indeterminate Beam***

Design the steel beam shown in Fig. E9.1.9. Assume that $F_y = 50$ ksi and $F_r = 10$ ksi. Also assume that the applied loads are the factored loads.

SOLUTION

Step 1: Initial guess.

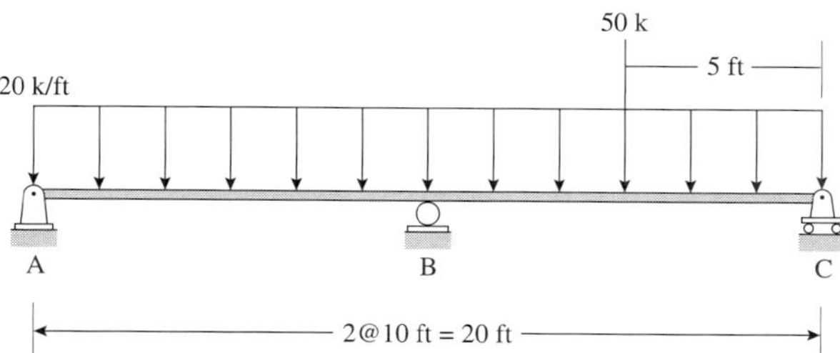

Fig. E9.1.9

We will assume that $I_{BC} = 1.5I_{AB}$ since the span BC is subjected to a larger load. Using the method for analysis of an indeterminate beam (see, for example, Example 5.1.3), we obtain the following response values. The largest bending moment and shear occurs at B and

$$M_u = M_{max} = 287.6 \text{ k-ft} = 3.45\left(10^6\right) \text{ lb-in}$$

$$V_u = V_{max} = 282.5 \text{ k}$$

Step 2: Design of beam AB.

Let us assume the required section to be W21 × 83. This section has the following properties.

$b_f = 8.35$ in $t_f = 0.835$ in $h = 18.25$ in $t_w = 0.515$ in

$X_1 = 2400$ ksi $X_2 = 0.00525/\text{ksi}^2$

$R_y = 1.83$ in $Z_x = 196 \text{ in}^3$ $S_x = 171 \text{ in}^3$ $r_x = 8.67$ in

Classification of the section

Flange:

$$\lambda = \frac{b_f}{2t_f} = \frac{8.355}{(2 \times 0.835)} = 5.0$$

$$\lambda_p = \frac{65}{\sqrt{F_y}} = 9.192$$

Since $\lambda < \lambda_p$, the section is compact.

Web:

$$\lambda = \frac{h}{t_w} = \frac{18.25}{0.515} = 35.43$$

$$\lambda_p = \frac{640}{\sqrt{F_y}} = 90.51$$

Since $\lambda < \lambda_p$, the section is compact.

Identification of the factor governing the design

$$L_p = \frac{300}{\sqrt{F_y}} r_y = 77.6 \text{ in} = 6.47 \text{ ft}$$

$$L_r = \frac{r_y X_1}{(F_y - F_r)} \sqrt{1 + \sqrt{1 + X_2 (F_y - F_r)^2}} = 221.4 \text{ in} = 18.4 \text{ ft}$$

Since $L_p < L_b < L_r$, the design is controlled by inelastic LTB.

Calculation of nominal strength

$$M_p = Z_x F_y = 9.8(10^6) \text{ lb-in}$$

$$M_r = S_x \left(F_y - F_r \right) = 6.84(10^6) \text{ lb-in}$$

$$M_n = M_p - (M_p - M_n) \left(\frac{L_b - L_p}{L_r - L_b} \right) = 8.556(10^6) \text{ lb-in}$$

$$\phi M_n = 7.7(10^6) \text{lb-in} > 3.45(10^6) \text{ lb-in}$$

Shear check

$$V_u = 282.5 \text{ k}$$

$$\frac{h}{t_w} = \frac{18.25}{0.515} = 35.43 < \frac{418}{\sqrt{F_y}} = 59.11$$

Hence, there is no web instability.

$$V_n = 0.6 F_y A_w = 331.1 \text{ k}$$

Also, $\phi V_n = 298 > V_{max}$

Hence, the section is adequate.

Step 3: Design of beam BC.

Let us assume the required section to be W27 × 84. This section has the following properties:

$b_f = 9.96$ in $\quad t_f = 0.64$ in $\quad h = 24.0$ in $\quad t_w = 0.46$ in

$X_1 = 1570$ ksi $\quad X_2 = 0.0311/\text{ksi}^2$

$R_y = 2.067$ in $\quad Z_x = 244 \text{ in}^3 \quad S_x = 213 \text{ in}^3 \quad r_x = 10.72$ in

Classification of the section

Flange:

$$\lambda = \frac{b_f}{2t_f} = \frac{9.96}{(2 \times 0.64)} = 7.78$$

$$\lambda_p = \frac{65}{\sqrt{F_y}} = 9.19$$

Since $\lambda < \lambda_p$, the section is compact.

Web:

$$\lambda = \frac{h}{t_w} = \frac{24.0}{0.46} = 52.2$$

$$\lambda_p = \frac{640}{\sqrt{F_y}} = 90.51$$

Since $\lambda < \lambda_p$, the section is compact.

Identification of the factor governing the design

$$L_p = \frac{300}{\sqrt{F_y}} r_y = 87.7 \text{ in} = 7.31 \text{ ft}$$

$$L_r = \frac{r_y X_1}{(F_y - F_r)} \sqrt{1 + \sqrt{1 + X_2 (F_y - F_r)^2}} = 231.2 \text{ in} = 19.3 \text{ ft}$$

Since $L_p < L_b < L_r$, the design is controlled by inelastic LTB.

Calculation of nominal strength

$$M_p = Z_x F_y = 12.2(10^6) \text{ lb-in}$$

$$M_r = S_x\left(F_y - F_r\right) = 8.52(10^6) \text{ lb-in}$$

$$M_n = M_p - (M_p - M_n)\left(\frac{L_b - L_p}{L_r - L_b}\right) = 11.13(10^6) \text{ lb-in}$$

$$\phi M_n = 10.0(10^6) \text{ lb-in} > 3.45(10^6) \text{ lb-in}$$

Shear check

$V_u = 282.5$ k

$$\frac{h}{t_w} = \frac{24}{0.46} = 52.2 < \frac{418}{\sqrt{F_y}} = 59.11$$

Hence, there is no web instability.

$V_n = 0.6 F_y A_w = 368.5$ k

Also, $\phi V_n = 331.6 > V_{max}$

Hence, the assumed section is adequate.

Serviceability check

Beam BC: Maximum displacement in the beam is under the point load (50 k) and can be computed using the Unit Load Method (or any other technique) as 0.026 in. The allowable deflection can be computed as

$$\frac{L}{360} = 0.33 \text{ in} > 0.026 \text{ in}$$

Hence, the assumed sections are adequate.

Step 4: Check original assumptions

a. I_{AB} = 1830 in^4; I_{BC} = 2850 in^4; I_{BC}/I_{AB} = 2850/1830 = 1.56. This is about 4% higher than the assumed ratio of 1.5. Another approach to solving the problem is to assume the actual sections in the first step. If we had assumed the sections as stated above, the ratio of 1.56 could have been used in the indeterminate structural analysis to obtain the internal force distribution in the beam.

b. Self-weight was not included in the analysis. The total load on the structure is 450 k. The weight of member AB is (83)(10) = 830 lb and the weight of member BC is (84)(10) = 840 lb for a total weight of 1.7 k. This is a tiny fraction of the total load on the beam, hence the assumption is justified.

Closure

The ideas explored in this section on design of steel structural components are introductory in nature. They are very useful in understanding not only steel as a structural material but also the important link between structural analysis, structural design, and optimal design. We have not covered a variety of topics necessary for a better understanding of steel design—connections, plates and plate girders, torsion, different forms of buckling, combined axial and bending, design of frames, to name a few. Clearly, a separate course on steel design is required by those interested in the design of steel structural components and systems.

EXERCISES

Whenever possible, also solve the following problems using the UCSD© program and compare the results.

Appetizers

9.1.1. Find the lightest W section for a tension member made of A36 steel. The member is 10 ft long. The axial tension in the member due to dead load is 70 kips and due to live load is 10 kips. The connections at the end of the member include 7/8 in diameter bolts, two per line.

9.1.2. Enhance the graph in Fig. 9.1.2.1 by including the plots for different grades of steel—50, 60, 70, and 100 ksi. Comment on the nature of the graph.

9.1.3. Find the lightest W section for a compression member made of A36 steel. The member is 15 ft long. The axial compression in the member due to dead load is 40 kips and due to live load is 70 kips. Assume that the member is pin-connected at both ends.

9.1.4. Design compression member CD in the frame in Fig. P9.1.4. Select the lightest W section for the member assuming that A572 Grade 60 steel is used. The member is subjected to a dead load of 80 kips and a live load of 140 kips. The frame is braced out-of-plane at the ends of the member.

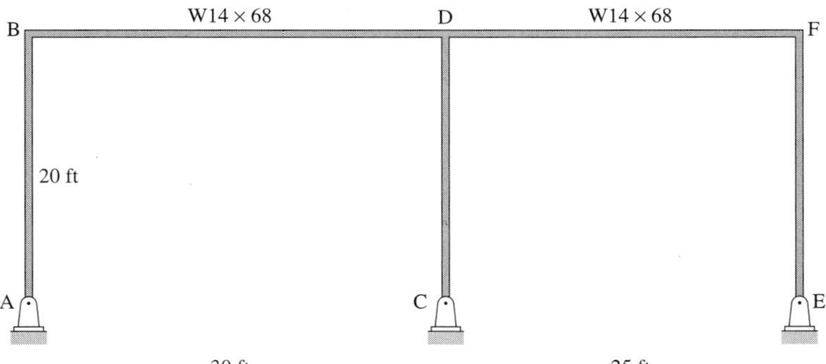

Fig. P9.1.4

9.1.5. Using the dimensions of a W18 × 65 member made of A36 steel, compute the plastic section modulus Z_x and the corresponding plastic moment capacity M_p. Classify this section when used as a beam.

Main Course

9.1.6. Redo Problem 9.1.1 but assume that a single angle section is used as the member and that one 7/8 in diameter bolt is used.

9.1.7. The truss in Problem 2.7.10 is made of A36 W shape steel members.

(a) The connections use a gusset plate on either side of the joint. Sketch the details of the connection at joint B.

(b) Design the tension member BC assuming that two bolts are used per line.

9.1.8. A 30 ft simply supported beam is made of W18 × 50 A36 section and is subjected to uniformly distributed factored loading w.

(a) The beam has continuous lateral support. What is the largest load can the beam carry?

(b) The beam is laterally supported at the ends. What is the largest load can the beam carry?

9.1.9. A 30 ft simply supported beam is made of W30 × 116 A36 section and is subjected to concentrated force of 200 k (live load) at the center of the beam. The beam is laterally supported at the ends. Is the design adequate?

9.1.10. A 30 ft simply supported A36 steel beam is to be designed to carry a concentrated force of 200 k (live load) at the center of the beam. The beam is laterally supported at the ends and at the center of the beam. Design the beam.

9.1.11. Compute the flexural design strength and the shear strength of the following sections that are used as continuously laterally supported beams.

(a) W12 × 96 A572 Grade 50 steel.

(b) W6 × 12 A 36 steel.

Structural Concepts

9.1.12. List the assumptions made for the derivation of the Euler buckling formula.

9.1.13. Rederive the all equations in which English/U.S. Customary Units are used, using SI units instead.

9.2 INTRODUCTION TO DESIGN OF CONCRETE STRUCTURES

Concrete is made from a combination of cement, water, air, aggregate, and sometimes admixtures. Once the components are mixed, it is placed in a formwork. When required, steel reinforcements are placed inside the formwork and concrete is poured around the reinforcements. Concrete has to cure before the formwork can be removed. The resulting material is structurally strong—concrete is strong in compression, and the steel reinforcements provide the required strength in tension and shear. The ACI Building Code called the *Building Code Requirements for Structural Concrete* (318-99) is usually used for the design of concrete structures. It contains performance standards and specifications that describe acceptable design and construction methods and what performance levels need to be met.

Concrete Composition

Portland cement is mainly composed of calcium and aluminum silicates (CaO, SiO_2, Al_2O_3). The cement reacts with water through a process of hydration. The resulting compound has a high compressive strength. It is the addition of water and the entrained air that make the mixture workable. The chemical reaction produces lime as a byproduct. In addition, heat is also generated during hydration. Both lime and heat must be controlled, otherwise they can have detrimental effects on the final properties of the concrete. ASTM-designated Portland Cement has Types I through V. Type I is the common cement that is used for most construction work. Type II has lower heat of hydration than Type I and can withstand some sulfate attack. Type III cement is used when curing time is an issue. Its strength in the first 24 hours is about twice that of Type I cement. Type IV is a low-heat cement in which the heat dissipates very slowly. Finally, Type V cement is used for structures that are exposed to high levels of sulfate.

Aggregates constitute the bulk of the finished product. Gravel and crushed stone typically make up the *coarse aggregates*, whereas sand makes up the *fine aggregate*. If the smallest size of the aggregate is greater than 0.25 in, it is classified as coarse aggregate. Sand gradation is a function of the sieve size—U.S. standard sieve sizes between No. 4 and No. 100.

Admixtures are compounds added before or during the mixing of the concrete. The motivation is to enhance the qualities of the concrete—reduce the curing time, increase workability, increase the strength, increase the setting time, etc. Examples include accelerating admixtures (e.g., calcium chloride), air-entraining admixtures, polymers, superplasticizers, retarding admixtures, silica fumes, etc.

Two phenomena distinguish concrete behavior from other materials—shrinkage and creep. As mentioned before, water reacts with cement. With time, the concrete volume decreases during hardening and drying. This phenomenon is termed *shrinkage*. Cracks develop and may lead to undesired results such as exposure of reinforcement to corrosion. In addition, one of the more important byproducts of shrinkage is the increase in deflection of concrete beams with time. *Creep*, on the other hand, is caused by sustained loads. When a concrete structure is loaded, the initial (or, instantaneous) strains are elastic strains. The total strain in the structure, however, continues to increase with time, even when the total load remains the same. Experience has shown that the creep strains can be two to three times the instantaneous strains. As we can see here, concrete structures continue to deform with time. The total strain $\varepsilon(t)$ at any given time is given as

$$\varepsilon(t) = \varepsilon_e + \varepsilon_c(t) + \varepsilon_s(t) + \varepsilon_T(t) \tag{9.2.1}$$

where ε_e is the (instantaneous) elastic strain, $\varepsilon_c(t)$ is the creep strain, $\varepsilon_s(t)$ is the shrinkage strain, and $\varepsilon_T(t)$ is the thermal strain. The last term represents the strain due to temperature change in the concrete. Good design practice requires the use of reinforcing steel to control the detrimental effects due to shrinkage and creep.

Reinforcing steel is used with concrete when additional strength in tension and shear is required. It is also used (a) when additional strength is required in the compression zone, and (b) to control long-term deflections. While reinforcements are available in different forms such as bars, wires, and welded mesh, we discuss only reinforcing bars in this section. The bars are made in accordance with ASTM Specifications. The properties of various types of steel used are shown in Table 9.2.1. Most of the design is carried out using billet steel and low-alloy steel. The price differential between the lower-strength (40 ksi) and higher-strength (60 ksi) steel is small enough that the latter is used when appropriate.

Table 9.2.1 Reinforcing Bar Steel Properties

Type	Yield strength, f_y(psi)	Ultimate strength, f_u(psi)
Billet steel (A615)		
Grade 40	40000	70000
Grade 60	60000	90000
Axle steel (A617)		
Grade 40	40000	70000
Grade 60	60000	90000
Low-alloy steel (A706)		
Grade 60	60000	80000

Reinforcements are available in various sizes ranging from bar #3 to bar #18. The outer surfaces of the bars have projections called *deformations* (see Fig. 9.2.1). The idea is to increase the bond between the reinforcement and concrete.

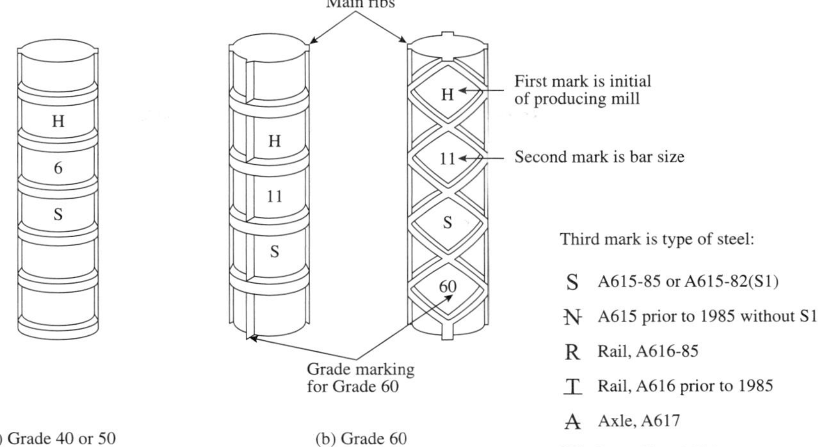

Fig. 9.2.1 Typical steel reinforcement bars.

Bar properties are shown in Table 9.2.2. Normally, the steel reinforcements are used as is as long as the initial state is free of rust and corrosion. When the steel is used in corrosive environment, special protection is required. As we will see later, regardless of the environment, the steel reinforcements must be adequately covered by the concrete on all sides to prevent these problems. Some details are presented in Table 9.2.3.

Table 9.2.2 Reinforcing Bar Properties

Bar size or designation	Weight lb/ft	Diameter in	Cross-sectional Area in^2
3	0.376	0.375	0.11
4	0.668	0.500	0.20
5	1.043	0.625	0.31
6	1.502	0.750	0.44
7	2.044	0.875	0.60
8	2.670	1.000	0.79
9	3.400	1.128	1.00
10	4.303	1.270	1.27
11	5.313	1.410	1.56
14	7.65	1.693	2.25
18	13.60	2.257	4.00

Table 9.2.3 Minimum Beam Width (in) per ACI Code*

Bar size	Number of bars in a single layer of reinforcement							Add for each additional bar
	2	3	4	5	6	7	8	
4	6.8	8.3	9.8	11.3	12.8	14.3	15.8	1.50
5	6.9	8.5	10.2	11.8	13.4	15.0	16.7	1.63
6	7.0	8.8	10.5	12.3	14.0	15.8	17.5	1.75
7	7.2	9.0	10.9	12.8	14.7	16.5	18.4	1.88
8	7.3	9.3	11.3	13.3	15.3	17.3	19.3	2.00
9	7.6	9.8	12.2	14.3	16.6	18.8	21.1	2.26
10	7.8	10.4	12.9	15.5	18.0	20.5	23.1	2.54
11	8.1	10.9	13.8	16.6	19.4	22.2	25.0	2.82
14	8.9	12.3	15.7	19.1	22.5	25.9	29.3	3.40
18	10.6	15.1	19.6	24.1	28.6	33.1	37.6	4.51

* Table shows minimum beam widths when #3 stirrups are used.

ACI also has established guidelines for minimum bar spacing and cover: (a) The clear distance between parallel bars in a layer cannot be less than the diameter of the bar or 1 in. (b) The clear distance between longitudinal bars in columns cannot be less than one and a half times the bar diameter or 1.5 in. (c) The minimum clear cover for cast-in-place beams and columns should not be less than 1.5 in. This is also the clear cover recommendation for stirrups and ties. A thicker cover may be necessary where the concrete is subjected to unusual weather conditions.

Concrete Properties

The properties of concrete can be determined using the appropriate testing procedure. These properties are either short-term or long-term. The former refers to the strength in compression, tension, and shear, and modulus of elasticity. Shrinkage and creep describe the long-term properties of concrete.

The *compressive strength* f_c' of concrete is found by taking a standard 6 in $\times$ 12 in cylinders that are cured and then tested at 28 days. The cylinders are subjected to compressive loads that are increased at a specified rate of loading until the cylinder fails. The compressive strength of most commonly used concrete varies between 2500 psi and 7000 psi. Concrete with strengths as high as about 20000 psi has been used for special high-rise construction. For design purposes (where the tensile property of concrete is required) the *modulus of rupture* f_r is more important than the tensile strength f_t. The modulus of rupture (stress at which concrete begins to crack) is found by testing plain concrete, and ACI Code recommends

$$f_r = 7.5\sqrt{f_c'} \text{ psi} \tag{9.2.2}$$

While the *shear strength* of concrete is difficult to determine, from a design perspective the values are not important because the design code stipulates a conservative approach.

The stress-strain curve for concrete is shown in Fig. 9.2.2. The relationship is nonlinear even for low values of stress and strain. The initial slope of the tangent to the curve is shown as line A and is generally known as the *initial modulus*. The straight line B that connects the origin to a point about $0.4f_c'$ is called the *secant modulus of elasticity*. The line c–d that is tangent to the curve at about $0.5f_c'$ is called the *tangent modulus*.

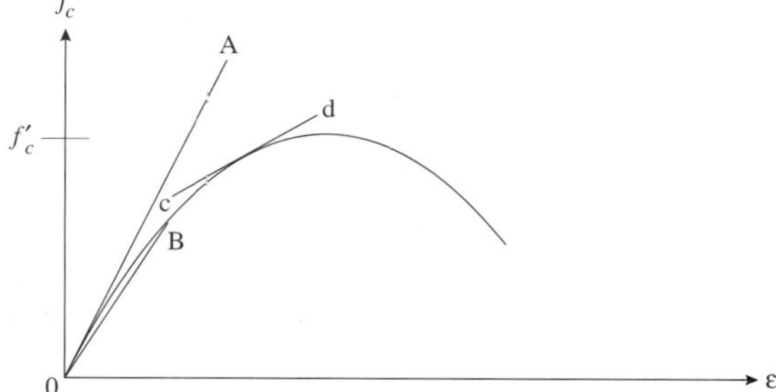

Fig. 9.2.2 Typical stress-strain diagram for concrete.

The Poisson's ratio of concrete varies depending on the composition of the mix and has a value between 0.11 for high-strength concrete to 0.21 for weaker concrete.

We will revisit the two design philosophies that we first saw in Chapter 3, working stress design (WSD) and strength design (SD). The philosophy behind working stress design is to investigate the conditions of the structure under the action of service loads. On the other hand, strength design deals with loads that are greater than service loads. The motivation is to investigate the state of the structure as it approaches one or more failure states. From a design perspective, where preventing failures is extremely important, the strength design methodology is more realistic.

In Chapter 3 we saw ASCE-specified load factors. Recall that the *factored* loads are used to take into account the worst effect on the structure under the action of the different types of loads. The ACI Code defines a different set of load factors. The ultimate load-carrying capacity of a member U under the action of dead load D and live load L must be at least equal to

$$U = 1.4D + 1.7L \tag{9.2.3}$$

Similarly, when wind load W is present,

$$U = 0.75(1.4D + 1.7L + 1.7W) \tag{9.2.4}$$

or

$$U = 0.9D + 1.3W \tag{9.2.5}$$

ACI also brings into play a second type of safety factor through the use of the *strength reduction factor* ϕ (see Table 9.2.4). This factor specifies by what percent the nominal strength of a member is to be reduced. The reason for this reduction is the uncertainty or variation in material properties and workmanship, approximations in analysis, etc.

Table 9.2.4 ACI Strength Reduction Factors, ϕ

Category	ϕ
Flexure, with or without axial tension	0.90
Axial tension	0.90
Axial compression, with or without flexure	
(i) Members with spiral reinforcement	0.75
(ii) Other reinforced members	0.70
Shear and torsion	0.85
Bearing on concrete	0.70

The ideas and concepts presented in this introductory section are used later for the analysis and design of concrete beams and columns.

9.2.1 Beam Design

It is appropriate to discuss first the analysis of reinforced concrete structures—beams and frames. The material is neither elastic nor homogeneous. The stress-strain curve is nonlinear even for small values of strain. However, as we will see, for the analysis of commonly used systems, experience has shown that the linear approximations made throughout this text are valid. The design, however, will be guided by strength design philosophy.

Flexural Analysis of Singly Reinforced Rectangular Beams

Consider a simply supported beam that is subjected to gradually increasing loads. Initially the concrete is uncracked. Both the stresses and strains are small. The tensile stresses are smaller than the modulus of rupture. As the loads are increased, the concrete begins to crack in the tensile zone (bottom of the beam). The moment at which cracks begin to form is called the *cracking moment* M_{cr}. The stresses are still, however, in the elastic regime (Fig. 9.2.1.1). ACI defines the cracking moment as

$$M_{cr} = \frac{f_r I_g}{y_t} \tag{9.2.1.1}$$

where I_g is the moment of inertia of the gross cross-section and y_t is the distance from the centroidal axis to the extreme tensile fiber.

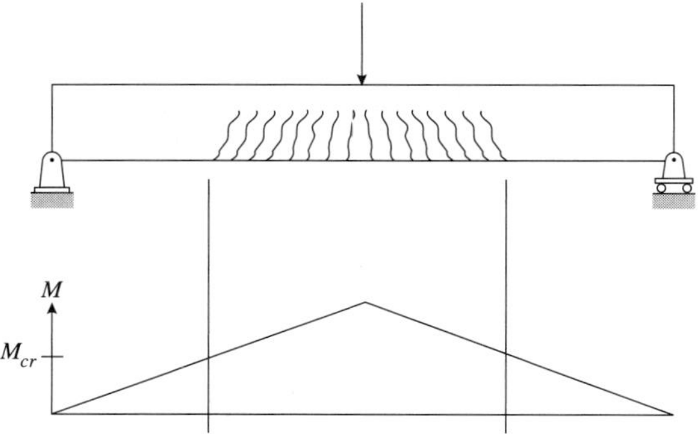

Fig. 9.2.1.1 Simply supported beam under flexure (elastic stresses).

With the concrete cracked and ineffective in the tensile zone, the steel becomes the primary load-carrying component. The neutral axis moves upwards where the concrete has cracked. The stress-strain relationship for concrete is linear as long as the maximum compressive stress in the concrete is less than $0.5f_c'$ and the strain in the steel reinforcements is less than the yield value, ε_y. Finally, as the loads are increased further closer to the ultimate strength (see Fig. 9.2.1.2), (i) the cracks move up, (ii) the neutral axis shifts up, and (iii) depending upon the amount of reinforcing steel, the bars yield, and (iv) the stress in the concrete is in the nonlinear regime. The concrete in the compressive zone may crush and/or the steel reinforcement may yield and fail.

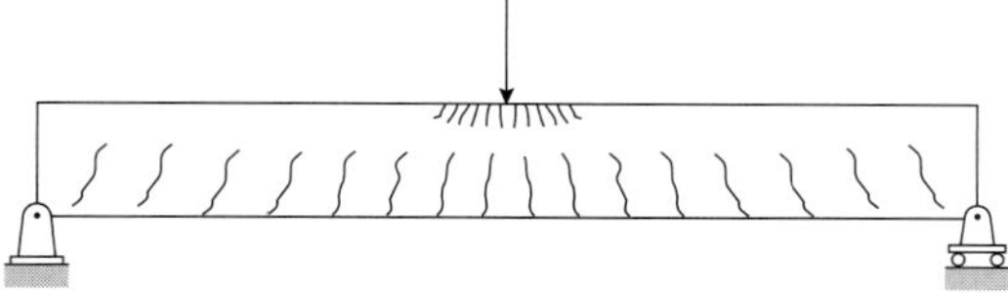

Fig. 9.2.1.2 Simply supported beam under flexure (ultimate strength).

Care must be taken during the structural analysis procedure to take this behavior of the reinforced concrete structural elements into account.

Modulus of elasticity E_c. The ACI code suggests the use of the following formula for concrete from 90 lb/ft^3 to 155 lb/ft^3:

$$E_c = w_c^{1.5}\,33\sqrt{f_c'}\ \text{psi} \tag{9.2.1.2a}$$

where w_c is the weight of concrete in lb/ft^3 and f_c' is the 28-day compressive strength in psi. ACI also provides a simpler form of the above equation for concrete weighing about 145 lb/ft^3:

$$E_c = 57000\sqrt{f_c'}\ \text{psi} \tag{9.2.1.2b}$$

Moment of Inertia. The computation of the moment of inertia is complicated by the fact that (a) reinforced concrete is made up of both concrete and steel, and (b) concrete's tensile strength is very small, implying that concrete section in tension is usually cracked, hence ineffective. Several methods have been proposed to compute the appropriate moment of inertia.

Gross moment of inertia I_g: The basic assumption is that the entire concrete cross-section contributes to the moment of inertia, and the steel's contribution is small and can be neglected. For example, the moment of inertia of a rectangular section of height h and width b is given as $I_g = bh^3/12$.

Transformed moment of inertia I_{gt}: The cross-section is transformed into an equivalent (homogenous) concrete cross-section (see Fig. 9.2.1.3). The steel is replaced with a fictitious but equivalent concrete area. Let A_s and f_s be the area and tensile stress in the actual steel. Let A_{eq} and f_{eq} be the area and tensile stress in the equivalent concrete. The transformation is valid if two conditions are met. First, equilibrium dictates that the two tensile forces be equal:

$$A_s f_s = A_{eq} f_{eq} \tag{9.2.1.3}$$

Second, compatibility requires that the two strains be equal:

$$\frac{f_s}{E_s} = \frac{f_{eq}}{E_c} \tag{9.2.1.4}$$

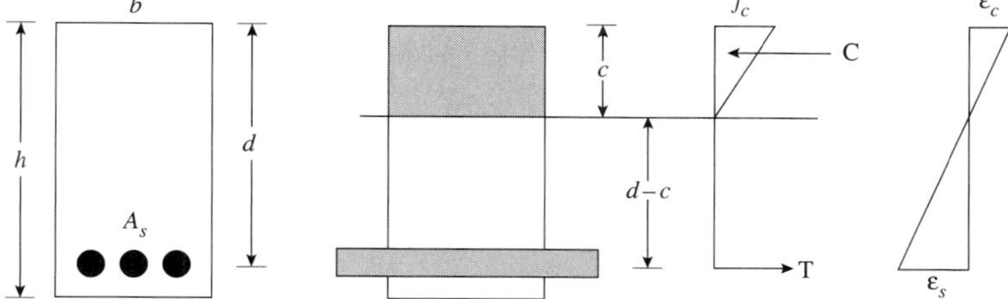

Fig. 9.2.1.3 Transformed section.

Let $n = E_s/E_c$. Using Eqs. (9.2.1.3) and (9.2.1.4), we have

$$A_{eq} = nA_s \tag{9.2.1.5}$$

$$f_{eq} = \frac{f_s}{n} \tag{9.2.1.6}$$

In other words, the equivalent concrete area is n times the actual steel area. To compute the moment of inertia, consider the following. From the stress diagram, we have

$$C = \frac{1}{2} b\, c\, f_c \tag{9.2.1.7}$$

$$T = A_s f_s \tag{9.2.1.8}$$

Note that C represents the volume of the compressive zone that is a prism or wedge. From equilibrium we have

$$\frac{1}{2} bcf_c = A_s f_s \tag{9.2.1.9}$$

From the stress-strain relationship, $f_s = E_s\varepsilon_s$ and $f_c = E_c\varepsilon_c$. Substituting in the above equation yields

$$\frac{bc}{2} E_c\varepsilon_c = A_s E_s \varepsilon_s \tag{9.2.1.10}$$

Now from the strain distribution we find

$$\frac{\varepsilon_c}{c} = \frac{\varepsilon_s}{d-c} \Rightarrow \varepsilon_s = \varepsilon_c\left(\frac{d-c}{c}\right) \tag{9.2.1.11}$$

Combining Eqs. (9.2.1.10) and (9.2.1.11) gives us

$$\frac{A_s \varepsilon_s}{E_c}\left(\frac{d-c}{c}\right) = \frac{bc}{2} \tag{9.2.1.12}$$

Since $n = E_s/E_c$, we have

$$\frac{b}{2}c^2 + nA_s c - nA_s d = 0 \tag{9.2.1.13}$$

This quadratic equation can be solved for c and the moment of inertia about the neutral axis can be computed as follows:

$$I_{gt} = I_{\text{concrete}} + I_{eq} \tag{9.2.1.14a}$$

or

$$I_{gt} = \frac{bc^3}{3} + nA_s(d-c)^2 \tag{9.2.1.14b}$$

I_{gt} is also called the cracking moment of inertia I_{cr}. As we can see, in this approach, the basic assumption is that the concrete behavior is linear.

Effective moment of inertia I_e: Both the above methodologies are based on assumptions that are not entirely valid. The actual moment of inertia is somewhere between I_g and I_{gt}. Branson[5] is credited with developing the expression for the effective moment of inertia as

$$I_e = I_{gt} + \left(\frac{M_{cr}}{M_a}\right)^3 \left(I_g - I_{gt}\right) \tag{9.2.1.15}$$

where M_a is the maximum moment in the beam and M_{cr} is the cracking moment capacity.

We will answer the question "Which moment of inertia expression should one use?" later in this section.

In the rest of this section we design beams that are subjected primarily to bending moment (flexure). To begin our discussions, we first assume that the beam cross-section and properties are such that the concrete crushes and the steel reinforcements fail at the same time. This state represents the ultimate strength of the beam. The state of the cross-section is shown in Fig. 9.2.1.4. The strain distribution is shown in Fig. 9.2.1.4(b). Since it is assumed that the cracked concrete is not effective, the stress state is shown in Fig. 9.2.1.4(c). The compressive stress zone is a parabola. Assuming that the cross-section is in equilibrium, we have

$$T = C \tag{9.2.1.16}$$

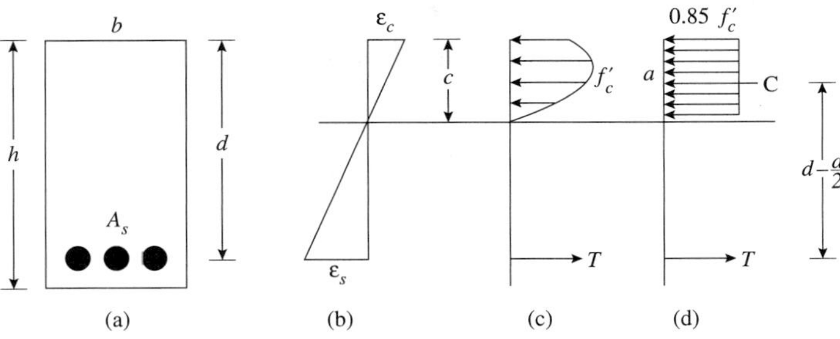

Fig. 9.2.1.4 Singly reinforced beam. (a) Beam cross-section. (b) Strain distribution. (c) Actual stress distribution. (d) Equivalent stress distribution.

If the concrete and steel fail at the same time, the largest stress in the concrete is f_c' and the strain in the steel is $\varepsilon_s = \varepsilon_y$. Instead of computing the volume of the parabolic compressive zone, it has been suggested by Whitney (see Bibliography) that a rectangular block can be used without undue approximation. The rectangular stress block has a uniform stress value of $0.85f_c'$ (based on experimental data) and a depth a. This depth is computed so that the volume of the stress blocks in (c) and (d) are approximately the same. We can now simplify Eq. (9.2.1.16) as

[5] D. E. Branson, Deformation of Concrete Structures, McGraw-Hill, New York, 1977.

$$A_s f_y = 0.85 f'_c ba \tag{9.2.1.17}$$

or

$$a = \beta_1 c = \frac{A_s f_y}{0.85 f'_c b} \tag{9.2.1.18}$$

where the β_1 parameter represents the fraction of the distance from the extreme compressive fiber to the neutral axis. The nominal strength M_n of the beam represents the moment at the cross-section as

$$M_n = T\left(d - \frac{a}{2}\right) = C\left(d - \frac{a}{2}\right) \tag{9.2.1.19a}$$

or

$$M_n = A_s f_y\left(d - \frac{a}{2}\right) = 0.85 f'_c ba\left(d - \frac{a}{2}\right) \tag{9.2.1.19b}$$

If we take $\rho = \dfrac{A_s}{bd}$ as the reinforcement ratio, then Eq. (9.2.1.18) can be written as

$$a = \frac{\rho d f_y}{0.85 f'_c} \tag{9.2.1.20}$$

$$M_n = \rho b d f_y\left(d - \frac{a}{2}\right) \tag{9.2.1.21}$$

Where can we use this procedure and these equations? Given the beam dimensions, concrete and reinforcement properties, and assuming that the concrete and steel fail simultaneously, we can compute the nominal moment that the beam can withstand. However, the fundamental question remains—Do we want a balanced section? The answer is "No!" Let's see why.

Based on experimental results, ACI recommends the maximum strain in concrete as ε_c = 0.003. A *balanced section* as we have just seen, is one in which the steel reinforcement yields ($\varepsilon_y = f_y/E_s$) just as the concrete starts crushing (the maximum compressive strain is 0.003 in/in). In Fig. 9.2.1.5, this is represented by line OB. On the other hand, an *underreinforced section* is one where the steel yields first. As the load increases, the steel yields and continues to deform. In such a section, the amount of tensile steel A_s is less than the amount of steel in a balanced section. The line OA represents the strain distribution in the cross-section. An *overreinforced section* is one where the concrete fails first through crushing. The corresponding strain in the steel is less than yield. In such a section, the amount of tensile steel A_s is greater than the amount of steel in a balanced section. The line OC represents the strain distribution in the cross-section. For all practical purposes, barring extreme values of the parameters, the location of the neutral axis is about the same for the three cases, since the strain values are very small to begin with.

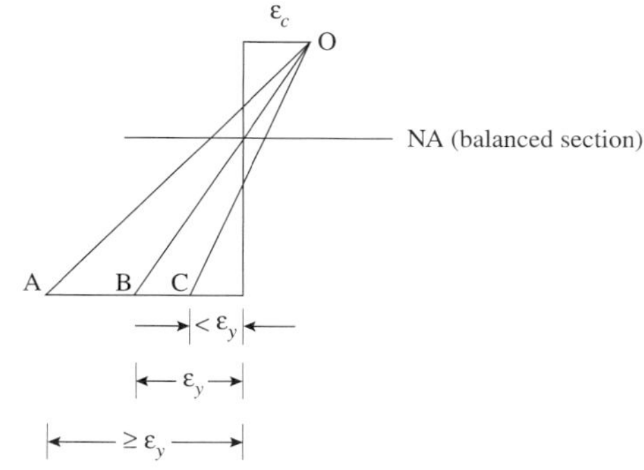

Fig. 9.2.1.5 Strain distribution for underreinforced, balanced, and overreinforced sections.

The reason that overreinforced or balanced sections are not desirable is that concrete is a brittle material. Failure due to crushing concrete can be sudden, without warning. Hence, ACI recommends that underreinforced beams be designed. In such a situation, if failure does occur, it will be preceded by large strains in the reinforcement, leading to large deflections. It is hoped that large deflections will serve as sufficient warning of impending failure. ACI Code recommends that the maximum steel reinforcement be limited to 75% of that required for a balanced section. In other words,

$$\rho \leq 0.75\rho_b \tag{9.2.1.22}$$

where $\rho = \dfrac{A_s}{bd}$ is the actual reinforcement ratio and $\rho_b = \dfrac{A_{sb}}{bd}$ is the balanced ratio. The ACI code also specifies the minimum amount of steel reinforcement as

$$\left(A_s\right)_{\min} = \frac{3\sqrt{f_c'}}{f_y}bd \geq \frac{200bd}{f_y} \tag{9.2.1.23}$$

While we have expressions for the minimum and the maximum amount of steel, from a practical viewpoint, the amount is dictated by how the steel reinforcements can be placed within the formwork and allow for proper distribution of concrete, cover, etc.

Going back to Fig. 9.2.1.4, the location of the neutral axis can be found as follows for the balanced section. From the strain diagram we have $\varepsilon_c = 0.003$ (ACI requirement). Hence, from similar triangles, we find

$$\frac{c_b}{d - c_b} = \frac{0.003}{f_y/E_s} \tag{9.2.1.24}$$

where c_b is the distance for the balanced section. Since, $E_s = 29(10^6)$ psi, we have

$$\frac{c_b}{d} = \frac{87000}{87000 + f_y} \tag{9.2.1.25}$$

For the equivalent rectangular block section (see Eq. 9.2.1.18), we have

$$a_b = \beta_1 c_b \tag{9.2.1.26}$$

where ACI recommends

for $0 < f_c' \leq 4000$ psi $\quad \beta_1 = 0.85 \quad$ (9.2.1.27a)

for $4000 < f_c' \leq 8000$ psi $\quad \beta_1 = 0.85 - 0.05\left(\dfrac{f_c' - 4000}{1000}\right) \quad$ (9.2.1.27b)

for $f_c' > 8000$ psi $\quad \beta_1 = 0.65 \quad$ (9.2.1.27c)

As we have seen and used before, the tensile force T is equal to the compressive force C, or

$$A_{sb} f_y = 0.85 f_c' b a_b \tag{9.2.1.28}$$

Using Eq. (9.2.1.25) yields

$$\rho_b = \beta_1 \left(\frac{0.85 f_c'}{f_y}\right)\frac{87000}{87000 + f_y} \tag{9.2.1.29}$$

We are now ready to develop the algorithm for the analysis of a singly reinforced concrete beam subjected primarily to bending.

General Procedure for Analysis. Given A_s, f_c', b, d, f_y in in, lb. Obtain M_n in lb–in.

Step 1: Use Eq. (9.2.1.23) to establish $\rho_{\min}$. Compute $\rho = \dfrac{A_s}{bd}$. Is $\rho > \rho_{\min}$? If the answer is no, the section is unacceptable.

Step 2: Find the value of β_1 using Eq. (9.2.1.27).

Step 3: Compute the balanced reinforcement ratio ρ_b using Eq. (9.2.1.29).

Step 4: Is $\rho \le 0.75\rho_b$? If the answer is no, the section is inadequate. The cross-sectional dimensions may have to be increased, or the amount of steel may have to be decreased, or both actions may be desirable.

Step 5: Compute the equivalent stress block dimension $a = \rho d f_y / 0.85 f_c'$ (Eq. (9.2.1.20)). Finally, compute the nominal strength using Eq. (9.2.1.21).

EXAMPLE 9.2.1 ***Moment Capacity***

For the concrete beam in Fig. E9.2.1, compute the ultimate moment capacity. The material properties are $f_c' = 4000$ psi and $f_y = 60000$ psi.

SOLUTION

Step 1: Compute the minimum reinforcement required:

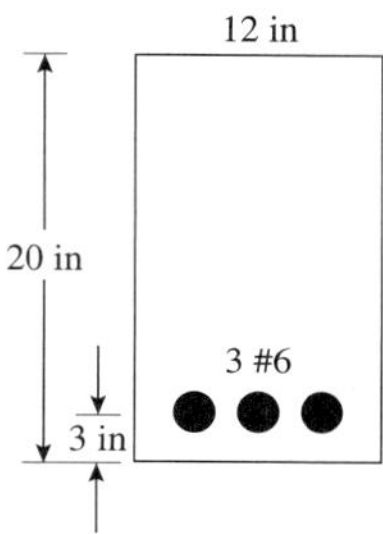

Fig. E9.2.1

$$\rho_{min} = \max\left(\frac{3\sqrt{4000}}{60000}, \frac{200}{60000}\right) = 0.0033$$

The actual reinforcement is $A_s = 3(0.44) = 1.32$ in^2 and

$$\rho = \frac{1.32}{(17)(12)} = 0.0065 > 0.0033. \qquad \text{OK.}$$

Step 2: For the given concrete, $\beta_1 = 0.85$.

Step 3: Compute the balanced reinforcement ratio, Eq. (9.2.1.29):

$$\rho_b = 0.85\frac{0.85(4000)}{60000}\frac{87000}{87000+60000} = 0.029$$

$$\rho < 0.75\rho_b. \qquad \text{OK.}$$

Step 4: Compute the stress block dimension:

$$a = \frac{\rho d f_y}{0.85 f_c'} = \frac{(0.0065)(17)(60000)}{0.85(4000)} = 1.95 \text{ in}$$

$$M_n = \rho b d f_y\left(d - \frac{a}{2}\right) = 0.0065(12)(17)(60000)\left(17 - \frac{1.95}{2}\right) = 1.275(10^6) \text{ lb-in}$$

Answer: $M_n = 106250$ lb-ft

Flexural Design of Singly Reinforced Rectangular Beams

The design of singly reinforced concrete beams involves finding the beam cross-section dimensions and the amount of reinforcement given the beam nominal strength. As we saw repeatedly in Chapter 3, such problems do not have a unique solution. It is possible to bring in additional considerations that narrow the available choices. One obvious consideration is cost. Experience has shown that the lowest cost can be obtained when the amount of reinforcement is at its upper bound and the ratio of the width to the height of the beam is at its lower bound. The bounds are function of construction practices. We outline several procedures for design—select one that meets your requirements.

General Procedure (Trial and Error). Given beam geometry, loading, f_c', and f_y, obtain b, d and A_s. (An improved method is shown in the examples).

Step 1: Compute the factored load on the beam. Assume initial guess for b, d, and A_s. Include the self-weight. Analyze the beam to find the largest bending moment $M_{\max}$.

Step 2: With the assumed values in step 1, analyze the beam as in the previous section find the nominal strength M_u.

Step 3: Is $M_u > M_{\max}$? If not, the beam needs to be redesigned—increase the cross-sectional dimensions, increase the amount of steel reinforcement, or both. If $M_u > M_{\max}$, can the cross-sectional dimensions be decreased, can the amount of steel reinforcement be decreased, or both? If the answer is yes, repeat steps 1 and 2.

General Procedure for Simple Design (as implemented in the UCSD program). Given f_c', b, h, f_y, d_c, M_u in in, lb, obtain A_s in in^2.

Step 1: Compute the nominal strength, $M_n = M_u/\phi$.

Step 2: Find β_1 based on the given concrete strength f_c' using Eq. (9.2.1.27).

Step 3: Compute the balanced reinforcement ratio ρ_b using Eq. (9.2.1.29).

Step 4: Compute $M_n = \rho bdf_y(d - a/2)$. Substitute $a = \rho df_y/0.85f_c'$ in the above equation and solve the quadratic equation to obtain ρ.

Step 5: Is $\rho \le 0.75\rho_b$? If the answer is no, the section is inadequate. The cross-sectional dimensions may have to be increased, or the amount of steel may have to be decreased, or both actions may be desirable.

Step 6: If the answer is yes, then compute $A_s = \rho bd$.

Step 7: If $A_s \le (A_s)_{\min}$ then the required steel is $\left(A_s\right)_{\text{reqd}} = \dfrac{3\sqrt{f_c'}}{f_y} bd \ge \dfrac{200bd}{f_y}$ else $(A_s)_{\text{reqd}} = A_s$.

Step 8: Select the reinforcement bars and decide the spacing accordingly.

Optimal Design Procedure (as implemented in the UCSD program). The design problem described above can be posed as an optimal design problem as follows. Let us consider a problem where the largest factored bending moment $M_{\max}$ can be written as

$$M_{\max} = M + cb(d + d_s) \tag{9.2.1.30}$$

where M is the largest bending moment due to live load, c is a constant (the second term accounts for the moment due to dead load), and d_s represents the total concrete cover. With reference to Fig. 9.2.1.4, $h = d + d_s$. In addition, we assume that the beam cross-section is uniform along the entire length of the beam. The optimal design problem can be stated as follows.

$$\text{Find} \qquad \mathbf{x} = b, d, A_s \tag{9.2.1.31}$$

$$\text{to minimize} \qquad f(\mathbf{x}) = w_1 A_s + w_2 b(d + d_s) \tag{9.2.1.32}$$

$$\text{subject to} \qquad g_1(\mathbf{x}) \equiv \frac{A_s}{bd} \ge \frac{3\sqrt{f_c'}}{f_y} \tag{9.2.1.33}$$

$$g_2(\mathbf{x}) \equiv \frac{A_s}{bd} \geq \frac{200}{f_y} \tag{9.2.1.34}$$

$$g_3(\mathbf{x}) \equiv \frac{A_s}{bd} \leq 0.75\rho_b \tag{9.2.1.35}$$

$$g_4(\mathbf{x}) \equiv M_n > M_{\max} \tag{9.2.1.36}$$

$$g_5(\mathbf{x}) \equiv b > b_{\min} \tag{9.2.1.37}$$

$$g_6(\mathbf{x}) \equiv d_s > (d_s)_{\min} \tag{9.2.1.38}$$

$$(A_s)_L \leq A_s \leq (A_s)_U \tag{9.2.1.39a}$$

$$b_L \leq b \leq b_U \tag{9.2.1.39b}$$

$$d_L \leq d \leq d_U \tag{9.2.1.39c}$$

The objective function, Eq. (9.2.1.31), is a weighted sum of the steel reinforcement and the cross-sectional area; w_1 and w_2 represent the weights. In other words, we are minimizing the amount of material used. If the weighting constants represent the unit cost of steel and concrete, then the objective function is a measure of the cost of the beam.[6] Two requirements need special handling. First, the value of d_s must somehow be estimated. Second, the minimum width b should be such that the reinforcements can be placed with no congestion (see Table 9.2.3). In other words, b_L is not a constant. Such constraints are easy to incorporate into a custom design program.

EXAMPLE 9.2.2

Design of a Singly Reinforced, Simply Supported Beam

A simply supported concrete beam is shown in Fig. E9.2.2(a). The live load on the beam is 1500 lb/ft. Design the beam. The material properties are $f_c' = 4000$ psi and $f_y = 60000$ psi.

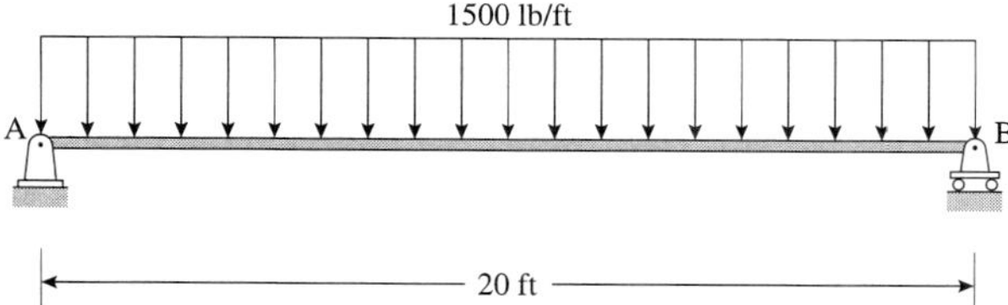

Fig. E9.2.2(a)

SOLUTION

We solve this problem using the trial-and-error approach.

Approach 1

Step 1: Compute factored load on the beam. Let us assume that the beam is 20 in × 12 in. Hence the dead load is (12 × 20/144)(150) = 250 lb/ft. The factored load can now be computed as

$$w_u = 1.4D + 1.7L = 1.4(250) + 1.7(1500) = 2900 \text{ lb/ft}$$

Step 2: Maximum moment in the beam. For a simply supported beam,

$$M_u = w_u L^2/8 = 2900(20^2)/8 = 145000 \text{ lb-ft} = 1.74(10^6) \text{ lb-in}$$

The required nominal moment capacity is

[6] This expression represents the material cost only. One could modify the expression to include the labor cost.

$$M_n = M_u/\phi = 1.74(10^6)/0.9 = 1.933(10^6) \text{ lb-in}$$

Step 3: Beam analysis. The balanced reinforcement ρ_b for this section is

$$\rho_b = \beta_1 \frac{0.85 f'_c}{f_y} \frac{87000}{87000 + f_y} = 0.85 \frac{0.85(4000)}{60000} \frac{87000}{87000 + 60000} = 0.029$$

Let us assume that the actual reinforcement is $\rho = 0.5\rho_b = 0.0145$. Hence,

$$a = \frac{\rho d f_y}{0.85 f'_c} = \frac{(0.0145)(20)(60000)}{0.85(4000)} = 5.12 \text{ in}$$

$$M_n = \rho b d f_y \left(d - \frac{a}{2} \right) = 0.0145(12 \times 20)(60000)(20 - 2.56) = 3.64(10^6) \text{ lb-in}$$

The required moment capacity is $1.933(10^6)$ lb–in. The design is acceptable but the beam is overdesigned.

Approach 2. There are two problems with the above approach: we did not include the concrete cover (hence its weight) and, as we will see, it is not necessary to assume a value for the width *b*.

Step 1: Compute factored load on the beam. Let us assume that the beam is 20 in × *b* with a clear cover of 1.5 in, as in Fig. E9.2.2(b). From the figure, d_e is the sum of the diameter of the stirrup and half the diameter of the longitudinal reinforcement. Let the stirrup be made of #3 bars. Since we do not know the size of the reinforcement, let $d_e = 1$ in (or $d_s = 2.5$ in). Hence, $d = 17.5$ in The dead load is $(b \times 20/144)(150) = 20.83b$ lb/ft. The factored load can now be computed as

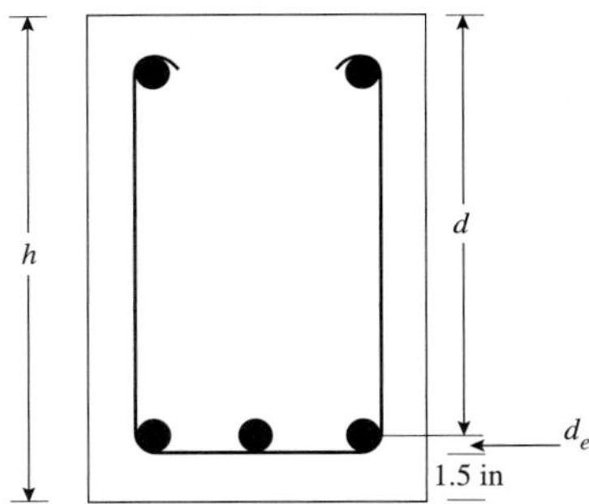

Fig. E9.2.2(b) Detailing (note that the top reinforcements are nonstructural and are needed to keep the stirrups in place).

$$w_u = 1.4D + 1.7L = 1.4(20.83b) + 1.7(1500) = (29.2b + 2550) \text{ lb/ft}$$

Step 2: Maximum moment in the beam. For a simply supported beam,

$$M_u = \frac{w_u L^2}{8} = 17520b + 1.53(10^6) \text{ lb-in}$$

The required nominal moment capacity

$$M_n = M_u/\phi = 19467b + 1.7(10^6) \text{ lb-in}$$

Step 3: Beam analysis. The balanced reinforcement ρ_b for this section is

$$\rho_b = \beta_1 \frac{0.85 f'_c}{f_y} \frac{87000}{87000 + f_y} = 0.85 \frac{0.85(4000)}{60000} \frac{87000}{87000 + 60000} = 0.029$$

Let us assume that the actual reinforcement is $\rho = 0.5\rho_b = 0.0145$. Hence,

$$a = \frac{\rho d f_y}{0.85 f_c'} = \frac{(0.0145)(17.5)(60000)}{0.85(4000)} = 4.48 \text{ in}$$

$$M_n = \rho b d f_y \left(d - \frac{a}{2} \right) = 0.0145(b \times 17.5)(60000)(17.5 - 2.24) = 232335b \text{ lb-in}$$

Step 4: Obtain the value of the beam width. Equating the required nominal moment capacity to the actual moment capacity yields

$$19467b + 1.7(10^6) = 232335b$$

Solving, we find $b = 8$ in. Hence, $A_s = 0.0145(17.5)(8) = 2.03$ in^2. We can use 2 #9 bars ($A = 2$ in^2) for which the minimum required beam width is 7.6 in < 8 in.

Doubly Reinforced Rectangular Beams

Sometimes it is necessary to place reinforcements in both the tension and compression sides. Such a situation arises when, for example, the depth of the beam is restricted for architectural reasons. The addition of the steel, $A_{s'}$ on the compressive side increases the moment capacity of the beam (see Fig. 9.2.1.6).

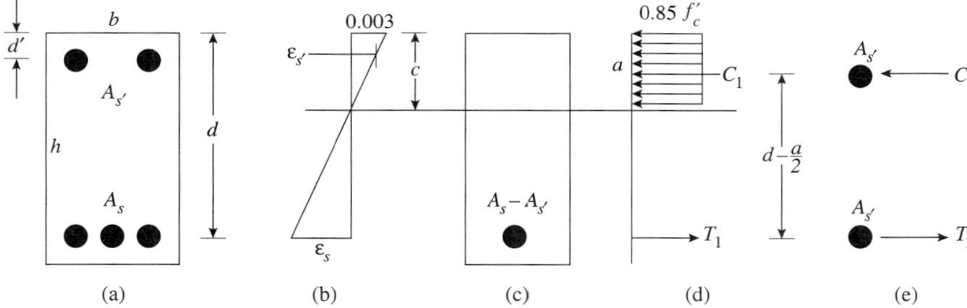

Fig. 9.2.1.6 Beam subjected to flexure. (a) Reinforcements. (b) Strain. (c) Equivalent. (d) FBD. (e) Compressive and tensile forces.

Figure 9.2.1.6(b) shows the strain distribution. We assume that the compressive steel yields. Once again, as per ACI code, we assume that the maximum strain on the compressive side is 0.003. The solution can be obtained as the superposition of (i) singly reinforced beam as shown in Fig. 9.2.1.6(c), where $A_s - A_{s'}$ is the amount of steel reinforcement, and (b) two areas of steel $A_{s'}$ on the compressive and tensile sides, as shown in Fig. 9.2.1.6(e). With the singly reinforced section,

$$T_1 = C_1 = (A_s - A_{s'}) f_y \tag{9.2.1.40}$$

Hence, the nominal moment is given as

$$M_{n1} = T_1 \left(d - \frac{a}{2} \right) = (A_s - A_{s'}) \left(d - \frac{a}{2} \right) \tag{9.2.1.41}$$

$$a = \frac{(A_s - A_{s'}) f_y}{0.85 f_c' b} \tag{9.2.1.42}$$

Now, using Fig. 9.2.1.6(e), we have from equilibrium considerations

$$T_2 = C_2 = A_{s'} f_y \tag{9.2.1.43}$$

$$M_{n2} = T_2 (d - d') = A_{s'} f_y (d - d') \tag{9.2.1.44}$$

Hence, the total nominal moment of the section is

$$M_n = M_{n1} + M_{n2} = (A_s - A_{s'}) f_y \left(d - \frac{a}{2} \right) + A_{s'} f_y (d - d') \tag{9.2.1.45a}$$

The factored design strength M_u can then be computed as

$$M_u = \phi M_n \tag{9.2.1.45b}$$

It is necessary to verify the original assumption about yielding of the compressive-side steel. Using Fig. 9.2.1.6(b), we have

$$\varepsilon_{s'} = \frac{0.003(c - d')}{c} \tag{9.2.1.46}$$

But,

$$c = \frac{a}{\beta_1} = \frac{(A_s - A_{s'})f_y}{\beta_1(0.85 f_c' b)} \tag{9.2.1.47}$$

Substituting Eq. (9.2.1.47) into (9.2.1.46), we have

$$\varepsilon_{s'} = 0.003\left(1 - \frac{0.85\beta_1 f_c' b d'}{(A_s - A_{s'})f_y}\right) \tag{9.2.1.48}$$

The compressive steel will yield if $\varepsilon_y \geq f_y/E_s = f_y/29(10^6)$, or

$$0.003\left(1 - \frac{0.85\beta_1 f_c' b d'}{(A_s - A_{s'})f_y}\right) \geq \frac{f_y}{29(10^6)} \tag{9.2.1.49}$$

If the compression steel has not yielded, then (a) the stress in the compressive steel is given by $f_{s'} = E_s \varepsilon_{s'}$, and (b) the neutral axis can still be located by noting that the total compressive force is equal to the total tensile force. In other words,

$$0.85 f_c'(\beta_1 bc) + A_{s'}\left[\frac{E_s 0.003(c - d')}{c}\right] = A_s f_y \tag{9.2.1.50a}$$

or

$$(283.833\beta_1 f_c' b)c^2 + (A_{s'}E_s - 333.33A_s f_y)c - (A_{s'}E_{s'}d') = 0 \tag{9.2.1.50b}$$

This is a quadratic equation in c with c as the sole unknown. One should also note that the maximum permissible steel reinforcement is given by

$$A_s = 0.75\rho_b bd + A_{s'} \tag{9.2.1.51}$$

General Procedure for Analysis. Given A_s, $A_{s'}$, f_c', b, d, d', f_y in in, lb, obtain M_n in lb-in.

Step 1: Use Eq. (9.2.1.23) to establish $\rho_{\min}$. Compute $\rho = \frac{A_s}{bd}$ and $\rho' = \frac{A_{s'}}{bd}$. Is $\rho > \rho_{\min}$? If the answer is no, the section is unacceptable.

Step 2: Use Eq. (9.2.1.50b) to compute the value of c.

Step 3: Compute strain using Eq. (9.2.1.46) and hence the stress in compressive steel $f_{s'}$.

Step 4: If $f_{s'} \geq f_y$, then $f_{s'} = f_y$. Otherwise use the above computed stress to obtain M_n in lb-in.

Step 5: Compute the balanced reinforcement ratio ρ_b using Eq. (9.2.1.29).

Step 6: Is $\rho - \rho' f_{s'}/f_y \leq 0.75\rho_b$ (modified form of Eq. (9.2.1.51))? If the answer is no, the section is inadequate. The cross-sectional dimensions may have to be increased, or the amount of steel may have to be decreased, or both actions may be desirable. Repeat the same procedure to compute the nominal strength M_n in lb-in.

EXAMPLE 9.2.3 ***Analysis of a Doubly Reinforced Rectangular Beam***

Given the material properties $f_c' = 4000$ psi and $f_y = 60000$ psi, compute the nominal moment capacity M_n in Fig. E9.2.3(a).

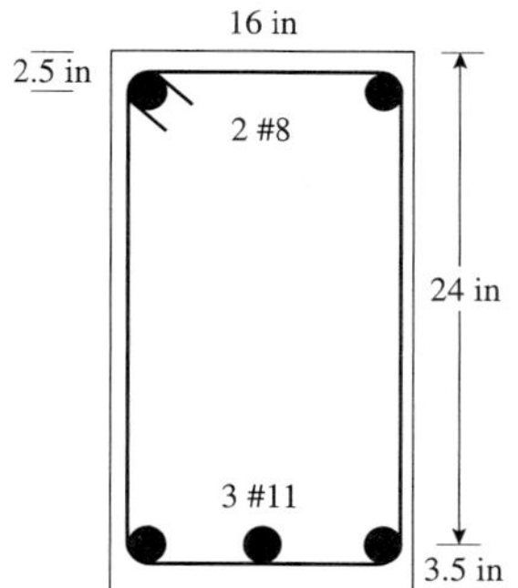

Fig. E9.2.3(a)

SOLUTION

Step 1: Beam properties:

$$\rho = \frac{4.68}{(16)(24)} = 0.012 \qquad \rho' = \frac{1.58}{(16)(24)} = 0.004$$

$$\rho_{\min} = \max\left(\frac{3\sqrt{4000}}{60000}, \frac{200}{60000}\right) = 0.0033$$

Since $\rho > \rho_{\min}$, the section meets the minimum reinforcement requirement.

Step 2: Use the quadratic equation 9.2.1.50b to solve for c. Using $\beta_1 = 0.85$, $f_c' = 4000$, $d' = 2.5$, we have

$$c^2 - 3.099c - 7.4312 = 0 \qquad c = 4.68513 \qquad c = -1.58613$$

We select $c = 4.68513$ as the lowest positive root.

Step 3: Compute strain using Eq. (9.2.1.43) and hence the stress in compressive steel $f_{s'}$.

$$\varepsilon_{s'} = \frac{0.003(c - d')}{c} = 0.0014 \text{ and hence}$$

$$f_{s'} = \varepsilon_{s'} E_s = 0.0014 \times 29(10^6) = 40600 \text{ psi} < f_y$$

Using Eq. (9.2.1.42a), where $a = \beta_1 c$ yields

$$M_n = (A_s - A_{s'}) f_y \left(d - \frac{a}{2}\right) + A_{s'} f_{s'} (d - d')$$

$$M_n = (4.68 - 1.58)60000\left(24 - \frac{3.9823}{2}\right) + 1.58 \times 40600(24 - 2.5)$$

$$M_n = 5472828 \text{ lb-in}$$

Step 4: Compute the balanced reinforcement ratio and show the check as in step 6 above:

$$\rho - \frac{\rho' f_{s'}}{f_y} = 0.012 - \frac{0.004(32929)}{60000} = 0.001 \le 0.75\rho_b. \qquad \text{OK.}$$

General Design Procedure for Doubly Reinforced Rectangular Beams (as implemented in the UCSD program)

Given b, d, d', M_u, ϕ, $f_{c'}$, f_y, f (compressive steel $A_{s'}$ is expressed as a fraction f of the tension steel A_s), compute the required reinforcement.

Step 1: Substituting $A_{s'} = fA_s$ in Eq. (9.2.1.50a) , we get

$$A_s = \left(\frac{0.85 f_{c'} b \beta c^2}{f_y c - f E_s 0.003(c - d')} \right)$$

Step 2: Substitute A_s and $f_{s'} = E_s\left(0.003(c-d')/c\right)$ in Eq. (9.2.1.45a) to get

$$M_n = \left(A_s - A_{s'}\right) f_y \left(d - \frac{a}{2}\right) + A_{s'} f_{s'} (d - d')$$

This simplifies to a cubic equation in c as follows:

$$\frac{B}{2} A\beta c^3 - B(Ad + C)c^2 + (BCd' + M_n f_y - D)c + (Dd') = 0$$

where $A = (1 - f) f_y$

$B = 0.85 f_c' b\beta$

$C = 87000 f (d - d')$

$D = 87000 f\, M_n$

Take the lowest positive root of the above cubic equation as a value of c.

Step 3: Compute strain using Eq. (9.2.1.46) and hence the stress in compressive steel $f_{s'}$.

Step 4: If $f_{s'} \geq f_y$, then $f_{s'} = f_y$. Otherwise use the above computed stress to obtain A_s using

$$M_n = \left(A_s - A_{s'}\right) f_y \left(d - \frac{a}{2}\right) + A_{s'} f_{s'} (d - d')$$

where $A_{s'} = fA_s$ and $a = \beta c$.

Step 5: If, $A_s \leq (A_s)_{\min}$ then the required steel is $\left(A_s\right)_{reqd} = \left(3\sqrt{f_c'}/f_y\right)bd \geq 200bd/f_y$. Else $(A_s)_{reqd} = A_{s'}$.

Step 6: Compute the balanced reinforcement ratio, ρ_b using Eq. (9.2.1.29).

Step 7: Is $\rho - \left(\rho' f_{s'}/f_y\right) \leq 0.75\rho_b$ (modified form of Eq. (9.2.1.51))? If the answer is no, the section is inadequate. The cross-sectional dimensions may have to be increased, or the amount of steel may have to be decreased or both actions may be desirable. Repeat the same procedure to compute the steel A_s in in^2.

Optimal Design Procedure (as implemented in the UCSD program). The above-mentioned design problem can be posed as an optimal design problem as follows. Let us consider a problem where the largest factored bending moment $M_{\max}$ can be written as

$$M_{\max} = M + cb(d + d_s) \tag{9.2.1.52}$$

where M is the largest bending moment due to live load, c is a constant (the second term accounts for the moment due to dead load) and d_s represents the tension-side concrete cover. With reference to Fig. 9.2.1.4, $h = d + d_s$. In addition, we assume that the beam cross-section is uniform along the entire length of the beam. The optimal design problem can be stated as follows.

Find $$\mathbf{x} = \left(b,\ d,\ A_s,\ A_{s'}'\right) \tag{9.2.1.53}$$

to minimize $$f(\mathbf{x}) = w_1 A_s + w_1 A_{s'} + w_2 b(d + d_s) \tag{9.2.1.54}$$

subject to $$g_1(\mathbf{x}) \equiv \frac{A_s}{bd} \geq \frac{3\sqrt{f_c'}}{f_y} \tag{9.2.1.55}$$

$$g_2(\mathbf{x}) \equiv \frac{A_s}{bd} \geq \frac{200}{f_y} \tag{9.2.1.56}$$

$$g_3(\mathbf{x}) \equiv \frac{A_s}{bd} - \frac{A_{s'} f_{s'}}{bdf_y} \leq 0.75\rho_b \tag{9.2.1.57}$$

$$g_4(\mathbf{x}) \equiv M_n > M_{\max} \tag{9.2.1.58}$$

$$g_5(\mathbf{x}) \equiv b > b_{\min} \tag{9.2.1.59}$$

$$g_6(\mathbf{x}) \equiv d_s > (d_s)_{\min} \tag{9.2.1.60}$$

$$g_7(\mathbf{x}) \equiv f_{s'} \leq f_y \tag{9.2.1.61}$$

$$(A_s)_L \leq A_s \leq (A_s)_U \tag{9.2.1.62a}$$

$$b_L \leq b \leq b_U \tag{9.2.1.63b}$$

$$d_L \leq d \leq d_U \tag{9.2.1.64c}$$

As one can see here, the problem formulation is very similar to the optimal design of a singly reinforced beam.

Design for Shear and Diagonal Tension

In Fig. 9.2.1.2 we saw a simply supported beam that was subjected to a concentrated force. On the left in Fig. 9.2.1.7 are two differential elements that represent the state of stress above and below the neutral axis. The top differential element is subjected to compression and shear, while the bottom element is subjected to tension and shear. Recall from Chapter 3 that $\sigma = \pm(Mc/I)$ and $\tau = VQ/It$.

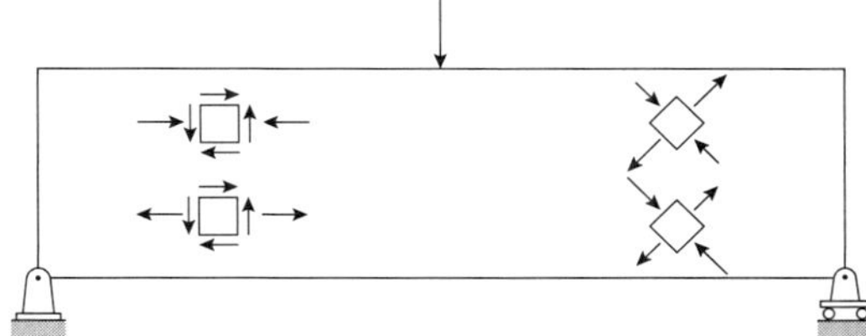

Fig. 9.2.1.7 Stress differential elements.

At both these locations, we can also compute the principal stresses. The differential elements showing the principal planes are shown on the right in Fig. 9.2.1.7. Below the neutral axis, the element is subjected to the largest tensile stress at an inclination with respect to the longitudinal axis given by

$$(\sigma_{\max})_t = \frac{\sigma}{2} + \sqrt{\left(\frac{\sigma}{2}\right)^2 + \tau^2} \tag{9.2.1.65}$$

Figure 9.2.1.8 shows some of the crack pattern as being similar to the plot above. Since concrete's strength in tension is small, diagonal cracks usually develop along the

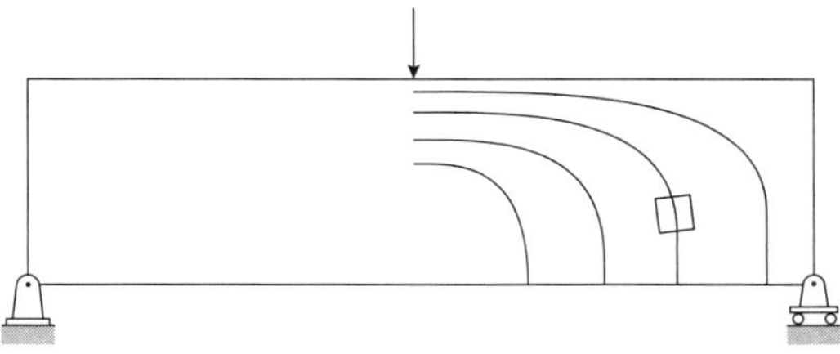

Fig. 9.2.1.8 Principal compressive stress plot and a typical stress differential element (similar plot for left half).

lines perpendicular to the principal tensile stress as if these tensile stresses are pulling the concrete apart. These cracks are called *diagonal tension cracks* or *shear cracks*.

With the limited discussion of failure modes so far, it should be noted that there are three types of failure—flexural failure, diagonal tension failure, and shear compression failure. Earlier we discussed flexural failure. With underreinforced beams, there is ample warning before the beam fails. Diagonal tension failure occurs when the beam is weaker in diagonal tension than strength in flexure. This is a brittle type of failure. The shear compression failure occurs when flexural cracks suddenly combine with the zone where concrete has crushed. This type of failure is also brittle and must be avoided. Once again, steel reinforcement can be used to control these other types of failures.

Figure 9.2.1.9 shows the reinforcement arrangement with the use of stirrups spaced a distance s apart. A truss analogy describes how the reinforcement arrangement arrests cracks due to diagonal tension. However, a detailed analysis of the diagonal tension phenomenon is outside the scope of this text. We will discuss ACI recommendations for design. Let V_u be the factored shear load, V_n the nominal shear resistance, V_c the nominal shear resistance of plain concrete, and V_s the shear strength provided by the shear reinforcements. The nominal strength of the section should at least be equal to

$$V_n = \phi V_c + \phi V_s \tag{9.2.1.66}$$

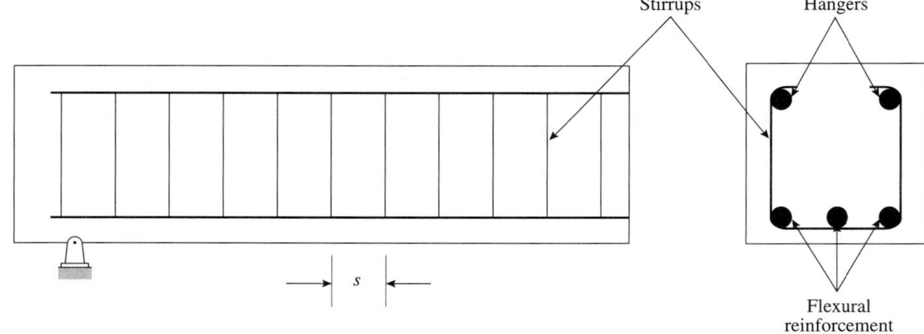

Fig. 9.2.1.9 Shear reinforcement for a rectangular beam in the form of vertical stirrups.

A conservative approach is to define

$$V_c = \lambda\left(2.0\sqrt{f_c'}bd\right) \tag{9.2.1.67}$$

where the factor λ is 1.0 for normal-weight concrete, 0.85 for sand–lightweight concrete, and 0.75 for all lightweight concrete. Equation (9.2.1.67) needs to be modified when axial compression and axial tension are present as

$$V_c = 2\lambda\left(1 + \frac{N_u}{2000A_g}\right)\sqrt{f_c'}bd \text{ (axial compression)} \tag{9.2.1.68}$$

$$V_c = 2\lambda\left(1 + \frac{N_u}{500A_g}\right)\sqrt{f_c'}bd \text{ (axial tension)} \tag{9.2.1.69}$$

where N_u is the axial force (positive in compression; negative in tension) and A_g is the gross cross-sectional area.

General Design Procedure. Given f_c', b, d, f_y, V_u in in, lb. Obtain stirrup details—bar size and spacing.

Step 1: Calculate V_u at a distance d from the support. When the beam is such that the ends of the member are in compression, the critical section is assumed to be at a distance d from the support.

Step 2: Calculate $V_c = \lambda\left(2.0\sqrt{f_c'}bd\right)$. If $V_u > (1/2)\phi V_c$, then stirrups are usually needed.

Step 3: If

$$V_u > \phi\left(V_c + 8\sqrt{f_c'}\,bd\right) \tag{9.2.1.70}$$

then the section dimensions must be increased.

Step 4: If $V_u < \phi V_c$, then the minimum amount of shear reinforcement must be provided as follows:

$$A_v = \frac{50bs}{f_y} \tag{9.2.1.71}$$

$$s \le \frac{d}{2} \le 24 \text{ in} \tag{9.2.1.72}$$

where A_v is the minimum area of all the vertical stirrup legs in the cross-section.

Step 5: If $V_u \ge \phi V_c$, then

$$A_v = \frac{\left(V_u/\phi - V_c\right)s}{f_y d} \tag{9.2.1.73}$$

Also, $s \le d/2 \le 24$ in unless $\left(V_u/\phi\right) - V_c > 4\sqrt{f_c'}\,bd$, then $s \le d/4 \le 12$ in.

EXAMPLE 9.2.4 ***Design of Shear Reinforcement (Example 9.2.2)***

A simply supported concrete beam is shown in Fig. E9.2.4. The live load on the beam is 1500 lb/ft. Design the shear reinforcements. The material properties are $f_c' = 4000$ psi and $f_y = 60000$ psi.

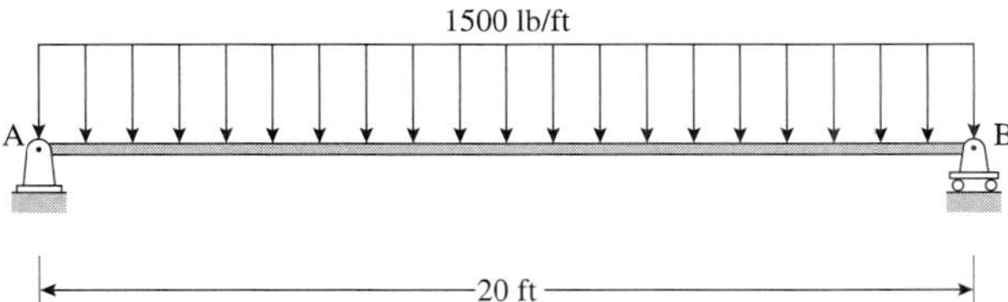

Fig. E9.2.4

SOLUTION

In Example 9.2.2 we designed the beam to meeting the flexural requirements. In this example we design the beam to meet the shear requirements. In that regard, note that the beam dimensions are 20 in × 8 in. We will assume that the beam is made of normal-weight concrete.

Step 1: Does the beam need stirrups? The computed factored load is

$$w_u = 1.4\left[\frac{(20)(8)}{144}(150)\right] + 1.7(1500) = 2785 \text{ lb/ft}$$

Note that $V(x) = 27850 - 2785x$. Hence the shear at a distance $d = 17.5$ in $= 1.458$ ft from the support is $V_u = 23450$ lb. Hence

$$V_c = 1.0(2.0\sqrt{4000})(8)(17.5) = 17709 \text{ lb} \qquad \phi V_c = 15053 \text{ lb}$$

Stirrups are needed since $V_u > 1/2\phi V_C$ (with $\phi = 0.85$).

Step 2: Are the section dimensions OK?

$$\phi\left(V_c + 8\sqrt{f_c'}\,bd\right) = (0.85)\left[17709 + 8\sqrt{4000}(8)(17.5)\right] = 75262 \text{ lb}$$

Since V_u is not greater than 75262 lb, the beam dimensions need not be changed.

Step 3: Stirrup design. *Let us assume that the stirrups are #3 bars*. Hence $A_v = 2(0.11) = 0.22$ in^2. Since $V_u \geq \phi V_c$, the theoretical spacing required is

$$s = \frac{A_v f_y d}{\left(V_u/\phi - V_c\right)} = \frac{(0.22)(60000)(17.5)}{(23450/0.85 - 15053)} = 18.4 \text{ in}$$

Note that since $\left(V_u/\phi\right) - V_c = \left(23450/0.85\right) - 17709 = 9879 < 4\sqrt{f_c'}\,bd$, $s \leq d/2 = 8.75$ in.

Answer: #3 bars at a spacing of 8.5 in should be adequate. Note that we could obtain a more economical design by varying the spacing over the length of the beam, since the shear force decreases towards the center of the span of the beam.

Serviceability Considerations

So far in this section we have looked at the strength aspects of concrete beams. What we have not addressed are the issues dealing with *serviceability*—how does the structure deflect, vibrate, etc., under service loads? The ACI code has recommendations on the maximum permissible deflections.

There are two types of deflections, short-term and long-term. The former can be computed assuming that these deflections are elastic in nature. The long-term deflections are due to shrinkage and creep, and can only be estimated. The ACI Code recommends that the additional long-term deflection resulting from creep and shrinkage be determined by multiplying the immediate deflection caused by sustained load by the factor

$$\lambda = \frac{\xi}{1 + 50\rho'} \tag{9.2.1.74}$$

where ρ' is the compressive reinforcement ratio at midspan for simple and continuous beams, and at support for cantilevers. The time-dependent factor ξ for sustained loads is given by

5 years or more (Eq. 9.2.1.75)	2.0
12 months	1.4
6 months	1.2
3 months	1.0

Furthermore, the code also specifies the maximum permissible computed deflections listed in Table 9.2.5.

EXAMPLE 9.2.5 ***Deflection of a Simply Supported Beam***

A simply-supported concrete beam is shown in Fig. E9.2.5(a). The total dead load on the beam is 1000 lb/ft. The live load on the beam is 1500 lb/ft. The material properties are $f_c' = 4000$ psi and f_y = 60000 psi. Calculate (a) the instantaneous deflection, and (b) the deflection after one year (assume that sustained load is dead load plus 50% of the live load).

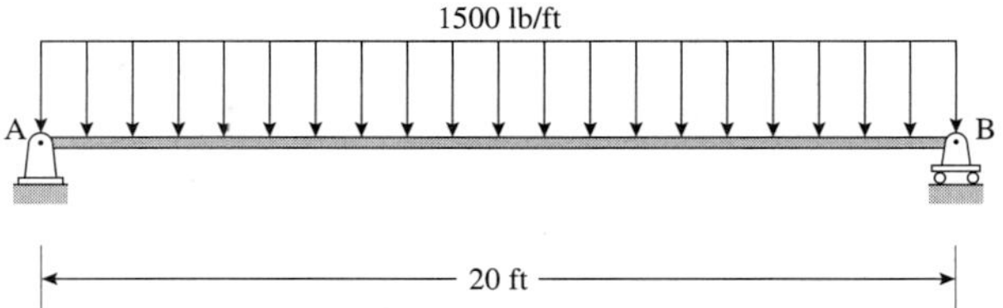

Fig. E9.2.5(a)

Table 9.2.5 Maximum Permissible Computed Deflections (ACI Table 9.5(b))

Type of member	Deflection to be considered	Deflection limitation
Flat roofs not supporting or attached to nonstructural elements likely to be damaged by large deflections	Immediate deflection due to live load **L**	$\frac{\ell}{180}$ (see 1)
Floors not supporting or attached to nonstructural elements likely to be damaged by large deflections	Immediate deflection due to live load **L**	$\frac{\ell}{360}$
Roof or floor construction supporting or attached to nonstructural elements likely to be damaged by large deflections	That part of the total deflection occurring after attachment of nonstructural elements (sum of the long-term deflection due to all sustained loads and the immediate deflection due to any additional live load) (see 2)	$\frac{\ell}{480}$ (see 3)
Roof or floor construction supporting or attached to nonstructural elements not likely to be damaged by large deflections		$\frac{\ell}{240}$ (see 4)

(1) Limit not intended to safeguard against ponding. Ponding should be checked by suitable calculations of deflection, including added deflections due to ponded water, and considering long-term effects of all sustained loads, camber, construction tolerances, and reliability of provisions for drainage.
(2) Long-term deflection shall be determined in accordance with 9.5.2.5 or 9.5.4.2, but may be reduced by amount of deflection calculated to occur before attachment of nonstructural elements. This amount shall be determined on basis of accepted engineering data relating to time-deflection characteristics of members similar to those being considered.
(3) Limit may be exceeded if adequate measures are taken to prevent damage to supported or attached elements.
(4) Limit shall not be greater than tolerance provided for nonstructural elements. Limit may be exceeded if camber is provided so that total deflection minus camber does not exceed limit.

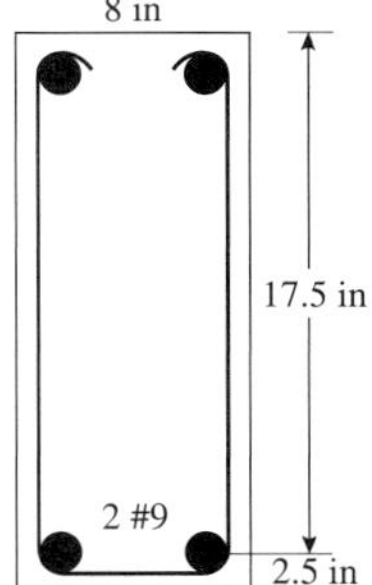

Fig. E9.2.5(b)

SOLUTION We use lb, in as the problem units.

Step 1: Instantaneous deflection. For a simply supported beam, the maximum deflection at the center of the beam is

$$\Delta_{max} = \frac{5wl^4}{384EI}$$

With the given data we can compute the following:

$$w = \frac{1000}{12} + \frac{1500}{12} = 208.3 \text{ lb/in} \qquad E = 57000\sqrt{4000} = 3.605(10^6) \text{ psi}$$

We will compute the different forms of moment of inertia discussed in Section 9.2.1.

$$I_g = \frac{bh^3}{12} = \frac{(8)(20)^3}{12} = 5333 \text{ in}^4$$

From Example 9.2.2, we have

$$a = 4.48 \text{ in},\ c = \frac{a}{\beta_1} = \frac{4.48}{0.85} = 5.27 \text{ in},\ d = 17.5 \text{ in},\ A_s = 2 \text{ in}^2,\ n = \frac{E_s}{E_c} = \frac{29(10^6)}{3.605(10^6)} = 8.04$$

$$I_{gt} = \frac{bc^3}{3} + nA_s(d-c)^2 = \frac{(8)(5.27^3)}{3} + 8.04(2)(17.5 - 5.27)^2 = 2795 \text{ in}^4$$

$$f_r = 7.5\sqrt{f_c'} = 7.5\sqrt{4000} = 474 \text{ psi}$$

$$M_{cr} = \frac{f_r I_g}{y_t} = \frac{(474)(5333)}{10} = 252784 \text{ lb-in}$$

$$M_a = \frac{wl^2}{8} = \frac{(208.3)(20\times 12)^2}{8} = 1.5(10^6) \text{ lb/in}$$

$$I_e = I_{gt} + \left(\frac{M_{cr}}{M_a}\right)^3 \left(I_g - I_{gt}\right) = 2795 + \left(\frac{252784}{1.5\times 10^6}\right)^3 (5333 - 2795) = 2807 \text{ in}^4$$

$$\text{Hence, } \Delta_{\max} = \frac{5wl^4}{384EI} = \frac{5(208.3)(240)^4}{384(3.605\times 10^6)(2807)} = 0.89 \text{ in}$$

Step 2: Long-term deflection. The initial deflection due to sustained loads is (1000 + 0.5(1500)/1000 + 1500)(0.89) = 0.623 in. Using λ (ξ/1 + 50ρ') = (1.4/1 + 50(0)) = 1.4, we have the total long-term deflection as

$$\Delta = 0.89 + (1.4)(0.623) = 1.76 \text{ in}$$

9.2.2 Column Design

As we have seen before, columns are members that are usually vertical and subjected to primarily compressive forces. Columns can be built in different sizes and shapes and loaded differently (see Fig. 9.2.2.1). As per ACI code, a *pedestal*, a column whose height is less than three times its smallest lateral dimension, may be designed with plain concrete. On the other hand, reinforced columns may be short or long. The former is a stocky column that exhibits material failure as the primary mode of failure. The latter have a larger slenderness ratio—the lateral dimensions are small compared to the length of the column. Consequently, the deformations cause the secondary moments to play a more important role in increasing the axial stress in the column.

In this section we will learn more about the analysis and design of short columns. Figure 9.2.2.2 shows a tied column, where the main reinforcements run longitudinally along the column and are held in place by *ties*. Tied columns can also have other cross-sectional shapes (much like steel columns) and are restricted by construction limitations.

Assuming that a column is subjected to primarily an axial force with the bending moments being negligible, the nominal strength of the column is given by

$$P_n = 0.85 f_c'(A_g - A_{st}) + f_y A_{st} \tag{9.2.2.1}$$

where A_g is the gross concrete area and A_{st} is the total area of the longitudinal steel reinforcement. To calculate the axial load capacity, the ACI code reduces the theoretical load capacity to take into account the presence of bending moments (arising from a variety of factors that do not show up in typical calculations) such that

$$P_u = \alpha \phi P_n \tag{9.2.2.2}$$

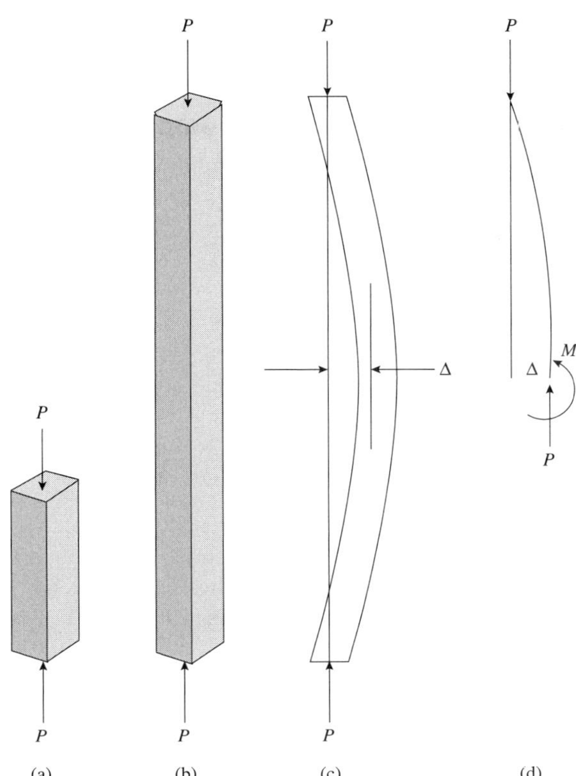

Fig. 9.2.2.1
(a) Pedestal. (b) Long column. (c) Deformation of a long column. (d) FBD of a long column showing the P–Δ effect.

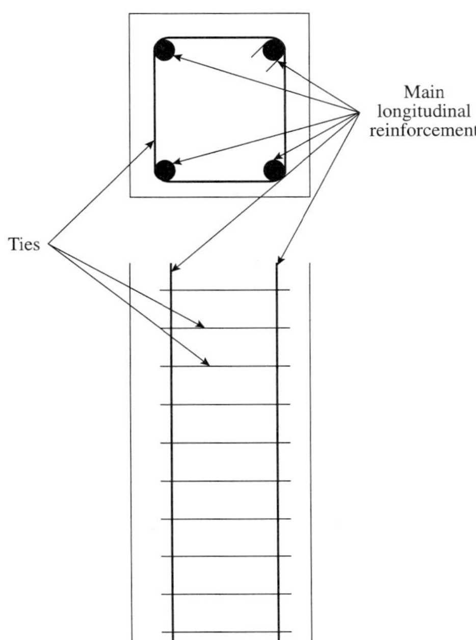

Fig. 9.2.2.2
Tied rectangular column.

where $\phi = 0.70$, $\alpha = 0.80$ for tied columns and $\phi = 0.70$, $\alpha = 0.80$ for spiral columns (columns with ties in the form of spirals). ACI has requirements for cast-in-place concrete columns, some of which are:

1. The minimum percentage of steel should not be less than 1% of the gross cross-sectional area and the maximum percentage should not be greater than 8%. The former provision is stated to prevent a sudden brittle failure and lessen the effects due to creep and shrinkage. The latter is intended to ensure that there is adequate space between the bars for proper bonding to take place between the bars and concrete. ACI also specifies the minimum number of longitudinal bars as four for rectangular or circular shaped columns. The minimum clear spacing between parallel bars in a layer should be greater of (i) bar diameter and (ii) 1 in.
2. The ties in the columns must be at least #3 bars when the longitudinal bars are #10 bars or smaller. For longitudinal bars that are greater than #10 bars, the ties must be at least #4 bars. The tie spacing cannot be greater than 16 times the diameter of the longitudinal bars, or the smallest lateral dimension of the column.

In reality, the columns are subjected to bending moments either because the axial load has an eccentricity *e* or because the loading introduces bending moment in the column. Figure 9.2.2.3 shows these two cases. Note that the two load cases are equivalent to each other. In the rest of this section, we look at an axial load with various degrees of eccentricity.

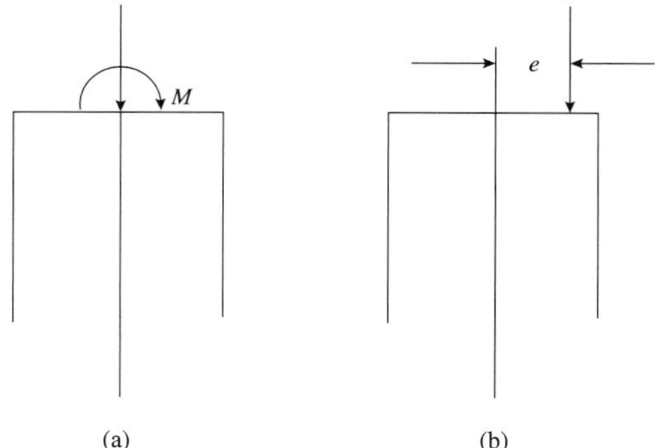

Fig. 9.2.2.3 Column with (a) axial load with no eccentricity and bending moment, (b) axial load with eccentricity *e*.

The beam analysis in the previous section can be used here to analyze and design an eccentrically loaded column. There are two modes of material failure that we must consider—tension failure with the steel on the tensile side yielding, and compression failure with the crushing of concrete. Figure 9.2.2.4 shows the state of stress and strain in the column cross-section. This diagram is very similar to Fig. 9.2.1.6 used for the analysis of doubly-reinforced beams. The notable differences are as follows. There is a special point inside the cross-section called the *plastic centroid* (PC). This is the location such that, if an axial load passes through this point it produces a constant state of strain in the cross-section. For symmetrical sections, PC coincides with the centroid of the section. We assume that the steel on the tension side is in fact in tension and the neutral axis lies within the depth of the beam. These assumptions may not be valid for small values of eccentricity (for which the entire cross-section may be in compression).

From the section properties, we have

$$C_c = 0.85 f_c' ba \tag{9.2.2.3a}$$

$$C_s = A_{s'} f_s' \tag{9.2.2.3b}$$

$$T = A_s f_s \tag{9.2.2.3c}$$

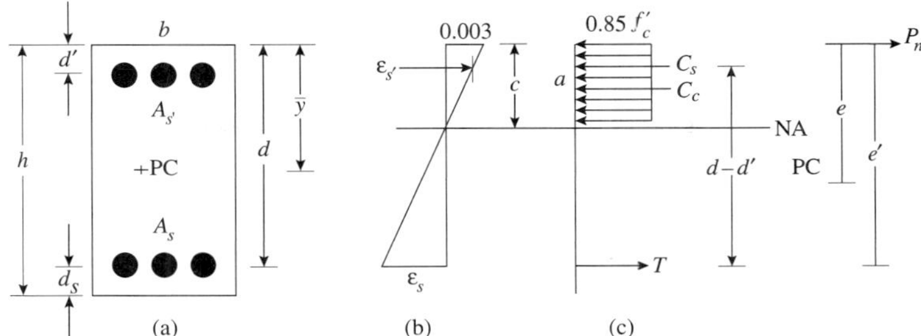

Fig. 9.2.2.4 Column analysis (compression failure).

From equilibrium considerations, we have

$$P_n = C_c + C_s - T \tag{9.2.2.4}$$

$$M_n = P_n e = C_c\left(\bar{y} - \frac{a}{2}\right) + C_s(\bar{y} - d') + T(d - \bar{y}) \tag{9.2.2.5}$$

Substituting Eq. (9.2.2.3) into (9.2.2.4) and (9.2.2.5), we have

$$P_n = 0.85 f_c' ba + A_{s'} f_{s'} - A_s f_s \tag{9.2.2.6}$$

$$M_n = 0.85 f_c' ba\left(\bar{y} - \frac{a}{2}\right) + A_{s'} f_{s'}(\bar{y} - d') + A_s f_s (d - \bar{y}) \tag{9.2.2.7}$$

A few points to note about this analysis. P_n cannot exceed the value given by Eq. (9.2.2.1). The compression steel $A_{s'}$ reaches the yield value f_y when concrete crushes. From Fig. 9.2.2.4, we see that

$$f_{s'} = E_s \varepsilon_{s'} = E_s \frac{0.003(c - d')}{c} \le f_y \tag{9.2.2.8a}$$

Similarly, the tension steel A_s reaches the yield value f_y when

$$f_s = E_s \varepsilon_s = E_s \frac{0.003(d - c)}{c} \le f_y \tag{9.2.2.8b}$$

Balanced Failure. As we have seen before, the balanced failure condition occurs when failure develops simultaneously in tension and in compression. Consider the same rectangular section as shown in Fig. 9.2.2.4. The balanced failure condition is when the strain in the tension steel is ε_y and the compressive strain in the concrete is 0.003. From Fig. 9.2.2.3(b), with units of lb, in, we find

$$\frac{c_b}{d} = \frac{0.003}{0.003 + f_y/E_s} \Rightarrow c_b = \frac{87000}{87000 + f_y} d \tag{9.2.2.9}$$

$$a_b = \beta_1 c_b \tag{9.2.2.10}$$

Similarly, we have

$$P_{nb} = 0.85 f_c' ba_b + A_{s'} f_{s'} - A_s f_s \tag{9.2.2.11}$$

$$M_{nb} = 0.85 f_c' ba_b\left(\bar{y} - \frac{a_b}{2}\right) + A_{s'} f_{s'}(\bar{y} - d') + A_s f_s (d - \bar{y}) \tag{9.2.2.12}$$

$$f_{s'} = 0.003 E_s \frac{c_b - d'}{c_b} \le f_y \tag{9.2.2.13}$$

Hence, tension failure occurs when $P_n < P_{nb}$, balanced failure occurs when $P_n = P_{nb}$ and compression failure occurs when $P_n > P_{nb}$.

General Procedure for Analysis. We will describe a trial-and-error procedure. Given b, d, A_s, $A_{s'}$, f_c', f_y, e. Compute P_n, M_n.

1. Assume a value for c. Compute $a = \beta_1 c$.
2. Compute the stresses in the compression and tension steel using Eq. (9.2.2.8) and $P_n = 0.85 f_c' ba + A_{s'} f_{s'} - A_s f_s$.
3. Calculate the eccentricity e using Eq. (9.2.2.5). Does it match the given eccentricity? If yes, stop the iterations.
4. If the calculated eccentricity is greater than the given eccentricity, then assume a larger value of c. Otherwise decrease c. Restart the process.

EXAMPLE 9.2.6 ***Analysis of a Short Column***

Compute the nominal load capacity of the beam in Fig. E9.2.6. Take the eccentricity as 8 in. The material properties are $f_c' = 3000$ psi and $f_y = 50000$ psi.

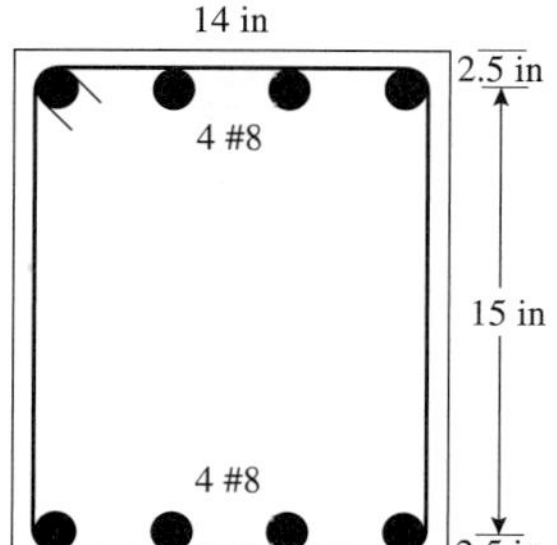

Fig. E9.2.6

SOLUTION

Step 1: With the given data, we have

$$A_s = A_{s'} = 3.16 \text{ in}^2$$

Let us assume c= 10 in. Then $a = \beta_1 c = (0.85)(10) = 8.5$ in. The stress in the tensile and compressive steel are

$$f_{s'} = E_s \frac{0.003(c-d')}{c} = (29\times 10^6)\left(\frac{0.003(10-2.5)}{10}\right) = 65250 \text{ psi} > f_y, \text{ hence } f_{s'} = 50000 \text{ psi}$$

$$f_s = E_s \frac{0.003(d-c)}{c} = (29\times 10^6)\left(\frac{0.003(17.5-10)}{10}\right) = 65250 \text{ psi} > f_y, \text{ hence } f_s = 50000 \text{ psi}$$

Hence,

$$P_n = 0.85(3000)(14)(8.5) = 303450 \text{ lb}$$

$$e = \left[C_c\left(\bar{y} - \frac{a}{2}\right) + C_s(\bar{y} - d') + T_s(d - \bar{y})\right] \Big/ P_n$$

or $$e = \frac{303450(10-4.25) + 3.16(50000)(10-2.5) + 3.16(50000)(17.5-10)}{303450} = 13.56 \text{ in}$$

Step 2: The rest of the iterations are shown next as spreadsheet output.

Iteration	1	2	3
Es	2.90E+07	2.90E+07	2.90E+07
As	3.16	3.16	3.16
Asp	3.16	3.16	3.16
fc	3000	3000	3000
fy	50000	50000	50000
b	14	14	14
d	17.5	17.5	17.5
dp	2.5	2.5	2.5
yb	10	10	10
beta1	0.85	0.85	0.85
c	10	15	13
a	8.5	12.75	11.05
fsp	50000	50000	50000
fs	50000	14500	30115.38
Cc	303450	455175	394485
Cs	158000	158000	158000
Ts	158000	45820	95164.62
Pn	303450	567355	457320.4
e	13.5601829	5.602593	8.012009

EXAMPLE 9.2.7 ***Analysis of a Short Column***

Find the balanced failure condition for the column in Fig. E9.2.7 that is subjected to axial force and bending moment. The material properties are $f_c' = 3000$ psi and $f_y = 50000$ psi.

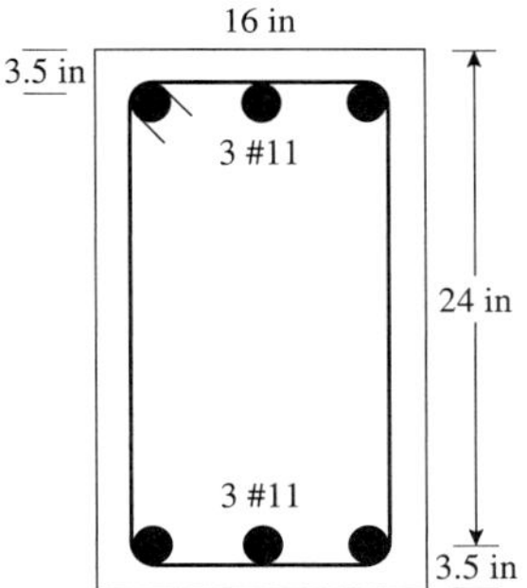

Fig. E9.2.7

SOLUTION

Step 1: Given data: b = 16 in, d = 24 in, h = 27.5 in, d' = 3.5 in, $A_s = 4.68\ \text{in}^2 = A_{s'}$, $f_c' = 3000$ psi, and $f_y = 50000$ psi.

Hence, using Eqs. (9.2.2.9) and (9.2.2.10), we have

$$c_b = \frac{87000}{87000 + f_y} d = \frac{87000}{87000 + 50000}(24) = 15.24 \text{ in}$$

$$a_b = \beta_1 c_b = (0.85)(15.24) = 12.95 \text{ in}$$

From Eq. (9.2.2.13) we can compute the stress in the compression steel:

$$f_{s'} = 0.003\left(29 \times 10^6\right)\frac{15.24 - 3.5}{15.24} = 67020 \text{ psi} > f_y$$

Hence, $f_{s'} = 50000$ psi.

Step 2: Compute axial load and bending moment. Using Eqs. (9.2.2.11) and (9.2.2.12)

$$P_{nb} = 0.85(3000)(16)(12.95) + (4.68)(50000 - 50000) = 528360 \text{ lb-in}$$

$$M_{nb} = 0.85(3000)(16)(12.95)\left(\frac{27.5}{2} - \frac{12.95}{2}\right) + 4.68(50000)\left(\frac{27.5}{2} - 3.5\right)$$

$$+4.68(50000)\left(24 - \frac{27.5}{2}\right) = 3.84\left(10^6\right) + 2.4\left(10^6\right) + 2.4(10^6) = 8.64\left(10^6\right) \text{ lb-in}$$

Hence the eccentricity is

$$e_b = \frac{M_{nb}}{P_{nb}} = \frac{8.64(10^6)}{528360} = 16.4 \text{ in}$$

In other words, a balanced failure is caused if an axial force of 528360 lb is applied at an eccentricity of 16.4". For the same eccentricity, if the axial load $P < 528360$ lb, then tension failure occurs. On the other hand, if $P > 528360$ lb, then compression failure occurs.

Analysis and Design of Short Columns via P–M *or Interaction Diagram*

We will obtain a graphical view of our previous discussion. The conventional process uses the *P–M* or interaction diagram in the analysis and design of the columns. Figure 9.2.2.5 shows the different components. The important curve is formed by the two lines 2–3–4 and 4–5. In other words, each point on the graph represents a set of nominal (P_n, M_n) values for a given cross-section. Point 4 represents the balanced condition. As discussed before, tension failure occurs when $P_n < P_{nb}$, as represented by the curve 4–5. Similarly, compression failure occurs when $P_n > P_{nb}$, as represented by the curve 2–3–4. The pure axial loading case is denoted by point 1, while point 2 shows the pure bending case.

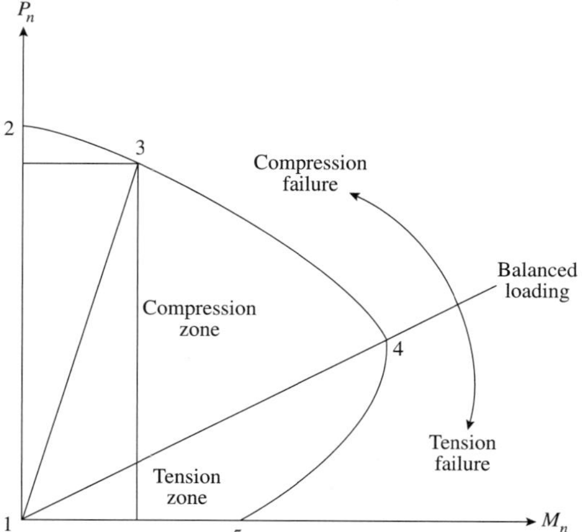

Fig. 9.2.2.5 *P–M* or Interaction diagram

Two modifications are necessary for the diagram to be a useful tool. First, the above diagram does not include the strength reduction factor, ϕ. The ACI code specifies that ϕ is 0.70 for tied columns and 0.75 for spiral columns. Hence the design values are ϕP_n and ϕM_n. The top part of the curve is for columns with small eccentricities or small moments. As mentioned earlier, the maximum axial load that can act on the column is given by Eqs. (9.2.2.1) and (9.2.2.2). The modified diagram is shown in Fig. 9.2.2.6 where 2–3 accounts for the above equations. On the lower part of the curve, the axial loads are small. Hence even if the axial load is small, should the reduction factors of 0.70 (tied column) and 0.75 (spiral column) be applicable? Some relief is provided by the code. If f_y does not exceed 60 ksi, the reinforcements are placed symmetrically, and $(h - d' - d_s)/h$ is not less than 0.7, the reduction factor may be increased from 0.7 or 0.75 to 0.9 as ϕP_n decreases from $0.10 f_c' A_g$ to 0, or as ϕP_n decreases from $0.10 f_c' A_g$ to ϕP_b to 0. This region is shown as 4–5 on the *P–M* diagram.

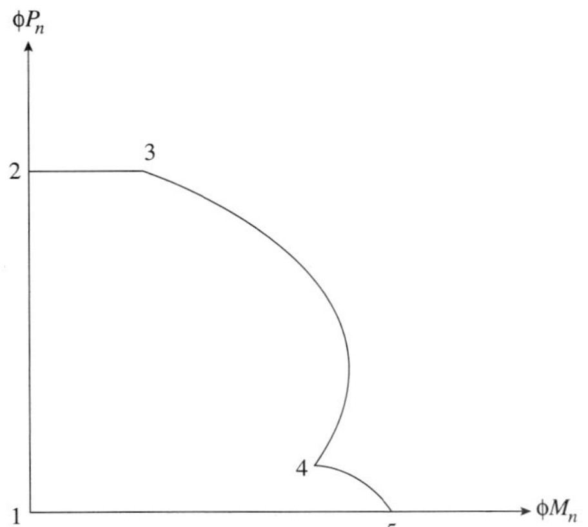

Fig. 9.2.2.6 Modified *P–M* or interaction diagram.

The interaction diagram is prepared for a specific cross-section—dimensions, steel reinforcements and their placement, and concrete grade. Clearly this will lead to hundreds of diagrams if one tries to vary the values of these parameters. The ACI code contains diagrams as shown in Fig. 9.2.2.7 in which the ordinate is $\dfrac{\phi P_n}{A_g}$ and the abscissa is $\dfrac{\phi P_n e}{A_g h}$. Note that A_g is the gross cross-sectional area. Such diagrams are available for different grades of steel and concrete, and different bar arrangements. The parameter $\gamma = \dfrac{h - d' - d_s}{h}$ represents the distance between the two sets of steel reinforcements as a fraction of the height of the column. We illustrate the analysis and design of columns using the interaction diagram.

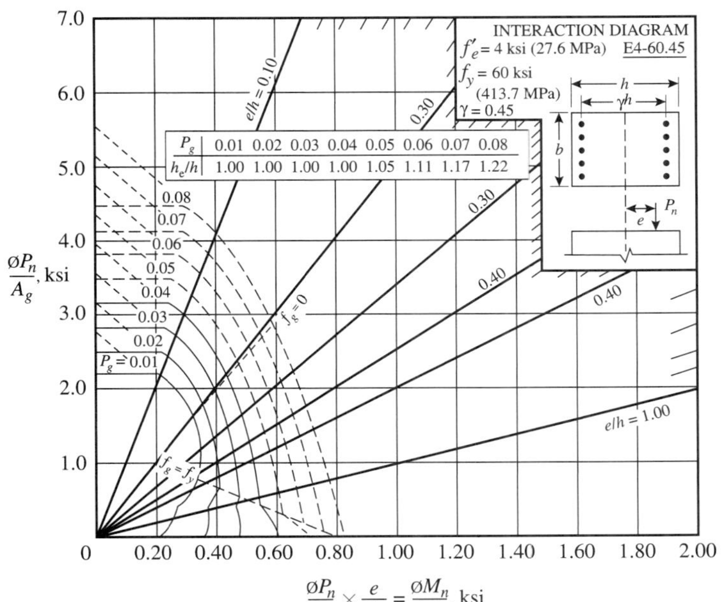

Fig. 9.2.2.7 ACI Design Interaction Diagram E4-60.45 for reinforcements on end faces.

Design Procedure. There are several approaches to designing a column. Two approaches are presented here.

Approach 1: Given b, d, f_c', f_y, P_n, M_n, compute reinforcement details.

1. Compute $\gamma = (h - d' - d_s)/h \qquad e = \dfrac{M_n}{P_n} \qquad \dfrac{\phi P_n}{A_g} \qquad \left(\dfrac{\phi P_n}{A_g}\right)\left(\dfrac{e}{h}\right)$
2. Locate the appropriate ACI Design Interaction Diagram. Note that this is based on f_c', f_y, γ and the bar arrangement. Interpolation may be necessary if γ is not one of the values for which a diagram is available. On the graph, locate the points $\phi P_n/A_g$ and $(\phi P_n/A_g)(e/h)$. Estimate the value of ρ_g that is applicable for this point. Using this value, compute the total amount of reinforcement A_{st}. Find an even number of bars for this area. Distribute the bars equally between the face(s).
3. Check whether the column satisfies the other requirements—maximum axial force, bar clear spacing, minimum and maximum percent reinforcement. If the requirements are not satisfied, then redesign is necessary.

Approach 2: Given f_c', f_y, P_n, M_n. Compute b, d, and reinforcement details.

1. One can estimate the initial column dimensions by assuming that the primary load is the axial force. Recall that a rectangular tied column must satisfy

   $$\phi P_n = 0.80\phi\left[0.85 f_c'(A_g - A_{st}) + f_y A_{st}\right]$$

   This equation can be rearranged to yield

   $$A_g \geq \frac{P_u}{0.45\left(f_c' + f_y \rho_t\right)}$$

   where $\rho_t = A_{st}/A_g$. Assuming a value for ρ_t (say, between 0.01 and 0.04), we can compute A_g. Since $A_g = bh$, we can assume suitable values for b, d.
2. The rest of the steps are similar to Approach 1. Again, it may be necessary to redesign if the required conditions are not met.

General Procedure for Column Design (as implemented in the UCSD program)

Given b, h, d, d', M_u, P_u, f_y, f_c'. Find $A_s, A_{s'}$.

Note: We will design the column so that the steel area in tension and compression is the same, i.e., $A_s = A_{s'}$. Also in the balanced failure condition, we will assume that both tension and compression steel yields.

Step 1: Calculate the nominal capacities M_n, P_n and the plastic centroid $\bar{y}$ as $\bar{y} = h/2$.

Step 2: Find β_1 based on the given concrete strength f_c' using Eq. (9.2.1.27).

Step 3: Calculate P_{nb} using Eq. (9.2.2.11), which reduces to $P_{nb} = 0.85 f_c' b a_b$ as per our assumptions.

Step 4a. Compression failure occurs if $P_n > P_{nb}$. Compute c by solving the cubic equation.

$$p_0 c^3 + p_1 c^2 + p_2 c + p_3 = 0$$

where

$$A = 0.85 f_c' b \beta_1$$

$$D = A f_y$$

$$G = 87000A$$

$$E = 87000 P_n (d - \bar{y})$$

$$p_0 = -(D + G)\frac{\beta_1}{2}$$

$$p_1 = (G+D)Y_P + \left(\frac{G\beta_1 d}{2}\right) - D(\bar{y}-d') + G(d-\bar{y})$$

$$p_2 = -(G\bar{y}d) + P_n f_y(\bar{y}-d') - E - G(d-\bar{y})d - M_n(f_y + 87000)$$

$$p_3 = (E + 87000M_n)d$$

Compute $f_s = 87000(d - c/c)$, and the tension and compression steel as follows:

$$A_s = A_{s'} = \left(\frac{P_n - 0.85f_c{}'\beta_1 bc}{f_y - f_s}\right)$$

Step 4b. Tension failure occurs if $P_n < P_{nb}$. Compute c by solving the cubic equation.

$$p_0c^3 + p_1c^2 + p_2c + p_3 = 0$$

where $A = 0.85f_c'b\beta_1$

$$D = Af_y$$

$$G = 87000A$$

$$E = 87000P_n(\bar{y} - d')$$

$$H = 87000M_n$$

$$p_0 = \frac{(D-G)\beta_1}{2}$$

$$p_1 = (G-D)Y_P + \left(\frac{G\beta_1 d'}{2}\right) - G(\bar{y}-d') - D(d-Y_P)$$

$$p_2 = -(G\bar{y}d') + P_n f_y(d-\bar{y}) + E + G(\bar{y}-d')d' + M_n f_y - H$$

$$p_3 = (-E+H)d'$$

Compute $f_{s'} = 87000\left(c - \frac{d'}{c}\right)$, and compute tension and compression steel as follows.

$$A_s = A_{s'} = \left(\frac{P_n - 0.85f_c{}'\beta_1 bc}{f_{s'} - f_y}\right)$$. If $f_{s'} > f_y$, go to Step 4c.

Step 4c. Balanced failure occurs if $P_n = P_{nb}$.

$$c_b = \left(\frac{87000}{87000 + f_y}\right)d \qquad a_b = \beta_1 c_b$$

$$P_{nb} = 0.85f_c'ba_b \qquad A_s = A_{s'} = \left(\frac{M_n - P_{nb}(\bar{y} - 0.5a_b)}{(d-d')f_y}\right)$$

Step 5: Compute the total steel and check for minimum required steel:

$$A_{\text{total}} = A_s + A_{s'} \qquad A_{\min} = 0.01bh$$

If $A_{\text{total}} < A_{\min}$, $A_{\text{total}} = A_{\min}$ and $A_s = A_{s'} = A_{\min}/2$.

Optimal Design Procedure. The above-mentioned design problem can be posed as an optimal design problem as follows. Let us consider a problem where the largest factored bending moment, $M_{\max}$ can be written as

$$M_{\max} = M + cb(d + d_s) \tag{9.2.2.14}$$

where M is the largest bending moment due to live load, c is a constant (the second term accounts for the moment due to dead load), and d_s represents the concrete cover. In addition, we assume that the cross-section is uniform along the entire length of the column. The optimal design problem can be stated as follows:

$$\text{Find} \quad \mathbf{x} = (b, d, A_s, A_{s'}) \tag{9.2.2.15}$$

$$\text{to minimize} \quad f(\mathbf{x}) = w_1 A_s + w_1 A_{s'} + w_2 b(d + d_c) \tag{9.2.2.16}$$

$$\text{subject to} \quad g_1(\mathbf{x}) \equiv (A_s + A_{s'}) > 0.01b(d + d_c) \tag{9.2.2.17}$$

$$g_2(\mathbf{x}) \equiv (A_s + A_{s'}) < 0.08b(d + d_c) \tag{9.2.2.18}$$

$$g_3(\mathbf{x}) \equiv M_n > M_{\max} \tag{9.2.2.19}$$

$$g_4(\mathbf{x}) \equiv b > b_{\min} \tag{9.2.2.20}$$

$$g_5(\mathbf{x}) \equiv d_s > (d_s)_{\min} \tag{9.2.2.21}$$

$$g_6(\mathbf{x}) \equiv f_{s'} \le f_y \tag{9.2.2.22}$$

$$g_7(\mathbf{x}) \equiv f_s \le f_y \tag{9.2.2.23}$$

$$(A_{\text{total}})_L \le A_{\text{total}} \le (A_{\text{total}})_U \tag{9.2.2.24a}$$

$$b_L \le b \le b_U \tag{9.2.2.24b}$$

$$d_L \le d \le d_U \tag{9.2.2.25c}$$

The objective function, Eq. (9.2.2.16) is a weighted sum of the steel reinforcement and the cross-sectional area—w_1 and w_2 represent the weights. In other words, we are minimizing the amount of material used. If the weighting constants represent the unit cost of steel and concrete, then the objective function is a measure of the cost of the column.

EXAMPLE 9.2.8 ***Design of a Short Column***

Design the reinforcements of a short rectangular tied column to support $P_u = 500$ k and $M_u = 250$ k-ft. The material properties are $f_c' = 4000$ psi and $f_y = 60000$ psi. The cross-section dimensions are shown in Fig. E9.2.8(a).

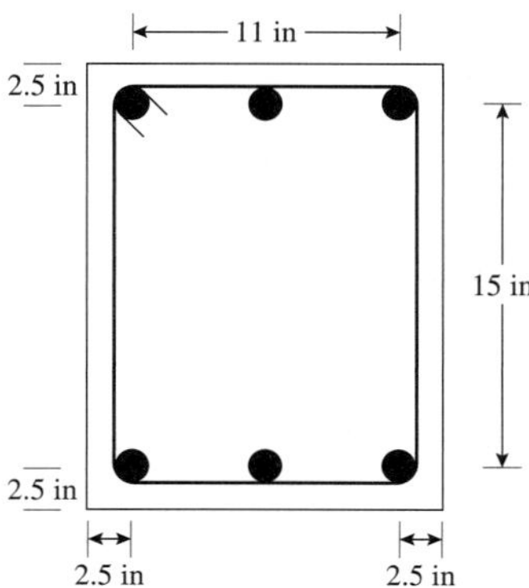

Fig. E9.2.8(a)

SOLUTION We will solve the design problem using the ACI Design Interaction Diagram.

Step 1: Compute the nominal loads and cross-sectional properties. For the given column,

$$P_n = \frac{P_u}{\phi} = (500/0.7) = 714 \text{ k} \qquad M_n = \frac{M_u}{\phi} = (250/0.7) = 357 \text{ k-ft.}$$

Also,

$$\gamma = \frac{h - d' - d_s}{h} = (20 - 2.5 - 2.5)/20 = 0.75 \qquad e = (357)(12)/714 = 6'' \qquad \frac{e}{h} = 6/20 = 0.3$$

Hence,

$$\frac{\phi P_n}{A_g} = (0.7)(714)/(16)(20) = 1.56 \qquad \left(\frac{\phi P_n}{A_g}\right)\left(\frac{e}{h}\right) = ((0.7)(714)/(16)(20))0.3 = 0.47$$

Step 2: Refer to the ACI Diagram (E4-60.75). Once we locate the point (0.47, 1.56) on the graph, we can estimate to find $\rho = 0.02$. Hence

$$A_s = (0.02)(16)(20) = 6.4 \text{ in}^2$$

We can use eight #8 bars = 6.32 in^2 (four bars in each face), as in Fig.E9.2.8(b).

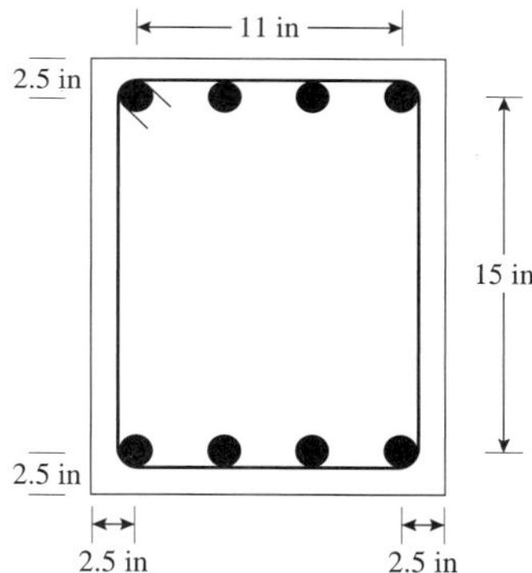

Fig. E9.2.8(b)

Step 3: With this design we should check the rest of the requirements.

Maximum axial force: $\phi P_n = 0.80\phi\left[0.85 f_c'(A_g - A_{st}) + f_y A_{st}\right]$

$\phi P_n = 0.80(0.8)[0.85(4)(320 - 6.32) + (60)(6.32)] = 925 \text{ k} > 500 \text{ k}$. OK.

Minimum reinforcement percentage: (0.01)(16)(20) = 3.2 in^2 < 6.4 in^2. OK
Maximum reinforcement percentage: (0.08)(16)(20) = 25.6 in^2 > 6.4 in^2. OK.
Minimum clear spacing of reinforcements: (11/3) – 1 = 2.7 in > 1.5 in. OK.
Number of bars: 8 > 4. OK.

EXAMPLE 9.2.9 ***Design of a Short Column***

Design a short rectangular tied column to support $P_y = 500$ k and $M_u = 250$ k-ft. The material properties are $f_c' = 4000$ psi and $f_y = 60000$ psi.

SOLUTION

Step 1: Initial dimensions. Let us assume that $\rho_t = 0.015$. Then

$$A_g = \frac{P_u}{0.45\left(f_c' + f_y \rho_t\right)} = \frac{500}{0.45(4 + 60 \times 0.015)} = 227 \text{ in}^2$$

Let us assume a square section of sides 17 in. A larger dimension is needed to take into account the presence of bending moment. Let us also assume that the cover on all sides is 2.5 in.

Step 2: Compute the nominal loads and cross-sectional properties. For the given column,

$$P_n = \frac{P_u}{\phi} = \frac{500}{0.7} = 714 \text{ k} \qquad M_n = \frac{M_u}{\phi} = \frac{250}{0.7} = (250/0.7) = 357 \text{ k-ft.}$$

Also,

$$\gamma = \frac{h - d' - d_s}{h} = \frac{17 - 2.5 - 2.5}{17} = 0.71 \qquad e = \frac{(357)(12)}{714} = 6 \text{ in} \qquad \frac{e}{h} = \frac{6}{17} = 0.35$$

Hence,

$$\frac{\phi P_n}{A_g} = \frac{(0.7)(714)}{(17)(17)} = 1.73 \qquad \left(\frac{\phi P_n}{A_g}\right)\left(\frac{e}{h}\right) = \frac{(0.7)(714)}{(17)(17)} 0.35 = 0.61$$

Step 3: Refer to the ACI Diagram (E4-60.75). Once we locate the point (0.61, 1.73) on the graph, we can estimate to find $\rho = 0.03$. Hence

$$A_s = (0.03)(17)(17) = 8.67 \text{ in}^2$$

We can use six #11 bars = 9.36 in^2 (three bars in each face), as in Fig. E9.2.9.

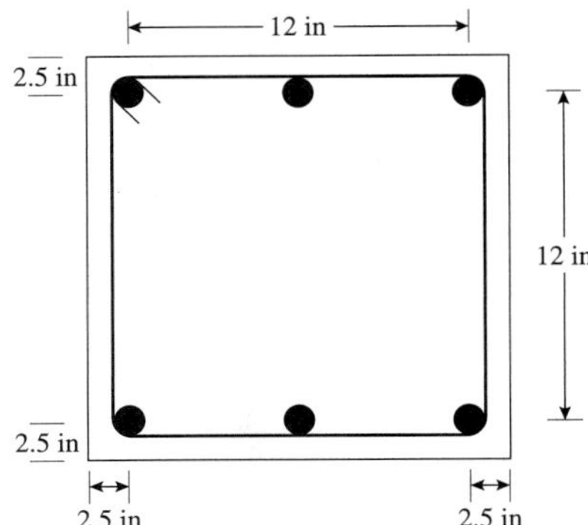

Fig. E9.2.9

Step 4: With this design we should check the rest of the requirements.
Minimum reinforcement percentage: (0.01)(17)(17) = 2.9 in^2 < 9.36 in^2. OK
Maximum reinforcement percentage: (0.08)(17)(17) = 23.1 in^2 > 9.36 in^2. OK.
Minimum clear spacing of reinforcements: (12/2) – 1.56 = 4.44 in > 2.1 in. OK.
Number of bars: 6 > 4. OK.

Design of Indeterminate Reinforced Concrete Structures

In Chapter 4 we examined in some detail the design process for simple determinate beams. In Chapter 8 we formulated the design problem as a mathematical programming problem. The process was applicable to both determinate and indeterminate structures. In this section, we will look in some detail at the trial-and-error procedure for the (sizing) design of indeterminate reinforced structures.

Typically, the detailed design process starts with the structural geometry (material, members, joints, support conditions) and loading defined. The designer must guess the initial size of the members in the structure. For concrete members, one must choose a shape (rectangular, circular, T-beam), the dimensions for that shape, and the amount of reinforcement. Designers rely on their experience to make this choice. The experience may take into account factors such as cost and constructability. Having established the member sizes, one must carry out a structural analysis to compute the following response quantities: (a) support reactions, (b) the internal force distribution in the form of the shear force and bending moment diagrams, and (c) key deflections.

The results from the structural analysis are then used to carry out the design checks as we have seen above in some of the examples. It is possible that (a) the checks are satisfied with a large margin of safety, (b) the checks are just about satisfied, or (c) one or more checks are not satisfied. In case (c), a redesign is required. In cases (a) and (b), the designer must then decide whether the savings offsets the redesign cost. Note that in the case of indeterminate structures, changing the cross-sectional properties of one member can lead to a redistribution of the internal forces.

EXAMPLE 9.2.10 ***Design of an Indeterminate Concrete Frame***

Design the concrete frame shown in Fig. E9.2.10.

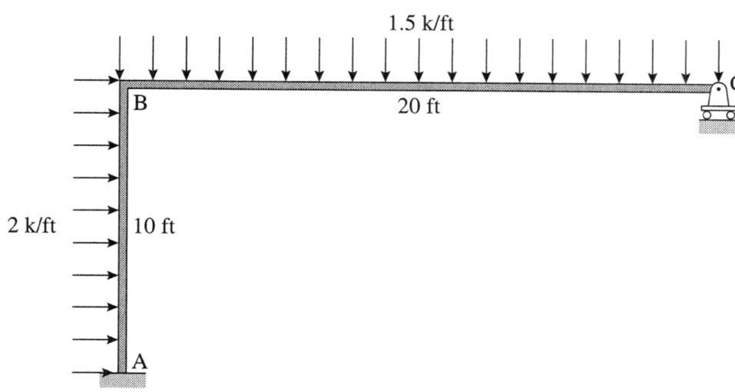

Fig. E9.2.10

SOLUTION

Step 1: Initial guess.

We will assume that $I_{BC} = 2I_{AB}$. Using the method for analysis for an indeterminate frame (see, for example, Example 5.1.6), we obtain the following response values.

Beam BC: Maximum bending moment is $M_{max} = 252$ k-ft with tension in the top fiber.
Maximum shear force is $V_{max} = 27.5$ k at joint B.

Column AB: Maximum bending moment is $M_{max} = 350$ k-ft with tension on left side.
Maximum axial force is $N_{max} = 27.5$ k (compression)

Step 2: Design of beam BC.

Let us assume the beam dimensions to be $b = 12$ in and $d = 18.5$ in. The material properties are $f'_c = 4000$ psi and $f_y = 60000$ psi. We can now compute the following quantities.

$$M_u = 252 \text{ k-ft} = 3.024(10^6) \text{ lb-in}$$

$$M_n = \frac{M_u}{\phi} = 3.36(10^6) \text{ lb-in}$$

$$\beta_1 = 0.85$$

$$\rho_b = \frac{0.85\beta_1 f'_c(87000)}{f_y(f_y + 87000)} = 0.0285$$

Substituting into Eqs. (9.2.1.20) and (9.2.1.21) we obtain $a^2 - 37a + 164.7 = 0$. Solving, $a = 5.175$ in. Furthermore, $c = a/\beta_1 = 6.59$ in. But, $\rho d f_y / 0.85 f'_c = a$. Hence, $\rho = 0.0158$. Finally,

$$A_s = \rho b d = 3.52 \text{ in}^2$$

We can use three #10 bars than have a total area of 3.81 in^2. On the compression side, we provide a minimum steel reinforcement $\rho_{min} = 0.792$ in^2.

Step 3: Shear reinforcements.

We know V_{max} = 27500 lb. Using Eq. (9.2.1-68), we obtain V_c = 28100 lb. Since V_{max} > (1/2)ϕV_c, stirrups are needed. Since $V_{max} < \phi V_c + 8\sqrt{f_c'}bd = 119300$ lb, beam dimensions are adequate. Let us assume two #3 bars have a total area of A_v = 0.22 in^2. Since $V_u \geq \phi V_c$,

$$s = \frac{A_v f_y d}{\frac{V_u}{\phi} - V_c} = 57.2 \text{ in}$$

Now $(V_u/\phi) - V_c < 4\sqrt{f_c'}bd = 56200$ lb. Hence $s \leq d/2$ = 9.25 in. Therefore provide A_v at a spacing of 9.25 in. We can obtain a more economical design by increasing the spacing over the length of the beam as shear force decreases from B to C.

Step 4: Design of column AB.

Let us assume the column dimensions to be b = 15 in and d = 12.5 in. The material properties are as follows f_c' = 5000 psi and f_y = 60000 psi. We will provide the same amount of steel in compression and in tension, i.e., $A_s = A_{s'}$. We can now compute the following quantities.

$$M_u = 350 \text{ k-ft} = 4.2(10^6) \text{ lb-in}$$

$$M_n = \frac{M_u}{\phi} = 4.666(10^6) \text{ lb-in}$$

$$P_u = 27500 \text{ lb}$$

$$P_n = 30550 \text{ lb}$$

$$\beta_1 = 0.85$$

$$\rho_b = \frac{0.85\beta_1 f_c'(87000)}{f_y(f_y + 87000)} = 0.0335$$

$$c_b = \frac{87000d}{(f_y + 87000)} = 11.836 \text{ in}$$

$$a_b = \beta c_b = 10.06 \text{ in}$$

$$f_s = f_y = 60000 \text{ psi}$$

$$f_s' = \frac{87000(c - d')}{c} = 66619.67 > 60000 \text{ psi}. \text{ Hence,}$$

$$f_s' = 60000 \text{ psi}$$

$$P_{nb} = 0.85 f_c' b a_b = 641325 \text{ lb} > P_n$$

Tension failure governs the design. We must now solve Eq. (9.2.2.18) to obtain the value of c as follows.

$A = 51000$ $D = 3.06(10^9)$ $G = 4.437(10^9)$ $E = 1.329(10^9)$

$H = 4.06(10^{11})$

Substituting, $c^3 + 60c^2 + 238.408c - 1782.456 = 0$. Solving, c = 3.91 in. Hence,

$$f_s' = \frac{87000(3.91 - 2.5)}{3.91} = 28828 \text{ psi}$$

$$A_s' = A_s = \frac{(P_n - 0.85 f_c' \beta_1 bc)}{(f_s' - f_y)} = 3.91 \text{ in}^2$$

Provide three #11 bars that have a total area of 4.68 in^2.

Step 5: Lateral ties.

Since the main reinforcements are #11 bars, provide ties as #4 bars with spacing of (16)1.56 = 25 in.

Step 6: Check assumptions

1. Moment of inertia

For beam: $$I_g = \frac{bh^3}{12} = 9261 \text{ in}^4 \qquad I_{gt} = \frac{bc^3}{3} + nA_s(d-c)^2 = 5490 \text{ in}^4$$

$$M_{cr} = \frac{f_r I_g}{\bar{y}} = 381714 \text{ lb-in} \qquad M_a = 3.024(10^6) \text{ lb-in}$$

$$I_e = I_{gt} + \left[\frac{M_{cr}}{M_a}\right]^3 (I_g - I_{gt}) = 5495 \text{ in}^4$$

For column: $$I_g = \frac{bh^3}{12} = 4219 \text{ in}^4$$

$$I_{gt} = \frac{bc^3}{3} + nA_s(d-c)^2 - nA_s'(c-d')^2 = 2715 \text{ in}^4$$

$$M_{cr} = \frac{f_r I_g}{\bar{y}} = 298311 \text{ lb-in}$$

$$M_a = 4.2(10^6) \text{ lb-in}$$

$$I_e = I_{gt} + \left[\frac{M_{cr}}{M_a}\right]^3 (I_g - I_{gt}) = 2715 \text{ in}^4$$

Hence, I_{BC}/I_{AB} = 5495/2715 = 2.02. This is close to the original assumption.

2. Self-weight: Taking the weight of concrete as 150 lb/ft^3, the weight of beam BC is 4625 lb and the weight of column AB is 1950 lb. The total weight of the structure is 6.6 k compared to a total applied load of 50 k. Redesign is perhaps necessary.

Closure

The ideas explored in this section on design of concrete structural components are introductory in nature. They are very useful in understanding not only concrete as a structural material but also the important link between structural analysis, structural design, and optimal design. In other words, the link between code-based concrete design and the rest of the book is established. We have not covered a variety of topics necessary for a better understanding of concrete design—establishing properties of concrete via tests, beams and columns that have other shapes (T-beams, circular columns etc.), one and two-way slabs, torsion, biaxial bending of columns, slender columns, footings, and indeterminate concrete systems, to name a few. Clearly, a separate course on concrete design is required for those interested in the design of reinforced and prestressed concrete systems.

EXERCISES

Whenever possible, also solve the following problems using the UCSD program and compare the results.

Appetizers

9.2.1. For the beam cross-section in Fig. P9.2.1, (i) determine what controls the design (crushing of concrete or yielding of steel), and (ii) whether ACI Code requirements are satisfied. Take $f_y = 60000$ psi.

(a) Assume that $f_c' = 4000$ psi, $A_s = 1.2\ \text{in}^2$.

(b) Assume that $f_c' = 6000$ psi, $A_s = 0.8\ \text{in}^2$.

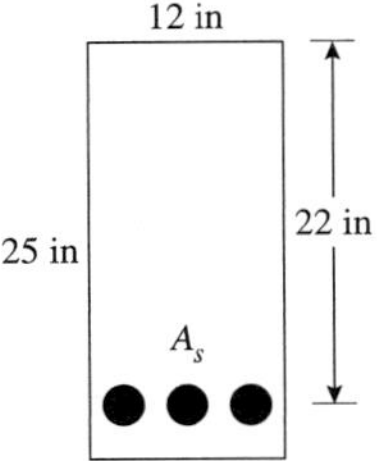

Fig. P9.2.1

9.2.2. For the beam dimensions in Fig. P9.2.1, determine whether the beam is underreinforced or overreinforced, if

(a) $f_y = 60000$ psi, $f_c' = 4000$ psi, and three #8 bars are used

(b) $f_y = 50000$ psi, $f_c' = 4000$ psi, and three #11 bars are used

(c) $f_y = 60000$ psi, $f_c' = 4000$ psi, and three #11 bars are used

9.2.3. Compute the nominal moment capacity of the beam in Fig. P9.2.3. Take $f_c' = 3000$ psi and $f_y =$ 50000 psi. Also compute its moment of inertia.

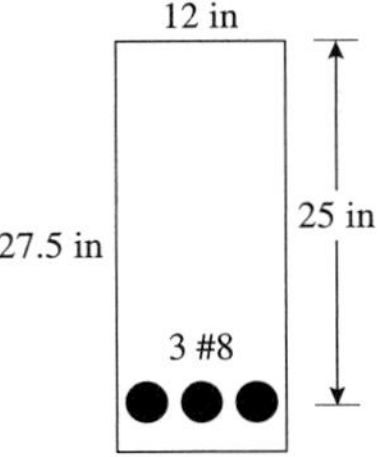

Fig. P9.2.3

9.2.4. What is the design moment capacity of the beam in Fig. P9.2.4? Take $f_c' = 3000$ psi and $f_y =$ 50000 psi.

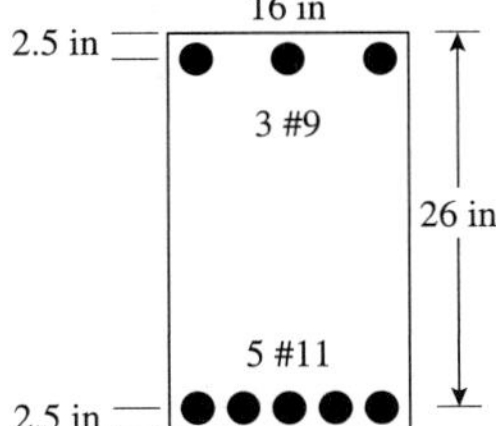

Fig. P9.2.4

9.2.5. For the column section in Fig. P9.2.5, use the interaction diagram to determine P_n given that $f_c' = 4000$ psi and $f_y = 60000$ psi.

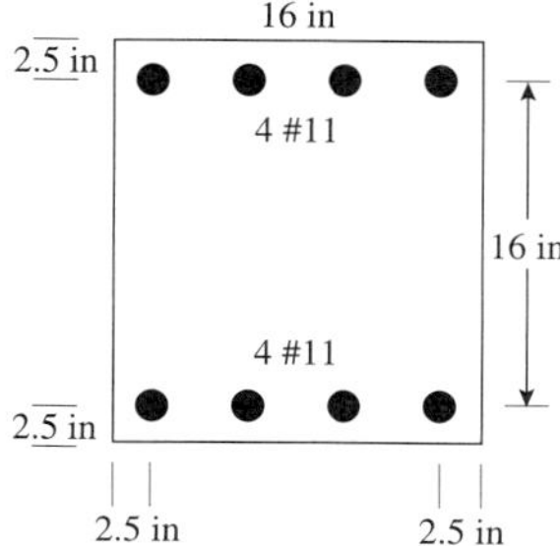

Fig. P9.2.5

Main Course

9.2.6. Design a rectangular simply supported beam of span 25 ft loaded by a (factored) uniformly distributed load of 3000 lb/ft. Take $f_c' = 4000$ psi and $f_y = 60000$ psi.

9.2.7. Design a rectangular simply supported beam of span 18 ft loaded by a (factored) uniformly distributed load of 10 k/ft. Also find the shear reinforcement requirement. Take $f_c' = 4000$ psi and $f_y =$ 60000 psi.

9.2.8. Design a rectangular beam with $b = 15$ in and $h = 30$ in for a dead load moment of 150 k-ft and a live load moment of 400 k-ft. Take $f_c' = 3000$ psi and $f_y = 60000$ psi.

9.2.9. Design a rectangular beam shown (including shear reinforcement) as in Fig. P9.2.9 to carry a dead load of 2 k/ft and a live load of 4 k/ft. Take $f_c' = 3000$ psi and $f_y = 60000$ psi. Compute the maximum deflection.

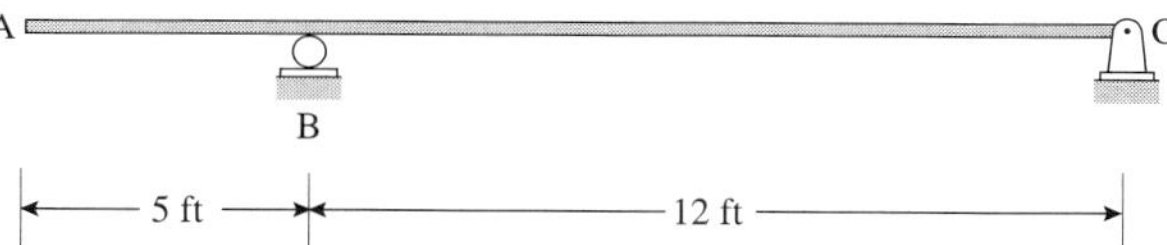

Fig. P9.2.9

9.2.10. Design a rectangular column to support $P_n = 390$ k and $M_n = 125$ k-ft. Take $f_c' = 3000$ psi and $f_y = 50000$ psi. Check the design using the interaction diagram.

9.2.11. Compute points in the Interaction Diagram that represent tension failure, compression failure and balanced failure conditions for a square column 20 in × 20 in, $f_c' = 3000$ psi and $f_y = 50000$ psi. The steel provided is three #11 bars on the compressive and on the tensile sides.

Structural Concepts

9.2.12. Rewrite the equations that implicitly use English units, using SI units instead.

SUMMARY

In this chapter we looked at design of steel and concrete structures: the design of steel structures as per the AISC Design Code, and the design of concrete structures as per ACI Design Code. These are the two most commonly used structural materials that are typically covered in separate follow-on courses to a course on structural analysis. It is hoped that the user has a better understanding and appreciation of the role structural analysis and structural design play in the engineering community.

We have come to the end of our journey as far as this book will take us. However, this should be just a chapter in our quest for life-long learning. Happy "structural engineering!"

Bibliography

Structural Analysis

J. M. Biggs, *Introduction to Structural Engineering—Analysis and Design*, Prentice–Hall, 1986.

M. Botwin and G. Murnen, *The Basics of Structural Analysis*, Engineering Press, Inc., 1986.

L. P. Felton and R. B. Nelson, *Matrix Structural Analysis*, Wiley, 1997.

S. E. French, *Determinate Structures*, Delmar Publishers, 1996.

R. C. Hibbeler, *Structural Analysis*, Prentice–Hall, 1999.

A. Kassimali, *Structural Analysis*, ITP, 1999.

J. C. McCormac and J. K. Nelson, *Structural Analysis—A Classical and Matrix Approach*, Addison–Wesley, 1997.

E. C. Rossow, *Analysis and Behavior of Structures*, Prentice–Hall, 1996.

R. L. Sack, *Structural Analysis*, McGraw–Hill, 1984.

D. L. Schodek, *Structures*, Prentice–Hall, 1998.

H. H. West, *Fundamentals of Structural Analysis*, Wiley, 1993.

W. Zalewski and E. Allen, *Shaping Structures: Statics*, Wiley, 1998.

Design

H. Allison, *Design Guide for Low- and Medium-Rise Steel Buildings*, AISC, 1991.

J. S. Arora, *Introduction to Optimum Design*, McGraw–Hill, 1989.

A. D. Belegundu and T. C. Chandrupatla, *Optimization Concepts and Applications in Engineering*, Prentice–Hall, 1999.

N. H. Cook, *Mechanics and Materials for Design*, McGraw–Hill, 1984.

D. E. Goldberg, *Genetic Algorithms*, Addison–Wesley, 1989.

W. F. Stoecker, *Design of Thermal Systems*, McGraw–Hill, 1989.

Structural Design

J. Ambrose, *Simplified Design of Building Structures*, Wiley, 1995.

J. Ambrose and D. Vergun, *Simplified Building Design for Wind and Earthquake Forces*, Wiley, 1995.

D. E. Breyer, *Design of Wood Structures*, McGraw–Hill, 1993.

S. Cooper and A. Chen, *Designing Steel Structures—Methods and Cases*, Prentice–Hall, 1985.

S. W. Crawley and R. M. Dillon, *Steel Buildings—Analysis and Design*, Wiley, 1993.

J. MacGregor, *Reinforced Concrete—Mechanics and Design*, Prentice–Hall, 1997.

J. McCormac, *Design of Reinforced Concrete*, Addison–Wesley, 1998.

C. Meyer, *Design of Concrete Structures*, Prentice–Hall, 1996.

E. G. Nawy, *Prestressed Concrete*, Prentice–Hall, 1996.

C-K. Wang and C. G. Salmon, *Reinforced Concrete Design*, Addison–Wesley, 1998.

T. V. Galambos, ed., *Guide to Stability Design Criteria for Metal Structures*, 4th ed., New York, Wiley, 1988.

D. E. Branson, *Deformation of Concrete Structures*, McGraw–Hill, New York, 1977.

Numerical Analysis

R. L. Burden and J. D. Faires, *Numerical Analysis*, PWS–Kent, 1988.

B. Carnahan, H. A. Luther and J. O. Wilkes, *Applied Numerical Methods*, Wiley, 1969.

S. C. Chapra and R. P. Canale, *Numerical Methods for Engineers*, McGraw–Hill, 1998.

J. Tuma, *Engineering Mathematics Handbook*, McGraw–Hill, 1987.

History of Engineering

D. P. Billington, *The Innovators—The Engineering Pioneers Who Made America Modern*, Wiley, 1996.

L. P. Grayson, *The Making of an Engineer*, Wiley, 1993.

R. S. Kirby, S. Withington, A. B. Darling and F. G. Kilgour, *Engineering in History*, Dover, 1990.

S. P. Timoshenko, *History of Strength of Materials*, Dover, 1952.

Mechanics of Materials

J. M. Gere and S. P. Timoshenko, *Mechanics of Materials*, PWS Publishing Company, 1997.

E. P. Popov, *Introduction to Mechanics of Solids*, Prentice–Hall, 1968.

W. F. Riley and L. W. Zachary, *Introduction to Mechanics of Materials*, Wiley, 1989.

Other References

Y. Jaluria, *Design and Optimization of Thermal Systems*, McGraw–Hill, 1998.

S. D. Rajan, B. Mobasher, S-Y. Chen and C. Young, "Cost-based design of residential steel roof systems: A case study," *Structural Engineering and Mechanics*, vol. 8(2):165–180, 1995.

V. H. Cochrane, "Rules for rivet hole deductions in tension members," *Engineering News Record*, 89:847–848, 1922.

F. Engesser, "Über die Knickfestigkeit gerader Stabe," *Zietschrift des Architeketen-und Ingenieur-Vereins zu Hannover*, 35:455–462, 1989.

A. Considere, "Résistance de pièces comprimées," *Congrès International des Procèdes de Construction*, Paris, 3, 371, 1891.

C. S. Whitney, "Plastic Theory of Reinforced Concrete Design," *Transactions of the ASCE*, 107:251–326, 1942.

Answers to Selected Problems

CHAPTER 1 SUMMARY EXERCISES

1.1. (a) 1 lb = 4.448 N (b) 1 m^3 = 35.3 ft^3 (c) $F_{\mathrm{CD}} = 8.33$ kN(C)

(d) 1 N/m = 0.0057 lb/in (e) 1 m / m -°C = 0.56 in / in -° F (f) 1 N-m = 8.85 lb-in

(g) 1 MPa = 145.05 psi

1.2. (a) (i) About 4–5 lb (ii) About 25–30 lb (iii) About 15 lb (iv) 8.3 lb

(b) About 10 ft to 12 ft

(c) About 12 ft

CHAPTER 2 EXERCISES

2.2.7. Resultant, R = 365 lb acting at the center of the beam.

2.2.8. The resultant and the location of the trapezoidal loading can be checked against the expressions for (a) uniformly distributed loading with $q_L = q_R = q$ and (b) triangular loading with $q_L = 0$ and $q_R = q$.

2.5.1. Beam is determinate. $A_x = 20$ kN(←), $A_y = 17.3$ kN(↑), and $B_y = 17.3$ kN(↑).

2.5.2. The beam is statically indeterminate to degree one.

2.5.4. Frame is determinate. $B_x = 2.01$ k(←), $B_y = 25.98$ k(↑), and $C_x = 12.99$ k(←).

2.5.5. $A_x = 0$, $A_y = 25$ k(↑), $M_A = 166.75$ k-ft(↻), $B_x = 0$, $B_y = 0$, and $C_y = 0$.

2.5.6. The beam is statically indeterminate to degree one.

2.5.8. Structure is determinate. $R_A = 21.93$ k, $B_x = 15$ k(→), and $B_y = 16$ k(↑).

2.5.9. $A_x = 0$, $A_y = 29.2$ k(↑), and $C_y = 45.8$ k(↑).

2.5.11. Structure is determinate. $A_x = 66.7$ k(→), $A_y = 50$ k(↑), $B_x = 66.7$ k, and $B_y = 0$.

2.7.1. $F_{\mathrm{BC}} = 16.7$ kN(T) $F_{\mathrm{BA}} = 13.3$ kN(C)

2.7.2. $F_{\mathrm{AC}} = 7800$ lb(C) $F_{\mathrm{AB}} = 7200$ lb(T) $F_{\mathrm{BC}} = 6000$ lb(C)

$F_{\mathrm{BD}} = 7200$ lb(T) $F_{\mathrm{CD}} = 7800$ lb(T) $F_{\mathrm{CE}} = 15600$ lb(C)

$F_{\mathrm{ED}} = 6000$ lb(T)

2.7.3. The truss is unstable.

2.7.6. $F_{AC} = F_{IG} = 0$ $F_{AB} = F_{IJ} = 0$ $F_{BC} = F_{JG} = 4242.6\text{ lb}(C)$

$F_{BD} = F_{JH} = 3000\text{ lb}(T)$ $F_{CD} = F_{GH} = 3000\text{ lb}(C)$ $F_{CE} = F_{GE} = 3000\text{ lb}(C)$

$F_{DE} = F_{HE} = 1220.8\text{ lb}(T)$ $F_{DF} = F_{HF} = 2300\text{ lb}(T)$ $F_{FE} = 2000\text{ lb}(C)$

2.7.8. $F_{AB} = F_{ED} = 3606.5\text{ lb}(C)$ $F_{AH} = F_{EF} = 3001.1\text{ lb}(T)$ $F_{BH} = F_{DF} = 838.6\text{ lb}(C)$

$F_{BC} = F_{DC} = 3155.6\text{ lb}(C)$ $F_{HC} = F_{FC} = 838.6\text{ lb}(T)$ $F_{HC} = F_{FC} = 2250.6\text{ lb}(T)$

$F_{GC} = 0$

2.7.10. $F_{AB} = F_{IJ} = 24000\text{ lb}(C)$ $F_{AC} = F_{IG} = 0$ $F_{BC} = F_{JG} = 25456\text{ lb}(T)$

$F_{BD} = F_{JH} = 18000\text{ lb}(C)$ $F_{DC} = F_{HG} = 12000\text{ lb}(C)$ $F_{CF} = F_{GF} = 8485\text{ lb}(C)$

$F_{CE} = F_{GE} = 24000\text{ lb}(T)$ $F_{DF} = F_{HF} = 18000\text{ lb}(C)$ $F_{EF} = 0$

2.7.11. It should be clear that we *cannot* solve the two equations of equilibrium since they are linearly dependent.

2.7.13. $F_{GH} = 25.6\text{ k}(C)$ $F_{CD} = 16\text{ k}(T)$ $F_{GD} = 10\text{ k}(T)$

2.7.15. $F_{DF} = 2300\text{ lb}(T)$ $F_{DE} = 1220.8\text{ lb}(T)$

2.7.17. $F_{GB} = F_{GD} = 15.8\text{ k}(C)$ $F_{GF} = F_{GH} = 45\text{ k}(T)$ $F_{GC} = 20\text{ k}(T)$

2.7.18. $F_{KJ} = 4.8\text{ k}(C)$ $F_{KI} = 63.8\text{ k}(C)$ $F_{JM} = 67.5\text{ k}(T)$

$F_{ST} = 24\text{ k}(T)$

2.7.20. Yes (with some modifications). When placing a bridge pier is extremely difficult or costly, the two halves of the bridge can be cantilevered from the ends.

2.7.21. (a) Unstable (b) Unstable

2.8.1. Center of beam: $V = 2.78\text{ kN}(\downarrow)$ $M = 27.5\text{ kN-m}(\curvearrowleft)$

2.8.4. Left of Point B: $V = 1.86\text{ kN}(\downarrow)$ $M = 34.3\text{ kN-m}(\curvearrowleft)$

Right of Point B: $V = 8.14\text{ kN}(\uparrow)$ $M = 4.3\text{ kN-m}(\curvearrowleft)$

Left of Point C: $V = 12.14\text{ kN}(\uparrow)$ $M = 15.98\text{ kN-m}(\curvearrowright)$

Right of Point C: $V = 8\text{ kN}(\downarrow)$ $M = 15.98\text{ kN-m}(\curvearrowright)$

2.8.5. Left of Point C: $V = 13.08\text{ kN}(\downarrow)$ $M = 43.74\text{ kN-m}(\curvearrowleft)$

Right of Point C: $V = 3.08\text{ kN}(\downarrow)$ $M = 43.74\text{ kN-m}(\curvearrowleft)$

2.8.7. $(V_B)_{\text{left}} = 1500\text{ lb}(\downarrow) = (V_B)_{\text{right}}$

$(M_B)_{\text{left}} = (M_B)_{\text{right}} = 0$

2.8.8. $V = 0$ $M = 10\text{ kN - m}(\curvearrowleft)$

2.8.14. $V(x) = -\dfrac{3x^2}{2} - 4x$ $M(x) = -\dfrac{x^3}{2} - 2x^2$

2.8.15. (a) True (b) True (c) True

2.8.16.

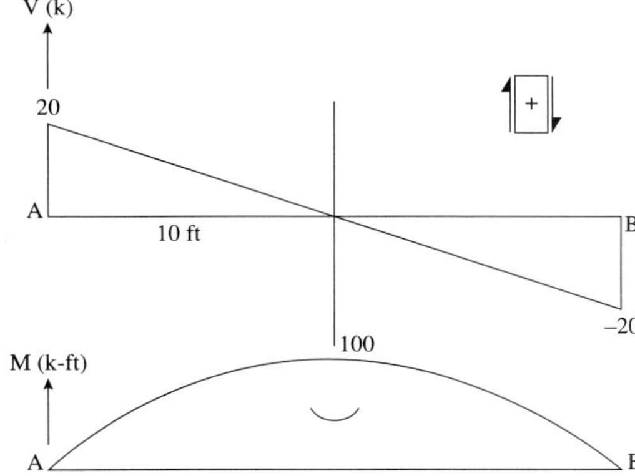

2.8.18.

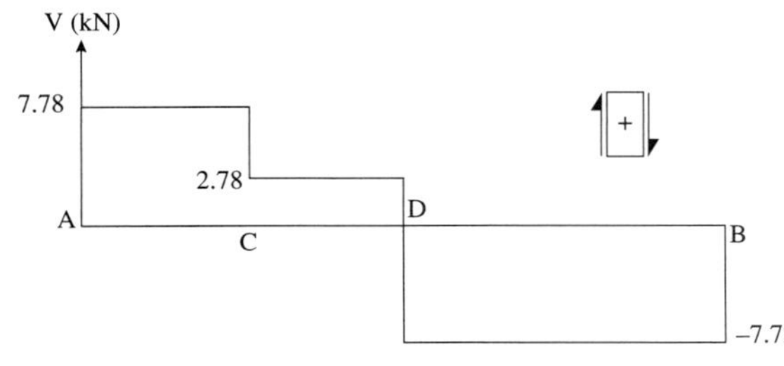

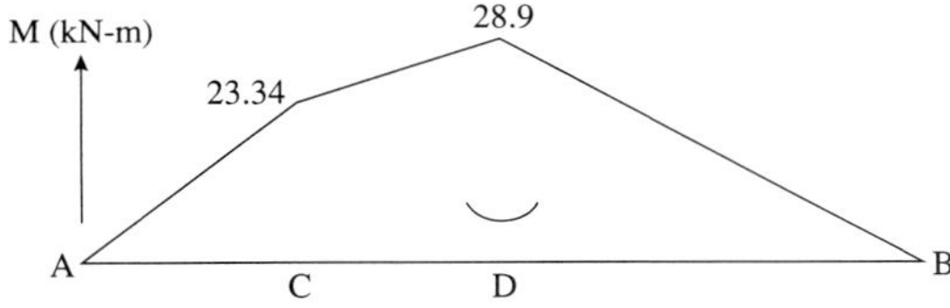

2.8.21.

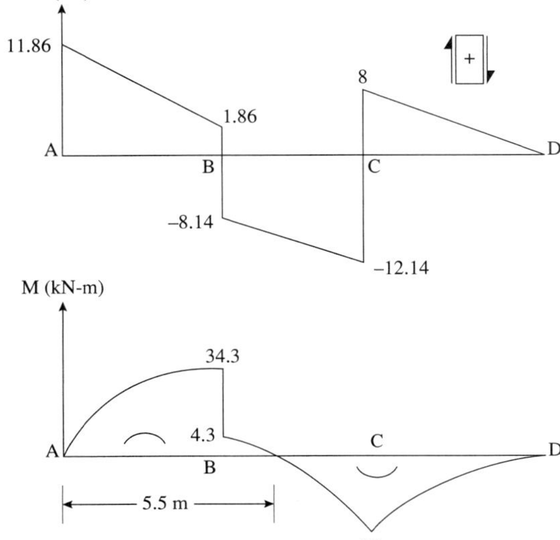

2.8.22.

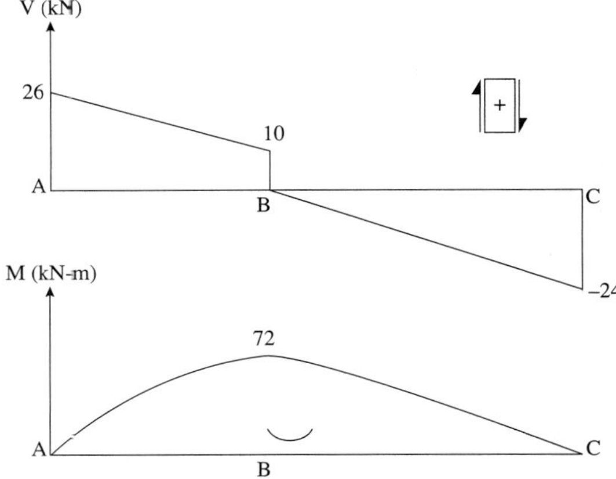

2.8.23.

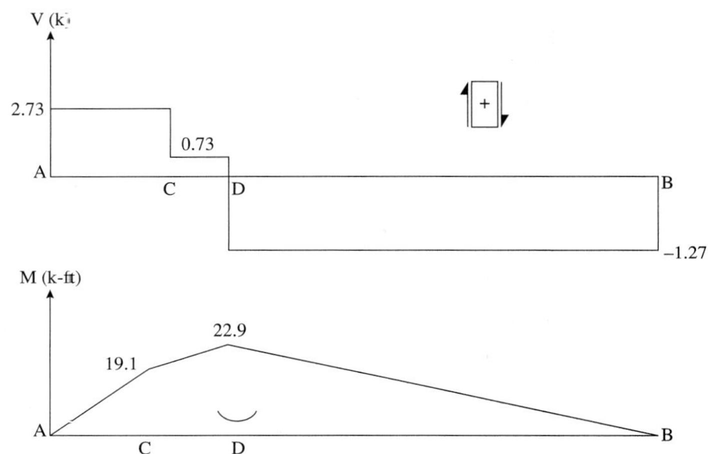

2.8.26. We will assume that the 3000 lb force transmitted to the foundation slab is approximated as a point load.

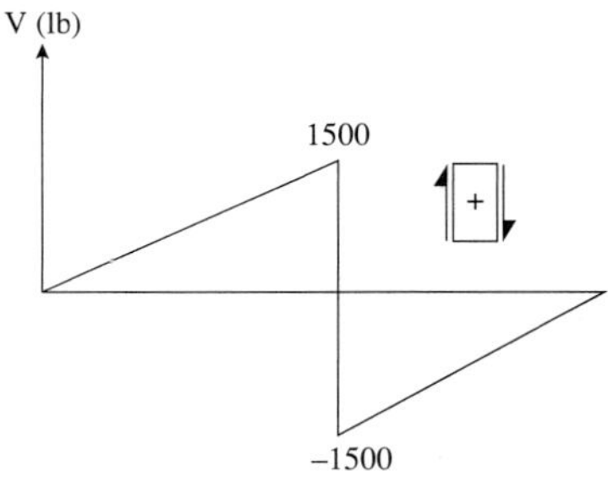

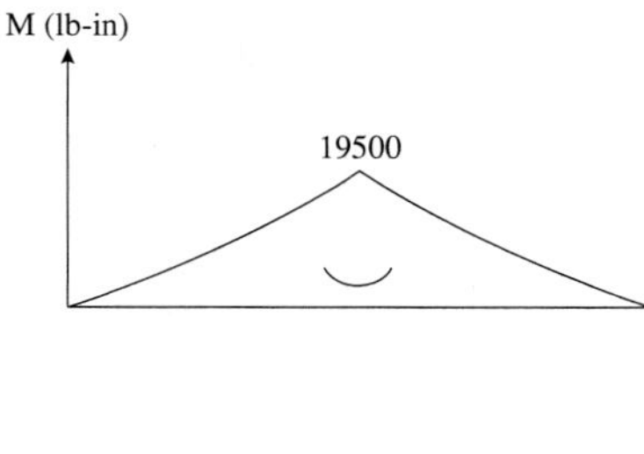

2.8.28.

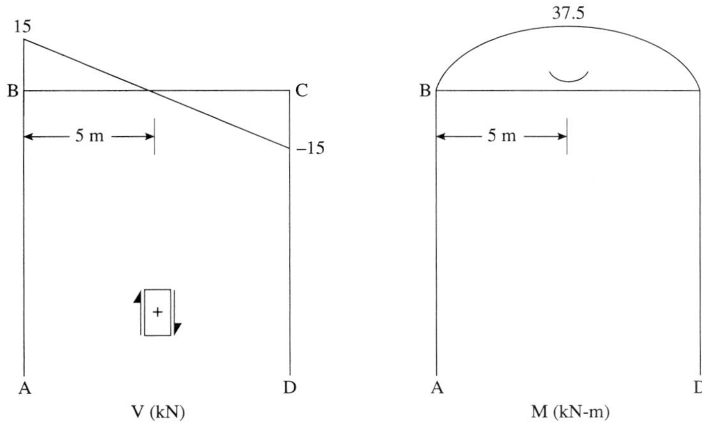

2.8.29.

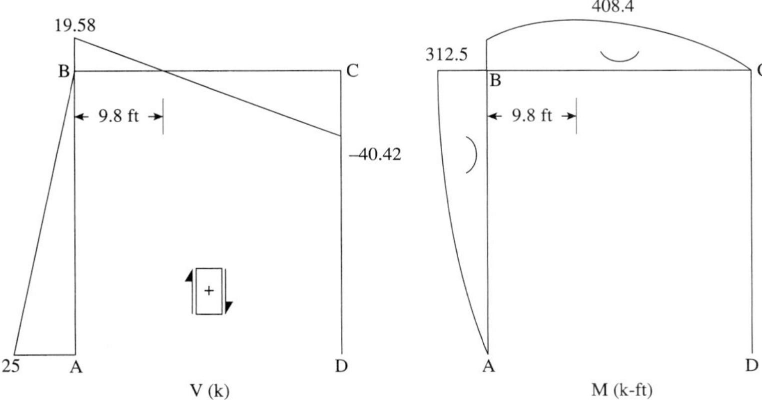

2.8.32.

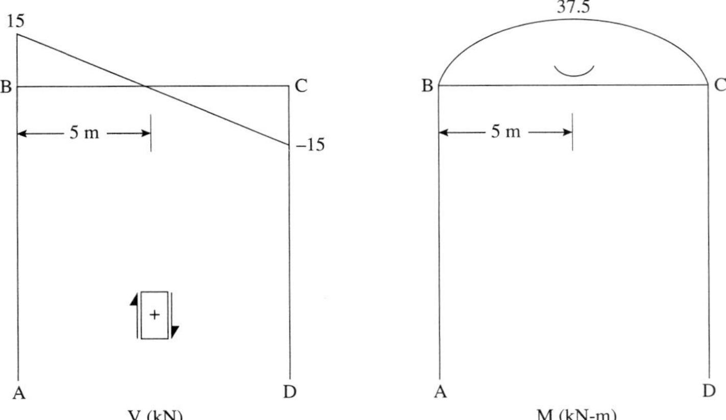

2.8.33. The frame is unstable.

2.8.36. There is at least one reason why axial force diagrams are not drawn. It is unusual to find problems where the axial force varies within a member. In other words, for most problems, it is safe to assume that the axial force (whether compression or tension) is constant along the length of a member.

CHAPTER 2 SUMMARY EXERCISES

2.1. $F_{GH} = F_{GF} = 4942\ \text{lb}(C)$ $F_{GB} = F_{GC} = 2099\ \text{lb}(T)$

2.3. $F_{AG} = 24.79\ \text{kN}(C)$ $F_{GI} = 22.5\ \text{kN}(C)$ $F_{IM} = 20.4\ \text{kN}(C)$ $F_{MO} = 18\ \text{kN}(C)$

$F_{FH} = 11.27\ \text{kN}(C)$ $F_{HJ} = 11.27\ \text{kN}(C)$ $F_{JN} = 11.27\ \text{kN}(C)$ $F_{NO} = 11.27\ \text{kN}(C)$

$F_{AB} = 20.63\ \text{kN}(T)$ $F_{BC} = 16.9\ \text{kN}(T)$ $F_{FE} = 9.38\ \text{kN}(T)$ $F_{ED} = 9.38\ \text{kN}(T)$

$F_{DC} = 9.38\ \text{kN}(T)$ $F_{KO} = 12.6\ \text{kN}(T)$ $F_{KC} = 8.44\ \text{kN}(T)$ $F_{LO} = 0$

$F_{LD} = 0$ $F_{IK} = 3.8\ \text{kN}(T)$ $F_{LJ} = 0$ $F_{CI} = 8.44\ \text{kN}(C)$

$F_{DJ} = 0$ $F_{BI} = 4.19\ \text{kN}(T)$ $F_{EJ} = 0$ $F_{MK} = 4.19\ \text{kN}(C)$

$F_{NL} = 0$ $F_{BG} = 4.19\ \text{kN}(C)$ $F_{EH} = 0$

2.7. One could assume that the structure is either a truss or a beam/frame. However, with either assumption, one would find that the structure is *unstable*. Drawing the FBDs will show that the equations of equilibrium are not satisfied.

2.8. This problem violates the small-displacement assumption. The joint is not in equilibrium in the displaced or deformed position.

CHAPTER 3 EXERCISES

3.1.1. The average normal strain $\varepsilon = \dfrac{\Delta L}{L} = \dfrac{0.4\,m}{10\,m} = 0.04$.

3.1.3. Shear modulus, $G = \dfrac{E}{2(1+\upsilon)} = \dfrac{200}{2(1+0.3)} = 76.9\ \text{GPa}$

3.1.4. $\sigma_{max} = -585\ \text{psi}$ located at an angle of 29.5° with respect to the x-axis.

$\sigma_{min} = -6415\ \text{psi}$ located at an angle of 119.5° with respect to the x-axis.

$\tau_{max} = 3208\ \text{psi}$ and the corresponding normal stress is $\sigma = -3208\ \text{psi}$.

3.1.5. (a) True. Strain has units of length/length.

(b) False. Isotropic implies that the properties at a point are the same in every direction. Homogenous implies that the properties are the same at every point in the body.

(c) False. There are permanent deformations and strain in a specimen that is loaded beyond the elastic limit.

3.2.2. $I = 739.6\ \text{in}^4$

3.2.4. $\sigma_{max} = 6680\ \text{psi}$ $\tau_{max} = 939\ \text{psi}$

3.2.5. Normal stress due to bending moment and axial force: $\sigma_{max} = 1.41(10^6)\ \text{psf}$ in member AB

Shear stress due to shear force: $\tau_{max} = 208632\ \text{psf}$ in member BC

3.2.6. The spacing of the screws must be less than 1.66 in.

3.6.1. Dead load is 90 psf.

3.6.2. Beam AB: $w = 2$ kN/m

Beam CD: $w = 4$ kN/m

Beam EF: First 5 m, $w = 4$ kN/m. Last 3 m, $w = 2$ kN/m

3.6.3. $R_{\text{beam}} = 11714$ lb $\quad R_{\text{girder}} = 37156$ lb $\quad N_{\text{column}} = 97740$ lb

3.6.8. Since the structure is indeterminate, the results are obtained by analyzing the beam using the GS-USA Frame program.

Member	Largest Shear Force (lb)	Largest Bending Moment on the Top Fiber (lb-ft)	Largest Bending Moment on the Bottom Fiber (lb-ft)
AB	12400	40500	46700
BC	11700	30000	46700
CD	12400	40500	46700

CHAPTER 3 SUMMARY EXERCISES

3.1. The largest load is 72 k.

3.2. $\sigma = 3.03(10^9)$ Pa $> 2(10^6)$ Pa

$\tau = 2.14(10^{10})$ Pa $> (10^6)$ Pa

3.5. W12 × 14 can be used for member ABC and W24 × 68 for member CD.

3.6. The required inner radius is 0.39 in with a wall thickness of 0.06 in.

CHAPTER 4 EXERCISES

4.2.1. (a) $\theta_B = \dfrac{a^2}{2EI}(\curvearrowright)$ $\quad \Delta_B = \dfrac{a^3}{3EI}(\downarrow)$

(b) $\theta_C = \dfrac{a^2}{2EI}(\curvearrowright)$ $\quad \Delta_C = \dfrac{a^2}{2EI}\left(L - \dfrac{a}{3}\right)(\downarrow)$

4.2.2. $\theta_B = \dfrac{wL^3}{6EI}(\curvearrowright)$ $\quad \Delta_B = \dfrac{wL^4}{8EI}(\downarrow)$

4.2.4. (b) $I > 0.00125$ m^4

4.2.5. (a) $\theta_C = \dfrac{5a^2}{8EI}(\curvearrowleft)$ $\quad \Delta_C = \dfrac{5a^2 b}{8EI}(\uparrow)$

(b) $\Delta_D = \dfrac{10a^3}{48EI}(\downarrow)$

4.3.7. $\theta_{BL} = \dfrac{166.67}{EI}(\curvearrowright)$ $\theta_{BR} = \dfrac{26.04}{EI}(\curvearrowright)$

4.3.10. (a) $\theta_A = \dfrac{450}{EI}(\curvearrowright)$ (b) $\Delta = \dfrac{1977}{EI}(\rightarrow)$

4.4.1. (a) $U = 62.5$ N-m (b) $\Pi = U + WP = -62.5$ kN-m

4.4.3. $U = 0.525$ N-m

4.4.4. $U = \dfrac{w^2 L^5}{120EI}$ assuming constant EI

4.5.10. (a) $\theta_A = \dfrac{45}{EI}(\curvearrowright)$ (b) $\Delta_C = \dfrac{114.75}{EI}(\rightarrow)$

4.5.12. (a) $\Delta_C = \dfrac{22917}{EI}(\rightarrow)$ (b) $\theta_A = \dfrac{2667}{EI}(\curvearrowright)$

4.5.14. $\Delta_C = \dfrac{9188}{EI}(\downarrow)$

4.5.17. (a) $(\Delta_C)_y = 0.00108$ in$(\downarrow)$ (b) $(\Delta_D)_x = 0$

4.5.19. $(\Delta_D)_y = 6.6(10^{-5})$ m$(\downarrow)$ $(\Delta_E)_y = 1.35(10^{-4})$ m$(\downarrow)$

CHAPTER 5 EXERCISES

5.1.1. (a) The possible redundants are A_y, M_A, and C_y.

(b) $C_y = \dfrac{a^2(3L - a)}{2L^3} = 0.148$ k$(\uparrow)$

5.1.2. $M_A = \dfrac{wL^2}{8}(\circlearrowleft)$

5.1.3. (a) The possible redundants are A_y, B_y, and C_y.

(b) $A_y = 3.75$ kN$(\uparrow)$, $B_y = 8.75$ kN$(\uparrow)$, and $C_y = 2.5$ kN$(\downarrow)$

5.1.6. $A_y = 2604$ lb$(\uparrow)$, $M_A = 52080$ lb-ft$(\curvearrowright)$, and $B_y = 2604$ lb$(\downarrow)$

5.1.7. $A_y = 15104$ lb$(\uparrow)$, $M_A = 102080$ lb-ft$(\curvearrowright)$, and $B_y = 4895$ lb$(\uparrow)$

5.1.10. $\theta_C = \dfrac{27.3}{EI}(\curvearrowleft)$

5.1.11. $A_y = 9937.5$ lb$(\uparrow)$, $A_x = 5875$ lb$(\leftarrow)$, $C_y = 5062.5$ lb$(\uparrow)$, and $C_x = 9125$ lb$(\leftarrow)$

5.1.13. $A_x = 9375$ lb$(\leftarrow)$, $A_y = 7500$ lb$(\uparrow)$, $M_A = 28125$ lb-ft$(\circlearrowleft)$, $C_x = 5625$ lb$(\leftarrow)$, and $C_y = 7500$ lb$(\uparrow)$

5.1.14. $A_x = 6\ 64$ kN$(\leftarrow)$, $A_y = 6.67$ kN$(\downarrow)$, $D_y = 16.67$ kN$(\uparrow)$, and $D_x = 3.356$ kN$(\leftarrow)$

5.1.16. $A_y =$ 1[illegible]477 lb$(\uparrow)$, $A_x = 0$, $M_A = 230$ lb-ft$(\curvearrowright)$, and $C_y = 5523$ lb$(\uparrow)$

5.1.18. $A_x = 10$ kN$(\leftarrow)$, $A_y = 0.89$ kN$(\uparrow)$, $M_A = 58.93$ kN-m$(\curvearrowleft)$, and $D_y = 9.11$ kN$(\uparrow)$

5.1.19. $A_x = 0.108$ k(←), $A_y = 14.1$ k(↑), $M_A = 13.52$ k-ft(↶), and $C_x = 15.023$ k(←)

5.1.21. (a) All the three members can be used as a redundant.

(b) $F_{AD} = 8.33$ kN(T), $F_{BD} = 0$, and $F_{CD} = 8.33$ kN(C)

5.1.22. $F_{AB} = 6$ kN(C), $F_{AC} = 0$, $F_{BC} = 6.71$ kN(T), $F_{BD} = 6$ kN(C), and $F_{CD} = 6$ kN(C). The rest of the members can be found by symmetry.

5.1.25. $F_{AB} = 225$ lb(T), $F_{AC} = 1344$ lb(T), $F_{BC} = 201$ lb(T), $F_{BD} = 226$ lb(T), $F_{CD} = 1622$ lb(C), and

$F_{AD} = 890$ lb(T)

5.1.26. $F_{AB} = 0.5$ k(T), $F_{BC} = 0.5$ k(T), $F_{DE} = 0$, $F_{GH} = 2$ k(C), $F_{JK} = 1.5$ k(C), $F_{IJ} = 1.5$ k(C), $F_{BD} = 0$,

$F_{DG} = 1.0$ k(C), $F_{GJ} = 2.0$ k(C), $F_{CE} = 0.5$ k(T), $F_{EF} = 0.5$ k(T), $F_{FH} = 1.5$ k(C), and $F_{HK} = 1.5$ k(C)

5.1.28. $A_x = 0$, $A_y = 5.177$ kN(↑), $M_A = 4.46$ kN-m(↶), $B_y = 8.037$ kN(↑), and $C_y = 1.79$ kN(↑)

5.1.30. $F_{AB} = 1716$ lb(C), $F_{AC} = 3116$ lb(C), $F_{BC} = 2427$ lb(T), $F_{AD} = 3230$ lb(C), $F_{CD} = 460$ lb(C),

$F_{CF} = 1534$ lb(C), $F_{DE} = 908$ lb(T), $F_{CE} = 521$ lb(C), $F_{EF} = 744$ lb(C), $F_{DF} = 880$ lb(T), and

$F_{BD} = 3684$ lb(T)

5.2.8. $M_{AB} = -93.78$ k-ft, $M_{BA} = 91.43$ k-ft, $M_{BC} = -91.43$ k-ft, $M_{CB} = 165.83$ k-ft, $M_{CD} = -165.83$ k-ft,

and $M_{DC} = -144.3$ k-ft

5.2.9. $M_{AB} = 0$, $M_{BA} = 15$ kN-m, $M_{BC} = -15$ kN-m, $M_{CB} = 15$ kN-m, $M_{CD} = -15$ kN-m, and $M_{DC} = 0$

5.2.11. $M_{AB} = -37.5$ k-ft, $M_{BA} = 37.5$ k-ft, $M_{BC} = -37.5$ k-ft, $M_{CB} = 37.5$ k-ft, $M_{CD} = -37.5$ k-ft, and

$M_{DC} = 37.5$ k-ft

5.2.12. $M_{AB} = -210.3$ k-ft, $M_{BA} = -123$ k-ft, $M_{BC} = 123$ k-ft, $M_{CB} = 136.3$ k-ft, $M_{CD} = 136.3$ k-ft, and

$M_{DC} = -11.9$ k-ft

CHAPTER 5 SUMMARY EXERCISES

5.2. $A_y = 3.75$ k(↑), $B_y = 61.875$ k(↑), and $C_y = 24.375$ k(↑)

5.3. $C_x = 0$, $C_y = 14.86$ kN(↑), $M_C = 0.3$ kN-m(↶), and $D_y = 6.14$ kN(↑)

5.5. $I = 0.00418 \text{ m}^4$

5.9. $A_x = 0$, $A_y = 1.25$ kN(↓), $B_y = 12.5$ kN(↑), and $C_y = 8.75$ kN(↑)

CHAPTER 6 EXERCISES

6.1.1. (a) **A** is symmetric, but is not an upper triangular matrix.

(b) $\mathbf{A} - \mathbf{B} = \begin{bmatrix} -4 & -1 \\ -27 & -23 \end{bmatrix}$

(c) $\mathbf{A}-\mathbf{B}=\begin{bmatrix} -4 & -1 \\ -27 & -23 \end{bmatrix}$

(d) $\mathbf{A}_{2\times 2}\,\mathbf{a}_{1\times 2}$ cannot be computed. $\mathbf{a}_{1\times 2}\,\mathbf{A}_{2\times 2}$ can be computed.

6.1.2. (a) $\mathbf{A}_{3\times 2}\mathbf{C}_{2\times 3}=\begin{bmatrix} -12 & 104 & -54 \\ 0 & -48 & 27 \\ -60 & -200 & 135 \end{bmatrix}$

(b) $\mathbf{B}_{3\times 2}\mathbf{C}_{2\times 3}=\begin{bmatrix} 0 & -80 & 45 \\ -132 & 200 & -63 \\ 228 & -88 & -36 \end{bmatrix}$

(c) $\mathbf{A}^T_{2\times 3}=\begin{bmatrix} 1 & 0 & 5 \\ 6 & -3 & -15 \end{bmatrix}$ and $\mathbf{C}^T_{3\times 2}=\begin{bmatrix} -12 & 0 \\ 8 & 16 \\ 0 & -9 \end{bmatrix}$

(d) $\mathbf{C}^T_{3\times 2}\mathbf{A}_{3\times 2}$ cannot be computed.

(e) $\mathbf{d}_{1\times 2}\mathbf{C}_{2\times 3}-\mathbf{a}^T_{1\times 3}=\begin{bmatrix} -42 & -45 & 38 \end{bmatrix}$

6.2.2, 6.2.6, and 6.2.11.

Element 1: $(D_3, D_4, D_1, D_2)\!:\mathbf{k}_{4\times 4}=10^6\begin{bmatrix} 1.6116 & 0 & -1.6116 & 0 \\ 0 & 0 & 0 & 0 \\ -1.6116 & 0 & 1.6116 & 0 \\ 0 & 0 & 0 & 0 \end{bmatrix}$

Element 2: $(D_5, D_6, D_1, D_2)\!:\mathbf{k}_{4\times 4}=10^6\begin{bmatrix} 0.348 & 0.464 & -0.348 & -0.464 \\ 0.464 & 0.6187 & -0.464 & -0.6187 \\ -0.348 & -0.464 & 0.348 & 0.464 \\ -0.464 & -0.6187 & 0.464 & 0.6187 \end{bmatrix}$

System Equations: $10^6\begin{bmatrix} 1.9591 & 0.464 \\ 0.464 & 0.6187 \end{bmatrix}\begin{Bmatrix} D_1 \\ D_2 \end{Bmatrix}=\begin{Bmatrix} 0 \\ -4000 \end{Bmatrix}$

$\{D_1, D_2\}=10^{-3}\{1.86207, 7.86207\}$ in

Element 1: $\begin{Bmatrix} f_1' \\ f_2' \end{Bmatrix}=\begin{Bmatrix} -3000 \\ 3000 \end{Bmatrix}$ lb

Element 2: $\begin{Bmatrix} f_1' \\ f_2' \end{Bmatrix}=\begin{Bmatrix} 5000 \\ -5000 \end{Bmatrix}$ lb

Node 2: X-reaction = –3000 lb   Y-reaction = 0

Node 3: X-reaction = 3000 lb   Y-reaction = 4000 lb

6.2.5, 6.2.9, and 6.2.12.

Element 1: $(D_3, D_4, D_7, D_8)\!:\mathbf{k}_{4\times 4}=10^6\begin{bmatrix} 4 & 0 & -4 & 0 \\ 0 & 0 & 0 & 0 \\ -4 & 0 & 4 & 0 \\ 0 & 0 & 0 & 0 \end{bmatrix}$

Element 2: $(D_1, D_2, D_5, D_6)\!:\mathbf{k}_{4\times 4}=10^6\begin{bmatrix} 4 & 0 & -4 & 0 \\ 0 & 0 & 0 & 0 \\ -4 & 0 & 4 & 0 \\ 0 & 0 & 0 & 0 \end{bmatrix}$

Element 3: (D_1, D_2, D_3, D_4): $\mathbf{k}_{4\times4} = 10^6 \begin{bmatrix} 0 & 0 & 0 & 0 \\ 0 & 8 & 0 & -8 \\ 0 & 0 & 0 & 0 \\ 0 & -8 & 0 & 8 \end{bmatrix}$

Element 4: (D_5, D_6, D_7, D_8): $\mathbf{k}_{4\times4} = 10^6 \begin{bmatrix} 0 & 0 & 0 & 0 \\ 0 & 8 & 0 & -8 \\ 0 & 0 & 0 & 0 \\ 0 & -8 & 0 & 8 \end{bmatrix}$

Element 5: (D_1, D_2, D_7, D_8): $\mathbf{k}_{4\times4} = 10^6 \begin{bmatrix} 2.828 & 2.828 & -2.828 & -2.828 \\ 2.828 & 2.828 & -2.828 & -2.828 \\ -2.828 & -2.828 & 2.828 & 2.828 \\ -2.828 & -2.828 & 2.828 & 2.828 \end{bmatrix}$

System Equations: $10^6 \begin{bmatrix} 6.828 & 0 & 0 & -2.828 & -2.828 \\ & 4 & 0 & -4 & 0 \\ & & 8 & 0 & 0 \\ & \text{Sym} & & 6.828 & 2.828 \\ & & & & 10.828 \end{bmatrix} \begin{Bmatrix} D_1 \\ D_3 \\ D_4 \\ D_7 \\ D_8 \end{Bmatrix} = \begin{Bmatrix} 0 \\ -4330 \\ 2500 \\ 2000 \\ 0 \end{Bmatrix}$

$\{D_1, D_3, D_4, D_7, D_8\} = 10^{-4}\{-5.825, -27.8, 3.125, -16.976, 2.9127\}$ m

Element 1: $\begin{Bmatrix} f_1' \\ f_2' \end{Bmatrix} = \begin{Bmatrix} -4330 \\ 4330 \end{Bmatrix}$ N

Element 2: $\begin{Bmatrix} f_1' \\ f_2' \end{Bmatrix} = \begin{Bmatrix} -2330 \\ 2330 \end{Bmatrix}$ N

Element 3: $\begin{Bmatrix} f_1' \\ f_2' \end{Bmatrix} = \begin{Bmatrix} -2500 \\ 2500 \end{Bmatrix}$ N

Element 4: $\begin{Bmatrix} f_1' \\ f_2' \end{Bmatrix} = \begin{Bmatrix} -2330 \\ 2330 \end{Bmatrix}$ N

Element 5: $\begin{Bmatrix} f_1' \\ f_2' \end{Bmatrix} = \begin{Bmatrix} 3295 \\ -3295 \end{Bmatrix}$ N

Node 1: Y-reaction = –169.87 N

Node 3: X-reaction = 2330 N Y-reaction = –2330 N

6.2.13. We will take the cross-sectional area as 10^4 m^2.

System Equations: $10^6 \begin{bmatrix} 83333 & 0 \\ 0 & 1.333 \end{bmatrix} \begin{Bmatrix} D_4 \\ D_6 \end{Bmatrix} = \begin{Bmatrix} 0 \\ 5000 \end{Bmatrix}$

$\{D_4, D_6\} = \{0, 0.00375 \text{ rad}\}$

Element 1: $\mathbf{f}'_{6\times1} = \{0, 9500\text{ N}, 10000\text{ N-m}, 0, 8500\text{ N}, -7000\text{ N-m}\}$

Element 2: $\mathbf{f}'_{6\times1} = \{0, 7125\text{ N}, 7000\text{ N-m}, 0, 4875\text{ N}, -2500\text{ N-m}\}$

6.2.15. We will take the cross-sectional area as 10^4 m^2.

System Equations: $10^7 \begin{bmatrix} 4(10^7) & 0 & 1.2 \\ & 2(10^7) & 4.8 \\ \text{Sym} & & 24 \end{bmatrix} \begin{Bmatrix} D_4 \\ D_5 \\ D_6 \end{Bmatrix} = \begin{Bmatrix} -10000 \\ 0 \\ -5000 \end{Bmatrix}$

$\{D_4, D_5, D_6\} = \{-2.4375(10^{-11})\,\text{m}, 5(10^{-12})\,\text{m}, -2.083(10^{-5})\text{ rad}\}$

Element 1: $\mathbf{f}'_{6\times1} = \{-1000\text{ N}, -250\text{ N}, -833\text{ N-m}, 1000\text{ N}, 250\text{ N}, -1667\text{ N-m}\}$

Element 2: $\mathbf{f}'_{6\times1} = \{-9750\text{ N}, -1000\text{ N}, -3333\text{ N-m}, 9750\,N, 1000\text{ N}, -1667\text{ N-m}\}$

6.2.16. System Equations: $10^7\begin{bmatrix} 4.8333 & 0 & 2.4167 & 0 & 0 \\ & 0.3625 & 0 & -0.1208 & 0 \\ & & 7.25 & 0 & 1.2083 \\ & & & 0.1208 & 0 \\ & & & & 2.4167 \end{bmatrix}\begin{Bmatrix} D_3 \\ D_4 \\ D_6 \\ D_7 \\ D_9 \end{Bmatrix} = \begin{Bmatrix} -3(10^5) \\ 0 \\ -9(10^5) \\ 0 \\ 1.2(10^6) \end{Bmatrix}$

$\{D_3, D_4, D_6, D_7, D_9\} = \{6.2069(10^{-3})\text{ rad}, 0, -2.483(10^{-2})\text{ rad}, 0, 6.207(10^{-2})\text{ rad}\}$

Element 1: $\mathbf{f}'_{6\times1} = \{0, 3750\text{ lb}, 0, 0, 26250\text{ lb}, -1.35(10^6)\text{ lb-in}\}$

Element 2: $\mathbf{f}'_{6\times1} = \{0, 35625\text{ lb}, 1.35(10^6)\text{ lb-in}, 0, 24375\text{ lb}, 0\}$

6.2.18. System Equations: $10^7\begin{bmatrix} 0.12 & 0 \\ 0 & 5.76 \end{bmatrix}\begin{Bmatrix} D_4 \\ D_6 \end{Bmatrix} = \begin{Bmatrix} 0 \\ 61340 \end{Bmatrix}$

$\{D_4, D_6\} = \{0, 1.0649(10^{-3})\text{ rad}\}$

Element 1: $\mathbf{f}'_{6\times1} = \{0, 4443\text{ lb}, 107130\text{ lb-in}, 0, 2057\text{ lb}, 0\}$

6.2.19. System Equations: $10^6\begin{bmatrix} 4.6703 & 0 & 4.53 \\ & 2.2912 & -7.65 \\ \text{Sym} & & 1766 \end{bmatrix}\begin{Bmatrix} D_1 \\ D_2 \\ D_3 \end{Bmatrix} = \begin{Bmatrix} 0 \\ -4000 \\ 1.2(10^5) \end{Bmatrix}$

$\{D_1, D_2, D_3\} = \{-5.8454(10^{-5})\text{ in}, -1.54(10^{-3})\text{ in}, 6.1426(10^{-5})\text{ rad}\}$

Element 1: $\mathbf{f}'_{6\times1} = \{184\text{ lb}, 887\text{ lb}, 40687\text{ lb-in}, -184\text{ lb}, -887\text{ lb}, 65733\text{ lb-in}\}$

Element 2: $\mathbf{f}'_{6\times1} = \{3237\text{ lb}, 275\text{ lb}, 16434\text{ lb-in}, -3237\text{ lb}, -275\text{ lb}, 33131\text{ lb-in}\}$

Element 3: $\mathbf{f}'_{6\times1} = \{-92\text{ lb}, 124\text{ lb}, 8612\text{ lb-in}, 92\text{ lb}, -124\text{ lb}, 21135\text{ lb-in}\}$

6.4.3. A-B-C are nodes 1-2-3, and AB and BC are elements 1 and 2. Take $A = 1\text{ in}^2$.

System Equations: $10^2\begin{bmatrix} 4.834 & 0 & 0 \\ & 1.2084 & 24.16 \\ \text{Sym} & & 580.18 \end{bmatrix}\begin{Bmatrix} D_4 \\ D_5 \\ D_6 \end{Bmatrix} = \begin{Bmatrix} 0 \\ -15 \\ 0 \end{Bmatrix}$

$\{D_4, D_5, D_6\} = \{0, -0.1354\text{ in}, 5.64(10^{-4})\text{ rad}\}$

Element 1: $\mathbf{f}'_{6\times1} = \{0, 14318\text{ lb}, 531818\text{ lb-in}, 0, 682\text{ lb}, 286364\text{ lb-in}\}$

Element 2: $\mathbf{f}'_{6\times1} = \{0, -682\text{ lb}, -286364\text{ lb-in}, 0, 15682\text{ lb}, -695455\text{ lb-in}\}$

6.4.9. $\{D_1, D_2, D_3\} = \{0.0412, -0.7501, -4.07\}$ in

Element 1: $\begin{Bmatrix} f'_1 \\ f'_2 \end{Bmatrix} = \begin{Bmatrix} 134 \\ -134 \end{Bmatrix}$ lb

Element 3: $\begin{Bmatrix} f'_1 \\ f'_2 \end{Bmatrix} = \begin{Bmatrix} -220 \\ 220 \end{Bmatrix}$ lb

Element 2: $\begin{Bmatrix} f_1' \\ f_2' \end{Bmatrix} = \begin{Bmatrix} 100 \\ -100 \end{Bmatrix}$ lb

6.5.1. System Equations: $10^8\begin{bmatrix} 10.12 & 0 \\ 0 & 2.88 \end{bmatrix}\begin{Bmatrix} D_7 \\ D_8 \end{Bmatrix} = \begin{Bmatrix} -26(10^6) \\ -10^4 \end{Bmatrix}$

$\{D_7, D_8\} = \{-2.569(10^{-3}), -3.472(10^{-5})\}$ m

Element 1: $\begin{Bmatrix} f_1' \\ f_2' \end{Bmatrix} = \begin{Bmatrix} -186200 \\ 186200 \end{Bmatrix}$ N

Element 2: $\begin{Bmatrix} f_1' \\ f_2' \end{Bmatrix} = \begin{Bmatrix} -169530 \\ 169530 \end{Bmatrix}$ N

Element 3: $\begin{Bmatrix} f_1' \\ f_2' \end{Bmatrix} = \begin{Bmatrix} 284585 \\ -284585 \end{Bmatrix}$ N

6.5.2. System Equations:

$$\begin{bmatrix} 6.8284(10^6) & 0 & 0 & 0 & -2.8284(10^6) & -2.8284(10^6) \\ & 1 & 0 & 0 & 0 & 0 \\ & & 4(10^6) & 0 & -4(10^6) & 0 \\ & & & 8(10^6) & 0 & 0 \\ & \text{Sym} & & & 6.8284(10^6) & 2.8284(10^6) \\ & & & & & 10.828(10^6) \end{bmatrix}\begin{Bmatrix} D_1 \\ D_2 \\ D_3 \\ D_4 \\ D_7 \\ D_8 \end{Bmatrix} = \begin{Bmatrix} 2.8284(10^4) \\ -0.01 \\ -4.3301(10^3) \\ -7.75(10^4) \\ -2.6284(10^4) \\ -2.8284(10^4) \end{Bmatrix}$$

$\{D_1, D_3, D_4, D_7, D_8\} = 10^{-4}\{-5.825, -100, -127.8, -96.875, -116.98, 2.9127\}$ m

Element 1: $\begin{Bmatrix} f_1' \\ f_2' \end{Bmatrix} = \begin{Bmatrix} -4330 \\ 4330 \end{Bmatrix}$ N

Element 2: $\begin{Bmatrix} f_1' \\ f_2' \end{Bmatrix} = \begin{Bmatrix} -2330 \\ 2330 \end{Bmatrix}$ N

Element 3: $\begin{Bmatrix} f_1' \\ f_2' \end{Bmatrix} = \begin{Bmatrix} -2500 \\ 2500 \end{Bmatrix}$ N

Element 4: $\begin{Bmatrix} f_1' \\ f_2' \end{Bmatrix} = \begin{Bmatrix} -2330 \\ 2330 \end{Bmatrix}$ N

Element 5: $\begin{Bmatrix} f_1' \\ f_2' \end{Bmatrix} = \begin{Bmatrix} 3295 \\ -3295 \end{Bmatrix}$ N

6.5.4. System Equations: $10^3\begin{bmatrix} 1208.7 & 0 \\ 0 & 6.7151 \end{bmatrix}\begin{Bmatrix} D_4 \\ D_5 \end{Bmatrix} = \begin{Bmatrix} 0 \\ -11250 \end{Bmatrix}$

$\{D_4, D_5\} = \{0, -1.675 \text{ in}\}$ $\theta_{BL} = -0.00698$ rad $\theta_{BR} = 0.012216$ rad

Element 1: $\mathbf{f}'_{6\times 1} = \{0, 1250 \text{ lb}, 450000 \text{ lb-in}, 0, -1250 \text{ lb}, 0\}$

Element 2: $\mathbf{f}'_{6\times 1} = \{0, 1250 \text{ lb}, 0, 28750 \text{ lb}, -2.475(10^6) \text{ lb-in}\}$

6.5.5. System Equations:

$$\begin{bmatrix} 4.2078(10^9) & 0 & -1.7533(10^7) & 2.1039(10^9) \\ & 3.787(10^{14}) & -5.0494(10^{14}) & 0 \\ & & 6.7325(10^{14}) & -1.7533(10^7) \\ \text{Sym} & & & 4.2078(10^9) \end{bmatrix} \begin{Bmatrix} D_3 \\ D_4 \\ D_5 \\ D_6 \end{Bmatrix} = \begin{Bmatrix} -3.6(10^6) \\ 0 \\ -6(10^4) \\ 3.6(10^6) \end{Bmatrix}$$

$$\{D_3, D_4, D_5, D_6\} = \{-1.73(10^{-3}) \text{ rad}, -0.09106 \text{ in}, -0.00683 \text{ in}, 1.6921(10^{-3}) \text{ rad}\}$$

Element 1: $\mathbf{f}'_{6\times1} = \{45000 \text{ lb}, 60000 \text{ lb}, 0, -45000 \text{ lb}, 60000 \text{ lb}, 0\}$

CHAPTER 6 SUMMARY EXERCISES

6.1. (a) System Equations: $10^7 \begin{bmatrix} 120 & 0 \\ 0 & 1.6 \end{bmatrix} \begin{Bmatrix} D_4 \\ D_6 \end{Bmatrix} = \begin{Bmatrix} 0 \\ 0 \end{Bmatrix}$

$\{D_4, D_6\} = \{0, 0 \text{ rad}\}$

Element 1: $\mathbf{f}'_{6\times1} = \{0, 10000 \text{ N}, 16667 \text{ N-m}, 0, 10000 \text{ N}, -16667 \text{ N-m}\}$

Element 2: $\mathbf{f}'_{6\times1} = \{0, 10000 \text{ N}, 16667 \text{ N-m}, 0, 10000 \text{ N}, -16667 \text{ N-m}\}$

(b) System Equations: $\begin{bmatrix} 120(10^7) & 0 & 0 \\ 0 & 1 & 0 \\ 0 & 0 & 1.6(10^7) \end{bmatrix} \begin{Bmatrix} D_4 \\ D_5 \\ D_6 \end{Bmatrix} = \begin{Bmatrix} 0 \\ -0.05 \\ 0 \end{Bmatrix}$

$\{D_4, D_5, D_6\} = \{0, -0.05 \text{ m}, 0 \text{ rad}\}$

Element 1: $\mathbf{f}'_{6\times1} = \{0, 22000 \text{ N}, 76666 \text{ N-m}, 0, -2000 \text{ N}, 43333 \text{ N-m}\}$

Element 2: $\mathbf{f}'_{6\times1} = \{0, -2000 \text{ N}, -43333 \text{ N-m}, 0, 22000 \text{ N}, -76666 \text{ N-m}\}$

6.3. *Due to External Loading*

System Equations: $$10^5 \begin{bmatrix} 4.6041 & 0 & 0 & 0 \\ & 6.7239 & 0 & 3.7772 \\ & & 6.0436 & 0 \\ Sym & & & 3.7772 \end{bmatrix} \begin{Bmatrix} D_3 \\ D_4 \\ D_5 \\ D_6 \end{Bmatrix} = \begin{Bmatrix} 0 \\ 0 \\ 0 \\ -5000 \end{Bmatrix}$$

$\{D_3, D_4, D_5, D_6\} = \{0, -0.01697 \text{ in}, 0, -0.030256 \text{ in}\}$

Element 1 (AC): $\begin{Bmatrix} f'_1 \\ f'_2 \end{Bmatrix} = \begin{Bmatrix} 0 \\ 0 \end{Bmatrix}$ lb

Element 2 (CD): $\begin{Bmatrix} f'_1 \\ f'_2 \end{Bmatrix} = \begin{Bmatrix} 0 \\ 0 \end{Bmatrix}$ lb

Element 3 (AB): $\begin{Bmatrix} f'_1 \\ f'_2 \end{Bmatrix} = \begin{Bmatrix} 4002 \\ -4002 \end{Bmatrix}$ lb

Element 4 (BD): $\begin{Bmatrix} f'_1 \\ f'_2 \end{Bmatrix} = \begin{Bmatrix} 4002 \\ -4002 \end{Bmatrix}$ lb

Element 5 (CB): $\begin{Bmatrix} f'_1 \\ f'_2 \end{Bmatrix} = \begin{Bmatrix} -5000 \\ 5000 \end{Bmatrix}$ lb

Support Settlement

System Equations: $$10^5 \begin{bmatrix} 4.6041 & 0 & 0 & 0 & 0 \\ & 6.7239 & 0 & -3.7772 & 0 \\ & & 6.0436 & 0 & 0 \\ \text{Sym} & & & 3.7772 & 0 \\ & & & & 1 \end{bmatrix} \begin{Bmatrix} D_3 \\ D_4 \\ D_5 \\ D_6 \\ D_8 \end{Bmatrix} = \begin{Bmatrix} 92083 \\ -73666 \\ 0 \\ 0 \\ -0.5(10^5) \end{Bmatrix}$$

$\{D_3, D_4, D_5, D_6, D_8\} = \{0.2 \text{ in}, -0.25 \text{ in}, 0, -0.25 \text{ in}, -0.5 \text{ in}\}$

Element 1 (AC): $\begin{Bmatrix} f_1' \\ f_2' \end{Bmatrix} = \begin{Bmatrix} 0 \\ 0 \end{Bmatrix}$ lb

Element 2 (CD): $\begin{Bmatrix} f_1' \\ f_2' \end{Bmatrix} = \begin{Bmatrix} 0 \\ 0 \end{Bmatrix}$ lb

Element 3 (AB): $\begin{Bmatrix} f_1' \\ f_2' \end{Bmatrix} = \begin{Bmatrix} 0 \\ 0 \end{Bmatrix}$ lb

Element 4 (BD): $\begin{Bmatrix} f_1' \\ f_2' \end{Bmatrix} = \begin{Bmatrix} 0 \\ 0 \end{Bmatrix}$ lb

Element 5 (CB): $\begin{Bmatrix} f_1' \\ f_2' \end{Bmatrix} = \begin{Bmatrix} 0 \\ 0 \end{Bmatrix}$ lb

Fabrication Error

System Equations: $$10^5 \begin{bmatrix} 4.6041 & 0 & 0 & 0 \\ & 6.7239 & 0 & -3.7772 \\ & & 6.0436 & 0 \\ \text{Sym} & & & 3.7772 \end{bmatrix} \begin{Bmatrix} D_3 \\ D_4 \\ D_5 \\ D_6 \end{Bmatrix} = \begin{Bmatrix} 0 \\ 1.5109(10^5) \\ 0 \\ -1.5109(10^5) \end{Bmatrix}$$

$\{D_3, D_4, D_5, D_6\} = \{0, 0, 0, -38.4 \text{ in}\}$

Element 1 (AC): $\begin{Bmatrix} f_1' \\ f_2' \end{Bmatrix} = \begin{Bmatrix} 0 \\ 0 \end{Bmatrix}$ lb

Element 2 (CD): $\begin{Bmatrix} f_1' \\ f_2' \end{Bmatrix} = \begin{Bmatrix} 0 \\ 0 \end{Bmatrix}$ lb

Element 3 (AB): $\begin{Bmatrix} f_1' \\ f_2' \end{Bmatrix} = \begin{Bmatrix} 0 \\ 0 \end{Bmatrix}$ lb

Element 4 (BD): $\begin{Bmatrix} f_1' \\ f_2' \end{Bmatrix} = \begin{Bmatrix} 0 \\ 0 \end{Bmatrix}$ lb

Element 5 (CB): $\begin{Bmatrix} f_1' \\ f_2' \end{Bmatrix} = \begin{Bmatrix} 0 \\ 0 \end{Bmatrix}$ lb

CHAPTER 7 EXERCISES

7.5.1. (a) Assume that the origin is at node 1.

Node	X Coordinate (ft)	Y Coordinate (ft)	X Displacement Fixity	X Displacement Fixity	Z Rotation Fixity
1	0	0	Fixed	Fixed	Free
2	36	12	Free	Fixed	Free
3	16	12	Free	Free	Free

(b) Assume that the origin is at node 1.

Node	X Coordinate (ft)	Y Coordinate (ft)	X Displacement Fixity	X Displacement Fixity	Z Rotation Fixity
1	0	0	Fixed	Fixed	Free
2	36	12	Free	Fixed	Free
3	16	12	Free	Free	Hinge

(c) Assume that the origin is at node 1.

Node	X Coordinate (m)	Y Coordinate (m)	X Displacement Fixity	X Displacement Fixity	Z Rotation Fixity
1	0	0	Fixed	Fixed	Fixed
2	5	0	Free	Fixed	Hinge
3	10	8.66	Free	Free	Free
4	17	8.66	Fixed	Fixed	Fixed

7.5.6. The truss is stable when the support at node 6 is changed from a Y-Roller (free to move in the Y-direction) to a X-Roller (free to move in the X-direction). Generally speaking, a collection of triangles leads to a stable truss. However, as this problem shows, the truss is not entirely made up of triangles.

7.5.8.

a (ft)	F_{CE} (lb)	F_{FD} (lb)	F_{DE} (lb)
4	27000	36000	10817
5	21000	28800	9372
6	18000	24000	8485
7	15429	20571	7902
8	13500	18000	7500
9	12000	16000	7210
10	10800	14400	6997

CHAPTER 7 SUMMARY EXERCISES

7.1. $A_x = 16\ \text{k}(\rightarrow)$ $\qquad A_y = 36\ \text{k}(\uparrow)$ $\qquad R_C = 20\ \text{k}$

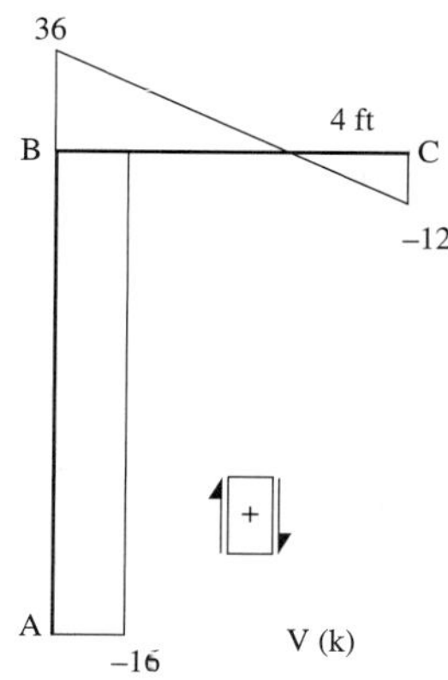

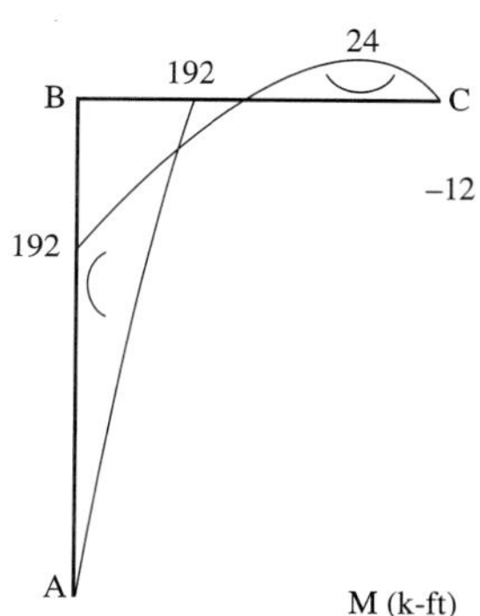

7.5.

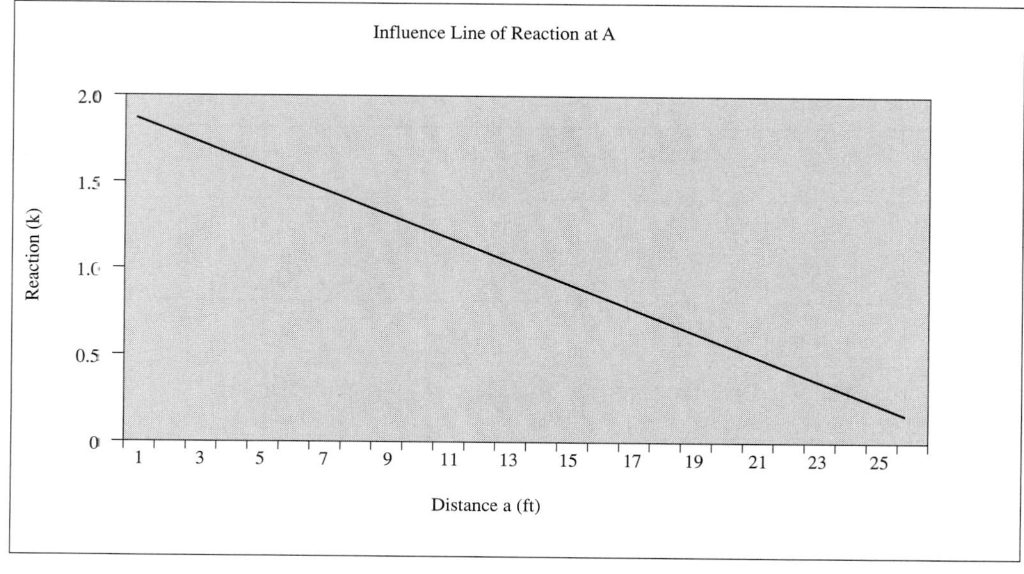

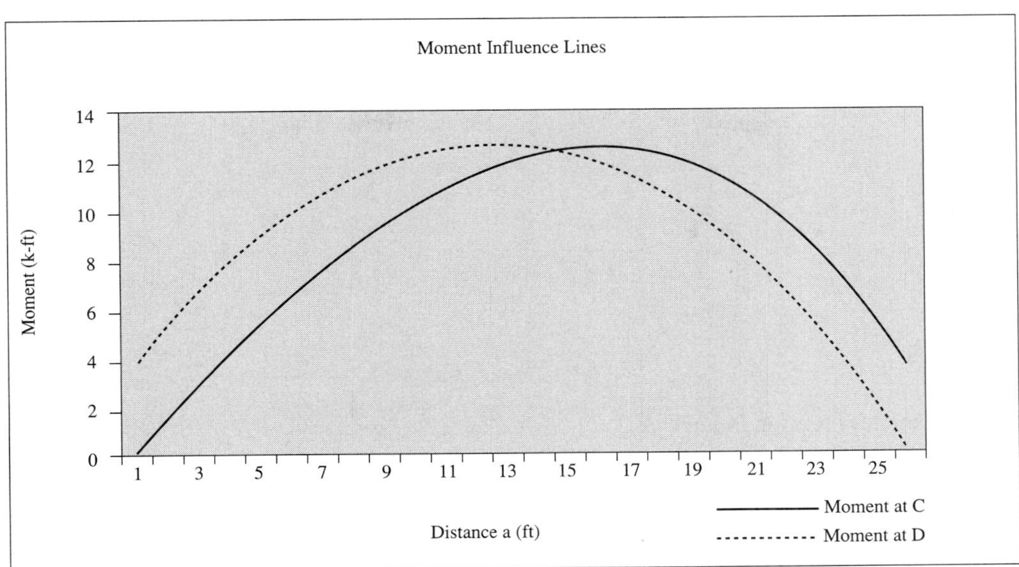

CHAPTER 8 EXERCISES

8.3.1 and 8.3.4.

Find $\mathbf{x} = (r, h)$

to minimize $f(\mathbf{x}) = 2\pi r(r + h)$

subject to

$$g_1(\mathbf{x}) \equiv h - 0.25 \le 0$$

$$g_2(\mathbf{x}) \equiv h - 6r \le 0$$

$$g_3(\mathbf{x}) \equiv 4r - h \le 0$$

$$h_1(\mathbf{x}) \equiv \pi r^2 h - 0.0004 = 0$$

$$r > 0, \ h > 0$$

Solution: $r = 0.03$ m and $h = 0.13$ m

8.3.7. (a) False. A feasible design is one that satisfies all the constraints.

(b) False. An optimal design problem can be unconstrained. However, most engineering problems are constrained optimization problems.

(c) False. The number of inequality constraints can be either less than, equal to, or greater than the number of design variables.

(d) False. An optimal design problem can be ill-posed so that there is no solution to the problem.

(e) True.

8.5.1. The minimum volume is $V = 509(10^5)$ cm^3, and $b = 14$ cm, and $h = 28$ cm.

8.5.4. The minimum volume is $V = 6.93(10^{-4})$ m^3, and $A_{AB} = 5.77(10^{-5})$ m^2, $A_{BC} = 5.77(10^{-5})$ m^2, and $x = 6.928$ m.

8.7.5. The optimized solution obtained by optimizing in stages is likely to be inferior to the solution obtained by simultaneously considering the entire design space (e.g., sizing and shape design variables). In any case, the simultaneous consideration of sizing and shape-optimal design variables will not be inferior to either sizing-optimal design problem or shape-optimal design problem considered separately.

CHAPTER 9 EXERCISES

9.1.1. W6 × 12

9.1.3. W8 × 31

9.1.4. W18 × 40

9.1.7. W6 × 9 can be used for tension member BC.

9.1.9. The section is inadequate since it does not pass the elastic lateral torsional buckling requirement.

9.1.10. W36 × 260

9.2.1. (a) Yielding of steel (b) Yielding of steel

9.2.2. (a) Section is overreinforced.

(b) Section is underreinforced.

(c) Section is underreinforced.

9.2.3. $M_n = 227.8$ k-ft

9.2.5. $P_n = 1196$ k

9.2.6. $b = 16$ in, $d = 22.5$ in, and $A_s = 2.46$ in^2

9.2.9. $b = 16$ in, $d = 20$ in, and $A_s = 1.41$ in^2

9.2.10. $b = 12$ in, $d = 17.5$ in, and $A_s = A_{s'} = 4.41$ in^2

Appendix A

Material Properties

The following table shows commonly used material property values. In reality, the range of values can be quite large and is a function of several factors such as the manufacturer, manufacturing process, test procedure used, etc.

Material	Modulus of elasticity (ksi)	Density (lb/in^3)	Shear modulus (ksi)	Poisson's ratio	Coeff. of thermal expansion ($\mu/°F$)	Ultimate stress (ksi)	Yield stress (ksi)
Douglas fir	1700	0.020					7.2
Western hemlock	1400	0.017					6.1
Eastern spruce	1200	0.016					5.5
Structural steel, low-strength	29000	0.283	11000	0.3	6.5	55	33
Structural steel, medium-strength	29000	0.283	11000	0.3	6.5	65	45
Structural steel, high-strength	29000	0.283	11000	0.3	6.5	90	60
Wrought iron AISI, 1020	29000	0.278	11000	0.3	6.5	65	36
Wrought iron AISI, 1090	29000	0.278	11000	0.3	6.5	122	67
Wrought iron AISI, 4063	29000	0.278	11000	0.3	6.5	180	160
Wrought iron AISI, 8760	29000	0.278	11000	0.3	6.5	218	200
Reinforcing steel, Grade 40	29000	0.278	11000	0.3	6.5	60	40
Reinforcing steel, Grade 50	29000	0.278	11000	0.3	6.5	80	50
Reinforcing steel, Grade 60	29000	0.278	11000	0.3	6.5	90	60
Aluminum, 6061-T6	10000	0.101	4000	0.33	12.8	45	39
Aluminum, 7075-T6	10000	0.101	4000	0.33	12.8	81	71
Concrete, low-strength	3000	0.087		0.15	6.1	2.9	
Concrete, high-strength	3000	0.087		0.15	6.1	5.1	

Appendix B

Cross-Sectional Properties

Table B.1 Properties of Common Shapes

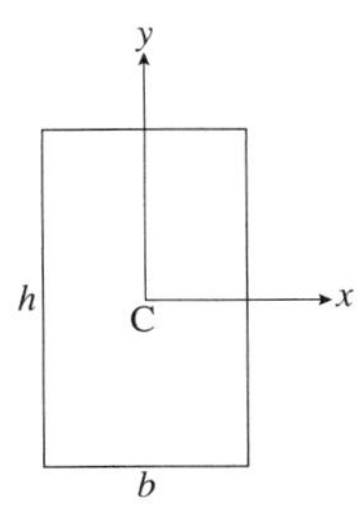

$$A = bh$$

$$I_x = \frac{1}{12}bh^3 \qquad S_x = \frac{1}{6}bh^2$$

$$I_y = \frac{1}{12}hb^3 \qquad S_y = \frac{1}{6}hb^2$$

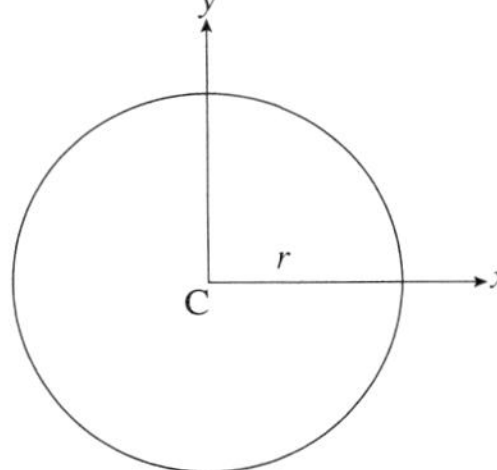

$$A = \pi r^2$$

$$I_x = \frac{\pi r^4}{4} \qquad S_x = \frac{\pi r^3}{4}$$

$$I_y = \frac{\pi r^4}{4} \qquad S_y = \frac{\pi r^3}{4}$$

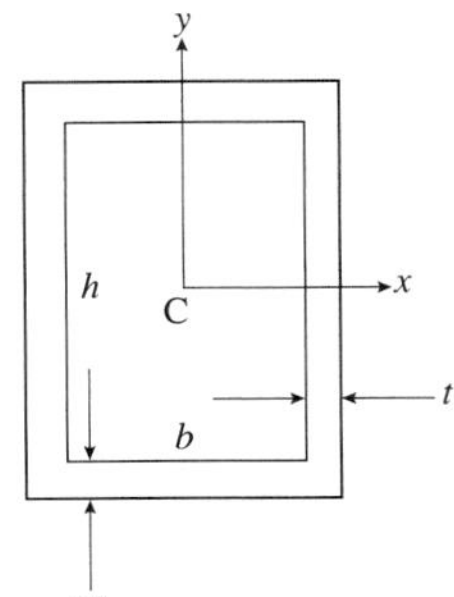

$$A = 2(ht + bw + 2wt)$$

$$I_x = \frac{(b+2t)(h+2w)^3}{12} \qquad S_x = \frac{(b+2t)(h+2w)^2}{6}$$

$$I_y = \frac{(h+2w)(b+2t)^3}{12} \qquad S_y = \frac{(h+2w)(b+2t)^2}{6}$$

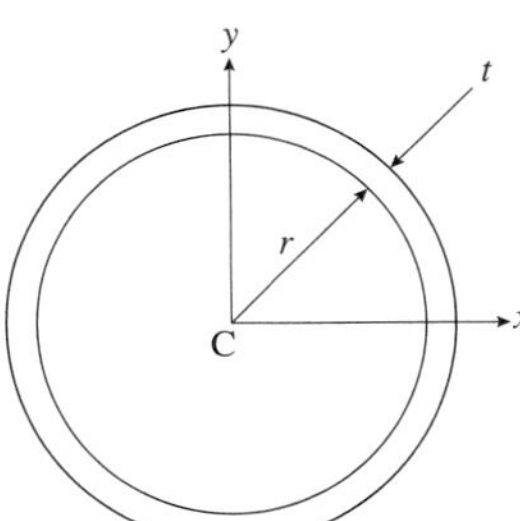

$$A = \pi t(t + 2)$$

$$I_x = \frac{\pi\left[(r+t)^4 - r^4\right]}{4} \qquad S_x = \frac{I_x}{(r+t)}$$

$$I_y = \frac{\pi\left[(r+t)^4 - r^4\right]}{4} \qquad S_y = \frac{I_y}{(r+t)}$$

Table B.2 Cross-Sectional Properties of Structural Lumber

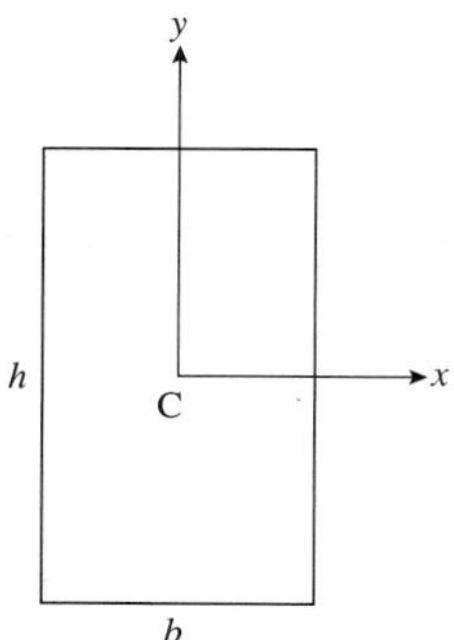

Dimensions				Area	Moment of inertia
Nominal		Actual		A	I
b	h	b	h	(in^3)	(in^4)
2	3	1.5	2.5	3.75	1.953
2	4	1.5	3.5	5.25	5.359
2	6	1.5	5.5	8.25	20.797
2	8	1.5	7.25	10.875	47.635
2	10	1.5	9.25	13.875	98.932
2	12	1.5	11.25	16.875	177.979
2	14	1.5	13.25	19.875	290.775
3	2	2.5	1.5	3.75	0.703
3	4	2.5	3.5	8.75	8.932
3	6	2.5	5.5	13.75	34.661
3	8	2.5	7.25	18.125	79.391
3	10	2.5	9.25	23.125	164.886
3	12	2.5	11.25	28.125	296.631
3	14	2.5	13.25	33.125	484.626
3	16	2.5	15.25	38.125	738.870
4	2	3.5	1.5	5.25	0.984
4	3	3.5	2.5	8.75	4.557
4	4	3.5	3.5	12.25	12.505
4	6	3.5	5.5	19.25	48.526
4	8	3.5	7.25	25.375	111.148
4	10	3.5	9.25	32.375	230.840
4	12	3.5	11.25	39.375	415.283
4	14	3.5	13.25	46.375	678.476
4	16	3.5	15.25	53.375	1034.419
6	2	5.5	1.5	8.25	1.547
6	3	5.5	2.5	13.75	7.161
6	4	5.5	3.5	19.25	19.651
6	6	5.5	5.5	30.25	76.255
6	8	5.5	7.5	41.25	193.359
6	10	5.5	9.5	52.25	392.964
6	12	5.5	11.5	63.25	697.068
6	14	5.5	13.5	74.25	1127.672
6	16	5.5	15.5	85.25	1706.776
8	2	7.5	1.5	11.25	2.109
8	3	7.5	2.5	18.75	9.766
8	4	7.5	3.5	26.25	26.797

Dimensions				Area	Moment of inertia
Nominal		Actual		A	I
b	h	b	h	(in^3)	(in^4)
8	6	7.5	5.5	41.25	103.984
8	8	7.5	7.5	56.25	263.672
8	10	7.5	9.5	71.25	535.859
8	12	7.5	11.5	86.25	950.547
8	14	7.5	13.5	101.25	1537.734
8	16	7.5	15.5	116.25	2327.422
8	18	7.5	17.5	131.25	3349.609
8	20	7.5	19.5	146.25	4634.297
10	10	9.5	9.5	90.25	678.755
10	12	9.5	11.5	109.25	1204.026
10	14	9.5	13.5	128.25	1947.797
10	16	9.5	15.5	147.25	2948.068
10	18	9.5	17.5	166.25	4242.839
10	20	9.5	19.5	185.25	5870.109
12	12	11.5	11.5	132.25	1457.505
12	14	11.5	13.5	155.25	2357.859
12	16	11.5	15.5	178.25	3568.714
12	18	11.5	17.5	201.25	5136.068
12	20	11.5	19.5	224.25	7105.922
12	22	11.5	21.5	247.25	9524.276
12	24	11.5	23.5	270.25	12437.130
14	14	13.5	13.5	182.25	2767.922
16	16	15.5	15.5	240.25	4810.005

Table B.3 Properties of AISC Cross-Sections

These properties are available through the GS-USA© program or the SlideTray© program.

Appendix C

Commonly Used Structural Values

Three tables are shown here that discuss the data associated with commonly encountered dead and live loads. All the tables are courtesy of ASCE.

TABLE C.1 Minimum Design Dead Loads*

Component	Load (psf)
CEILINGS	
Acoustical Fiber Board	1
Gypsum board (per 1/8-in. thickness)	0.55
Mechanical duct allowance	4
Plaster on tile or concrete	5
Plaster on wood lath	8
Suspended steel channel system	2
Suspended metal lath and cement plaster	15
Suspended metal lath and gypsum plaster	10
Wood furring suspension system	2.5
COVERINGS, ROOF, AND WALL	
Asbestos-cement shingles	4
Asphalt shingles	2
Cement tile	16
Clay tile (for mortar add 10 lb.):	
Book tile, 2-in.	12
Book tile, 3-in.	20
Ludowici	10
Roman	12
Spanish	19
Composition:	
Three-ply ready roofing	1
Four-ply felt and gravel	5.5
Five-ply felt and gravel	6
Copper or tin	1
Corrugated asbestos-cement roofing	4
Deck, metal, 20 gage	2.5
Deck, metal, 18 gage	3
Decking, 2-in. wood (Douglas fir)	5
Decking, 3-in. wood (Douglas fir)	8
Fiberboard, 1/2-in.	0.75
Gypsum sheathing, 1/2-in	2
Insulation, roof boards (per inch thickness)	
Cellular glass	0.7
Fibrous glass	1.1
Fiberboard	1.5
Perlite	0.8
Polystyrene foam	0.2
Urethane foam with skin	0.5
Plywood (per 1/8-in. thickness)	0.4
Rigid insulation, 1/2-in.	0.75
Skylight, metal frame, 3/8-in. wire glass	8
Slate, 3/16-in.	7
Slate, 1/4-in.	10
Waterproofing membranes:	
Bituminous, gravel-covered	5.5
Bituminous, smooth surface	1.5
Liquid applied	1
Single-ply, sheet	0.7
Wood sheathing (per inch thickness)	3
Wood shingles	3

Component	Load (psf)
FLOOR FILL	
Cinder concrete, per inch	9
Lightweight concrete, per inch	8
Sand, per inch	8
Stone concrete, per inch	12
FLOORS AND FLOOR FINISHES	
Asphalt block (2-in.), 1/2-in. mortar	30
Cement finish (1-in.) on stone-concrete fill	32
Ceramic or quarry tile (3/4-in.) on 1/2-in. mortar bed	16
Ceramic or quarry tile (3/4-in.) on 1-in. mortar bed	23
Concrete fill finish (per inch thickness)	12
Hardwood flooring, 7/8-in.	4
Linoleum or asphalt tile, 1/4-in.	1
Marble and mortar on stone-concrete fill	33
Slate (per inch thickness)	15
Solid flat tile on 1-in. mortar base	23
Subflooring, 3/4-in.	3
Terrazzo (1-1/2-in.) directly on slab	19
Terrazzo (1-in.) on stone-concrete fill	32
Terrazzo (1-in.), 2-in. stone concrete	32
Wood block (3-in.) on mastic, no fill	10
Wood block (3-in.) on 1/2-in. mortar base	16
FLOORS, WOOD-JOIST (NO PLASTER)	
DOUBLE WOOD FLOOR	

Joist sizes (inches):	12-in. spacing (lb/ft^2)	16-in. spacing (lb/ft^2)	24-in. spacing (lb/ft^2)
2 × 6	6	5	5
2 × 8	6	6	5
2 × 10	7	6	6
2 × 12	8	7	6

Component	Load (psf)
FRAME PARTITIONS	
Movable steel partitions	4
Wood or steel studs, 1/2-in. gypsum board each side	8
Wood studs, 2 × 4, unplastered	4
Wood studs, 2 × 4, plastered one side	12
Wood studs, 2 × 4, plastered two sides	20
FRAME WALLS	
Exterior stud walls:	
2 × 4 @ 16-in., 5/8-in. gypsum, insulated, 3/8-in. siding	11
2 × 6 @ 16-in., 5/8-in. gypsum, insulated, 3/8-in. siding	12
Exterior stud walls with brick veneer	48
Windows, glass, frame and sash	8

Component					Load (psf)
Clay brick wythes:					
4 in.					39
8 in.					79
12 in.					115
16 in.					155
Hollow concrete masonry unit wythes:					
Wythe thickness (in inches)	4	6	8	10	12
Density of unit (105 pcf):					
No grout	22	24	31	37	43
48″ o.c.		29	38	47	55
40″ o.c. grout		30	40	49	57
32″ o.c. spacing		32	42	52	61
24″ o.c.		34	46	57	67
16″ o.c.		40	53	66	79
Full Grout		55	75	95	115
Density of unit (125 pcf):					
No grout	26	28	36	44	50
48″ o.c.		33	44	54	62
40″ o.c. grout		34	45	56	65
32″ o.c. spacing		36	47	58	68
24″ o.c.		39	51	63	75
16″ o.c.		44	59	73	87
Full Grout		59	81	102	123
Density of unit (135 pcf):					
No grout	29	30	39	47	54
48″ o.c.		36	47	57	66
40″ o.c. grout		37	48	59	69
32″ o.c. spacing		38	50	62	72
24″ o.c.		41	54	67	78
16″ o.c.		46	61	76	90
Full Grout		62	83	105	127
Solid concrete masonry unit wythes					
Wythe thickness (in inches)	4	6	8	10	12
Density of unit (105 pcf):	32	51	69	87	105
Density of unit (125 pcf):	38	60	81	102	124
Density of unit (135 pcf):	41	64	87	110	133

*Weights of masonry include mortar but not plaster. For plaster, add 5 lb/ft^2 for each face plastered. Values given represent averages. In some cases there is a considerable range of weight for the same construction.

TABLE C.2 Minimum Densities for Design Loads from Materials

Material	Load (lb/cu ft)	Material	Load (lb/cu ft)
Aluminum	170	Lead	710
Bituminous products		Lime	
Asphaltum	81	Hydrated, loose	32
Graphite	135	Hydrated, compacted	45
Parafin	56	Masonry, Ashlar Stone	
Petroleum, crude	55	Granite	165
Petroleum, refined	50	Limestone, crystalline	165
Petroleum, benzine	46	Limestone, oolitic	135
Petroleum, gasoline	42	Marble	173
Pitch	69	Sandstone	144
Tar	75	Masonry, Brick	
Brass	526	Hard (low absorbtion)	130
Bronze	552	Medium (medium absorbtion)	115
Cast-stone masonry (cement, stone, sand)	144	Soft (high absorbtion)	100
Cement, portland, loose	90	Masonry, Concrete*	
Ceramic tile	150	Lightweight units	105
Charcoal	12	Medium weight units	125
Cinder fill	57	Normal weight units	135
Cinders, dry, in bulk	45	Masonry Grout	140
Coal		Masonry, Rubble Stone	
Anthracite, piled	52	Granite	153
Bituminous, piled	47	Limestone, crystalline	147
Lignite, piled	47	Limestone, oolitic	138
Peat, dry, piled	23	Marble	156
Concrete, plain		Sandstone	137
Cinder	108	Mortar, cement or lime	130
Expanded-slag aggregate	100	Particleboard	45
Haydite (burned-clay aggregate)	90	Plywood	36
Slag	132	Riprap (Not submerged)	
Stone (including gravel)	144	Limestone	83
Vermiculite and perlite aggregate, nonload-bearing	25-50	Sandstone	90
Other light aggregate, load-bearing	70-105	Sand	
Concrete, Reinforced		Clean and dry	90
Cinder	111	River, dry	106
Slag	138	Slag	
Stone (including gravel)	150	Bank	70
Copper	556	Bank screenings	108
Cork, compressed	14	Machine	96
Earth (not submerged)		Sand	52
Clay, dry	63	Slate	172
Clay, damp	110	Steel, cold-drawn	492
Clay and gravel, dry	100	Stone, Quarried, Piled	
Silt, moist, loose	78	Basalt, granite, gneiss	96
Silt, moist, packed	96	Limestone, marble, quartz	95
Silt, flowing	108	Sandstone	82
Sand and gravel, dry, loose	100	Shale	92
Sand and gravel, dry, packed	110	Greenstone, hornblende	107
Sand and gravel, wet	120	Terra Cotta, Architectural	
Earth (submerged)		Voids filled	120
Clay	80	Voids unfilled	72
Soil	70	Tin	459
River mud	90	Water	
Sand or gravel	60	Fresh	62
Sand or gravel and clay	65	Sea	64
Glass	160	Wood, Seasoned	
Gravel, dry	104	Ash, commercial white	41
Gypsum, loose	70	Cypress, southern	34
Gypsum, wallboard	50	Fir, Douglas, coast region	34
Ice	57	Hem fir	28
Iron		Oak, commercial reds and whites	47
Cast	450	Pine, southern yellow	37
Wrought	48	Redwood	28
		Spruce, red, white, and Stika	29
		Western hemlock	32
		Zinc, rolled sheet	449

* Tabulated values apply to solid masonry and to the solid portion of hollow masonry.

TABLE C.3 Minimum Uniformly Distributed Live Loads

Occupancy or use	Live Load lb/ft² (kN/m²)	Occupancy or use	Live Load lb/ft² (kN/m²)
Air-conditioning (machine space)	200* (9.58)	Laboratories, scientific	100 (4.79)
Amusement park structure	100* (4.79)	Laundries	150* (7.18)
Attic, Nonresidential		Libraries, corridors	80* (3.83)
Nonstorage	25 (1.20)	Manufacturing, ice	300 (14.36)
Storage	80* (3.83)	Morgue	125 (6.00)
Bakery	150 (7.18)	Office Buildings	
Exterior	100 (4.79)	Business machine equipment	100* (4.79)
Interior (fixed seats)	60 (2.87)	Files (see file room)	
Interior (movable seats	100 (4.79)	Printing Plants	
Boathouse, floors	100* (4.79)	Composing rooms	100 (4.79)
Boiler room, framed	300* (14.36)	Linotype rooms	100 (4.79)
Broadcasting studio	100 (4.79)	Paper storage	**
Catwalks	25 (1.20)	Press rooms	150* (7.18)
Ceiling, accessible furred	10# (0.48)	Public rooms	100 (4.79)
Cold Storage		Railroad tracks	‡
No overhead system	250† (11.97)	Ramps	
Overhead system		Driveway (see garages)	
Floor	150 (7.18)	Pedestrian (see sidewalks and corridors in Table	
Roof	250 (11.97)	Seaplane (see hangars)	
Computer equipment	150* (7.18)	Rest rooms	60 (2.87)
Courtrooms	50 - 100 (2.40 - 4.79)	Rinks	
Dormitories		Ice Skating	250 (11.97)
Nonpartitioned	80 (3.83)	Roller skating	100 (4.79)
Partitioned	40 (1.92)	Storage, hay or grain	300* (14.36)
Elevator machine room	150* (7.18)	Telephone exchange	150* (7.18)
Fan room	150* (7.18)	Theaters:	
File room		Dressing rooms	40 (1.92)
Duplicating equipment	150* (7.18)	Grid-iron floor or fly gallery:	
Card	125* (6.00)	Grating	60 (2.87)
Letter	80* (3.83)	Well beams, 250 lb/ft per pair (373 kg/m)	
Foundries	600* (28.73)	Header beams, 1,000 lb/ft (1,490 kg/m)	
Fuel rooms, framed	400 (19.15)	Pin rail, 250 lb/ft (373 kg/m)	
Garages - trucks	§	Projection room	100 (4.79)
Greenhouses	150 (7.18)	Toilet rooms	60 (2.87)
Hangars	150§ (7.18)	Transformer rooms	200* (9.58)
Incinerator charging floor	100 (4.79)	Vaults, in offices	250* (11.97)
Kitchens, other than domestic	150* (7.18)		

*Use weight of actual equipment or stored material when greater.

†Plus 150 lb/ft² (7.18 kN/m²) for trucks.

§Use American Association of State Highway and Transportation Officials lane loads. Also subject to not less than 100% maximum axle load.

**Paper storage 50 lb/ft (2.40 kN/m²) of clear story height.

‡As required by railroad company.

#Accessible ceilings normally are not designed to support persons. The value in this table is intended to account for occasional light storage or suspension of items. If it may be necessary to support the weight of maintenance personnel, this shall be provided for.

TABLE C.4 Typical Live Load Statistics

	Survey Load		Transient Load		Temporal Constants			Mean maximum load*
Occupancy or use	m_s lb/ft² (kN/m²)	σ_s* lb/ft² (kN/m²)	m_t* lb/ft² (kN/m²)	σ_t* lb/ft² (kN/m²)	τ_s† (years)	ν_e‡ (per year)	T§ (years)	lb/ft² (kN/m²)
Office buildings								
offices	10.9 (0.52)	5.9 (0.28)	8.0 (0.38)	8.2 (0.39)	8	1	50	55 (2.63)
Residential								
renter occupied	6.0 (0.29)	2.6 (0.12)	6.0 (0.29)	6.6 (0.32)	2	1	50	36 (1.72)
owner occupied	6.0 (0.29)	2.6 (0.12)	6.0 (0.29)	6.6 (0.32)	10	1	50	38 (1.82)
Hotels								
guest rooms	4.5 (0.22)	1.2 (0.06)	6.0 (0.29)	5.8 (0.28)	5	20	50	46 (2.2)
Schools								
classrooms	12.0 (0.57)	2.7 (0.13)	6.9 (0.33)	3.4 (0.16)	1	1	100	34 (1.63)

*For 200-ft² (18.58 m²) area, except 1000 ft² (92.9 m²) for schools.

†Duration of average sustained load occupancy.

‡Mean rate of occurrence of transient load.

§Reference period.

Appendix D

Catalog of Structural Solutions

Table D.1 Fixed-End Moments

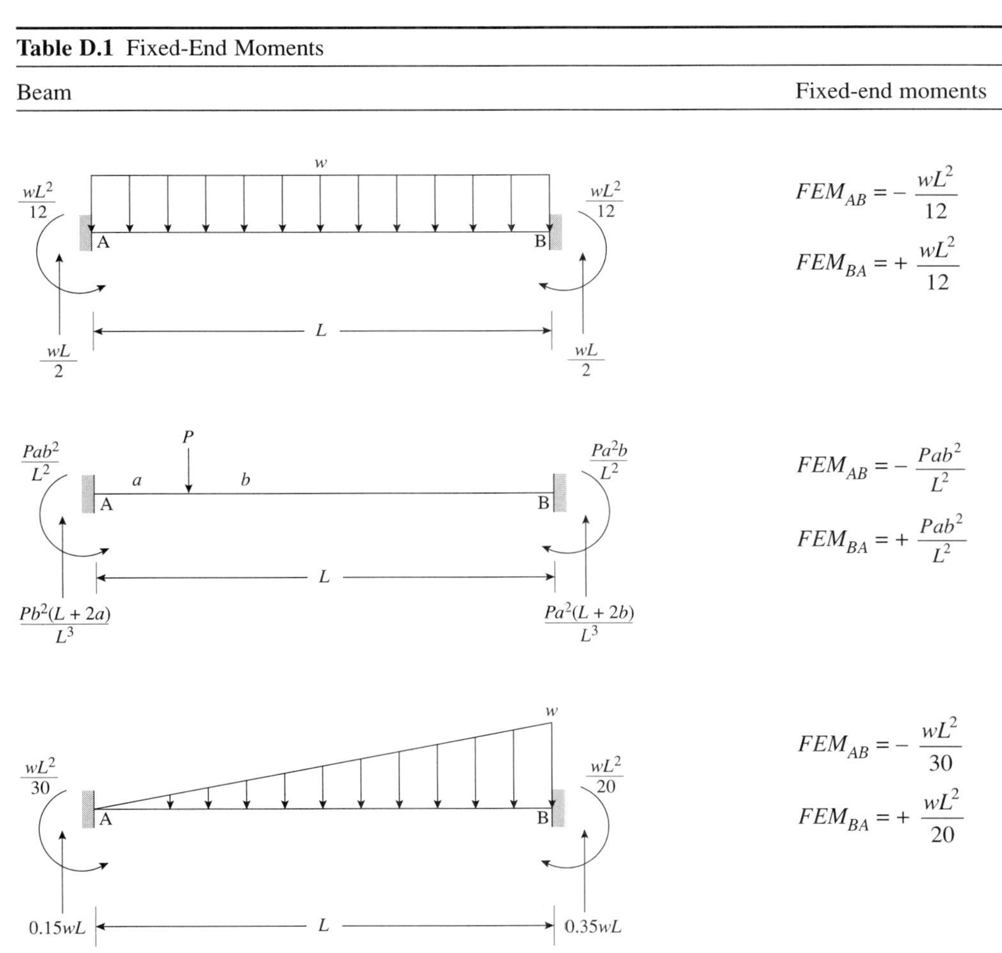

Table D.2 Equivalent Nodal Forces

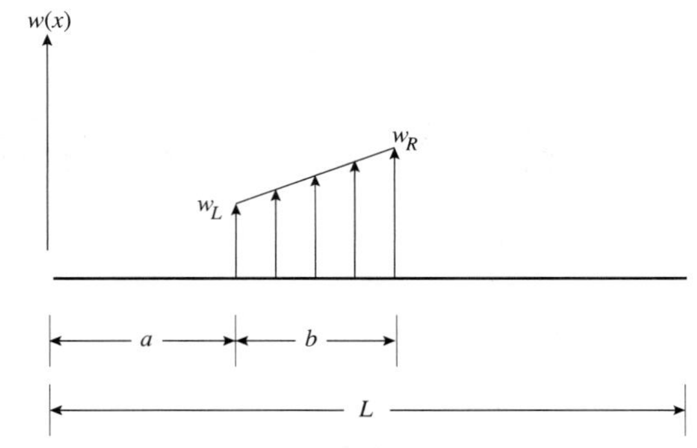

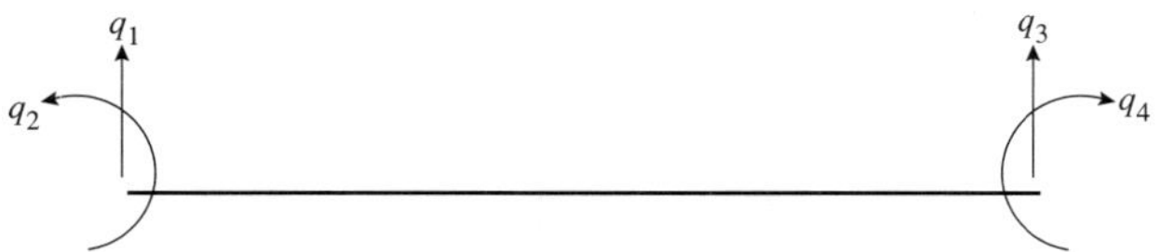

$$q_1 = \frac{2(w_R - w_L)}{5L^3 b}\left(b^5 - a^5\right) + \left[\frac{-3(w_R - w_L)}{4L^2 b} + \frac{w_L}{2L^3}\right]\left(b^4 - a^4\right)$$

$$- \frac{w_L}{L^2}\left(b^3 - a^3\right) + \frac{(w_R - w_L)}{2b}\left(b^2 - a^2\right) + w_L(b - a)$$

$$q_2 = \frac{(w_R - w_L)}{5L^2 b}\left(b^5 - a^5\right) + \left[\frac{-(w_R - w_L)}{2Lb} + \frac{w_L}{4L^2}\right]\left(b^4 - a^4\right)$$

$$+\left(\frac{w_R - w_L}{3b} - \frac{2w_L}{3L}\right)\left(b^3 - a^3\right) + \frac{w_L}{2}\left(b^2 - a^2\right)$$

$$q_3 = -\frac{2(w_R - w_L)}{5L^3 b}\left(b^5 - a^5\right) + \left[\frac{3(w_R - w_L)}{4L^2 b} - \frac{w_L}{2L^3}\right]\left(b^4 - a^4\right) + \frac{w_L}{L^2}\left(b^3 - a^3\right)$$

$$q_4 = \frac{(w_R - w_L)}{5L^2 b}\left(b^5 - a^5\right) + \left[\frac{-(w_R - w_L)}{4Lb} + \frac{w_L}{4L^2}\right]\left(b^4 - a^4\right) - \frac{w_L}{3L}\left(b^3 - a^3\right)$$

Table D.3 Deflections

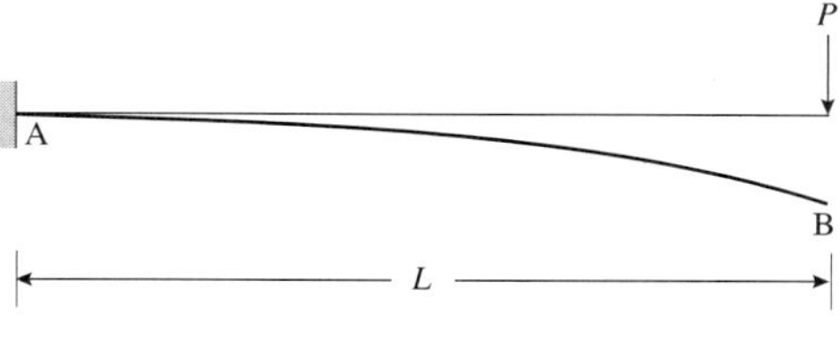

$$\theta_B = \frac{PL^2}{2EI}\ (\curvearrowright)$$

$$\Delta_B = \frac{PL^3}{3EI}(\downarrow)$$

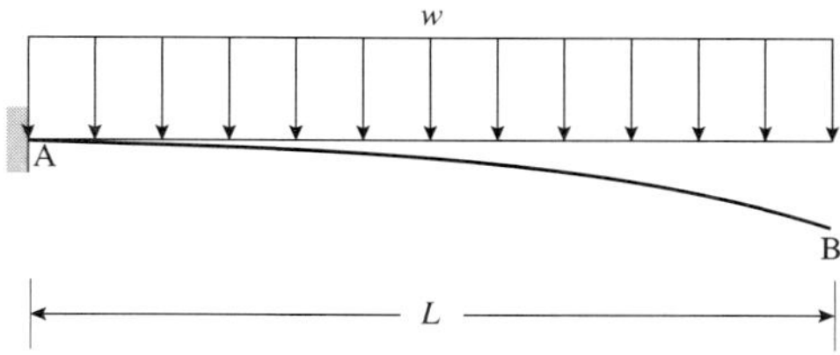

$$\theta_B = \frac{wL^2}{6EI}\ (\curvearrowright)$$

$$\Delta_B = \frac{wL^4}{8EI}(\downarrow)$$

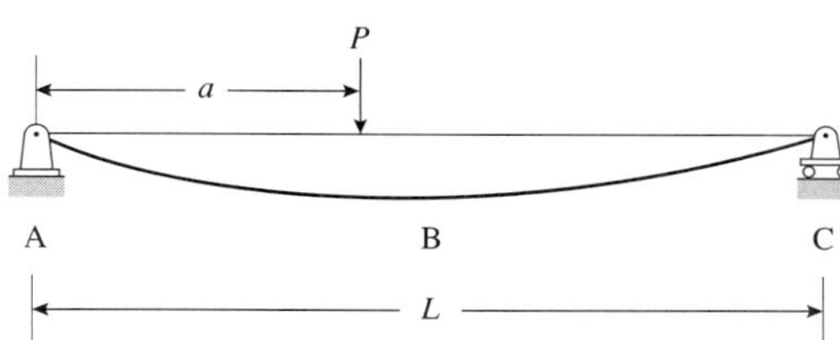

$$\theta_A = \frac{Pa(L-a)(2L-a)}{6EI}\ (\curvearrowright)$$

$$\theta_B = \frac{Pa(L-a)(2L+a)}{6EI}\ (\curvearrowleft)$$

$$v(x) = \frac{Pa(L-a)x}{6LEI}\left(-a^2+2aL-x^2\right)(\downarrow)$$

When $a = L/2$

$$\theta_{max} = \frac{PL^2}{16EI} \text{ at A and C}$$

$$\Delta_{max} = \frac{PL^3}{48EI}(\downarrow)$$

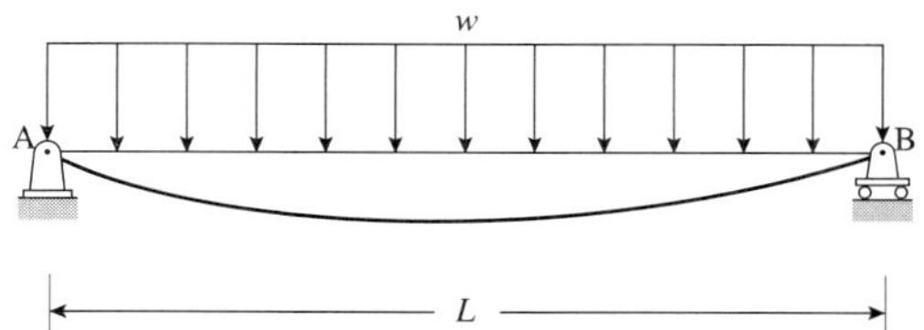

$$\theta_{max} = \frac{wL^3}{24EI} \text{ at A and B}$$

$$\Delta_{max} = \frac{5wL^4}{384EI}(\downarrow)$$

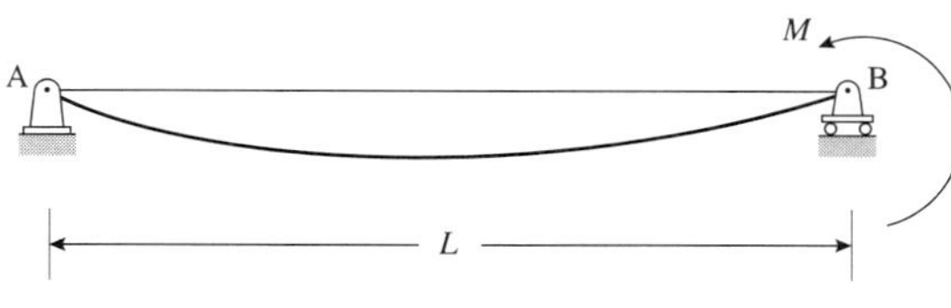

$$v(x) = \frac{Mx}{6EIL}\left(x^2-3Lx+2L^2\right)(\downarrow)$$

$$\theta_A = \frac{ML}{6EI}\ (\curvearrowright)$$

$$\theta_B = \frac{ML}{3EI}\ (\curvearrowleft)$$

Appendix E

Mathematical Background

Chapter 2

Definite Integration

$$\int_a^b c\,dx = [cx]_a^b = c(b-a) \tag{2.1}$$

$$\int_a^b x^n\,dx = \left[\frac{x^{n+1}}{n+1}\right]_a^b = \frac{1}{(n+1)}\left[b^{n+1} - a^{n+1}\right] \tag{2.2}$$

Examples: a. $\displaystyle\int_0^5 (x^2 - 3x + 10)\,dx = \left[\frac{x^3}{3} - 3\frac{x^2}{2} + 10x\right]_0^5 = \left[\frac{5^3}{3} - 3\frac{5^2}{2} + 10(5)\right] = \frac{325}{6}$

b. $\displaystyle\int_5^{10} (x^2 - 3x + 10)\,dx = \left[\frac{x^3}{3} - 3\frac{x^2}{2} + 10x\right]_5^{10}$

$$= \left[\frac{10^3}{3} - 3\frac{10^2}{2} + 10(10)\right] - \left[\frac{5^3}{3} - 3\frac{5^2}{2} + 10(5)\right] = \frac{1375}{6}$$

System of Linear Algebraic Equations

1. Solution of linear algebraic equation with a single unknown: The equations are usually for the form $ax + b = 0$, from which $x = -(b/a)$.

Example: $2.57x - 357.8 = 0 \Rightarrow x = \dfrac{357.8}{2.57} = 139.222$

2. Solution of linear algebraic equations with two unknowns: The equations usually are of the form

$$a_{11}x + a_{12}y = b_1 \tag{1}$$

$$a_{21}x + a_{22}y = b_2 \tag{2}$$

We use the elimination technique to solve for the unknowns. From (1), we have

$$x = \frac{b_1 - a_{12}y}{a_{11}} \tag{3}$$

Substituting (3) in (2) yields

$$a_{21}\left(\frac{b_1 - a_{12}y}{a_{11}}\right) + a_{22}y = b_2$$

or $y(a_{22} - a_{21}a_{12}/a_{11}) = b_2 - (a_{21}b_1/a_{11})$, which is an equation with a single unknown.

Simplifying, we have

$$y = \frac{b_2a_{11} - b_1a_{21}}{a_{11}a_{22} - a_{21}a_{12}} \tag{2.3}$$

Once y is determined, it can be substituted in (3) to obtain x as follows:

$$x = \frac{b_1a_{22} - b_2a_{12}}{a_{11}a_{22} - a_{21}a_{12}} \tag{2.4}$$

Example: $2.3x + 5.8y = 13.3$

$-4.4x + 0.5y = 96.5$

From the first equation, $x = 5.783 - 2.522y$. Substituting in the second equation gives us $-4.4(5.783 - 2.522y) + 0.5y = 96.5$. This can be reduced to

$$y(0.5 + 11.1) = 96.5 + 25.45 \Rightarrow y = \frac{121.95}{11.6} = 10.51$$

Substituting in the equation for x yields $x = 5.783 - 2.522(10.51) = -20.72$.

Check: $2.3(-20.72) + 5.8(10.51) \approx 13.3$ OK

$-4.4(-20.72) + 0.5(10.51) \approx 96.5$ OK

Or, using Eqs. (2.5.1.1) and (2.5.1.2) directly, we have

$$x = \frac{13.3(0.5) - 96.5(5.8)}{2.3(0.5) - (-4.4)(5.8)} = -20.74$$

$$y = \frac{96.5(2.3) - 13.3(-4.4}{2.3(0.5) - (-4.4)(5.8)} = 10.52$$

Example: $2.3x + 5.8y = 13.3$

$4.6x + 11.6y = 96.5$

There is no solution since the equations are linearly dependent. Mathematically, this can be determined by computing the denominator in Eqs. (2.3) or (2.4). If it is zero, the equations are linearly dependent and no solution exists.

Geometry and Trigonometry. A significant portion of the solutions involving truss analysis deals with trigonometry and geometry. Using most problem data, it is possible to construct several right-angle triangles (see Fig. 2.1). These triangles can then be used to determine the length and orientation of members using the above equations. In those few situations where this procedure may not be possible, the following can be used instead (see Fig. 2.2):

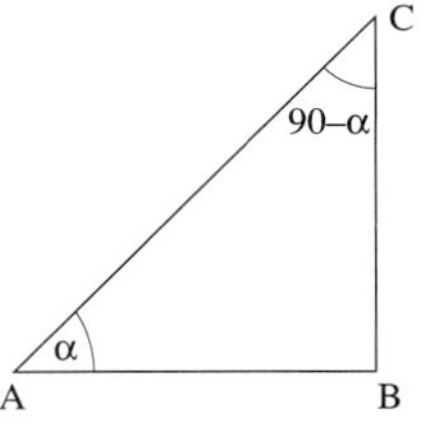

Fig. 2.1

$$\sin\alpha = \cos(90-\alpha) = \frac{CB}{AC} \qquad \cos\alpha = \frac{AB}{AC} \qquad \tan\alpha = \frac{CB}{AB} \qquad AC = \sqrt{AB^2 + CB^2} \quad (2.5)$$

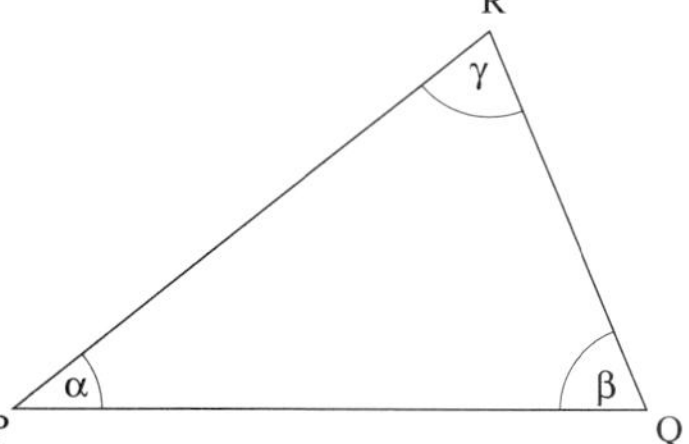

Fig. 2.2

Law of sines: $\dfrac{QR}{\sin\alpha} = \dfrac{PR}{\sin\beta} = \dfrac{PQ}{\sin\gamma}$ (2.6)

Law of cosines: $QR^2 = PQ^2 + PR^2 - 2(PQ)(PR)\cos\alpha$ (2.7)

Graphing

1. The derivative of a function $f(x)$ is denoted $f'(x)$ or $df(x)/dx$:

   $$\frac{d}{dx}(c) = 0 \qquad \frac{d}{dx}(x^n) = nx^{n-1}$$

2. The necessary condition for finding the maximum of a function $f(x)$ is $\dfrac{df(x)}{dx} = 0$.

   Solving this equation yields points $x = x^*$ at which the function has an extreme value.

   A sufficient condition for maximum of a function, $f(x)$ is $\dfrac{d^2f(x = x^*)}{dx^2} < 0$.

3. The roots of a quadratic equation $ax^2 + bx + c = 0$ are $x_{1,2} = \dfrac{-b \pm \sqrt{b^2 - 4ac}}{2a}$

   While we could list a rather involved set of conditions and equations (Tuma, 1987), roots of a cubic equation can be found by trial and error.
4. Slope of a function: The slope of a function $f(x)$ is denoted $f'(x)$. Consider the function in Fig. 2.3(a). The slope of the function is positive at A, zero at B, and negative at C. Now, consider the function shown in Fig. 2.3(b) shown with a left-handed coordinate system. The slope of the function is positive at R, zero at Q, and negative at P.

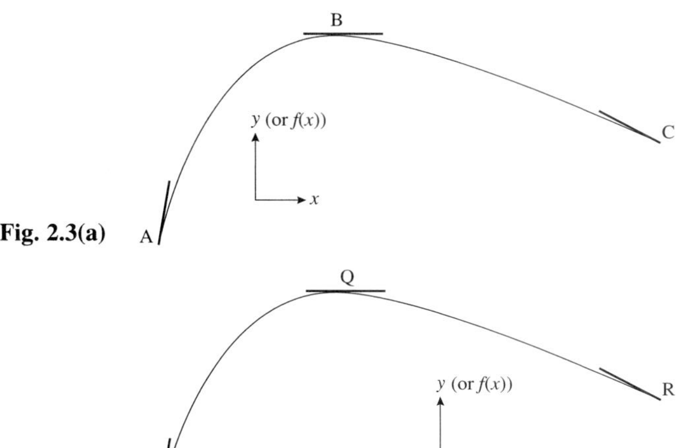

Fig. 2.3(a)

Fig. 2.3(b)

5. Graphing a polynomial: There are basically two approaches to graphing a function $f(x)$ in the interval $x_L \leq x \leq x_R$. The brute-force approach is to compute the function value for several values of x in the given interval and then connect the points. This approach is typically used in computer programs. Since our intent is to sketch the function, we will use a more cerebral approach.

   a. Constant function: $f(x) = d$ and $f'(x) = 0$. A single point is needed to define a constant function (see Fig. 2.4).

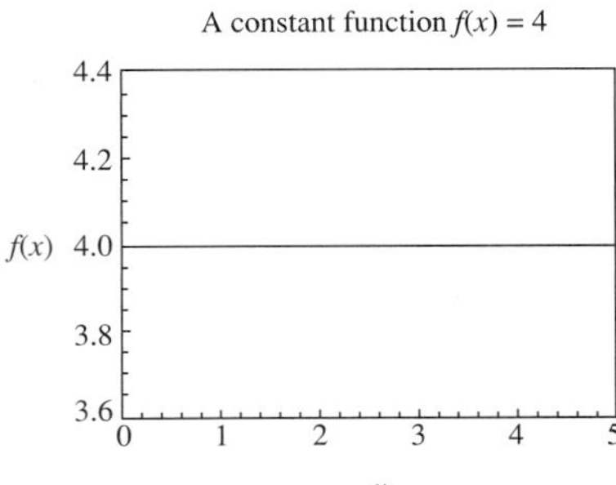

Fig. 2.4

   b. Linear function: $f(x) = cx + d$ and $f'(x) = c$. Two points are needed to define a linear function. The graphs of two functions are shown in Fig. 2.5 below in the range $0 \leq x \leq 5$. The function $f(x) = 4x - 10$ starts at (0, –10), ends at (5, 10), and has a positive slope (c value). The function $f(x) = 5x + 10$ starts at (0, 10), ends at (5, –15), and has a negative slope (c value). The graph will intersect the x axis if the values at the ends of the interval have opposite signs. The intercept can be found by solving $f(x) = 0$.

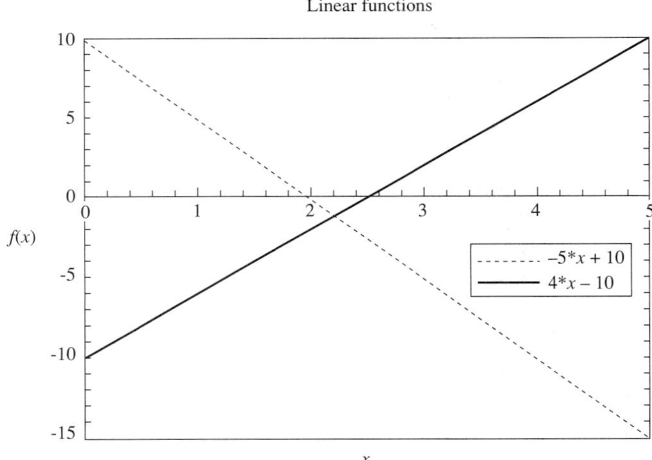

Fig. 2.5

   c. Quadratic function: $f(x) = bx^2 + cx + d$ and $f'(x) = 2bx + c$. We use the values of the functions at the end points and monitor the slope of the function when graphing. Figure 2.6 shows three functions in the interval $0 \leq x \leq 10$.

   Graphing $f(x) = -1.25x^2 + 10x + 10$: The derivative of the function is $f'(x) = -2.5x + 10$. First, we find the function values at the end points: $f(x = 0) = 10$ and $f(x = 10) = -15$. Next we look at the slope of the function. The slope is positive at $x = 0$, decreases and is zero at $x = 4$, and is increasingly negative thereafter.

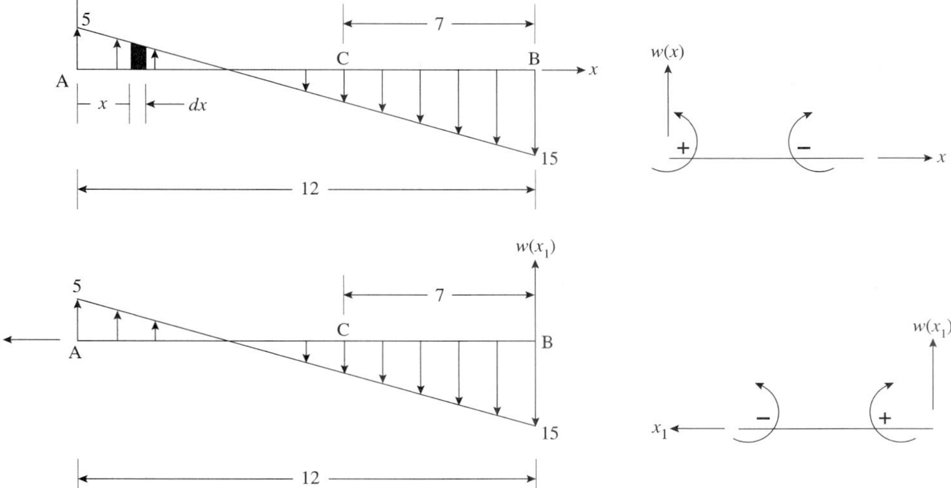

Fig. 4.1
Identical loading on two beams with different coordinate systems. (a) Righthanded coordinate system. (b) Left-handed coordinate system.

where $w(x)dx$ is the differential force and x is the lever arm of the differential force about A. *The moment is written with a positive sign since a positive force causes a positive moment (counterclockwise or about the positive z axis) about A*. Hence,

$$M_A = \int_0^{12} (w(x)\,dx)(x)$$

$$\text{or} \quad M_A = \int_0^{12} (w(x)\,dx)(x) = \int_0^{12} \left(-\frac{5x}{3} + 5 \right) x dx = \left[-\frac{5x^3}{9} + \frac{5x^2}{2} \right]_0^{12} = -600 \Rightarrow 600\,(CW)$$

The negative sign implies that in reality the moment is in the negative z direction or clockwise. For the beam in Fig. 4.1(b), $w(x_1) = (5x_1/3) - 15$. The moment of the loading about A is given by

$$M_A = -\int_0^{12} (w(x_1)\,dx_1)(12 - x_1) = -\int_0^{12} \left(\frac{5x_1}{3} - 15 \right)(12 - x_1)\,dx_1 = -\left[-\frac{5x_1^3}{9} + \frac{35x_1^2}{2} - 180x_1 \right]_0^{12} = 600$$

Note that the lever arm of the differential force about A is $(12 - x_1)$, not x_1. The positive sign once again is used since a positive force causes a positive moment (clockwise) about A. As a final check, we compute the moment using the resultants of the two triangular loading on the beam. The zero point of the loading is at 3 units from A. Assuming that counterclockwise moments are positive, we have

$$M_A = R_1\bar{x}_1 + R_2\bar{x}_2 = \left(\frac{1}{2}(3)(5) \right)\left(\frac{1}{3}(3) \right) - \left(\frac{1}{2}(9)(15) \right)\left(3 + \frac{2}{3}(9) \right) = -600$$

The last method can be tedious or even difficult unless one recalls the formulae for the areas and the locations of the centroids for arbitrary shapes. To illustrate how simple the integration approach is, let us compute the moment about C. For the beam on the left, the moment of the loading about C is given by

$$M_C = -\int_0^{12} (w(x)\,dx)(5 - x)$$

The negative sign is due to the differential moment *about C* being in the negative z direction for a positive $w(x)$. Hence,

$$M_C = -\int_0^{12} (w(x)\,dx)(5-x) = -\int_0^{12}\left(-\frac{5x}{3}+5\right)(5-x)dx = \left[-\frac{5x^3}{9}+\frac{20x^2}{3}-25x\right]_0^{12} = -300$$

As mentioned before the negative sign implies that the moment is along the negative z direction, or clockwise.

Index

Fixed-End Moments

Beam	Fixed-End Moments
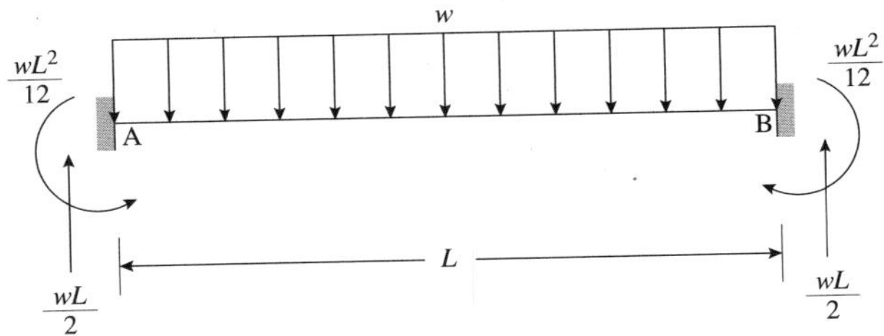	$FEM_{AB} = -\frac{wL^2}{12}$ $FEM_{BA} = +\frac{wL^2}{12}$
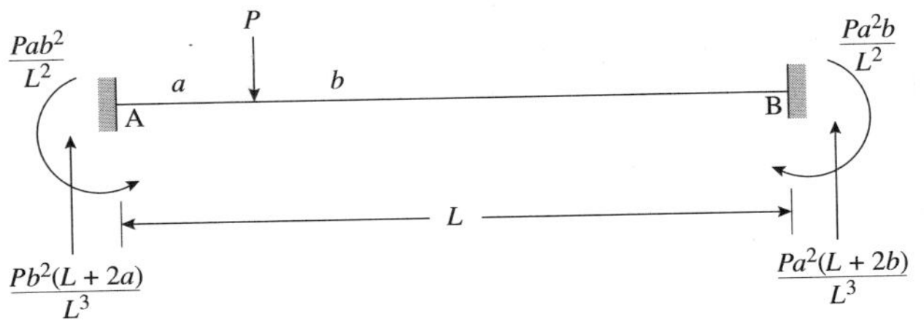	$FEM_{AB} = -\frac{Pab^2}{L^2}$ $FEM_{BA} = +\frac{Pab^2}{L^2}$
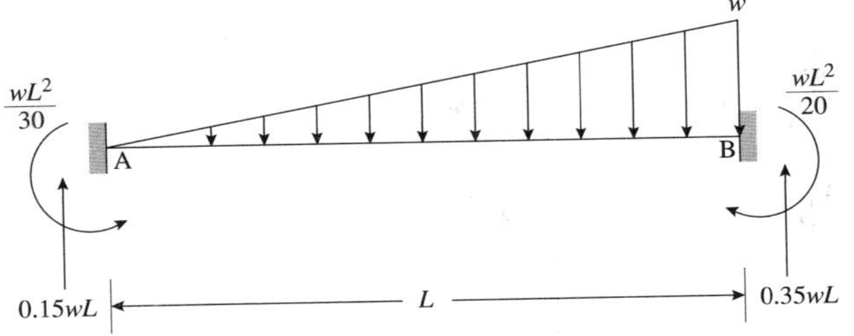	$FEM_{AB} = -\frac{wL^2}{30}$ $FEM_{BA} = +\frac{wL^2}{20}$